Bayesian Models of Cognition

Bayesian Models of Cognition

Reverse Engineering the Mind

Thomas L. Griffiths, Nick Chater,
and Joshua B. Tenenbaum

The MIT Press
Cambridge, Massachusetts
London, England

The open access edition of this book was made possible by generous funding and support from MIT Libraries.

This book was set in Times New Roman by Westchester Publishing Services. Printed and bound in the United States of America.

Library of Congress Cataloging-in-Publication Data

Names: Griffiths, Thomas L., author. | Chater, Nick, author. | Tenenbaum, Joshua, author.
Title: Bayesian models of cognition : reverse engineering the mind /
Thomas L. Griffiths, Nick Chater, and Joshua Tenenbaum.
Description: Cambridge, Massachusetts : The MIT Press, [2024] | Includes
bibliographical references and index.
Identifiers: LCCN 2023058931 (print) | LCCN 2023058932 (ebook) |
ISBN 9780262049412 (hardcover) | ISBN 9780262381048 (epub) |
ISBN 9780262381055 (pdf)
Subjects: LCSH: Cognitive science. | Cognition. | Cognitive psychology.
Classification: LCC BF311 .G758 2024 (print) | LCC BF311 (ebook) |
DDC 153—dc23/eng/20240209
LC record available at https://lccn.loc.gov/2023058931
LC ebook record available at https://lccn.loc.gov/2023058932

10 9 8 7 6 5 4 3 2 1

To Roger Shepard (1929–2022)

Contents

Preface

This book is about how the human mind comes to understand the world—and ultimately, perhaps, how we humans may come to understand ourselves. Many disciplines, ranging from neuroscience to anthropology, share this goal—but the approach that we adopt here is quite specific. We adopt the framework of *cognitive science*, which aims to create such an understanding through *reverse-engineering*: using the mathematical and computational tools from the engineering project of creating *artificial intelligence (AI)* systems to better understand the operation of human thought. AI generates a rich and hugely diverse stream of hypotheses about how the human mind might work. But cognitive science does not just take AI as a source of inspiration. What we have learned about the mathematical and computational underpinnings of human cognition can also help to build more human like intelligence in machines.

The fields of AI and cognitive science were born together in the late 1950s, and grew up together over their first decades. From the beginning, these fields' twin goals of engineering and reverse-engineering human intelligence were understood to be distinct, yet deeply related through the lens of computation. The rise of the digital computer and the possibility of computer programming simultaneously made it plausible to think that, at least in principle, a machine could be programmed to produce the input-output behavior of the human mind. So it was a natural step to suggest that the human mind itself could be understood as having been programmed, through some mixture of evolution, development, and maybe even its own reflection, to produce the behaviors we call "intelligent." In these early days, AI researchers and cognitive scientists shared their biggest questions: What kind of computer was the brain, and what kind of program could the mind be? What model of computation could possibly underlie human intelligence—both its inner workings and its outwardly observable effects?

Now, almost 70 years later, these two fields have matured and (as often happens to siblings) grown apart to some extent. Cognitive science has become a thriving, occasionally hot, but still relatively small interdisciplinary field of academic study and research. AI has become a dominant societal force, intellectually, culturally, and economically. It is no exaggeration to say that we are living in the first "AI era," in the sense that we are surrounded by genuinely useful AI technologies. We have machines that appear able to do things we used to think only humans could do—driving a car, having a conversation, or playing a game like

Go or chess—yet we still have no real AI, in the sense that the founders of the field originally envisioned. We have no general-purpose machine intelligence that does everything a human being can or thinks about everything a human can, and it's not even close. The AI technologies we have today are built by large, dedicated teams of human engineers, at great cost. They do not learn for themselves how to drive, converse, or play games, or want to do these things for themselves, the way any human does. Rather, they are trained on vast data sets, with far more data than any human being ever encounters, and those data are carefully curated by human engineers. Each system does just one thing: the machine that plays Go doesn't also play chess or tic-tac-toe or bridge or football, let alone know how to see the stones on the Go board or pick up a piece if it accidentally falls on the floor. It doesn't drive a car to the Go tournament, engage in a conversation about what makes Go so fascinating, make a plan for when and how it should practice to improve its game, or decide if practicing more is the best use of its time. The human mind, of course, can do all these things and more—independently learning and thinking for itself to operate in a hugely complex physical, social, technological, and intellectual world. And the human mind spontaneously learns to figure all this out without a team of data scientists curating the data on which it learns, but instead through growing up interacting with that complex and chaotic world, albeit with crucial help with caregivers, teachers, and textbooks.

To be sure, recent and remarkable developments in deep learning have created AI models which, with the right prompting, can be used to perform a surprisingly diverse range of tasks, from writing computer code, academic essays, and poems, and even to creating images. But, by contrast, humans autonomously create their own objectives and plans and are variously curious, bored, or inspired to explore, create, play, and work together in ways that are open-ended and self-directed. AI is smart; but as yet it is only a faint echo of human intelligence.

What's missing? Why is there such a gap between what we call AI today and the general computational model of human intelligence that the first computer scientists and cognitive psychologists envisioned? And how did AI and cognitive science lose, as has become increasingly evident, their original sense of common purpose? The pressures and opportunities arising from market forces and larger technological developments in computing, along with familiar patterns of academic fads and trends, have all surely played a role. Some of today's AI technologies are often described as inspired by the mind or brain, most notably those based on artificial neural networks or reinforcement learning, but the analogies, although they have historically been crucial in inspiring modern AI methods, are loose at best. And most cognitive scientists would say that while their field has make real progress, its biggest questions remain open. What are the basic principles that govern how the human mind works? If pressed to answer that question honestly, many cognitive scientists would say either that we don't know or that at least there is no scientific consensus or broadly shared paradigm for the field yet.

Together with our many coauthors, we have written this book in an attempt to close these gaps: to bring the fields of cognitive science and AI back together and to establish a firmly grounded and deeply shared mathematical and computational foundation for the understanding of human intelligence. These are ambitious goals, and we will not pretend otherwise. We are sure that this attempt falls short in many ways and leaves many questions still unanswered or unaddressed. But we have been working on this project for more than two decades now, developing tools, ideas, and insights that are hard to appreciate if one reads only individual papers—and especially if one reads only papers written in the last few years, as today's students, overwhelmed by the continually rising floodtide of AI developments, mostly do. It is well past time to share what we have collectively learned so far, however incomplete.

Our goal is to offer something to our readers, regardless of their academic background and interests. The book can be read from start to finish; but many will pick and choose their own path through the chapters. The book is organized into two parts. Part I, consisting of chapters 1–7, provides an introduction to the key mathematical ideas and illustrations with examples from the psychological literature. These chapters provide detailed derivations of specific models and references that can be used to learn more about the underlying principles. Part II, consisting of the remaining chapters, introduces more advanced topics and goes into depth in areas where these ideas have been applied. That part ends with an imagined dialogue in which we consider and respond to some of the concerns that we often hear about the reverse-engineering approach to understanding the mind.

To readers coming from *AI and machine learning*, the mathematical ideas in part I of the book may be familiar, but the applications are hopefully novel. Reading through these chapters will be useful for seeing how models that might have appeared in a class in computer science or statistics—such as the Bayesian Occam's razor (chapter 3) or identifying causal relationships (chapter 4)—are connected to the problems that human minds have to solve. For these readers, part II of the book offers insights into how Bayesian models relate to neural networks (chapter 12); how people make efficient use of their limited cognitive resources (chapter 13); and classic challenges for AI systems, such as reasoning about the mental states of others (chapter 14); engaging in common-sense reasoning about the physical world (chapter 15); and learning language from limited data (chapter 16). Structured representations—such as logic, grammars, and programs—play a key role in advanced applications of Bayesian models of cognition and are discussed in detail in chapters 17–19. Chapter 20 discusses developmental psychology, of particular relevance to those seeking to train intelligent systems from data, and chapter 21 considers the limits of learning.

To *cognitive psychologists*, the Bayesian approach offers a different way of thinking about the mind from the more traditional focus on dissecting the mind into distinct cognitive mechanisms. As we've stressed, Bayesian models of cognition explicitly adopt a reverse-engineering approach, trying to understand human behavior in terms of the abstract computational problems that human minds have to solve. The constraint that one's model of the mind should actually *solve* the information-processing problems that the mind faces is a powerful one—and it radically reshapes what it means to create an adequate psychological theory. This perspective is introduced in chapter 1, contextualized in chapter 2, and illustrated through the examples presented in chapters 3–7. Chapters 8 and 9 take on classic problems in cognitive psychology—learning how to learn and forming the representations that support learning—using sophisticated tools from Bayesian statistics. Chapter 10 presents some traditional experimental methods in a novel light, showing how having a sequence of people transmit a piece of information can reveal systematic biases in learning and memory. Chapters 11–13 begin to make contact with more traditional approaches to computational modeling in psychology, showing how exemplar models, neural networks, and heuristics and biases can be reconciled with Bayesian inference. Chapters 14–19 present applications to important topics in higher-level cognition, such as language, concept learning, and intuitive physics.

To *developmental psychologists*, the Bayesian approach offers a way to engage with fundamental questions about what humans can learn through experience. Bayesian models of cognition make explicit assumptions about the prior knowledge that learners bring to bear on a problem, and hence they provide a tool for exploring the limits of learning. Chapter 20 explicitly discusses how these ideas can be applied in the context of developmental psychology, so it might be a good place to go after reading part I to structure further reading. Chapter 8 discusses learning to learn, which allows learners to develop more informative priors about the world over time. Chapter 9 describes models that gradually form increasingly complex representations. Chapter 10 shows how cultural transmission of information can make that information easier to learn—an idea that has been applied in the context of language learning. Chapters 11–13 introduce mechanisms that support Bayesian inference and have room to accommodate developmental change. Chapter 14 considers Theory of Mind, chapter 15 intuitive physics, and chapter 16 language learning, all of which have played important roles in developmental psychology. For the more theoretically inclined, chapter 21 provides a theoretical analysis of the limits of learning that is relevant to debates about the poverty of the stimulus and related arguments concerning the nativist-empiricist debates in the study of language acquistion and throughout cognitive development. *Comparative psychologists* might enjoy a similar path through the book. While we do not explicitly discuss comparative research, many of these topics are ripe for exploration across different species, regarding both mechanisms that seem likely to be highly conserved across species (e.g., basic principles of perception, learning, and action) and those that are often presumed to be distinctively human (e.g., aspects of Theory of Mind and social cognition).

To *neuroscientists*, the work presented here primarily offers a challenge. While we discuss cognitive processes that could be used to implement Bayesian inference (chapters 11 and 13) and make connections to neural networks (chapter 12), the neural mechanisms that might underlie the most sophisticated models that we present are still deeply mysterious. How neurons could execute computations that result in inferences over the kind of structured representations used in these models remains an open question. To the neuroscientist hungry for such a challenge, we recommend chapter 8 as an appetizer and chapters 14–19 as the main course.

We have written this book to be accessible to as broad an audience as possible, from undergraduates who have discovered cognitive science and have a basic familiarity with probability theory to established researchers who want to broaden their perspective. In doing so, we have benefited from feedback from our students at Stanford University, the Massachusetts Institute of Technology (MIT), Brown University, the University of California, Berkeley, Princeton University, the University of Oxford, and the University of Warwick, as well as our colleagues in various fields.[1] The seed that grew into this book was planted by a summer school we held at the Institute for Pure and Applied Mathematics at the University

1. Seasoned professors will already have noticed that the number of chapters in the book is a perfect match for the number of sessions in a biweekly semester-long course. You don't need to master the Bayesian methods presented in chapter 3 to be able to recognize that this is no coincidence. We think the current order of the chapters works

of California, Los Angeles, which in turn built on tutorials we presented at the Annual Conference of the Cognitive Science Society and the Neural Information Processing Systems conference. We are grateful for all these opportunities to discuss these ideas over the last two decades, and we look forward to continuing this conversation with our readers in years to come.

well for such a course, but the chapters in part II are relatively independent of one another and can be shifted to suit the interests of your students.

I THE BASICS

1

Introducing the Bayesian Approach to Cognitive Science

Joshua B. Tenenbaum, Thomas L. Griffiths, and Nick Chater

Our goal in this first chapter is to introduce the Bayesian approach to cognitive science in terms of the big questions that it lets us ask and the distinctive kinds of answers that it offers. How does human intelligence work, in engineering terms? What mathematical principles and computational building blocks can explain our capacities to make sense of the world and our place in it? These are the biggest questions, the ones that originally brought us and many others to cognitive science. They go back to the earliest attempts to model human thinking as a species of computation (e.g., Newell & Simon, 1956), with links to philosophical ideas that go back much further (e.g., Hume, 1739/1978; Kant, 1781/1964; Mill, 1843). Everything in the pages to follow could be seen as a response to them, and we have written this book because we think that the Bayesian approach offers the most valuable tools we know to address them.

But such big questions may also feel impossible to confront at first. It may not be clear what would count as satisfying answers, or how we would know when we are right—or at least on the right track. So let us instead start with a question that is easy to state and perhaps easier to appreciate possible answers to: How do our minds get so much from so little?

1.1 Generalization and Induction

In every domain where we see the mind at work, we see a mismatch between the information coming in through our senses and the ouputs of cognition. We build rich causal models of the world, make strong generalizations and construct powerful abstractions, and even invent whole new worlds and ways of understanding them, while the input data are sparse, noisy, incomplete, and ambiguous—in every way far too limited. How do we do it?

This is the question that motivated the earliest research described in this book, and the challenge that it presents—along with what would count as a satisfying solution—can be made concrete and precise with a few familiar examples. Consider first the situation of a child learning the meanings of words. A typical two-year-old can learn how to use a new word such as "horse" or "hairbrush" from seeing just a few examples—and sometimes even a single example suffices (Carey & Bartlett, 1978; Bloom, 2000; Smith, 2000; Xu & Tenenbaum, 2007). We know that they grasp the meaning of these words, not just the sounds, because they can generalize: they can use the words appropriately (if not always

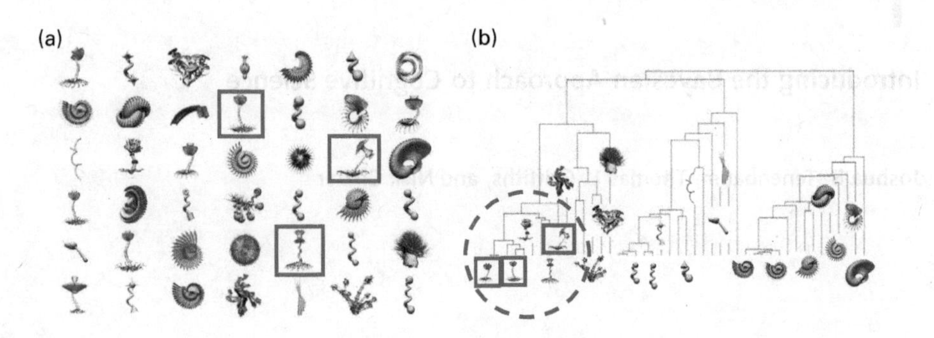

Figure 1.1
Human children learning names for object concepts routinely make strong generalizations from just a few examples. The same processes of rapid generalization can be studied in adults learning names for novel objects created with computer graphics. (a) Given these alien objects, and three examples (boxed in red) of "tufas"—a word in the alien language—which other objects are tufas? (b) Learning names for categories can be modeled as Bayesian inference over a tree-structured domain representation (Xu & Tenenbaum, 2007). Objects are placed at the leaves of the tree, and hypotheses about categories that words could label correspond to different branches. Branches at different depths pick out hypotheses at different levels of generality. By assuming that more distinctive hypotheses—those with longer branches—are more probable, and that the observed data are more likely to be generated from hypotheses that cover the examples tightly, Bayesian inference favors generalizing across the lowest distinctive branch that spans all the observed examples (shown here with a dashed circle). Figure adapted from Tenenbaum, Kemp, Griffiths, and Goodman (2011).

perfectly) in new situations. They come to use these words *categorically* to refer to any entity of the appropriate kind. This is a remarkable feat, viewed in terms of computation or statistical inference. Within the infinite (and probably infinite-dimensional) space of all possible objects, there is an infinite, but much smaller and highly constrained, subset that can be called "horses," and another for "hairbrushes." How does the child grasp the boundaries of these subsets from seeing just one or a few examples of each? And how could a machine learn to do the same?

Conventional algorithmic approaches from statistics and machine learning typically require tens or hundreds of labeled examples to classify objects into categories, and do not generalize nearly as reliably or robustly. How do children do so much better? Adults less often face the challenge of learning entirely novel object concepts, but they can be just as good at it: see for yourself with the computer-generated objects in figure 1.1.

Fundamentally the same challenge arises across cognition. Consider as a second case the many puzzles and games that children and adults make for themselves and learn from each other. Take any simple board game with just a few rules, such as tic-tac-toe, Connect Four, checkers, or Othello, and imagine that you are encountering it for the first time, seeing two people playing. The rules have not been explained to you—you are just watching the players' actions. Merely by observing their play, you would surely not be able to learn everything you need to know to play these games with skill. But seeing just a few rounds of action would be sufficient in many cases to infer much of how the game is played: what moves are allowed, what their effects are on the state of the game, and perhaps even what goal the moves appear to be aiming toward. Your inferences might be only guesses, with some degree of uncertainty, but even uncertain and even if slightly off, your guesses about the rules of the game would be far better than you could have made before seeing any play at all. You could likewise infer the rules and goals behind many popular children's playground games, watching just a minute of play for tag, hopscotch, Red Light, Green Light, or many others.

In every one of these cases, even when the concepts and rules that we infer strike us as clearly the right ones, and *are* the right ones, there always remains an infinite set of alternative possibilities that would be consistent with all the same data for any finite sequence of play. We have studied this phenomenon in the lab with even simpler abstract tasks, such as the following number game (Tenenbaum, 1999; Tenenbaum, 2000): Imagine that we are thinking of a rule that picks out a subset of numbers between 1 and 100—any subset at all, in principle. Now we give you four examples of numbers that are in the target subset—60, 80, 10, and 30—and we ask which other numbers are in that same subset. Almost everyone will say that the rule likely picks out "multiples of 10" and respond with the other numbers in that subset (i.e., 20, 40, 50, and so on). But why? What makes you so sure, when logically there are many other possibilities—a vast number. The examples given are just as consistent with a rule picking out "even numbers," "multiples of 5," or "numbers between 10 and 90," not to mention more arcane but just as logically consistent possibilities, such as "numbers between 7 and 97," "multiples of 10 except 50," or "multiples of 20, and also numbers less than 40." Why is one particular way of going beyond the data given here so much more compelling than others? And why would that same rule, "multiples of 10," not have been nearly as compelling if we had presented to you only the first one or two examples (60, or 60 and 80) and asked you to generalize at that point—even though in both those cases, all the same possible alternative rules would have been in play?

This in essence is the problem of generalization from sparse data. It is the same problem faced by a child learning how to use words or how to play a new playground game from observing what others do and say. And once you know how to recognize the challenge, you will see it showing up everywhere in cognition. It arises in learning every aspect of language, not just the meanings of words, and plays an especially central role in the study of how children acquire the syntax or morphology of their native language (Pinker, 1999; Marcus, Vijayan, Bandi Rao, & Vishton, 1999; Gomez & Gerken, 1999). It presents starkly in causal learning: every statistics class teaches that correlation does not imply causation, yet under the right circumstances, even young children routinely and reliably infer causal links from just a handful of events (Gopnik et al., 2004)—far too small a sample to compute even a reliable correlation by traditional statistical means.

Perhaps the deepest forms of generalization in human learning occur over cognitive development, in the construction of larger-scale systems of knowledge: intuitive theories of physics, psychology, or biology, or rule systems for social structure or moral judgment. Building these systems can take years, much longer than learning a single new word, a novel causal connection, or the rules of a new game, but on this scale too the final product of learning far outstrips the observed data—if only because the core concepts of these theories, such as weight and density, beliefs and preferences, or life and death, are not observable at all (Gopnik & Meltzoff, 1997; Carey, 1985, 2009). An intuitive theory can also be transformed in an instant, modified to accommodate the possibility of new causal laws given only a sliver of sensory evidence—if it's the right sliver. Consider a child seeing magnets for the first time, or a remote control, a touch-screen device, or an anti-gravity machine. A single demonstration is enough to suggest that the world could work differently than they had previously thought; one or two more experiments, playing around with the new system in question for themselves, confirm that suspicion. Of course, human scientists have yet to invent an anti-gravity machine, but you can bet that if they do, and it becomes available for anyone to purchase and play with around the home, human children will be learning about its dramatic effects in no time—and from the same minimal patterns of data.

Trying to solve the problem of how we get so much from so little is by no means the only interesting place one could start in building models of the mind, but it is surely one of the most central, with many other interesting questions growing out of attempts to answer it in a compelling way. It lays bare the severe challenges facing any attempt to understand intelligence as the solution to a set of mathematical and computational problems. It is timely, as today's artificial intelligence (AI) technologies are increasingly driven by machine learning frameworks that are endlessly data-hungry and hopelessly data-dependent. Generalizing *beyond* the data observed is at the heart of the difficulty in building machines with anything resembling human like intelligence. And it is timeless.

Philosophers have puzzled over versions of this question for 2,000 years—most famously as "the problem of induction," as David Hume's version has come to be known (Henderson, 2022), but with roots going back to Plato and Aristotle, and spanning the entire Western philosophical tradition. The nineteenth-century philosopher John Stuart Mill inspired generations of thinking about thinking when he wrote in *A System of Logic*: "Why is a single instance, in some cases, sufficient for a complete induction, while in others, myriads of concurring instances, without a single exception known or presumed, go such a very little way toward establishing a universal proposition? Whoever can answer this question knows more of the philosophy of logic than the wisest of the ancients, and has solved the problem of induction." (Mill, 1843, p. 380).

Mill was writing about the problem of generalization as it confronts scientists observing their world, but he could just as well have been talking about our modern problem of generalization in cognitive science: Out of the vast array of possible ways that human learners could go beyond the data given to them, how do we know—sometimes from only a single example—which of these generalizations should hold, and which should not? William Whewell, William Stanley Jevons, Bertrand Russell, Rudolf Carnap, Hans Reichenbach, Carl Hempel, Karl Popper, Willard Van Orman Quine, Nelson Goodman, and many others in the nineteenth and twentieth centuries deepened these questions and continue to drive debates about the origins of knowledge in contemporary epistemology and philosophy of science (Godfrey-Smith, 2003): How can we come to grasp the laws of the universe, or other enduring properties of the world? What does it mean to be a law of nature or an enduring property? What is our true basis of confidence for believing that a law or universal concept that has held everywhere and always will not be falsified tomorrow or next year?

In 1959, as AI and cognitive science were only just beginning to emerge as fields of study, Popper wrote that he chose to study the origins of scientific knowledge because it was essentially the problem of everyday epistemology—the problem of the origins of common-sense knowledge—writ large (Popper, 1959/1990). In Popper's time, as in Mill's, scientific knowledge was perhaps easier and more compelling to study than common-sense knowledge. We see the same analogy now, but in reverse. We choose to study the human mind's commonsense understanding of the world, and how it operates through everyday inductive leaps, because it exposes the deepest problems of the origins and growth of knowledge in their most tractable and urgent form. At this moment in time, with the mathematical, computational, and experimental tools available to us, this is the form of the classic problem of induction most amenable to rigorous scientific explanation and practical engineering progress.

At some level, our answer to these questions must be the same kind of answer that philosophers have given to problems of induction since Plato. If the mind goes beyond the data given, some other source of information must make up the difference. There can be no free lunch, no magic, no something-for-nothing in a computational or information-processing account of cognition (an idea formalized in "no free lunch" theorems for machine learning; e.g., Wolpert & Macready, 1997). Some more abstract background knowledge must generate and delimit the hypotheses that learners consider, or meaningful generalization would be impossible (e.g., McAllester, 1998). Developmental psychologists and linguists speak of "constraints" (e.g., Carey, 2009; Spelke & Kinzler, 2007; Chomsky, 1965); machine learning and AI researchers, "inductive bias" (e.g., Mitchell, 1997; Russell & Norvig, 2021); statisticians, "priors" (e.g., Jaynes, 2003; Robert, 2007). Neuroscientists asking analogous questions about how perception interprets noisy sensory inputs talk about the brain's "internal models" (e.g., Wolpert, Miall, & Kawato, 1998). The terminology varies but always describes some kind of world-modeling that fills in the gaps of experience, letting our minds rationally turn data into reliable, reasonable beliefs.

This "answer" is not really an answer, of course, but rather an invitation to ask more questions and a guide to where they should focus. For a reverse-engineering account, the key questions now center on explaining the nature, functions, and origins of this abstract knowledge in computational terms: How exactly does it solve the problem of induction, what is its specific form and content, and how does it arise in human minds and brains? Unpacking these questions a bit, we want to understand: How does abstract knowledge guide inference from impoverished data, to yield reliably valuable generalizations, predictions, and decisions? What form does that knowledge take, and what informational content does it have across various domains of thought? How does knowledge grow over a lifetime, balancing the need for strong constraint with the flexibility to continually learn about new domains? How does the knowledge that we build guide our planning and acting in the world—how do we put it to use? How are the algorithms and data structures needed to represent, acquire, and deploy human-scale knoweldge actually implemented in human minds, or the physical subtrate of human brains? How can learning and inference with complex world models be implemented efficiently in minds with finite—often quite limited—resources? And how do our deepest notions of abstract knowledge ultimately arise in cognitive development—must they all be innate in some sense, wired into a brain or a mind's machinery from birth, or could the fundamental forms and contents of abstract knowledge themselves be learned or constructed, in whole or in part?

This rest of this book represents our attempt to answer these questions in reverse-engineering terms: to develop a coherent, unifying framework in the language of mathematics and computation that addresses all these questions in an integrated fashion, at a level of precision and explanatory depth that is sufficient to build machines with these same capacities and to be tested rigorously in quantitative behavioral experiments. The rest of this chapter offers a more concrete overview of the ground to be covered in the book as a whole, along with some highlights and key insights. Before we begin, however, we should signal several points about the approach and how we write about it that will be helpful in orienting readers.

First, if forced to sum up our approach to these questions in a single word, we would naturally have to say: "Bayes." As explained later in this chapter and expanded upon in chapter 3, the principle of probability theory known as Bayes' rule provides a guide to how rational agents should solve problems of induction, integrating prior knowledge with the information provided by data. If granted a whole sentence, we might say: "Algorithms and data structures that efficiently approximate hierarchical Bayesian inference over richly and flexibly structured representations of the world." Of course, to properly answer all of these questions requires a whole book—this book—not a one-word label or one-sentence slogan. But these catchphrases are also useful to orient us. The work that we will describe over the next few hundred pages has fairly and unquestionably come to be known within cognitive science as the Bayesian or probabilistic approach. We ourselves have featured these words prominently in talks and paper titles, and proudly embrace the ways that our view of reverse-engineering the mind have been influenced by the engineering successes of Bayesian and probabilistic approaches in machine learning and AI (Pearl, 1988; Thrun, Burgard, & Fox, 2005; Russell & Norvig, 2021; Murphy, 2012)—along with the ways that our own work has inspired more powerful and more humanlike Bayesian approaches to machine learning, reasoning, and perception (Griffiths & Steyvers, 2004; Griffiths & Ghahramani, 2006; Kemp, Perfors, & Tenenbaum, 2007; Goodman, Mansinghka et al., 2008a; Shafto, Kemp, Mansinghka, & Tenenbaum, 2011; Kulkarni, Kohli, Tenenbaum, & Mansinghka, 2015; Lake, Salakhutdinov, & Tenenbaum, 2015; Grant, Finn, Levine, Darrell, & Griffiths, 2018; Ellis et al., 2023).

At the same time, there is much more to the Bayesian approach in cognitive science than might be suggested at first by that one word. Indeed, no view of the mind that aspires to accurate reverse-engineering could possibly be encapsulated in a single idea or principle. Consider as an example the "connectionist" or "neural network" paradigm (Rumelhart & McClelland, 1986). These models crucially do involve claims about connections: for instance, that knowledge is stored in the network of connections between neuron like processing units, and learning consists of adjusting the strengths of those connections. But they also typically involve other core claims as well, such as the primacy of distributed representations, error-driven learning, and graded activation (O'Reilly & Munakata, 2000). Hence the term "connectionism" has come to stand for a cluster of ideas that these models collectively embodied and developed over several decades. So too should "Bayesian" or "probabilistic" models of cognition be taken as labels for a body of mutually reinforcing and supporting concepts, principles, and tools that work together in powerful ways to explain how intelligence works in natural minds and how we might build more intelligent artificial minds. The goal of this book is to introduce, survey, and explain this landscape. Bayes provides a starting point, but only the start.

Second, while we have talked so far and will continue to talk about human knowledge as our subject, our focus will be on one kind of abstract knowledge—what we could call "world models" or, more precisely, "probabilistic generative world models." By "generative," we mean to refer to the human mind's models of the causal processes at work in the world: the mechanisms that make things happen, that give rise to (or "generate") the events that we observe and also those we do not directly observe but know are there, latent, behind the scenes. By "probabilistic," we are referring to all the ways that these models must inherently represent and work with uncertainty: even when the generative mechanisms they posit are deterministic, probabilistic modeling and inference are essential because of all we do not capture and we do not know about the world's latent causal structure.

These models span both the inanimate physical world and the animate world of agents, and the causal processes that go on inside those other agents' minds to generate *their* behavior. They may often be unconscious, although some surely have conscious aspects as well. They reach to domains well beyond our direct experience, that we come to think about only from others' testimony or our own imaginations. And they even extend to (or, some speculate, start with) our mind's model of its own internal processes, our own subjective world.

Structurally, these generative world models operate over multiple scales and levels of abstraction, ranging from each of our very specific models of what is directly in front of us right now, as we work on this chapter at our desks, or your model of what is in front of you as you read these words, to our much more general notions of how the world works, and likely has always worked and will always work. They also represent many worlds besides the actual one: hypothesis spaces of ways that the world could possibly or probably be, which support our inferences about how the actual world probably is; and spaces of hypothetical and counterfactual possibility, to support our reasoning about what we should probably do next to achieve our goals, or what could have been or what we should have done before when things didn't unfold as we hoped.

Why focus on this particular kind of knowledge as the basis for how our minds get so much from so little? Since the earliest days of both cognitive psychology and AI, the human mind's expansive capacity for modeling the world has been central to every serious account of human intelligence. It is frequently cited as the essence of distinctively human "common sense," as well as the cognitive basis upon which science is founded, the difference between intelligence in humans and nonhuman animals, and the difference between human intelligence and data-driven AI systems. We will not rehearse most of those arguments here, but we refer interested readers to the many versions that appear in Koffka (1925), Craik (1943), Heider (1958), Newell, Shaw, and Simon (1959), McCarthy (1959), Neisser (1967), Minsky (1982), Norman (1972), Gentner and Stevens (1983), Johnson-Laird (1983), Rumelhart, Smolensky, McClelland, and Hinton (1986b), Pearl (1988), Shepard (1994), Gopnik and Meltzoff (1997), Carey (2009), Levesque (2012), Davis (2014), Kohler (2018), and LeCun (2022). More on our own thinking about the centrality of people's models of the world can be found throughout this book.

We will say, however, that even with so much prior work on the subject of human mental models, the approach and tools presented in this book are new in the scope and technical depth with which they attempt to engage all that we humans do with our world models: our capacities to not only recognize the things we see, but to understand and explain them; to not only predict what is likely to happen next, but to judge what might have happened had circumstances or own actions been different; to imagine events that might occur but are extremely unlikely, or things that have never existed but could possibly; and then to set those outcomes as goals, making plans to achieve them effectively and efficiently, and solving problems that come up along the way; and last but hardly least, to learn by building models not only from our own experiences in the world, but by sharing our models with others, to learn socially and build knowledge culturally across individual lifetimes and generations. Our goal over the last two decades of work on Bayesian models of cognition has been to develop a unifying mathematical and computational appproach for explaining how human minds can do all these things, grounded in rigorous behavioral experiments testing how far those explanations can reach. This approach, and the tools needed to work with it and extend it, is what we hope you will take away from reading this book.

Finally, because throughout the book (and especially in this introduction) we move freely between speaking intuitively of people's "world models" and more general but philosophically laden terms such as "knowledge" or "abstract knowledge," along with more specifically Bayesian technical notions of "generative models" or "probabilistic generative models," we want to note to readers concerned about the important distinctions between these terms that we are also not insensitive to them. A probabilistic generative model is not the only kind of world model an agent can have—humans surely have others. And just because a person has a certain world model in their head does not mean that epistemologists should genuinely credit them with "knowledge"—surely many aspects of human models are false or otherwise fail to qualify.

However, we do want to defend the claim that the mind's world models as we study them here, if they indeed follow roughly the form and function that we lay out in this book, do constitute genuine knowledge. Consider the classical definition of knowledge as "justified true belief" (which is not without its own problems; see Gettier, 1963). World models are mental representations, or beliefs. Built the way that people build them, we argue, they should come out to be true, or true enough. And they will do so in virtue of both their form and their function, as hierarchical probabilistic generative models brought to bear on a world of facts by learning and inference procedures that are rational and reasonably justified. So it seems permissible to call these models "knowledge." Like any realistic knowledge that any realistic agent has, they will be incomplete and imperfect in many ways. In particular they are approximate, probabilistic, and at best only probably, approximately true. They are designed only to support good guesses. But they are the best guesses we can make, and this book offers the best guesses we currently have about how they work.

1.2 From One Question to Many

The material in this book can be organized around a set of foundational questions about human beings' abstract knowledge of their world—its structure, use, and origins. We introduced these questions informally earlier in this chapter, but we will now lay them out more systematically.

We start with the three most basic questions that originally guided our own work on everyday inductive leaps, trying to account for how our minds get so much from so little:

Q1 How does abstract knowledge of the world guide learning and inference from sparse data?

Q2 What forms does our abstract world knowledge take across different domains and tasks?

Q3 How are world models themselves acquired or constructed?

Initially, we embraced and developed the Bayesian approach to learning and inference because it appeared to offer the best prospects for giving principled and unifying answers to these questions in ways that promised to generalize across many domains of cognition.[1] We focused on modeling archetypal inductive problems for human beings that are also classic problems that cognitive psychologists have studied in some form since the field's beginnings: learning concepts (or categories), learning causal relations, inductive reasoning, making predictions, and the like. We have already introduced some of these problems, and next we will give a preview of how the Bayesian framework has developed to explain them, specifically by answering questions Q1–Q3. This represents the core of our approach to reverse-engineering the mind and the focus of part I of this book.

Part II of the book represents more recent work, aimed at a number of more advanced but equally important questions:

Q4 How do we use our world models to make decisions and act in the world successfully?

Q5 How can learning and inference with complex world models be implemented efficiently in minds with bounded computational resources?

Q6 How are complex world models implemented in a physical machine, brain, or computer?

Q7 What are the origins of our world models in evolution and development—what is built into a baby's mind, and how do children learn within and beyond that starting point?

Q8 What is needed to scale up learning to all the knowledge that a human being acquires over their lifetime and human cultures have built over generations?

Extensions of the Bayesian framework to address these questions arose organically in response to our early efforts focused on Q1–Q3. In some cases, these were attempts to respond to pointed critiques or objections; in others, they came from our own thinking about what was most missing in our work. We recognized that having made some progress motivated by Q1–Q3, our tentative answers also raised new questions of their own, or suggested new approaches to classic questions of cognition that we hadn't seen how to take on before. We see it as a strength of the Bayesian paradigm (or any productive approach to reverse-engineering the mind) if in response to both external criticism and internal reflection, it does not remain stuck with a fixed set of questions to answer, but rather opens up new lines of inquiry that build on and enrich earlier ones. In trying to understand a subject as deep and as vast as the human mind, at this comparatively early stage of the inquiry, we make the most progress not when our initial questions are definitively and finally settled, but when the set of questions we can ask productively grows larger, more challenging, and more interesting.

We also came to believe that any approach to computational cognitive science (or cognitive AI) makes the most progress when it bridges concrete, quantitatively testable models of specific aspects of cognition with a broader account of how the mind works as a whole, as well as how it sometimes fails to work. Our earlier work on Q1–Q3 prioritized the specific

1. Some readers may notice a resemblance between our questions about world knowledge and the questions that Noam Chomsky has often used to frame research on language (e.g., Chomsky, 1986). This is no coincidence: language involves structured representations and challenging inductive problems, factors that we see as being key to understanding human cognition more broadly. Other work has drawn similar inspiration from Chomsky's questions in studying aspects of higher-level cognition (e.g., Mikhail, 2008, on morality).

and the value of elegant, precise, quantitatively testable models, but Q4–Q8 seemed to us the most urgent questions to address to achieve this larger integration. We know that we are still very far from the full account of the mind we all seek, but nonetheless, the last decade has seen great progress on these more advanced questions and made it possible to take seriously the idea that such an account might be in reach. These developments also represent some of the most active areas of current research and are the focus of part II.

In the remainder of this first chapter, we want to give you a taste of the Bayesian approach by way of a preview of how it approaches all these questions. We focus on Q1–Q3, as the foundation of our approach, and then move more quickly through Q4–Q8. Along the way, we will also include pointers to specific later chapters where we take on each question in detail, as a guide to the reader.

We will illustrate the approach with examples drawn chiefly from two archetypal inductive problem domains: learning and reasoning about natural kind concepts and learning and reasoning about causal relations. We will also contrast the Bayesian approach developed in this book with two other classic ways of thinking about the nature and origins of human knowledge: the modern cognitive inheritors of the rationalism of René Descartes and Immanuel Kant and the empiricism of John Locke and David Hume. The history of these ideas and their debates with each other is one of philosophy's great rivalries. Cognitive scientists and AI researchers have forcefully joined both sides of this debate, including, on the rationalist side, various versions of linguistic, conceptual, and evolutionary nativism (Pinker, 1997; Fodor, 1998; Spelke, 1990; Leslie, 1994; Spelke & Kinzler, 2007; Chomsky, 2015; Marcus & Davis, 2019); and, on the empiricist side, both the associationist streak in classic connectionist models (McClelland & Rumelhart, 1986; Elman et al., 1996; McClelland et al., 2010) as well as contemporary AI's deep reinforcement learning systems and very large sequence-learning learning models (Silver et al., 2016; Silver, Singh, Precup, & Sutton, 2021; LeCun, 2022; Brown et al., 2020; Alayrac et al., 2022). Bayesian models of cognition have appealed to many by offering an alternative to these two poles, with genuinely different ways to think about the foundational issues at stake. We discuss the history of these various approaches, and how they relate to the Bayesian approach, in chapter 2.

At its heart, the approach that we present in this book combines richly structured, expressive representations of the world with powerful statistical inference mechanisms, arguing that only a synthesis of sophisticated approaches to both knowledge representation and inductive inference can account for human intelligence. Until recently, it was not understood how this fusion could work computationally. Cognitive modelers were forced to choose between two alternatives (Pinker, 1997): powerful statistical learning operating over the simplest, unstructured forms of knowledge, such as matrices of associative weights in connectionist accounts of semantic cognition (McClelland & Rumelhart, 1986; Rogers & McClelland, 2004), or richly structured symbolic knowledge equipped with only the simplest, nonstatistical forms of learning, checks for logical inconsistency between hypotheses and observed data, as in nativist accounts of language acquisition (Niyogi & Berwick, 1996). It appeared necessary to accept either that people's abstract knowledge is not learned or induced in a nontrivial sense from experience (hence essentially innate), or that it is not nearly as abstract or structured—as "knowledge like"—as it seems (hence simply associations). Many developmental researchers rejected this choice altogether and pursued less formal approaches to describing the growing minds of children, under the headings of "constructivism" or the "theory theory" (Gopnik & Meltzoff, 1997). The potential to explain how people can genuinely learn with abstract structured knowledge may be the most salient feature of Bayesian cognitive models—the biggest reason for their popularity in some developmental circles (Gopnik & Tenenbaum, 2007; Griffiths, Chater, Kemp, Perfors, & Tenenbaum, 2010; Perfors, Tenenbaum, & Wonnacott, 2010; Griffiths, Sobel, Tenenbaum, & Gopnik, 2011b; Xu, 2019; Spelke, 2022), and the biggest target of skepticism from others, in both the traditional nativist and empiricist camps (Berwick, Pietroski, Yankama, & Chomsky, 2011; McClelland et al., 2010).

1.2.1 The Role of Abstract Knowledge

Over the last two decades, many aspects of higher-level cognition have been illuminated by the mathematics of Bayesian statistics: our sense of similarity (Tenenbaum & Griffiths, 2001a), representativeness (Tenenbaum & Griffiths, 2001b), and randomness (Griffiths & Tenenbaum, 2001; Griffiths, Daniels, Austerweil, & Tenenbaum, 2018); coincidences as a cue to hidden causes (Griffiths & Tenenbaum, 2007a), judgments of causal strength (Lu, Yuille, Liljeholm, Cheng, & Holyoak, 2008) and evidential support (Griffiths & Tenenbaum, 2005); diagnostic and conditional reasoning (Krynski & Tenenbaum, 2007; Oaksford & Chater, 2001); and predictions about the future of everyday events (Griffiths & Tenenbaum, 2006).

The claim that human minds learn and reason according to Bayesian principles does not mean that the mind can implement any Bayesian inference. Only those inductive computations that the mind is designed to perform well, where biology has had time and cause to engineer effective and efficient mechanisms, are likely to be understood in Bayesian terms. In addition to the general cognitive abilities just mentioned, Bayesian analyses have shed light on many specific capacities that result from rapid, reliable, and unconscious processing, including perception (Yuille & Kersten, 2006), language (Chater & Manning, 2006), memory (Shiffrin & Steyvers, 1997; Steyvers, Griffiths, & Dennis, 2006), and sensorimotor systems (Körding & Wolpert, 2004). In contrast, in tasks that require explicit conscious manipulations of probabilities as numerical quantities—a recent cultural invention that few people become fluent with, and only then after sophisticated training—judgments can be notoriously biased away from Bayesian norms (Tversky & Kahneman, 1974).

In essence, *Bayes' rule* is simply a tool for answering Q1: How does abstract knowledge guide learning and inference from incomplete data? Abstract knowledge is encoded in a *probabilistic generative model*, a kind of mental model that describes the causal processes in the world that give rise to the learner's observations, as well as unobserved *latent variables* that support effective prediction and action—if the learner can infer their hidden state. Generative models must be probabilistic to handle the learner's uncertainty about the true states of latent variables and the true causal processes at work. A generative model is abstract in two senses: it describes not only the specific situation at hand but also a broader class of situations over which learning should generalize, and it captures in parsimonious form the essential world structure that causes learners' observations and makes generalization possible.

Bayesian inference gives a rational framework for updating beliefs about latent variables in generative models given observed data (Jaynes, 2003; Mackay, 2003). Background knowledge is encoded through a constrained space of hypotheses $\mathcal{H}$ about possible values for the latent variables—candidate world structures that could explain the observed data. Finer-grained knowledge comes in the *prior probabilities* $P(h)$ that specify the learner's degree of belief in each hypothesis h prior to (or independent of) the observations. Bayes' rule updates these prior probabilities to *posterior probabilities* $P(h|d)$ conditional on the observed data d:

$$P(h|d) = \frac{P(d|h)P(h)}{\sum_{h' \in \mathcal{H}} P(d|h')P(h')} \propto P(d|h)P(h). \tag{1.1}$$

The posterior probability is proportional to the product of the prior probability and the *likelihood* $P(d|h)$, measuring how probable the data are under hypothesis h, relative to all other hypotheses h' in $\mathcal{H}$.

To illustrate Bayes' rule in action, suppose that we observe John coughing (d) and we consider three hypotheses as explanations: h_1, John has a cold; h_2, lung disease; or h_3, heartburn. Intuitively, only h_1 seems compelling. Bayes' rule explains why. The likelihood favors h_1 and h_2 over h_3: only colds and lung disease cause coughing, and thus elevate the probability of the data above baseline. The prior, in contrast, favors h_1 and h_3 over h_2: colds and heartburn are much more common than lung disease. Bayes' rule weighs hypotheses according to the product of priors and likelihoods and thus yields only explanations like h_1, which score highly on both terms. We provide a more technical introduction to Bayes' rule—and the range of problems it can be used to solve—in chapter 3.

Bayesian inference can be used to explain how people learn from sparse data. In concept learning, the data might correspond to several example objects (as in figure 1.1) and the hypotheses to possible extensions of the concept. Why, given three examples of different kinds of horses, would a child generalize the word "horse" to all, and only, horses (h_1)? Why not h_2, "all horses except Clydesdales," h_3, "all animals," or any other rule consistent with the data? Likelihoods favor the more specific patterns, h_1 and h_2; it would be a highly suspicious coincidence to draw three random examples that all fall within the smaller sets h_1 or h_2 if they were actually drawn from the much larger h_3 (Tenenbaum & Griffiths, 2001a). The prior favors h_1 and h_3, because as more coherent and distinctive categories, they are more likely to be the referents of common words in language (Bloom, 2000). Only h_1 scores highly on both terms. Likewise, in causal learning, the data could be cooccurences between events; and the hypotheses, possible causal relations linking the events. Likelihoods favor causal links that make the co-occurence more probable, while priors favor links that fit our background knowledge of what kinds of events are likely to cause which others; for example, a disease (e.g., *cold*) is more likely to cause a symptom (e.g., *coughing*) than the other way around.

Viewed in these terms, Bayes' rule provides a simple quantitative framework for answering questions about inductive inference. In particular, it indicates how prior knowledge should be combined with data to yield new conclusions. This then sets up a research program where we can work backwards from the data that people observe and the conclusions that they reach to the knowledge that must have informed their inferences. This project of characterizing people's inductive biases, expressed in the form of prior distributions, characterized much of our early work on Bayesian models of cognition (e.g., Tenenbaum, 2000; Tenenbaum & Griffiths, 2001a; Griffiths & Tenenbaum, 2001, 2006). Along the way, we also developed sophisticated experimental methods for revealing human prior distributions based on algorithms used in statistics and computer science (see chapter 10). For example, under certain assumptions, information that is passed from person to person—as in the game of Telephone—will converge to a form that reflects the inductive biases of the people involved (Griffiths & Kalish, 2007). These methods make it possible to quantify people's abstract knowledge across a wide range of problems, giving us the ingredients for forming generalizations about its form and origins.

1.2.2 The Form of Abstract Knowledge

Abstract knowledge provides essential constraints for learning, but in what form? This brings us to the next question, Q2: What forms does our abstract world knowledge take across different domains and tasks? For complex cognitive tasks such as concept learning or causal reasoning, it is impossible to simply list every logically possible hypothesis along with its prior and likelihood. Some more sophisticated forms of knowledge representation must underlie the probabilistic generative models needed for Bayesian cognition.

In traditional associative or connectionist approaches, statistical models of learning were defined over large numerical vectors. Learning was seen as estimating strengths in an associative memory, weights in a neural network or parameters of a high-dimensional, nonlinear function (McClelland & Rumelhart, 1986; Rogers & McClelland, 2004). Bayesian cognitive models, in contrast, have had the most success with defining probabilities over more structured symbolic forms of knowledge representations used in computer science and AI, such as graphs, grammars, predicate logic, relational schemas, and functional programs. Different forms of representation are used to capture people's knowledge in different domains and tasks, as well as at different levels of abstraction.

In learning words and concepts from examples, the knowledge that guides both children's and adults generalizations has been well described using probabilistic models defined over tree-structured representations (as in figure 1.1b) (Xu & Tenenbaum, 2007). Reasoning about other biological concepts for natural kinds—for example, given that cows and rhinos have protein X in their muscles, how likely is it that horses or squirrels do?—is also well described by Bayesian models that assume nearby objects in the tree are likely to share properties (Kemp & Tenenbaum, 2009). However, trees are by no means a universal representation. Inferences about other kinds of categories or properties are best captured using probabilistic models with different forms (figure 1.2): two-dimensional (2D) spaces or grids for reasoning about geographic properties of cities, one-dimensional (1D) orders for reasoning about values or abilities, or directed networks for causally transmitted properties of species (e.g., diseases) (Kemp & Tenenbaum, 2009).

Knowledge about causes and effects more generally can be expressed in a *directed graphical model* (Pearl, 1988): a graph structure where nodes represent variables and directed edges between nodes represent probabilistic causal links (see chapter 4). In a medical setting, for instance (see figure 1.3a), nodes might represent whether a patient has a cold, a cough, a fever, or another condition, and the presence or absence of edges show that colds tend to cause coughing and fever, but not chest pain; lung disease tends to cause coughing and chest pain, but not fever; and so on.

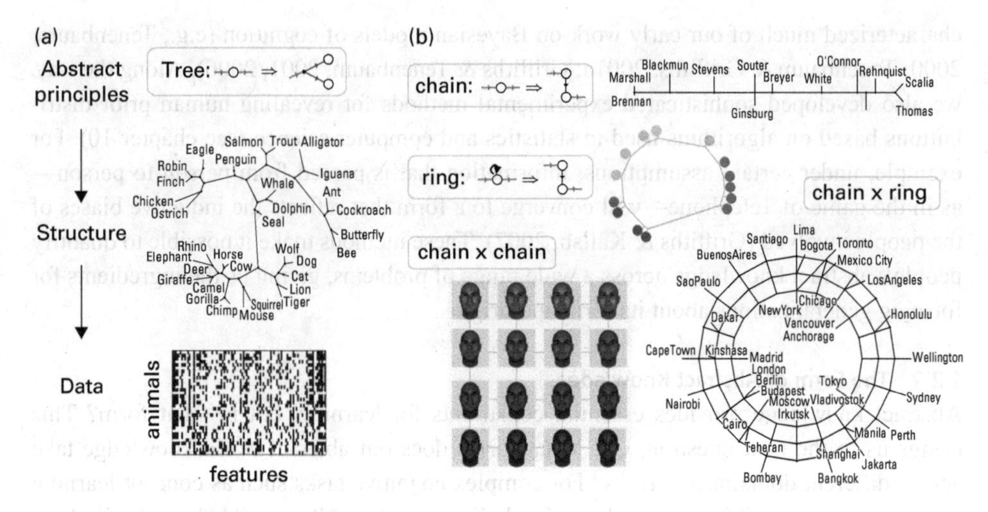

Figure 1.2
Kemp and Tenenbaum (2008) showed how the form of structure in a domain can be discovered using a hierarchical Bayesian model defined over graph grammars. At the bottom level of the model is a matrix of objects and their properties, or similarities between pairs of objects. Each cell of the matrix represents whether a given feature (column) is observed for a given object (row). One level up is a graph describing how properties are distributed over objects. Intuitively, objects nearby in the graph are expected to share properties. At the highest level, grammatical rules specify the form of structure in the domain—rules for growing graphs of a constrained form out of an initial seed node. A search algorithm attempts to find the combination of a form grammar and graph generated by that grammar that jointly receive highest probability. (a) Given observations about the features of animals, the algorithm infers that a tree structure best explains the data. The best tree found captures intuitively sensible categories at multiple scales. (b) The same algorithm discovers that the voting patterns of historical US Supreme Court justices are best explained by a linear "left-right" spectrum. (c) Subjective similarities among colors are best explained by a circular ring. (d) Given proximities between cities on the globe, the algorithm discovers a cylindrical representation analogous to latitude and longitude: the cross product of a ring and a ring. (e) Given images of realistically synthesized faces varying in two dimensions, race and masculinity, the algorithm successfully recovers the underlying 2D grid structure: a cross-product of two chains. Figure adapted from Tenenbaum et al. (2011).

Such a "causal map" represents a simple kind of intuitive theory (Gopnik et al., 2004), but learning causal networks from limited data depends on the constraints of more abstract knowledge (Griffiths & Tenenbaum, 2009). For example, learning causal dependencies between medical conditions is enabled by a higher-level *framework theory* (Wellman & Gelman, 1992) specifying two classes of variables (or nodes)—*diseases* (D) and *symptoms* (S)—and the tendency for causal relations (or graph edges) to run from D to S, rather than within these classes, or from S to D (figure 1.3a–c). This abstract framework can be represented using probabilistic models defined over relational data structures such as *graph schemas* (Kemp, Tenenbaum, Niyogi, & Griffiths, 2010b), templates for graphs based on types of nodes, or probabilistic *graph grammars* (Griffiths & Tenenbaum, 2007b), similar in spirit to the probabilistic grammars for strings that have become standard for representing linguistic knowledge (Chater & Manning, 2006) (see chapters 16 and 17). At the most abstract level, the very concept of causality itself, in the sense of a directed relationship that supports intervention or manipulation by an external agent (Woodward, 2003), can be formulated as a set of logical laws expressing constraints on the structure of directed graphs relating actions and observable events (figure 1.3d).

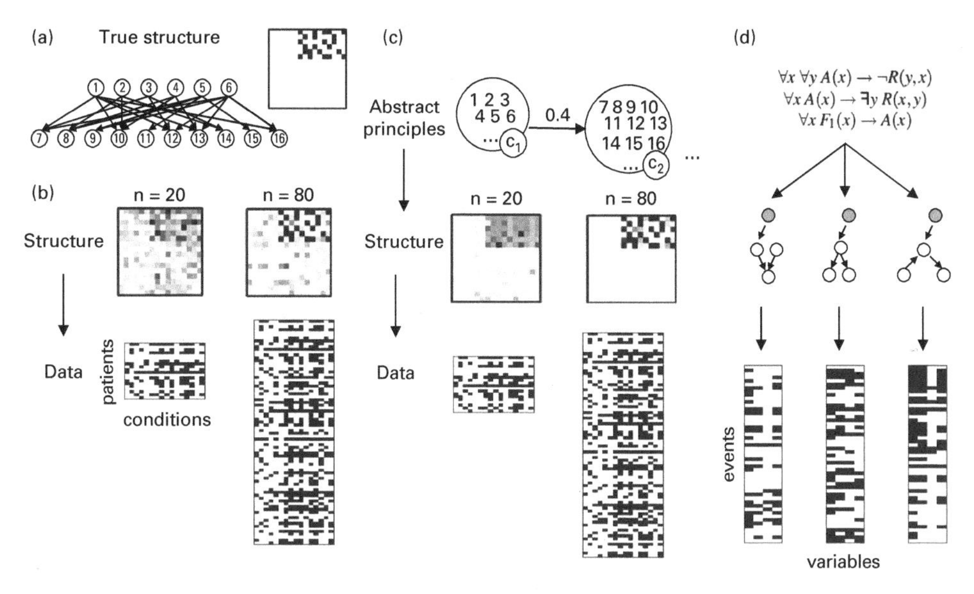

Figure 1.3
Hierarchical Bayesian models (HBMs) defined over graph schemas can explain how intuitive theories are acquired and used to learn about specific causal relations from limited data (Kemp et al., 2010b). (a) A simple medical reasoning domain might be described by relations among 16 variables: the first 6 encode presence or absence of "diseases" (upper row), with causal links to the next 10 "symptoms" (lower row). This network can also be visualized as a matrix (upper right, links shown in black). The causal learning task is to reconstruct this network based on observing data on the states of these variables in a set of patients. (b) A two-level HBM formalizes bottom-up causal learning, or learning with an uninformative prior on networks. The bottom level is the data matrix. The second level (structure) encodes hypothesized causal networks: a grayscale matrix visualizes the posterior probability that each pairwise causal link exists, conditioned on observing n patients; compare this matrix with the black-and-white ground truth matrix shown in (a). With $n = 80$, spurious links (gray squares) are inferred, and with $n = 20$, almost none of the true structure is detected. (c) A three-level nonparametric HBM (Mansinghka, Kemp, Tenenbaum, & Griffiths, 2006) adds a level of abstract principles, a graph schema. The schema encodes a prior on the level below (causal network structure) that constrains and thereby accelerates causal learning. Both schema and the network structure are learned from the same data observed in (b). The schema discovers the disease-symptom framework theory by assigning variables 1–6 to class C_1, variables 7–16 to class C_2, and a prior favoring only $C_1 \rightarrow C_2$ links. These assignments, along with the effective number of classes (here, two), are inferred automatically. Although this three-level model has many more degrees of freedom than the model in (b), learning is faster and more accurate. With $n = 80$ patients, the causal network is identified nearly perfectly. Even $n = 20$ patients are sufficient to learn the high-level $C1 \rightarrow C2$ schema, and thereby to limit uncertainty at the network level to just the question of which diseases cause which symptoms. (d) An HBM for learning an abstract theory of causality (Goodman et al., 2011). At the highest level are laws expressed in first-order logic representing the abstract properties of causal relationships, the role of exogenous interventions in defining the direction of causality, and features that may mark an event as an exogenous intervention. These laws place constraints on possible directed graphical models at the level below, which in turn are used to explain patterns of observed events over variables. Given observed events from several causal systems and a hypothesis space of possible laws at the highest level, the model converges quickly on a correct theory of intervention-based causality and uses that theory to constrain inferences about specific causal networks. Figure adapted from Tenenbaum et al. (2011).

Each of these forms of knowledge makes different kinds of prior distributions natural to define and therefore imposes different constraints on induction. Successful generalization depends on getting these constraints right. While inductive constraints are often graded, it is easiest to appreciate the effects of qualitative constraints that simply restrict the hypotheses that learners can consider (i.e., setting priors for many logical possible hypotheses to zero). For instance, in learning concepts over a domain of n objects, there are 2^n subsets, and hence 2^n logically possible hypotheses for the extension of a novel concept. Assuming that concepts correspond to the branches of a specific binary tree over the objects, as in figure 1.1b, restricts this space to only $n-1$ hypotheses. In learning a causal network over 16 variables, there are roughly 10^{46} logically possible hypotheses (directed acyclic graphs), but a framework theory restricting hypotheses to bipartite disease-symptom graphs reduces this to roughly 10^{23} hypotheses. Knowing which variables belong to the disease and symptom classes further restricts this to roughly 10^{18} networks. The smaller the hypothesis space, the more accurately a learner can be expected to generalize, but only so long as the true structure to be learned remains within or near (in a probabilistic sense) the learner's hypothesis space (McAllester, 1998). It is no coincidence, then, that our best accounts of people's mental representations often resemble simpler versions of how scientists represent the same domains, such as tree structures for biological species. A compact description that approximates how the grain of the world actually runs offers the most useful form of constraint on inductive learning.

1.2.3 The Origins of Abstract Knowledge

The need for abstract knowledge—and the need to get it right—brings us to Q3: How do learners learn what they need to know to make learning possible? How does a child know which tree structure is the right way to organize hypotheses for word learning? At a deeper level, how can a learner know that a given domain of entities and concepts should be represented using a tree at all, as opposed to a low-dimensional space or some other form? Or in causal learning, how do people come to correct framework theories such as knowledge of abstract *disease* and *symptom* classes of variables with causal links from diseases to symptoms?

The acquisition of abstract knowledge or new inductive constraints is primarily the province of cognitive development (Gopnik & Meltzoff, 1997; Carey, 2009). For instance, children learning words initially assume a flat, mutually exclusive division of objects into nameable clusters; only later do they discover that categories should be organized into tree-structured hierarchies (as in figure 1.1) (Markman, 1989). Such discoveries are also pivotal in scientific progress: Dmitri Mendeleev launched modern chemistry with his proposal of a periodic structure for the elements, and Carl Linnaeus famously proposed that relationships between biological species are best explained by a tree structure rather than a simpler linear order (premodern Europe's "great chain of being") or some other form.

Such structural insights have long been viewed by psychologists and philosophers of science as deeply mysterious in their mechanisms—more magical than computational. Conventional algorithms for unsupervised structure discovery in statistics and machine learning—including hierarchical clustering, principal component analysis, multidimensional scaling, and clique detection—assume a single fixed form of structure (Shepard, 1980). Unlike human children or scientists, they cannot learn multiple forms of structure or discover new forms in novel data. Neither traditional approach to cognitive development has a fully satisfying response: nativists have assumed that if different domains of cognition are represented in qualitatively different ways, those forms must be innate (Chomsky, 1980; Atran, 1998); on the other hand, connectionists have suggested these representations may be learned, but in a generic system of associative weights that at best only approximates

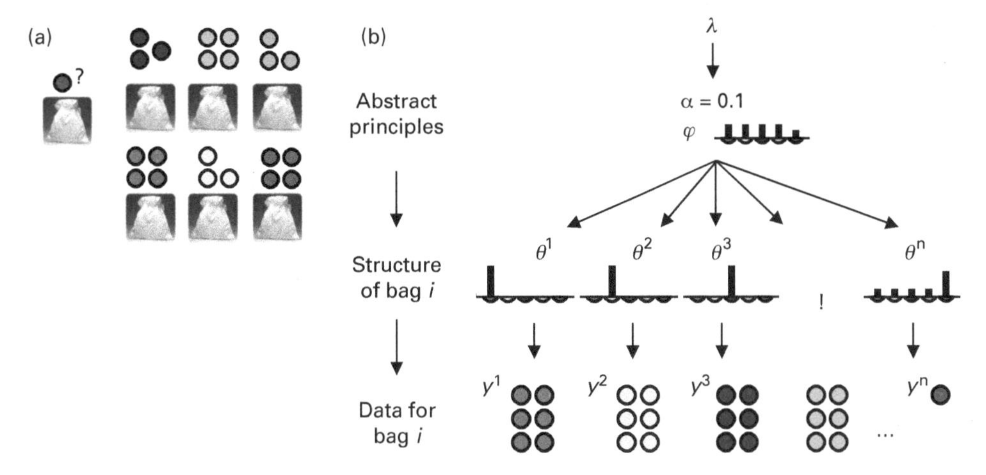

Figure 1.4
A thought experiment to illustrate transfer learning. (a) On its own, observing a single blue marble drawn from a bag tells us little about the colors of other marbles that we are likely to draw. But if we have also observed several draws from several other bags of the same type, an *overhypothesis* can arise: bags are homogeneous in color, although color varies across bags. (b) A hierarchical Bayesian model explains how this overhypothesis can be acquired and suggests that strong generalizations can be made about a new bag's distribution from just one example.

trees, causal networks, and other forms of structure that people appear to know explicitly (Rogers & McClelland, 2004).

Bayesian cognitive modelers have answered these challenges by combining the structured knowledge representations described in this chapter with advanced methods from Bayesian statistics known as *hierarchical Bayesian models* (HBMs; Gelman, Carlin, Stern, & Rubin, 1995). HBMs address the origins of hypothesis spaces and priors by positing not just a single level of hypotheses to explain the data, but multiple levels: hypothesis spaces of hypothesis spaces, with priors on priors. Bayesian inference across all levels allows hypotheses and priors needed for a specific learning task to themselves be learned at larger or longer-time scales, at the same time as they constrain lower-level learning (see chapter 8).

In machine learning and AI, HBMs have primarily been used for *transfer learning* or *learning to learn*: the acquisition of inductive constraints from experience in previous related tasks (Kemp et al., 2007). This idea is reflected in the contemporary machine learning literature on *metalearning* (see chapter 12). Learning to learn is central to human cognition as well, and was the first prominent application of HBMs in cognitive modeling. Figure 1.4 illustrates the basic mathematics, while figure 1.5 illustrates a classic example from the study of human cognitive development that has been modeled with this approach, the acquisition of the shape bias in early word learning.

HBMs have been especially valuable for cognitive scientists seeking to explain how people can acquire the different forms of abstract knowledge appropriate to different domains. Kemp and Tenenbaum (2008, 2009) showed how HBMs defined over graph- and grammar-based representations can discover the form of structure governing similarity in a domain. Structures of various forms—trees, clusters, spaces, rings, orders, and others—can all be represented as graphs, while the abstract principles underlying each form are expressed

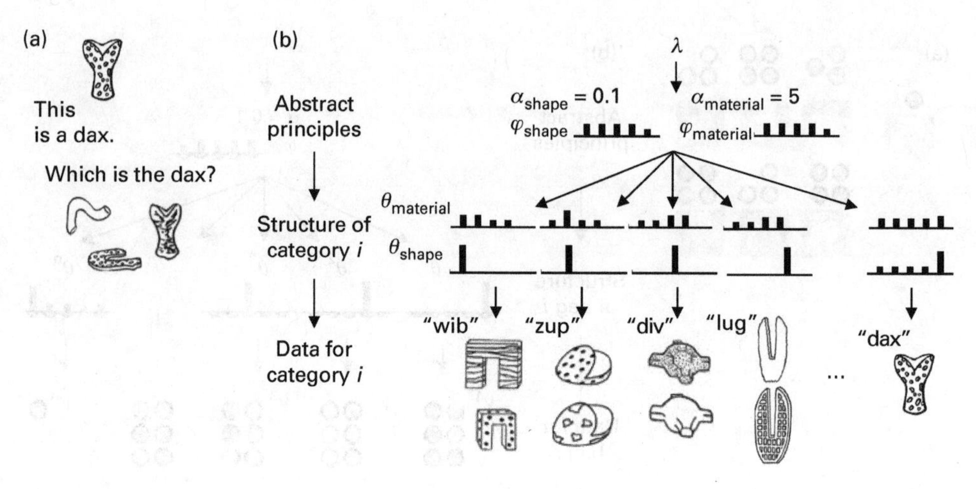

Figure 1.5
Shape bias and hierarchical Bayesian inference. (a) A standard paradigm for demonstrating the shape bias in two-year-olds (Jones & Smith, 2002). Given one example of a "dax," with a novel shape, material, and texture, children asked to find another dax choose an object matching based on shape over other objects matching along other dimensions. (b) A hierarchical Bayesian model can explain how the shape bias emerges as a learned overhypothesis from very little data (Kemp et al., 2007): just two examples from each of four named categories of novel objects ("wibs," "zups," "divs," and "lugs"). The HBM makes inferences about separate parameters describing the shape and material properties of objects. Both dimensions are inferred to vary broadly over the objects as a whole, but shape is inferred to be homogeneous within categories while material is inferred to be heterogeneous within categories. Hence strong one-shot generalization for novel categories is expected only along the shape dimension.

as simple grammatical rules for growing graphs of that form. Embedded in a hierarchical Bayesian framework, this approach can discover the correct forms of structure (the grammars) for many real-world domains, along with the best structure (the graph) of the appropriate form, as shown in figure 1.2. In particular, it can infer that a hierarchical organization for the novel objects in figure 1.1a (as in figure 1.1b) better fits the similarities that people see in these objects, compared to alternative representations such as a 2D space.

Hierarchical Bayesian models can also be used to learn abstract causal knowledge, such as the framework theory of diseases and symptoms (figure 1.3) and other simple forms of intuitive theories (Kemp et al., 2010b). Mansinghka et al. (2006) showed how a *graph schema* representing two classes of variables, *diseases* (*D*) and *symptoms* (*S*), and a preference for causal links running from *D* to *S* variables can be learned from the same data that support learning causal links between specific diseases and symptoms—and learned just as fast or faster (figure 1.3b,c). The learned schema in turn dramatically accelerates the learning of specific causal relations (the directed graph structure) at the level below. Getting the big picture first—discovering that *diseases* cause *symptoms* before pinning down any specific disease-symptom links—and then using that framework to fill in the gaps of specific knowledge is a distinctively human mode of learning. That mode figures prominently in children's development and scientific progress but has not previously fit into the landscape of rational or statistical learning models.

While this HBM imposes strong and valuable constraints on the hypothesis space of causal networks, it is also extremely flexible: it can discover framework theories defined by any number of variable classes and any pattern of pairwise regularities on how variables in these classes tend to be connected. Not even the number of variable classes (two for the disease-symptom theory) need be known in advance. This is enabled by another state-of-the-art Bayesian tool, known as *infinite models* or *nonparametric Bayesian models* (see chapter 9). These models posit an unbounded amount of structure, but only finitely many degrees of freedom are actively engaged for a given data set (Rasmussen, 2000). An automatic preference for simplicity embodied in Bayesian inference trades off model complexity and fit to ensure that new structure—in this case, a new class of variables—is introduced only when the data truly require it.

The specific nonparametric distribution on node classes in figure 1.3c is a *Chinese restaurant process* (CRP), which has been much more broadly influential across many areas of machine learning and cognitive modeling. CRP models have given the first principled account of how people form new categories without direct supervision—including the number of categories when that is unknown (Anderson, 1991a; Sanborn, Griffiths, & Navarro, 2010a; Griffiths, Sanborn, Canini, & Navarro, 2008c). As each object is observed, CRP models infer whether that object is best explained by assimilation to an existing category or by positing a previously unseen category, as shown in figure 1.6. This is the basic problem that any lifelong learner faces in making sense of a world of unknown complexity. Seeing the first instance of a new kind of thing, how can you know if it is really a new kind, or merely a strange instance of a familiar kind? For example, upon seeing a zebra or a camel for the first time, you could try to understand it as a horse with black and white stripes or a horse with two humps. The trade-off in different probabilistic pressures captured by the CRP enables a principled and computationally efficient way of answering this question.

The fundamental principle behind nonparametric Bayesian models—assuming that you have seen only a fraction of the infinite complexity of the world—can be applied to other aspects of learning. One of the strengths of Bayesian models is that the underlying generative models can be combined to create new models. If we assume that not just objects but the properties of those objects form clusters, we obtain a new model: the *CrossCat* model (Shafto, Kemp, Mansinghka, & Tenenbaum, 2011), illustrated in figure 1.7. In the real world, with all its complexity, a single way of categorizing objects is rarely sufficient. How different kinds of properties arise may be best explained by different ways of categorizing the objects. Earlier categorization models typically ignored this complexity, but CrossCat captures it by discovering clusters that capture diferent views of these different subclasses of properties. A nonparametric Bayesian formulation allows CrossCat to automatically discover both the number of views and the number of categories within each view, and also allows structure at both levels to grow as new objects, properties, or both are encountered.

CRPs can be embedded in probabilistic models for many other settings. For instance, in language, they have been invoked to explain how children discover words in unsegmented speech (Goldwater, Griffiths, & Johnson, 2009), learn morphological rules (Johnson, Griffiths, & Goldwater, 2007b), and organize word meanings into hierarchical semantic networks (Griffiths, Steyvers, & Tenenbaum, 2007; Blei, Griffiths, & Jordan, 2010) (figure 1.8). Such tree-structured semantic networks have long been posited by cognitive psychologists studying the structure and dynamics of human memory. Only with the advent of sophisticated nonparametric Bayesian models, however, have we been able to explain how people

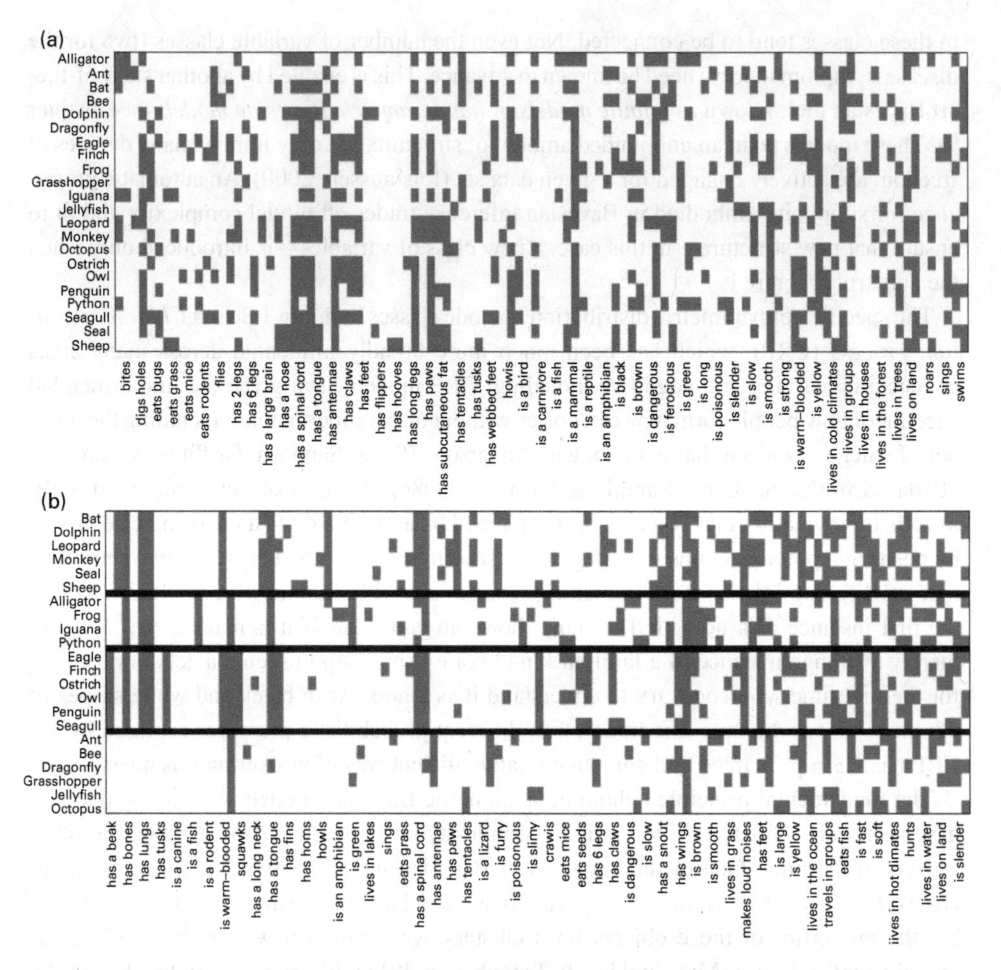

Figure 1.6
Illustration of unsupervised categorization with a Chinese restaurant process (CRP). (a) The data consist of a matrix of animal species (rows) and their anatomical, physiological, behavioral, and ecological attributes (columns). Only every other column is labeled. (b) A Bayesian model using a CRP as a prior infers that the animals are best divided into four broad taxonomic categories corresponding roughly to mammals, reptiles and amphibians, birds, and insects and invertebrates. Both the number of categories and the specific assignments of animals to categories are inferred automatically. Relative to (a), rows have been reordered to reflect these four categories, and columns have been reordered according to how well they are explained by the model—how "clean" they are within categories. Figure adapted from Shafto et al. (2011).

can discover such a mental model from unsupervised linguistic experience. A related but novel nonparametric construction, the *Indian buffet process* (IBP) explains how new perceptual features can be constructed during object categorization (Griffiths & Ghahramani, 2006; Austerweil & Griffiths, 2011).

More generally, nonparametric Bayesian models address the principal challenge that human learners face as knowledge grows over a lifetime: balancing constraint and flexibility, or the need to restrict the hypotheses available for generalization at any moment, with the capacity to expand one's hypothesis spaces, to learn new ways that the world could work. Placing nonparametric distributions at higher levels of a hierarchical model yields flexible

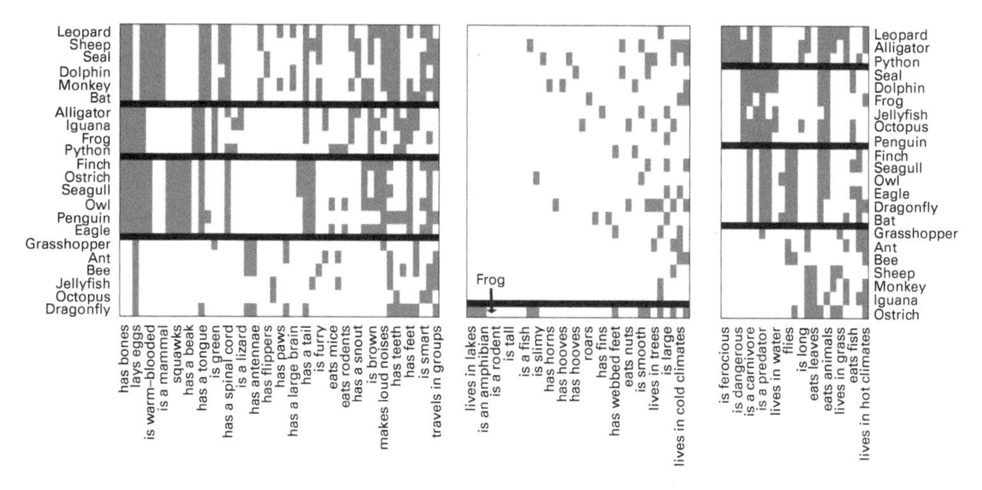

Figure 1.7
Illustration of unsupervised discovery of multiple categorization "views" using the CrossCat model (Shafto et al., 2011). Given the animal-property data from figure 1.6, CrossCat finds three views, shown in three panels here. The left view corresponds exactly to the taxonomic categorization found by the Chinese restaurant process in figure 1.3b. Primarily core biological properties are associated with this view: anatomical and physiological features such as "has bones," "lays eggs," "is warm-blooded," "has a beak," and so on. The middle view is composed of a number of idiosyncratic properties—features which in this data set are true of few species and do not covary in a coherent way with many other properties. The right view is most interesting, as it captures an intuitively compelling and coherent structure that is completely missed by the CRP and is likely to be important for many natural reasoning tasks. Here, the species are categorized according to ecological niches and predator/prey roles: from top to bottom, groups correspond to land predators, water species, air species, and land prey. Associated properties likewise are mostly ecologically relevant ones—"is ferocious," "is a carnivore," "lives in water," "flies," "eats animals" and so on—which are not well explained by the dominant taxonomic category structure of this domain but are well fit by this alternative way of organizing the species. As in figure 1.6, only every other column is labeled. Figure adapted from Shafto et al. (2011).

inductive biases for lower levels, while the Bayesian preference for simplicity ensures the proper balance of constraint and flexibility as knowledge grows.

Across several case studies of learning abstract knowledge—discovering structural forms, causal framework theories, and other inductive constraints acquired through transfer learning—it has been found that abstractions in HBMs can be learned remarkably fast, and from relatively little data compared to what is needed for learning at lower levels. This is because each degree of freedom at a higher level of the HBM influences—and pools evidence from—many variables at the levels below. This property of HBMs has been called the the *blessing of abstraction* (Goodman, Ullman, & Tenenbaum, 2011). It offers a top-down route to the origins of knowledge that contrasts sharply with the two classic approaches: nativism (Chomsky, 1986; Spelke, Breinlinger, Macomber, & Jacobson, 1992), in which abstract concepts are assumed to be present from birth; and empiricism or associationism (Rogers & McClelland, 2004), in which abstractions are constructed but only approximately, and only slowly, in a bottom-up fashion, by layering many experiences on top of each other and filtering out their common elements. Only HBMs thus seem suited to explaining the two most striking features of abstract knowledge in humans: that it can be learned from experience and that it can be engaged remarkably early in life, serving to constrain more specific learning tasks.

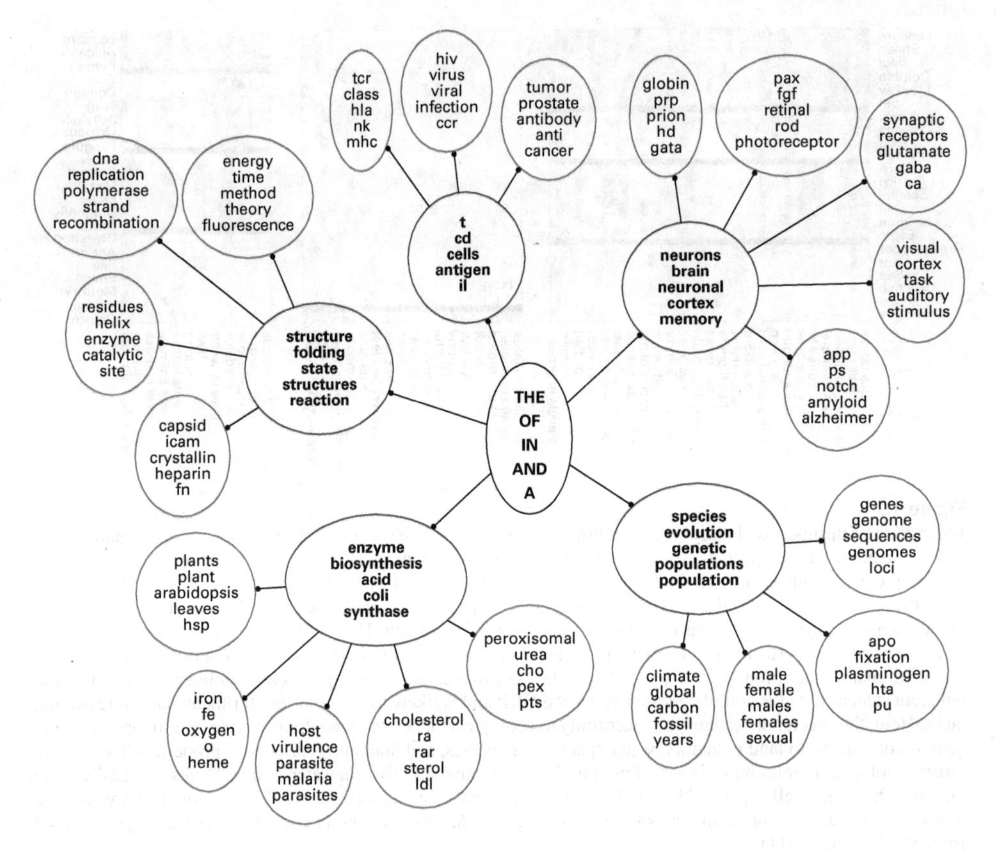

Figure 1.8
The *nested* Chinese restaurant process is the basis for a nonparametric hierarchical model that can automatically discover humanlike semantic networks from text corpora (Blei et al., 2010). The model's input is a large set of documents represented as the number of times that each word in the language occurs in that document. The output is a rooted tree of "topics," and a pointer for each document to one leaf node of the tree. Here, the text corpus consists of abstracts from the *Proceedings of the National Academy of Sciences*. A topic is a distribution over characteristic words with a common semantic theme, such as "neuroscience" or "evolutionary biology" here or, at a lower level of the tree, more fine-grained themes such as "synaptic plasticity" or "perception." Several most characteristic words for each topic are shown. The model attempts to explain the corpus data by fitting the observed distribution of words in each document with a combination of characteristic words from the topics lying on a path from the root to the leaf node containing that document. Figure adapted from Blei et al. (2010).

1.2.4 Using Knowledge to Inform Action

So far, we have been focused on the form that knowledge takes and how it is acquired. However, for any organism, the true value of knowledge is that it makes it possible to take better-informed actions. So how should we choose what to do? This brings us to Q4: How do we use our world models to make decisions and act in the world successfully?

One nice property of Bayesian models of cognition is that their grounding in Bayesian statistics provides an immediate answer to this question. For almost a century, statisticians and probabilists have been exploring the relationship between beliefs and actions, resulting in an extensive body of literature on *statistical decision theory* (see chapter 7). If people's preferences about the outcomes of their actions satisfy some simple rules, then those preferences can be captured by assigning a *utility* to each possible outcome (where higher utility indicates that the outcome is more desirable). A rational agent, with beliefs about the world

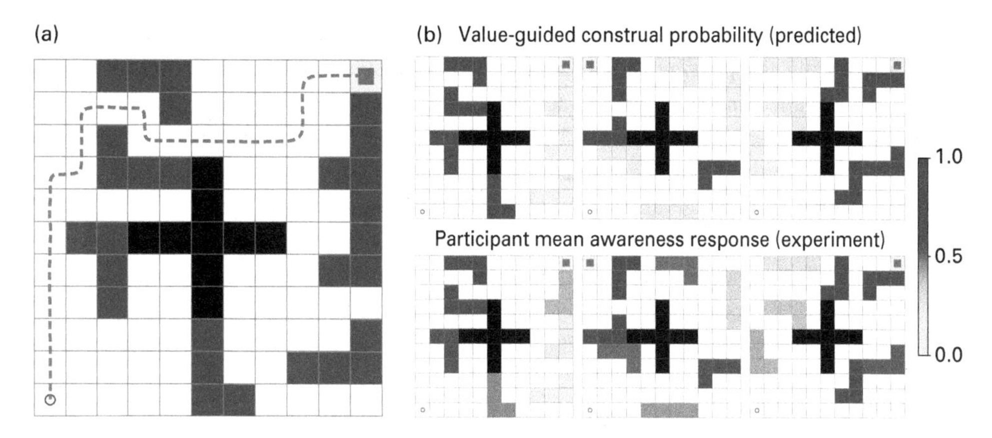

Figure 1.9
A sequential decision problem. (a) How can we get from the starting point (the bottom-left corner) to the goal (the green and yellow square in the top-right corner) as efficiently as possible while avoiding the obstacles shown in blue and black? This problem can be formalized as an MDP, where states correpond to the location of the agent, actions are the places the agent can move to, and rewards are a positive payoff for reaching the goal but there is a small cost for each step along the way. (b) People's representations of these problems can be explained in terms of the efficient use of cognitive resources. The top row shows which objects are represented by a model that tries to find a path while minimizing the number of objects considered. The bottom row shows how aware human participants were of those objects when asked questions about them after solving the navigation task. Figure adapted from Ho et al. (2022).

expressed in a probability distribution over hypotheses, seeks to maximize the expected utility of the outcomes resulting from their actions.

The principles of statistical decision theory extend beyond individual decisions, making it possible to characterize how an agent should carry out a sequence of interdependent actions. For example, consider the problem shown in figure 1.9a. The goal is to get to the yellow square in the top-right corner, starting at the square marked with a circle in the bottom-left corner. From each square, the available actions are to move to the square above, below, left, or right. However, the obstacles shown in blue and black cannot be traversed—trying to move through them results in staying in the same square. Each move that you make changes your position and hence changes the options available to you.

This kind of sequential decision problem can be described in terms of a *Markov decision process* (MDP), where the current position of the agent is encoded as the state of the system and actions (probabilistically) modify that state (described in more detail in chapter 7). Reaching the goal provides some reward, but each step that it takes to get there has a cost. The goal is to maximize expected long-term reward, summed across all the actions taken by the agent. Expressed in these terms, it is possible to use sophisticated algorithms such as dynamic programming to compute the optimal action in every state—the *policy* that the agent should follow in this situation.

MDPs provide a link between Bayesian models and two classic topics in psychology: problem solving and reinforcement learning. Problem solving has been a core topic in cognitive psychology since the pioneering work of Herbert Simon and Allen Newell, who characterized it in terms of search in an abstract "problem space": each possible action results in a new state that has another set of possible actions, resulting in a branching tree of decisions (Newell & Simon, 1972). For deterministic tasks, this formulation of the problem maps precisely onto the structure of an MDP. However, MDPs expand upon this idea by allowing the possibility that actions influence states only probabilistically—that there's some chance that an attempted action fails, for example. Even more uncertainty can be introduced in *partially observable Markov decision processes* (POMDPs), which allow the possibility that the agent doesn't actually know what state they are in but needs to infer that information from observable features. In the example shown in figure 1.9a, instead of starting in the bottom-left corner, the agent might start in a randomly selected square and have to identify that square based on limited perceptual information about nearby obstacles. This kind of uncertainty is intrinsic to many real-world problem-solving situations, and methods for solving MDPs and POMDPs thus provide us with a rich source of ideas about human cognition.

Reinforcement learning introduces another kind of uncertainty: so far, we have been assuming that agents know the consequences of their actions, but what if this too has to be learned? The reinforcement learning problem is one of trying to find the optimal policy for an MDP in the absence of information about how actions modify states or what costs or rewards might result. The agent has to acquire this information through experience. This can be done by following two quite different strategies: the *model-free* approach focuses on simply learning the best action to take in a given state, perhaps by estimating the rewards associated with each action; meanwhile, the *model-based* approach tries to estimate the probability distributions over states and rewards associated with each action and then find an optimal policy by solving the resulting MDP (Daw, Niv, & Dayan, 2005). These two strategies map to different aspects of human cognition. Model-free learning resembles the associative learning mechanisms that humans share with many other animals, forming an association between a state and an action. Model-based learning is more deliberative and provides an opportunity to use the kind of sophisticated tools for acquiring and representing knowledge discussed in this chapter.

At present, the reinforcement learning algorithms that are dominant in AI applications fall into the model-free category (broadly construed). These algorithms are based on using large neural networks to learn the relationship between a state and the reward associated with an action (Mnih et al., 2015) or to learn the policy directly (Sutton, Precup, & Singh, 1999). This approach has been influential in part because modern machine learning has focused on settings where it is possible to generate the large amounts of data that these systems need to learn to make good decisions. For example, for computer games, board games, or simulated environments, it is easy to get as much data as might be required by running the game or simulation as many times as needed. Human learners are able to learn new tasks from orders of magnitude less data because they are able to build explicit models of their environment and thus significantly constrain the learning problem. Being able to learn structured representations, such as simple programs directly from data, makes it possible to learn complex policies from limited experience (Tsividis et al., 2021).

1.2.5 Making the Most of Limited Cognitive Resources

Up to now, we have focused on Bayesian inference as the ideal solution to inductive problems. Used in this way, the Bayesian approach lets us make predictions about how any intelligent agent should go about solving these problems. Indeed, much of our work on Bayesian models of cognition was inspired by Roger Shepard's idea that we should be able to identify universal laws of cognition by thinking about the ideal solutions to the abstract problems that all intelligent agents need to solve (Shepard, 1987).

However, such ideal solutions are ideal not just in the sense of being optimal, but in the sense of being unachievable. The high computational cost of exact inference in large probabilistic models implies that human minds and brains can at best approximate Bayesian computations, just as in any working Bayesian AI system (Russell & Norvig, 2021). This sets up Q5: How can learning and inference with complex world models be implemented efficiently in minds with bounded computational resources?

One way to answer this question is to look to the algorithms used to approximate Bayesian inference in computer science and statistics (see chapter 6). From this perspective, the key research questions are: What approximate algorithms does the mind use? How do they relate to the engineering approximations in AI? And how are they implemented in neural circuits? Two distinct answers to these questions have been pursued. One line of research focuses on *Monte Carlo* or stochastic sampling-based approximations as a unifying framework for understanding how Bayesian inference may work practically across all these levels, in minds, brains, and machines (Sanborn, Griffiths, & Navarro, 2006; Brown & Steyvers, 2009; Fiser, Berkes, Orbán, & Lengyel, 2010; Vul, Goodman, Griffiths, & Tenenbaum, 2014). This approach is reviewed in chapter 11. Monte Carlo inference in richly structured models is possible (e.g., Goodman, Mansinghka, et al., 2008a) but very slow; constructing more efficient samplers is a major focus of current work.

Another line of research focuses on methods that approximate complex probabilistic computations by substituting probability distributions that are easier to work with. In particular, recent advances in methods for training artificial neural networks make it possible to train such networks to approximate arbitrary probability distributions. By training a neural network to approximate a posterior distribution over hypotheses given observed data, we can create an *amortized inference* system that pays a significant up-front cost (training) to reduce the cost of inference for new observations (Dasgupta & Gershman, 2021). This approach is summarized in chapter 12.

The availability of different algorithms for approximating Bayesian inference creates a new problem: How do we know which algorithm to use? Even if we commit to an algorithm—say, one that samples from the posterior distribution—how do we know how many samples to draw? Considering these questions requires thinking about the trade-off between the quality of an approximation and the time required to compute it—quantities that may be critical to any human decision. An early investigation of this tradeoff produced the perhaps surprising result that in many cases, drawing a single sample strikes the right balance (Vul et al., 2014). This *one and done* strategy has the side effect of explaining the curious observation that when people perform tasks that are modeled as Bayesian inference, the probabilities with which they select hypotheses often correspond closely to the posterior probabilities of those hypotheses. This kind of *probability matching* is exactly what we would expect if they are making a decision based on a single sample.

Considering this trade-off suggests an answer to another question about how we get so much from so little: How do we solve such a wide range of inductive problems in real time, with limited computational resources? The answer has to be that we use those limited resources well, making intelligent decisions about how and when to approximate. This idea has led to a wave of research exploring how well we can explain human behavior in terms of the rational use of cognitive resources (for an overview, see Lieder & Griffiths, 2020, and chapter 13 of this book). When applied to decision-making, this perspective provides a way to reconcile the *heuristics and biases* research program of Kahneman and Tversky (e.g., Tversky & Kahneman, 1974) with Bayesian models of cognition, defining a good heuristic as one that strikes the right balance between approximation quality and computational cost.

Making intelligent use of our cognitive resources doesn't just mean using efficient *algorithms*—the heuristics and strategies that we follow when making decisions. It also means using efficient *representations*. We can apply the same lens to another set of classic questions from the problem-solving literature: How do people choose what aspects of a problem to represent? How do they select meaningful subgoals? These questions can be analyzed in terms of how well a particular representation supports planning. For example, when people solve sequential decision problems like the one shown in figure 1.9a, they seem to minimize the number of objects they include in their representation of the problem. Figure 1.9b shows the predictions of a model assuming that each object is associated with a representational cost, and people trade off this cost against the quality of the plan that they are able to form (Ho et al., 2022).

1.2.6 From Abstract Models to Universal Languages and Their Physical Implementation

Methods like sampling and amortized inference provide us with a way to see how it may be possible for human minds and brains to perform the challenging computations involved in Bayesian inference. However, even these algorithms struggle when we consider problems that involve the full complexity of people's models of the world. This challenge is at the heart of Q6: How are world models implemented in a physical machine, brain, or computer?

This challenge results in part from our discovery that formalizing the full content of intuitive theories appears to require Turing-complete compositional representations, going even beyond probabilistic first-order logics (Milch, Marthi, & Russell, 2004; Kemp, Goodman, & Tenenbaum, 2008b), as we develop in chapter 17, to *universal probabilistic programming languages* (Goodman, Mansinghka, Roy, Bonawitz, & Tenenbaum, 2012) (chapter 18). Remarkably, the class of all computable probabilistic models and conditional distributions is easily defined in these languages, using a formalism known as the *stochastic λ-calculus*. It spans in principle any Bayesian inference that any computational agent could possibly perform. This includes Bayesian inferences about the programs that best explain (are most likely to have generated) an observed data set—in other words, *Bayesian learning of probabilistic programs*. Effectively constraining learning with such flexible representations is a challenge, and effectively implementing or even roughly approximating probabilistic inferences and learning in such expressive languages is even more so. We explore these issues in chapters 18 and 19, but the hardest challenges of efficient and scalable probabilistic inference over richly structured program like representations remain mostly open for future work. This is also an area where serious engagement with the neural computational basis of cognition might be useful, or even essential.

Indeed, the biggest remaining challenge for our program of reverse-engineering the mind may be to understand how not just probabilistic inference but also structured symbolic knowledge, as well as inference over symbolic structures, can be represented in neural circuits. Connectionist models traditionally have sidestepped these challenges by denying the proposition that—or avoiding the question of how—brains actually encode such rich knowledge. The Bayesian approach, in contrast, attempts to engage directly with the cognitive science and AI traditions showing all the ways that explicit symbolic representations and languages are valuable for thought, but then the challenge of neurally implementing these structures looms large. Uncovering their neural basis is arguably the greatest computational challenge in cognitive neuroscience more generally—our modern mind-body problem.

1.2.7 Capturing the Contents of the Minds of Infants

Our focus so far in this chapter has been on how to build a mind—identifying the pieces, such as hierarchical and nonparametric Bayesian inference, structured representations, and intelligent policies for action and deliberation, that we view as explaining different aspects of human cognition. But minds aren't just built—they grow. How is it possible to grow a mind? How does it all start? This is Q7: What are the origins of our world models in evolution and development—what is built into a baby's mind, and how do children learn within and beyond that starting point?

As discussed previously, philosophy has offered two perspectives on these questions: the nativist view, which emphasizes innate knowledge, and the empiricist view, which emphasizes learning. Modern developmental psychology offers more nuanced versions of these positions. Developmentalists in the core knowledge tradition have argued that learning can get off the ground only with some innate stock of abstract concepts such as *agent*, *object*, and *cause* to provide the basic ontology for carving up experience (Carey, 2009; Spelke, 2022). Surely some aspects of mental representation are innate, but without disputing this, Bayesian modelers have recently argued that even the most abstract concepts may in principle be learned. For instance, an abstract concept of causality expressed as logical constraints on the structure of directed graphs can be learned from experience in an HBM that generalizes across the network structures of many specific causal systems, as shown in figure 1.3d. Following the "blessing of abstraction," these constraints can be induced from only small samples of each network's behavior, and in turn enable more efficient causal learning for new systems (Goodman et al., 2011). How this analysis might extend to other abstract concepts such as *agent* or *object*, and whether children actually acquire their deepest abstract concepts in this way, remain substantially open questions.

While HBMs have addressed the acquisition of simple forms of abstract knowledge, they have only touched on the hardest subjects of cognitive development: framework theories for core common-sense domains such as intuitive physics, psychology, and biology (Carey, 1985, 2009; Gopnik & Meltzoff, 1997). First steps have been taken to explain developing theories of mind—how children come to understand explicit false beliefs (Goodman et al., 2006) and individual differences in preferences (Lucas, Griffiths et al., 2014)—as well as the origins of essentialist theories in intuitive biology, and forming beliefs about magnetism and other novel forces in intuitive physics (Griffiths & Tenenbaum, 2007b; Kemp et al., 2010b; Bonawitz, Ullman, Bridgers, Gopnik, & Tenenbaum, 2019; Ullman, Stuhlmüller, Goodman, & Tenenbaum, 2018).

We will present an overview of some of these developmental models in chapter 20, along with more recent work aiming to capture some of the earliest core knoweldge capacities of young children in terms of an object system, a place system, and an agent system (figure 1.10). Together, these can be modeled by Bayesian inference over probabilistic generative simulators, with representations structured around a set of approximations and

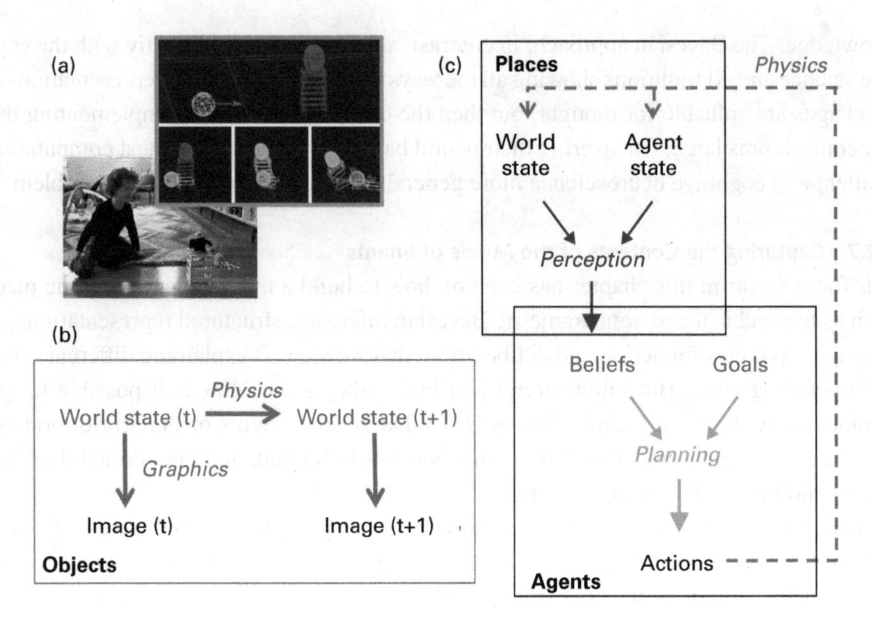

Figure 1.10
Core common-sense capacities in young children for intuitive physics and intuitive psychology, inspired by Spelke and Kinzler (2007) and Spelke (2022), can be modeled in terms of probabilistic inference over structured generative models of the world in some ways analogous to the simplifications and approximations of a video game engine—hence these models are often referred to with the slogan "the game engine in the head" (Ullman, Spelke, Battaglia, & Tenenbaum, 2017). These models capture the child's ability to probabilistically simulate likely or possible future outcomes, on short temporal and spatial scales, under a range of possible actions—and thus to choose actions that could produce a desired outcome (a). Core systems modeled in this way include (b) an object system for predicting dynamic physical events, such as the rolling of a ball and its collision with the stack of blocks (in (a)), and (c) a place system for navigating and orienting in space, integrated with an agent system for reasoning about how others' beliefs give rise to their actions through goal-directed planning and update their informational content through perceptual inference and reasoning processes.

simplifications that we have called "the game engine in the head" (Ullman et al., 2017; Lake, Ullman, Tenenbaum, & Gershman, 2017). The resulting architectures serve as candidate hypotheses for what is innate in cognitive development, as well as the computational substrate of early learning (Ullman & Tenenbaum, 2020). In chapters 14 and 15, we introduce these models and some of the ways that they have been tested rigorously and quantitatively as models of intuitive physics and intuitive psychology in adults. We also sketch how these models can be integrated into a unified "probabilistic language of thought (PLoT)" view of cognitive architecture, taking advantage of the fundamental simulation-based nature of knowledge representation and inference in probabilistic programming languages (chapter 18), and introduce program induction methods by which these simulators might be extended beyond their initial core capacities over the course of learning and development (chapter 19).

1.2.8 Learning from and About Other People

The standard formulation of Bayes' rule, as in equation (1.1), imagines a learner drawing inferences from data. But where do these data come from? In many of the models that we have discussed so far, these data are generated by the world, being sampled from the process that the learner is trying to build a world model to capture. But in reality, much of human learning relies on a different kind of data: data generated by people. It is impossible for us to directly experience enough of the world to build rich models of all of its aspects that are relevant to guiding intelligent action. As a consequence, we rely on other people to tell us about their experiences, to give us data structured in a way that supports learning, and to effectively accumulate and distill the insights of the people that they in turn have learned from. Understanding these sources of data is the key to answering our final question, Q8: What is needed to scale up learning to all the knowledge that a human being acquires over their lifetime and human cultures have built over generations?

One distinctive aspect of Bayesian models of cognition, compared with other learning paradigms such as artificial neural networks, is that they require precisely specifying assumptions about how data are generated. The $P(d|h)$ term in Bayes' rule needs to incorporate a model of the data-generating process to be able to assign a probability to each d that we might observe. This sensitivity to the way that data are sampled was explored even in early Bayesian models of concept learning (Tenenbaum, 2000) and word learning (Xu & Tenenbaum, 2007).

That early work on sampling assumptions grew into a rich body of literature using Bayesian models to understand the principles of pedagogy and communication. Analyzing pedagogical thinking in terms of teachers considering what data will be most beneficial to learners, as well as learners assuming that teachers are engaging in such a thought process, resulted in models that show how it is possible for complex concepts to be communicated with very small amounts of data (Shafto, Goodman, & Griffiths, 2014).

Language more broadly offers rich opportunities for applying Bayesian models of cognition, not just in understanding how aspects of language are learned, but also for seeing how people learn from and reason about what other people say. Indeed, the same principles are at work in the influential *Rational Speech Act* framework for pragmatic linguistic inference (Goodman & Frank, 2016), which provides a way to make sense of the classic finding that the interpretation of people's utterances depends just as much on what they didn't say as what they did (see chapter 16). Here, the trick is that the listener is not confined to reading the literal meaning of what is said—the listener also knows that the speaker chose to use that particular utterance (rather than the huge variety of alternative possibilities)—and hence, assuming that the speaker is trying to be as informative as possible, this allows the listener to infer more than is actually said. So, for example, hearing "*the play has many good qualities*," the listener will likely infer that the speaker believes that the play has at least as many bad qualities, and probably overall the speaker dislikes it, or they would have been more positive. And the inferential picture is more complex still, because speakers choose their words knowing that listeners will make such inferences (i.e., that the speaker will be using Bayesian inference to infer the listener's interpretation). And the listener also may take this fact into account, so ascending to yet a further level of inference; and this process can continue without obvious limit, except presumably by the limits of our computational abilities.

Bayesian pragmatics in language can also be seen as a special case of the more general Bayesian treatment of "theory of mind" or "mentalizing": the core human capacity to observe other agents' actions and infer the beliefs, desires, and intentions that gave rise to them, assuming that those actions were the result of approximately rational goal-directed planning and decision-making processes in the agent's mind. Chapter 14 develops the

Bayesian approach to modeling theory of mind with a focus on understanding the actions of agents moving around us and in our local environment, on relatively short spatial and temporal scales, as developed in seminal work by Baker, Saxe, and Tenenbaum (2009), Lucas, Griffiths, et al. (2014), Jern, Lucas, and Kemp (2017), Jara-Ettinger, Gweon, Schulz, and Tenenbaum (2016), Baker, Jara-Ettinger, Saxe, and Tenenbaum (2017), and Ullman et al. (2009). At its core, this framework posits that individuals infer the intentions and beliefs of others by observing their actions and then "inverting" the agents' planning process. Instead of predicting future actions based on given intentions (as in forward planning and rational decision-making; see chapter 7), *Bayesian inverse planning* starts with the observed actions and works backwards to infer the most likely intentions and beliefs that could have produced those actions. To achieve this, the model uses Bayes' rule, combining prior beliefs about the actor's preferences and intentions with the likelihood of the observed actions, given that those intentions assume an approximately rationally planner, to produce a posterior distribution over possible intentions. This method provides a structured way to interpret the behavior of others in terms of underlying mental states, giving us not only a candidate mechanism behind the human ability to understand and predict the actions of others, but also opening up a vast landscape of insights and modeling opportunities for human social cognition and the cultural basis of cognition more generally.

Finally, the Bayesian approach can be used to shed light on some aspects of cultural learning as well. By considering how information changes when passed along a sequence of Bayesian learners, we can explore how the inductive biases of those learners influence the outcome of this process of cultural transmission (see chapter 10). Such processes of *iterated learning* can change the information being transmitted to become more consistent with the priors of the learners, making it easier for subsequent learners to learn (Griffiths & Kalish, 2007). This provides a way to understand how cultural objects passed from generation to generation are shaped by human minds. Under this view, languages—passed from speaker to speaker—should be expected to change to take on a form that is more easily learned, providing another way to understand how human learners are able to acquire these languages from limited data.

1.3 A Reverse-Engineering View of the Mind and Brain

In presenting the Bayesian approach in this book, we have chosen to emphasize the fact that the underlying philosophy is one of reverse-engineering. To create a Bayesian model of cognition, we begin by thinking about the computational problem that the mind is solving. This involves identifying the data available to learn from, specifying hypotheses about how those data are being generated, and assigning prior probabilities to those hypotheses. Bayes' rule then indicates the ideal solution to that problem, generating predictions that can be compared against human behavior. If we see a correspondence between model and behavior—and particularly if it holds up as we run further experiments designed to test the model's predictions—then we have a way to understand why people might be doing what they are doing, and a model that we can use to make machines that perform similarly.

This reverse-engineering approach instantiates a view of how to make progress in cognitive science that was first articulated by David Marr—the notion of *levels of analysis*.

Marr (1982) argued that information-processing systems can be analyzed at three levels: the *computational level* characterizes the problem that a system solves and the principles by which its solution can be computed from the available inputs in natural environments; *algorithmic-level* analysis describes the procedures executed to produce this solution and the representations or data structures over which the algorithms operate; and the *implementation level* specifies how these algorithms and data structures are instantiated in the circuits of a brain or machine. Reverse-engineering means beginning at the computational level, trying to understand the function of a system before diving into algorithms and implementation. For this reason, we have referred to it as a *top-down* or *function-first* approach (Griffiths et al., 2010). This idea was made explicit in Anderson's (1990) framework for *rational analysis*, which focuses on analyzing cognition in terms of adaptive solutions to problems posed by the environment, and resulted in several groundbreaking Bayesian models of cognition.

Many early Bayesian models addressed only the computational level, characterizing cognition in purely functional terms as approximately optimal statistical inference in a given environment, without reference to how the computations are carried out (Oaksford & Chater, 2001; Tenenbaum & Griffiths, 2001b; Griffiths & Tenenbaum, 2007b). Subsequent work, such as that summarized in our response to Q5, has begun to explore how the same principle of optimization might be used to guide investigation at the lower levels. *Resource-rational analysis* (Griffiths, Lieder, & Goodman, 2015) explicitly aims to make sense of people's cognitive processes—an algorithmic-level question—in terms of optimal solutions to the trade-off between the quality of a solution and its computational cost.

The focus on optimal solutions as a source of insight into human cognition is part of what connects Bayesian models in cognitive science with research in statistics, machine learning, and AI. These disciplines focus on solving problems of inference and optimization, providing a rich source of solutions that can be applied to understanding human cognition. To the extent that machines face similar computational problems to people, we can also hope to export the human solutions back to these various disciplines.

This reverse-engineering perspective puts Bayesian models of cognition in a slightly different class from traditional computational modeling efforts in psychology. While a cognitive psychologist might typically be hoping to elucidate the *mechanisms* behind human behavior, the models that we present in this book operate at another level of abstraction. Taking these models as claims about cognitive processes is a potential source of confusion (a point discussed further in chapter 22). For example, offering an explanation of concept learning in terms of Bayesian inference does not mean that we expect that people are explicitly calculating Bayes' rule in their head. It means that the assumptions of this model—particularly the assumptions about inductive biases—are aligned with people's behavior, helping us to understand that behavior in terms of the ideal response to an abstract formulation of the concept-learning task.

When we find that a Bayesian model captures people's behavior, there are several possibilities for what could be going on inside their heads. In rare cases, they might actually be explicitly performing Bayesian inference. More typically, they are following a strategy that implicitly allows them to approximate the ideal solution indicated by Bayes' rule. This allows us to make predictions about how they will behave in other situations, and gives us clues about what kind of representations and algorithms might be at work. It also highlights the implicit assumptions behind people's behavior that can provide deeper insight or allow us to build machines that make similar assumptions.

One concern that this might raise is whether it results in a falsifiable theory. Many scientists evaluate theories using the criterion highlighted by Popper (1959/1990), considering a theory to be valuable only if it makes predictions that can be tested and potentially proven false. The approach that we have outlined—in which we need to assume that people are producing something close to an optimal solution to a problem for reverse-engineering to be effective—doesn't sound particularly falsifiable, as we can make a lot of choices about how we define problems and their solutions. We agree with this—the basic assumption of optimality is not falsifiable. But it's also not intended as a theoretical claim. We aren't asserting that people are optimal, are rational, or live up to some particular ideal. Rather, we are suggesting that this assumption can be useful as a means of generating explanations of human cognition—a methodological framework, rather than a theory intended to be falsified. Individual theories generated within this theory can be tested and shown to be false, but the broader idea that we can make progress in understanding human cognition by considering ideal solutions to computational problems is a framework that should be evaluated based on its utility in producing meaningful explanations. This perspective is in line with positions taken by more recent philosophers of science (e.g., Lakatos, 1970; Godfrey-Smith, 2003), providing an update to the narrow focus on falsification advocated by Popper.

1.4 The Promise of the Bayesian Approach to Cognitive Science

We have outlined an approach to understanding cognition and its origins in terms of Bayesian inference over richly strucutured, hierarchical generative models. While we are far from having a complete understanding of how human minds work and develop, the Bayesian approach brings us closer in several ways. The first is the promise of a unifying mathematical language for framing cognition as the solution to inductive problems and building principled quantitative models of thought with a minimum of free parameters and ad hoc assumptions. Another, deeper contribution is a framework for understanding why the mind works in the way it does, in terms of rational inference adapted to the structure of real-world environments, and what the mind knows about the world—abstract schemas and intuitive theories revealed only indirectly through how they constrain generalizations.

Most important, the Bayesian approach lets us move beyond classic "either-or" dichotomies that have long shaped and limited debates in cognitive science: "empiricism versus nativism," "domain-general versus domain-specific," "logic versus probability," and "symbols versus statistics." Instead, we can ask harder questions of reverse-engineering, and get answers that are potentially rich enough to help us build more human like AI. How can domain-general mechanisms of learning and representation build domain-specific systems of knowledge? How can structured symbolic knowledge be acquired by statistical learning? The answers that are emerging suggest new ways to think about the development of a cognitive system. Powerful abstractions can be learned surprisingly quickly, together with or prior to learning the more concrete knowledge they constrain. Structured symbolic representations need not be rigid, static, hard-wired, or brittle. Embedded in a probabilistic framework, they can grow dynamically and robustly in response to the sparse, noisy data of experience.

Taken together, the approach that we present in this book provides a unifying mathematical language that can be used across all the cognitive sciences. Inductive problems occur in all the domains that have traditionally been studied by cognitive scientists: not just learning and reasoning, but vision, language, social interaction, and even basic cognitive processes such as memory and attention. It also creates a bridge between AI and cognitive science, between engineering and reverse-engineering. Even as artificial neural networks currently dominate the methods used in AI, we believe that the principles laid out here are going to be essential for understanding any kind of intelligent system. They can also help us understand how people learn from small amounts of data and get the most out of their limited cognitive resources—two abilities that form a striking contrast with the massive increase in amounts of data and computation being used when training AI systems. As a consequence, they can provide a guide to where to go next when those AI systems start to reach the limits of what is possible in terms of scaling data and computation, and where to go now if we want to build AI systems that learn and think more like people.

2

Probabilistic Models of Cognition in Historical Context

Nick Chater, Thomas L. Griffiths, and Joshua B. Tenenbaum

The founding objective of cognitive science is to understand the mind and brain in computational terms. And if the brain is to be understood as a kind of biological computer, then ideas about how computation can work in principle will be a crucial source of inspiration when creating hypotheses about how the mind/brain works. So to reverse engineer the mind, we should look for inspiration from the engineering disciplines concerned with building intelligent machines. Thus, as we described in chapter 1, the starting point for cognitive science is its twin discipline of artificial intelligence (AI) and, more broadly, computer science.

This picture is oversimplified in various ways, though. First, even if the mind is a kind of biological computer, we should expect it to look, and perhaps to operate, very differently from any computers we have actually built or computer programs we have written. This is true with reverse engineering in general, of course—the heart may be a pump, but it is a very different, and far more sophisticated, than any pump of human design. Brains certainly don't look much like silicon chips, so if the brain is a biological computer, any relationship with real computers is likely to be rather abstract, and by no means obvious. Second, reverse engineering does not merely need to passively select from tools that computer scientists happen to have created for other purposes (such as doing arithmetic or building data bases). Indeed, by addressing the question of how human intelligence works, cognitive science has itself helped to inspire new computational ideas and methods. Just as in other areas of science, nature can provide inspiration for new engineering solutions. So insights flow both ways between the engineering disciplines underpinning AI and the reverse engineering project of cognitive science.

Inevitably, though, if we view the project of cognitive science through the lens of reverse engineering, we should expect the development of the field to be strongly influenced by the available stock of engineering, computational, and mathematical tools and techniques. In this chapter, we describe three influential traditions in cognitive science that have been built on different styles of engineering ideas: symbolic AI, connectionism, and rational approaches to cognition. The viewpoint explored in this book can be viewed as a synthesis of key insights from each of these approaches. The possibility of such a synthesis has itself depended on fresh engineering developments and in particular the technology of using and learning sophisticated probabilistic models, that is, probabilistic models that are defined over structured representations, such as graphs, grammars, or programs. Exploring these models, and their application in helping to understand a diverse range of areas in human cognition, is a core theme of this book.

This chapter outlines some of the history of cognitive science, and its interactions with its twin discipline of AI, from this perspective. The first three sections describe in turn the symbolic, connectionist, and rational research traditions. Building on chapter 1, we then note how engineering developments, from fields including machine learning, computational linguistics, and computational vision have made it possible to synthesize these approaches, by developing inference methods over sophisticated probabilistic models that can be defined over complex *symbolic* representations which may ultimately be implemented in *connectionist* networks and have a *rational* justification. This integrated perspective provides a starting point for the narrative of the rest of the book, which introduces and explains probabilistic models of cognition of successively increasing complexity. We will also stress that a reverse-engineering approach suggests a top-down approach to the mind, focusing on understanding the computational problems that the mind solves, before turning to the neural hardware on which its computations are implemented. Despite this top-down focus, the question of how sophisticated probabilistic models can efficiently be embedded in the biological machinery of the brain remains a major and crucial challenge for future research.

2.1 Symbolic Cognitive Science

We have noted that the objective of cognitive science is to understand the mind in computational terms. Thus, it was natural that, from the beginning of the discipline in the late 1950s, cognitive scientists took as their starting point the theory and practice of computation that had emerged over the preceding decade or so. This was done very deliberately and self-consciously: the core characteristics of early digital computation were therefore taken up by early cognitive science as hypotheses about the operation of the mind.

Indeed, two of the pioneers of both early cognitive science and AI, Allen Newell and Herbert Simon, proposed the *physical symbol system* hypothesis: that human intelligence is a system for the manipulation of symbols, physically instantiated in the hardware of the brain, just as a digital computer operates by manipulating symbols in a silicon chip (Newell & Simon, 1976).

What does this mean in practice? Let us start with a simple information-processing challenge, such as sorting a list of words into alphabetical order. First, we need some way of representing the individual words; and we need some data structure to represent the current order that they are in—typically a data structure known as a *list*. A list is defined by the information-processing operations that can be carried out on it. For example, given a list, we can append a new item to the beginning so that *pear* can be added to the list *[banana, orange, blueberry]* to create a new list: *[pear, banana, orange, blueberry]*. By contrast, in this technical sense of a list (unlike the everyday "shopping list" sense), an item can't be directly appended to the far end of the list. We can also directly remove the first item (or "head") of the list (stripping off *pear*) to leave *[banana, orange, blueberry]* (but again, for lists, we can't directly strip off the last item, *blueberry*).

So putting a list in alphabetical order will involve a complex dance of symbolic manipulations in which items are stripped off the list, compared with each other to see which is earlier in the alphabet, and ultimately reassembled into a final list. This symbolic dance

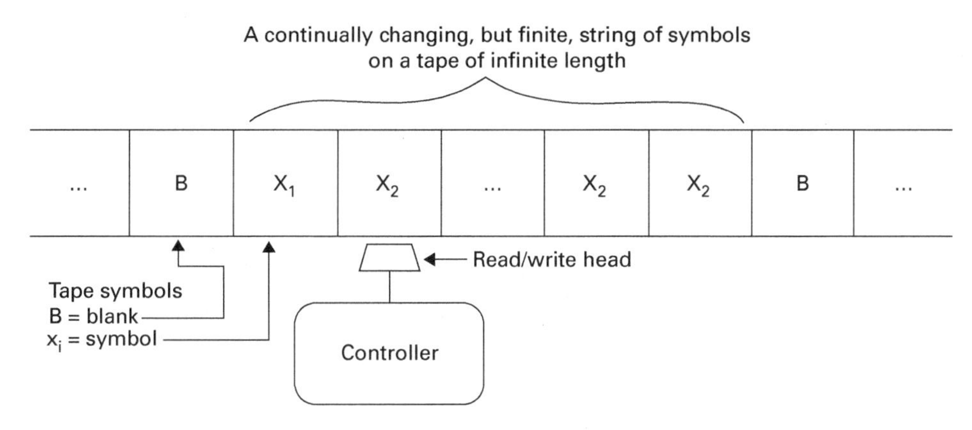

Figure 2.1
The Turing machine: The simplest possible physical symbol system. A Turing machine consists of an infinitely long tape on which a finite repertoire of symbols can be written and read (this repertoire can be just two basic symbols, which we can label "0" and "1," and other symbols can be encoded as strings of 0s and 1s). A very simple "controller" system moves up and down the tape one step at a time. At each step, it can read only the symbol at its current location on the tape, and, depending on that symbol and the controller's current state (one of a finite number of possible states), it may rewrite the current symbol on the tape, and/or move one step left or right along the tape. Over time, the string of symbols on the tape will gradually change, representing the steps of the computation, and finally giving the output of the computation when the machine halts. A simple but crucial further step is to see the symbols of the tape as divided into two blocks, one block of which is viewed as an algorithm that should be carried out on the data encoded by the other block. Remarkably, this incredibly simple "programmable" computer is capable of carrying out any computation, although very slowly. The physical symbol systems that are embodied in today's digital computers, and in Newell and Simon's proposals about the operation of the human machine, can be viewed as incredibly sophisticated and efficient elaborations of the Turing machine.

is governed by an algorithm—a procedure that spells out the steps that take us from the original list, step by step, to the sorted list *[banana, blueberry, orange, pear]*.

Actually, working with lists is pretty awkward for putting items in order because items at the end of the list can't be accessed quickly without decomposing the rest of the list. So we might choose a different data structure, like a one-dimensional array, in which items can be directly accessed via a number (0, 1, 2. . .), which corresponds to a location in the array.[1] But even so, sorting is surprisingly slow and time-consuming, particularly with many items (e.g., if we are putting an entire dictionary in alphabetical order). One strategy, for example, is to start by picking an item in the array (by whatever method), and then to successively relocate all the items in the original array to one side of this "pivot" item if they are before it in the alphabet, and to the other side if they come after it in the alphabet. Then we apply the same approach to the left portion of the array (finding a pivot item in that array and sorting according to that item); and next we do the same to the right portion of the array. And we can keep doing this recursively until each subarray contains just one item (so there is nothing to sort). This procedure is called *quicksort*, but there are many other methods for sorting a list. And despite its apparent triviality, the problem of sorting turns out to be a perennial one in computer science, and it has been the subject of considerable theoretical and practical research.

1. This is the difference between sequential access and random access memory.

How do we persuade the computer to follow the particular dance of symbolic operations that we intend? The answer is that we need to specify an *algorithm* that spells out the sequence of symbolic operations that are required. The algorithm will depend on the data structure that is being manipulated, of course; and it will typically itself be encoded in symbolic form, written in a computer programming language. So we might write a quicksort algorithm over arrays to put words in alphabetical order in Python, C++, or any other programming language.

All the same—this is just sorting: not a very rich or interesting task for cognitive science. Chapter 1 has already introduced many much more interesting data structures, and far richer and more interesting information-processing problems than this. If we want to represent the complexity of visual images, objects, categories, and natural language sentences, to say nothing of the abstract realms of science and mathematics, we will need a gamut of rich representations, including not just lists and arrays, but graphs, grammars, logical formalisms, and more. But, from the point of view of the symbol-processing perspective on cognitive science, this is ultimately just more of the same. Symbolic computation involves algorithms coordinating the intricate manipulation of the contents of data structures of whatever kind.

This emphasis on the manipulation of complex symbolic representations was a dramatic theoretical break from the previous decades of psychological theory, in which the doctrine of *behaviorism* had been in the ascendancy (e.g., Skinner, 1953). The behaviorists wanted to build a psychological theory purely based on externally measurable stimuli and observable responses. The very idea of internal representations inside the mind or brain was rejected as hopelessly unscientific. The computational revolution in thinking about the mind broke free of these concerns. The example of digital computers showed how information processing could be carried out through symbolic manipulation over data structures, capturing information of arbitrary richness and complexity, and implementing it in a physical system. In particular, while the behaviorists restricted themselves to a view of the mind as consisting of associations between various stimuli (so a bell might, for example, evoke the thought of the food that typically followed) or between responses and subsequent stimuli (so that pressing a lever might somehow be reinforced when consistently rewarded with food), symbolic cognitive science saw the observable inputs and outputs to the mind as merely the peripheral disturbances on the surface of a vastly complex information-processing system.

Language provided a particularly potent illustration of the radical shift in thinking that the view of the mind as a symbol-processing system represented. Behaviorist views of language viewed words as associated with aspects of the environment (actual dogs becoming associated with the word *dog*, for example) and chained together in associations with each other, supposedly leading to the sequential structure of language (Skinner, 1957). But this story never really worked because, among other things, language depends on complex structural relationships between linguistic units of varying sizes (morphemes, whole words, noun and verb phrases, and so on), rather than associations between successive words (Chomsky, 1959). And language refers to the world (and to conjectures about the world, thoughts about how the world might have been, or entirely fictional realms) in a highly complex and indirect fashion. Indeed, it is hard to imagine how we might even begin explaining the meaning of *the quick brown fox* on the basis of associations between occurrences of the words *quick*, *brown*, and *fox* and presentations of actual quick things, brown patches, and foxes. The particular relationships between the words (the fox is both quick and brown) and the function of the word *the* (which tells us that there is one and only one fox that is both quick and brown, in whatever context we happen to be talking about) are equally mystifying in a behaviorist account.

The linguist Noam Chomsky, one of the key figures in the computational revolution in thinking about the mind, argued that sentences have a tree like grammatical structure (Chomsky, 1957). He drew directly from, and extended, a branch of theoretical computer science, *automata theory*, to argue that the underlying machinery required to generate all and only the sentences of human language must be rather complex, and certainly much more sophisticated than can be captured by pure association between words or any simple generalization of association (we will meet some of these issues, and in particular, the Chomsky hierarchy of languages, in chapter 16).

Chomsky also proposed, along with many other cognitive scientists in the symbolic tradition, that the mind translates to and from the natural languages (Chinese, Hausa, Finnish, etc.) into a single internal logical representation. This internal representation was presumed to capture the logical form of the sentence—clarifying that, for example, there is a unique fox that is both quick and brown, and allowing inferences such as that the fox is brown, that there is at least one thing that is both brown and quick, and so on. The goal of understanding the logical form of natural-language sentences was central to the analytic tradition in philosophy in the first part of the twentieth century. The novel cognitive science angle, though, was to put the logical form—and the logical system of representation out of which it is constructed—into the head of the speaker and listener. That is, the proposal is that the mind represents and reasons over a logical *language of thought* (Fodor, 1975). Indeed, this language of thought can be viewed as a rich, abstract, and highly flexible system for representing the world. Moreover, it can be viewed as providing not merely an inert repository of knowledge but also a high-level *programming language*, which allows algorithms to be defined through guided chains of logical inferences over these representations (corresponding to the logic programming paradigm in computer science [Kowalski, 1974] and most famously embodied in the programming language Prolog [Clocksin & Mellish, 2003]). And the proposed language of thought is assumed not just to underpin the meaning of sentences but also to be the format in which knowledge of all kinds is encoded in memory and used as the basis for thought. This is an idea to which we will return, with a Bayesian twist, with the *probabilistic language of thought* hypothesis in chapter 18.

Returning to our question at the beginning of chapter 1, Chomsky also considered how languages could be learned: how children are able to learn so much (i.e., gaining the ability to create and express an infinite number of potentially novel thoughts) from so little (the scrappy and incomplete corpus of utterances of the people around them). Chomsky's conclusion was striking: he argued that the complexity of human language was not really learnable at all (Chomsky, 1980). Instead, he argued that a blueprint for the supposedly common core of all human languages must be represented innately, in the form of a *universal grammar*—and that children could merely tweak this universal grammar when exposed to actual speech, allowing them to home in on the specific language or languages around them (whether Chinese, Hausa, or Finnish). And in this tradition, the presumed process of tweaking the universal grammar into the specific language to be acquired was itself viewed as inherently a matter of symbolic computation, for which an algorithm could be specified (Pinker, 1979; Gibson & Wexler, 1994).

The symbolic approach to computation was initially associated with a more general shift toward nativism: the proposal that knowledge of many kinds is not primarily learned from experience at all. Instead, it was assumed that the child possesses an innate knowledge of language, morality, mathematics, an everyday understanding of physics or psychology, and perhaps even of categories themselves (Fodor, 1975). The answer to the question of how the child can apparently get so much out of so little was, in broad terms, that a great deal of knowledge is built in, rather than learned through experience. While the behaviorists had assumed the mind to be a blank slate on which associations of any and every kind could be overlaid, the mind was now viewed instead as containing a wealth of prepackaged information and algorithms, perhaps gradually accumulated through millions of years of biological evolution (Fodor, 1983; Pinker, 2003). One of the main themes of this book, as will become clear across many chapters (and especially chapter 20, on cognitive development), is that a Bayesian reverse-engineering viewpoint allows us to provide a more nuanced perspective on the perennial debate between the nativist viewpoint and its contrary, empiricism. We will see how putting a quite specific, but limited, innate structure into a computational model can, in some cases at least, allow remarkably rich and powerful understanding to be learned from surprisingly little experience. But this is getting ahead of ourselves.

Stepping back from specific debates, it is worth reflecting on what is distinctive about the symbolic approach to understanding cognition. What is a symbol anyway? And what does it mean to do *computation* with symbols? In line with the discussion so far, for Newell and Simon, a physical symbol system consists of expressions of a symbolic representation (e.g., lists, graphs, mathematical and logical formulas, and words), together with a set of processes for manipulating these representations (e.g., adding, combining, duplicating, or deleting symbols; linking symbols together in a list or a matrix; and so on). Crucially, both the symbols and processes are embodied in physical form, ultimately in the binary 0/1 switches in a silicon chip in the case of a conventional digital computer (indeed, it is this discrete binary substrate, rather than working with continuous values, that makes conventional computers *digital*). And as we have seen, these symbols are not merely causally inert, like marks of ink on a piece of paper. Rather, they (and ultimately the binary switches in which they are encoded) shape and take part in the complex choreography that is the process of computation.

The symbols are not, of course, merely meaningless physical patterns. Crucially, they can be viewed as having an *interpretation*, either as representing aspects of the world (so the symbolic structures can be viewed as encoding *knowledge*) or as specifying sequences of symbolic manipulations (so they can be viewed as representing *programs*).

Newell and Simon (1976) noted that digital computers are working exemplars of the physical symbol system hypothesis: digital computers show how it is possible to solve problems, prove theorems, or engage in simple linguistic behavior via the physical manipulation of symbols. Thus, adopting the perspective of reverse engineering, they proposed that human thought has the same basis. The idea that cognition is computation is taken to be the proposal that computation should be construed as symbol manipulation.

These and related ideas were embodied in the research program of early AI and cognitive science. For example, the generative approach to linguistics (e.g., Chomsky, 1957) aimed to specify the knowledge of the syntax of natural language in terms of a formal grammar, specified as a set of symbols and processes of symbol manipulation (e.g., rewrite rules, transformations) that could generate all and only the acceptable sentences of each language. From this perspective, the task of parsing a sentence becomes a problem of mapping a symbolic representation of a string of words into a hierarchical tree structure, indicating the syntactic derivation of that sequence of words according to the grammar of the language concerned. Similarly, the problem of language production involves, among other things, creating such a tree and reading off the linear order of the words in that tree to produce a sentence. And learning a language involves inferring the formal grammatical rules of the language, given a set of example sentences, perhaps together with other constraints, such as innate prior knowledge (see chapter 16).

Everyday knowledge was, similarly, assumed to be encoded in symbolic representational formalisms, such as the *situation calculus* (McCarthy & Hayes, 1969), *semantic networks* (Collins & Quillian, 1969), *frames* (Minsky, 1977), and *scripts* (Schank & Abelson, 1977). Reasoning was then modeled as logical inference over such representations (or slight departures from strict logical inference). In this framework, perceptual representations and processes were also viewed as fundamentally symbolic in character (e.g., Marr, 1982).

Some aspects of this symbolic approach to cognition have, moreover, been influenced by further engineering developments, such as the project of constructing logical foundations for aspects of computer science and AI (e.g., Genesereth & Nilsson, 1987). Thus, many aspects of symbolic computation have been productively interpreted as logical inference—and there has been an important strand of research to reconstruct particular programming languages, and by extension specific programs, in logical terms. According to this perspective, symbol manipulations can be viewed as correct or incorrect, depending on whether they correspond to valid inferences under some logical interpretation.

Moreover, developments in *declarative* approaches to computer programming have made this logical standpoint concrete (i.e., relevant to the programmer, rather than merely the theoretical computer scientist). A declarative computer program can be viewed, to an approximation at least, as involving a statement of various facts, constraints, or relationships, rather than outlining a procedure specifying a sequence of symbol manipulations (like a computer program, written, say, in C, BASIC, or Pascal). The pattern of symbol manipulation arises indirectly through automatically deriving logical consequences of those facts, constraints, or relationships.

The increasing scope of logical methods in computer science was paralleled within AI, where there were increasingly vigorous attempts to build a logical framework for representation and inference, both to tidy up, and to extend, the rather ad hoc procedures embodied in early AI. Similarly, the philosophy of cognitive science began to shift from the core idea that cognition is symbol manipulation to the more specific notion that cognition is mechanized logical inference over a logical language of thought (Fodor & Pylyshyn, 1988), a theme that we will pick up shortly.

The symbolic approach to cognitive science has been spectacularly successful in many ways. Indeed, modern engineering approaches to a wide range of problems in AI are based on the idea that knowledge can be formulated and manipulated using complex symbolic representations, many of which can usefully be understood in logical terms. Nonetheless, the symbolic approach to cognitive science appeared, to many, to be inadequate to explain

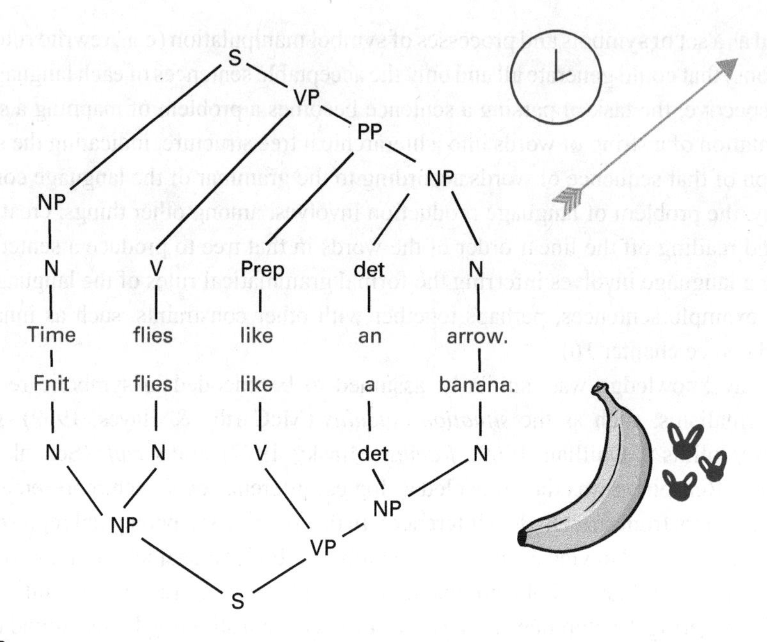

Figure 2.2
The challenge of syntactic ambiguity. Apparently similar strings of words can have very different grammatical structures. "Fruit flies" could refer to the way in which fruit projects through the air; "time flies" could be a rare type of fly; but these are improbable. Similarly, "like a banana" seems more naturally to be interpreted as expressing a preference for bananas, rather than a comparison to a banana. And, of course, putting the fragments together doesn't entirely resolve the ambiguity (those mysterious "time flies" could have a penchant for arrows) but makes one interpretation overwhelmingly likely. Early symbolic approaches to explaining how people process sentences typically avoided the use of probabilistic information and tried to guide the "parsing" process purely by structural features of the parses (roughly, preferring some shapes of trees to others, e.g., Frazier & Fodor, 1978). The sheer prevalence of local linguistic ambiguity generates a forbidding combinatorial explosion of possible parses—which probabilistic constraints can help to control.

many central cognitive phenomena; and researchers began to search for alternative, or perhaps complementary, engineering approaches to provide a starting point for understanding cognition.

The reasons for the perceived limitations of symbolic approaches to cognition are complex and numerous. From the point of view outlined in this book, the most important missing ingredient was *probability*: almost all problems faced by the cognitive system involve uncertainty; and probability, the calculus for dealing with uncertainty, is critical to understanding how such problems can be solved (see the discussion later in this chapter and, of course, throughout this book). For example, as we have noted, according to the traditional symbolic perspective on language understanding, a key goal in parsing a sentence is to recover the underlying tree structure that expresses its grammatical derivation. Finding this structure involves analyzing parts of the sentence (and then attempting to link possible analyses of these parts to produce a globally coherent analysis, just as we may attempt to fit together the pieces of a puzzle). Yet attempts to carry out this program soon ran into the problem that natural language is spectacularly locally ambiguous—each part of a sentence can be parsed in a variety of ways, leading to a combinatorial explosion of possible global parses to consider. Crucially, though, not all possible local parses are equally probable: focusing on the small number of *probable* parses may be crucial to cut down the search space (we will return to these issues in more detail in chapter 16). If this is right, then a critical question is how to determine, given current linguistic context and/or background knowledge, which local parses *are* the most probable; and, equally crucially, how the probabilities of each local parse are modified by the probabilities of the others as the puzzle is pieced together.

The same point applies not just to parsing but throughout perception. So, for example, visual input is spectacularly locally ambiguous—any part or aspect of an image can typically be interpreted in a myriad of ways, so there is an explosive number of combinations of local interpretations that might potentially be explored. But, again, some local interpretations are much more probable than others, and, as the puzzle of the interpretation of the image is gradually assembled, probabilities for each local interpretation will substantially constrain one other. But because they left out probabilistic reasoning, early symbolic approaches to language and vision did not allow the exploitation of these powerful constraints.

While, from the present standpoint, the absence of probability was perhaps the most crucial lacuna in the symbolic approach to cognitive science, the drive to explore alternative models of computation was, historically, somewhat different. Dissatisfaction with the symbolic paradigm arose, at least in part, from the intuition that biological computation is graded, tolerant of noise, and able to learn gradually from experience and to integrate together many "soft" constraints—while symbolic methods were viewed as excessively rigid, all-or-nothing, and not well suited to modeling learning and degradation under damage (e.g., Hinton & Anderson, 1981; Rumelhart, Smolensky, McClelland, & Hinton, 1986b). Moreover, theorists began to wonder how symbolic methods, which are typically implemented in very fast serial computers, could be implemented in the slow and highly parallel machinery of the brain (Feldman & Ballard, 1982). This raised the possibility that the view that the mind is a symbol-processing system was either fundamentally flawed or at least in need of reconstruction. Advocates of this viewpoint attempted to create a computational paradigm that appears more compatible with the operation of the brain: *connectionism*.

2.2 Connectionism

Symbolic computation turned out to be astonishingly successful as the basis for the computer industry. But other engineering approaches to computation had an equally long history—and might perhaps provide an equally or even more fruitful starting point for modeling the mind. The alternative family of approaches was based, at its core, on the continuous mathematics of the calculus rather than the discrete mathematics of 0s and 1s, symbol manipulation, and logic. One strand of this approach was *analog computation*—the attempt to solve differential equations or to model specific physical or economic systems using electrical or hydraulic circuits that embody the relevant mathematical relationships. Another source of metaphors came from analog telecommunications: networks of copper wires across the world provided a natural parallel with the networks of nerve fibers running through the body and making up the brain.

How far, then, might it be possible to build a model of brain function from the bottom up—that is, by directly modeling the properties of neural circuits? McCulloch and Pitts (1943) influentially suggested that neural circuits should be viewed as computing logical functions such as AND or OR—just the type of computational component that underlies digital, symbolic computation. Later models, however, focused on the ability of neurons to

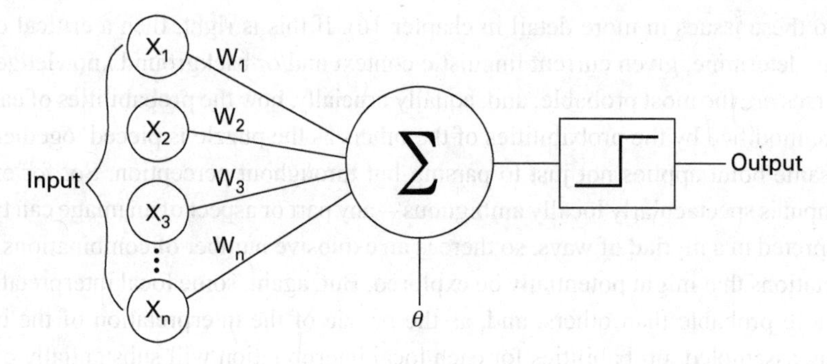

Figure 2.3
The perceptron (Rosenblatt, 1958)—the very simplest connectionist network, where inputs x_i are multiplied by weights w_i, and the result is summed (also adding in a bias term θ). If the sum exceeds a threshold value, the output is 1; otherwise, it is 0. We can think of an output of 1 as indicating that an input pattern is a member of a category and an output 0 indicating that it is not. The network learns by repeated exposure to pairs of inputs and outputs (a training set), and gradually adjusting its weights to minimize its classification errors on that training set. The single-layer perceptron, as shown here, has strict representational limitations—for example, it can only learn to split the space of inputs by a hyperplane, on one side of which are category members and on the other side of which are category nonmembers. By discovering how to train multilayered perceptrons, using soft rather than sharp thresholds (as here), most famously using the back-propagation algorithm (Rumelhart et al., 1985), connectionist researchers were able to show that much more complex concepts and mappings could be learned.

compute weighted functions of their inputs, where the weights might be viewed as analogous to synaptic strengths between neurons. One notable model, the *perceptron* (Rosenblatt, 1958), computed a linear weighted function of its inputs and produced an output of 1 if that function exceeded a threshold, and an output of 0 otherwise. The use of weighted sums rather than logical functions provided a simple model of the "soft" integration of many constraints. Moreover, the intuitive appeal of the perceptron was enhanced by the fact that the network need not be specified by a designer but rather could *learn* from exposure to prior data. Specifically, the perceptron could learn to classify a set of new inputs into two types based on exposure to prior input examples, labeled as category members and nonmembers. Indeed, such learning can also be implemented *locally*. For example, each link in the network need merely observe the inputs and outputs that it connects and can update its weight accordingly. Thus, in principle, learning can be carried out, in parallel, through local adjustments of the components of the network. This feature—that both processing and learning occur from the bottom up and without needing external intervention—is typical of most connectionist models.

A perceptron, with its single layer of trainable connections, has very limited classification abilities. Because it relies on a linear sum, it can only divide the input items by a plane (or, in many dimensions, a hyperplane), where the two categories correspond to the different sides of the plane. However, more complex neural networks can be constructed by adding multiple layers of nodes and weights between the inputs and the outputs. Each of these nodes still computes the weighted sum of its inputs, in a similar fashion to the perceptron (for a more detailed introduction to neural networks, see chapter 12). Many-layered networks can represent more general functions; but initially, it was not clear how such networks could be trained. One approach (Rosenblatt, 1958) is to train only one layer of the weights, setting the others randomly. The representational limitations of this approach, however, were viewed as severe by advocates of symbolic computation (Minsky & Papert, 1969).

Interest in increasingly sophisticated connectionist networks continued, however (e.g., Grossberg, 1987; Hinton & Anderson, 1981); and the invention of learning methods that could train multilayer networks (e.g., Ackley, Hinton, & Sejnowski, 1985; Rumelhart, Hinton, & Wilson, 1986a) gave fresh impetus to the connectionist approach and led to an impressive range of cognitive models of memory, reading, aspects of language processing and acquisition, and so on. Many connectionist networks focused on apparently elementary, but powerful, cognitive processes such as pattern completion. Others were concerned with mappings from one domain to another, some of which had traditionally been viewed as requiring explicit symbol manipulation (e.g., between stems and past tense forms; Rumelhart & McClelland, 1986), according to symbolic models in linguistics (Pinker & Prince, 1988). Moreover, it turned out that introducing a feedback loop into a one-directional feedforward network (Elman, 1990) appeared to be a promising avenue for finding sequential structure in linguistic input. This development raised the possibility that at least some apparently symbolic aspects of syntax might usefully be approximated by a learning system without explicit representations of syntactic categories or regularities, at least for very simple languages (Christiansen & Chater, 1999; Elman, 1990).

Significant progress has been made in developing neural network models of language, including new insights into neural network architectures and new data sets and technologies for training these systems at scale. The large language models of today show that such neural networks can learn to deal fluently with much of the complexity of human language. These methods have recently become widely adopted across a variety of applications in machine learning, with spectacular results (Bubeck et al., 2023; Wei et al., 2022; Ramesh, Dhariwal, Nichol, Chu, & Chen, 2022). The relationship between these models and the types of sophisticated probabilistic cognitive models that are the primary focus of this book is an area of active debate—and a topic that we revisit in later chapters.

But does this mean that cognitive science can do away with symbolic representations altogether? Not so fast! First, note that, as a matter of pure engineering, connectionist networks are built on a foundation of symbolic computation, of course: they run on digital computers that not only encode the complex structure of the network, propagate activity through the network, run the learning algorithm, and so on (which might perhaps ultimately be implemented in specialized neuron-like hardware), but also depend on training data that is assembled and encoded in symbolic form. Thus, the input to large language models is a series of discrete words, each mapped to a single node of the network, gleaned from symbolic representations of language on the web, rather than as a raw sensory stimulus (e.g., a representation of the raw acoustic waveform, as might be recorded by the neurons attached to the hair cells in the inner ear, for example). Similarly, training a network to link images to descriptions requires symbolic encodings of those descriptions and, apparently at least, some way of representing which images are paired with which descriptions. It is conceivable that this symbolic "machinery" is, as it were, merely a ladder that can discarded in later and purer neural network models—but this is by no means clear. But, as we touched on in chapter 1, there may also be a deep reason why symbolic models are crucial in cognitive science: that rich symbolic representations may be crucial to explaining how the mind can get so much from so little.

For many, this was certainly the lesson from the early phase of the connectionist movement in cognitive science which, starting in the 1980s, provided a remarkable range of rich cognitive models. Yet, like the symbolic paradigm, it rapidly ran up against what appeared at the time to be fundamental limitations (although many of these limitations have since been overcome by the latest generation of connectionist models). The most notable, and most discussed, limitation concerned the lack of representational power. As we have seen, symbolic computation can define highly structured languages in terms of which any number of messages can be encoded, each with a clearly specifiable interpretation. Thus, symbolic models can embody rich representations such as graphs, grammars, or logical formulas—they can embody the language of thought that can elegantly and succinctly capture abstract information. And symbol manipulation, often following the rules of such a logic-like system, can allow flexible and powerful reasoning about the implications that follow from the information that is encoded in those representations.

There are two rather different connectionist responses to the apparent need for rich symbolic representations to explain human language, reasoning, planning, categorization, and so on. One approach is that the problem can be sidestepped—either because sufficiently powerful connectionist models will be able to learn to mimic cognition without such representations, or perhaps by the connectionist network building such representations in an ad hoc way during learning. The second approach accepts the centrality of symbolic computation in cognitive science and explores how symbolic computations can be *implemented* in connectionist units (Rumelhart et al., 1986b; Smolensky, 1990; Shastri & Ajjanagadde, 1993). These two approaches to the relationship between connectionism and symbolic representations can be viewed as illustrative of two broader interpretations of the connectionist project: connectionism as *cognitive science*: that is, as a replacement for, or at least a substantial augmentation of, the symbolic approach; or connectionism as *computational neuroscience* (e.g., Dayan & Abbott, 2001): that is, as the project of explaining the computational properties of neural structures in the brain, in terms of which cognitive processes are implemented.

The connectionism-as-cognitive-science viewpoint, that bottom-up connectionist learning could carry out the kind of rich computations for which symbol manipulation has been thought to be essential seemed largely speculative until recently. But striking advances in connectionist AI have shown not only that connectionist networks are able to carry out computations over externally defined symbolic representations (e.g., engaging in complex linguistic interactions, writing stories, or writing computer programs), but that they may do so by learning to create and manipulate internal symbolic representations to some limited degree—although understanding the internal workings of vast connectionist networks remains extremely difficult (Pavlick, 2023).

The perspective adopted here, as will become clear, is that rich symbolic representations are essential for accounting for the complexity and variety of cognitive processes, from perception to planning and from language understanding to reasoning. From an engineering perspective, symbolic representations appear indispensible for addressing such problems; and good reverse engineering must at minimum be good engineering. Hence, we shall assume that symbolic representations are required, too, for reverse-engineering the mind. Nonetheless, the connectionist approach has advanced cognitive science in a number of key respects: highlighting the centrality of learning and the relationship between learning and probabilistic inference (Mackay, 2003); stressing the importance of understanding how multiple soft constraints are traded off, rather than focusing on rigid all-or-nothing rules; and focusing attention on how cognitive processes can be implemented in real neural hardware (Dayan & Abbott, 2001).

The recent advances in connectionist AI also raise interesting questions regarding the crucial question of how the brain learns so much from so little. Large-scale connectionist systems are trained on gigantic quantities of data (e.g., a corpus of text not far short of the entire contents of the internet), while human children work with much smaller quantities of input (e.g., of the order of millions of words of verbal input). Nonetheless, once trained, large-scale connectionist models can be surprisingly flexible in responding to new input. Indeed, they can be given a small number of examples of a task (e.g., an English sentence and its translation into French) and then can successfully infer that the task is to translate an additional English sentence into French (Brown et al., 2020). How far extensions of such "few-shot" learning in connectionist networks can mirror human-level flexibility in responding to novel problems or whether the addition of built-in symbolic machinery is required (Zhang et al., 2022) is an important open question for future research.

2.3 Rational Approaches

A third long tradition for explaining cognition has its roots in economics, mathematics, and philosophy, rather than theories of computation. Here, the starting point is trying to understand intelligence, whether human or artificial, by attempting to construct formal theories of *rationality*.

In the most general terms, theories of rationality typically start with some conception of what it means for an agent to be coherent or incoherent in its thoughts and/or actions. So, for example, a rational theory of mathematical, logical, physical, or social reasoning might start from the assumption that no intelligent agent should believe that any proposition (say, Fermat's last theorem) is *both* true and false—or indeed, have any set of beliefs that imply (whether the agent has yet noticed this or not) that Fermat's last theorem is both true and false. As this example indicates, achieving this consistency can be arbitrarily difficult—before the mathematician Andrew Wiles constructed his vastly complex and subtle proof of the theorem, no one knew whether the basic truths of number theory were compatible with the theorem or not. Indeed, one could see the whole of mathematics as a formal rational theory helping us understand what we can coherently believe. Thus, formal rational theories automatically provide constraints for theories of intelligence, whether human or artificial. If an agent is reasoning about numbers, its thoughts should be constrained by the rules of arithmetic. If the agent is reasoning about objects in space, its reasoning might be guided by the principles of the relevant aspects of geometry and physics (themselves perhaps captured in some mathematical form), as discussed in chapter 15. To do this, of course, these rational principles need to be embodied in the computations that the agent performs, just as the principles of arithmetic inform the operations of a pocket calculator.

The same line of thought applies to understanding how we make decisions (see chapter 7). We might require that an agent does not simulantaeously choose to engage in, and refrain from, the very same action A; that it does not simultaneously decide to take two incompatible routes or plans; and that it otherwise avoids combinations of choices that will lead it into deadlock. Fairly modest constraints of this kind rapidly generate a rich mathematical theory, which has been developed in economics, engineering, and other disciplines, concerning what it means to be a rational agent (see chapter 7). But figuring out how to decide and plan, particularly when dealing with a complex world, which may contain other intelligent agents, can require both a complex rational theory, but also very subtle computations to take account of that theory (think, for example, of the essentially limitless complexity of deciding the best plan in competitive games like chess or Go).

A particularly central aspect of reverse-engineering the mind is dealing with uncertainty.[2] Here, as we have seen in chapter 1, the aim is to capture degrees of belief with probabilities—and coherence between degrees of belief is maintained by following the laws of probability theory, of which Bayes' rule is a particularly important consequence (we discuss this more in chapter 3).

Rational theories of all kinds can be viewed as primarily matters of coherence (Chater & Oaksford, 2012; Nau & McCardle, 1991). But, in practice, rational approaches to understanding cognition do not typically attempt simply to ensure that an intelligent agent has coherent beliefs, degree of beliefs, or decisions. Rather, they attempt to solve a specific computational problems, in an optimal way. Indeed, the approach typically fixes some goal (finding the best interpretation of the environmental layout from sensory input, choosing actions to maximize expect reward, and so on) and asks how this goal can best be achieved. Theories of rationality play a crucial role in providing constraints on the types of reasoning that are legitimate (e.g., that the agent follows the rules of probability theory when carrying out Bayesian updating, given some sensory or linguistic input).

Indeed, in line with the spirit of reverse engineering, the aim is to work out which engineering problems the mind is solving; and to figure out the optimal way in which the problem can be solved, in the light of the constraints of rationality.[3] Within the cognitive sciences, then, rational approaches to cognition typically abstract away from the question of what calculations the mind performs, but focus instead on the nature of the cognitive problem being solved. The question of interest is then: What is the optimal solution for the specified cognitive problem? It is then assumed that the cognitive system will, to a greater or sometimes lesser extent, accord with that solution. It is typically, though not universally, assumed that finding an optimal solution requires following rational principles, though perhaps implicitly. In particular, dealing optimally with uncertainty might involve behaving in line with the rules of probability theory—hence the focus on Bayesian probabilistic methods throughout this book.

Working out the optimal solution to a cognitive problem may itself, of course, require substantial calculation. But this does not imply that the agent need necessarily carry out such calculation—merely that it adopts, at least to some approximation, the resulting solution. To borrow an example from Marr (1982), complex aerodynamic calculations will be required

2. In behavioral economics, finance and psychology, *uncertainty* is sometimes used to describe unquantifiable unknowns, and *risk* is used to describe cases in which events can be assigned probabilities (e.g., derived from a statistical model of some kind). In a Bayesian framework, the distinction is less hard-and-fast because the scope of probabilistic models is expanded considerably. In any case, we use uncertainty in the informal sense here, so in our usage, probability theory is potentially useful for modelling uncertainty, rather than uncertainty being "out of scope" by definition.

3. The connection between rationality and optimality is not, though, uncontroversial (Chater et al., 2018). The approach taken here is that rational constraints provide a framework within which optimal, or approximately optimal, solutions can most readily be formulated.

to determine the optimal shape of a bird's wing or the optimal movements of the wing that lead to efficient flight. The bird need not, of course, have carried out such calculations—its wing shape, or patterns of movements, are instead presumably the product of natural selection. Indeed, this type of adaptationist explanation of biological structures, and of the basis of behavior (e.g., optimal foraging theory; see Stephens & Krebs, 1986), are ubiquitous through the biological sciences.

Or consider the economic explanation of the distribution of patterns of production and trade across nations, according to the principle of comparative advantage, popularized by David Ricardo (1817). The principle is, roughly, that each nation will specialize in producing goods or services in which it is comparatively efficient in relation to other nations, even it is absolutely more or less efficient than other nations in producing all goods. The optimal equilibrium distribution of production that is calculated by economic theory need not be calculated by farmers and factory owners. Instead, in conditions of free trade, this solution will arise simply if those farmers or factory owners switch production to other goods if they can make more profit by doing so. More generally, the explanation of economic phenomena requires calculation of equilibria, and such calculations may in general be complex; but no assumption that economic agents find these equilibria by carrying out any such calculations is needed.

From the perspective of reverse engineering, the rational approach operates by specifying aspects of the nature of the solution that must be found—without specifying the details of how the solution is achieved. So the shape and weight of the bird's wing might be specified, by virtue of the nature of the problem—without specifying precisely what such a structure might be made of or how it might grow.

Marr's (1982) computational level of explanation, which we introduced in chapter 1, is perhaps the paradigm of rational explanation in cognitive science. Such explanation focuses on the nature of problems that an information-processing system confronts, considers how the problem may be optimally solved in principle, and assumes that the system adopts a solution close to this optimum. Crucially, for many interesting problems in cognitive science, this involves a careful analysis of the *environment*. In the case of vision, on which Marr was focused, understanding the visual environment requires understanding the relationship between the natural world and the images to which it gives rise. Only then can the inference problem of reconstructing the structure of the world from those images properly be characterized (Richards, 1988).

Marr drew attention to Gibson's (1979) ecological analyses of visual perception for some elegant examples of computational-level explanation. So, for example, the rate of expansion of the optic array may indicate time-to-contact with an object in the array (e.g., when a walking observer will hit a wall in front of her if she continues at a constant speed, or when a moving object will reach a static observer, and hence how an action to catch or avoid that object should be timed). Similarly, for an agent in forward motion, the point from which optic flow radiates indicates the direction in which the agent is heading (Gibson, 1950). If people and animals are sensitive to such information, then it may be possible to explain aspects of their perceptuo-motor behavior without reference to any knowledge of the specific calculations, or neural circuits, involved in estimating optic flow. Nonetheless, the task of processing the image itself (e.g., finding edges, surfaces and objects from raw pixel-level input) was, for Marr, also a problem that required computational-level analysis, and hence required understanding the underlying natural constraints from which images are generated by the world. This was a problem that Gibson's ecological approach ignored; and that symbolic approaches to vision in AI (e.g., Winston, Horn, Minsky, Shirai, & Waltz, 1975) typically presumed already to have been solved.

In the same spirit, Shepard (1987, 1994) provided a number of important rational explanations, without direct reference to computational mechanisms. For example, Shepard (1987) famously argued that a universal law of generalization between pairs of objects should apply in inductive inference—a law that aims to relate the nearness of those objects in an internal psychological space and the probability that they will share arbitrary properties. The scope and justification of the universal law have been extensively analyzed (e.g., Tenenbaum & Griffiths, 2001a; Myung & Shepard, 1996; Chater & Vitányi, 2003a; Chater & Brown, 2008)—and the argument has proceeded, with some exceptions (Ennis, 1988), without direct reference to the calculations involved in such generalization.

Anderson (1990, 1991b) extended this standpoint to a wide range of phenomena in higher-level cognition. His methodology of *rational analysis* made general assumptions about the cognitive environment, the task to be achieved, and (where necessary) relevant cognitive constraints (e.g., concerning memory or processing limitations). The aim was to explain detailed empirical data as an optimal (or near-optimal) solution to the task, given an environment with a particular structure and subject to the relevant cognitive limitations. Anderson's (1991b) account of categorization elegantly captures the process of expanding the set of categories, potentially indefinitely, as a stream of new items comes in (here, one aspect of the environment is that it can contain indefinitely many different types of objects, rather than being limited to some finite number of categories); and he attempts to explain data on human categorization in the light of the structure of natural categories. We shall consider Anderson's model of categorization in chapter 9, as an early and important example of a nonparametric Bayesian mixture model.

Anderson (1990) and his colleagues' work on rational models of memory provides an elegant illustration of the power of the rational approach. Traditional theories of memory in cognitive psychology had typically viewed limitations on memory performance as arising from the operation of particular cognitive *mechanisms*; but Anderson argued that memory may be carefully adapted to the demands of information retrieval in natural environments. Items should be easy to retrieve in proportion to their probability of being needed by the cognitive system. And this probability, he noted, depends on the structure of the informational environment. So, in a world in which the same topics recur over short periods, recently encountered information is more likely to be required in the current context than information encountered in the distant past. Thus, the greater memorability of recent information may result not from some mechanistic aspect of memory (e.g., a putatively inexorable process of decay of memory traces over time), but rather because such information is more likely to be required in the present context. Anderson and Milson (1989) studied models of library borrowing of specific books and repeated access to computer files, indicating temporal "clumping" of information-retrieval in these domains. They argued that estimates concerning how the probability of needing a piece of information reduces over time accurately reflect the degree to which information remains available in human memory. Anderson and Schooler (1991) directly collected information about the recurrence of topics in newspaper headlines over time and used this to model a variety of experimental memory data. Schooler and Anderson (1997) extended this work further to capture the observation that the probability that an item will recur depends not only on its history of past occurrence but also on the occurrence of associated items—and provides an integrated model of the operation of both factors in human memory. This implies that "priming" a word by presenting its associates may have a rational basis, in sharp contrast to the accounts of traditional theories in cognitive psychology, which had typically assumed that priming was no more than a side effect of processing mechanisms, such as spreading activation. A more general rational account of word association effects has been developed more recently in the form of Bayesian topic models (Griffiths, Canini, Sanborn, & Navarro, 2007), which will be discussed in chapter 5.

Rational approaches to cognition do not necessarily require the application of probabilistic ideas. But when the problem under consideration is a problem of uncertain inference, the application of probability is often very natural. So, while Marr and Gibson focused on the structure of the natural environment and the structure of the visual input to which that environment gives rise, the process of inferring one of many possible structures from an ambiguous visual image is naturally interpreted as a problem of Bayesian inference. Indeed, the Bayesian perspective has been highly influential in the engineering arena of computer vision (e.g., Geman & Geman, 1984; Weiss, 1996), and, in parallel, has been rapidly gaining ground in theories of human perception (Feldman, 2001; Feldman & Singh, 2005; Yuille & Kersten, 2006; Knill & Richards, 1996). This recent work can be seen as continuous with Helmholtz's (1866/1962) statement of the *likelihood principle*—namely, that the perceptual system seeks the most probable interpretation given its input data.

A similar probabilistic revolution has been underway in the domain of language. Parsing a sentence (i.e., constructing a representation of the sentence's syntactic structure) could, in the symbolic tradition, be viewed as analogous to logical inference. To take the simplest possible approach, we can set the goal to be to infer *S* (denoting a complete sentence) from a concatenation of words (e.g., *the dog barked*), together with a grammar. This grammar specifies rules, such as $S \rightarrow NP\ VP$, $NP \rightarrow Det\ N$, or $N \rightarrow dog$, which may be interpreted as meaning that the sequence *NP VP* allows *S* to be inferred; the sequence *Det N* allows *NP* to be inferred; or the word *dog* allows *N* to be inferred (for a recent example of the symbolic grammar-as-logic viewpoint, see Morrill, 2019). But the spectacularly local, and sometimes global, ambiguity of natural language means that there are typically very large numbers of parses that must be pursued, creating severe problems of computational complexity. A probabilistic alternative is to associate probabilities with grammar rules so that the grammar can be viewed as a probabilistic sentence generator. From this viewpoint, parsing is an inverse problem, that of inferring the most likely parse (i.e., application of the grammatical rules) that might have generated the sentence. By investigating the statistical structure of corpora of real language, it becomes possible to "tune" such a parser to have probabilistic "biases" in resolving local syntactic ambiguity that are most likely to lead to successful parsing. A range of work in psycholinguistics has shown that such models explain a good deal of empirical data (e.g., Chater & Manning, 2006; Jurafsky, 2003). Intriguingly, similar Bayesian models in other domains, such as spoken word recognition and reading (e.g., Feldman, Griffiths, & Morgan, 2009; Norris, 2006), have been able to capture many of the phenomena explained by connectionist models in these domains, but more simply and transparently.

How far has it been possible to apply rational approaches to what might be expected to be its core domain: human reasoning? Experimental tasks in reasoning typically involve presenting a verbally stated set of premises, and asking participants to draw a verbally stated conclusion. One popular standard against which performance can be assessed has been deductive logic, according to which the conclusion must follow with certainty, if the premises are true. According to this standard, human reasoning performance has consistently been judged to be significantly and systematically error-prone (e.g., Evans, Newstead, & Byrne, 1993).

But is logic necessarily the right standard? The ubiquity of uncertainty in cognition applies just as much to reasoning and argument as it does to vision or language processing. Thus, most everyday inference, indeed perhaps almost all inference outside mathematics, is *defeasible*: conclusions follow only tentatively from the given premises and can be overturned in the light of new information. This applies even to apparently canonical logical inferences such as *modus ponens*, such as the inference from *if the weather is good, then I will play tennis*, and *the weather is good*, to *I will play tennis*, which can be overturned by any number of additional premises such as *my tennis partner is sick*, *my car won't start*, or *the court is unexpectedly closed*. In symbolic AI, one approach to handling defeasability was to extend logical methods to create so-called nonmonotonic logics (McCarthy, 1980; Reiter, 1980), a project that ran into considerable technical difficulties. But an alternative approach is to assume that knowledge and reasoning is probabilistic through-and-through, an approach pioneered in AI by Pearl (1988, 2000).

Applying this perspective to the psychology of reasoning, human reasoning should not be viewed as a matter of deriving new information that follows by logical necessity from what one already knows. Rather, reasoning serves to draw tentative conclusions about what is *probable* given current knowledge. From this point of view, many of the logical blunders and fallacies to which people succumb can be reinterpreted as good probabilistic arguments. Suppose, for example, that we know that *if Mac the Knife is the murderer, then his DNA will be present at the scene of the crime*; and Mac's DNA is duly found, so that we come to know that *Mac's DNA is present at the scene of the crime*. Then we may naturally draw the conclusion that Mac is indeed the murderer. But notice that this line of argument, from a logical point of view, corresponds to an elementary logical fallacy sometimes known as *modus morons*: to infer *p* from statements of the form *if p, then q* and *q*. This line of reasoning is, indeed, logically invalid. There is no logical necessity that Mac is the murderer; his DNA might, for example, be at the scene of the crime because he was innocently present before (or indeed after) the time of the murder; or perhaps some of Mac's possessions were planted at the crime scene by the police to incriminate him. But, given our background knowledge (including, perhaps, our knowledge of Mac's appalling character and record of misdeeds), we might reasonably suppose that, without further information to the contrary, overwhelmingly the most *probable* explanation of the DNA is Mac's guilt. Indeed, this style of reasoning, known as *inference to the best explanation* appears central to both everyday and scientific thinking (e.g., Harman, 1965).

So, from a Bayesian standpoint, our natural inference that the discovery of the DNA evidence counts against Mac's innocence is entirely reasonable. Yet from a strictly logical point of view, such an argument is fallacious. Thus, switching from a logical to probabilistic standard helps make sense of human reasoning and restores our faith in human rationality

(Oaksford & Chater, 1998a, 1994). It turns out, moreover, that a probabilistic approach to human reasoning also provides the foundation for a productive program of explaining detailed empirical data about patterns of human reasoning and argument (e.g., Chater & Oaksford, 1999; Evans & Over, 2004; Hahn & Oaksford, 2007; Oaksford & Chater, 1994), in contrast to accounts of reasoning that take logical validity as their starting point (e.g., Johnson-Laird, 1983; Rips, 1994).

2.4 The Bayesian Synthesis

We have briefly outlined three important approaches to cognitive science: the symbolic approach, which emphasizes the importance of structured representation; connectionist approaches, which focus on learning and neural implementation; and rational approaches, which aim to explain *why* the cognitive processes involved in vision, language processing, and reasoning provide a reliable guide to thought and behavior.

The approach outlined in this book attempts to synthesize the key insights from all three approaches. We shall emphasize the importance of structured representations throughout, running from relatively restricted, but nonetheless powerful, structures such as graphs to grammars, relations, logics, and programs. The patterns of inference defined over such structured representations, however, will not be restricted to standard logical operations, but will follow the probability calculus. We shall define Bayesian inference over a range of sophisticated structured representations. The connectionist emphasis on learning can be captured by viewing learning as a particular case of Bayesian inference, a viewpoint that has long been mainstream within the foundations of connectionism and has indeed been the inspiration for the development of many connectionist learning methods (Ackley, Hinton, & Sejnowski, 1985; Hinton & Sejnowski, 1986; Hinton & Salakhutdinov, 2006; Hinton & Sejnowski, 1983; MacKay, 1992a). Moreover, the connectionist emphasis on computations that do not require a central controller, but can be distributed over a large number of simple processing units, operating cooperatively and in parallel, also aligns naturally with many Bayesian methods, and especially with those using graphical models (see chapter 4).

Our aim, then, is to combine the complementary insights of the symbolic, connectionist, and rational traditions in cognitive science rather than to outline a competing approach. The contribution of each tradition is, we believe, essential; and each alone is incomplete. Given that reverse-engineering the mind can draw only on the range of engineering techniques that have so far been developed to create intelligent systems, the fact that modern computer science relies centrally on symbolic representations and their manipulation provides a powerful prima facie argument for imputing such representations to the cognitive system. To appreciate the power of this style of argument, consider the following: Suppose that we want to build an engineering system for processing language, recognizing images, controlling a robot arm, or reasoning about the everyday world; but suppose that we do not permit ourselves the use of any symbolic representations (i.e., no lists, graphs, trees, relations, grammars, and so on). We would not know even how to begin—indeed, no such engineering approaches exist. To be sure, there are elements of these problems that may be analyzed nonsymbolically. For example, perhaps certain aspects of morphology can be analyzed in connectionist terms (consider the controversial case of the English past tense; e.g., Rumelhart & McClelland, 1986); but even if this is right, such a connectionist model presupposes a symbolic representation of speech in terms of phonemes, or at least bundles of phonetic features—the inputs and outputs of a connectionist system are not raw acoustic waveforms, but rather symbolic representations of language. Indeed, creating such representations requires that speech be split from nonspeech in the acoustic input; and that speech be encoded in highly abstract symbolic terms (abstracting away, for example, from details of the voice, intonation, distance of the speaker, head angle of the listener, amount and nature of echoes in the acoustic environment, and so on). Moreover, to apply a connectionist model to map present to past tenses of English verbs, it is necessary to first represent some inputs as *verbs* and to represent which verbs are in the present or the past tense. Moreover, the learner needs to be able to store and retrieve these representations to attempt to learn the mapping between them (i.e., the learner needs to be able to pick out corresponding forms, such as *pass* and *passed*, *go* and *went*, or *sing* and *sang*, and to shunt a symbolic representation of present-tense forms into the "input" buffer in the connectionist network, and to shunt symbolic representation of corresponding past-tense forms to the "output" buffer. Only then can the connectionist system even begin to learn the mapping between present and past tenses. Thus, even to set up a learning problem so that a connectionist system can begin to learn appears to presuppose a vast amount of preprocessing. And this preprocessing seems to require the liberal use of symbolic representations.

An advocate for a purely connectionist approach to cognition might respond that, ultimately, all such preprocessing might itself be implemented in some nonsymbolic, connectionist fashion. Indeed, it is possible that connectionist engineering may one day be developed to show how this can be done, without merely re-implementing symbolic operations in connectionist machinery (which would itself be a hugely important achievement). But at present, this remains a mere theoretical possibility; for now, we simply have no idea how to reverse-engineer the mind without harnessing the enormous power of structured symbolic representations.

We provisionally assume, then, that structured, symbolic representations are crucial to reverse-engineering cognition. Considerations, outlined in this chapter, at the level of rational explanation require that it must be possible to carry out probabilistic calculations over such structured representations. And the driving motivations for connectionism, the emphasis on learning and on implementation in parallel, cooperative computational hardware defined over simple processing units, further require that such representations can be learned; and that inference and learning can themselves be implemented in parallel over simple processing units. This book aims to sketch how such a Bayesian synthesis may be developed; we suggest that this is the most promising and powerful approach currently available for reverse-engineering the mind.

2.5 Explaining the Mind, from the Top Down

Mental phenomena can be explained at a number of levels of explanation, ranging from phenomenology to neurochemistry. As mentioned in chapter 1, Marr (1982) highlighted three key levels: the computational level (mentioned previously), which captures the problem faced by the agent and what counts as a solution to that problem; the algorithmic level, describing the representations and computational processes operating on those representations, which allow the problems faced at the computational level to be solved successfully;

and the implementational level (i.e., the specification of how the algorithmic-level computations are physically instantiated in neural activity).

The very project of attempting to reverse-engineer the mind suggests a particular order of priority for generating explanations at each of these levels, as we noted in chapter 1. In reverse engineering, as in conventional engineering, the primary questions are: What is the problem to be solved? And what would count as a solution? These are questions at the computational level. Only then is it possible to ask what representations and processes might embody a solution (i.e., to formulate possible explanations at the algorithmic level). And only once those representations and processes are specified can we determine how to build such representations and processes into physical machinery. Thus, to the degree that reverse-engineering parallels conventional engineering, we would expect that a "top-down" approach to building cognitive models is likely to be most productive (Griffiths, Chater, Kemp, Perfors, & Tenenbaum, 2010).

We stress that this top-down focus is no more than a methodological recommendation. Ultimately, explanations at each level must be mutually compatible, and hence insights concerning any level can help constrain theorizing at any of the others. For example, insights about neural architecture may lead to powerful constraints at the algorithmic level—the class of algorithms that can run quickly on slow, highly interconnected, and very simple, neural units—is likely to be small. Or the observation of apparently distinct neural systems underpinning some aspect of behavior might reasonably strengthen the suggestion that the brain may be implementing distinct algorithms, perhaps solving distinct computational problems (see Dayan & Daw, 2008).

The reverse-engineering approach suggests, though, that the most productive research strategy is that of the engineer: start with a function; find a design that can work to achieve that function; and figure out how to instantiate that design in physical (or, in the case of reverse engineering, biological) machinery.

Our approach in this book is therefore initially to sketch classes of probabilistic inference problems faced by the cognitive system; we then consider how such problems can be solved (or, more typically, approximated) using specific representations and algorithms using methods originally developed in optimization and machine learning. Many of these methods can naturally be implemented in parallel hardwire over simple processing units. This gives room for some optimism that such algorithms may ultimately be implemented in neural hardware. There have been some promising inroads toward understanding how interesting computations, including Bayesian calculations, can be implemented neurally (e.g., Dayan & Abbott, 2001; Ma, Beck, Latham, & Pouget, 2006); but these issues lie outside the scope of this book.

2.6 Summary and Prospectus

In this chapter, we have outlined three important traditions in cognitive science: the symbolic approach, connectionism, and rational explanation. We have argued that successfully reverse engineering the mind is likely to require synthesizing the key insights from all three, carried out in a top-down manner: focusing initially on the nature of the problems that the cognitive system faces, then moving to questions of representations and algorithms, and finally ending up with neural implementation.

The rest of this book focuses on spelling out what this research program leads to in practice; and, we hope, inspiring others to help push this approach forward over the coming decades. We are going to recapitulate some of the history of this research program in the following chapters, beginning with fundamental concepts and examples from some of the earlier work on Bayesian models of cognition and working toward advanced ideas and more contemporary results.

3

Bayesian Inference

Thomas L. Griffiths and Joshua B. Tenenbaum

People solve a vast array of challenging inductive problems, learning words, categories, rules, causal relationships, and more from limited data. Solving these problems requires dealing with uncertainty. Inductive inferences go beyond the information given, seeking a reasonable answer even though we know we can never be absolutely certain that it is correct. In this chapter, we provide a more formal introduction to the ideas behind Bayesian inference, which gives us a powerful set of tools for solving these problems.

Bayesian inference starts from the idea that we need a mathematical system that lets us represent uncertainty—that we have different hypotheses about the state of the world, and those hypotheses are assigned different degrees of belief. These degrees of belief quantify how strongly we would endorse each hypothesis, indicating how likely we think it to be true and what kinds of bets we might be willing to place on it. The core assumption behind Bayesian models of cognition is that the appropriate way to represent this uncertainty is via *probability theory*, which immediately gives us a set of principles to use to answer questions about how we should change our beliefs in light of evidence.

We spell out this idea in detail in the next section, and then turn from a consideration of what Bayes' rule means to a demonstration of how it can be applied in a wide range of contexts. As we introduce new mathematical concepts, we provide examples of how those concepts can be used to make sense of various aspects of human cognition, drawing on the existing literature on Bayesian models of cognition. Our focus here is on providing an introduction to key ideas from Bayesian statistics—deeper and more mathematical treatments can be found in Berger (1993), Bernardo and Smith (1994), and Robert (2007).

3.1 What Is Bayes' Rule, and Why Be Bayesian?

Being Bayesian requires a leap of faith. You need to be willing to assume that people's degrees of belief can be represented by probability distributions. This means that for each hypothesis h that you have about the state of the world, there is some probability that h is true, $P(h)$, that obeys the laws of probability theory. More precisely, if we use $\mathcal{H}$ to denote the set of all hypotheses, $P(h)$ should obey the following *axioms*, which define the mathematical object that is a probability:

- $P(h) \geq 0$ for all $h \in \mathcal{H}$.
- $P(\text{some } h \in \mathcal{H} \text{ is true}) = 1$.
- For any countable set of mutually exclusive hypotheses $\mathcal{H}_n = \{h_1, \ldots, h_n\}$, $P(\text{some } h \in \mathcal{H}_n \text{ is true}) = \sum_{i=1}^{n} P(h_i)$.

From these axioms, it is possible to derive all the properties that we associate with probabilities, such as $P(\text{some } h \in \mathcal{H} \text{ that is not } h' \text{ is true}) = 1 - P(h')$.

One important consequence of representing degrees of belief with probabilities is that believing more in one hypothesis means believing less in another (assuming that those hypotheses are mutually exclusive). In physics, it is common to talk about physical laws, such as the conservation of momentum. This law asserts that the total momentum of a closed system remains constant. For example, the total momentum of two objects that collide with one another remains the same after the collision, even if their velocities change. Since the total probability that some hypothesis is true is 1 (via the second axiom given in this chapter), we only have one unit of probability to share among all our hypotheses. As a consequence, increasing the probability of one hypothesis has to be accompanied by decreasing the probability of another. This can be thought of as a law of *conservation of belief*—whatever happens, we only have so much belief to distribute among our hypotheses.

Thinking that degrees of belief can be represented by probabilities (a notion known as *subjective probability*) is particularly important in the context of making probabilistic models of *cognition*. If we just wanted to make probabilistic models of *behavior*, we would have no need to talk about internal states or degrees of belief, and we could survive with another interpretation of probability (such as *frequentism*, where probabilities are taken to reflect the long-run relative frequency of an event occurring if the circumstances in which it occurs are repeated many times). If you accept subjective probability, then it follows immediately that Bayes' rule (which we will get to in a moment) is the way that you should update your beliefs, because Bayes' rule is just a simple consequences of the axioms of probability theory. So we will start by considering some of the arguments for subjective probability and then turn to the consequences of making this assumption. A historical perspective on these different views of probability is given in Hacking (1975), Stigler (1986), and Gigerenzer et al. (1989), and a detailed treatment of arguments for subjective probability appears in Jaynes (2003).

3.1.1 Why Accept Subjective Probability?

There are two traditional forms of argument that degrees of belief can be represented by probabilities.[1] One form shows that this is inevitable for any agent who obeys some simple rules of common sense. The other argues that not following the probability axioms will result in being able to be exploited (and would thus not be adaptive).

Cox's Theorem The classic "inevitability" argument is *Cox's theorem*. Richard Cox (1946, 1961) showed that if you accept some simple axioms about degrees of belief—he called these degrees of belief "plausibilities"—you find that these degrees of belief end up being

1. It is worth noting that the idea that rational agents should encode their degrees of belief via probability theory remains controversial; for example, see Machina and Siniscalchi (2014).

equivalent to probabilities. Cox wanted plausibilities to be real numbers that behaved in a way that everybody could agree looked like common sense. He showed, for a simple set of axioms, that plausibilities following those axioms would be equivalent to probabilities. An example of such an axiom is that the plausibility of a proposition and its negation are inversely related. Since a double negative is a positive, the function that maps the plausibility of a statement to the plausibility of its negation must be its own inverse. This is satisfied by probability theory. If we use $\neg h$ ("not" h) to indicate that some hypothesis other than h is true, then $P(\neg h) = 1 - P(h)$. Since h is the same as $\neg\neg h$, it follows that $P(h) = 1 - P(\neg h) = 1 - (1 - P(h)))$. Starting with a wider set of axioms along these lines, Cox showed that the standard laws describing probabilities could be derived from his axioms. Under this view, probability theory is just common sense appropriately formalized.

Dutch Books A *Dutch book argument* (Ramsey, 1926/1931; de Finetti, 1937) shows that violating the laws of probability will cost you money. The basic form of the argument relies on the mechanism of buying and selling tickets. You get to set the price of a ticket that pays the owner $1 if a particular event occurs, and then another player gets to choose whether they will buy that ticket from you, or sell it to you (and you have to buy a ticket that they sell you). It is possible to show that the prices of the tickets should correspond to the probabilities of the events, and consequently, that violating the principles of probability theory in setting these tickets will lose you money. For example, if there are only two possible hypotheses that could be true, h_1 and h_2, then the prices of the tickets that pay on the truth of these hypotheses should sum to $1. If the sum is less than $1, then your opponent can buy both tickets and be guaranteed to make the difference between the total price and $1. If the sum is greater than $1, then your opponent can sell these tickets to you and again make the difference between the total price and $1. Similar tricks work for all the probability axioms, establishing that an agent that doesn't behave in accordance with these axioms will lose money in any situation analogous to this betting game. In practice, we take bets with not just our money but our lives, as we choose actions based on our beliefs, so maintaining beliefs in a way that follows the axioms of probability theory might be expected to be adaptive behavior.

A More Pragmatic Bayesianism The other two arguments given in this section are about why people *should* be Bayesian. However, to make probabilistic models of cognition, we only need to convince ourselves that it is reasonable to *assume* that people are Bayesian. If they really are (or really should be), then this assumption seems justified. This is the essence of the argument behind Anderson's (1990) approach of rational analysis—try to explain people's behavior in terms of adaptive solutions to the problems that they face, which naturally leads to Bayesian models. But there can also be other reasons for assuming that people are Bayesian.

One good reason for assuming that people are Bayesian is that doing so is useful. To the extent that Bayesian models of cognition are able to predict and explain behavior, it seems that this might be a worthwhile working hypothesis. As mentioned in chapter 2, there are some advantages to rational analysis, such as forming connections to other disciplines, and having a language for describing human learning that is expressed in terms of problems and translates naturally into implementation in machines. These are all ways in which

Bayesianism can be a useful assumption in analyzing cognition: if it turns out to be an effective source of models of human behavior, the payoffs are significant.

Another, perhaps worse reason for assuming that people are Bayesian is that we would like it to be true. If it really were the case that people are Bayesian, our lives as cognitive scientists would be much easier. We would have a complete theory of learning—Bayesian inference—and we would just need to work out the assumptions and representations that people use, with this framework providing a guide to the predictions about behavior that result from using specific assumptions and representations. Since this would be valuable, it might be worth seeing how far this framework can take us.

3.1.2 Probabilistic Preliminaries

Bayesian inference grows out of the simple formula known as *Bayes' rule*. When stated in terms of abstract random variables, Bayes' rule is no more than an elementary result of probability theory. In this section, we provide a brief introduction to the key ideas behind probability theory and some of the language that we are going to use to talk about them. For a more detailed introduction to probability theory, see Pitman (1993).

Previously, we stated the axioms of probability theory as applied to hypotheses. More generally, we can apply these axioms to mutually exclusive events that could occur in the world. For example, if we flipped a coin twice, there are four events that could occur, corresponding to all possible sequences of heads $\mathtt{H}$ and tails $\mathtt{T}$: $\mathtt{TT}$, $\mathtt{TH}$, $\mathtt{HT}$, and $\mathtt{HH}$. Each of these events is assigned a probability that is greater than or equal to 0, such that the probabilities of all the events sum to 1.

Often, rather than talking about these *atomic events*, we want to talk about variables that describe sets of events. For example, we could define the *random variable* A to correspond to the outcome of the first coin flip. The probability that A takes on value $\mathtt{H}$ is the sum of the probabilities of all sequences where the first coin flip comes up heads (i.e., $P(\mathtt{HT}) + P(\mathtt{HH})$). The probability that A takes on the value $\mathtt{T}$ is the sum of the probabilities of all sequences where the first coin flip comes up tails (i.e., $P(\mathtt{TT}) + P(\mathtt{TH})$) or, equivalently, $P(A = \mathtt{T}) = 1 - P(A = \mathtt{H})$. We will use uppercase letters to indicate random variables, and matching lowercase variables to indicate the values that those variables take on when those values are unspecified, as in $P(A = a)$. When we write probabilities, the random variables will often be implicit. For example, $P(a)$ refers to the probability that variable A takes on value a.

When we have multiple random variables defined over the same set of events, we can ask questions about their *joint probability*. For example, we could use the random variable B to describe the outcome of the second coin flip. We can now ask about the joint probability distribution $P(A = a, B = b)$, which we can also write as $P(a, b)$. Other notations for joint probabilities include $P(a\&b)$ and $P(a \cap b)$, but we won't use those in this book.

When we have multiple random variables, it's common to refer to the probability of one variable alone, such as $P(A = a)$ or $P(a)$, as a *marginal probability*. These marginal probabilities can be obtained from the joint probabilities by summing over all the values of the other variables (e.g., $P(A = a) = \sum_b P(A = a, B = b)$). As with all the properties of probability theory presented in this section, this can be derived from the axioms—in this case by confirming that this sum correctly identifies the set of atomic events corresponding to the random variable A taking value a. In our example with coin flips, we already showed that $P(A=\text{H})=P(\text{HT})+P(\text{HH})$, which corresponds to summing over all the values of b in $P(A=\text{H}, B=b)$. This process of "summing out" a variable is called *marginalization*.

With two or more random variables, we can also ask questions about *conditional probability*. For example, we might want to ask what the probability is that the second coin flip comes up heads after the first coin flip came up heads. Intuitively, seeing heads once might make you think that it is more likely that you will see it again, particularly if you didn't get a good look at the coin before it was flipped. We can write this conditional probability as $P(B=b|A=a)$ or $P(b|a)$, where | is read as "given." So $P(b|a)$ is the probability of b given a, telling us the probability that b occurs if a is known to have occurred. We can also say $P(b|a)$ is the probability of b "conditioned on" a.

The conditional probability is calculated by restricting our world to the atomic events consistent with a having occurred and reallocating the probability among those events. After the first coin flip comes up heads, there are only two possible events: HT and HH. However, the probabilities of these events no longer sum to 1. To correct for this, we take the new probability of each event to be proportional to its original probability, but multiply it by a number that results in the probabilities summing to 1. In the case of calculating $P(b|a)$, this corresponds to multiplying each probability by $1/P(a)$, since $P(a)$ is the sum of the probabilities of all the events that remain as possibilities. Finally, we calculate the probability of b occurring in this world. This is proportional to $P(a, b)$, and thus it becomes

$$P(b|a)=\frac{P(a,b)}{P(a)}, \tag{3.1}$$

which is equivalent to the sum of the relevant set of atomic events with their new probabilities after conditioning on a having occurred.

Conditional probabilities give us a way to talk about the relationship between random variables. For example, if we were absolutely convinced that the coin being flipped was fair, we might think that the probability of seeing heads on the second flip is just 0.5, regardless of the outcome of the first flip. In this case, the conditional probability of getting heads on the second flip is just the same as the marginal probability. We can capture this by saying that two random variables A and B are independent of one another. *Independence* means that the conditional probabilities of the random variables are not affected by the values of one another, so $P(b|a)=P(b)$ and $P(a|b)=P(a)$.

We can re-arrange equation (3.1) to derive an important principle of probability theory—the factorization of joint distributions, sometimes called the *chain rule*. This allows us to write the joint probability of the two random variables A and B taking on particular values a and b, $P(a, b)$, as the product of the conditional probability that B will take on value b given that A takes on value a, $P(b|a)$, and the marginal probability that A takes on value a, $P(a)$:

$$P(a,b)=P(b|a)P(a). \tag{3.2}$$

Intuitively, we are getting the probability that a happened and b happened by multiplying the probability that a happened by the probability that b happened, given that a happened. In our example with coins, we get the probability that the first coin came up heads and the second coin came up tails by multiplying the probability that the first coin came up heads by the probability that the second coin came up tails, given that the first coin came up heads.

We can now derive Bayes' rule. There was nothing special about the choice of B rather than A in factorizing the joint probability in equation (3.2), so we can also write

$$P(a,b) = P(a|b)P(b). \tag{3.3}$$

It follows from equations (3.2) and (3.3) that $P(a|b)P(b) = P(b|a)P(a)$, which can be rearranged to give

$$P(b|a) = \frac{P(a|b)P(b)}{P(a)}. \tag{3.4}$$

This expression is Bayes' rule, which indicates how we can compute the conditional probability of b given a from the conditional probability of a given b.

3.1.3 Making Sense of Bayes' Rule

In a world where the events that we are concerned with are simply flips of a coin, Bayes' rule doesn't seem like a very dramatic result.[2] It gives us a simple way to calculate some probabilistic quantities (conditional probabilities) from others (conditional and marginal probabilities). But if we now consider a richer world, in which the events that we are reasoning about correspond to the truth or falsehood of hypotheses and the data that might inform such inferences, the assumption that degrees of belief should be represented as subjective probabilities transforms Bayes' rule from a simple tautology of the probability calculus into a powerful description of how those degrees of belief should be changed by experience (i.e., a mathematical theory of learning).

Assume that a learner assigns probabilities $P(h)$ to hypotheses h belonging to set $\mathcal{H}$. We will call these *prior probabilities* since they are the probabilities assigned prior to observing the data. The learner then observes data d and seeks to calculate the degrees of belief that should be assigned to hypotheses in light of these data, $P(h|d)$. If degrees of belief follow the probability axioms, then these *posterior probabilities* should be computed by applying the formula for conditional probabilities, substituting a with d and b with h in equation (3.4) to obtain

$$P(h|d) = \frac{P(d|h)P(h)}{P(d)}, \tag{3.5}$$

where $P(d|h)$ is known as the *likelihood*, and specifying the probability of observing d if h were true.[3]

It is easier to understand Bayes' rule if we expand the denominator to give

$$P(h|d) = \frac{P(d|h)P(h)}{\sum_{h' \in \mathcal{H}} P(d|h')P(h')}, \tag{3.6}$$

2. Although it's worth noting that in his original paper (Bayes, 1763/1958), the Reverend Thomas Bayes was actually trying to solve a problem quite similar to calculating the probability of sequential coin flips—one that was equivalent to the problem analyzed in section 3.3.2.

3. For historical reasons, the likelihood is a property of hypotheses rather than data. So $P(d|h)$ is the likelihood of h, not the likelihood of d given h (which is sometimes how people talk about it). This is due to the way the likelihood is used in maximum-likelihood estimation, which we talk about later in this chapter: in maximum-likelihood estimation, you are choosing the hypothesis with the highest likelihood. This use of likelihood is similar to the way that a Bayesian would use the posterior probability of h and was part of the attempt by frequentist statisticians to avoid having to do things like define subjective priors. For an example, see Fisher (1930).

where $P(d) = \sum_h P(d, h)$ by the marginalization principle, and $P(d, h) = P(d|h)P(h)$ by the chain rule. This makes it clear that the main role of the denominator is to serve as a *normalizing constant*, ensuring that the resulting probabilities sum to 1. Consequently, it is also common to see Bayes' rule written as

$$P(h|d) \propto P(d|h)P(h), \tag{3.7}$$

where the normalizing constant is absorbed into the relation of proportionality. This makes it very clear that the posterior probabilities are really just the prior probabilities modified by the extent to which each hypothesis predicts the observed data.

The two terms—likelihood and prior—that determine the posterior probability have quite natural interpretations in many problems. The likelihood reflects the fit between hypothesis and data, and the prior indicates the a priori plausibility of a hypothesis (which might decrease for hypotheses that have low frequency, are complicated, or seem otherwise improbable). The contribution of these two factors to the conclusions that we should draw is fairly natural and makes intuitive sense in a variety of contexts. Returning to an example introduced in chapter 1, if you see John coughing (your data d), you might consider three hypotheses about the cause of the cough: a cold (h_1), lung disease (h_2), or heartburn (h_3). You might rule out heartburn on the basis of fit, since it might only slightly increase the chance of coughing. A cold and lung disease both fit well with the cough—they increase the probability of coughing—but they differ in plausibility. Normally, a cold is far more common than lung disease, and thus might be the hypothesis that you would select to explain the coughing. However, the plausibility of these two hypotheses might change if you were passing by a hospital and saw John coughing inside. All inductive inferences require considering fit and plausibility. Bayes' rule just tells you how they should be combined to reach a conclusion, using the common language of probability theory to determine the impact of each factor.

In the context of cognitive science, priors become a useful way of describing the *inductive biases* of learners—those factors that make them choose one hypothesis over another when both hypotheses are equally consistent with the data (Mitchell, 1997). It should be clear from equation (3.7) that if $P(d|h)$ is the same for two hypotheses, then the hypothesis that receives greater posterior probability is that which has the highest prior probability. Many questions in cognitive science reduce to questions about inductive biases, including the following:

- What constraints on learning are necessary for children to acquire language?
- What kinds of category structures are easy or hard to learn?
- Do people prefer simpler hypotheses?
- What expectations do people have about the strength of causal relationships?
- What kinds of structures will people tend to identify in random data?

Bayesian models of cognition are thus particularly effective for answering these questions. The assumption of rationality is useful here too, because we can say things like "Given these data, and these assumptions about the interpretation of the data (i.e., the likelihood, $P(d|h)$), an ideal learner can only reach this conclusion if they have inductive biases with these properties."

When talking about priors in cognitive science, it is tempting to equate priors with innate constraints on learning. This is certainly one way of using Bayesian models, but it is not the only way of thinking about priors. Really, the prior probability of a hypothesis reflects everything that a learner knows, except the current piece of data d. So priors can be learned through experience and can capture the influence of knowledge that the learner might have acquired in other domains. A simple demonstration of this is the idea that *today's prior is yesterday's posterior.* Assume that you want to evaluate hypothesis h, and you see two pieces of data, d_1 and d_2. If the two pieces of data are independent conditioned on h, then we can write

$$P(h|d_1, d_2) \propto P(d_1|h)P(d_2|h)P(h) \tag{3.8}$$

since $P(d_1, d_2|h) = P(d_1|d_2, h)P(d_2|h)$ by factorization, and the assumption of independence means that the conditional probability of d_1 does not change based on the value of d_2, so $P(d_1|d_2, h) = P(d_1|h)$. Since we also know that $P(h|d_1) \propto P(d_1|h)P(h)$, we can rewrite this as

$$P(h|d_1, d_2) \propto P(d_2|h)P(h|d_1), \tag{3.9}$$

where we have just changed the constant of proportionality. So if you saw d_1 yesterday and updated y our distribution over hypotheses from $P(h)$ to $P(h|d_1)$, and then you saw d_2 today, you could just apply Bayes' rule using $P(h|d_1)$ as y our prior and compute the new posterior $P(h|d_1, d_2)$. Any prior distribution thus assumes a body of background knowledge that is implicitly being conditioned upon—the data accumulated in all our yesterdays.

3.2 Bayesian Inference with a Discrete Set of Hypotheses

Bayes' rule applies for any form of data d and hypotheses h. However, it is easiest to understand in the case where the hypotheses form a discrete set. We will explore this case in detail next before considering Bayesian inference with continuous hypotheses. We will start with two hypotheses and then consider how this generalizes to multiple hypotheses.

3.2.1 Comparing Two Discrete Hypotheses

The mathematics of Bayesian inference is most easily introduced in the context of comparing two hypotheses. For example, imagine that you are told that a box contains two coins: one that produces heads 50 percent of the time, and one that produces heads 90 percent of the time. You choose a coin and then flip it ten times, producing the sequence HHHHHHHHHH. Which coin did you pick? How would your answer change if you had obtained HHTHTHTTHT instead?

To formalize this problem in Bayesian terms, we need to identify the hypothesis space, $\mathcal{H}$, the prior probability of each hypothesis, $P(h)$, and the probability of the data under each hypothesis, $P(d|h)$. We have two coins, and thus two hypotheses. If we use θ to denote the probability that a coin produces heads, then h_0 is the hypothesis that $\theta = 0.5$ and h_1 is the hypothesis that $\theta = 0.9$. Since we have no reason to believe that one coin is more likely to be picked than the other, it is reasonable to assume equal prior probabilities: $P(h_0) = P(h_1) = 0.5$. The probability of a particular sequence of coin flips containing n_H heads and n_T tails being generated by a coin that produces heads with probability θ is

$$P(d|\theta) = \theta^{n_H}(1-\theta)^{n_T}. \tag{3.10}$$

Formally, this expression follows from assuming that each flip is drawn independently from a *Bernoulli distribution* with parameter θ; stated less formally, heads occurs with probability θ and tails with probability $1-\theta$ on each flip. The likelihoods associated with h_0 and h_1 can thus be obtained by substituting the appropriate value of θ into equation (3.10).

We can take the priors and likelihoods defined in the previous paragraph and plug them directly into equation (3.6) to compute the posterior probabilities for both hypotheses, $P(h_0|d)$ and $P(h_1|d)$. However, when we have just two hypotheses, it is often easier to work with the *posterior odds*, or the ratio of these two posterior probabilities. The posterior odds in favor of h_1 is

$$\frac{P(h_1|d)}{P(h_0|d)} = \frac{P(d|h_1)}{P(d|h_0)}\frac{P(h_1)}{P(h_0)}, \tag{3.11}$$

where we have used the fact that the denominator of equation (3.6), $P(d)$, is the same whether we are computing $P(h_1|d)$ or $P(h_0|d)$. The first and second terms on the right side are called the *likelihood ratio* and the *prior odds*, respectively. We can use equation (3.11) (and the priors and likelihoods defined previously) to compute the posterior odds of our two hypotheses for any observed sequence of heads and tails: for the sequence HHHHHHHHHH, the odds are approximately 357:1 in favor of h_1; for the sequence HHTHTHTTHT, they are approximately 165:1 in favor of h_0.

The form of equation (3.11) helps to clarify how prior knowledge and new data are combined in Bayesian inference. The two terms on the right side each express the influence of one of these factors: the prior odds are determined entirely by the prior beliefs of the agent, while the likelihood ratio expresses how these odds should be modified in light of data d. This relationship is made even more transparent if we examine the expression for the log posterior odds:

$$\log\frac{P(h_1|d)}{P(h_0|d)} = \log\frac{P(d|h_1)}{P(d|h_0)} + \log\frac{P(h_1)}{P(h_0)}, \tag{3.12}$$

in which the extent to which one should favor h_1 over h_0 reduces to an additive combination of a term reflecting prior beliefs (the log prior odds) and a term reflecting the contribution of the data (the log likelihood ratio).[4] Based upon this decomposition, the log likelihood ratio in favor of h_1 is often used as a measure of the evidence that d provides for h_1.

It is straightforward to translate log posterior odds back into posterior probabilities by using the sigmoid function

$$s(x) = \frac{1}{1+\exp\{-x\}}, \tag{3.13}$$

which squashes $x \in (-\infty, \infty)$ down to $s(x) \in (0, 1)$. You can check that $P(h_1|d) = s\left(\log\frac{P(h_1|d)}{P(h_0|d)}\right)$.[5]

4. This expression uses the principle that $\log xy = \log x + \log y$. When we take logarithms, products become sums. We will use this principle extensively throughout the book.

5. The fact that Bayesian inference can be expressed as a nonlinear transformation of a linear function of the data (plus a term for the prior) provides a way to connect this approach to artificial neural networks, in which nonlinear transformations of linear functions are the basic elements of computation (McClelland, 1998). We revisit this theme in chapter 12.

3.2.2 An Example: Psychic Priors

Some of the earliest uses of Bayes' rule as a model of human inductive inference focused on the case of two hypotheses (for a review, see Peterson & Beach, 1967). The focus of much of this literature was on whether people update their beliefs as rapidly as Bayes' rule suggests they should. This was studied using tasks in which people have to infer which of two urns balls are being drawn from, where the urns differ in the probability of producing balls of different colors and those probabilities are known. In this setting, people tend to update their beliefs more slowly than Bayes' rule would indicate—a phenomenon that has been called "conservatism." However, the two-hypothesis setting doesn't just allow us to measure the *rate* of belief updating, but its starting point—the prior probabilities that people assign to different hypotheses. This is particularly important in the context of reverse-engineering the mind, as it illustrates how we can use Bayesian models to identify people's inductive biases.

As part of a Bayesian analysis of people's sense of what makes something a coincidence, Griffiths and Tenenbaum (2007a) ran a study where people were asked to make judgments about one of two scenarios. In one scenario, scientists studying genetic engineering are testing drugs that affect the sex of rats. In the other, scientists studying paranormal phenomena are testing people who claim to possess psychokinetic abilities. People were shown putative data from these studies, consisting of tables that showed how many rats out of a sample of 100 treated with the drug were male, or how many of 100 coin flips came up heads in the presence of a potential psychic. They saw results corresponding to trials on which 47, 51, 55, 59, 63, 70, 87, and 99 males (heads) were observed. They were then asked to make one of two judgments about each of these trials. One group of people (in the "Coincidence" condition) evaluated whether these results were "just a coincidence" or whether they presented "compelling evidence" for the phenomenon. The other group of people (in the "Posterior" condition) used a rating scale from 1–10 to evaluate whether they thought the drug worked or the test subject had psychic powers.

The critical question for Griffiths and Tenenbaum (2007a) was whether these two judgments were correlated—they were testing the theory that our sense of coincidence reflects an inference about an underlying causal relationship in a context where such a relationship is a prior unlikely. If this is the case, coincidence judgments should track the posterior probability of a relationship. As you can see in figure 3.1, these two kinds of judgments were closely (inversely) related.

However, these data can also be used to illustrate a key property of Bayesian modeling: if we know the data that people see and the conclusions that they reach, we can work backwards to infer their priors. In this case, there are two hypotheses: either the data arise by chance (with male/female or heads/tails occurring with probability 0.5) or there is something else at work (and the probability is something other than 0.5). Later in the chapter, we will show how to compute the probability of the observed data under these different hypotheses, but for now, the critical point is that we can actually calculate these probabilities and hence compute a log-likelihood ratio. Using the sigmoid function $s(\cdot)$ introduced in equation (3.13), we can write the posterior probability of a specific hypothesis h_1 as

$$P(h_1|d) = s\left(\log\frac{P(h_1|d)}{P(h_2|d)}\right) \tag{3.14}$$

$$= s\left(\log\frac{P(d|h_1)}{P(d|h_2)} + \log\frac{P(h_1)}{P(h_2)}\right), \tag{3.15}$$

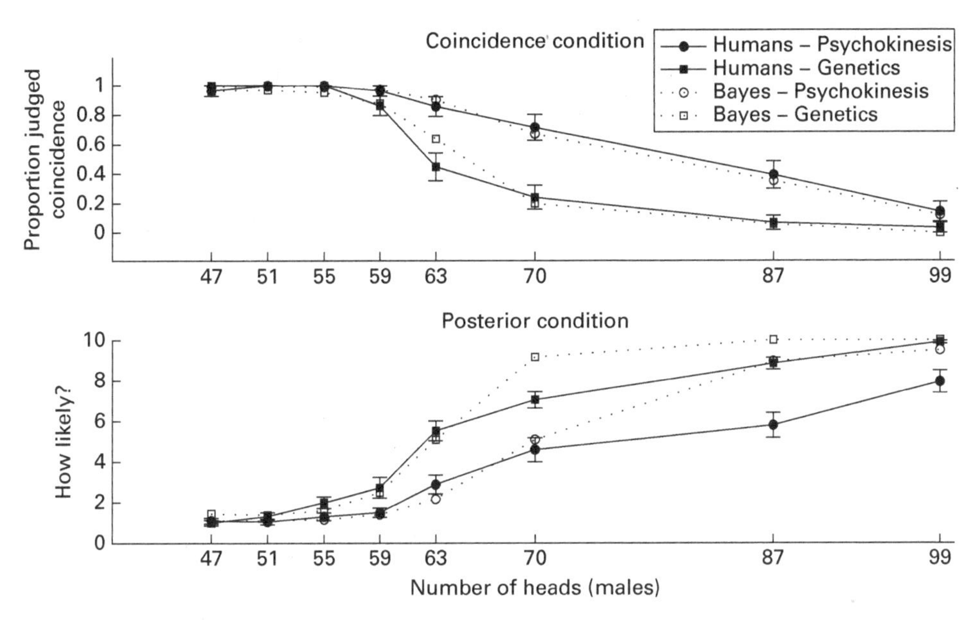

Figure 3.1
Different beliefs in the efficacy of genetic engineering and psychokinesis. Participants made judgments about the results of a set of eight experiments. Under one cover story, these experiments evaluated whether drugs influenced the sex of rats, and under another, whether psychics were able to influence the outcome of a coin flip. One group of participants judged whether these data were just a coincidence (as opposed to compelling evidence), and the other decided whether they thought a causal relationship existed based on these results. People are more willing to judge events involving psychics as coincidences and less willing to conclude that psychokinetic powers exist, consistent with assigning a lower prior probability to this hypothesis. Using Bayesian models, we can obtain quantitative estimates of the difference in prior probability. Figure reproduced with permission from Griffiths and Tenenbaum (2007a).

where we know the log likelihood ratio, $\log \frac{P(d|h_1)}{P(d|h_2)}$ but want to estimate the log prior odds, $\log \frac{P(h_1)}{P(h_2)}$.

Applying this procedure to the judgments of each participant in the Posterior condition, Griffiths and Tenenbaum (2007a) estimated the median prior probabilities assigned to the efficacy of genetic engineering and psychic powers were 0.20 and 0.0011, respectively. The judgments from the Coincidence condition yielded similar estimates. These numbers seem very reasonable as indicators of people's beliefs about science and the paranormal, revealing the inductive biases that people use when reasoning about these kinds of causal relationships. These results also show some evidence of the kind of conservatism observed in classic studies of belief updating (Peterson & Beach, 1967): in the Posterior condition, people's ratings increase more slowly than the Bayesian model would suggest they should, corresponding to an under weighting of the evidence. Interestingly, this does not seem to be the case in the Coincidence condition.

3.2.3 Comparing Multiple Discrete Hypotheses

Bayes' rule naturally extends to comparing multiple discrete hypotheses. Imagine if our goal is not to evaluate just two hypotheses about the nature of a coin, but four: h_1 is "A fair coin," h_2 is "A coin that always alternates heads and tails," h_3 is "A coin that mostly comes up heads," and h_4 is "A coin that always comes up heads." Upon observing a particular sequence of heads and tails d, we compute the posterior distribution over these hypotheses by applying equation (3.5). For example, we might observe the sequence HHTHTTTH. Under h_1, the probability of this sequence is $P(d|h_1)=0.5^8$. Taking "always" to be a probability of 0.99, and assuming that an alternating sequence can start with heads or tails with equal probability, the probability of this sequence under h_2 is $P(d|h_2)=0.5\times 0.01\times 0.99^3\times 0.01^2\times 0.99=0.5\times 0.01^3\times 0.99^4$. If "mostly" means a probability of 0.85, then $P(d|h_3)=0.85^4\times 0.15^4$. Finally, $P(d|h_4)=0.99^4\times 0.01^4$. If all these hypotheses have equal prior probability, then Bayes' rule simply says to divide each of these likelihoods by their sum, giving a posterior probability to h_1, the fair coin, of $P(h_1|d)=0.937$—a pretty compelling case for believing that the coin is fair. Of course, other sequences would yield different inferences.

Bayesian inference isn't just about identifying the most probable hypotheses. It is also a tool for making informed predictions. Having observed sequence d_1, we can make a prediction about what sequence d_2 we are likely to observe next. Applying the rules of the probability calculus, we can write this probability as

$$P(d_2|d_1)=\sum_i P(d_2|h_i,d_1)P(h_i|d_1). \tag{3.16}$$

If d_2 and d_1 are independent conditioned on h (as is the case for coin flips—once you know the probability with which the coin produces heads, it doesn't matter what the results of past coin flips were), then this simplifies to

$$P(d_2|d_1)=\sum_i P(d_2|h_i)P(h_i|d_1), \tag{3.17}$$

where the first term on the right side is the likelihood for that hypothesis and the new data d_2, and the second term on the right side is the posterior probability of h given d_1. The distribution $P(d_2|d_1)$ is known as the *posterior predictive distribution*: what we predict about new data given the beliefs expressed in the posterior distribution. In our coin-flipping example, having observed HHTHTTTH, h_1 predicts heads with probability 0.5, h_2 predicts heads with probability 0.01, h_3 predicts heads with probability 0.85, and h_4 predicts heads with probability 0.99. Applying equation (3.17) gives a posterior predictive probability of heads of 0.522—a smidge more likely than what would be expected from a fair coin.

The posterior predictive distribution has an intuitive interpretation: it's the distribution that we get by averaging the predictions made by each hypothesis, weighted by the posterior probabilities of those hypotheses. For this reason, the calculation shown in equations (3.16) and (3.17) is known as *hypothesis averaging*. In cognitive science, hypothesis averaging is an interesting idea: it shows how a set of hypotheses that each makes deterministic predictions can be averaged into a set of continuous predictions. This is relevant to debates about whether people form hypotheses that are expressed as discrete rules or instead represent information in terms of more continuous quantities: if they do use discrete rules but have uncertainty about which rule applies, patterns of behavior that look continuous can emerge.

3.2.4 An Example: The Number Game

We can now revisit one of the examples introduced in chapter 1 in more detail. Imagine that you are trying to figure out how a computer program works. Unfortunately, you don't have access to the code. The computer program responds either "yes" or "no" to numbers according to whether they satisfy a simple concept. Some possible concepts might be "is odd," "is between 30 and 45," "is a power of 3," or "is less than 10." For simplicity, you can assume that only numbers under 100 are under consideration.

When faced with this task, people's behavior shows some interesting patterns (Tenenbaum, 1999; Tenenbaum, 2000). Given a single example—say, that the program says "yes" to the number 60—people are pretty willing to guess that the computer will say "yes" to both nearby numbers (say 66) and distant numbers (say 90). However, they become much more certain about their generalizations as they get more information, and the resulting patterns of generalization can look like deterministic rules (consider how you would generalize if you discovered that the program says "yes" to 60, 80, 10, and 30) or a more continuous gradient favoring nearby numbers (if, say, the machine accepts 60, 52, 57, and 55). Importantly, both these patterns of generalization can be explained in terms of Bayesian inference.

In this scenario, which is known as the *number game*, learners can entertain many different hypotheses about the numbers accepted by the program. For simplicity, we consider just two kinds of hypotheses: simple mathematical properties, such as being a multiple of 3 or a power of 2, and magnitude properties, such as being a number between 20 and 40 or a number between 57 and 67. Each of these hypotheses can be expressed as a subset of the integers between 1 and 100. In Tenenbaum's model, there were a relatively large number of magnitude hypotheses (5,050, corresponding to all intervals with lower and upper bounds in the range 1–100), and a much smaller number of mathematical hypotheses (corresponding to the multiples and powers of numbers up to 12, odd numbers, even numbers, square numbers, cubes, primes, and numbers ending in the same digit). To apply Bayes' rule to these hypotheses, we need to specify two ingredients: the likelihood and the prior.

Defining a Likelihood via Sampling Tenenbaum presented a Bayesian model of human behavior in the number game, making specific choices for both likelihood and prior. The likelihood was defined by assuming that learners treat the examples of numbers that the program accepts as independent random samples from the set of numbers consistent with that hypothesis. This means that the likelihood for any hypothesis will be inversely proportional to the size of the subset of consistent numbers—because each number is equally likely to be drawn. For example, under the hypothesis that the program accepts all numbers between 57 and 67 (of which there are 11), the probability of observing any one of those numbers as an example would be 1/11. Formally, we can write the likelihood for an arbitrary hypothesis h_i given a sequence X of $n \geq 1$ example numbers accepted by the program as

$$P(X|h_i) = \begin{cases} \frac{1}{|h_i|^n} & X \subseteq h_i \\ 0 & \text{otherwise,} \end{cases} \tag{3.18}$$

where $|h_i|$ is the size of the set of numbers associated with h_i and the exponentiation with n reflects the fact that we are multiplying a probability of $1/|h_i|$ for each of the n examples based on the assumption that they are independent samples. For any hypothesis that does not contain all the examples (i.e., for which the numbers in X are not a subset of those picked by h_i), the likelihood is zero because the examples could not possibly have been drawn from that set.

Equation (3.18) embodies an idea that Tenenbaum dubbed the *size principle*: smaller hypotheses receive more likelihood, and exponentially more as n increases. This principle is just a consequence of the assumption that observations are randomly sampled, and it is an instance of the principle of conservation of belief: we have only one unit of probability to spread over all the observations that we might make, so a hypothesis that includes fewer possibilities assigns each of those possibilities a higher probability.

The size principle provides a way to capture the rapid clarification of our beliefs as we obtain more evidence. For example, on discovering that the program accepts the number 60, we might give some weight to both the hypothesis that the program accepts multiples of 10 and the hypothesis that the program accepts even numbers. There are 10 multiples of 10 between 1 and 100, so the likelihood of this hypothesis is 1/10. There are 50 even numbers between 1 and 100, resulting in a slightly lower likelihood of 1/50 (and perhaps consistent with the intuition that "multiples of 10" seems a little more compelling). However, observing that the program also accepts 80, 10, and 30 makes "multiples of 10" seem much more plausible than "even numbers," despite the fact that these observations are still consistent with both hypotheses. It just seems that it would be a big coincidence to see exactly these examples—all of which are multiples of 10—if the true hypothesis is "even numbers." The size principle explains why: the likelihood for "multiples of 10" is now $1/10^4$, while the likelihood of "even numbers" is $1/50^4$. "Multiples of 10" has gone from being 5 times as likely to $5^4 = 625$ times as likely.

A Prior Distribution on Sets of Numbers Prior probabilities also play an important role in explaining the concepts that people consider. Under the size principle, all that matters when considering a hypothesis is the size of the set of numbers that it picks out. This means that a hypothesis such as "multiples of 10 except 50 and 20," which picks out 8 numbers, would potentially be prefered to "multiples of 10." What makes this seem wrong? The answer has to be located in the prior distribution, which expresses what kinds of concepts people are willing to entertain. Complex hypotheses, such as "multiples of 10 except 50 and 20," should be given very low prior probability (although perhaps not zero—after observing every multiple of 10 except 50 and 20 a few times, you might be convinced!).

Tenenbaum specified a prior distribution $P(h_i)$ by assuming that the two types of hypotheses—mathematical properties and magnitude intervals—are assigned different prior probabilities. This can be captured in a *hierarchical prior*, which breaks down the problem of specifying the probability of a hypothesis into a sequence of choices at different levels of abstraction. In the simplest prior used by Tenenbaum, some probability (call it λ) is assigned to the concept being a mathematical property and some probability $(1 - \lambda)$ is assigned to the concept being a magnitude interval. The probability of specific hypotheses in each of those classes—we might call them "hypothesis subspaces"—is then calculated by distributing the probability of that subspace over all hypotheses within it. Tenenbaum assumed that this was done uniformly for mathematical properties, assigning the same probability to each property, and based on size for magnitude intervals (with a one-parameter Erlang prior over sizes, inspired by Shepard (1987)). Because there are many more magnitude hypotheses than mathematical hypotheses in this model (on the order of 100 times more), the prior probability assigned to any one magnitude hypothesis is typically much less than the prior assigned to any mathematical hypothesis. This can also be seen as a version of the size principle, but now applied at the level of the prior: all other things being equal, hypotheses belonging to smaller, more constraining hypothesis subsapces will be more probable a priori and will tend to receive correspondingly higher posterior probability as well.

Evaluating the Model Having defined likelihoods $P(X|h_i)$ and priors $P(h_i)$, calculating the predictions of this model proceeds in a straightforward manner in two steps. We first multiply priors and likelihoods and renormalize to obtain the posterior probabilities $P(h_i|X)$. Then, to model judgments about the probability that any other number y could be accepted by the program given the examples observed so far, we use hypothesis averaging—averaging the votes of each hypothesis h_i about how to generalize to new instances, weighted by its posterior $P(h_i|X)$ computed by Bayes' rule:

$$P(y \in C|X) = \sum_i P(y \in C|h_i)P(h_i|X). \tag{3.19}$$

This is equivalent to using equation (3.17) to compute the posterior predictive distribution $P(d_2|d_1)$, where d_1 is now the sequence of positive example numbers X observed, and d_2 is the binary proposition $y \in C$ encoding whether an arbitrary number y is in the concept C picked by the computer program. Crucially, the prediction that each number game hypothesis makes about y, or $P(y \in C|h_i)$ in equation (3.19), is just 1 or 0 depending on whether y is in the subset of numbers picked out by h_i.[6]

Figure 3.2 shows human judgments in the number game and model predictions $P(y \in C|X)$ for several sequences of observations. The model captures human generalization quite well, both qualitatively and quantitatively: the Bayesian model's posterior predictive distribution (equation (3.19)) accounts for more than 90 percent of the variance in the average judgments of people. The qualitative behavior is especially revealing, both about the ways that people can generalize concepts from one or a few examples and the way that a single unifying Bayesian model can capture these behavioral differences in a principled fashion. Sequences in class I feature just a single example, either 16 or 60: here, generalization for both people and the Bayesian model is diffuse, representing many possible hypotheses that could explain the one observed data point and the relatively flat posterior that results. In class II sequences, three additional examples are presented after the first to suggest a clear regularity, either 16, 8, 2, and 64 or 60, 80, 10, and 30: here, both people and the Bayesian model converge to a sharp rule like pattern of generalization. Finally in class III sequences, different sets of three examples are presented to suggest more of a magnitude-based concept, either 16, 23, 19, and 20 or 60, 52, 57, and 55: now, as predicted, generalization for both people and the Bayesian model looks more like a graded function of magnitude, smoothly decreasing with increasing distance from the range spanned by the examples.

6. Unlike in computing the likelihood $P(X|h_i)$, the conditional probability $P(y \in C|h_i)$ is not based on the size principle because we are not asking about whether y will be observed from a random sampling process; we are just asking about the probability that y is in the subset picked out by the program or not, yielding a probability of 1 or 0, respectively.

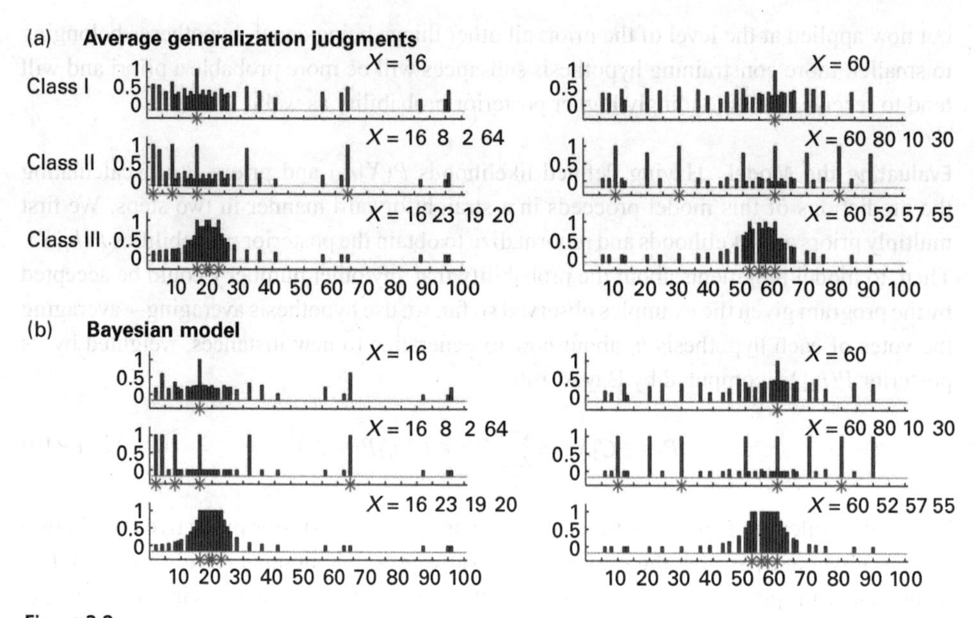

Figure 3.2
Human judgments and model predictions in the number game. The rows show three classes of example sets, indicated by X: class I consists of a single example, class II provides multiple examples consistent with a simple mathematical rule, and class III provides multiple examples clustering in magnitude. (a) People rated 30 probe numbers according to their probability of belonging to the concept, given the observed examples X. Average probability of generalizing the concept to each number is indicated by vertical bar height. (No bar showing for a given number means that the number was not probed.) People's generalization patterns tightened and reshaped when moving from class I to class II or class III, focusing on the most relevant hypotheses and following either a sharp all-or-none rule like pattern or a more graded similarity gradient (classes II and III, respectively). (b) The probability of belonging to the concept as predicted by the Bayesian model. The tightening of generalization, as well as the possibility of convergence to either an all-or-none rule or a similarity gradient, are captured by the interaction of the size principle and hypothesis averaging. Reproduced with permission from Tenenbaum (2000).

Figure 3.3 illustrates how these patterns of rulelike or similarity-like generaliation all fall out of the Bayesian model, reflecting the qualitatively different shapes of the posterior distributions that these short observation sequences induce over hypotheses. Initially, with just a single example (16 or 60), many hypotheses are consistent with the data and have nonnegligible posterior probabilities, including both mathematical and magnitude properties (figure 3.3a). The mathematical hypotheses score higher because their priors are higher (since there are many fewer hypotheses of this class); within each class, the smaller hypotheses also score higher because of the use of the size principle in the likelihood. After several more examples have been observed, generalization tightens up depending on how the examples are distributed. Given examples consistent with a regularity such as "powers of 2," which is much more specific (i.e., smaller) than any other hypothesis consistent with the same data, the posterior probability is much higher for that hypothesis and dominates the posterior predictive distribution. Generalization then appears all-or-none, following that rule (figure 3.3b). By contrast, many similar hypotheses—intervals with very similar ranges and sizes—are consistent with any set of numbers that fall into a small magnitude interval such as the range 16–23, but they are not consistent with any simple mathematical property. These hypotheses all have similar priors and size-based likelihoods, leading to very similar posterior probabilities. Applying hypothesis averaging results in a

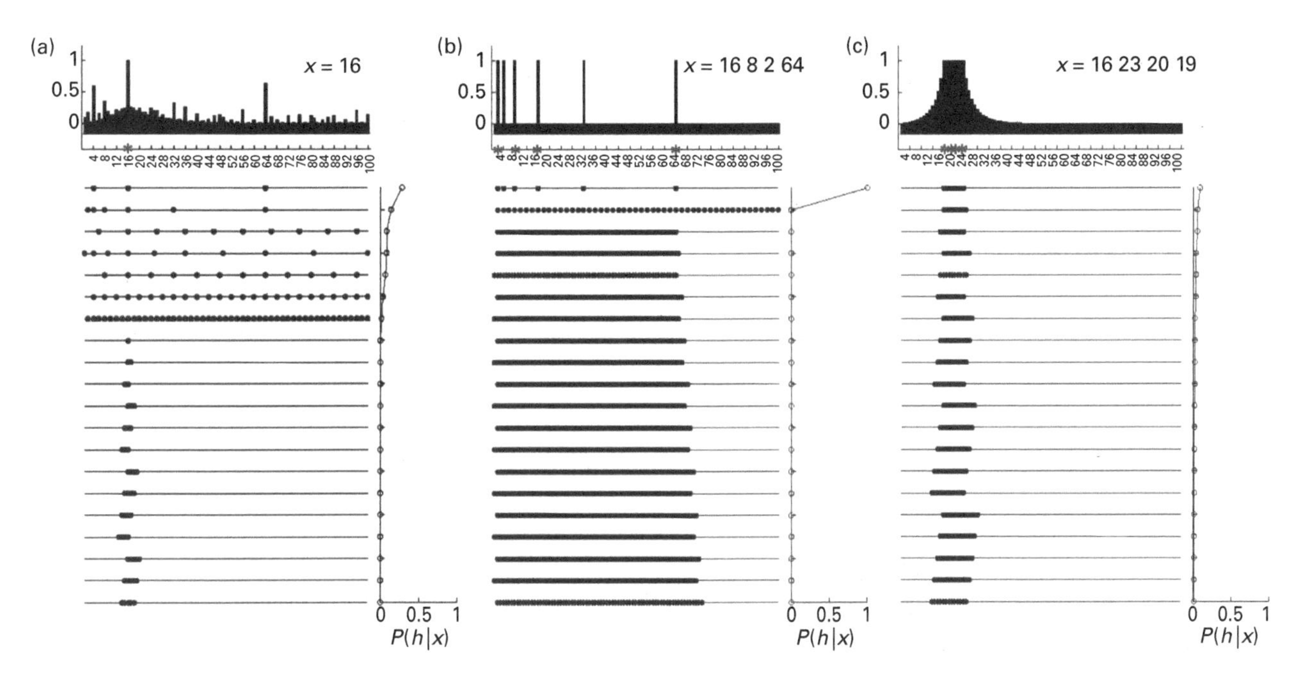

Figure 3.3
A single computational process in the Bayesian model for the number game is able to produce very different patterns of generalization depending on the sequence X of examples observed and the posterior probabilities $P(h_i|X)$ induced by those examples. For each of three example sets (a) through (c), the top panel shows the posterior predictive distribution $P(y \in C|X)$, computed via hypothesis averaging, which represents the probability that each number y between 1 and 100 is in the concept C given the observed examples X (blue stars). Below that are shown the hypotheses with highest posterior probability in order of decreasing posterior $P(h_i|X)$, with each hypothesis plotted using black dots on a number line to indicate the numbers consistent with that hypothesis. Each hypothesis contributes a vote of 1 or 0 to $P(y \in C|h_i)$ for every number y, depending on whether y is in the set of numbers picked out by the hypothesis h_i. These votes are weighted by the posterior $P(h_i|X)$ in hypothesis averaging, yielding graded generalization in the posterior predictive when the posterior is broadly spread out ((a) and (c)), but all-or-none rule like generalization ((b)) when the posterior is concentrated on a single hypothesis that dominates the average. Adapted from Tenenbaum (1999).

smooth gradient of prediction that falls away from the observed examples as the intervals in question become larger (figure 3.3c). This continuous pattern of generalization emerges despite the fact that each hypothesis is discrete, due to the broad posterior distribution and the posterior-weighted voting behavior of hypothesis averaging.

While we are rarely faced with the problem of deciphering the preferences of a computer program, the structure of the number game parallels that of many real inductive problems. Perhaps the most obvious example is word learning: when a child learns a new word, she often does so from positive examples ("That's a *dog*"). Figuring out what set of objects can be given this label is analogous to probing the responses of our hypothetical computer program and can be approached in the same way: hypotheses are sets of objects and observations are sampled from those sets. Xu and Tenenbaum (2007) showed that a similar Bayesian model can be used to explain the inferences that adults and children make in laboratory word-learning tasks, including analogous patterns of graded similarity-like generalization, given one example and convergence to more all-or-none generalization—based on the most specific consistent hypothesis—after just a few examples. This example is discussed in more detail in chapter 1.

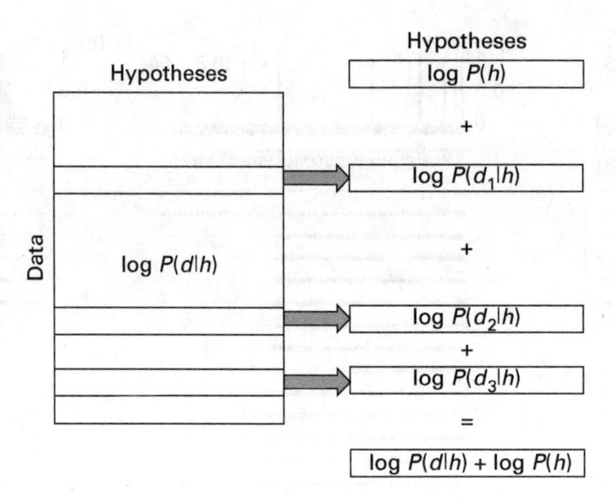

Figure 3.4
Implementing discrete Bayesian inference with linear algebra. With a discrete set of hypotheses and possible observations, the likelihood can be expressed in the form of a matrix. The prior is a vector indicating the probability of each hypothesis. Using log probabilities, Bayesian inference becomes a matter of summing the prior with the vectors of likelihoods corresponding to each d_i. The resulting vector can be exponentiated and renormalized to yield the posterior.

3.2.5 Implementing Discrete Bayesian Inference with Linear Algebra

Bayesian inference with multiple discrete hypotheses and discrete observations—as in the number game—can be implemented efficiently using a little linear algebra. In this section, we will briefly outline how this can be done, as it can also be a useful strategy in other settings where we approximate a more complicated problem by reducing it to a discrete set of hypotheses. It also provides a chance to introduce some of the notation that we will use for linear algebra later in the book. For an accessible introduction to the aspects of linear algebra used in this book, see Jordan (1986).

We will use bold capital letters (e.g., $\mathbf{M}$) to denote matrices, bold lowercase letters (e.g., $\mathbf{v}$) to denote vectors (with the default being a column vector), and $\mathbf{M}^T$ and $\mathbf{v}^T$ to denote the transpose of those matrices and vectors. $\mathbf{M}^{-1}$ is the matrix inverse of $\mathbf{M}$. $\mathbf{Mv}$ is the matrix product of $\mathbf{M}$ and $\mathbf{v}$, $\mathbf{v}^T\mathbf{w}$ the inner product of $\mathbf{v}$ and $\mathbf{w}$, and $\mathbf{vw}^T$ the outer product. $\mathbf{v} \otimes \mathbf{w}$ is the Hadamard (element wise) product, being the vector that has entries corresponding to $v_i w_i$.

If we have a discrete hypothesis space and a discrete set of possible observations, the likelihood can be specified in the form of matrix $\mathbf{L}$, where each row corresponds to a possible observation (in the number game, a number) and each column corresponds to a hypothesis. Each entry in this matrix L_{ji} indicates the probability of observation j under hypothesis i, $P(d = j|h_i)$. In the case of the number game, this can be done by initially making matrix $\mathbf{H}$, which indicates whether the number j belongs to hypothesis i, with $H_{ji} = 1$ if $j \in h_i$ and 0 otherwise, and then summing the columns to get the size of each hypothesis. The likelihood is then obtained by dividing each entry of $\mathbf{H}$ by the size of h_i, so $L_{ji} = H_{ji}/|h_i|$.

The prior can be specified in a row vector $\mathbf{p}$, where $p_i = P(h_i)$. Bayesian inference is then a procedure for updating this vector. Upon observing the data containing observation j, the posterior distribution $\mathbf{p}$ is updated to $\mathbf{p} = \mathbf{L}^{(j)} \otimes \mathbf{p}$, where $\mathbf{L}^{(j)}$ is the jth row of $\mathbf{L}$.

This makes each entry of $\mathbf{p}$ into $L_{ji}p_i$, which is equivalent to $P(d=j|h_i)P(h_i)$. To complete applying Bayes' rule, we renormalize $\mathbf{p}$ by dividing its entries by their sum. This process can be repeated for each observation. Normalizing after each observation is not necessary—it can be done at the end, after all observations have been made. The posterior predictive distribution is then given by the matrix product $\mathbf{L}\mathbf{p}^T$.

This implementation will work if there are only a few observations, but with more observations, the product of the L_{ji} terms can approach the numerical limits of a digital computer. For this reason, it is typically safer to implement probabilistic models using log probabilities (see figure 3.4). In this case, we take $L_{ji} = \log P(d=j|h_i)$ and $p_i = \log P(h_i)$. After observing j, we take the update $\mathbf{p} = \mathbf{L}^{(j)} + \mathbf{p}$. When we are ready to compute the posterior distribution, we need to convert back to probabilities from log probabilities and normalize the results. The first step is to subtract the maximum value from the entries of $\mathbf{p}$ to avoid numerical underflow when we convert back to probabilities from log probabilities. Then, we then exponentiate and renormalize, which gives us $P(h_i|d) \propto \exp\{p_i\}$. This exponentiate-and-renormalize procedure is also known as the *softmax* function, and it generalizes the sigmoid to the case where there are more than two hypotheses.

3.3 Bayesian Inference with a Continuous Hypothesis Space

When we want to infer continuous quantities, we can still use Bayes' rule, representing the relevant distributions with *probability densities*. With a continuous random variable, it no longer makes sense to refer to the probability of a particular value of that variable because the probability of any such value becomes infinitesimally small. Instead, we talk about the *density* of probability at a particular value. We will explain this concept and then turn to some applications of Bayesian inference with continuous hypothesis spaces.

3.3.1 Probability Densities

To get an intuition for why we need to use probability densitities to describe distributions on continuous quantities, imagine dividing the line between 0 and 1 into 10 bins: one from 0 to 0.1, one from 0.1 to 0.2, and so on. We could assign equal probability to each of these bins—a *uniform distribution*—so each has a probability of 0.1. Now do the same thing with 100 bins, from 0 to 0.01, 0.01 to 0.02, and so on. The probability of each of these bins would be 0.01. If we keep increasing the number of bins by a factor of 10 each time, the probability of each of those bins gets smaller and smaller.

Now, instead of focusing on the probability of each bin (the "mass" in each bin), let's focus on the density. Divide the probability by the size of the bin. When our bins are 0.1 units wide and have 0.1 units of probability mass, the density is 1. When they are 0.01 units wide with 0.01 units of mass, the density is still 1. The probability density remains the same even as the bins get smaller and smaller.

We can define a uniform distribution over the continuous interval from 0 to 1 by assigning a probability density of 1 to each point $x \in [0, 1]$. We would write this as $p(x) = 1$, using lowercase to distinguish probability density functions from the uppercase probability mass function $P(x)$ that we have used so far. The probability density function can also be used to work backwards and calculate the probability of different events. For example, if we want to calculate the probability that x falls into a bin that ranges from 0 to 0.1, we can do so by integrating the probability density over that interval:

$$P(x \in [0, 0.1]) = \int_0^{0.1} p(x)\, dx = \int_0^{0.1} 1\, dx = x|_0^{0.1} = 0.1. \tag{3.20}$$

The same kind of calculation can be performed for any interval—the probability density function tells us how to compute the probability mass. For more on probability densities and their properties, see Pitman (1993).

In the contexts where we will be using them, probability densities will obey the same rules of joint, conditional, and marginal probabilities that we have used so far in this chapter, being subject to marginalization, the chain rule, and others. We will just be replacing sums over random variables with integrals. As a consequence, Bayes' rule for a continuous set of hypotheses θ becomes

$$p(\theta|x) = \frac{p(x|\theta)p(\theta)}{\int p(x|\theta')p(\theta')\, d\theta'}, \tag{3.21}$$

where x is some continuous data.

The next couple of sections are going to work through examples of using Bayes' rule with continuous quantities, which illustrate some general principles behind Bayesian inference. The first example is the case of inferring a proportion.

3.3.2 Estimating the Bias of a Coin

Imagine that you see a coin flipped N times, producing a sequence x containing n_H heads and n_T tails. Letting θ represent the probability of the coin producing heads, what should you infer for the value of θ? More generally, this is the problem of estimating the parameter of a Bernoulli distribution (i.e., a distribution where a desired outcome occurs with probability θ).

Maximum-Likelihood Estimation Under one classical approach, inferring θ is treated as a problem of estimating a fixed parameter of a probabilistic model, to which the standard solution is *maximum-likelihood estimation*. Maximum-likelihood estimation is simple and often sensible, but it can also be problematic—particularly as a way to think about human inference. Our coin-flipping example illustrates some of these problems. The maximum-likelihood estimate of θ is the value $\hat{\theta}$ that maximizes the probability of the data. In our case, this probability is

$$p(x|\theta) = \theta^{n_H}(1-\theta)^{n_T} \tag{3.22}$$

which, as noted previously, is known as the "likelihood." With a little calculus, we can show that $\hat{\theta} = \frac{n_H}{n}$. The first step is to take the logarithm of $p(x|\theta)$, which will not affect the maximum of the function since it is a monotonic transformation. This is called the log-likelihood, sometimes written as $\mathcal{L}(\theta)$. We can then differentiate to obtain

$$\frac{d}{d\theta}\mathcal{L}(\theta) = \frac{d}{d\theta}(n_H \log\theta + n_T)\log(1-\theta) \tag{3.23}$$

$$= \frac{n_H}{\theta} + \frac{n_T}{1-\theta}, \tag{3.24}$$

which allows us to find extrema by setting the derivative to zero. This gives

$$\frac{n_H}{n_T} = \frac{\theta}{1-\theta}, \tag{3.25}$$

which implies $cn_H = \theta$ and $cn_T = 1 - \theta$ for some constant c. Since we know $cn_H + cn_T = 1$, it follows that $c = 1/(n_H + n_T) = 1/n$.

Drawbacks of Maximum-Likelihood Estimation It should be immediately clear that the single value of θ, which maximizes the probability of the data, might not provide the best basis for making predictions about future data. For example, if we see the sequence HHHHHHHHH, the maximum-likelihood estimate of θ is 1. Inferring that θ is 1 implies that we should predict that the coin will never produce tails. This might seem reasonable after observing a long sequence consisting solely of heads, but the same conclusion follows for all-heads sequences of *any* length (because n_T is always 0, so $\frac{n_H}{n_H+n_T}$ is always 1). Would you really predict that a coin would produce only heads after seeing it produce heads on just one or two flips?

A second problem with maximum-likelihood estimation is that it does not take into account other knowledge that we might have about θ. This is largely by design: maximum-likelihood estimation and other classical statistical techniques have historically been promoted as "objective" procedures that do not require prior probabilities, which were seen as inherently and irremediably subjective. While such a goal of objectivity might be desirable in certain scientific contexts, intelligent agents typically do have access to relevant and powerful prior knowledge, and they use that knowledge to make stronger inferences from sparse and ambiguous data than could be rationally supported by the data alone. For example, given the sequence HHH produced by flipping an apparently normal, randomly chosen coin, many people would say that the coin's probability of producing heads is nonetheless around 0.5—perhaps because we have strong prior expectations that most coins are (at least nearly) fair.

A Bayesian Approach Both of the problems with maximum-likelihood estimation are addressed by a Bayesian approach to inferring θ. If we assume that θ is a random variable, then we can apply Bayes' rule to obtain

$$p(\theta|x) = \frac{P(x|\theta)p(\theta)}{P(x)}, \tag{3.26}$$

where

$$P(x) = \int_0^1 P(x|\theta)p(\theta)\,d\theta, \tag{3.27}$$

since the only valid values of θ are in $[0, 1]$. This integral is intuitively exactly the same as the denominator of Bayes' rule with a discrete set of hypotheses, adding up the product of the prior and likelihood across all the hypotheses.

The posterior distribution over θ contains more information than a single estimate: it indicates not just which values of θ are probable, but also how much uncertainty there is about those values. Collapsing this distribution down to a single number discards information, so Bayesians prefer to maintain distributions wherever possible. However, there are two methods that are commonly used to obtain a point estimate (i.e., a single value) from a posterior distribution. The first method is *maximum a posteriori (MAP) estimation*: choosing the value of θ that maximizes the posterior probability, as given by equation (3.26). The second method is computing the *posterior mean* of the quantity in question: a weighted average of all possible values of the quantity, where the weights are given by the posterior distribution. For example, the posterior mean value of the coin weight θ is computed as follows:

$$\bar{\theta} = \int_0^1 \theta\, p(\theta|x)\, d\theta. \tag{3.28}$$

In the case of coin-flipping, the posterior mean also corresponds to the *posterior predictive distribution*: the probability that the next flip of the coin will produce heads, given the observed sequence of outcomes.

So which of these estimates—the MAP or the mean—should we use? The right estimate to use depends on the problem being solved, a point that we return to in chapter 7. If an incorrect estimate is penalized by the squared difference between the estimate and the true value, the mean is the optimal estimator. If all that matters is maximizing the probability of getting the right answer, the MAP estimate is optimal.[7]

3.3.3 Setting a Prior

Different choices of the prior, $p(\theta)$, will lead to different inferences about the value of θ. A first step might be to assume a uniform prior over θ, with $p(\theta)$ being equal for all values of θ between 0 and 1 (our familiar probability density, where $p(\theta) = 1$ for $\theta \in [0, 1]$). With this choice of $p(\theta)$ and the Bernoulli likelihood from equation (3.10), equation (3.26) becomes

$$p(\theta|x) = \frac{\theta^{n_H}(1-\theta)^{n_T}}{\int_0^1 \theta^{n_H}(1-\theta)^{n_T}\, d\theta}, \tag{3.30}$$

where the denominator is just the integral from equation (3.27). We also need a little calculus to compute this integral.

The key step in evaluating the integral is recognizing that it takes a standard form, being the definition of what is called the *beta function*. The beta function is written as $B(r, s)$, where r and s are two arguments, and defined as

$$B(r, s) = \int_0^1 \theta^{r-1}(1-\theta)^{s-1}\, d\theta, \tag{3.31}$$

7. If given a choice of point estimates, many Bayesians prefer the posterior mean over the MAP estimate. The reason is that the MAP estimate is not invariant to reparameterization of the posterior. When you change variables in a probability density, you can change the maximum of that density. If we specify a new variable $\phi = f(\theta)$ to be the quantity that we care about, then any distribution on θ implies a distribution on ϕ. Taking the prior for the sake of simplicity, the distribution on ϕ is

$$p(\phi) = p(\theta)/|f'(\theta)|, \tag{3.29}$$

where $\theta = f^{-1}(\phi)$, $f'(\cdot)$ is the derivative of $f(\cdot)$ with respect to θ, and we assume that $f(\cdot)$ is one-to-one. The derivative of f appears because a transformation can locally stretch or shrink the scale on which the distribution is defined. It should be clear that for any $p(\theta)$, we can choose a one-to-one function $f(\theta)$, such that the resulting $p(\phi)$ has a maximum at a value of ϕ that does not correspond to the value of θ that maximizes $p(\theta)$.

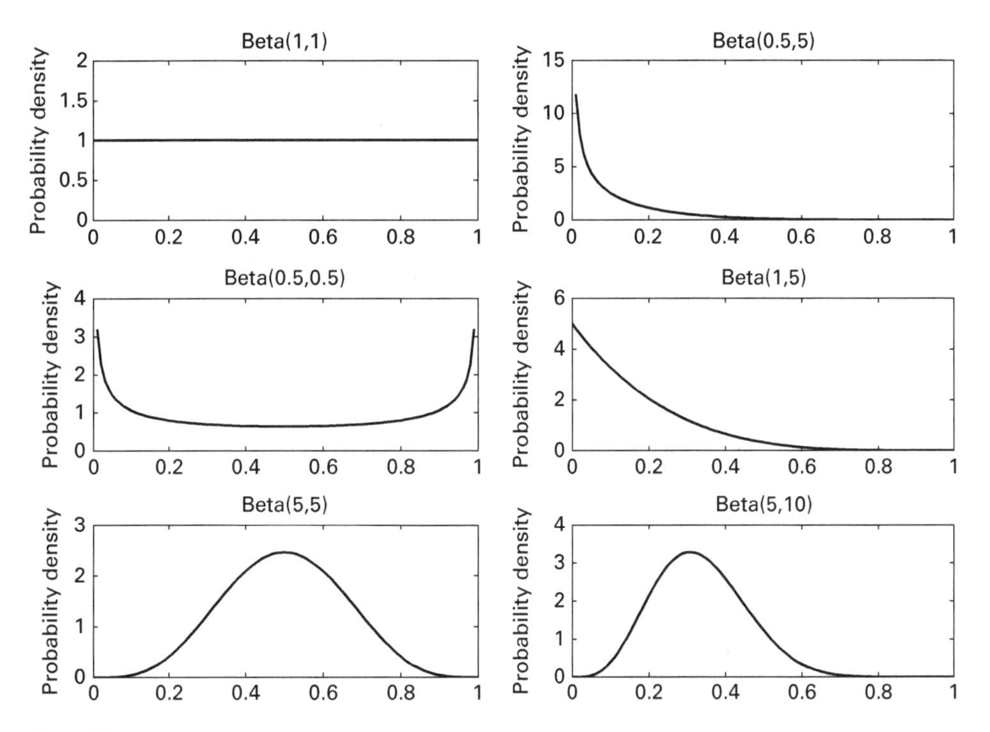

Figure 3.5
Beta distributions. Depending on the values of the two parameters of the distribution, it can be uniform, favor particular values, or favor extreme values.

although there are also several other integrals that are equivalent (being the result of substitution of variables). The beta function can be written in terms of other functions, with one of the most useful forms being that using the *gamma function*

$$B(r,s) = \frac{\Gamma(r)\Gamma(s)}{\Gamma(r+s)}, \tag{3.32}$$

where $\Gamma(\cdot)$ is the gamma function. The gamma function is defined as

$$\Gamma(r) = \int_0^\infty y^{r-1} \exp\{-y\}\, dy \tag{3.33}$$

and is known as the "generalized factorial function" because when r is an integer, $\Gamma(r) = (r-1)!$ (you can prove this using integration by parts and induction). More generally, for any value of r, $\Gamma(r) = (r-1)\Gamma(r-1)$.

Using these results to simplify the beta function, the posterior distribution over θ produced by a sequence x with n_H heads and n_T tails is

$$p(\theta|d) = \frac{(n_H + n_T + 1)!}{n_H!\, n_T!}\, \theta^{n_H}(1-\theta)^{n_T}. \tag{3.34}$$

This is actually a distribution of a well-known form: a *beta distribution* with parameters $n_H + 1$ and $n_T + 1$, denoted as $\text{Beta}(n_H + 1, n_T + 1)$ (see figure 3.5). Using the uniform prior, the MAP estimate for θ is the same as the maximum-likelihood estimate, $\frac{n_H}{n_H+n_T}$, which you can check by differentiating $p(\theta|d)$ in the same way as before.[8] The posterior mean (or posterior predictive distribution) is

$$\bar{\theta} = \int_0^1 \theta \frac{(n_H + n_T + 1)!}{n_H!\, n_T!} \theta^{n_H} (1-\theta)^{n_T}\, d\theta \tag{3.35}$$

$$= \frac{(n_H + n_T + 1)!}{n_H!\, n_T!} \int_0^1 \theta^{n_H+1} (1-\theta)^{n_T}\, d\theta \tag{3.36}$$

$$= \frac{(n_H + n_T + 1)!}{n_H!\, n_T!} B(n_H + 2, n_T + 1) \tag{3.37}$$

$$= \frac{(n_H + n_T + 1)!}{n_H!\, n_T!} \frac{(n_H + 1)!\, n_T!}{(n_H + n_T + 2)!} \tag{3.38}$$

$$= \frac{n_H + 1}{n_H + n_T + 2}. \tag{3.39}$$

The posterior mean is thus sensitive to the consideration that we might not want to put as much evidential weight on seeing a single head as on a sequence of 10 heads in a row: on seeing a single head, the posterior mean predicts that the next flip will produce a head with probability $\frac{2}{3}$, while a sequence of 10 heads leads to the prediction that the next flip will produce a head with probability $\frac{11}{12}$.

We can also use priors that encode stronger beliefs about the value of θ. For example, we can take a Beta$(\nu_H + 1, \nu_T + 1)$ distribution for $p(\theta)$, where ν_H and ν_T are positive integers. This distribution gives

$$p(\theta) = \frac{(\nu_H + \nu_T + 1)!}{\nu_H!\nu_T!} \theta^{\nu_H} (1-\theta)^{\nu_T}, \tag{3.40}$$

having a mean at $\frac{\nu_H+1}{\nu_H+\nu_T+2}$, and gradually becoming more concentrated around that mean as $\nu_H + \nu_T$ becomes large. For instance, taking $\nu_H = \nu_T = 1{,}000$ would give a distribution that strongly favors values of θ close to 0.5. Using such a prior with the Bernoulli likelihood from equation (3.10) and applying the same kind of calculations as before, we obtain the posterior distribution

$$p(\theta|d) = \frac{(n_H + n_T + \nu_H + \nu_T + 1)!}{(n_H + \nu_H)!\, (n_T + \nu_T)!} \theta^{n_H+\nu_H} (1-\theta)^{n_T+\nu_T}, \tag{3.41}$$

which is Beta$(n_H + \nu_H + 1, n_T + \nu_T + 1)$. Under this posterior distribution, the MAP estimate of θ is $\frac{n_H+\nu_H}{n_H+n_T+\nu_H+\nu_T}$, and the posterior mean is $\frac{n_H+\nu_H+1}{n_H+n_T+\nu_H+\nu_T+2}$. Thus, if $\nu_H = \nu_T =$

8. It is tempting to look at this and say, "Maximum-likelihood estimation is exactly the same as MAP estimation with a uniform prior!" but remember that the MAP estimate is not invariant to reparameterization, while the maximum-likelihood estimate is (since reparameterization of θ does not change $P(x|\theta)$). If you really want to, you can say "Assuming a fixed parameterization of θ, maximum-likelihood estimation is the same as MAP estimation with a uniform prior!," although this seems a little bit less exciting. In fact, it is the kind of thing that both Bayesians and frequentists see as proving their point—Bayesians can say that frequentists are using procedures that end up being Bayesian anyway, and frequentists can point to the problem with reparameterization as a reason why they want nothing to do with prior distributions.

1000, seeing a sequence of 10 heads in a row would induce a posterior distribution over θ with a mean of $\frac{1011}{2012} \approx 0.5025$. In this case, the observed data matter hardly at all. A prior that is much weaker but still biased toward approximately fair coins might take $v_H = v_T = 5$. Then an observation of 10 heads in a row would lead to a posterior mean of $\frac{16}{22} \approx .727$, significantly tilted toward heads but still closer to a fair coin than the observed data would suggest on their own. We can say that such a prior acts to *smooth* or *regularize* the observed data, damping what might be misleading fluctuations when the data are far from the learner's initial expectations.

Finally, we can also define a prior distribution that favors very small or very large values of θ. In the most general case, we can take a Beta(α, β) prior, where α and β are real numbers in $(0, \infty)$. The result of observing n_H heads and n_T tails is a posterior distribution that is Beta$(n_H + \alpha, n_T + \beta)$. For values of α and β less than 1, the prior favors values of θ close to 0 and 1 (you can check that the density goes to infinity at these values). This prior is harder to interpret in terms of "virtual examples." If you calculate the posterior mean or posterior predictive distribution, α and β still add fractional amounts to n_H and n_T, but computing the MAP estimate with the standard parameterization of the posterior requires taking into consideration the boundary conditions (i.e., the values of $p(\theta|x)$ at 0 and 1) and can be complicated and just as degenerate as maximum-likelihood estimation.

3.3.4 Conjugate Priors

Our analysis of coin-flipping with informative priors has two features of more general interest. First, the prior and posterior are specified using distributions of the same form (both being beta distributions). Second, the parameters of the prior, v_H and v_T, act as "virtual examples" of heads and tails, which are simply added to the real examples tallied in n_H and n_T to produce the posterior, as if both the real and virtual examples had been observed in the same data set. These two properties are not accidental: they define a class of priors called *conjugate priors*. The likelihood determines whether a conjugate prior exists for a given problem, and the form that the prior will take. The results that we have given in this section exploit the fact that the beta distribution is the conjugate prior for the Bernoulli or binomial likelihood (equation (3.10))—the uniform distribution on $[0, 1]$ is also a beta distribution, being Beta$(1, 1)$. Conjugate priors exist for many of the distributions commonly used in probabilistic models, such as Gaussian, Poisson, and multinomial distributions, and greatly simplify many Bayesian calculations. Using conjugate priors, posterior distributions can be computed analytically, and the interpretation of the prior as contributing virtual examples is intuitive. We discuss conjugate priors further when we consider Bayesian inference for other distributions later in this chapter.

Of course, conjugate priors can't capture all the prior knowledge that people have, even in cases as simple as flipping a coin. Imagine seeing a coin flipped 10 times, coming up heads 4 times and tails 6 times. It seems reasonable to guess that the probability of heads is around 0.5, consistent with a conjugate prior with large and equal numbers of virtual examples of heads and tails. But if you see a coin flipped 10 times and come up heads each time, you're probably not going to have the same estimate of the probability of heads, despite the fact that a strong conjugate prior would still favor something close to 0.5. If you see the same coin produce heads another 10 times, you can be pretty confident that it is severely biased—perhaps even a two-headed coin.

This example illustrates that when we propose to use a conjugate prior—and interpret it in terms of virtual examples—we are often trying to approximate some other, richer, prior knowledge. It's not really the case that you've seen thousands of coins flipped and that's what induces your prior. Rather, you have knowledge of the physical dynamics of a coin flip—that if a coin spins enough, it is essentially random—and the physical structure of a coin—heads on one side, tails on the other—and base your expectations on that. The possibility that the coin might have heads on both sides changes these expectations in a way that can't be captured by a simple conjugate prior.

In practice, Bayesian modelers often try to get around the limits of conjugate priors by specifying priors in terms of more complex generative processes. For example, we could have a prior that is a *mixture distribution*. This prior combines three distributions—one that always produces heads, one that always produces tails, and one that is a beta distribution concentrated around 0.5. Each of these three distributions is assigned some probability—say a 1 in 100 chance of always heads, a 1 in 100 chance of always tails, and a 98 in 100 chance of being something close to a fair coin. Specifying the distribution in this way lets us use the simplicity of the conjugate prior, while capturing some of the complexity of people's prior knowledge. Bayesian inference then becomes a combination of evaluating these three discrete hypotheses and evaluating continuous hypotheses about bias. We discuss how to work with these distributions in chapter 4.

3.3.5 Learning Multinomial Distributions

When we flip a coin, we have only two alternatives. However, many learning problems require inferring a distribution over multiple alternatives. For example, we might want to estimate the distribution over words in a document, or the probability of seeing different objects in a category. This is the problem of estimating a *multinomial distribution*. As before, we will use θ to indicate the unknown parameters of this distribution, with $(\theta_1, \ldots, \theta_k)$ corresponding to the probabilities of k outcomes.

As you might expect, there is a conjugate prior for multinomials, which is known as the *Dirichlet distribution*. The Dirichlet is the multivariate generalization of the beta distribution, with hyperparameters $\alpha_1, \ldots, \alpha_k$ corresponding to the k different outcomes. The probability of a vector θ under this distribution is

$$p(\theta) = \frac{\Gamma(\sum_i \alpha_i)}{\prod_i \Gamma(\alpha_i)} \prod_i \theta_i^{\alpha_i - 1}, \tag{3.42}$$

with the mean corresponding to $\theta_i = \frac{\alpha_i}{\sum_i \alpha_i}$. All the results for the beta-Bernoulli model given in this chapter extend to the Dirichlet-multinomial model: the posterior distribution on θ given a sequence of n events in which outcome i occurs with frequency n_i is Dirichlet with parameters $n_i + \alpha_i$, and the posterior predictive distribution indicates that outcome i occurs with probability $\frac{n_i + \alpha_i}{n + \sum_i \alpha_i}$. In the context of estimating a multinomial, taking $\alpha_i < 1$ indicates a preference for *sparsity* in θ, giving the highest probability to distributions that assign a probability of zero to a large number of outcomes. The strength of this preference for sparsity increases as α_i approaches 0. This property will turn out to be important in some of the models discussed later in the book.

3.3.6 An Example: Estimating Distributions over Words

When making probabilistic models of natural language, researchers have to make a decision about how to assign probabilities to events that never occurred in the linguistic data they have access to. For example, if the researcher has access to only the first paragraph of a document, how should they calculate the probability that a particular word occurs in the second paragraph of the document? If that word didn't appear in the first paragraph, its frequency is zero. However, assigning it a probability of zero seems hasty.

Estimating a distribution over a known set of words is just a problem of estimating a multinomial, and the approach outlined in section 3.3.5 provides a solution. Taking a Dirichlet(α) prior over the multinomial parameters θ (i.e., setting α_i to the same value α for all words) results in an estimate of $\frac{\alpha}{n+\alpha}$ for the probability of any word that hasn't been seen so far, where n is the sum of the frequencies of all observed words.

In the natural language processing community, this is known as the problem of "smoothing" a probability distribution, and a smoothing scheme of exactly this form was widely used for a long time before its connections to Bayesian inference were articulated (e.g., Chen & Goodman, 1996). A variety of other, more complex smoothing techniques exist, some of which have recently been shown to also correspond to sensible Bayesian estimation methods (e.g., Goldwater, Griffiths, & Johnson, 2006b; Teh, 2006; Favaro, Nipoti, & Teh, 2016).

3.4 Bayesian Inference for Gaussians

Another common case where one needs to do Bayesian inference over a continuous quantity is estimating the parameters of a Gaussian distribution. The Gaussian is parameterized by a mean, μ, and a standard deviation, σ, with likelihood

$$p(x|\mu,\sigma) = \frac{1}{\sqrt{2\pi}\sigma} \exp\left\{-\frac{(x-\mu)^2}{2\sigma^2}\right\}, \tag{3.43}$$

where $\frac{1}{\sqrt{2\pi}\sigma}$ is the normalizing constant of the distribution, with

$$\int_{-\infty}^{\infty} \exp\left\{-\frac{(x-\mu)^2}{2\sigma^2}\right\} = \sqrt{2\pi}\sigma. \tag{3.44}$$

We will focus on the problem of estimating μ from an observation x that holds σ fixed.

3.4.1 Inferring the Mean of a Gaussian

As a first step, we can consider using maximum-likelihood estimation to find the value of μ. We want to maximize the likelihood, $p(x|\mu,\sigma)$, with respect to μ. Since the logarithm is a monotonic function, we can also choose to maximize the log-likelihood, which is

$$\log p(x|\mu,\sigma) = -\frac{1}{2}\log 2\pi\sigma^2 - \frac{(x-\mu)^2}{2\sigma^2}, \tag{3.45}$$

in which only the second term depends on μ. Differentiating, setting the resulting expression to zero, and solving for μ give the maximum-likelihood estimate $\hat{\mu} = x$. This makes intuitive sense since $(x-\mu)^2$ increases as the distance between x and μ increases, meaning that the probability of x is maximized when $\mu = x$.

We can now explore the consequences of incorporating prior knowledge about μ by considering Bayesian solutions to this estimation problem. Applying Bayes' rule, we want to compute

$$p(\mu|x,\sigma) \propto p(x|\mu,\sigma)p(\mu|\sigma), \tag{3.46}$$

which we will simplify by assuming that μ and σ are independent, so $p(\mu|\sigma) = p(\mu)$. If we look at the likelihood in equation (3.43) carefully, we can see that the part that involves μ takes the form $\exp\left\{-\frac{(x-\mu)^2}{2\sigma^2}\right\}$, which is an exponentiated quadratic function in μ. It follows that the conjugate prior should also be an exponentiated quadratic function in μ, and the distribution on μ that satisfies this requirement is a Gaussian. Consequently, we will take the prior on μ to be Gaussian with mean μ_0 and standard deviation σ_0. Substituting the expressions for our Gaussian likelihood and prior into equation (3.46), we obtain

$$p(\mu|x,\sigma) \propto \frac{1}{\sqrt{2\pi}\sigma} \exp\left\{-\frac{(x-\mu)^2}{2\sigma^2}\right\} \frac{1}{\sqrt{2\pi}\sigma_0} \exp\left\{-\frac{(\mu-\mu_0)^2}{2\sigma_0^2}\right\} \tag{3.47}$$

$$\propto \exp\left\{-\frac{1}{2}\left[\frac{(x-\mu)^2}{\sigma^2} + \frac{(\mu-\mu_0)^2}{\sigma_0^2}\right]\right\}, \tag{3.48}$$

where the second step removes the constants and combines the exponents. We can now expand the bracketed terms to obtain

$$\frac{(x-\mu)^2}{\sigma^2} + \frac{(\mu-\mu_0)^2}{\sigma_0^2} = \mu^2\left(\frac{1}{\sigma^2} + \frac{1}{\sigma_0^2}\right) - 2\mu\left(\frac{x}{\sigma^2} + \frac{\mu_0}{\sigma_0^2}\right) + \frac{x^2}{\sigma^2} + \frac{\mu_o^2}{\sigma_0^2}. \tag{3.49}$$

The last two terms are constant in μ, and thus will not affect $p(\mu|x)$. By a similar argument, we can add or subtract any other term constant in μ. We thus have

$$p(\mu|x,\sigma) \propto \exp\left\{-\frac{1}{2}\left[\frac{\mu^2}{\sigma_1^2} - 2\frac{\mu\mu_1}{\sigma_1^2} + c\right]\right\} \tag{3.50}$$

where

$$\frac{1}{\sigma_1^2} = \frac{1}{\sigma^2} + \frac{1}{\sigma_0^2} \tag{3.51}$$

$$\mu_1 = \frac{\frac{x}{\sigma^2} + \frac{\mu_0}{\sigma_0^2}}{\frac{1}{\sigma^2} + \frac{1}{\sigma_0^2}} \tag{3.52}$$

and c is a constant. Taking $c = \frac{\mu_1^2}{\sigma_1^2}$ allows us to complete the square, with

$$p(\mu|x,\sigma) \propto \exp\left\{-\frac{(\mu-\mu_1)^2}{2\sigma_1^2}\right\}, \tag{3.53}$$

which we should not be surprised to discover is a Gaussian with mean μ_1 and standard deviation σ_1:

$$p(\mu|x,\sigma) = \frac{1}{\sqrt{2\pi}\sigma_1} \exp\left\{-\frac{(\mu-\mu_1)^2}{2\sigma_1^2}\right\}, \tag{3.54}$$

consistent with our use of a conjugate prior. The posterior mean is thus μ_1, given by equation (3.52), and the posterior predictive distribution is also Gaussian with mean μ_1 and variance $\sigma_1^2 + \sigma^2$. We can use the knowledge that the mean of a Gaussian is also its mode, or differentation of the log posterior, to recognize that the MAP estimate of μ is also μ_1.

The interaction of the prior and the data in equation (3.52) is reminiscent of the result obtained by using a conjugate prior when estimating a proportion. The mean of the posterior (or the MAP estimate) linearly interpolates between the observed value, x, and the mean of the prior, μ_0, with the weight assigned to each being inversely proportional to the associated variance. Again, this represents a simple compromise between prior and data, with the terms of the compromise being set by the reliability of each of these sources of information, and acts as a way of smoothing or regularizing the empirical data x (in this case, the movement toward μ_0 is often described as "shrinking" toward the value predicted by the prior).

3.4.2 Uninformative and Improper Priors

Using a Gaussian prior on μ implies that we have enough knowledge about the value of μ to specify the mean and variance of the prior. However, sometimes we might feel that we really know nothing about μ, and any value might seem equally plausible. In this case, we could want to use an *uninformative prior*—one that expresses our ignorance. In particular, if we believe that any value between $-\infty$ and ∞ is equally plausible, we should use a uniform prior on μ.

One problem with using a uniform prior over $(-\infty, \infty)$ is that it will not satisfy one of our axioms of probability, since the integral of a constant over this range diverges. However, many Bayesians are happy to use such "improper" priors, provided that the posterior is well defined. This is the case for inferring the mean of a Gaussian with a known standard deviation: you can repeat the analysis given in section 3.4.1 with a uniform prior and find that the posterior has mean x and standard deviation σ (providing an equivalent estimate of μ to that produced by maximum-likelihood estimation). If the posterior distribution is undefined—for example, because you cannot compute the integral of the function in the denominator—then there is not enough information in the data to overcome the extreme uncertainty implied by the prior.

An easier way to evaluate the consequences of using an uninformative prior is to consider the posterior in the limit as $\sigma_0 \to \infty$. As the standard deviation of the prior increases, it eventually approaches a uniform distribution over $(-\infty, \infty)$. Taking this limit for the posterior distribution given previously gives us the same result: with no information from the prior, the posterior has a mean corresponding to the observed value of x and a standard deviation determined by the uncertainty in that observation.

3.4.3 An Example: Categories and Memory

Gaussian distributions provide a natural way to capture the knowledge that people might have about a category of objects. For example, if we were to ask you to think about how large an adult trout is, you would probably characterize the distribution in terms of an average value and the variance around that, as could be captured by a Gaussian distribution. So how does this knowledge of the structure of categories affect the way that we perceive the world?

Huttenlocher, Hedges, and Vevea (2000) took one approach to answering this question, examining how category knowledge influenced people's reconstructions of stimuli from memory. In their experiments, people saw a stimulus (such as a schematic image of a fish) very briefly and then had to adjust another stimulus along a single dimension (such as the fatness of the fish) until they thought that it matched the original stimulus. This was done many times, allowing the participants to learn something about the distribution of stimuli and allowing the experimenters to examine how this distribution affected people's reconstructions.

Reconstructing a stimulus from memory is an interesting example of a problem that we might not initially think of as inductive. However, it has exactly the form of the kind of problems we have been considering, with the data being the noisy recollection of the stimulus supplied by memory and the hypotheses concerning the true value of the stimulus. Consequently, we should expect people's reconstructions from memory to be influenced by their prior expectations about which stimuli they are likely to see. In the case where those prior expectations take the form of a Gaussian, reconstructions should be biased toward the mean of the Gaussian, with the amount of bias (the degree of error in reconstruction) increasing with the distance between the stimulus and the mean.

We can express the relationship between the bias and the stimulus more formally by using the notation introduced in this discussion. Here, x is the stimulus and μ_1 the reconstruction. Following equations (3.51) and (3.52), the difference between μ_1 and x is

$$\mu_1 - x = \frac{\frac{x}{\sigma^2} + \frac{\mu_0}{\sigma_0^2}}{\frac{1}{\sigma^2} + \frac{1}{\sigma_0^2}} - x$$

$$= x\left[\frac{\frac{1}{\sigma^2}}{\frac{1}{\sigma^2} + \frac{1}{\sigma_0^2}} - 1\right] + \frac{\frac{\mu_0}{\sigma_0^2}}{\frac{1}{\sigma^2} + \frac{1}{\sigma_0^2}},$$

which is linear in x. Thus, the prediction is that the bias in the reconstruction of x should depend linearly on the value of x, decreasing as x increases and taking the value 0 only when x is equal to the mean of the prior, μ_0.

Huttenlocher et al. (2000) conducted several experiments, and their results supported this Bayesian account of reconstruction from memory.[9] Figure 3.6 shows the results of a condition from experiment 1 of Huttenlocher et al. (2000). In this condition, participants saw fish whose fatness was drawn from a normal distribution. The bias in reconstruction was measured by comparing the size of people's reconstructions to the actual size of the fish. This bias shows the predicted linear effect in the actual size of the fish. Similar effects were found when fish were drawn from uniform and bimodal distributions, and for other one-dimensional (1D) stimuli such as the brightness of a gray patch.

9. Their original analysis was not explicitly Bayesian; rather, it focused on a strategy for increasing accuracy in reconstructions that results in the same kind of estimator.

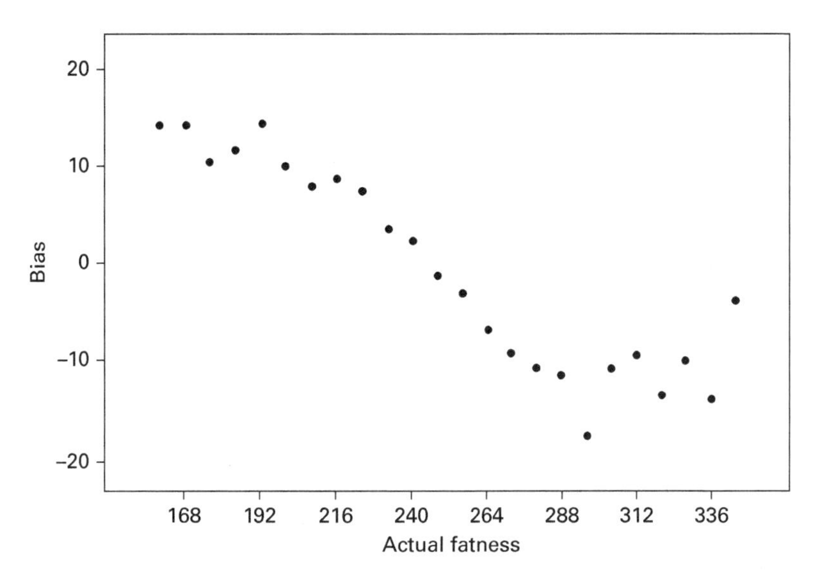

Figure 3.6
Bias as a function of stimulus value in the *normal* condition of experiment 1 of Huttenlocher et al. (2000). Consistent with estimating the mean of a Gaussian with a Gaussian prior, bias is a decreasing linear function of the stimulus value. Figure adapted from Huttenlocher et al. (2000).

3.4.4 Multiple Observations

So far, we have focused on the case where the observed data consist of a single value drawn from a Gaussian distribution. However, the same analysis extends naturally to the case of multiple observations. If we have multiple independent samples from a Gaussian distribution, the likelihood becomes

$$p(x|\mu,\sigma)=\prod_{i=1}^{n}p(x_i|\mu,\sigma) \tag{3.55}$$

$$=\prod_{i=1}^{n}\frac{1}{\sqrt{2\pi}\sigma}\exp\left\{-\frac{(x-\mu)^2}{2\sigma^2}\right\} \tag{3.56}$$

$$=\left(\frac{1}{\sqrt{2\pi}\sigma}\right)^n\exp\left\{-\sum_{i=1}^{n}\frac{(x_i-\mu)^2}{2\sigma^2}\right\}, \tag{3.57}$$

where $x=(x_1,\ldots,x_n)$ are n samples. We can go through the same math as for the case of a single observation, but a quick shortcut is to observe that

$$\sum_{i=1}^{n}(x_i-\mu)^2=\sum_{i=1}^{n}((x_i-\bar{x})+(\bar{x}-\mu))^2 \tag{3.58}$$

$$=\sum_{i=1}^{n}(x_i-\bar{x})^2-2(\bar{x}-\mu)\sum_{i=1}^{n}(x_i-\bar{x})+n(\bar{x}-\mu)^2 \tag{3.59}$$

$$=\sum_{i=1}^{n}(x_i-\bar{x})^2+n(\bar{x}-\mu)^2, \tag{3.60}$$

where $\bar{x} = \frac{1}{n}\sum_{i=1}^{n} x_i$ is the mean of x, and the last line follows from the fact that $\sum_{i=1}^{n} x_i = \sum_{i=1}^{n} \bar{x} = n\bar{x}$. Since $\sum_{i=1}^{n}(x_i - \bar{x})^2$ is constant in μ, it plays no role in determining the posterior, and the relevant part of the likelihood is

$$p(x|\mu, \sigma) \propto \exp\left\{-\frac{n(\bar{x}-\mu)^2}{2\sigma}\right\} \tag{3.61}$$

which means that the posterior is the same as that given previously, replacing x with $\bar{x}$ and σ with σ/n. Likewise, the maximum-likelihood estimate of μ is $\hat{\mu} = \bar{x}$.

3.4.5 Multivariate Gaussians

While we have focused on the 1D case, Gaussian distributions can be defined in multiple dimensions. In this setting, the basic intuition still holds: we assign points x probabilities that decrease with the square of the distance to the mean μ. What differs is how we measure that distance, taking into account the fact that there are now multiple directions in which points can move from the mean and some of them might be penalized more than others.

The multivariate Gaussian distribution in d dimensions assigns the vector $\mathbf{x} = (x_1, \ldots, x_d)$ the probability

$$p(\mathbf{x}|\boldsymbol{\mu}, \boldsymbol{\Sigma}) = \frac{1}{(2\pi)^{d/2}|\boldsymbol{\Sigma}|^{1/2}} \exp\left\{-\frac{1}{2}(\mathbf{x}-\boldsymbol{\mu})^T \boldsymbol{\Sigma}^{-1}(\mathbf{x}-\boldsymbol{\mu})\right\}, \tag{3.62}$$

where $\boldsymbol{\mu}$ is a d-dimensional vector representing the mean and $\boldsymbol{\Sigma}$ is the $d \times d$ covariance matrix indicating the covariance between dimensions. The $(\mathbf{x}-\boldsymbol{\mu})^T \boldsymbol{\Sigma}^{-1}(\mathbf{x}-\boldsymbol{\mu})$ is a quadratic form for vectors, equivalent to $(x-\mu)^2/\sigma^2$. The conjugate prior for $\boldsymbol{\mu}$ is also multivariate Gaussian, with mean $\boldsymbol{\mu}_0$ and covariance $\boldsymbol{\Sigma}_0$. By an analysis similar to that given here,[10] the posterior is multivariate Gaussian, with mean $\boldsymbol{\mu}_1$ and covariance matrix $\boldsymbol{\Sigma}_1$ equal to

$$\boldsymbol{\Sigma}_1^{-1} = \boldsymbol{\Sigma}^{-1} + \boldsymbol{\Sigma}_0^{-1} \tag{3.63}$$

$$\boldsymbol{\mu}_1 = (\boldsymbol{\Sigma}^{-1} + \boldsymbol{\Sigma}_0^{-1})(\boldsymbol{\Sigma}^{-1}\mathbf{x} + \boldsymbol{\Sigma}_0^{-1}\boldsymbol{\mu}_0), \tag{3.64}$$

where $\boldsymbol{\mu}_1$ has the same interpretation as linearly interpolating between the data $\mathbf{x}$ and the mean of the prior $\boldsymbol{\mu}_0$. The analysis for multiple observations proceeds in the same way.

3.5 Bayesian Inference for Other Distributions

The basic framework for Bayesian inference with continuous variables that we have illustrated for multinomial and Gaussian distributions generalizes naturally to other distributions. The main challenge is that the integral in the denominator of equation (3.21) may not be possible to evaluate analytically. In this case, MAP estimates may still be possible to obtain by simply maximizing the product of the prior and likelihood. Alternatively, numerical techniques can be used to approximate the integral. The simplest approach for a

10. The tricky part is completing the square for vectors. If you have $\mathbf{x}^T\mathbf{A}\mathbf{x} + 2\mathbf{b}^T\mathbf{x}$, then adding $\mathbf{b}^T\mathbf{A}^{-1}\mathbf{b}$ gets you to the factorized form $(\mathbf{x} - \mathbf{A}^{-1}\mathbf{b})^T\mathbf{A}(\mathbf{x} - \mathbf{A}^{-1}\mathbf{b})$.

small number of parameters defined over a fixed range is to grid the space of parameters and sum the product of the prior and likelihood over the grid, reducing the continuous problem to a discrete one. More sophisticated approaches based on Monte Carlo methods are discussed in chapter 6.

Analytic results for the posterior distribution can always be obtained for a class of distributions known as *exponential family distributions*, provided that conjugate priors are used (and conjugate priors always exist for these distributions). Exponential family distributions can be written in the form $p(\mathbf{x}|\boldsymbol{\eta}) = f(\mathbf{x}) \exp\{\boldsymbol{\eta}^T \mathbf{s}(\mathbf{x}) + \psi(\boldsymbol{\eta})\}$, where $\boldsymbol{\eta}$ is a vector of *natural parameters*, $f(\mathbf{x})$ is an arbitrary function that depends only on $\mathbf{x}$, $\mathbf{s}(\mathbf{x})$ is a vector containing the *sufficient statistics* of the data $\mathbf{x}$, and $\psi(\boldsymbol{\eta})$ is the logarithm of the normalizing constant. This might seem like an arbitrary form for a distribution, but many of the distributions used in Bayesian models can be cast in this form, including the multinomial and Gaussian distributions. For example, the Bernoulli distribution where x takes the value 1 with probability θ and 0 otherwise can be written as $p(x|\eta) = \exp\{x\eta + \psi(\eta)\}$, where $\eta = \log \frac{\theta}{1-\theta}$ and $\psi(\eta) = \log(1 + \exp\{\eta\}) = -\log(1-\theta)$.

For any exponential family distribution, there is a corresponding family of conjugate priors with $p(\boldsymbol{\eta}|\boldsymbol{\nu}, \lambda) \propto \exp\{\boldsymbol{\eta}^T \boldsymbol{\nu} + \lambda\psi(\boldsymbol{\eta})\}$ (you can check that this corresponds to the Beta$(\nu + 1, \lambda + 1)$ prior for the Bernoulli distribution). Conjugacy can be demonstrated by observing that the posterior is given by

$$p(\boldsymbol{\eta}|\mathbf{x}) \propto p(\mathbf{x}|\boldsymbol{\eta})p(\boldsymbol{\eta}) \tag{3.65}$$

$$= f(\mathbf{x}) \exp\{\boldsymbol{\eta}^T \mathbf{s}(\mathbf{x}) + \psi(\boldsymbol{\eta})\} \exp\{\boldsymbol{\eta}^T \boldsymbol{\nu} + \lambda\psi(\eta)\} \tag{3.66}$$

$$\propto \exp\{\boldsymbol{\eta}^T \mathbf{s}(\mathbf{x}) + \psi(\boldsymbol{\eta}) + \boldsymbol{\eta}^T \boldsymbol{\nu} + \lambda\psi(\boldsymbol{\eta})\} \tag{3.67}$$

$$= \exp\{\boldsymbol{\eta}^T (\mathbf{s}(\mathbf{x}) + \boldsymbol{\nu}) + (1 + \lambda)\psi(\boldsymbol{\eta})\}, \tag{3.68}$$

which takes the same form as the prior.

3.5.1 An Example: Predicting the Future

If you were assessing the prospects of a 60-year-old man, how much longer would you expect him to live? If you were an executive evaluating the performance of a movie that had made $40 million at the box office so far, what would you estimate for its total gross? These are examples of prediction problems, where we get some information about the current extent or duration of a phenomenon but want to infer its total extent or duration. Such problems can be formulated as a case of Bayesian inference, with the data being the current value and the hypotheses being the total.

Defining the Problem Assume that point t is sampled uniformly at random from the interval $[0, t_{total})$. What should we guess for the value of t_{total}? A Bayesian solution to this problem involves computing the posterior distribution over t_{total} given t. Applying Bayes' rule, this is

$$p(t_{total}|t) = \frac{p(t|t_{total})p(t_{total})}{p(t)}, \tag{3.69}$$

where

$$p(t) = \int_0^\infty p(t|t_{total}) p(t_{total})\, dt_{total}. \tag{3.70}$$

By the assumption that t is sampled uniformly at random, $p(t|t_{total}) = 1/t_{total}$ for $t_{total} \geq t$ and 0 otherwise. Equation (3.70) thus simplifies to

$$p(t) = \int_t^\infty \frac{p(t_{total})}{t_{total}}\, dt_{total}. \tag{3.71}$$

The form of the posterior distribution for any given value of t is thus determined entirely by the prior, $p(t_{total})$.

This approach to prediction is quite general, applicable to any problem that requires estimating the upper limit of a duration, extent, or other numerical quantity given a sample from that interval (Buch, 1994; Garrett & Coles, 1993; Gott, 1993, 1994; Jaynes, 2003; Jeffreys, 1961; Maddox, 1994; Shepard, 1987; Griffiths & Tenenbaum, 2000). However, different priors will be appropriate for different kinds of phenomena, and the predictions produced will vary substantially as a result. For example, the total gross of movies follows a power-law distribution, with $p(t_{total}) \propto t_{total}^{-\gamma}$ for some $\gamma > 0$.[11] This distribution has a highly non-Gaussian shape, with most movies taking in only modest amounts but occasional blockbusters making huge amounts of money. Other phenomena, such as human life spans, are better described by Gaussian priors, which fall off rapidly on either side of the mean, or Erlang priors, which rise to a peak and then fall off exponentially. Figure 3.7 illustrates these three kinds of prior distribution.

Bayesian Inference with Different Priors We can derive an analytic form for the posterior distribution obtained with power-law and Erlang priors. The posterior distribution resulting from the Gaussian prior has no simple analytic form, but it can be approximated using the method of discrete approximation mentioned the beginning of section 3.5. With the power-law prior, $p(t_{total}) \propto t_{total}^{-\gamma}$ for $\gamma > 0$. This prior is improper if $\gamma \leq 1$, since the integral over t_{total} diverges, but the posterior remains a proper probability distribution regardless. Applying equation (3.71), we have

$$\begin{aligned} p(t) &\propto \int_t^\infty t_{total}^{-(\gamma+1)}\, dt_{total} \\ &= -\frac{1}{\gamma} t_{total}^{-\gamma}\Big|_t^\infty \\ &= \frac{1}{\gamma} t^{-\gamma}, \end{aligned}$$

where the constant of proportionality remains the same as in the original prior. We can substitute this result into Bayes' rule (equation (3.69)) to obtain

$$p(t_{total}|t) = \frac{t_{total}^{-(\gamma+1)}}{\frac{1}{\gamma} t^{-\gamma}}$$

11. When $\gamma > 1$, a power-law distribution is often referred to in statistics and economics as a "Pareto distribution." With $\gamma = 1$, this is the "uninformative" prior for a variable that ranges from 0 to ∞, since it is invariant to multiplicative rescaling and thus does not impose an intrinsic scale on the variable.

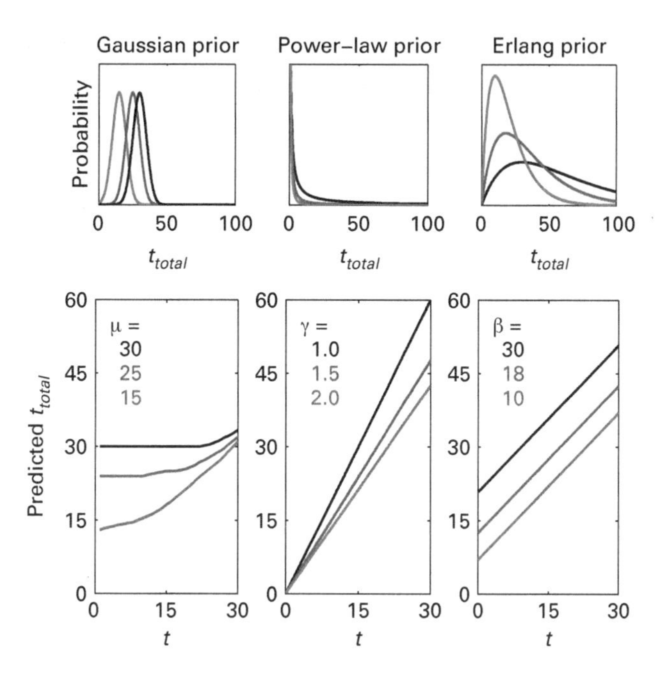

Figure 3.7
Bayesian predictions depend on the form of the prior distribution. The three columns represent qualitatively different statistical models appropriate for different kinds of events. The top row shows three parametric families of prior distributions for the total duration or extent, t_{total}, that could describe events in a particular class. Lines of different colors represent different parameter values (e.g., different mean durations) within each family. The bottom row shows the optimal predictions for t_{total} as a function of t, the observed duration or extent of an event so far, assuming the prior distributions shown in the top panel. For Gaussian priors (left column), parameterized by mean μ and standard deviation σ, the prediction rule always has a slope less than 1 and an intercept near the mean μ: predictions are never much smaller than the mean of the prior distribution, nor much larger than the observed duration. Power-law priors (middle column), with $p(t_{total}) \propto t_{total}^{-\gamma}$, result in linear prediction rules with slope always greater than 1: Predicted $t_{total} = 2^{1/\gamma} t$. The uninformative prior is a power-law with $\gamma = 1$. Erlang priors (right column), with $p(t_{total}) \propto t_{total} \exp\{-t_{total}/\beta\}$, yield a linear prediction rule that always has slope equal to 1 and a nonzero intercept: Predicted $t_{total} = t + \beta \log 2$. Figure adapted from Griffiths and Tenenbaum (2006).

$$= \frac{\gamma\, t^{\gamma}}{t_{total}^{\gamma+1}}, \tag{3.72}$$

for all values of $t_{total} \geq t$. Under the Erlang prior, $p(t_{total}) \propto t_{total} \exp\{-t_{total}/\beta\}$, we have

$$\begin{aligned} p(t) &\propto \int_t^{\infty} \exp\{-t_{total}/\beta\} \\ &= -\beta \exp\{-t_{total}/\beta\} \,|_t^{\infty} \\ &= \beta \exp\{-t/\beta\}, \end{aligned}$$

where the constant of proportionality remains the same as in the original prior. Again, we can substitute this result into Bayes' rule (equation (3.69)) to obtain

$$p(t_{total}|t) = \frac{\exp\{-t_{total}/\beta\}}{\beta \exp\{-t/\beta\}}$$

$$= \frac{1}{\beta} \exp\{-(t_{total} - t)/\beta\}, \tag{3.73}$$

for all values of $t_{total} \geq t$.

Reducing the Posterior to an Estimate We take the predicted value of t_{total}, which we will denote as t^*, to be the posterior median. This is the point t^* such that $P(t_{total} > t^* | t) = 0.5$: a Bayesian predictor believes that there is a 50 percent chance that the true value of t_{total} is greater than t^* and a 50 percent chance that the true value of t_{total} is less than t^*. The median makes sense as a point estimate in this setting, as the MAP is often the observed value t and the posterior for power-law distributions is heavy-tailed, meaning that the mean can be very skewed toward high values. The posterior median can be computed from the posterior, using the fact that

$$P(t_{total} > t^* | t) = \int_{t^*}^{\infty} p(t_{total} | t) \, dt_{total}. \tag{3.74}$$

We can derive t^* analytically in the case of a power-law or Erlang prior. For the power-law prior, we can use equation (3.72) to rewrite equation (3.74) as

$$\begin{aligned} P(t_{total} > t^* | t) &= \int_{t^*}^{\infty} \frac{\gamma \, t^{\gamma}}{t_{total}^{\gamma+1}} \, dt_{total} \\ &= -\left(\frac{t}{t_{total}}\right)^{\gamma} \Big|_{t^*}^{\infty} \\ &= \left(\frac{t}{t^*}\right)^{\gamma}. \end{aligned} \tag{3.75}$$

We can now solve for t^* such that $P(t_{total} > t^* | t) = 0.5$, obtaining $t^* = 2^{1/\gamma} t$. For the Erlang prior, we can use equation (3.73) to rewrite equation (3.74) as

$$\begin{aligned} P(t_{total} > t^* | t) &= \int_{t^*}^{\infty} \frac{1}{\beta} \exp\{-(t_{total} - t)/\beta\} dt_{total} \\ &= -\exp\{-(t_{total} - t)/\beta\}|_{t^*}^{\infty} \\ &= \exp\{-(t^* - t)/\beta\}. \end{aligned} \tag{3.76}$$

Again, we can solve for t^* such that $P(t_{total} > t^* | t) = 0.5$, obtaining $t^* = t + \beta \log 2$. For the Gaussian prior, we can find values of t^* by numerical optimization.

Evaluating the Predictions Figure 3.7 shows the results of making predictions using these three priors. For power-law priors, the Bayesian prediction function picks a value for t_{total} that is a multiple of the observed sample t. The exact multiple depends on parameter γ. For the particular power law that best fits the actual distribution of movie grosses, an optimal Bayesian observer would estimate the total gross to be approximately 50 percent greater than the current gross: if we observe a movie has made \$40 million to date, we should guess a total gross of around \$60 million; if we had observed a current gross of only \$6 million, we should guess about \$9 million for the total. While such "constant multiple" prediction rules

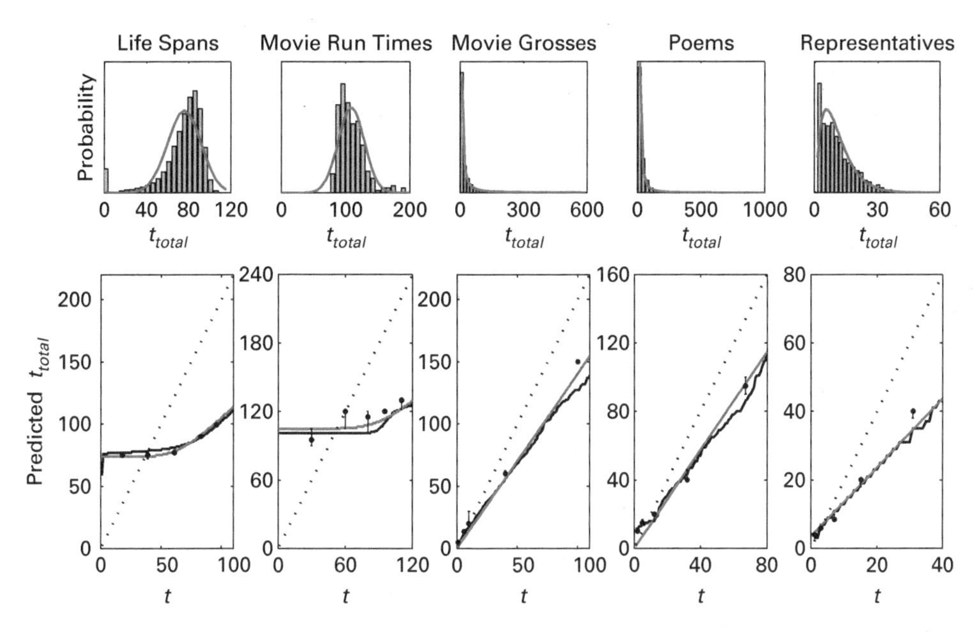

Figure 3.8
People integrate the evidence from a single observation with the appropriate prior distributions. The upper panels show the empirical distribution of the total duration or extent, t_{total}, for five everyday phenomena. The values of t_{total} are the hypotheses h to be evaluated, and these distributions are the appropriate priors. The first two distributions are approximately Gaussian, the next two are approximately power-law, and the last is approximately Erlang. Best-fitting parametric distributions are plotted in red. In the lower panels, black dots show subjects' median predictions for t_{total} when given a single observed sample t of a duration or extent in each of five domains (the data d used when applying Bayes' rule). Judgments are consistent with Bayesian predictions using the empirical prior distribution shown in the upper panel (black lines), and the best-fitting parametric prior (red lines). Predictions based on a single uninformative prior (dotted lines) are not consistent with these judgments. Figure adapted from Griffiths and Tenenbaum (2006).

are optimal for event classes that follow power-law priors, they are clearly inappropriate for other phenomena, such as human life spans (consider predicting the life span of a 6-year-old boy and a 90-year-old man by applying a multiplicative rule!). The Gaussian and Erlang priors produce functions that do not have the same multiplicative character. For the Gaussian, the best prediction is the mean until the observed value approaches that mean, at which point it increases with a slope close to 1. For the Erlang distribution, the best guess of t_{total} is simply t plus a constant determined by parameter β, which is always a little more than the observed value.

The fact that different priors produce such different prediction functions provides an opportunity to explore whether people are sensitive to the distributions of everyday quantities in forming predictions. To address this question, Griffiths and Tenenbaum (2006) conducted an experiment in which participants were asked to produce predictions of the total duration or extent of a range of everyday phenomena. Some of the results of this experiment are shown in figure 3.8. The empirical distributions of the everyday phenomena were estimated using information taken from online databases, and they were used to determine the posterior median of t_{total} as a function of t. People's predictions showed a surprisingly close correspondence to the posterior median, suggesting that people are sensitive to these distributions and can combine this information with evidence when making predictions.

This study also provides another example of being able to estimate people's prior distributions from their behavior: when we know the data that they see and the conclusion that they reach, we can infer the prior that captures their inference.

3.6 Bayesian Model Selection

Many problems of statistical inference require comparing hypotheses that differ in their complexity. For example, the problem of inferring whether a coin is fair or biased based upon an observed sequence of heads and tails requires comparing a hypothesis that gives a single value for θ—if the coin is fair, then $\theta = 0.5$—with a hypothesis that allows θ to take on any value between 0 and 1. A similar problem arises when evaluating whether a causal relationship exists, where the hypothesis that it exists also allows that relationship to vary in strength. Likewise, when we try to decide whether data make more sense organized in a hierarchy or laid out in a spatial representation, it seems that we need to take into account the fact that these representations differ in flexibility.

Using observed data to choose between two probabilistic models that differ in complexity is often called the problem of *model selection* (Myung & Pitt, 1997; Myung, Forster, & Browne, 2000). One familiar statistical approach to this problem uses hypothesis testing, but this approach is often complex and counter intuitive. In contrast, the Bayesian approach to model selection is a seamless application of the methods discussed so far. Hypotheses that differ in their complexity can be compared directly using Bayes' rule, once they are reduced to probability distributions over the observable data (see Kass & Raftery, 1995). We will illustrate this principle using the example of determining whether a coin is biased.

3.6.1 Looking for Biased Coins

Assume that we flip a coin n times to obtain n_H heads and n_T tails, and we want to determine whether the coin is biased. We have two hypotheses: h_0 is the hypothesis that $\theta = 0.5$, and h_1 is the hypothesis that θ takes a value drawn from a uniform distribution on [0, 1]. If we have no a priori reason to favor one hypothesis over the other, we can take $P(h_0) = P(h_1) = 0.5$. The probability of the data under h_0 is straightforward to compute, using the Bernoulli likelihood, giving $P(d|h_0) = 0.5^{n_H+n_T}$. But how should we compute the likelihood of the data under h_1, which does not make a commitment to a single value of θ?

One strategy would be to use the maximum-likelihood value of θ, taking that as our best estimate of the bias of the coin. However, pursuing this strategy always yields a probability of the data that is at least as high as $P(d|h_0)$ no matter what the data (intuitively, since $\hat{\theta}$ is the maximum-likelihood estimate, the likelihood using $\hat{\theta}$ must be at least as high as when using $\theta = 0.5$, which corresponds to h_0). So we need to have some way to correct for the flexibility of h_1. The standard approach to hypothesis testing in frequentist statistics consists of a method for implementing this correction, with the goal of minimizing the probability that we erroneously decide in favor of h_1. However, the problem has a simple solution in Bayesian statistics.

The Bayesian solution to this problem is to compute the marginal probability of the data under h_1. As discussed earlier in this chapter, given a joint distribution over a set of variables, we can always sum out variables until we obtain a distribution over just the variables that interest us. In this case, we define the joint distribution over d and θ given h_1, and then integrate over θ to obtain

$$P(d|h_1) = \int_0^1 P(d|\theta, h_1)p(\theta|h_1)\,d\theta, \tag{3.77}$$

where $p(\theta|h_1)$ is the distribution over θ assumed under h_1—in this case, a uniform distribution over [0, 1]. This does not require any new concepts—it is exactly the same kind of computation that we needed to perform to compute the denominator for the posterior distribution over θ. Performing this computation, we obtain $P(d|h_1) = \frac{n_H!\, n_T!}{(n_H+n_T+1)!}$, where again the fact that we have a conjugate prior provides a neat analytic result. Having computed this likelihood, we can apply Bayes' rule just as we did for two simple hypotheses. Figure 3.9a shows how the log posterior odds in favor of h_1 change as n_H and n_T vary for sequences of length 10.

This marginal probability approach was used by Griffiths and Tenenbaum (2007a) to calculate the log likelihood ratios used in the model of coincidences discussed in section 3.2.2. If you recall, participants were presented with data generated from an experiment investigating either genetic engineering or paranormal phenomena. The genetic engineers were trying to determine whether a drug influenced the sex of 100 rats, while the paranormal investigators were evaluating whether psychics could influence the outcome of 100 coin flips. Participants were told the frequency of male rats or coins coming up heads and asked to evaluate whether a relationship existed. In this case, if there is no relationship, then male rats or heads will occur with probability 0.5. If there is a relationship, that probability could be anything between 0 and 1. The results given in the previous paragraph provide us with everything that we need to calculate the probability of the observed data under this hypothesis, and hence the log likelihood ratio needed to infer people's priors about genetic engineering or psychic powers.

3.6.2 The Bayesian Occam's Razor

The ease with which hypotheses differing in complexity can be compared using Bayes' rule conceals the fact that this is actually a very challenging problem. Complex hypotheses have more degrees of freedom that can be adapted to the data, and thus they can always be made to fit the data better than simple hypotheses. It seems that a complex hypothesis would thus have an inherent unfair advantage over a simple hypothesis.

The Bayesian approach to comparing hypotheses that differ in their complexity takes this into account. More degrees of freedom provide the opportunity to find a better fit to the data, but this greater flexibility also makes a worse fit possible. To return to our coin-flipping example, for d consisting of the sequence HHTHTTHHHT, $P(d|\theta, h_1)$ is greater than $P(d|h_0)$ for $\theta \in (0.5, 0.694]$, but it is less than $P(d|h_0)$ outside that range. Marginalizing over θ averages these gains and losses: a more complex hypothesis will be favored only if its greater complexity consistently provides a better account of the data. The log posterior odds shown in figure 3.9a capture this: the more complex hypothesis h_1 is favored only when $P(d|\theta, h_1)$ is higher than $P(d|h_0)$ across a wide range of values of θ.

To phrase this principle in another way, a Bayesian learner judges the fit of a parameterized model not by how well it fits using the *best* parameter values, but by how well it fits using *randomly selected* parameters, where the parameters are drawn from a prior specified by the

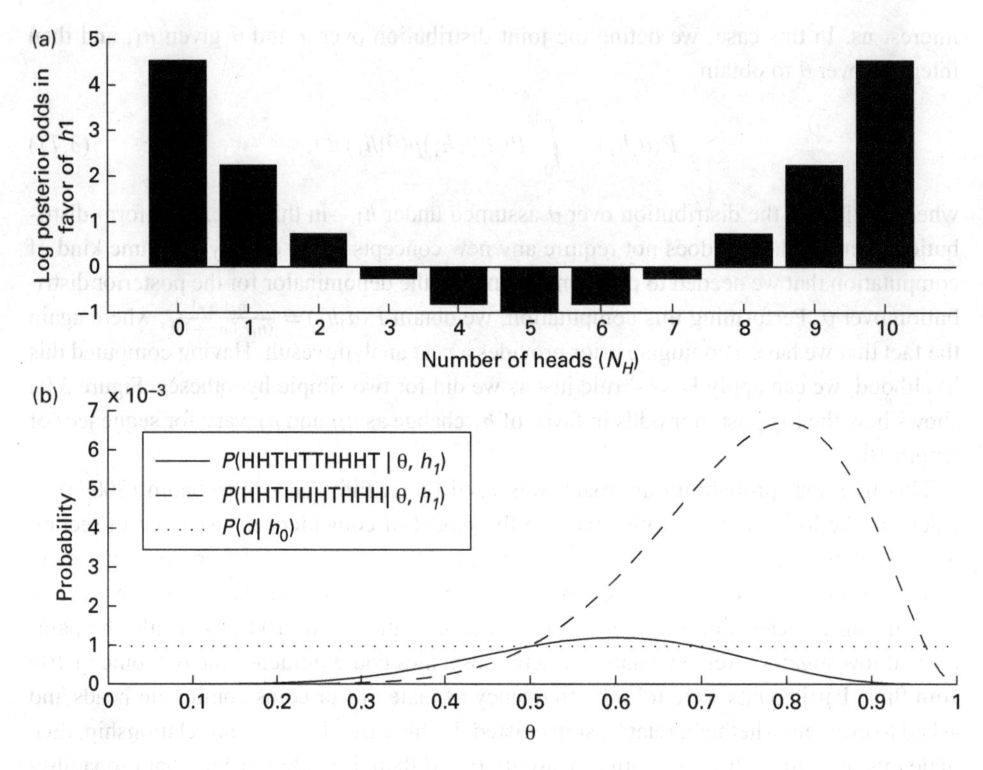

Figure 3.9
Comparing hypotheses about the weight of a coin. (a) The vertical axis shows log posterior odds in favor of h_1, the hypothesis that the probability of heads (θ) is drawn from a uniform distribution on $[0, 1]$, rather than h_0, the hypothesis that the probability of heads is 0.5. The horizontal axis shows the number of heads, n_H, in a sequence of 10 flips. As n_H deviates from 5, the posterior odds in favor of h_1 increase. (b) The posterior odds shown in (a) are computed by averaging over the values of θ with respect to the prior, $p(\theta)$, which in this case is the uniform distribution on $[0, 1]$. This averaging takes into account the fact that hypotheses with greater flexibility—such as the free-ranging θ parameter in h_1—can produce both better and worse predictions, implementing an automatic "Bayesian Occam's razor." The solid line shows the probability of the sequence HHTHTTHHHT for different values of θ, while the dotted line is the probability of any sequence of length 10 under h_0 (equivalent to $\theta = 0.5$). While there are some values of θ that result in a higher probability for the sequence, on average the greater flexibility of h_1 results in lower probabilities. Consequently, h_0 is favored over h_1 (this sequence has $n_H = 6$). In contrast, a wide range of values of θ result in higher probability for the sequence HHTHHHTHHH, as shown by the dashed line. Consequently, h_1 is favored over h_0 (this sequence has $n_H = 8$). Figure adapted from Griffiths et al. (2008c).

model $p(\theta|h_1)$ in equation (3.77) (Ghahramani, 2004). This penalization of more complex models (which leads the learner to favor simpler explanations) is known as the *Bayesian Occam's razor* (Jeffreys & Berger, 1992; Mackay, 2003) and is illustrated in figure 3.9b. Here, two sequences of coin flips differ in the extent to which $P(d|\theta, h_1)$ is greater than $P(d|h_0)$. Only for the sequence with greater imbalance in the number of heads and tails is $P(d|\theta, h_1)$ consistently greater, and hence it would be favored if we were to randomly sample values of θ.

3.6.3 An Example: Coincidences in Space

Inferring causal relationships provides a classic example of a model selection problem, requiring us to choose between a model of the world where the relationship exists and another where it does not. As mentioned in section 3.2.2, Griffiths and Tenenbaum (2007a)

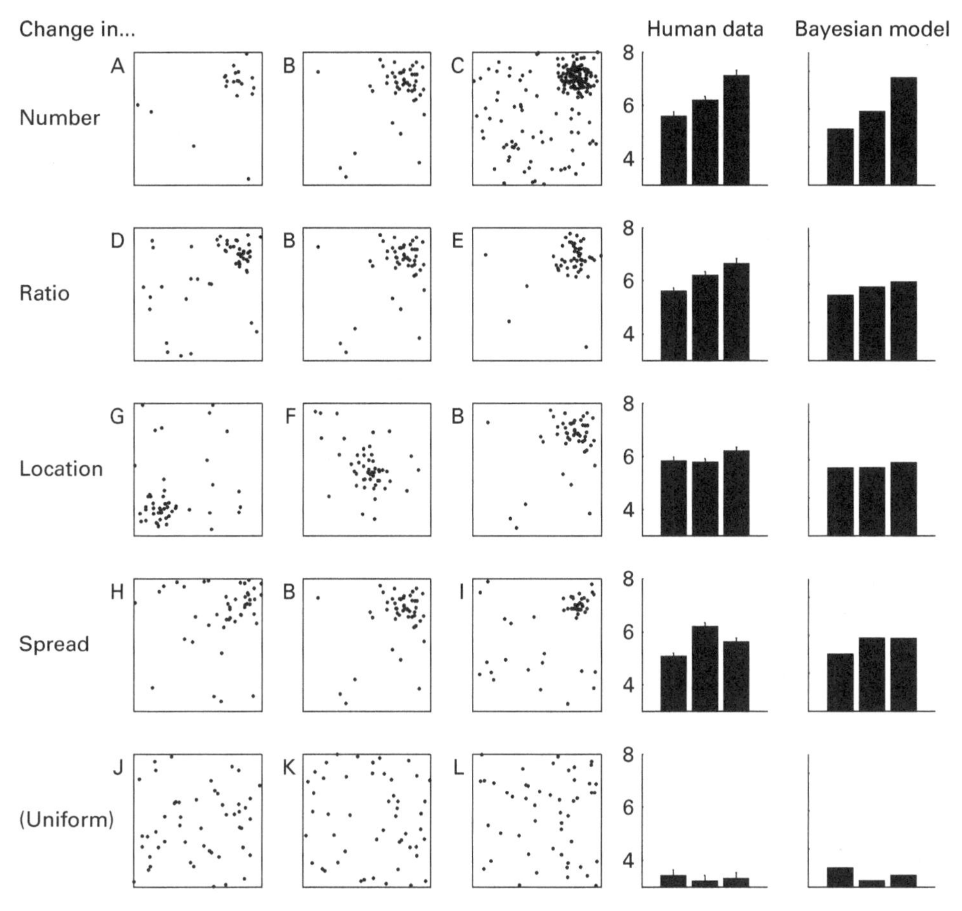

Figure 3.10
Results of experiment 2 from Griffiths and Tenenbaum (2007a). Each row shows the three stimuli used to test the effects of manipulating one of the statistical properties of the stimulus, together with the mean judgments of strength of coincidences from human participants and the predictions of the Bayesian model. Error bars show one standard error, and letters label the different stimuli. Figure adapted from Griffiths and Tenenbaum (2007a).

observed that our intuitive sense of coincidence seems to correspond well with cases where events provide strong evidence for an unexpected causal relationship. Such events generate a suspicion that a cause might be at work, despite our assumption that no such relationship exists.

Coincidences in the spatial location of events provide a good example of this phenomenon. Figure 3.10 shows several stimuli that were used in experiment 2 from Griffiths and Tenenbaum (2007a). Each stimulus consists of a set of dots inside a square, and participants were told that these dots represented hypothetical locations for the bombs that fell on London during World War II. A statistical analysis after the war revealed that the bombs apparently had fallen at random, but people in the city believed that there was a more systematic process at work. Participants in the experiment were asked to rate how big a coincidence it would have been for the bombs to have fallen in each pattern. The mean ratings are also shown in the figure, and they indicate clear effects of the total number of bombs and the ratio of the number of bombs that seemed to cluster together to the number that seemed to fall more widely.

People's judgments of the strength of coincidence in this experiment can be modeled using the Bayesian model selection framework outlined earlier in this chapter. Letting h_0 denote the hypothesis that the bombs fell at random and h_1 denote the hypothesis that there was a common cause behind the targets of at least some bombs, the evidence in favor of a causal relationship provided by the observed data d is given by the log likelihood ratio $\log \frac{p(d|h_1)}{p(d|h_0)}$. Under h_0, each bomb falls at a location in the square given by a uniform distribution. Under h_1, there is a region where the bombs are concentrated, represented by a multivariate Gaussian distribution. Evaluating the probability of the data under h_0 is straightforward, as we can assume that these are simply independent draws from the uniform distribution, each occurring with probability $1/|\mathcal{R}|$, where $\mathcal{R}$ is the region in which bombs are falling and $|\mathcal{R}|$ is its area. However, evaluating the probability of the data under h_1 requires integrating over all possible locations and sizes for the regularity, as well as all proportions of bombs falling within the regularity. This can be done using some ideas from chapters 5 and 6—mixture distributions and Monte Carlo methods, respectively—but we refer eager readers to Griffiths and Tenenbaum (2007a) for details. The results of this computation are shown in the rightmost column of figure 3.10. There is a close match between people's assessment of the strength of coincidences and the evidence in favor of a cause, as evaluated by Bayesian model selection.

3.7 Summary

Bayesian inference describes how a rational agent approaches the problem of induction. It indicates how to combine prior knowledge with the information provided by new data. As a result, it is a powerful tool for understanding human cognition. Using Bayesian models, we can precisely identify the prior knowledge that informs human inductive inferences in a wide range of settings. In the simplest cases, this can be done simply by presenting people with an inductive inference and comparing the conclusions that they reach with those that result from different prior distributions. The results can help to explain biases in human cognition that otherwise might be perplexing.

When applied to discrete hypotheses, Bayesian inference helps us understand how discrete, symbolic structure can result in continuous, graded patterns of generalization. Using this approach, it is possible to make meaningful inferences from very small amounts of data. Provided enough data, the resulting models can transition from graded generalization to categorical confidence in a way that captures human inductive inference.

Bayesian statistics also provides us with tools for characterizing more challenging inferences, including cases where we need to compare hypotheses that differ in their complexity. Rather than requiring the development of new methods, these inferences are naturally handled using the principles of probability theory. The result is an implicit preference for simpler explanations, which we will see playing an increasingly important role later in the book as we consider how people learn world models that potentially become arbitrarily complex.

4

Graphical Models

Thomas L. Griffiths and Alan Yuille

Our discussion of Bayesian inference so far has been formulated in the language of "hypotheses" and "data." However, the principles of Bayesian inference extend to much broader settings. Ultimately, we want to be able to describe rich models of the world that people build from experience and use to support rapid inferences from limited data. These world models can be expressed as probability distributions over large numbers of random variables. However, defining and working with such probability distributions have their own challenges.

In its most general form, a probabilistic model simply defines the joint distribution for a set of random variables. Representing and computing with these joint distributions become harder as the number of variables grows. Big joint distributions can also be hard to interpret, making it difficult to understand the assumptions that are expressed in a model. In this chapter, we introduce *graphical models*, which provide an efficient and intuitive framework for working with complex probability distributions.

A graphical model associates a probability distribution with a *graph*. Here, we use the word "graph" in the sense of graph theory—a set of nodes and edges, typically depicted by drawing nodes as circles and edges as lines that connect them. The nodes of the graph represent the variables on which the probability distribution is defined, and the edges between the nodes reflect their probabilistic dependencies. These dependencies are translated into a distribution via a set of functions that convert the values of nodes and their neighbors in the graph into probabilities. Together, the graph and these functions define a joint distribution over all the variables.

There are two kinds of graphical models, differing in the nature of the edges that connect the nodes. If the edges simply indicate a dependency between variables without specifying a direction, then the result is an *undirected graphical model*. Undirected graphical models have long been used in statistical physics, and many probabilistic neural network models, such as Boltzmann machines (Ackley, Hinton, & Sejnowski, 1985), can be interpreted as models of this kind. If the edges indicate the direction of a dependency, the result is a *directed graphical model*. Our focus here will be on directed graphical models, which are also known as *Bayesian networks* or Bayes nets (Pearl, 1988). Bayesian networks can often be given a causal interpretation (Pearl, 2000), where an edge between two nodes indicates that one node is a direct cause of the other, which makes them particularly appealing for modeling higher-level cognition.

In this chapter, we introduce Bayesian networks and their role in defining *generative models* that specify complex probability distributions. We demonstrate three ways in which Bayesian networks can be useful in developing computational models of human cognition. First, they simplify representing and computing with probability distributions. Second, they make it easy to define generative models, a key step in making probabilistic inferences that go beyond the simplest cases of hypotheses and data and take us in the direction of more comprehensive world models. Third, they can be augmented to provide a language for representing and reasoning about causal relationships, which can be extremely useful in developing models of higher-level cognition, where causality can play an important role in the structure of mental representations. For a more detailed mathematical treatment of graphical models and their properties, see Koller and Friedman (2009).

4.1 Bayesian Networks

Bayesian networks are powerful tools for representing and reasoning with probability distributions. Their value is both computational and conceptual: they can reduce the computational cost of working with large probability distributions, but they also clarify the relationships and assumptions that go into those distributions. In this section, we provide a more detailed definition of Bayesian networks, explain how they can be used to represent probability distributions, and illustrate their use in specifying generative models.

4.1.1 Defining a Distribution via the Markov Condition

A Bayesian network represents the probabilistic dependencies that relate a set of variables to one another. If an edge exists from node A to node B, then A is referred to as a *parent* of B, and B is a *child* of A. This genealogical relation is often extended to identify the *ancestors* and *descendants* of a node. The directed graph used in a Bayesian network has one node for each random variable in the associated probability distribution, and is constrained to be *acyclic*: one can never return to the same node by following a sequence of directed edges. Acyclic graphs are required for the mathematical definition of a Bayesian network to make sense, but the constraint is intuitive: if we ultimately want to think about the edges in the graph representing steps in a generative process or causal relationships, cycles would mean that two things simultaneously generate or cause one another.[1]

The edges between nodes express the probabilistic dependencies between the corresponding variables in a fashion consistent with the *Markov condition*: conditioned on its parents, each variable is independent of all other variables except its descendants (Pearl, 1988; Spirtes, Glymour, & Schienes, 1993). As a consequence of the Markov condition, any Bayesian network specifies a canonical factorization of a joint probability distribution

1. Genuine cases of cyclic causality can typically be "unrolled" over time. For example, if we say that wage growth causes inflation and inflation causes wage growth, we really mean that wage growth at time t causes inflation at time $t+1$ and inflation at time t causes wage growth at time $t+1$. If two variables are really coupled in a simultaneous relationship, then the solution is to represent those two variables with a single node in the Bayesian network, as if they were a single variable that takes on values corresponding to the joint distribution of the original variables.

into the product of local conditional distributions, one for each variable conditioned on its parents. That is, for a set of variables $X_1, X_2, \ldots, X_n$, we can write

$$P(x_1, x_2, \ldots, x_n) = \prod_i P(x_i | \mathrm{Pa}(X_i)), \tag{4.1}$$

where $\mathrm{Pa}(X_i)$ is the set of parents of X_i.

To understand how this works, it is helpful to assign a numerical ordering to the variables such that each node is assigned a higher number than its children (the requirement that the underlying graph be acyclic guarantees that such an ordering exists). Assume that our variables $X_1, X_2, \ldots, X_n$ are in such an order. We can always write their joint distribution as the product of a series of conditional distributions–namely,

$$P(x_1, x_2, \ldots, x_n) = P(x_1 | x_2, \ldots, x_n) P(x_2 | x_3, \ldots, x_n) \ldots P(x_{n-1} | x_n) P(x_n). \tag{4.2}$$

The Markov condition means that we can simplify these conditional distributions, replacing $P(x_i | x_{i+1}, \ldots, x_n)$ with $P(x_i | \mathrm{Pa}(X_i))$, as X_i is independent of all nondescendants (i.e., higher-numbered variables) conditioned on its parents. As a result, we obtain $P(x_1, x_2, \ldots, x_n) = \prod_i P(x_i | \mathrm{Pa}(X_i))$.

4.1.2 Identifying Independence

As we will see, part of the power of Bayesian networks derives from the way in which they characterize the relationships between random variables. A set of simple rules can be used to determine whether two random variables in a Bayesian network are independent (e.g., Schachter, 1998). Any two variables that are directly connected by an edge are clearly dependent on one another, with the value of one variable depending on the value of the other. However, variables that are not directly connected can also be dependent. Intuitively, dependency is about the flow of information along a path (i.e., a sequence of edges) connecting two variables. Information flows from X_i to X_j if observing X_i tells you anything about X_j, given your other observations. This flow of information doesn't have to follow the arrows in our directed graph: it doesn't matter whether X_j is the parent or the child of X_i; learning something about X_i still tells you something about X_j.

Every path between any two variables in a directed graph can be broken into sections of the four types shown in figure 4.1, with the whole path being the result of chaining these sections together (these possibilities are exhaustive because they include both possible directions of each of the two edges). If we interpret the edges in the graph as showing causal relationships, figures 4.1a and 4.1b depict *causal chains* that flow in different directions (an example of a causal chain might be anxiety, causing insomnia, causing tiredness). Figure 4.1c is a *common cause* structure, where X_2 causes both X_1 and X_3 (eating ice cream and swimming have the common cause of being at the seaside). Figure 4.1d is a *common effect* structure, with X_2 being caused by both X_1 and X_3 (oversleeping and bad traffic have the common effect of being late for work).

We can determine whether dependency can flow along a path by establishing whether it can flow through each section in that path. This can be done by applying a simple rule: for sections with a chain structure or common cause structure, 4.1a-c, X_1 and X_3 are independent when X_2 is observed (i.e., they are independent when conditioned on X_2); for sections with the structure shown in figure 4.1d, X_1 and X_3 are independent unless X_2 or one of

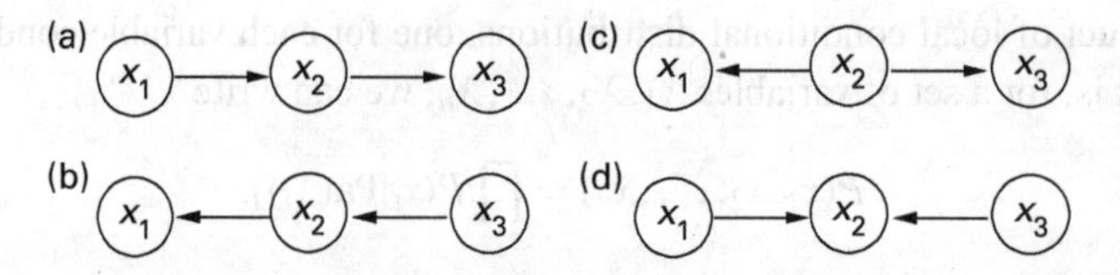

Figure 4.1
All possible sections of a path between two variables in a directed graph. (a)–(b) Causal chains. Dependency flows along chains of edges, unless x_2 is observed. (c) Common cause. Dependency flows between children of a common parent unless x_2 is observed. (d) Common effect. Dependency does not flow between parents of a common child unless x_2 is observed.

its descendants is observed. You can check that this is the case by writing down the joint distribution associated with each of these structures, and then computing the conditional distribution given X_2.

Patterns of dependency in common effect structures (figure 4.1d) can be counterintuitive, so it's worth looking at an example. Assume that oversleeping and bad traffic have the common effect of being late for work. Whether somebody oversleeps and whether the traffic is bad are presumably independent of one another. But when you observe that a person is late for work, both oversleeping and bad traffic become possible explanations. Subsequently finding out that the traffic is bad would reduce your estimate of the probability that person overslept. This means that oversleeping and bad traffic have become dependent on one another, conditioned on having observed somebody being late for work. We don't even have to observe them being late—observing that their co-worker is annoyed or that they don't show up for a meeting (both downstream consequences, and hence descendants of X_2) would be enough to induce this dependency. We discuss this pattern of reasoning, known as *explaining away*, in more detail in section 4.2.2.

4.1.3 Representing Probability Distributions over Propositions

Graphical models have been used in artificial intelligence (AI) systems, where the variables of interest represent the truth value of certain logical propositions (Russell & Norvig, 2021). To use an example that builds on the "psychic priors" study mentioned in chapter 3, imagine that a friend of yours claims to possess psychic powers—in particular, the power of psychokinesis. She proposes to demonstrate these powers by flipping a coin and influencing the outcome to produce heads. You suggest that a better test might be to see if she can levitate a pencil, since the coin producing heads could also be explained by some kind of sleight of hand, such as substituting a two-headed coin. We can express all possible outcomes of the proposed tests, as well as their causes, using the binary random variables X_1, X_2, X_3, and X_4 to represent (respectively) the truth of the coin being flipped and producing heads, the pencil levitating, your friend having psychic powers, and the use of a two-headed coin (these variables take the value 1 when true, 0 when false). Any set of beliefs about these outcomes can be encoded in a joint probability distribution, $P(x_1, x_2, x_3, x_4)$. For example, the probability that the coin comes up heads ($x_1 = 1$) should be higher if your friend actually does have psychic powers ($x_3 = 1$). Figure 4.2 shows a Bayesian network expressing a possible pattern of dependencies among these variables. For example, X_1 and X_2 are assumed to be independent given X_3, indicating that once it was known whether your friend was psychic,

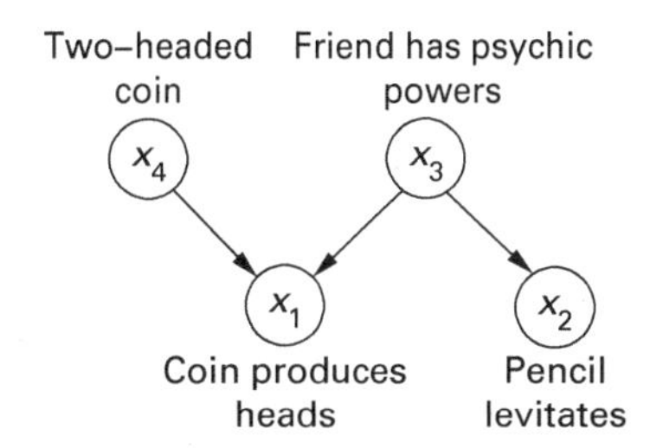

Figure 4.2
Directed graphical model (Bayesian network) showing the dependencies among variables in the "psychic friend" example. Figure reproduced with permission from Griffiths, Kemp, and Tenenbaum (2008).

the outcomes of the coin flip and the levitation experiments would be completely unrelated. By the Markov condition, we can write $P(x_1,x_2,x_3,x_4) = P(x_1|x_3,x_4)P(x_2|x_3)P(x_3)P(x_4)$.

In addition to clarifying the dependency structure of a set of random variables, Bayesian networks provide an efficient way to represent and compute with probability distributions. In general, a joint probability distribution on n binary variables requires $2^n - 1$ numbers to specify (one for each of the 2^n joint values taken by the variables, minus 1 because we know that probability distributions sum to 1, so we can derive the last number from the rest). In the case of the psychic friend example, there are four variables. Each of these variables takes the value 1 or 0, so there are $2^4 = 16$ possible joint values: 0000, 0001, 0010, etc. Specifying a probability distribution over these 16 values would require $2^4 - 1 = 15$ numbers.

Bayesian networks make it easier to specify complex joint distributions by factorizing those distributions into the product of conditional distributions. Each conditional distribution requires fewer numbers to specify. We only need one number for each variable conditioned on each possible set of values its parents can take, or $2^{|\mathrm{Pa}(X_i)|}$ numbers for each variable X_i (where $|\mathrm{Pa}(X_i)|$ is the size of the parent set of X_i). For our "psychic friend" network, this adds up to 8 numbers rather than 15 because X_3 and X_4 have no parents (contributing one number each), X_2 has one parent (contributing two numbers), and X_1 has two parents (contributing four numbers). The simplification of the joint distribution is a result of the conditional independence of these variables: if we just used the standard factorization of the joint distribution into conditionals, $P(x_1|x_2,x_3,x_4)P(x_2|x_3,x_4)P(x_3|x_4)P(x_4)$ without any assumed independence, we would need 8 numbers to specify $P(x_1|x_2,x_3,x_4)$, 4 numbers for $P(x_2|x_3,x_4)$, 2 numbers for $P(x_3|x_4)$, and 1 number for $P(x_4)$, for a total of 15 numbers.

4.1.4 An Example: Feature Inference

The Markov condition makes clear predictions about the circumstances under which variables should be independent of one another. Rehder and Burnett (2005) conducted a study intended to determine whether people are appropriately sensitive to these independence relations. The setting that they focused on was inferring the value of an unobserved feature of an object from its observed features. For example, if you were visiting a new country and saw an animal that had wings, built nests in trees, and ate bugs, you might make the reasonable assumption that it could also fly.

This problem of feature inference is just a kind of probabilistic inference: we want to infer the value of an unobserved feature F_i based on the values of a set of observed features (we will write the vector of the values of all features other than feature i as $\mathbf{F}_{-i}$). What we want is thus the conditional distribution $P(F_i|\mathbf{F}_{-i})$. However, the properties of this distribution—and the contributions that each of the features in $\mathbf{F}_{-i}$ makes to our beliefs about F_i—will be determined by the dependency structure on those features.

Rehder and Burnett explored a variety of dependency structures, but for simplicity we will focus on their experiment 1. In this experiment, participants were given a short vignette about the features of a new category of objects and the way that they relate to one another. For example, they might read:

On the volcanic island of Kehoe, in the western Pacific Ocean near Guam, there is a species of ant called Kehoe Ants. For food, Kehoe Ants consume vegetation rich in iron and sulfur.

With this backstory, the features could be specified, such as the following:

(F_1) Some Kehoe Ants have blood that is very high in iron sulfate. Others have blood that has low levels of iron sulfate.

(F_2) Some Kehoe Ants have an immune system that is hyperactive. Others have a suppressed immune system.

(F_3) Some Kehoe Ants have blood that is very thick. Others have blood that is very thin.

(F_4) Kehoe Ants build their nests by secreting a sticky fluid that then hardens. Some Kehoe Ants are able to build their nests quickly. Others build their nests slowly.

Participants learned that these features occurred either 75 percent or 25 percent of the time (with the exact assignments of probabilities to features varying across participants). One group of participants were then taught that features comprised a common cause structure, as shown in figure 4.3a, for example:

($F_1 \rightarrow F_2$). Blood high in iron sulfate causes a hyperactive immune system. The iron sulfate molecules are detected as foreign by the immune system, and the immune system is highly active as a result.

($F_1 \rightarrow F_3$). Blood high in iron sulfate causes thick blood. Iron sulfate provides the extra iron that the ant uses to produce extra red blood cells. The extra red blood cells thicken the blood.

($F_1 \rightarrow F_4$). Blood high in iron sulfate causes faster nest building. The iron sulfate stimulates the enzymes responsible for manufacturing the nest-building secretions, and an ant can build its nest faster with more secretions.

Participants were then tested to make sure that they understood the right structure of these relationships. The other group of participants were a control group that didn't learn about the causal relationships.

After being educated about this new category of objects, participants were asked to assess the probability of one feature given the observation of other features. Critically, some features were effects of the common cause (F_2, F_3, and F_4 in our example), while others were the cause itself. The Markov condition makes very different predictions about how these features should affect people's inferences.

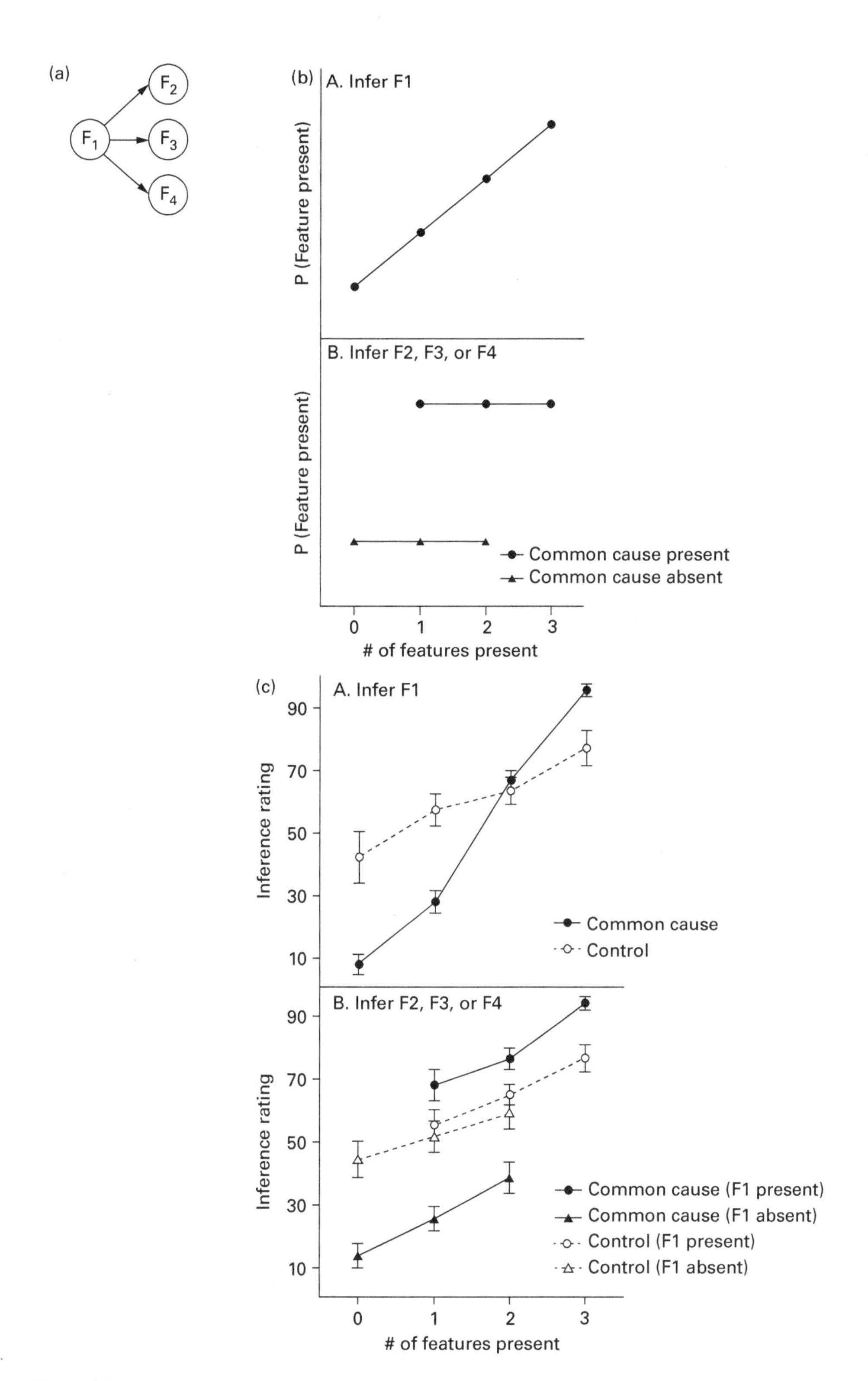

Figure 4.3
Feature inference (after Rehder & Burnett, 2005). (a) A common cause relationship between a set of features. (b) If the common cause (F_1) is unobserved, each additional feature that is observed provides more evidence about its value. However, once the common cause is observed, the other features become independent and do not provide evidence for one another's value. (c) People rated the strength of the inference that each feature was present based on the presence or absence of other features. People's judgments do not seem to obey the Markov condition in this case: they continue to treat features as mutually informative even when the common cause is observed. Figure adapted from Rehder and Burnett (2005).

As shown in figure 4.3b, if the unobserved feature is the common cause then observing each effect provides more evidence that the feature is present. By contrast, if the unobserved feature is one of the effects, then once the common cause is observed observing other effects provides no information. This is a consequence of the Markov condition: effects are independent of one another once the common cause is observed.

Rehder and Burnett found that people violated the Markov condition in this case. As shown in figure 4.3c, people continued to treat effects as providing evidence about other effects even when the common cause was observed. This suggests that people are allowing dependencies between these variables that go beyond those represented in the graphical model.

What could account for this result? One way to understand it is that people *do* follow the Markov condition, but they assume a more complex graphical model than the one they learned in the experiment. Rehder and Burnett performed a series of follow-up experiments to explore this possibility, finding a pattern of results consistent with a domain-general bias toward postulating a hidden mechanism behind category membership—an additional variable that is unobserved. This extra variable makes the features of objects that belong to a category dependent on one another, producing violations of the Markov condition in the original graphical model. This idea provides a connection to the literature on developmental psychology on essentialism (e.g., Gelman, 2003), which shows that children believe there is an unobservable "essence" that makes something a member of a category and is not altered by modifying its external features. Rehder has subsequently explored violations of the Markov condition that appear in people's judgments in a variety of other settings (Rehder, 2014; Rehder & Waldmann, 2017; Rehder, 2018).

4.1.5 Defining Generative Models

Beyond being useful for formalizing probabilistic reasoning, Bayesian networks provide an intuitive representation for the structure of many probabilistic models. By breaking the process of producing data into a sequence of simple steps—a generative model—it becomes much easier to define the corresponding probability distribution.

Examples of Simple Generative Models We previously discussed the problem of estimating the weight of a coin, θ, using a Beta(α, β) prior. One detail that we left implicit in that discussion was the assumption that successive coin flips are independent, given a value for θ. This conditional independence assumption is expressed in the graphical model shown in figure 4.4a, where $x_1, x_2, \ldots, x_n$ are the outcomes (heads or tails) of n successive tosses. Applying the Markov condition, this structure represents the probability distribution

$$P(x_1, x_2, \ldots, x_n, \theta, \alpha, \beta) = p(\alpha)p(\beta)p(\theta|\alpha, \beta) \prod_{i=1}^{n} P(x_i|\theta), \tag{4.3}$$

in which x_i is independent given the value of θ. Often, we would assume that the hyperparameters α and β are known, giving the conditional distribution

$$P(x_1, x_2, \ldots, x_n, \theta|\alpha, \beta) = p(\theta|\alpha, \beta) \prod_{i=1}^{n} P(x_i|\theta), \tag{4.4}$$

which can be used to infer θ when we condition on the values of $x_1, x_2, \ldots, x_n$.

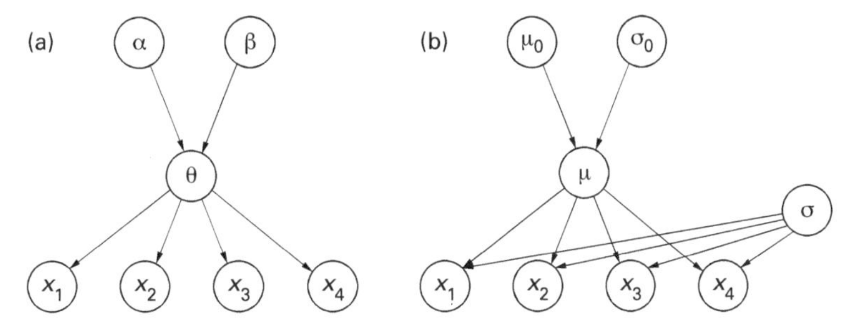

Figure 4.4
Graphical models for two simple examples of Bayesian inference. (a) Inferring a proportion, where θ is the weight of a coin, α and β are the hyperparameters of the prior, and $x_1, x_2, \ldots, x_n$ are the outcomes of n flips. (b) Inferring the mean of a Gaussian, where μ is the mean, μ_0 and σ_0 define the prior on the mean, σ is the standard deviation, and $x_1, x_2, \ldots, x_n$ are n observations.

Figure 4.4b shows a Bayesian network for the problem of estimating the mean of a Gaussian, μ, when that Gaussian has standard deviation σ and μ follows a Gaussian prior with mean μ_0 and standard deviation σ_0. The joint distribution is given by

$$p(x_1, x_2, \ldots, x_n, \mu, \sigma, \mu_0, \sigma_0) = p(\mu_0)p(\sigma_0)p(\mu|\mu_0, \sigma_0)p(\sigma) \prod_{i=1}^{n} p(x_i|\mu, \sigma), \tag{4.5}$$

and the conditional distribution when σ, μ_0, and σ_0 are known is

$$p(x_1, x_2, \ldots, x_n, \mu|\sigma, \mu_0, \sigma_0) = p(\mu|\mu_0, \sigma_0) \prod_{i=1}^{n} p(x_i|\mu, \sigma), \tag{4.6}$$

where we condition on $x_1, x_2, \ldots, x_n$ to infer μ.

These examples also illustrate that the "graphical" part of a graphical model is not itself sufficient to specify the associated probability distribution. We can derive equations (4.3)–(4.6) from the form of the associated graphs, but the graphs do not tell us what the actual distributions are that are used to generate the random variables. This information needs to be provided separately, defining a conditional probability distribution for each variable given its parents. The complete specification of a Bayesian network includes the graph and these conditional distributions, and together they define the associated joint distribution. For example, in figure 4.4a, we would specify the conditional distributions

$$\theta \mid \alpha, \beta \sim \text{Beta}(\alpha, \beta) \tag{4.7}$$

$$x_i \mid \theta \quad \sim \text{Bernoulli}(\theta), \tag{4.8}$$

while in figure 4.4b, we would have the conditional distributions

$$\mu \mid \mu_0, \sigma_0 \sim \text{Gaussian}(\mu_0, \sigma_0) \tag{4.9}$$

$$x_i \mid \mu, \sigma \quad \sim \text{Gaussian}(\mu, \sigma), \tag{4.10}$$

where $\sim$ should be read as "is distributed as." This is what makes these graphical models correspond to the examples presented earlier in this book.

Generative Models as a Sequence of Steps When introducing the basic ideas behind Bayesian inference, we emphasized the fact that hypotheses correspond to different assumptions about the process that could have generated the observed data. Bayesian networks help to make this idea transparent. Every Bayesian network indicates a sequence of steps that one could follow to generate samples from the joint distribution over the random variables in the network. First, one samples the values of all variables with no parents in the graph. Then, one samples the variables with parents taking known values, one after the other. For example, the graphical model shown in figure 4.4a corresponds to generating $x_1, x_2, \ldots, x_n$ by choosing α and β, sampling θ conditioned on those values, and then sampling each x_i conditioned on θ. Likewise, the graphical model shown in figure 4.4b corresponds to generating $x_1, x_2, \ldots, x_n$ by choosing μ_0, σ_0, and σ, sampling μ conditioned on μ_0 and σ_0, and then sampling each x_i conditioned on μ and σ. The directed graph associated with a probability distribution provides an intuitive representation for the steps involved in such a generative model.

The generative models shown in figure 4.4 both assume that observations are independent of one another, conditioned on the parameters θ or μ and σ. Other dependency structures are possible. For example, the flips could be generated in a *Markov chain*, a sequence of random variables in which each variable is independent of all its predecessors given the variable that immediately precedes it (e.g., Norris, 1997). We could use a Markov chain to represent a hypothesis space of coins that are particularly biased toward alternating or maintaining their last outcomes, letting the parameter θ be the probability that the outcome x_i takes the same value as x_{i-1} (and assuming that x_1 is heads with probability 0.5). This distribution would correspond to the graphical model shown in figure 4.5b. (Figure 4.5a reproduces the independent case for the sake of comparison, suppressing the dependence of θ on α and β.) Applying the Markov condition, this structure represents the probability distribution

$$P(x_1, x_2, \ldots, x_n, \theta) = p(\theta)P(x_1) \prod_{i=2}^{n} P(x_i | x_{i-1}, \theta), \tag{4.11}$$

in which each x_i depends only on x_{i-1}, given θ. More elaborate structures are also possible: any directed acyclic graph on $x_1, x_2, \ldots, x_n$ and θ corresponds to a valid set of assumptions about the dependencies among these variables.

Plate Notation and Latent Variables When dealing with graphical models involving large numbers of variables, it can be convenient to use a summary notation indicating repeated structure. The standard notation for doing this is using *plates*, which are rectangles enclosing a set of variables. A plate includes a number indicating how many times it should be replicated. All replicated variables have the same incoming and outgoing edges, and any edges between variables on the plate are replicated. Statistical independence is a common context in which structure is replicated, with the independent variables sharing common incoming edges from some parameters that they depend on, but not having edges between them. Figure 4.5d shows how the graphical model from figure 4.5a can be expressed more efficiently using plate notation.

For the generative models represented by figure 4.5a or 4.5b, we have assumed that all variables except θ are observed in each sample from the model or each data point. More

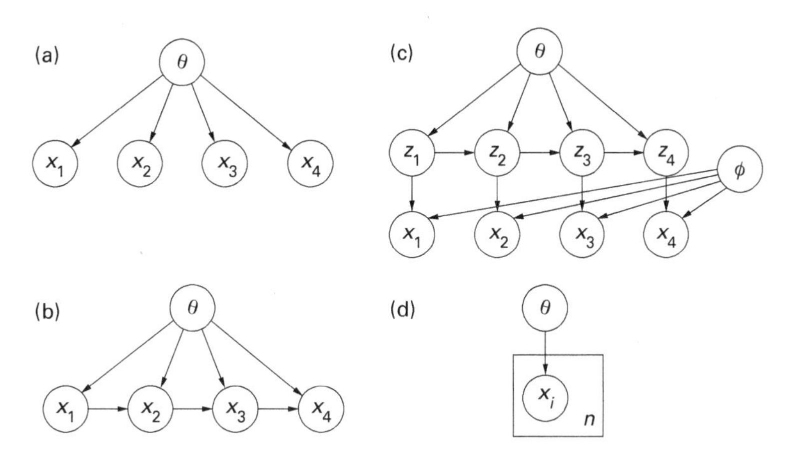

Figure 4.5
Graphical models showing different kinds of processes that could generate a sequence of coin flips. (a) Independent flips, with parameters θ determining the probability of heads. (b) A Markov chain, where the probability of heads depends on the result of the previous flip and the parameters θ defines the probability of heads after heads and after tails. (c) A hidden Markov model, in which the probability of heads depends on a latent state variable z_i. Transitions between values of the latent state are set by parameter θ, while an other parameter ϕ determines the probability of heads for each value of the latent state z_i. This kind of model is commonly used in computational linguistics, where the x_i might be the sequence of words in a document, and the z_i the syntactic classes from which they are generated. (d) A plate (the rectangle) can be used to indicate the replication of an element of a graphical model. Here, x_i is replicated n times. Taking $n = 4$ produces the graphical model shown in (a).

generally, generative models can include a number of steps that refer to unobserved or *latent variables*. Introducing latent variables can lead to apparently complicated dependency structures among the observable variables. For example, in the graphical model shown in figure 4.5c, a sequence of latent variables $z_1, z_2, \ldots, z_n$ influences the probability that each respective coin flip in a sequence $x_1, x_2, \ldots, x_n$ comes up heads (in conjunction with a set of parameters ϕ). The latent variables form a Markov chain, with the value of z_i depending only on the value of z_{i-1} (in conjunction with the parameters θ). This model, called a *hidden Markov model* (HMM), has historically been used in computational linguistics, where z_i might be the syntactic class (such as noun or verb) of a word, θ encodes the probability that a word of one class will appear after another (capturing simple syntactic constraints on the structure of sentences), and ϕ encodes the probability that each word will be generated from a particular syntactic class (e.g., Charniak, 1993; Jurafsky & Martin, 2000; Manning & Schütze, 1999). The dependencies among the latent variables induce dependencies among the observed variables—in the case of language, the constraints on transitions between syntactic classes impose constraints on which words can follow one another. We will return to HMMs in chapters.

4.2 Probabilistic Inference in Graphical Models

Recognizing the structure in a probability distribution can also greatly simplify the computations that we want to perform with that distribution. When variables are independent or conditionally independent of others, it reduces the number of terms that appear in sums over subsets of variables necessary to compute marginal beliefs about a variable or conditional beliefs about a variable, given the values of one or more other variables. A variety of algorithms have been developed to perform these probabilistic inferences efficiently on complex models by recognizing and exploiting conditional independence structures in Bayesian networks (Pearl, 1988; Mackay, 2003). These algorithms are used in AI systems, making it possible to reason efficiently under uncertainty (Korb & Nicholson, 2003; Russell & Norvig, 2021). In this section, we will illustrate the way in which knowing the dependencies between variables simplifies probabilistic computations and describe the particularly important kind of probabilistic inference, known as explaining away in more detail.

4.2.1 Simplifying Probabilistic Computations

While we will not go into detail about different inference algorithms here, we can provide an intuition for how knowing about patterns of statistical dependency can simplify inference by returning to the "psychic friend" example introduced earlier in the chapter. Let's say that we observed that the coin toss produced heads ($x_1 = 1$) and we wanted to infer the values of the other variables (x_2, x_3, and x_4). We then want to compute $P(x_2, x_3, x_4 | x_1 = 1)$, which we obtain via Bayes' rule, with

$$P(x_2, x_3, x_4 | x_1 = 1) = \frac{P(x_1 = 1, x_2, x_3, x_4)}{P(x_1 = 1)}. \tag{4.12}$$

Computing the denominator requires a sum of $P(x_1 = 1, x_2, x_3, x_4)$ over all values of x_2, x_3, and x_4, of which there are eight possible combinations. However, we can exploit the independence between variables to simplify this sum as

$$P(x_1 = 1) = \sum_{x_2, x_3, x_4} P(x_1 = 1, x_2, x_3, x_4) \tag{4.13}$$

$$= \sum_{x_2, x_3, x_4} P(x_1 = 1 | x_2, x_3, x_4) P(x_2 | x_3, x_4) P(x_3 | x_4) P(x_4) \tag{4.14}$$

$$= \sum_{x_2, x_3, x_4} P(x_1 = 1 | x_3, x_4) P(x_2 | x_3) P(x_3) P(x_4) \tag{4.15}$$

$$= \sum_{x_3, x_4} P(x_1 = 1 | x_3, x_4) P(x_3) P(x_4) \sum_{x_2} P(x_2 | x_3) \tag{4.16}$$

$$= \sum_{x_3, x_4} P(x_1 = 1 | x_3, x_4) P(x_3) P(x_4), \tag{4.17}$$

where equation (4.14) applies the chain rule to factorize the probability distribution, equation (4.15) uses the Markov condition for this graphical model, and equation (4.17) exploits the fact that the sum of x_2 over $P(x_2 | x_3)$ is just 1. Now our sum involves only four combinations rather than eight. Consequently, we are able to reduce the amount of computation we need to do by a factor of 2. This reduction in computation is purely a consequence of independence—if we know the value of X_3, then X_2 is independent of X_1. With many variables, knowing when different sets of variables are independent of one another can significantly speed up probabilistic computations.

4.2.2 Explaining Away

Part of the original motivation behind the development of Bayesian networks is that they naturally handle cases where there are multiple competing explanations for observed data. These cases posed a challenge for other systems for automated reasoning that were used in AI research before Bayesian networks, such as expert systems based on production rules (Pearl, 1988). The characteristic pattern of inference that allowed Bayesian networks to deal with these cases appropriately is known as "explaining away," which we briefly introduced earlier in the chapter.

We will illustrate explaining away with the "psychic friend" example. If you observe that the coin has come up heads, you know there are three possible explanations: your friend has psychic powers, a two-headed coin, or the coin came up heads by chance. If you had to assign a probability to whether your friend is psychic, the fact that the coin had come up heads would factor into that. However, if you inspect your friend's coin and discover that it is two-headed, you immediately think that it is less likely that your friend is psychic. The two-headed coin "explains away" the evidence that informed this inference.

If we work through the math, we come to the same conclusions. Assume that the probability that the coin toss produces heads (i.e., $x_1 = 1$) is 1 if either your friend has psychic powers or a two-headed coin, and otherwise is 0.5. Before observing the coin toss, the probability of your friend having psychic powers is $P(x_3 = 1) = \psi$, and the probability of a two-headed coin is $P(x_4 = 1) = \gamma$. After observing the coin being tossed and coming up heads, the probability of both of these variables being true is

$$P(x_3 = 1|x_1 = 1) = \frac{P(x_1 = 1|x_3 = 1)P(x_3 = 1)}{P(x_1 = 1)} \tag{4.18}$$

$$= \frac{P(x_1 = 1|x_3 = 1)P(x_3 = 1)}{\sum_{x_3,x_4} P(x_1 = 1|x_3, x_4)P(x_3)P(x_4)} \tag{4.19}$$

$$= \frac{\psi}{0.5(1-\psi)(1-\gamma) + \psi(1-\gamma) + (1-\psi)\gamma + \psi\gamma} \tag{4.20}$$

$$= \frac{\psi}{1 - 0.5(1-\psi)(1-\gamma)}, \tag{4.21}$$

where we use the result from equation (4.17) for $P(x_1 = 1)$ and then simplify. Alternatively, we can calculate $P(x_1 = 1)$ by observing that $x_1 = 0$ only when neither x_3 nor x_4 is true, and occurs only half the time even in this case. The analogous result for $P(x_4 = 1|x_1 = 1)$ can be obtained by substituting γ for ψ and vice versa. Since the denominator is less than 1, the evidence in favor of both x_3 and x_4 being true is increased by observing $x_1 = 1$.

Now, consider what would happen when you inspected your friend's coin and discovered that it actually did have two heads ($x_4 = 1$). The resulting distribution for x_3 is given by

$$P(x_3 = 1|x_1 = 1, x_4 = 1) = \frac{P(x_1 = 1|x_3 = 1, x_4 = 1)P(x_4 = 1)P(x_3 = 1)}{P(x_1 = 1, x_4 = 1)} \tag{4.22}$$

$$= \frac{\gamma\psi}{\gamma}, \tag{4.23}$$

where we obtain the denominator by observing that $x_1 = 1$ whenever $x_4 = 1$. Simplifying, $P(x_3 = 1|x_1 = 1, x_4 = 1)$ is just ψ, exactly the same probability as $P(x_3)$. The effect that observing the coin come up heads had on your beliefs about your friend's psychic powers has completely disappeared, with the outcome of the coin toss being fully explained away by the two-headed coin.

Explaining away may seem obvious when presented in this way, but it has several deep implications. First, it is a characteristic pattern of dependency that is observed only between variables that form this common effect structure. As a consequence, seeing this pattern of dependency can be a strong clue about the relationships between variables. Second, it instantiates a principle that is very relevant when thinking about psychological and neural mechanisms: causes of the same effect should inhibit one another.

4.2.3 An Example: Visual Inference

Common cause and common effect structures also arise in vision. We will illustrate this with examples from Kersten and Yuille (2003). As we have briefly mentioned previously, vision is naturally viewed as a problem of Bayesian inference. Figure 4.6a shows a simple graphical model depicting a visual inference problem (in this case, the nodes correspond to high-dimensional random variables, rather than the binary variables we have considered so far). A variety of three-dimensional (3D) objects project to the same two-dimensional (2D) image on the retina. Interpreting these data requires making an inference about the generative process, informed by prior distributions over 3D structures in the world. But it also requires making some assumptions about the dependencies between the various random variables involved.

It is often natural to think of multiple factors combining to generate an image, similar to a common effect structure. For example, in figure 4.6b the image of a bicycle is the common effect of the shape and reflectance properties of the bicycle and the viewpoint and illumination conditions. Often one of the causes is considered a nuisance variable and will be marginalized over. This will depend on the task. If we only want to identify the bicycle, we will not care about the viewpoint and illumination and will integrate them out. But if we want to grab the bicycle, then we need to estimate its shape and viewpoint, but the lighting can be discounted. In a few unusual situations, we may want to estimate the lighting and ignore the viewpoint or the shape and other properties of the bicycle.

Common cause structures also arise in vision, typically where there are multiple cues that could be generated from a single source. An example of a common cause structure occurs when there are two surfaces with one slightly above the other, as in figure 4.6c. There are two types of cues that indicate the relative depth of the surfaces. One cue is binocular stereo (viewing the surface with two eyes and estimating depth by trigonometry). Another cue arises from the shadow patterns thrown by the upper surface on the lower. The relationship between these cues can be captured by the graphical model shown in figure 4.6c. In this case, the random variables represent the positions in 3D space of all points on the two surfaces.

Visual inference can also demonstrate explaining away. Figure 4.6d shows a situation where two possible explanations exist for a percept: it could be a red diamond behind an occluding set of vertical lines, or four separate line segments. In this case, perceptual

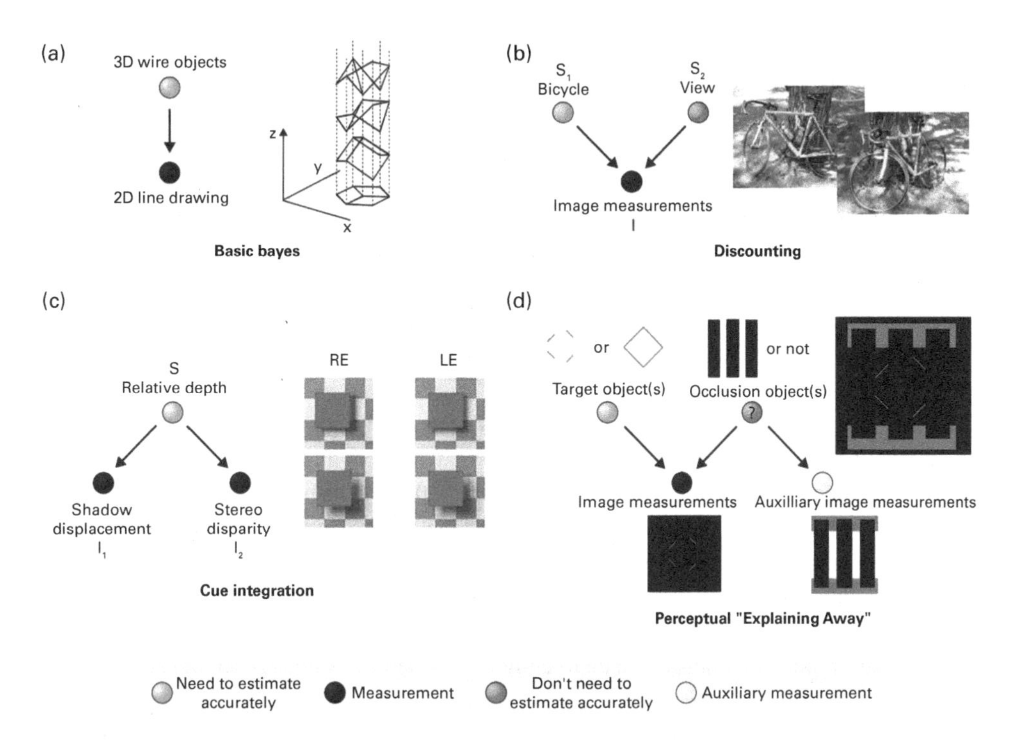

Figure 4.6
Graphical models for visual inference. (a) Formulation of the basic problem of visual inference. (b) The image of the bicycle is caused by the pose of the bicycle, the viewpoint of the camera, and the lighting conditions. (c) An example of common cause. The shading and binocular stereo cues are caused by the same event—two surfaces with one partially occluding the other. (d) Explaining away in vision. 3D = three-dimensional; 2D = two-dimensional. Figure reproduced with permission from Kersten and Yuille (2003).

evidence for the occluder will change the interpretation of the red lines: if the occluder seems to be present, then the percept is a diamond, while if it is absent, it is just four line segments. This structure is directly analogous to the "psychic friend" example that we have been using throughout the chapter: it's exactly the same graphical model, but with a different set of variables.

4.3 Causal Graphical Models

So far, we have been talking informally about causality in considering how graphical models can capture the structure of generative processes. However, recent work has resulted in a more precise specification of how graphical models can be used to represent causal processes, based on a calculus for understanding the consequences of actions, which are characterized as *interventions* on the values of variables (Pearl, 2000; Spirtes et al., 1993). In a standard Bayesian network, an edge between variables indicates only a statistical dependency between them. To reason about causality, we need to augment directed graphical models with a stronger assumption about the relationships indicated by edges: that they indicate direct causal relationships (Pearl, 2000; Spirtes et al., 1993).

4.3.1 From Graphical Models to Causal Graphical Models

The assumption that edges indicate causal relationships, not just dependency, allows causal graphical models to represent not just the probabilities of events that one might observe, but also the probabilities of events that one can produce through intervening on a system—reaching into the system and setting variables to particular values. The inferential implications of an event can differ strongly, depending on whether it was observed passively or under conditions of intervention. For example, observing that nothing happens when your friend attempts to levitate a pencil would provide evidence against her claim of having psychic powers; but secretly intervening to hold the pencil down while your friend attempts to levitate it would make the pencil's nonlevitation unsurprising and uninformative about her powers.

In causal graphical models, the consequences of intervening on a particular variable can be assessed by removing all incoming edges to that variable and performing probabilistic inference in the resulting "mutilated" model (Pearl, 2000). This procedure produces results that align with our intuitions in the psychic powers example: intervening on the pencil levitation (X_2) breaks its connection with your friend's psychic powers (X_3), rendering the two variables independent. As a consequence, X_2 cannot provide evidence about the value of X_3. Pearl defined a version of the Markov condition—the *causal Markov condition*—to describe how causal graphical models are used to compute probabilities under both observation and intervention.

Pearl (2000) formalized his notion of causality by introducing a "do" operator to indicate an intervention that fixes the value of a variable. Using this operator, we could write our query about our friend's psychic powers when we have intervened to prevent the pencil from levitating in terms of the conditional probability $P(X_3|\text{do}(X_2=0))$, where $\text{do}(X_2=0)$ indicates that X_2 has been fixed to take the value 0 rather than observed to have that value. This conditional probability is different from what we would have obtained if we had just observed $X_2=0$, which is $P(X_3|X_2=0)$—the two would be evaluated in different graphical models. In Pearl's formulation, a conditional probability that involves a "do" operator is computed by taking the causal graphical model that would normally be used to compute the relevant conditional probability and severing the incoming edges for all variables whose values are set by "do."

Extending graphical models to capture probability distributions under intervention makes it possible to reason about and potentially learn from a wider range of data. Relationships that might be impossible to identify from observational data alone are potentially identifiable when interventions are allowed. Several papers have investigated whether people are sensitive to the consequences of intervention, generally finding that people differentiate between observational and interventional evidence appropriately (Hagmayer, Sloman, Lagnado, & Waldmann, 2007; Lagnado & Sloman, 2004; Steyvers, Tenenbaum, Wagenmakers, & Blum, 2003). Introductions to causal graphical models that consider applications to human cognition are provided by Glymour (2001) and Sloman (2005).

4.3.2 An Example: Do We "Do"?

Pearl (2000) provided a way to use causal graphical models ot reason about the consequences of intervening on particular variables, in the form of his "do" operator. A natural question to ask is whether people make the same kind of inferences. Sloman and Lagnado (2005) explored this question in a series of experiments. The first of these presented participants with a simple scenario: "There are three billiard balls on a table that act in the following way: Ball 1's movement causes Ball 2 to move. Ball 2's movement causes Ball 3 to move." Participants were then asked to answer two questions:

1. Imagine that Ball 2 could not move; would Ball 1 still move? Circle one of the three options:
   It could. It could not. I don't know.
2. Imagine that Ball 2 could not move; would Ball 3 still move? Circle one of the three options:
   It could. It could not. I don't know.

The causal graphical model underlying this scenario is one in which variables denoting the notion of the three balls (call them B_1, B_2, and B_3) are linked in a causal chain: $B_1 \rightarrow B_2 \rightarrow B_3$. Intervening on B_2 breaks the link from B_1 to B_2, rendering B_1 and B_2 independent. Consequently, the use of the "do" operator predicts that participants should answer the first question with "It could" and the second question with "It could not."

Sloman and Lagnado (2005) found exactly this pattern of results—90 percent of participants answered each question in the way that Pearl's framework predicts. A second experiment showed that similar results hold when the underlying relationship is probabilistic (as opposed to the deterministic relationship between the balls described here). Interestingly, the causal formulation seems critical to producing these results—using a "logical" scenario ("Someone is showing off her logical abilities. She is moving balls without breaking the following rules: If Ball 1 moves, then Ball 2 moves. If Ball 2 moves, then Ball 3 moves.") did not produce the same pattern of results.

Pearl (2000) also outlined a way in which the "do" operator is also used in counterfactual reasoning—basically, in forming a counterfactual, we are imagining the consequence of intervening on a variable. Sloman and Lagnado (2005) also found that people's counterfactual reasoning was broadly consistent with Pearl's account, and subsequent work has built on this in investigating the formal structure of human counterfactual reasoning in more detail (for a review, see Sloman & Lagnado, 2015).

4.4 Learning Graphical Models

We have seen how graphical models can be used to specify complex generative models and capture causal relationships—key components of the intuitive models of the world around us that we build. But people need to be able to *learn* those world models from experience. We now turn to the question of how graphical models can be learned.

4.4.1 Different Approaches to Causal Learning

The prospect of using graphical models to express the probabilistic consequences of causal relationships has led researchers in several fields to ask whether these models could serve as the basis for learning causal relationships from data. Every introductory class in statistics teaches that "correlation does not imply causation." This is certainly the case, but it doesn't mean that correlation carries *no* information about causation. Patterns of causation imply patterns of correlation, meaning that a learner should be able to work backwards from observed correlations (or statistical dependencies) to make probabilistic inferences about the underlying causal structures likely to have generated those observed data.

Constructing a graphical model from a set of observed data involves two kinds of learning: structure learning and parameter estimation. *Structure learning* refers to identification of the topology of the underlying graph, while *parameter estimation* involves determining the parameters of the conditional probability distributions of the different variables. Structure learning is arguably more fundamental than parameter estimation, since the parameters can only be estimated once the structure is known. Learning the structure of a graph defined on many variables is a difficult computational problem, as the number of possible structures is a super exponential function of the number of variables (Koller & Friedman, 2009).

Solutions to the problem of parameter estimation take the general approach for inferring parameter values presented in chapter 3 (e.g., Heckerman, 1998). In particular, maximum-likelihood estimation or Bayesian methods can be used, with the likelihood function being based on the probability of the observed values of the variables in the graph. In typical machine-learning applications, Bayesian networks are learned from large databases that provide multiple observations of the values of these variables. For example, the parameters for a Bayesian network connecting diseases with symptoms could be estimated from a database of patients, each of whom has some symptoms and a diagnosis, providing multiple samples from the joint distribution implied by the network.

Structure learning attempts to identify the dependency structure underlying a set of observed data. There are two major approaches to structure learning: constraint-based learning and Bayesian inference. Constraint-based algorithms for structure learning (e.g., Pearl, 2000; Spirtes et al., 1993) proceed in two steps. First, standard statistical hypothesis tests such as Pearson's χ^2 test are used to identify which variables are dependent and independent. Since the Markov condition implies that different causal structures should result in different patterns of dependency among variables, the observed dependencies provide constraints on the set of possible structures. The second step of the algorithms identifies this set, reasoning deductively from the pattern of dependencies. The result is one or more structures that are consistent with the statistically significant dependencies exhibited by the data.

In contrast, the Bayesian approach to structure learning evaluates each graph structure in terms of the probability that it assigns to a data set. By integrating over the values that parameters could assume, it is possible to compute the probability of a data set given a graphical structure without committing to a particular choice of parameter values (e.g., Cooper & Herskovits, 1992). This computation is a form of model selection, as discussed in detail in chapter 3. Often, priors either are uniform (giving equal probability to all graphs) or give lower probability to more complex structures. Bayesian structure learning proceeds by either searching the space of structures to find the one with the highest posterior probability (Friedman, 1997; Heckerman, 1998), or evaluating particular causal relationships by integrating over the posterior distribution over graphs using sophisticated Monte Carlo methods (Friedman & Koller, 2000).

Constraint-based and Bayesian approaches to causal learning represent two philosophies, with constraint-based methods relying on the discrete results from frequentist statistical tests to turn an inductive problem into a deductive problem, and Bayesian methods casting structure learning as a special case of Bayesian inference. These two approaches have different advantages and disadvantages. Constraint-based methods are potentially more scalable, as the Bayesian approach is highly computationally intensive. However, the Bayesian approach makes it possible to integrate multiple pieces of weak evidence and incorporate prior knowledge—something that is important when modeling human cognition.

4.4.2 An Example: Structure and Strength in Causal Induction

Much psychological research on causal induction has focused on this simple causal learning problem: given a candidate cause, C, and a candidate effect, E, people are asked to give a numerical rating assessing the degree to which C causes E.[2] The exact wording of the judgment question varies, and until recently it was not the focus of much attention, although as we will see, it is potentially quite important. Most studies present information corresponding to the entries in a 2×2 contingency table, as in table 4.1. People are given information about the frequency with which the effect occurs in the presence and absence of the cause, represented by the numbers $N(e^+, c^+), N(e^-, c^-)$ and so forth. In a standard example, C might be injecting a chemical into a mouse and E the expression of a particular gene. $N(e^+, c^+)$ would be the number of injected mice expressing the gene, while $N(e^-, c^-)$ would be the number of uninjected mice not expressing the gene. We refer to tasks of this sort as *elemental causal induction* tasks.

Psychological Models of Elemental Causal Induction The leading psychological models of elemental causal induction are measures of association that can be computed from simple combinations of the frequencies in table 4.1. A classic model first suggested by Jenkins and Ward (1965) asserts that the degree of causation is best measured by the quantity ΔP, defined as

$$\Delta P = \frac{N(e^+, c^+)}{N(e^+, c^+) + N(e^-, c^+)} - \frac{N(e^+, c^-)}{N(e^+, c^-) + N(e^-, c^-)} = P(e^+|c^+) - P(e^+|c^-), \tag{4.24}$$

where $P(e^+|c^+)$ is the empirical conditional probability of the effect given the presence of the cause, estimated from the contingency table counts $N(\cdot)$. ΔP thus reflects the change in the probability of the effect occuring as a consequence of the occurence of the cause. An alternative model proposed by Cheng (1997) suggested that people's judgments are better

Table 4.1
Contingency table representation used in elemental causal induction

	Effect Present (e^+)	Effect Absent (e^-)
Cause Present (c^+)	$N(e^+, c^+)$	$N(e^-, c^+)$
Cause Absent (c^-)	$N(e^+, c^-)$	$N(e^-, c^-)$

2. As elsewhere in this book, we will represent variables such as C and E with capital letters, and their instantiations with lowercase letters, with c^+, e^+ indicating that the cause or effect is present, and c^-, e^- indicating that the cause or effect is absent.

captured by a measure called *causal power*:

$$\text{power} = \frac{\Delta P}{1 - P(e^+|c^-)}, \tag{4.25}$$

which takes ΔP as a component but predicts that ΔP will have a greater effect when $P(e^+|c^-)$ is large. Intuitively, dividing by $1 - P(e^+|c^-)$ normalizes ΔP by the maximum value that it could possibly take. It thus estimates the proportion of the times when E wouldn't occur on its own that it *does* occur when C is present—that is, the power of C to influence E.

Several experiments have been conducted with the aim of evaluating ΔP and causal power as models of human jugments. In one such study, Buehner and Cheng (1997, experiment 1B; this experiment also appears in Buehner, Cheng, & Clifford, 2003) asked people to evaluate causal relationships for 15 sets of contingencies expressing all possible combinations of $P(e^+|c^-)$ and ΔP in increments of 0.25. Specifically, they were asked to rate how strongly the cause produced the effect on a scale from 0 ("not at all") to 100 ("every time"). The results of this experiment are shown in figure 4.7, together with the predictions of ΔP and causal power. As can be seen from the figure, both ΔP and causal power capture some of the trends in the data, producing correlations of $r = 0.89$ and $r = 0.88$, respectively. However, since the trends predicted by the two models are essentially orthogonal, neither model provides a complete account of the data.[3]

Formulating the Problem Using Graphical Models ΔP and causal power seem to capture some important elements of human causal induction but miss others. We can gain some insight into the assumptions behind these models, and identify some possible alternative models, by considering the computational problem behind causal induction using the tools of causal graphical models and Bayesian inference. The task of elemental causal induction can be seen as trying to infer which causal graphical model best characterizes the relationship between variables C and E. Figure 4.8 shows two possible causal structures relating C, E, and another variable, B, which summarizes the influence of all of the other "background" causes of E (which are assumed to be constantly present). The problem of learning which causal graphical model is correct has two aspects: inferring the right causal structure, a problem of model selection; and determining the right parameters assuming a particular structure, a problem of parameter estimation.

To formulate the problems of model selection and parameter estimation more precisely, we need to make some further assumptions about the nature of the causal graphical models shown in figure 4.8. In particular, we need to define the form of the conditional probability distribution $P(E|B, C)$ for the different structures, often called the *parameterization* of the graphs. Sometimes the parameterization is trivial—for example, C and E are independent in Graph 0, so we just need to specify $P_0(E|B)$, where the subscript indicates that this probability is associated with graph 0. This can be done using a single numerical parameter w_0, which provides the probability that the effect will exist in the presence of the background

3. See Griffiths and Tenenbaum (2005) for the details of how these correlations were evaluated, using a power-law transformation to allow for nonlinearities in participants' judgment scales.

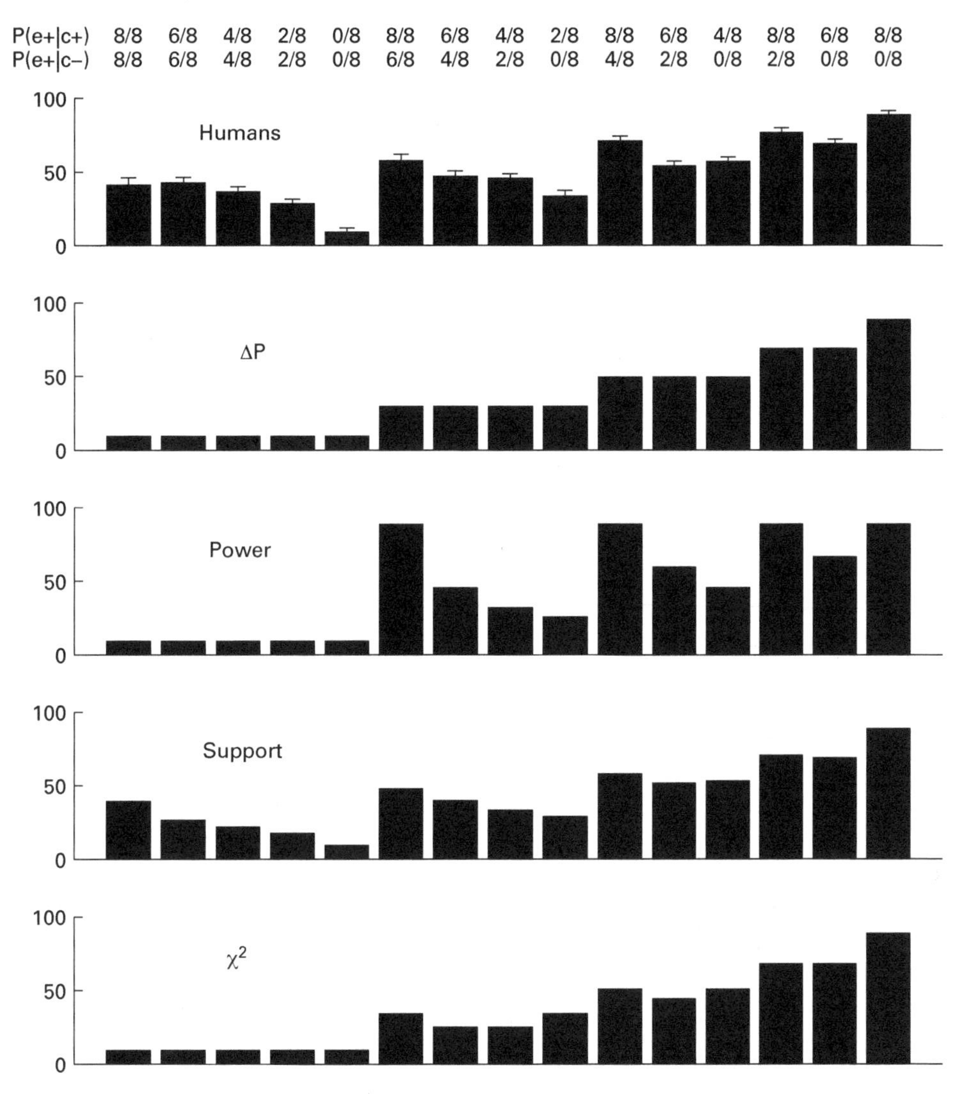

Figure 4.7
Predictions of models compared with the performance of human participants from Buehner and Cheng (1997, experiment 1B). The numbers across the top of the figure show stimulus contingencies. The bars show the magnitude of mean human ratings for how strongly the cause produced the effect, on a scale from 0–100, with error bars showing one standard error, and the corresponding model predictions. "Power" refers to causal power (Cheng, 1997), "Support" to causal support (Griffiths & Tenenbaum, 2005), and χ^2 to the Pearson χ^2 test statistic for the corresponding contingency table. Adapted from Griffiths and Tenenbaum (2005).

cause, $P_0(e^+|b^+; w_0) = w_0$. However, when a node has multiple parents, there are many different ways in which the functional relationship between causes and effects could be defined. For example, in graph 1, we need to account for how causes B and C interact to produce the effect E.

Different Parameterizations of Causal Relationships A simple and widely used parameterization for Bayesian networks of binary variables is the *noisy-OR distribution* (Pearl, 1988).

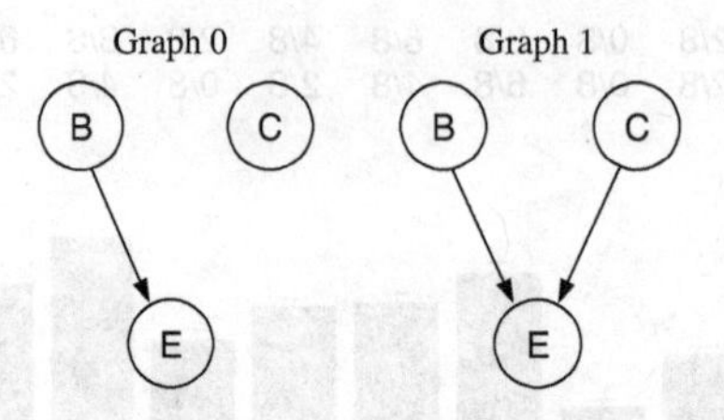

Figure 4.8
Directed graphs involving three variables, B, C, and E, relevant to elemental causal induction. B represents background variables, C a potential causal variable, and E the effect of interest. Graph 1 is assumed for computing ΔP and causal power. Computing causal support involves comparing the structure of graph 1 to that of graph 0, in which C and E are independent.

The noisy-OR can be given a natural interpretation in terms of causal relations between multiple causes and a single joint effect. For graph 1, these assumptions are that B and C are both generative causes, each having an independent opportunity to produce the effect. The probability of E in the presence of just B is w_0, and in the presence of just C, it is w_1. When both are present, the probability of E occurring is $w_0 + w_1 - w_0 w_1$, where the last term prevents double-counting cases where both B and C produce the effect. We can write out all the cases in a single equation as follows:

$$P_1(e^+|b, c; w_0, w_1) = 1 - (1 - w_0)^b (1 - w_1)^c, \tag{4.26}$$

where w_0 and w_1 are the parameters associated with the strength of B and C, respectively, and $b^+ = c^+ = 1$ and $b^- = c^- = 0$ for the purpose of arithmetic operations.

This parameterization is called a noisy-OR because if w_0 and w_1 are both 1, equation (4.26) reduces to the logical OR function: the effect occurs if and only if B or C is present, or both are. With w_0 and w_1 in the range $[0, 1]$, the noisy-OR softens this function but preserves its essentially disjunctive interaction: the effect occurs if and only if B causes it (which happens with probability w_0), C causes it (which happens with probability w_1), or both.

An alternative to the noisy-OR might be a linear parameterization of graph 1, asserting that the probability of E occuring is a linear function of B and C. This corresponds to assuming that the presence of a cause simply increases the probability of an effect by a constant amount, regardless of any other causes that might be present. The result is

$$P_1(e^+|b, c; w_0, w_1) = w_0 \cdot b + w_1 \cdot c. \tag{4.27}$$

This parameterization requires that we constrain $w_0 + w_1$ to lie between 0 and 1 to ensure that equation (4.27) results in a legal probability distribution. Because of this dependence between parameters that seem intuitively like they should be independent, such a linear parameterization is not normally used in Bayesian networks. However, it is relevant for understanding models of human causal induction.

Parameter Estimation versus Structure Learning Given a particular causal graph structure and a particular parameterization—for example, graph 1 parameterized with a noisy-OR function—inferring the strength parameters that best characterize the causal relationships in that model is straightforward. We can use any of the parameter-estimation methods discussed in chapter 3, such as maximum-likelihood or maximum a posteriori estimation, to find the values of the parameters (w_0 and w_1 in graph 1) that best fit a set of observed contingencies. Tenenbaum and Griffiths (2001b; Griffiths & Tenenbaum, 2005) showed that the two psychological models of causal induction introduced in this chapter—ΔP and causal power—both correspond to maximum-likelihood estimates of the causal strength parameter w_1, but under different assumptions about the parameterization of graph 1. ΔP results from assuming the linear parameterization, while causal power results from assuming the noisy-OR.

This view of ΔP and causal power helps to reveal their underlying similarities and differences: they are similar in being maximum-likelihood estimates of the strength parameter describing a causal relationship, but differ in the assumptions that they make about the form of that relationship. This analysis also suggests another class of models of causal induction: models of learning causal graph structure or causal model selection rather than parameter estimation. Recalling our discussion of model selection, we can express the evidence that a set of contingencies d provide in favor of the existence of a causal relationship (i.e., graph 1 over graph 0) as the log-likelihood ratio in favor of graph 1. Terming this quantity *causal support*, we have

$$\text{support} = \log \frac{P(d|\text{Graph } 1)}{P(d|\text{Graph } 0)}, \tag{4.28}$$

where $P(d|\text{Graph } 1)$ and $P(d|\text{Graph } 0)$ are computed by integrating over the parameters associated with the different structures:

$$P(d|\text{Graph } 1) = \int_0^1 \int_0^1 P_1(d|w_0, w_1, \text{Graph } 1)\, P(w_0, w_1|\text{Graph } 1)\, dw_0\, dw_1 \tag{4.29}$$

$$P(d|\text{Graph } 0) = \int_0^1 P_0(d|w_0, \text{Graph } 0)\, P(w_0|\text{Graph } 0)\, dw_0. \tag{4.30}$$

Tenenbaum and Griffiths (2001b; and also Griffiths & Tenenbaum, 2005) proposed this model, and specifically assumed a noisy-OR parameterization for graph 1 and uniform priors on w_0 and w_1. Equation (4.30) is related to the normalizing constant for the beta distribution and has an analytic solution (see chapter 3). Evaluating equation (4.29) is more of a challenge, but one that we will return to later in this chapter when we discuss Monte Carlo methods for approximate probabilistic inference.

The results of computing causal support for the stimuli used by Buehner and Cheng (1997) are shown in figure 4.7. Causal support provides an excellent fit to these data, with $r = 0.97$. The model captures the trends predicted by both ΔP and causal power, as well as trends that are predicted by neither model. These results suggest that when people evaluate contingency, they may be taking into account the evidence that those data provide for a causal relationship, as well as the strength of the relationship they suggest. The figure also shows the predictions obtained by applying Pearson's χ^2 test to these data, a standard hypothesis-testing method of assessing the evidence for a relationship (and a common ingredient in nonBayesian approaches to structure learning; e.g., Spirtes et al., 1993). These predictions miss several important trends in the human data, suggesting that the ability to assert expectations about the nature of a causal relationship that go beyond mere dependency (such as the assumption of a noisy-OR parameterization) is contributing to the success of this model. Causal support predicts human judgments on several other data sets that are problematic for ΔP and causal power, and also accommodates causal learning based upon the rate at which events occur (see Griffiths & Tenenbaum, 2005, for more details).

Learning More Complex Causal Relationships The Bayesian approach to causal induction can be extended to cover a variety of more complex cases, including learning and intervening in larger causal networks (e.g., Steyvers et al., 2003; Bramley, Dayan, Griffiths, & Lagnado, 2017), continuous causes and effects (e.g., Griffiths & Pacer, 2011; Davis, Bramley, & Rehder, 2020; Lu, Rojas, Beckers, & Yuille, 2016), and continuous time (e.g., Pacer & Griffiths, 2012; Pacer & Griffiths, 2015).

Modeling learning in these more complex cases often requires us to work with stronger and more structured prior distributions than were needed before to explain elemental causal induction. This prior knowledge can be usefully described in terms of intuitive domain theories (Carey, 1985; Wellman & Gelman, 1992; Gopnik & Meltzoff, 1997), systems of abstract concepts and principles that specify the kinds of entities that can exist in a domain, their properties and possible states, and the kinds of causal relations that can exist between them. These abstract *causal theories* can be formalized as probabilistic generators for hypothesis spaces of causal graphical models, using probabilistic forms of generative grammars, predicate logic, or other structured representations that we discuss in more detail later in the book (e.g., Griffiths & Tenenbaum, 2009; Kemp et al., 2010b). Given observations of causal events relating a set of objects, these probabilistic theories generate the relevant variables for representing those events, a constrained space of possible causal graphs over those variables, and the allowable parameterizations for those graphs. They also generate a prior distribution over this hypothesis space of candidate causal models, which provides the basis for Bayesian causal learning in the spirit of the methods described earlier in this chapter.

4.4.3 An Example: Multisensory Integration with Structural Uncertainty

Human observers are sensitive to both visual and auditory cues. Sometimes these cues have a common cause—for instance, you see a dog moving and hear it barking. In other situations the auditory and visual cues are due to different causes—for instance, a cat moves and a nearby dog barks. Ventriloquists are able to fake these interactions by making the audience think that a puppet is speaking by associating the sound (produced by the ventriloquist) with the movement of the puppet. The *ventriloquism* effect occurs when visual and auditory cues have different causes—and thus are in conflict—but the audience perceive them as having the same cause.

Körding et al. (2007) developed a Bayesian model intended to capture this ventriloquism effect. The model formulates this problem as one of determining whether two cues have a common cause. They formulated this using variable C to denote the causal structure, as shown in figure 4.9. When $C=1$, there is a common cause behind the sensory signals, so the positions of the auditory cue X_A and the visual cue X_V are generated from the same underlying location S. The resulting joint distribution is $p(x_A, x_V, s) = p(x_A|s)p(x_V|s)p(s)$. Here, $p(x_A|s)$ and $p(x_V|s)$ are Gaussian distributions with the same mean s and variances σ_A^2 and σ_V^2. It is assumed that the visual cues are more precise than the auditory cues so $\sigma_V^2 < \sigma_A^2$.

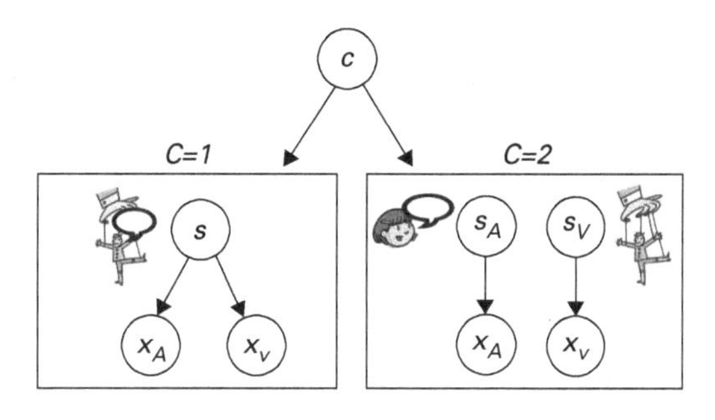

Figure 4.9
The ventriloquism effect. The participant is asked to estimate the position of the cues and to judge whether the cues are from a common cause—that is, at the same location—or not. In Bayesian terms, the task of judging whether there is a common cause can be formulated as model selection—that is, are the auditory and visual cues more likely to generated from a single cause ($C=1$) or by two independent causes ($C=2$)? Figure reproduced with permission from Körding et al. (2007).

The true position S is drawn from a probability distribution $p(s)$, which is assumed to be a Gaussian with mean 0 and variance σ_P^2.

By contrast, $C=2$ means that the cues are generated from two locations S_A and S_B, in which case we have $p(x_A|s_A)$ and $p(x_V|s_V)$, both Gaussian with mean and variance (s_A, σ_A^2) and (s_V, σ_V^2), respectively. We assume that S_A and S_V are independent samples from a Gaussian distribution with mean 0 and variance σ_P^2. The resulting joint distribution is $p(x_A, x_V, s_A, s_V) = p(x_A|s)p(x_V|s)p(s_A)p(s_V)$.

Deciding between $C=1$ and $C=2$ requires performing model selection. The posterior distribution $P(C|x_A, x_V)$ is calculated by summing out the estimated locations of the cues to obtain the likelihoods $p(x_A, x_V|C)$. For $C=1$, we have

$$p(x_A, x_V|C=1) = \int_{-\infty}^{\infty} p(x_A|s)p(x_V|s)p(s)\, ds, \tag{4.31}$$

while for $C=2$, we have

$$p(x_A, x_V|C=2) = \int_{-\infty}^{\infty}\int_{-\infty}^{\infty} p(x_A|s_A)p(x_V|s_V)p(s_A)p(s_V)\, ds_A\, ds_V. \tag{4.32}$$

There are two ways to combine the cues. The first is model selection. This estimates the most probable model $C^* = \arg\max P(C|x_V, x_A)$ from the input x_A, x_V and then uses this model to estimate the most likely positions s_A, s_V of the cues from the posterior distribution. The second way to combine the cues is by model averaging. This does not commit itself to choosing C^* but instead averages over both models:

$$P(s_V, s_A|x_V, x_A) = \sum_C P(s_V, s_A|x_V, x_A, C)P(C|x_V, x_A), \tag{4.33}$$

where $s_V = s_A = s$ when $C=1$.

This model was compared to experiments where brief auditory and visual stimuli were presented simultaneously with varying amounts of spatial disparity. Participants were asked to identify the spatial location of the cue and whether they perceive a common cause (Wallace et al., 2004). The closer the visual stimulus was to the audio stimulus, the more likely

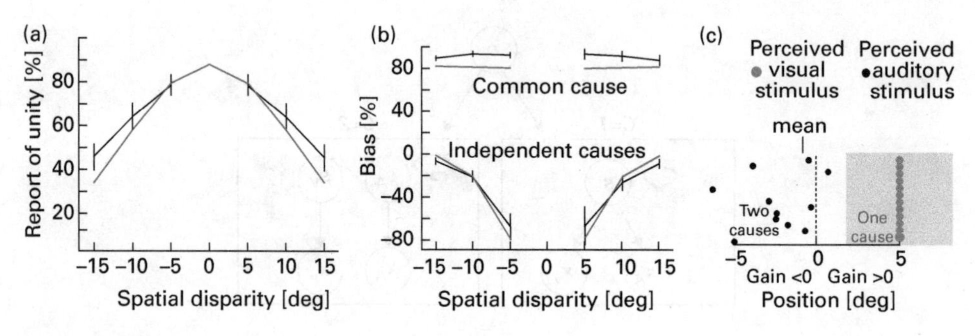

Figure 4.10
Reports of causal inference in the ventriloquism effect. (a) The relative frequency of subjects reporting one cause (black) is shown with the prediction of the causal inference model (red). (b) The bias, i.e., the influence of vision on the perceived auditory position is shown (gray and black). The predictions of the model are shown in red. (c) A schematic illustration explaining the finding of negative biases. Blue and black dots represent the perceived visual and auditory stimuli, respectively. In the pink area people perceive a common cause. Figure reproduced with permission from Körding et al. (2007).

people perceived a common cause. In this case, people's estimate of its position is strongly biased by the visual stimulus (because it is considered more precise with $\sigma_V^2 < \sigma_A^2$). But if people perceive *distinct* causes, then their estimate is pushed away from the visual stimulus and exhibits *negative bias*. Körding et al. (2007) argue that this bias is a selection bias stemming from restricting to trials in which causes are perceived as being distinct. For example, if the auditory stimulus is at the center and the visual stimulus at 5 degrees to the right of center, then sometimes the very noisy auditory cue will be close to the visual cue and hence judged to have a common cause, while on other cases, the auditory cue will be farther away (more than 5 degrees). Hence, the auditory cue will have a truncated Gaussian (if judged to be distinct) and will yield negative bias.

Natarajan, Murray, Shams, and Zemel (2009) investigated these issues further. In particular, they showed that human performance on these types of experiments could be better modeled by replacing the Gaussian distributions by a more robust alternative. It is well known that Gaussian distributions are not robust because the tails of their distributions fall off rapidly, which gives very low probability to rare events. Hence, in many real-world applications, distributions with heavier tails are preferred. Following this reasoning, Natarajan et al. (2009) assumed that the observations X_A and X_V were generated by distributions with heavier tails. More precisely, they assumed that the data is distributed by a mixture of a Gaussian distribution (as in the models described here) and a uniform distribution that yields heavier tails. The resulting model was able to better account for human behavior.

4.5 Summary

Graphical models are a valuable tool both for working with probability distributions and for characterizing the inferences that people make about causal relationships. The ideas covered in this chapter are a step toward letting us define more expressive world models, and give us new tools for understanding how people might learn those models through experience. In particular, they show how a level of abstraction can be valuable in defining generative models—a theme that will become increasingly important in later chapters as we begin to consider how intuitive theories might be formalized.

In the following chapters, we will make extensive use of graphical models as we begin to work with increasingly complex probability distributions. The formalism provides us with a language that we can use to think about and define these distributions. The first step in building a Bayesian model of some aspect of human cognition is often trying to figure out a way to express the hypotheses that a learner might use—a generative model for the data that the learner gets to observe. Graphical models make that process significantly easier, allowing us to explore richer generative models.

5

Building Complex Generative Models

Thomas L. Griffiths and Alan Yuille

As we start to build more expressive models of the world, we run into a new problem: we may not have some of the information that we need to make an inference. If we want to determine whether a chemical causes gene expression, we need to rule out other variables that could be involved. If we want to infer whether one object caused another object to move, it might help to know the masses of those objects. If we are going to take the fact that a friend ordered pasta as an indicator of his preferences, we should check whether he saw that there was a separate pizza menu.

In the models that we have discussed so far, all the variables we have needed to reason about have either been observable data or the hypotheses or parameters that we want to infer from those data. However, as our generative models become more complex, the steps that produce the observable data are likely to also reflect some kind of underlying structure that we do not have the opportunity to observe. This structure is captured using *latent variables*.

Clustering is a classic example of a latent variable problem. Imagine that you visited an animal rescue facility and were told that they currently housed three breeds of dogs. As you walked around and looked at the dogs, you might be trying to figure out which dog was of each breed. Even if you didn't have any information about what differentiates one breed from another, you could probably come up with some good guesses based on the observed similarities between the dogs, clustering some of the dogs together in your mind.

In this setting, the cluster assignments of the dogs are latent variables—variables that influence our observations but are not themselves observable. Likewise, we might imagine documents being organized by the topics that they discuss, the properties of plants resulting from their positions in a taxonomic hierarchy, or faces being characterized in terms of a small number of meaningful psychological dimensions. Each of these representations posits a different set of latent variables, corresponding to topic assignments, nodes in trees, or locations in space that we might seek to infer from our observations.

Understanding how people make inferences of this kind is a step toward understanding how we engage in *unsupervised learning*—learning about the structure of our world without explicit labels. Some of the most impressive scientific breakthroughs have involved postulating new representations in this way—think of Mendeleev's organization of the elements into the periodic table, or Darwin's insight that species should be organized into trees. However, latent variables are also key to building more complex generative models that are able to capture some of the structure that underlies everyday experience, and hence to providing potential explanations for how people make inferences that exploit that structure. In this chapter, we will explore how such models can be defined, how their parameters can be estimated, and how they can be used to explain aspects of human cognition.

5.1 Mixture Models and Density Estimation

In general, a latent variable model involves two kinds of variables: those that we can observe and those that we cannot. The latter are the latent variables. We will use X_i to denote the observable variables and Z_i to denote the latent variables. Typically, these variables would be related via a generative model that specifies how the X_i are generated via the latent structure Z_i. Such a model can ultimately be used to make inferences about the latent structure associated with new data points, as well as providing a more accurate model of the distribution of X_i.

Perhaps the most widely used latent variable model is the *mixture model*, in which the distribution of X_i is assumed to be a mixture of several other distributions. We have briefly discussed mixtures of distributions in previous chapters as a convenient way of specifying prior distributions without making the latent variables involved explicit. In a mixture model, the latent variable Z_i indicates which component of the mixture is used to generate each observed data point X_i.

More formally, a mixture model specifies the probability of X_i as a mixture of a set of k component distributions:

$$p(x_i) = \sum_{j=1}^{k} p(x_i|z_i = j)P(z_i = j), \tag{5.1}$$

where $p(x_i|z_i = j)$ is the distribution associated with component j and $P(z_i = j)$ is the probability that a data point would be generated from that component. The underlying generative model is one in which we first sample the value of the latent variable Z_i, determining the component that will be used to generate X_i, and then sample X_i from the distribution associated with that component. Figure 5.1 shows a simple graphical model expressing this structure, suppressing the parameters used to define the underlying distributions.

One common application of mixture models is to the problem of clustering. In this setting, the latent variable Z_i indicates which cluster an observation X_i comes from, and the distribution $p(x_i|z_i)$ characterizes the form of the clusters. For example, we might believe that our data were generated from two clusters, each associated with a different Gaussian distribution. $P(Z_i)$ would specify the prevalence of each Gaussian distribution in the data, and the parameters of each Gaussian would define $p(x_i|z_i)$. If we can estimate the parameters that characterize these distributions, we can infer the probable cluster membership (z_i) for any data point (x_i).

More generally, mixture models allow us to capture probability distributions that have a shape that is not a good match for any of the simple distributions discussed so far. For this reason, mixture models are often used for *density estimation*—estimating the form of a probability density function. This use of mixture models turns out to have an interesting connection to models of how humans represent categories.

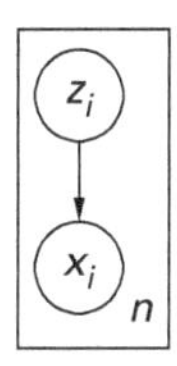

Figure 5.1
Basic graphical model for clustering with a mixture of Gaussians. Here, x_i is the ith data point and z_i its cluster assignment. The box around the variables is a plate indicating that this structure is replicated across n observations.

5.1.1 An Example: Mixture Models and Categorization

In chapter 3, we briefly discussed a model of memory in which categories were represented as Gaussian distributions (Huttenlocher, Hedges, & Vevea, 2000). This way of representing categories makes it easy not just to reconstruct objects from memory, but also to decide upon the category membership of those objects. The presentation here draws on that in Griffiths, Sanborn, Canini, Navarro, and Tenenbaum (2011a).

Categorization as Bayesian Inference A standard way to formalize the problem of categorization in psychology is to assume that people are given a set of $n-1$ stimuli with features $\mathbf{x}_{n-1}=(x_1,x_2,\ldots,x_{n-1})$ and category labels $\mathbf{c}_{n-1}=(c_1,c_2,\ldots,c_{n-1})$, and need to compute the probability that a new stimulus with features x_n is assigned to category c. We can calculate this probability by applying Bayes' rule, with

$$P(c_n=c|x_n,\mathbf{x}_{n-1},\mathbf{c}_{n-1})=\frac{p(x_n|c_n=c,\mathbf{x}_{n-1},\mathbf{c}_{n-1})P(c_n=c|\mathbf{c}_{n-1})}{\sum_c p(x_n|c_n=c,\mathbf{x}_{n-1},\mathbf{c}_{n-1})P(c_n=c|\mathbf{c}_{n-1})}, \quad (5.2)$$

where the posterior probability of category c is proportional to the product of the probability of an object with features x_n being produced from that category and the prior probability of choosing that category, taking into account the features and labels of the previous $n-1$ objects (assuming that only category labels influence the prior).

This formulation of the problem of categorization makes it clear that learning a category is a problem of determining the form of these probability distributions—a problem of density estimation. From the previous observations $\mathbf{x}_{n-1}$ and $\mathbf{c}_{n-1}$ we need to infer the probability density for the distribution of x_n given each value of c_n. One strategy for solving this problem is to assume that each category is a distribution of a standard form—such as a Gaussian—and estimate the parameters of these distributions from the previous observations. This can be done using the methods for maximum-likelihood estimation or maximum a posteriori (MAP) estimation introduced in chapter 3. Such an approach is called *parametric density estimation*, as it reduces the problem to parameter estimation (e.g., Rice, 1995).

Categories and Mixtures Mixture models provide a more flexible way to define probability distributions. Intuitively, it seems natural to think about certain kinds of categories as mixtures. To return to the example from the introduction to this chapter, the category of dogs contains many breeds that vary in their attributes and might each be captured by a distinct mixture component. We might thus try to solve our density estimation problem using mixture models.

Rosseel (2002) proposed an account of human category learning based on this idea. The Mixture Model of Categorization assumes that $P(x_n|c_n = c, \mathbf{x}_{n-1}, \mathbf{c}_{n-1})$ is a mixture distribution. Specifically, the model assumes that each object x_i comes from a cluster z_i, and each cluster is associated with a probability distribution over the features of the objects generated from that cluster. When evaluating the probability of a new object x_n, it is necessary to sum over all the clusters from which that object might have been drawn, with

$$p(x_n|c_n = c, \mathbf{x}_{n-1}, \mathbf{c}_{n-1}) = \sum_{j=1}^{k_c} p(x_n|z_n = j, \mathbf{x}_{n-1}, \mathbf{z}_{n-1})P(z_n = j|\mathbf{z}_{n-1}, c_n = c, \mathbf{c}_{n-1}), \quad (5.3)$$

where k_c is the total number of clusters for category c, $p(x_n|z_n = j, \mathbf{x}_{n-1}, \mathbf{z}_{n-1})$ is the probability of x_n under cluster j, and $P(z_n = j|\mathbf{z}_{n-1}, c_n = c, \mathbf{c}_{n-1})$ is the probability of generating a new object from cluster j in category c. The clusters can either be shared between categories or specific to a single category (in which case $P(z_n = j|\mathbf{z}_{n-1}, c_n = c, \mathbf{c}_{n-1})$ is 0 for all clusters not belonging to category c).

Psychological Models of Categorization While we have explicitly formulated the problem of categorization in terms of Bayesian inference and density estimation, psychological models of human categorization have traditionally been defined in different terms. The account presented here focuses on Marr's (1982) computational level, considering the abstract problem that human minds need to solve and its ideal solution. Previous psychological models have been expressed at the algorithmic level, focusing on the representations and processes that support categorization. Our presentation of these models in this section is based on Griffiths, Kemp, and Tenenbaum (2008c), which provides further details and information about ways in which these models can be extended.

Psychological models typically identify the problem of categorization as one of assigning stimuli to categories based on a subjective sense of similarity (e.g., Reed, 1972; Medin & Schaffer, 1978; Nosofsky, 1986). In this formulation, the probability that x_n is assigned to category c is given by

$$P(c_n = c|x_n, \mathbf{x}_{n-1}, \mathbf{c}_{n-1}) = \frac{\eta_{nc}\beta_c}{\sum_c \eta_{nc}\beta_c}, \quad (5.4)$$

where η_{nc} is the similarity of stimulus x_n to category c and β_c is the response bias for category c, capturing how much people are predisposed to producing that category label. Different models of categorization can be defined by making different assumptions about how η_{nc}, the similarity of a stimulus to a category, is computed. Three such strategies are illustrated in figure 5.2.

In a *prototype model* (e.g., Reed, 1972), category c is represented by a single prototypical instance. Under this account, your category of dogs would be represented by a single prototypical dog that captures the general properties of dogs. To formalize this, the similarity of stimulus n to category c is defined to be

$$\eta_{nc} = \eta_{np_c}, \quad (5.5)$$

where p_c is the prototypical instance of the category and η_{np_c} is a measure of the similarity between stimulus n and prototype p_c. One common way of defining the prototype is as the

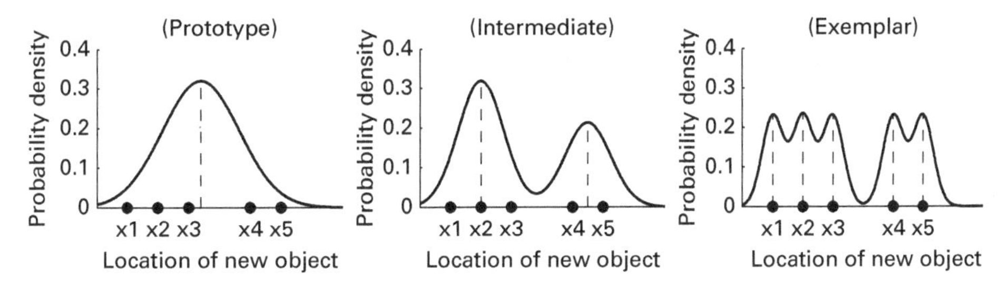

Figure 5.2
Different models of categorization can be expressed as different kinds of mixture models. In the prototype model (left), a category is represented as a single parametric probability distribution, such as a Gaussian. In an exemplar model (right), the category is represented as the sum of a set of kernels centered on the exemplars. This can be thought of as a mixture model with as many components as there are data points. Mixture models with fewer clusters (center) provide a way to interpolate between these extremes.

centroid of all instances of the category in some psychological space; that is,

$$p_c = \frac{1}{n_c} \sum_{i|c_i=c} x_i, \tag{5.6}$$

where n_c is the number of instances of the category (i.e., the number of stimuli for which $c_i = c$).

In an *exemplar model* (e.g., Medin & Schaffer, 1978; Nosofsky, 1986), a category is represented by all the stored instances of that category. Under this account, your representation of the category of dogs is simply all the dogs you have ever seen. The similarity of stimulus n to category c is calculated by summing the similarity of the stimulus to all stored instances of the category. That is,

$$\eta_{nc} = \sum_{i|c_i=c} \eta_{ni}, \tag{5.7}$$

where η_{ni} is a symmetric measure of the similarity between the two stimuli x_n and x_i. The similarity measure is typically either an exponential or a Gaussian function of the distance between the two stimuli.

Exemplar and prototype models represent two extreme solutions to the problem of defining the similarity between stimuli. Vanpaemel, Storms, and Ons (2005) observed that we can formalize a set of interpolating models by allowing the instances of each category to be partitioned into clusters, where the number of clusters k_c ranges from 1 to n_c. Then each cluster is represented by a prototype, and the similarity of stimulus n to category c is defined to be

$$\eta_{nc} = \sum_{j=1}^{k_c} \eta_{np_{jc}}, \tag{5.8}$$

where p_{jc} is the prototype of cluster j in category c. When $k_c = 1$ for all c, this is equivalent to the prototype model, and when $k_c = n_c$ for all c, this is equivalent to the exemplar model.

Connecting Levels of Analysis While they were originally proposed purely as an account of the psychological processes behind categorization, we can actually give a reasonable computational-level interpretation of prototype and exemplar models. Ashby and Alfonso-Reese (1995) observed a connection between the Bayesian solution to the problem of categorization presented in equation (5.2) and the way that the probabilities of category membership is computed in exemplar and prototype models (i.e., equation (5.4)). Specifically, η_{nc} can be identified with $p(x_n|c_n = c, \mathbf{x}_{n-1}, \mathbf{c}_{n-1})$, while β_c captures the prior probability of category c, $p(c_n = c|\mathbf{c}_{n-1})$. The difference between exemplar and prototype models thus comes down to different ways of estimating $p(x_n|c_n = c, \mathbf{x}_{n-1}, \mathbf{c}_{n-1})$.

The definition of η_{nc} used in an exemplar model (equation (5.7)) corresponds to estimating $P(x_n|c_n = c, \mathbf{x}_{n-1}, \mathbf{c}_{n-1})$ as the sum of a set of functions (known as *kernels*) centered on the x_i that are already labeled as belonging to category c, with

$$p(x_n|c_n = c, \mathbf{x}_{n-1}, \mathbf{c}_{n-1}) \propto \sum_{i|c_i=c} f(x_n, x_i), \tag{5.9}$$

where $f(x, x_i)$ is a probability distribution centered on x_i.[1] This is a method that is widely used for approximating distributions in statistics, being a form of *nonparametric density estimation* (meaning that it can be used to identify distributions without assuming that they come from an underlying parametric family) called *kernel density estimation* (e.g., Silverman, 1986).

The definition of η_{nc} used in a prototype model (equation (5.5)) corresponds to estimating $p(x_n|c_n = c, \mathbf{x}_{n-1}, \mathbf{c}_{n-1})$ by assuming that the distribution associated with each category comes from an underlying parametric family, and then finding the parameters that best characterize the instances labeled as belonging to that category. The prototype corresponds to these parameters—for a Gaussian distribution, it would be the mean. Again, this is a common method for estimating a probability distribution—it is the parametric density estimation strategy we introduced earlier.

The interpretation of exemplar and prototype models as different schemes for density estimation suggests that a similar interpretation might be found for interpolating models. Indeed, the corresponding solution is given by Rosseel's (2002) Mixture Model of Categorization. By a similar argument to that used for the exemplar model, we can connect equation (5.3) with the definition of η_{nc} in equation (5.8), providing a justification for adopting representations that interpolate between exemplars and prototypes. In fact, this model can produce all the kinds of representations that we have discussed: it reduces to kernel density estimation when each stimulus has its own cluster and the clusters are equally weighted, and parametric density estimation when each category is represented by a single cluster.

5.2 Mixture Models as Priors

Mixture models provide a good way to define complex distributions using simple parts. In chapter 3, we introduced the idea of using conjugate priors to simplify Bayesian inference for continuous variables, but many plausible prior distributions are not in the family of conjugate priors for a given likelihood function. Taking a prior that is a mixture of conjugate

1. The constant of proportionality is determined by $\int f(x, x_i)\, dx$, being $\frac{1}{n_c}$ if $\int f(x, x_i)\, dx = 1$ for all i, and is absorbed into β_j to produce direct equivalence to equation (5.7)).

priors retains much of the attractive tractability that conjugate priors provide, but also allows us to define more expressive prior distributions. As a simple example, we can return to the problem of estimating the mean of a Gaussian and see what happens when we use a mixture of Gaussians rather than a single Gaussian as a prior.

Assume that we observe data x generated from a Gaussian distribution with mean μ and standard deviation σ. Rather than simply assuming a Gaussian prior on μ as we did in chapter 3, assume that the prior on μ is a mixture distribution with $p(\mu) = \sum_z p(\mu|z)P(z)$, where z ranges over the components of the mixture and $p(\mu|z)$ is Gaussian. We can then calculate the joint posterior distribution on z and μ given an observation x, $p(\mu, z|x)$ by applying Bayes' rule. If we choose to factorize this joint distribution as $p(\mu|x,z)P(z|x)$, we obtain the following expression for the posterior mean:

$$\bar{\mu} = \int \mu \sum_z p(\mu|x,z)P(z|x)\, d\mu \tag{5.10}$$

$$= \sum_z \int \mu p(\mu|x,z)P(z|x)\, d\mu \tag{5.11}$$

$$= \sum_z P(z|x) \int \mu p(\mu|x,z)\, d\mu, \tag{5.12}$$

where we seize the opportunity to change the order of summation for probability distributions (in this case, switching around the order in which we evaluate the sum and the integral).

Since $p(\mu|x,z)$ is the posterior distribution on μ using the Gaussian associated with mixture component z as a prior, we can substitute our previous results from chapter 3 for the posterior mean of μ under a Gaussian prior for $\int \mu p(\mu|x,z)\, d\mu$. We can write this posterior mean as

$$\bar{\mu} = \frac{\frac{x}{\sigma^2} + \frac{\mu_0}{\sigma_0^2}}{\frac{1}{\sigma^2} + \frac{1}{\sigma_0^2}} = \frac{\sigma_0^2}{\sigma_0^2 + \sigma^2}x + \frac{\sigma^2}{\sigma_0^2 + \sigma^2}\mu_0, \tag{5.13}$$

where σ^2 is the variance of x given μ and μ_0 and σ_0^2 are the mean and variance of the prior. The posterior mean using the mixture is then the average of the posterior means obtained using each of the components as a prior, weighted by the probability of that component. In the case where all the components have the same variance σ_0^2, we obtain

$$\bar{\mu} = \frac{\sigma_0^2}{\sigma_0^2 + \sigma^2}x + \frac{\sigma^2}{\sigma_0^2 + \sigma^2} \sum_j p(z=j|x)\mu_0^{(z)}, \tag{5.14}$$

where $\mu_0^{(j)}$ denotes the mean of the jth component. The posterior mean thus linearly interpolates between the observed value of x and the weighted average of the means of the components of the prior.

5.2.1 An Example: The Perceptual Magnet Effect

The *perceptual magnet effect* is a phenomenon that has been documented in the perception of speech sounds, in which sounds that are close to the center of a phonetic category are perceived as being more similar to one another than sounds that are close to the boundary between two categories (Iverson & Kuhl, 1995). We can give a rational explanation for why such an effect might be observed by using a model based on a mixture of Gaussians (Feldman & Griffiths, 2007; Feldman, Griffiths, & Morgan, 2009).

When we perceive a speech sound, we receive a continuous signal x and try to reconstruct the sound produced by the speaker, μ. Assume that the noise in the transmission of the signal is Gaussian, with the distribution of x given μ having mean μ and standard deviation σ. If there is only one category of speech sounds, and we assume that the category corresponds to a Gaussian distribution over possible values of μ with mean μ_0 and standard devaition σ_0, then we are back in the familiar territory of estimating the mean of a Gaussian with a Gaussian prior. That is, the posterior distribution of μ given x will be Gaussian, with a mean that interpolates between x and μ_0. Taking the perceived speech sound to be the expectation of μ given x, the posterior mean, implies that perceptual space will be a linear transformation of the stimulus space, compressing stimuli into the region near the mean of the category. This is illustrated in the left panel of figure 5.3.

In reality, languages have multiple categories of speech sounds. The appropriate prior for μ is thus a mixture of Gaussians, with each component corresponding to a different speech sound. The reconstruction of a perceived speech sound will thus be affected by all the categories that the sound could belong to, with the mean of the posterior of μ given x being given by equation (5.14). The right panel of figure 5.3 illustrates the predictions that result from this equation for the case of two categories of speech sounds. When the stimulus is close to the mean of one category, reconstructions are drawn to the mean of that category. However, as they approach the region where their category membership is ambiguous, both categories exert an influence. As a consequence, the mean of the posterior is much closer to the original stimulus value. This produces a compression of perceptual space near the means of the categories and an expansion between the boundaries, exactly what occurs in the perceptual magnet effect.

5.3 Estimating Parameters in the Presence of Latent Variables

The presence of latent variables in a model poses two problems: inferring the values of the latent variables conditioned on observed data, and learning the probability distribution characterizing both the observable and latent variables. In the probabilistic framework, both these forms of inference reduce to inferring the values of unknown variables, conditioned on known variables. This is conceptually straightforward, but the computations involved are difficult and can require complex algorithms.

To understand the challenge, imagine that we have a model for data $\mathbf{x}$ that has parameters θ and latent variables $\mathbf{z}$. A mixture model is one example of such a model. The likelihood for this model is $p(\mathbf{x}|\theta) = \sum_{\mathbf{z}} p(\mathbf{x}, \mathbf{z}|\theta)$, where the latent variable $\mathbf{z}$ are unknown. To apply maximum-likelihood (or MAP) estimation, we would need to compute the derivative of the likelihood (or log-likelihood) with respect to θ. This can be a challenge since the likelihood involves a sum over $\mathbf{z}$. In particular, this approach is intractable when there are many possible values for $\mathbf{z}$ (e.g., with k clusters and n data points, we would have to sum over k^n possible values of $\mathbf{z}$ since each of the n cluster assignments z_i can take on k values).

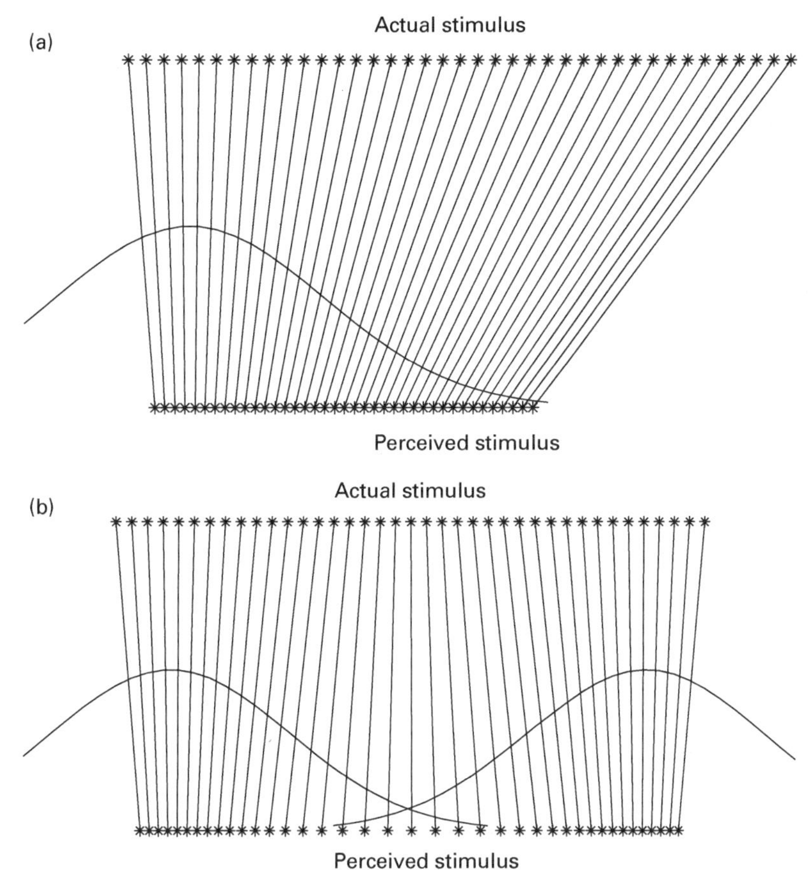

Figure 5.3
Different patterns of warping of perceptual space. With a single category, the perceived stimulus will be drawn to the mean of the category. With two categories, the perceived stimulus is drawn to the means of both categories, with the influence of each category determined by the posterior probability of having been generated from that category. As a consequence, perceptual space is compressed near the means of the categories and expanded at the boundary between categories. Figure reproduced with permission from Feldman and Griffiths (2007).

5.3.1 The Expectation-Maximization Algorithm

A standard approach to solving the problem of estimating probability distributions involving latent variables without needing to deal with this intractable sum is the *Expectation-Maximization (EM) algorithm* (Dempster, Laird, & Rubin, 1977). The EM algorithm is a procedure for obtaining a maximum-likelihood (or MAP) estimate for θ without resorting to differentiating $\log p(\mathbf{x}|\theta)$. The key idea is that we have two problems, each of which could be solved if we were able to solve the other problem: if we knew the value of the latent variable $\mathbf{z}$, then we could find θ by using the standard methods for estimation discussed in the previous chapters; on the other hand, if we knew θ, we could compute $P(\mathbf{z}|\mathbf{x}, \theta)$ and infer the values of the latent variable $\mathbf{z}$.

Surprisingly, we can make progress in inferring both θ and $\mathbf{z}$ by using of the partial knowledge we have about each. The EM algorithm for maximum-likelihood estimation proceeds by repeatedly alternating between assigning probabilities to $\mathbf{z}$ based on $\mathbf{x}$ and our current

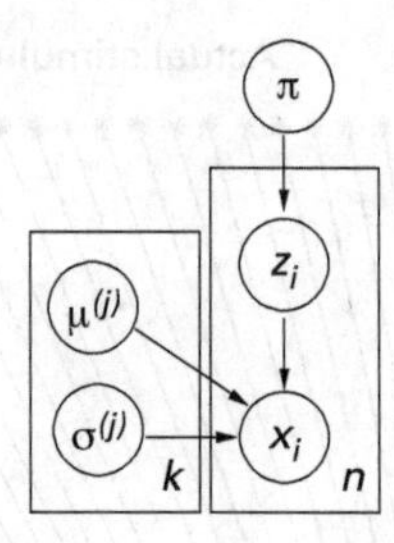

Figure 5.4
A more detailed graphical model for clustering with a mixture of Gaussians. Here, x_i is the ith data point and z_i its cluster assignment; π are the parameters of the discrete distibution on z_i with $P(z_i=j)=\pi_j$ and $\mu^{(j)}$ and $\sigma^{(j)}$ for the parameters of the Gaussian corresponding to the jth cluster. The distribution $p(x_i|z_i=j,\mu,\sigma)$ is Gaussian($\mu^{(j)},\sigma^{(j)}$). The boxes around the variables are plates that indicate that this structure is replicated across n observations and k clusters.

guess of θ, and using these probabilities to guess the values of $\mathbf{z}$ and hence estimate θ. More formally, this results in an iterative procedure with two steps: evaluating the expectation of the *complete log-likelihood* $\log p(\mathbf{x},\mathbf{z}|\theta)$ with respect to $P(\mathbf{z}|\mathbf{x},\theta)$ (the *E-step*, short for Expectation), and maximizing the resulting quantity with respect to θ (the *M-step*, short for Maximization). This algorithm is guaranteed to converge to a *local maximum* of $p(\mathbf{x}|\theta)$, finding different solutions depending on the value of θ used to initialize it (Dempster et al., 1977).

While the EM algorithm can be used for a variety of latent variable models, we will illustrate it for the case of clustering. Assume that our observations $\mathbf{x}=(x_1,\ldots x_n)$ are generated from a Gaussian mixture model where the two components have a known common standard deviation, σ, and unknown means $\mu^{(1)}$ and $\mu^{(2)}$. The cluster assignments $\mathbf{z}=(z_1,\ldots,z_n)$ are our latent variables, each sampled from a *discrete distribution* (i.e., a multinomial distribution with just two outcomes). The probability that an observation is drawn from the first cluster, π, is also unknown. The parameters that we want to estimate are thus $\theta=(\mu^{(1)},\mu^{(2)},\pi)$, corresponding to the parameters in the graphical model shown in figure 5.4. We start by choosing some arbitrary values for θ.

The E-step In the E-step of the algorithm, we evaluate the expectation of the complete log-likelihood with respect to $P(\mathbf{z}|\mathbf{x},\theta)$:

$$E_{P(\mathbf{z}|\mathbf{x},\theta)}[\log p(\mathbf{x},\mathbf{z}|\theta)]=\sum_{\mathbf{z}} P(\mathbf{z}|\mathbf{x},\theta)\log p(\mathbf{x},\mathbf{z}|\theta). \tag{5.15}$$

Since each x_i is independent of all other variables given z_i and θ, and the z_i are independent given θ, this simplifies to

$$E_{P(\mathbf{z}|\mathbf{x},\theta)}[\log p(\mathbf{x},\mathbf{z}|\theta)]=\sum_{\mathbf{z}} P(\mathbf{z}|\mathbf{x},\theta)\sum_{i=1}^{n}[\log p(x_i|z_i,\theta)+\log P(z_i|\theta)]. \tag{5.16}$$

We can then reverse the order in which we carry out the sums, obtaining

$$E_{P(\mathbf{z}|\mathbf{x},\theta)}[\log p(\mathbf{x},\mathbf{z}|\theta)]=\sum_{i=1}^{n}\sum_{\mathbf{z}} P(\mathbf{z}|\mathbf{x},\theta)\,[\log p(x_i|z_i,\theta)+\log p(z_i|\theta)] \tag{5.17}$$

$$= \sum_{i=1}^{n} \sum_{z_i} P(z_i|x_i, \theta) \left[\log p(x_i|z_i, \theta) + \log P(z_i|\theta)\right], \quad (5.18)$$

where the second line uses the fact that $p(x_i|z_i, \theta)$ and $P(z_i|\theta)$ are constant when we sum over $\mathbf{z}_{-i}$. Exploiting this conditional independence of the x_i and z_i is what allows us to overcome the intractable sum over all values for $\mathbf{z}$.

Substituting our Gaussian and discrete distributions for $p(x_i|z_i, \theta)$ and $P(z_i|\theta)$, respectively, we obtain

$$E_{P(\mathbf{z}|\mathbf{x},\theta)}[\log p(\mathbf{x}, \mathbf{z}|\theta)] = \sum_{i=1}^{n} \sum_{z_i} P(z_i|x_i, \theta) \left[-\frac{1}{2} \log 2\pi\sigma^2 - \frac{(x_i - \mu^{(z_i)})^2}{2\sigma^2} \right.$$

$$\left. + I(z_i = 1) \log \pi + I(z_i = 2) \log(1 - \pi) \right],$$

$$= -\frac{n}{2} \log 2\pi\sigma^2 - \sum_{i=1}^{n} \sum_{z_i} P(z_i|x_i, \theta) \frac{(x_i - \mu^{(z_i)})^2}{2\sigma^2}$$

$$+ \sum_{i=1}^{n} P(z_i = 1|x, \theta) \log \pi + \sum_{i=1}^{n} P(z_i = 2|x, \theta) \log(1 - \pi) \quad (5.19)$$

where $I(\cdot)$ is the indicator function, taking the value 1 when its argument is true and 0 otherwise.

The M-step In the M-step, we seek to maximize the expected complete log-likelihood with respect to θ. Since θ appears in this expression in two places, it is useful to denote the "old" values of θ (those used in computing $P(\mathbf{z}|\mathbf{x}, \theta)$) as θ^{old}, and the "new" values that we aim to find that maximize the expected complete log-likelihood as θ^{new}. Our goal is thus to find the value of θ^{new} that maximizes $E_{P(\mathbf{z}|\mathbf{x},\theta^{\text{old}})}[\log p(\mathbf{x}, \mathbf{z}|\theta^{\text{new}})]$. In the case of the mixture of Gaussians, this means maximizing the expression in equation (5.19) with respect to $\mu^{(1)}$, $\mu^{(2)}$, and π. Differentiating this expression with respect to these parameters, setting the derivative to zero, and solving the resulting equations give

$$\hat{\mu}^{(j)} = \frac{\sum_{i=1}^{n} P(z_i = j|x_i, \theta^{\text{old}}) x_i}{\sum_{i=1}^{n} P(z_i = j|x_i, \theta^{\text{old}})} \quad (5.20)$$

$$\hat{\pi} = \frac{\sum_{i=1}^{n} P(z_i = 1|x_i, \theta^{\text{old}})}{n}, \quad (5.21)$$

both of which have a very simple interpretation: the estimate of the mean of component z is the weighted average of the x_i, where the weights correspond to the probability that x_i belongs to component z, and the estimate of π is the expected number of the x_i belonging to the first component.

Iterating to Convergence In the M-step, the parameters used in computing the probability that observation x_i is assigned to component z_i are the old parameters, θ^{old}, that we chose when we initialized the algorithm. However, the algorithm now iterates back and forth between the E- and M-steps, using the new estimates of θ produced in the M-step to compute

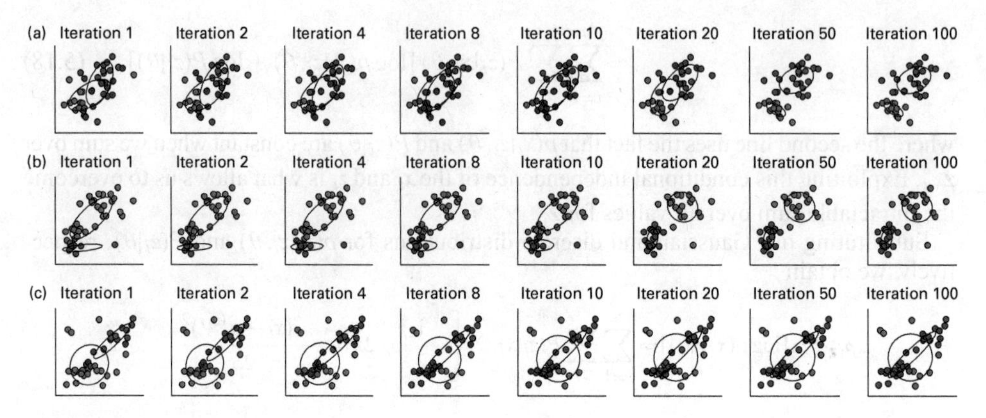

Figure 5.5
Three runs of the EM algorithm for a mixture of two Gaussians in two dimensions. Each row shows a single run of the algorithm, with different initial parameter values and data being used each time. The columns show the resulting parameter estimates after different numbers of iterations. The color of each point indicates its posterior probability of being assigned to each component, and the two Gaussians are indicated with equiprobability ellipses picking out a set of points that are equally probable under the inferred parameters. While EM often picks out natural clusters, as in (a) and (b), it only finds a local maximum of the likelihood, meaning that some initializations of the parameters result in other solutions, such as that shown in (c).

the expectation in the E-step. The algorithm thus proceeds to develop better parameter estimates and a better idea of which component each observation came from over time, and ultimately converges to a steady state that is a local maximum of $p(\mathbf{x}|\theta)$.

An illustration of EM for a two-dimensional (2D) mixture of Gaussians appears in figure 5.5. While the analysis given here focuses on maximum-likelihood estimation, MAP estimation can be done by including a prior $p(\theta)$ and computing the expected complete joint log probability, $\log p(\mathbf{x}, \mathbf{z}, \theta)$, in the E-step.

Alternating between updating the cluster assignments and updating the parameters that define the clusters is an intuitive way of solving the problem posed by clustering. It is also the strategy used in the classic *k-means clustering algorithm*, which alternates between assigning each data point to the cluster with the nearest mean and updating the cluster means based on those assignments. Indeed, the k-means algorithm is equivalent to the EM algorithm for a Gaussian mixture model where the standard deviation $\sigma^{(j)}$ are equal for all j and approach 0. In this case, $p(z_i = j|x_i, \theta)$ approaches 1 for the cluster with $\mu^{(j)}$ closest to x_i.

5.3.2 Analyzing the EM Algorithm

The EM algorithm works by reducing a problem that is hard to solve—finding the θ that maximizes $p(\mathbf{x}|\theta)$ when we have to marginalize over latent variables $\mathbf{z}$—to two problems that are easy to solve—computing the posterior distribution on $\mathbf{z}$ when θ is known, $P(\mathbf{z}|\mathbf{x}, \theta)$, and estimating θ when $\mathbf{z}$ is known by maximizing $p(\mathbf{x}, \mathbf{z}|\theta)$. Starting with an initial guess of θ and alternating between these two steps result in more accurate estimates of θ, which then provide more accurate guesses about the values of the latent variables, which again result in more accurate estimates of θ. Intuitively, maximizing the expected complete log-likelihood makes sense because if we knew the exact value of $\mathbf{z}$, we would just want to maximize the complete log-likelihood $\log p(\mathbf{x}, \mathbf{z}|\theta)$, so averaging over the possible values of $\mathbf{z}$ provides a way to take into account our uncertainty. However, it is also possible to give a more formal analysis of why the EM algorithm works.

One way of understanding the EM algorithm is to recognize that the expected complete log-likelihood is a lower-bound on the log-likehood. To see this, note that we can write

$$P(\mathbf{z}|\mathbf{x},\theta) = \frac{p(\mathbf{x},\mathbf{z}|\theta)}{p(\mathbf{x}|\theta)}. \tag{5.22}$$

Since $P(\mathbf{z}|\mathbf{x},\theta)$ is nonzero everywhere $P(\mathbf{x},\mathbf{z}|\theta)$ is nonzero, we can write

$$p(\mathbf{x}|\theta) = \frac{p(\mathbf{x},\mathbf{z}|\theta)}{P(\mathbf{z}|\mathbf{x},\theta)}. \tag{5.23}$$

Taking the logarithm of both sides, we have

$$\log p(\mathbf{x}|\theta) = \log p(\mathbf{x},\mathbf{z}|\theta) - \log P(\mathbf{z}|\mathbf{x},\theta). \tag{5.24}$$

If we now take the expectation of both sides with respect to $P(\mathbf{z}|\mathbf{x},\theta)$, we obtain

$$\log p(\mathbf{x}|\theta) = \sum_{\mathbf{z}} P(\mathbf{z}|\mathbf{x},\theta)\log p(\mathbf{x},\mathbf{z}|\theta) - \sum_{\mathbf{z}} P(\mathbf{z}|\mathbf{x},\theta)\log P(\mathbf{z}|\mathbf{x},\theta) \tag{5.25}$$

$$= E_{P(\mathbf{z}|\mathbf{x},\theta)}[\log p(\mathbf{x},\mathbf{z}|\theta)] + H[P(\mathbf{z}|\mathbf{x},\theta)] \tag{5.26}$$

where we use the fact that $\log p(\mathbf{x}|\theta)$ is constant in $\mathbf{z}$, and $H[P(\mathbf{z}|\mathbf{x},\theta)]$ is the *entropy* of $P(\mathbf{z}|\mathbf{x},\theta)$, which we introduce in more detail in chapter 7. Since the entropy is non-negative (see Cover & Thomas, 1991), $E_{P(\mathbf{z}|\mathbf{x},\theta)}[\log p(\mathbf{x},\mathbf{z}|\theta)]$—the expected complete log-likelihood—provides a lower bound on $\log p(\mathbf{x}|\theta)$.

The EM algorithm thus alternates between computing a function that provides a lower bound on $\log p(\mathbf{x}|\theta)$ in the E-step and maximizing this function with respect to θ in the M-step. Both steps can also be interpreted as performing hillclimbing on a single *free energy* function that has local maxima corresponding to the local maxima of $p(\mathbf{x}|\theta)$ (Neal & Hinton, 1998).

Applying EM is easiest in models where the latent variables are independent when conditioned on data and parameters (as in a mixture model), and the distributions for which we want to estimate parameters belong to exponential families (introduced in chapter 3). In this case, estimators typically take a form similar to that seen in equations (5.20) and (5.21), being based on observations weighted by the posterior probability of the values of the relevant latent variables. In models using more complex distributions, it is sufficient to improve the expected complete log-likelihood rather than maximizing it, so gradient descent or other optimization methods can be employed (Neal & Hinton, 1998).

5.3.3 An Example: Unsupervised Category Learning

Applying the EM algorithm to mixture models illustrates how it is possible to learn the parameters of the distributions that characterize clusters without needing any of the observations to be labeled. This kind of "unsupervised" learning characterizes much of human experience as well: while as a child, you probably had some observations labeled as "dogs" and others as "cats," your understanding of what dogs and cats are is just as dependent on all the unlabeled examples that you have seen throughout your life. In an even more extreme case, naturalists visiting a new continent are able to recognize that they are seeing animals from different species even without those animals being given verbal labels. So how are people able to learn categories without labels?

Fried and Holyoak (1984) set out to answer this question, comparing supervised and unsupervised learning for simple categories. They defined categories by choosing some simple 2D binary arrays as "standards" and then generating other arrays by randomly modifying the arrays, as shown in figure 5.6a. They told participants that they were going to see abstract designs by two artists—Smith and Wilson—and then compared how well people learned to categorize the images by artist. In one condition, people received feedback on their decisions. In the other, they did not.

Fried and Holyoak (1984) found that people were able to learn the categories either with or without labels, as shown in figure 5.6b. The best performance was seen in the condition where participants received feedback on their classification decisions, but in all conditions people were able to learn to generalize to new instances of the categories. This behavior was consistent with the predictions of a model that Fried and Holyoak (1984) proposed that is very similar to the EM algorithm for mixture models, updating estimates of the parameters of categories using a fractional allocation of observations based on the posterior probability of belonging to that category.

5.4 Topic Models

The basic idea behind mixture models can be used to define models that capture people's inferences in other settings. We now turn to *topic models*, which add a simple twist to mixture models to capture an important aspect of language. These models have also been used to account for aspects of human semantic memory, tracking the abstract structure behind the associations that we perceive between words.

Semantic Memory Several computational models have been proposed to account for the large-scale structure of semantic memory, including semantic networks (e.g., Collins & Loftus, 1975; Collins & Quillian, 1969) and semantic spaces (e.g., Landauer & Dumais, 1997; Lund & Burgess, 1996; Mikolov et al., 2013; Pennington, Socher, & Manning, 2014). These approaches embody different assumptions about the way that words are represented. In semantic networks, words are nodes in a graph where edges indicate semantic relationships, as shown in figure 5.7a. In semantic space models, words are represented as points in high-dimensional space, where the distance between two words reflects the extent to which they are semantically related, as shown in figure 5.7b.

Words and Topics Probabilistic models provide an opportunity to explore alternative representations for the meaning of words. Topic models represent words in terms of the set of topics to which they belong (Hofmann, 1999; Blei, Ng, & Jordan, 2003; Griffiths & Steyvers, 2004). Each topic is a probability distribution over words, and the content of the topic is reflected in the words to which it assigns high probability. For example, high probabilities for WOODS and STREAM would suggest that a topic refers to the countryside, while high probabilities for FEDERAL and RESERVE would suggest that a topic refers to finance. Each word will have a probability under each of these topics, as shown in figure 5.7c.

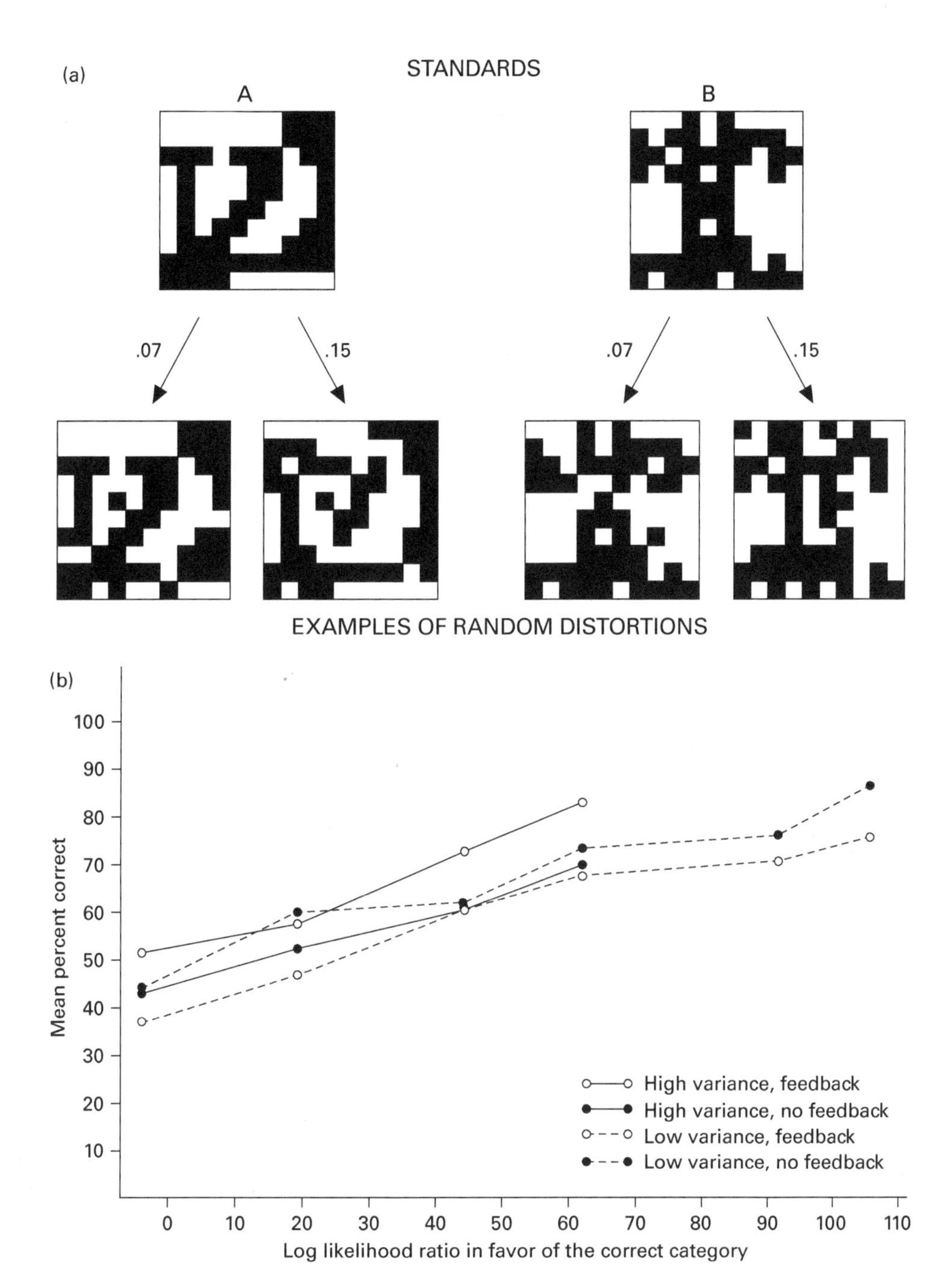

Figure 5.6
Fried and Holyoak (1984) examined whether people can learn categories without labels. (a) Their stimuli were binary arrays described as designs generated by different artists. Each category was defined by choosing a "standard" array (a prototype) and then randomly flipping bits (i.e., changing a black square to white or vice versa). The probability of flipping a bit was either .07 or .15, resulting in "low variability" and "high variability" categories, respectively. (b) People were able to learn the categories both with and without feedback on their categorization decisions. The results here show that people were increasingly correct in identifying the category membership of an array as that array provided more evidence of belonging to one of the categories, as reflected in the log-likelihood ratio. While the best performance was seen in the high variance categories with feedback, participants were still able to learn the categories without labels. Figure adapted from Fried and Holyoak (1984).

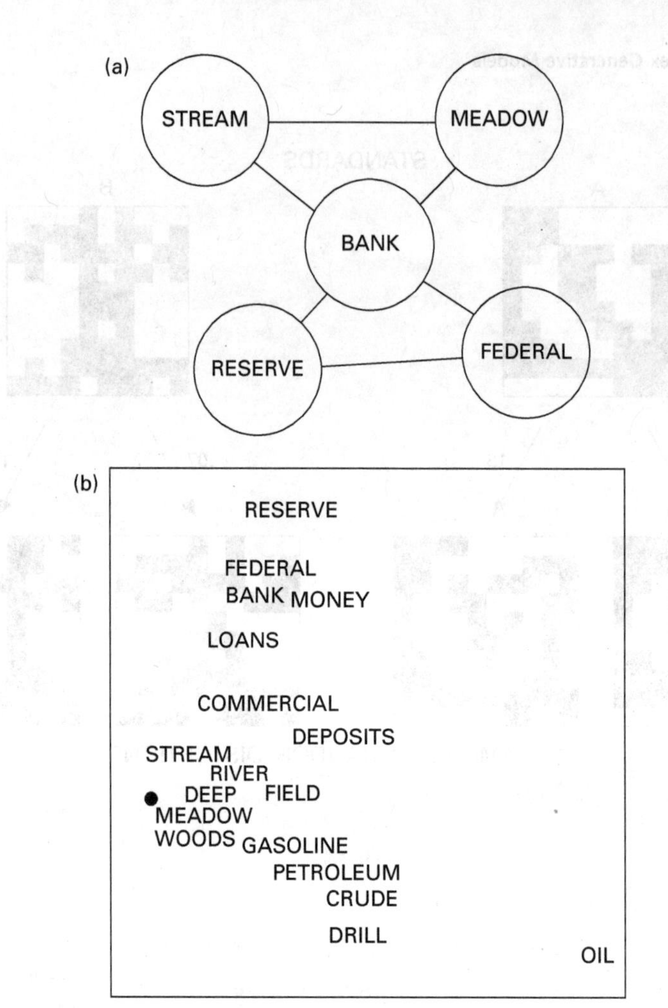

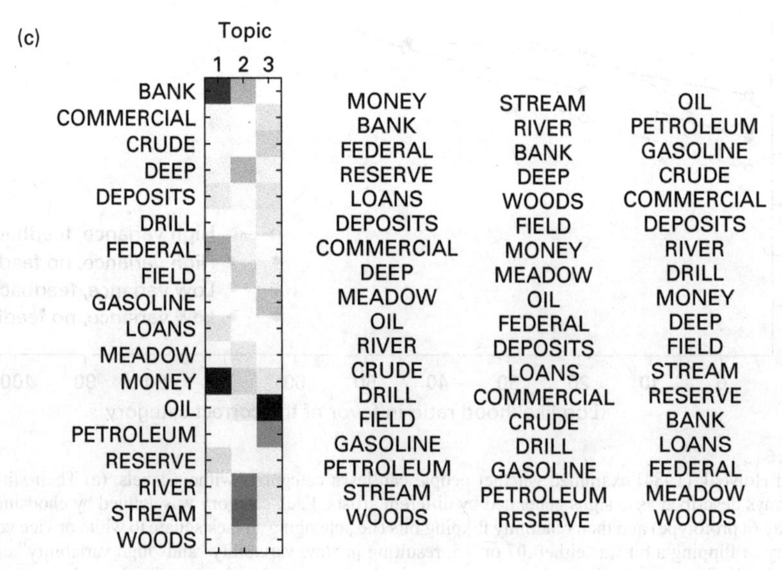

Figure 5.7
Approaches to semantic representation. (a) In a semantic network, words are represented as nodes, and edges indicate semantic relationships. (b) In a semantic space, words are represented as points, and proximity indicates semantic association. These are the first two dimensions of a solution produced by Latent Semantic Analysis (Landauer & Dumais, 1997). The black dot is the origin. (c) In the topic model, words are represented as belonging to a set of probabilistic topics. The matrix shown on the left indicates the probability of each word under each of three topics. The three columns on the right show the words that appear in those topics, ordered from highest to lowest probability. Figure reproduced with permission from Griffiths et al. (2007).

For example, MEADOW has a relatively high probability under the countryside topic, but a low probability under the finance topic, similar to WOODS and STREAM.

Representing word meanings using probabilistic topics makes it possible to use Bayesian inference to answer some of the critical problems that arise in processing language. In particular, we can make inferences about which semantically related concepts are likely to arise in the context of an observed set of words or sentences, in order to facilitate subsequent processing. Let z denote the dominant topic in a particular context, and w_1 and w_2 be two words that arise in that context. The semantic content of these words is encoded through a set of probability distributions that identify their probability under different topics: if there are k topics, then these are the distributions $P(w|z)$ for $z=\{1,\ldots,k\}$. Given w_1, we can infer which topic z was likely to have produced it by using Bayes' rule:

$$P(z|w_1) = \frac{P(w_1|z)P(z)}{\sum_{j=1}^{k} P(w_1|z'=j)P(z'=j)}, \tag{5.27}$$

where $P(z)$ is a prior distribution over topics. Having computed this distribution over topics, we can make a prediction about future words by summing over the possible topics:

$$P(w_2|w_1) = \sum_{j=1}^{k} P(w_2|z=j)P(z=j|w_1). \tag{5.28}$$

A topic-based representation can also be used to disambiguate words: if BANK occurs in the context of STREAM, it is more likely that it was generated from the bucolic topic than the financial topic.

Probabilistic topic models are an interesting alternative to traditional approaches to semantic representation, and in many cases, they actually provide better predictions of human behavior (Griffiths & Steyvers, 2003; Griffiths et al., 2007; Nematzadeh, Meylan, & Griffiths, 2017). However, one critical question in using this kind of representation is that of which topics should be used. Fortunately, work in machine learning and information retrieval has provided an answer to this question. As with popular semantic space models (Landauer & Dumais, 1997; Lund & Burgess, 1996), the representation of a set of words in terms of topics can be inferred automatically from the text contained in large document collections. The key to this process is viewing topic models as generative models for documents, making it possible to identify a set of topics that are likely to have generated an observed collection of documents. This can be done using the EM algorithm (Hofmann, 1999; Blei et al., 2003) or Monte Carlo methods, which we will introduce in chapter 6 (Griffiths & Steyvers, 2004). Figure 5.8 shows a sample of topics inferred from the TASA corpus (Landauer & Dumais, 1997), a collection of passages excerpted from educational texts used in curricula from the first year of school to the first year of college.

A Generative Model for Documents We can specify a generative model for documents by assuming that each document is a mixture of topics, with each word in that document being drawn from a particular topic and the topics varying in probability across documents. For any particular document, we write the probability of a word w in that document as

$$P(w) = \sum_{j=1}^{k} P(w|z=j)P(z=j), \tag{5.29}$$

PRINTING	**PLAY**	TEAM	JUDGE	HYPOTHESIS	STUDY	**CLASS**	ENGINE
PAPER	PLAYS	GAME	TRIAL	EXPERIMENT	**TEST**	MARX	FUEL
PRINT	STAGE	BASKETBALL	**COURT**	SCIENTIFIC	STUDYING	ECONOMIC	ENGINES
PRINTED	AUDIENCE	PLAYERS	CASE	OBSERVATIONS	HOMEWORK	CAPITALISM	STEAM
TYPE	THEATER	PLAYER	JURY	SCIENTISTS	NEED	CAPITALIST	GASOLINE
PROCESS	ACTORS	**PLAY**	ACCUSED	EXPERIMENTS	**CLASS**	SOCIALIST	AIR
INK	DRAMA	PLAYING	GUILTY	SCIENTIST	MATH	SOCIETY	**POWER**
PRESS	SHAKESPEARE	SOCCER	DEFENDANT	EXPERIMENTAL	TRY	SYSTEM	COMBUSTION
IMAGE	ACTOR	PLAYED	JUSTICE	**TEST**	TEACHER	**POWER**	DIESEL
PRINTER	THEATRE	BALL	**EVIDENCE**	METHOD	WRITE	RULING	EXHAUST
PRINTS	PLAYWRIGHT	TEAMS	WITNESSES	HYPOTHESES	PLAN	SOCIALISM	MIXTURE
PRINTERS	PERFORMANCE	BASKET	CRIME	TESTED	ARITHMETIC	HISTORY	GASES
COPY	DRAMATIC	FOOTBALL	LAWYER	**EVIDENCE**	ASSIGNMENT	POLITICAL	CARBURETOR
COPIES	COSTUMES	SCORE	WITNESS	BASED	PLACE	SOCIAL	GAS
FORM	COMEDY	**COURT**	ATTORNEY	OBSERVATION	STUDIED	STRUGGLE	COMPRESSION
OFFSET	TRAGEDY	GAMES	HEARING	SCIENCE	CAREFULLY	REVOLUTION	JET
GRAPHIC	**CHARACTERS**	TRY	INNOCENT	FACTS	DECIDE	WORKING	BURNING
SURFACE	SCENES	COACH	DEFENSE	DATA	IMPORTANT	PRODUCTION	AUTOMOBILE
PRODUCED	OPERA	GYM	CHARGE	RESULTS	NOTEBOOK	CLASSES	STROKE
CHARACTERS	PERFORMED	SHOT	CRIMINAL	EXPLANATION	REVIEW	BOURGEOIS	INTERNAL

Figure 5.8
A sample of topics from a 1,700-topic solution derived from the TASA corpus. Each column contains the 20 highest-probability words in a single topic, as indicated by $P(w|z)$. Words in boldface occur in different senses in neighboring topics, illustrating how the model deals with polysemy and homonymy. These topics were discovered in a completely unsupervised fashion, using just word-document cooccurrence frequencies. Figure reproduced with permission from Griffiths et al. (2007).

where $P(w|z)$ is the probability of word w under topic z, which remains constant across all documents, and $P(z=j)$ is the probability of topic j in this document. We can summarize these probabilities with two sets of parameters, taking $\phi_w^{(j)}$ to indicate $P(w|z=j)$, and $\theta_z^{(d)}$ to indicate $P(z)$ in a particular document d. The procedure for generating a collection of documents is then straightforward. First, we generate a set of topics, sampling $\phi^{(j)}$ from some prior distribution $p(\phi)$. Then for each document d, we generate the weights of those topics, sampling $\theta^{(d)}$ from a distribution $p(\theta)$. Assuming that we know in advance how many words will appear in the document, we then generate those words in turn. A topic z is chosen for each word that will be in the document by sampling from the distribution over topics implied by $\theta^{(d)}$. Finally, the identity of the word w is determined by sampling from the distribution over words $\phi^{(z)}$ associated with that topic.

To complete the specification of our generative model, we need to specify distributions for ϕ and θ so we can make inferences about these parameters from a corpus of documents. As in the case of coin-flipping, calculations can be simplified by using a conjugate prior. Both ϕ and θ are arbitrary distributions over a finite set of outcomes—multinomial distributions when we care about the counts of events, otherwise discrete distributions—and the conjugate prior for these distributions is the Dirichlet distribution. Just as the discrete distribution is a multivariate generalization of the Bernoulli distribution used in the coin-flipping example, the Dirichlet distribution is a multivariate generalization of the beta distribution. We assume that the number of "virtual examples" of instances of each topic appearing in each document is set by parameter α, and likewise use parameter β to represent the number of instances of each word in each topic. Figure 5.9 shows a graphical model depicting the dependencies among these variables. This model, known as *latent Dirichlet allocation*, was introduced in machine learning by Blei et al. (2003).

Modeling Human Semantic Associations Griffiths and Steyvers (2002, 2003) suggested that topic models might provide an alternative to traditional approaches to semantic representation, and showed that they can provide better predictions of human word association data than latent semantic analysis (LSA) (Landauer & Dumais, 1997). Topic models can also be applied to a range of other tasks that draw on semantic association, such as semantic priming and sentence comprehension (Griffiths et al., 2007).

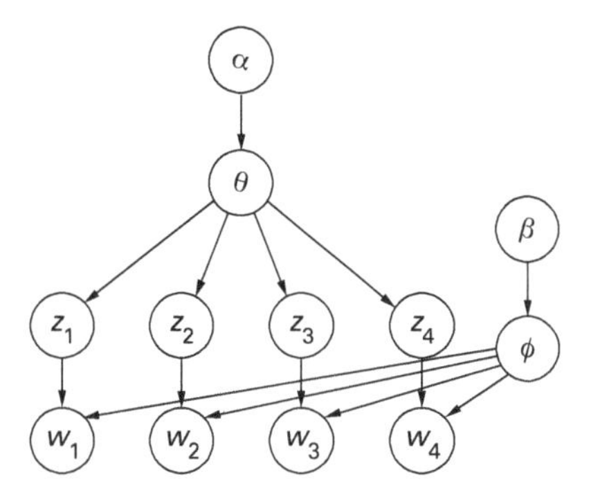

Figure 5.9
Graphical model for Latent Dirichlet Allocation (Blei, Ng, & Jordan, 2003). The distribution over words given topics, ϕ, and the distribution over topics in a document, θ, are generated from Dirichlet distributions with parameters β and α respectively. Each word in the document is generated by first choosing a topic z_i from θ, and then choosing a word according to $\phi^{(z_i)}$.

The key advantage that topic models have over semantic space models is postulating a more structured representation—different topics can capture different senses of words, allowing the model to deal with polysemy and homonymy in a way that is automatic and transparent. For instance, similarity in semantic space models must obey a version of the triangle inequality for distances: if there is high similarity between words w_1 and w_2, and between words w_2 and w_3, then w_1 and w_3 must be at least fairly similar. But word associations often violate this rule. For instance, ASTEROID is highly associated with BELT and BELT is highly associated with BUCKLE, but ASTEROID and BUCKLE have little association. LSA thus has trouble representing these associations. Out of approximately 4,500 words in a large-scale set of word association norms (Nelson, McEvoy, & Schreiber, 1998), LSA judges that BELT is the 13th most similar word to ASTEROID, that BUCKLE is the 2nd most similar word to BELT, and consequently BUCKLE is the 41st most similar word to ASTEROID–more similar than TAIL, IMPACT, or SHOWER. In contrast, using topics makes it possible to represent these associations faithfully because BELT belongs to multiple topics, one highly associated with ASTEROID but not BUCKLE, and another highly associated with BUCKLE but not ASTEROID. More recent semantic space models such as word2vec (Mikolov et al., 2013) or GloVe (Pennington et al., 2014), which can be trained on larger data sets than topic models, produce better performance in predicting human semantic associations but still have difficulty handling phenomena like the triangle inequality due to their representational assumptions (Nematzadeh et al., 2017).

5.5 Hidden Markov Models

So far, we have discussed some very simple models for sequential structure in data—Markov models. However, when we come to model inferences that involve reasoning about the dynamics of the environment or aspects of language, richer models of sequential dependency can be useful. The simplest of these models is the *hidden Markov model (HMM)*. We will start by reviewing Markov models and explaining how HMMs build on them, and then consider some of the basic ideas behind computing with HMMs. While these

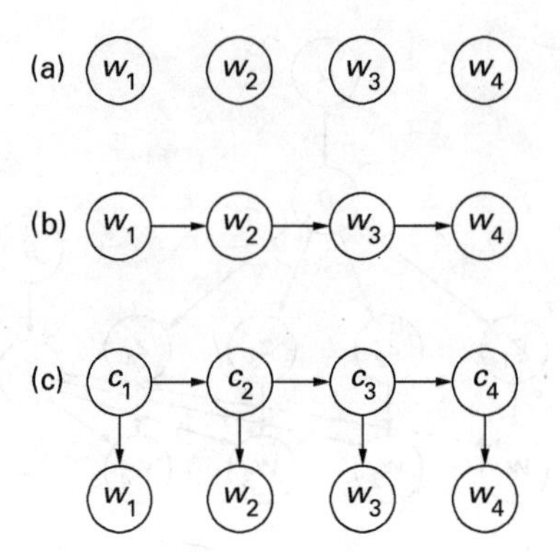

Figure 5.10
Graphical models illustrating the dependencies among variables in statistical models of language. (a) The simplest model is a unigram model, in which words are independently drawn from a common distribution. (b) A Markov model (in this case, a bigram model) adds dependencies between successive words. (c) In a hidden Markov model, the dependencies are among latent classes, with words being generated conditioned on the class. The variable c_i is the class associated with word w_i.

models can be applied to a range of problems, we will focus on the linguistic case in our description.

Simple Models of Language The simplest model of language assumes that each word is generated independently at random from a single fixed distribution over words. This model assumes no dependencies among words, predicting new words using the same distribution, $P(w)$, regardless of the other words that might appear in a sentence. In this model, the probability of a collection of words $\mathbf{w}$ depends only upon the frequency of each individual word in $\mathbf{w}$. If we define the distribution $P(w)$ using a vector of parameters θ such that $P(w)=\theta_w$, then we can write this model as

$$w_i \mid \theta \sim \text{Discrete}(\theta),$$

where $\sim$ should be read as "distributed as," and Discrete(θ) is the multivariate analog of the Bernoulli distribution. The dependency structure assumed by this model is illustrated in figure 5.10a, and the unknown parameters that have to be estimated from data are just those of the distribution over words, θ.

A slightly more complex model, a *Markov model*, allows dependencies among words. The dependency structure of one such model, a first-order Markov model, is shown in figure 5.10b. In this model, each word after the first is chosen from a distribution that is conditioned on the previous word: the ith word, w_i, is chosen from $P(w_i|w_{i-1})$, which specifies the transition probabilities between words. The probability of a collection of words $\mathbf{w}$ depends only upon the frequency with which one word precedes another. Associating $P(w_i|w_{i-1})$ with a vector of parameters $\theta^{(w_{i-1})}$, we can write

$$w_i \mid w_{i-1}, \theta \sim \text{Discrete}(\theta^{(w_{i-1})}).$$

This focus on pairs of words is the central innovation of this model over that shown in figure 5.10a, and this leads to the name *bigram model*. In general, a model in which each word is generated from a distribution that depends on the preceding $n-1$ words (a Markov model of order $n-1$) is referred to as an *n-gram model*, with the model shown in figure 5.10a being a *unigram model*. In any n-gram model, the parameters characterizing the distribution over words given the preceding words (for a bigram model, $\theta^{(w)}$ for all words w) need to be estimated from data.

The application of Markov models to language has a long history. Language modeling was one of the early examples explored by Markov (1913), and the approach was popularized by Shannon (1948). Research in language development has explored the sensitivity of learners to the transition probabilities of artificial languages, showing that infants can learn the kind of patterns captured by a Markov model (e.g., Saffran, Aslin, & Newport, 1996). However, the syntactic capabilities of Markov models have been roundly criticized, as they give low probability to grammatical sentences in which transitions between particular words are rare. Chomsky (1957) famously used this property of Markov models to argue against their adequacy as a model of language because people are able to judge sentences such as "colorless green ideas sleep furiously" to be grammatical even though no two successive words in that sentence are likely to have been heard together before.

Adding Latent Structure Some of the weaknesses of Markov models can be addressed by introducing latent structure into the model, associating each word w_i with an unobserved class c_i. Defining the transition probabilities in terms of the classes instead of the words makes it possible to give high probability to sentences that display appropriate transitions among classes, even though particular pairs of words may never have been seen before. Consequently, Chomsky's "colorless" sentence is not problematic since the transitions among word classes (nouns, adjectives, verbs, and adverbs) are consistent with the statistics of English (Pereira, 2000). This generative model is known as a hidden Markov model, since the classes of the words are "hidden" and has the dependency structure shown in figure 5.10c. In an HMM, each word is generated by choosing a class c_i from a distribution specified by c_{i-1}, $P(c_i|c_{i-1})$, then choosing a word w_i from the distribution associated with that class, $P(w_i|c_i)$. If we let $\phi^{(c)}$ be the parameters of the distribution over words associated with class c, and $\theta^{(c)}$ be the distribution over the next class selected from class c, we can write an HMM as

$$\begin{aligned} w_i \mid c_i, \phi &\sim \text{Discrete}(\phi^{(c_i)}) \\ c_i \mid c_{i-1}, \theta &\sim \text{Discrete}(\theta^{(c_{i-1})}), \end{aligned}$$

where ϕ and θ need to be estimated from data.

Hidden Markov models were first described by Baum and Petrie (1966). HMMs have been used for processing text and speech (e.g., Charniak, 1993; Jurafsky & Martin, 2000). One of their most successful applications was part-of-speech tagging (e.g., Charniak, Hendrickson, Jacobson, & Perkowitz, 1993), in which the word class $\mathbf{c}$ are trained to match the syntactic categories, such as nouns and verbs, that comprise the most basic level of linguistic structure. When these syntactic categories are learned from scratch (e.g., Goldwater & Griffiths, 2007), they are inferred based on the extent to which words tend to be similar in the distribution of neighboring words. These models thus provide a way to capture the idea of *distributional clustering* that cognitive scientists have claimed may play a role in the acquisition of syntactic categories by children (e.g., Redington, Chater, & Finch, 1998). We discuss other models of language in chapter 16.

5.5.1 Computing with Hidden Markov Models

Part of the attraction of HMMs is that they can capture latent structure but still remain computationally tractable. To a large extent, this tractability is the consequence of the simple dependency structure between the latent variables in the model. This dependency structure admits efficient procedures (based on *dynamic programming*) for computing probabilities of interest. Now we will consider a few of the questions that one might want to answer using HMMs and show how the resulting computations simplify.

What Is the Probability of a Sentence? A basic question that we might want to answer is how likely it is that a sentence $\mathbf{w} = (w_1, \ldots, w_n)$ would be generated from a particular HMM with known parameters θ. This is a marginal probability, calculated by summing over the latent variables $\mathbf{c}$,

$$P(\mathbf{w}) = \sum_{\mathbf{c}} P(\mathbf{w}, \mathbf{c}), \tag{5.30}$$

which will require summing over k^n values of $\mathbf{c}$ for an HMM with k states (since each of the n latent variables can take on k values). Fortunately, this can be simplified significantly because of the dependency structure of the variables involved.

The key is to define a simple recursive computation that allows us to make use of previously stored results. First, from the graphical model shown in figure 5.10c, we observe that $P(\mathbf{w})$ can also be written as

$$P(\mathbf{w}) = \sum_{c_n} P(w_n|c_n)P(c_n, \mathbf{w}_{n-1}), \tag{5.31}$$

where $\mathbf{w}_{n-1} = (w_1, \ldots, w_{n-1})$. This is easy to compute, provided that we can calculate $P(c_n, \mathbf{w}_{n-1})$. We can then observe that

$$P(c_n, \mathbf{w}_{n-1}) = \sum_{c_{n-1}} P(c_n|c_{n-1})P(w_{n-1}|c_{n-1})P(c_{n-1}, \mathbf{w}_{n-2}), \tag{5.32}$$

which is easy to compute if we know $P(c_{n-1}, \mathbf{w}_{n-2})$. This sets up a recursion that terminates at $P(c_1)$. Since we know $P(c_1)$, which is just the probability distribution over the first state, we can start there and compute $P(c_i, \mathbf{w}_{i-1})$ for $i = 2, \ldots, n$, and then apply equation (5.31). The resulting algorithm requires only n sums over k states, and it is thus much more efficient than summing over all states.

This procedure for computing $P(c_i, \mathbf{w}_{i-1})$ is known as the *forward procedure*. There is a corresponding procedure for computing $P(w_i, \ldots, w_n|c_i)$, which is known as the *backward procedure*. The basic idea is that

$$P(w_i, \ldots, w_n|c_i) = P(w_i|c_i) \sum_{c_{i+1}} P(w_{i+1}, \ldots, w_n|c_{i+1})P(c_{i+1}|c_i), \tag{5.33}$$

which establishes a similar kind of recursion. Here, the recursion grounds out at $P(w_n|c_n)$, which is also known. The backward procedure can also be used to calculate $P(\mathbf{w})$ since we can observe that $P(\mathbf{w}) = \sum_{c_1} P(\mathbf{w}|c_1)P(c_1)$.

Inferring a State Sequence Assume that we want to find the state sequence $\mathbf{c}$ that was used in generating a sentence $\mathbf{w}$. A natural way to formulate this problem is in terms of maximizing the posterior probability $P(\mathbf{c}|\mathbf{w})$. Naively, finding this sequence would require considering all k^n possibilities. However, the same kind of approach can yield an efficient procedure for identifying state sequences as well. If we want to find the posterior probability of a single state, c_i, given sentence $\mathbf{w}$, we can observe that $P(c_i|\mathbf{w}_i) \propto P(w_1, \ldots, w_{i-1}, c_i)P(w_i, \ldots, w_n|c_i)$, and use the forward and backward procedures to compute the probabilities that appear on the right side. However, finding the most likely sequence is a little more involved.

First, we observe that the $\mathbf{c}$ that maximizes $P(\mathbf{c}|\mathbf{w})$ also maximizes $P(\mathbf{w}, \mathbf{c})$. We thus need only maximize the latter. When computing $P(\mathbf{w})$, we needed to sum over $\mathbf{c}$, but now we want to maximize over $\mathbf{c}$. Fortunately, a similar recursion applies. We might hope that if we knew what $\mathbf{c}_{n-1}$ maximized $P(\mathbf{w}_{n-1}, \mathbf{c}_{n-1})$, we could use that information to find the $\mathbf{c}$ that maximizes $P(\mathbf{w}, \mathbf{c})$. However, it is a little more complicated than that. If we actually write the joint probability using this recursion, we see

$$P(\mathbf{w}, \mathbf{c}) = P(w_n|c_n)P(c_n|c_{n-1})P(\mathbf{w}_{n-1}, \mathbf{c}_{n-1}), \tag{5.34}$$

which makes it clear that the most likely $\mathbf{c}_{n-1}$ for $\mathbf{w}_{n-1}$ could have lower $P(c_n|c_{n-1})$ and $P(w_n|c_n)$ than another $\mathbf{c}_{n-1}$ and thus not maximize $P(\mathbf{w}, \mathbf{c})$. However, since the only part of $\mathbf{c}_{n-1}$ that is relevant to these new terms is c_{n-1}, this does suggest an algorithm: for every c_{n-1}, find the $\mathbf{c}_{n-2}$ that maximizes $P(\mathbf{w}_{n-1}, \mathbf{c}_{n-1})$, giving the most likely sequence of states that ends in c_{n-1}. The most likely sequence of states that ends in c_n must use one of these sequences of states. The algorithm that uses this observation is known as the *Viterbi algorithm* (Viterbi, 1967).

5.5.2 Estimating Parameters

While the preceding analysis assumes that the parameters of the HMM are known, it is more common that these parameters need to be estimated from data. This is often a key part of applying HMMs to problems in cognitive science and computational linguistics. Fortunately, the fact that probabilistic inference is straightforward in these models means that parameter estimation is also relatively easy.

Since the HMM involves latent variables, maximum-likelihood (or MAP) estimation can be done using the EM algorithm. The E-step takes the expectation of the complete log-likelihood over the posterior distribution on the latent classes. The distributions involved in this HMM are all discrete distributions and are thus estimated from the frequency with which events involving $\mathbf{c}$ and $\mathbf{w}$ occur. The transition probabilities are estimated from the frequency of transitions between states, and the emission probabilities are estimated from the frequency with which words are produced from a given state. When the expectation over the posterior distribution on the latent classes is taken, these frequencies are replaced by the expected number of transitions and the expected number of emissions, respectively.

Consequently, the only challenge in estimating the parameters of the model is computing these expectations.

The posterior probability that c_i is one state and c_{i+1} is another is given by

$$P(c_i, c_{i+1}|\mathbf{w}) \propto P(c_i, w_1, \ldots, w_{i-1})P(w_i|c_i)P(c_{i+1}|c_i)P(w_{i+1}, \ldots, w_n|c_{i+1}), \qquad (5.35)$$

where the first and last terms on the right side are given by the forward and backward procedures, respectively. The posterior probability that c_i is a particular state, needed to calculate the expected number of emissions, is computed directly from the forward and backward probabilities, as noted previously. Summing these quantities over all i yields the expected number of transitions and emissions required for either maximum-likelihood or MAP estimation of the parameters of the model. The resulting algorithm is known as the *Baum-Welch* algorithm or the *forward-backward* algorithm (for a more detailed tutorial, see Rabiner, 1989).

5.5.3 An Example: Changepoint Detection

While language is one of the prime applications of hidden Markov models, they also provide a simple tool for modeling other kinds of events that unfold over time. If the dynamics of a sequence of events can be captured by assuming that there are a fixed set of possible latent states and transitions take place between those states, then an HMM is an appropriate model. A simple example of a scenario fitting this description is a problem known in statistics as *changepoint detection*.

The most basic assumption used in defining a statistical model is that events are independent and identically distributed—that each event is drawn independently from the same distribution, with the distribution undergoing no change over time. However, this is a poor description of many real-world phenomena. In particular, it is possible for sudden changes to take place that completely change the distribution from which events are drawn. For example, the amount of road traffic increases significantly between 4 and 7p.m., so assuming that traffic levels are identically distributed throughout the day could cause significant problems for planning your evening. Consequently, a better assumption might be that the distribution from which events are drawn can change over time, perhaps moving between a few discrete states. This is exactly the assumption that is made in an HMM.

Brown and Steyvers (2009) explored the human capacity to detect and predict changes using a simple task in which people observed events taking place in a factory that produced cans of tomatoes. Tomato cans could fall from one of four chutes, and they tended to repeatedly fall from the same chute but sometimes switched chutes (see figure 5.11). Cans falling from each chute followed a Gaussian distribution. The task of inferring which chute produced a can and predicting which chute would be used for the next can can be optimally solved using an HMM, illustrating that these models can be applied to continuous observations (can locations), as well as discrete observations (words).

5.6 The Bayes-Kalman Filter and Linear Dynamical Systems

So far, we have focused on dynamical models for discrete random variables. However, the same approach can be applied to continuous variables. Defining dynamical models for continuous variables can be challenging because there are many ways in which continuous

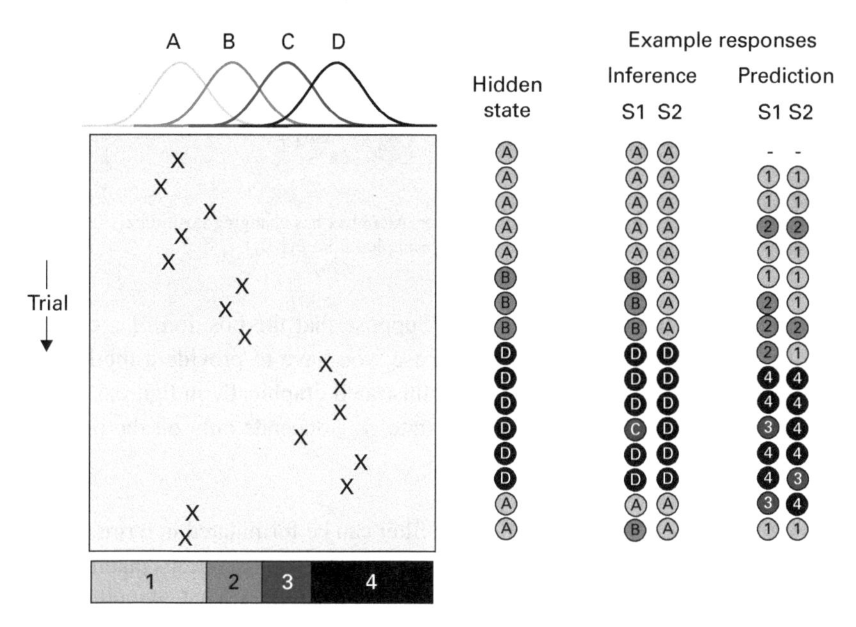

Figure 5.11
A changepoint detection task used by Brown and Steyvers (2009). In this experiment, people (here represented by two participants, S1 and S2) had to infer which of four chutes was releasing tomato cans onto a conveyer belt based on the past pattern of the locations of tomato cans ("Inference" responses) and predict which chute the next can would come from ("Prediction" responses). Here, the can locations are shown with X marks, and the big rectangle represents the conveyer belt. Each chute was associated with a Gaussian distribution over locations, and there was a small probability that there would be a change in chutes between cans. The dynamics of this system are well described by an HMM, which can be used to make optimal inferences and predictions in this task. Figure reproduced with permission from Brown and Steyvers (2009).

dynamics can be expressed. Perhaps the simplest assumption is that of linear dynamics—an assumption that leads to the model known as the *Bayes-Kalman filter* (or just *Kalman filter*).

A Generative Model for Dynamic Sequences The Bayes-Kalman filter was originally developed in the 1950s and early 1960s for tracking aircraft and spacecraft (Kalman, 1960). It is a technique commonly used in prediction and control. But it has also been applied to many problems in cognitive psychology, including classical conditioning (Daw, Courville, & Dayan, 2008) and causal learning (Lu, Rojas, Beckers, & Yuille, 2016). It was applied to vision to model human perception of a moving target dot over time in the presence of randomly moving background dots (Yuille, Burgi, & Grzywacz, 1998). It is applicable to any problem where you want to estimate a state z_t that changes over n discrete time steps t and you have some noisy observations x_t.

For the sake of concreteness, we will focus on the traditional task of estimating the position of an object over time, but the formulation presented here generalizes naturally to these other cases. Suppose that you want to estimate the object's position z and you have observations $x_1, \dots, x_n$ drawn from a distribution $p(x|z)$. If these observations are independent, and the object is static, then you can estimate z from them by maximum likelihood $z^*_{ML} = \arg\max_z \prod_{t=1}^n p(x_t|z)$. If you have prior knowledge $p(z)$ about the position of the object, then a better estimate is the MAP value $z^*_{MAP} = \arg\max_z p(z) \prod_{t=1}^n p(x_t|z)$.

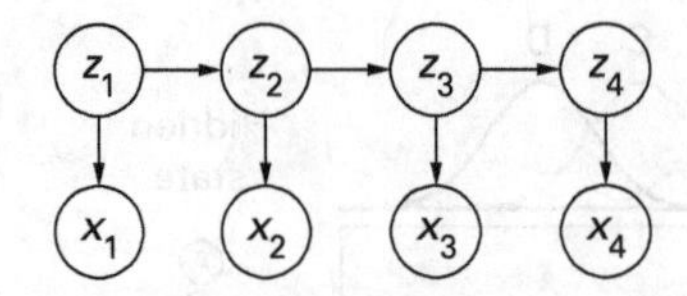

Figure 5.12
The graphical model assumed by the Bayes-Kalman filter. An object has changing position $z_1, \ldots, z_n$, generated by a distribution $p(z_{t+1}|z_t)$, and we have observations x_t sampled from $p(x_t|z_t)$.

But what happens if the object is moving? Suppose that the position of z changes with time, so it is a sequence $z_1, \ldots, z_n$. In this case, you have to provide a model $p(z_{t+1}|z_t)$ for how the object is moving. This model is illustrated graphically in figure 5.12. It makes conditional independence assumptions that state z_{t+1} depends only on the previous time state z_t.

Computing the Updates The Bayes-Kalman filter can be formulated in terms of an update of the distribution $p(z_t|\mathbf{x}_t)$, where $\mathbf{x}_t = (x_1, \ldots, x_t)$ is all the observations taken up to time t. We update this distribution to $p(z_{t+1}|\mathbf{x}_{t+1})$ in two stages. First, we calculate the probability $p(z_{t+1}|\mathbf{x}_t)$ of the position z_{t+1} of the object at time $t+1$ (i.e., where the object will be in the future). This is given by

$$p(z_{t+1}|\mathbf{x}_t) = \int p(z_{t+1}|z_t)p(z_t|\mathbf{x}_t)\, dz_t. \tag{5.36}$$

Next, we must take into account the new measurement x_{t+1} using $p(z_{t+1}|\mathbf{x}_t)$ as the new "prior." This uses Bayes' rule (and the fact that $\mathbf{x}_{t+1} = (\mathbf{x}_t, x_{t+1})$) to obtain

$$p(z_{t+1}|\mathbf{x}_{t+1}) = \frac{p(x_{t+1}|z_{t+1})p(z_{t+1}|\mathbf{x}_t)}{p(x_{t+1}|\mathbf{x}_t)}. \tag{5.37}$$

The Bayes-Kalman filter is characterized by equations (5.36) and (5.37), which *predict* z_{t+1} and then *correct* this prediction using the new measurement x_{t+1}, respectively. The filter is initialized by setting $p(z_1|x_1) = p(x_1|z_1)p(z_1)/p(x_1)$, where $p(z_1)$ is the prior for the original position of the object at the start of the sequence.

Observe that the Bayes-Kalman filter is really a procedure for updating the probability distribution $p(z_t|\mathbf{x}_t)$ representing the current beliefs about the location of the object. To estimate an exact position for the object, we need to estimate z_t from $p(z_t|\mathbf{x}_t)$. In the Bayes-Kalman filter, this can be done analytically because the prior $p(z_1)$, the dynamics $p(z_{t+1}|z_t)$, and the observation model $p(x_t|z_t)$ are all Gaussian distributions. In this case, $p(z_t|x_t)$, $p(z_{t+1}|\mathbf{x}_t)$, and $p(z_{t+1}|\mathbf{x}_{t+1})$ are also all Gaussian distributions (this shouldn't be a surprise—it happens because the priors and likelihoods for all the inferences that we need to make are conjugate).

The One-Dimensional Case Using $\mathcal{N}(\mu, \sigma^2)$ to denote the Gaussian distribution with mean μ and standard deviation σ, we can write the simplest 1D model as

$$p(x_t|z_t) = \mathcal{N}(x_t, \sigma_m^2),\ p(z_{t+1}|z_t) = \mathcal{N}(z_t + v, \sigma_p^2),\ p(z_1) = \mathcal{N}(\mu_1, \sigma_1^2), \tag{5.38}$$

where ν is the mean distance traveled by the object from t to $t+1$. If we write the distribution $p(z_t|\mathbf{x}_t)$ as a Gaussian with mean μ_t and standard deviation σ_t (i.e., $\mathcal{N}(\mu_t, \sigma_t^2)$) then we can reexpress the prediction and correction update equations (equations (5.36) and (5.37)) as

$$p(z_{t+1}|\mathbf{x}_t) = \mathcal{N}(\mu_t + \nu, \sigma_p^2 + \sigma_t^2) \tag{5.39}$$

$$p(z_{t+1}|\mathbf{x}_{t+1}) = \mathcal{N}(\mu_{t+1}, \sigma_{t+1}^2), \tag{5.40}$$

where

$$\mu_{t+1} = \mu_t + \nu - \frac{(\sigma_p^2 + \sigma_t^2)[(\mu_t + \nu) - x_{t+1}]}{\sigma_m^2 + (\sigma_p^2 + \sigma_t^2)},$$

$$\sigma_{t+1}^2 = \frac{\sigma_m^2(\sigma_p^2 + \sigma_t^2)}{\sigma_m^2 + (\sigma_p^2 + \sigma_t^2)}. \tag{5.41}$$

The update for μ_{t+1} includes a prediction part $(\mu_t + \nu)$ and a correction part (the rest). Observe that if the object is at the mean predicted position (i.e., $x_{t+1} = \mu_t + \nu$), then the correction part disappears. Also, note that the update combines the various sources of information—the observation x_{t+1} and the mean estimated position $\mu_t + \nu$ by a linear weighted average similar to that seen for other Gaussian models that we have considered.

We can get a better understanding of the Bayes-Kalman filter by considering special cases. If the observations are noiseless (i.e., $\sigma_m^2 = 0$), then it follows that $\mu_{t+1} = x_{t+1}$, so we should forget the history and just use the current observation as our estimate of z_{t+1}. If $\sigma_p^2 = 0$, then we have perfect prediction, and so $\mu_{t+1} = \frac{\sigma_t^2}{\sigma_m^2+\sigma_t^2} x_{t+1} + \frac{\sigma_m^2}{\sigma_m^2+\sigma_t^2}(\mu_t + \nu)$ with $\sigma_{t+1}^2 = \frac{\sigma_m^2 \sigma_t^2}{\sigma_m^2+\sigma_t^2}$, which corresponds to taking the weighted average of x_{t+1} with $\mu_t + \nu$. If we also require that $\nu = 0$ (i.e., the object does not move), then we obtain $\mu_{t+1} = \frac{\sigma_t^2}{\sigma_m^2+\sigma_t^2} x_{t+1} + \frac{\sigma_m^2}{\sigma_m^2+\sigma_t^2} \mu_t$, which is simply an online method for computing the MAP estimate of a static object at position x (as described in our first model earlier in this chapter).

5.6.1 An Example: Multiple Object Tracking

A common paradigm that is used in the psychophysical literature on visual attention is the Multiple Object Tracking (MOT) task (see figure 5.13). In this task, people are shown an array of objects (typically represented as geometric shapes such as circles) and a subset of the objects are indicated as the ones to pay attention to, either by flashing or being highlighted with a color. Then all the objects begin to move, and the participant has to try to keep track of which were the indicated objects. After a period of motion, the participant is asked to click on the objects that they think they were supposed to be tracking.

Research using the MOT task has revealed a number of interesting properties of human performance: people can track only a limited number of objects, but the exact number depends on other aspects of the task, such as the speed and spacing of the objects. A natural question to ask is whether these limitations are an intrinsic consequence of the structure of the computational problem being solved, or whether they reflect a fundamental limitation of the human visual system.

Vul, Alvarez, Tenenbaum, and Black (2009) set out to answer this question, formulating a probabilistic model of the MOT task that can be used to evaluate optimal performance.

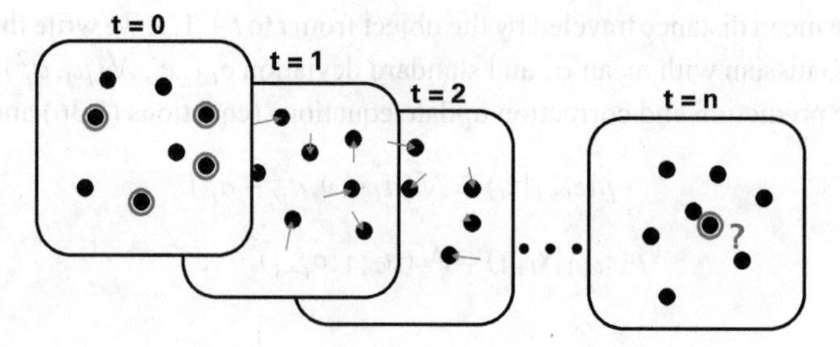

Figure 5.13
The Multiple Object Tracking (MOT) task. A set of objects are shown on the screen, and a subset are highlighted as targets to be tracked. After the objects move randomly for a while, participants have to indicate the ones that they were supposed to be tracking. Figure reproduced with permission from Vul et al. (2009).

If the goal is to model the motion of a single object, then the problem can be formulated in the terms outlined here—a linear dynamical system, which can be solved using the Bayes-Kalman filter. But with multiple objects, there is a new challenge: if two objects get close to one another, it is plausible that they could be confused for one another. To address this, Vul et al. (2009) introduced another variable into their model that identifies which object is which. Successfully solving the MOT problem requires maintaining correct assignments for the objects that are supposed to be tracked. The motion of the objects is modeled using a separate linear dynamical system for each object, and the assignment variable links these linear dynamical systems to the corresponding objects on the screen. If the assignment of objects to linear dynamical systems is known, inference in each system simply reduces to the Bayes-Kalman filter. Exploiting this, Vul et al. (2009) used a sampling algorithm to maintain estimates of the assignments, applying the Bayes-Kalman filter separately with each sample (for more details of this kind of approach, see the next chapter).

Vul et al. (2009) found that some of the basic results from the MOT literature could be captured by this model: easier tracking of objects moving at lower speeds or with greater separation is just a consequence of the statistical structure of the problem. However, this approach doesn't explain why people find it harder to track more objects. For example, people find it easier to track 4 out of 16 objects than 8 out of 16 objects, even though from the model's perspective both cases require modeling the dynamics of 16 objects. Vul et al. (2009) suggested that this phenomenon must reflect an additional constraint on the visual system, which they were able to capture effectively by assuming that people experience more uncertainty about the location of objects when they are asked to track more objects.

5.7 Combining Probabilistic Models

One of the strengths of probabilistic generative models is that they are easy to combine. Since a generative model is just a procedure for generating data, we can make one step in that procedure another generative model. For example, we can define a Markov model whose emission probabilities are mixture distributions, or a mixture model whose component distributions are Markov processes. This ability to compose generative models in various ways creates the capacity to define richly structured probabilistic models that can be applied to increasingly complex problems.

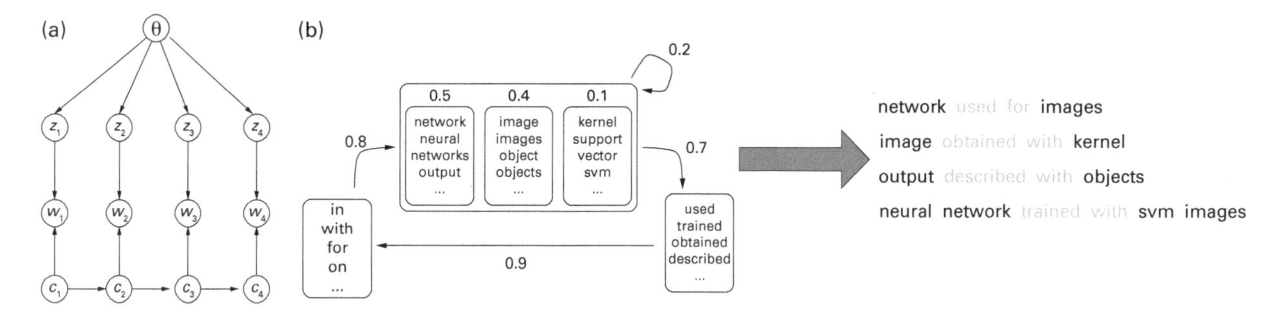

Figure 5.14
The LDA-HMM model combines an HMM with a topic model. (a) Graphical model, with θ being the mixture weights of different topics in a document, z_i being a topic, c_i a syntactic class, and w_i a word. (b) The distribution over words given classes and topics is a standard HMM, with one special state in which the distribution over words comes from a document-specific mixture of topics. As a result, words are naturally factorized into two kinds—those generated by the HMM "syntax" and those generated by the topic model "semantics." Figure reproduced with permission from Griffiths et al. (2004).

BLOOD	FOREST	FARMERS	GOVERNMENT	LIGHT	WATER	STORY	DRUGS	BALL
HEART	TREES	LAND	STATE	EYE	MATTER	STORIES	DRUG	GAME
PRESSURE	FORESTS	CROPS	FEDERAL	LENS	MOLECULES	POEM	ALCOHOL	TEAM
BODY	LAND	FARM	PUBLIC	IMAGE	LIQUID	CHARACTERS	PEOPLE	*
LUNGS	SOIL	FOOD	LOCAL	MIRROR	PARTICLES	POETRY	DRINKING	BASEBALL
OXYGEN	AREAS	PEOPLE	ACT	EYES	GAS	CHARACTER	PERSON	PLAYERS
VESSELS	PARK	FARMING	STATES	GLASS	SOLID	AUTHOR	EFFECTS	FOOTBALL
ARTERIES	WILDLIFE	WHEAT	NATIONAL	OBJECT	SUBSTANCE	POEMS	MARIJUANA	PLAYER
*	AREA	FARMS	LAWS	OBJECTS	TEMPERATURE	LIFE	BODY	FIELD
BREATHING	RAIN	CORN	DEPARTMENT	LENSES	CHANGES	POET	USE	BASKETBALL
THE	IN	HE	*	BE	SAID	CAN	TIME	,
A	FOR	IT	NEW	HAVE	MADE	WOULD	WAY	;
HIS	TO	YOU	OTHER	SEE	USED	WILL	YEARS	(
THIS	ON	THEY	FIRST	MAKE	CAME	COULD	DAY	:
THEIR	WITH	I	SAME	DO	WENT	MAY	PART	)
THESE	AT	SHE	GREAT	KNOW	FOUND	HAD	NUMBER	
YOUR	BY	WE	GOOD	GET	CALLED	MUST	KIND	
HER	FROM	THERE	SMALL	GO		DO	PLACE	
MY	AS	THIS	LITTLE	TAKE		HAVE		
SOME	INTO	WHO	OLD	FIND		DID		

Figure 5.15
Results of applying the LDA-HMM to the TASA corpus. The top part of the table shows topics from the LDA model, comprised of words that were inferred to be content words in the various documents. The bottom part shows syntactic classes corresponding to the states of the HMM, comprised of words that were inferred to be function words. Each column shows a separate distribution over words, ordered in decreasing probability. Figure reproduced with permission from Griffiths et al. (2004).

5.7.1 An Example: Combining Syntax and Semantics

We have discussed two simple models of language in this chapter: HMMs, which capture a rudimentary form of syntax, and topic models, which capture a rudimentary form of semantics. Since both are expressed as generative models, we can combine them into a single model that incorporates both "syntax" and "semantics" and makes it possible to identify the different roles that words play in documents.

Griffiths, Steyvers, Blei, and Tenenbaum (2005) defined a probabilistic generative model for text that combines latent Dirichlet allocation and a hidden Markov model—appropriately called the LDA-HMM (see figure 5.14). This *composite model* assumes that a sequence of words is generated by an HMM, but that one state in that HMM is a distribution over words that depends on the particular document. The emission probabilities for that state are determined by a topic model, where the weight given to each topic depends on the document in exactly the way specified in the latent Dirichlet allocation model introduced previously.

The LDA-HMM makes it possible to capture the two kinds of roles that words can play in documents—function words that make up part of the standard structure of sentences, regardless of the documents in which they appear, and content words that vary across documents. In the model, each word that appears in the document appears in one of these roles—either it comes from the general HMM states or from the special document-dependent state. In practice, this model "factorizes" the words that appear in sentences into these two parts, and makes it possible to simultaneously estimate both semantic representations and syntactic classes within a single model (see figure 5.15).

5.8 Summary

In chapter 4, we showed how graphical models provide a way to express the independence assumptions that make probabilistic inference tractable. In this chapter, we have explored how we can define increasingly complex probabilistic models while continuing to exploit conditional independence to make it possible to estimate parameters and perform inference. In particular, by postulating latent variables that play a role in generating the observed data, we can define models that begin to capture the richer structure that underlies the observable world. By making careful choices about how those latent variables relate to one another and to observed data, we can define expressive models that nonetheless remain tractable. Mixture models, topic models, HMMs, and the Bayes-Kalman filter all extend the kinds of observed data that we can hope to understand via Bayesian inference. However, even with these models—and increasingly as we try to go beyond them—we can run into challenges that make exact probabilistic inference difficult. In the next chapter, we consider methods of addressing these challenges.

6

Approximate Probabilistic Inference

Thomas L. Griffiths and Adam N. Sanborn

The probability distributions that we need to evaluate when applying Bayesian inference can quickly become very complicated. Graphical models provide some tools for speeding up probabilistic inference, but these tools tend to work best when most variables are directly dependent on a relatively small number of other variables. There are still many cases where calculating posterior distributions or conditional probabilities is a computational challenge. For example, how can we perform Bayesian inference over continuous variables when we cannot evaluate the integral required to calculate the denominator of Bayes' rule? Alternatively, how can we compute posterior distributions over large discrete hypothesis spaces, such as the space of all directed graphs defined on a set of variables?

The computational complexity of probabilistic inference leads to two kinds of problems. The first is an engineering problem: how can we as modelers calculate the predictions of our probabilistic models? The second is a reverse-engineering problem: how could human behavior be consistent with the predictions of these models if probabilistic inference is so hard to do? In this chapter, we will present methods for approximating probabilistic computations that are relevant to both of these problems, but we will focus on the engineering problem. We will examine to the reverse-engineering problem in detail in chapter 11.

There are various approaches to approximating probabilistic inference, including generic methods for performing numerical integration or optimization and methods developed specifically for probabilistic inference, such as loopy belief propagation (Pearl, 1988) and variational inference (Jordan, Ghahramani, Jaakkola, & Saul, 1999). All these methods are relevant to solving the engineering problem of working with probabilistic models, and many of them are likely to shed light on how human minds and brains might perform probabilistic inference. Our focus in this chapter will be on *Monte Carlo methods*, in which probabilistic computations are approximated by replacing probability distributions with samples from those distributions. Monte Carlo methods are one of the basic tools for approximate inference, and they can be adapted to work with a wide range of probabilistic models. They can also provide a clear source of hypotheses about how human minds and brains might deal with the severe computational challenges involved in probabilistic inference. At the end of the chapter, we will briefly discuss variational inference, which has become increasingly popular in the machine learning community with the development of automated tools for optimization. We recommend Neal (1993), Mackay (1998), and Robert and Casella (1999) for more details on Monte Carlo methods, and Blei, Kucukelbir, and McAuliffe (2017) as an introduction to variational inference.

6.1 Simple Monte Carlo

The basic idea behind Monte Carlo methods is to represent a probability distribution by a set of samples from that distribution. Those samples provide an indication of which values have high probability (since high probability values are more likely to be produced as samples), and they can be used in place of the distribution itself when performing various computations. When working with probabilistic models of cognition, we are typically interested in understanding the posterior distribution over a parameterized model—such as a causal network with its causal strength parameters—or over a class of models—such as the space of all causal network structures on a set of variables or all taxonomic tree structures on a set of objects. Samples from the posterior distribution can be useful in discovering the best parameter values for a model or the best models in a model class, and for estimating how concentrated the posterior is on those best hypotheses (i.e., how confident a learner should be in those hypotheses).

Sampling can also be used to approximate averages over the posterior distribution. For example, in computing the posterior probability of a parameterized model given data, it is necessary to compute the model's marginal likelihood, or the average probability of the data over all parameter settings of the model. Averaging over all parameter settings is also necessary for ideal Bayesian prediction about future data points (as in computing the posterior predictive distribution for a weighted coin). Finally, we could be interested in averaging over a space of model structures, making predictions about model features that are likely to hold regardless of which structure is correct. For example, we could estimate how likely it is that one variable A causes another variable B in a complex causal network of unknown structure by computing the probability that a link $A \to B$ exists in a high-probability sample from the posterior over network structures (e.g., Friedman & Koller, 2000).

Monte Carlo methods were originally developed primarily for approximating these sophisticated averages—that is, approximating a sum over all of the values taken on by a random variable with a sum over a random sample of those values. Assume that we want to evaluate the average (also called the *expected value*) of a function $f(\mathbf{x})$ over a probability distribution $p(\mathbf{x})$ defined on a set of k random variables taking on values $\mathbf{x} = (x_1, x_2, \ldots, x_k)$. This can be done by taking the integral of $f(\mathbf{x})$ over all values of $\mathbf{x}$, weighted by their probability $p(\mathbf{x})$, with

$$E_{p(\mathbf{x})}\left[f(\mathbf{x})\right] = \int f(\mathbf{x})p(\mathbf{x})\,d\mathbf{x}, \tag{6.1}$$

where $E[\cdot]$ denotes the expectation. For convenience, we will denote this expected value μ, with $\mu = E_{p(\mathbf{x})}\left[f(\mathbf{x})\right]$.

Monte Carlo provides an alternative to explicitly evaluating an expectation. The *law of large numbers*, a standard result in probability theory, indicates that the average of a set of samples from a probability distribution will approximate the expectation with respect to that distribution increasingly well as the number of samples increases. This is exactly what we do in Monte Carlo, giving us the approximation

$$\int f(\mathbf{x})p(\mathbf{x})\,d\mathbf{x} \approx \frac{1}{m}\sum_{i=1}^{m} f(\mathbf{x}^{(i)}) = \hat{\mu}_{MC}, \tag{6.2}$$

where the $\mathbf{x}^{(i)}$ are a set of m samples from the distribution $p(\mathbf{x})$. This procedure—replacing $p(\mathbf{x})$ with samples from $p(\mathbf{x})$—is called *simple Monte Carlo.*

The estimator $\hat{\mu}_{MC}$ has several desirable properties. Under weak conditions on the functions $f(\mathbf{x})$ and $p(\mathbf{x})$, it is *consistent*, with $(\hat{\mu}_{MC} - \mu) \to 0$ almost surely as $m \to \infty$. It is also *unbiased*, with $E\left[\hat{\mu}_{MC}\right] = \mu$. Finally, it is *asymptotically normal*, with

$$\sqrt{m}(\hat{\mu}_{MC} - \mu) \to \mathcal{N}(0, \sigma_{MC}^2), \tag{6.3}$$

in distribution, where $\mathcal{N}(0, \sigma_{MC}^2)$ is the Gaussian with mean 0 and variance $\sigma_{MC}^2 = E_{p(\mathbf{x})}\left[(f(\mathbf{x}) - \mu)^2\right]$. These properties make simple Monte Carlo the method of choice in cases where it is straightforward to sample from $p(\mathbf{x})$ and the fluctuations in $f(\mathbf{x})$ are not correlated with $p(\mathbf{x})$.

6.1.1 An Example: Computing Causal Support

To show how the Monte Carlo approach to approximate numerical integration is useful for evaluating Bayesian models, recall the model of causal structure-learning introduced in chapter 4, causal support (Griffiths & Tenenbaum, 2005). To compute the evidence that a set of contingencies d provides in favor of a causal relationship, we needed to evaluate the integral

$$P(d|\text{Graph 1}) = \int_0^1 \int_0^1 P_1(d|w_0, w_1, \text{Graph 1})\, P(w_0, w_1|\text{Graph 1})\, dw_0\, dw_1, \tag{6.4}$$

where $P(w_0, w_1|\text{Graph 1})$ is assumed to be uniform over all values of w_0 and w_1 between 0 and 1 and $P_1(d|w_0, w_1, \text{Graph 1})$ is derived from the noisy-OR parameterization

$$P_1(e^+|b, c; w_0, w_1) = 1 - (1 - w_0)^b(1 - w_1)^c, \tag{6.5}$$

with e^+ being the occurrence of the effect, and b and c indicating the presence (1) or absence (0) of the background b and cause c on a given trial. If we view $P_1(d|w_0, w_1, \text{Graph 1})$ simply as a function of w_0 and w_1, it is clear that we can approximate this integral using Monte Carlo. The analog of equation (6.2) is

$$P(d|\text{Graph 1}) \approx \frac{1}{m}\sum_{i=1}^{m} P_1(d|w_0^{(i)}, w_1^{(i)}, \text{Graph 1}), \tag{6.6}$$

where the $w_0^{(i)}$ and $w_1^{(i)}$ are m samples from the distribution $P(w_0, w_1|\text{Graph 1})$. This is the approach to computing causal support taken in Griffiths and Tenenbaum (2005).

6.2 When Does Simple Monte Carlo Fail?

One limitation of the simple Monte Carlo method is that it is not easy to automatically generate samples from most probability distributions. Using the Monte Carlo method requires that we are able to generate samples from distribution $p(\mathbf{x})$. However, this is often not the case. For distributions like the uniform, Gaussian, exponential, Poisson, gamma, and beta, there are straightforward algorithms for generating random samples, but these are particularly well behaved distributions (in fact, they are all exponential family distributions, as introduced in chapter 3). Other distributions can require ingenuity to sample from, and developing a method for sampling from such distributions can be a research topic in its own right. When we want to use Monte Carlo to sample from posterior distributions computed by applying Bayes' rule, we often run into distributions with a very specific problem that makes it hard to apply simple Monte Carlo methods.

When dealing with posterior distributions, we can end up in a situation where we don't actually know the probability of a particular hypothesis—we just know the value of that probability up to a constant. If we have a prior distribution $P(h)$, a likelihood function $p(d|h)$, and an observed data point d, the posterior distribution is given by

$$P(h|d) = \frac{p(d|h)P(h)}{\sum_{h'} p(d|h')P(h')}, \tag{6.7}$$

where the sum in the denominator ranges over all hypotheses. Since the denominator is what ensures that $P(h|d)$ is "normalized"—that it is a well-defined probability distribution, with the probabilities of all the hypotheses summing to 1—this sum is often referred to as the *normalizing constant*. With a large discrete hypothesis space—say, all partitions of a set of n objects or all directed graphs on n variables—calculating the normalizing constant can quickly become intractable. If it's easy to compute the numerator, we end up in a situation where we can calculate something that is *proportional* to $P(h|d)$, and we can compare the relative values of these probabilities for different hypotheses, but we know their actual probability only up to a constant.

The same issue arises when we apply Bayes' rule to continuous variables. In this case, we have

$$p(\theta|d) = \frac{p(d|\theta)p(\theta)}{\int p(d|\theta)(\theta)\, d\theta}, \tag{6.8}$$

where the integral in the denominator may not have an analytic solution. In this case, we can try approximating that integral numerically—something that becomes increasingly difficult as the dimensionality of θ increases—or we can rely on methods that require us to know $p(\theta|d)$ only up to a constant.

In these situations, simple Monte Carlo is not a viable solution—we don't even know the probability distribution we want to sample from, let alone how to sample from it. However, there are a number of more sophisticated Monte Carlo methods that can be applied even when we know a probability distribution only up to a constant.

6.3 Rejection Sampling

Assume that we have a distribution $p(\mathbf{x})$ that is difficult to sample from—perhaps because we know its value only up to a constant. If another distribution is close to $p(\mathbf{x})$ but easy to sample from, *rejection sampling* can often be used to generate samples from $p(\mathbf{x})$.

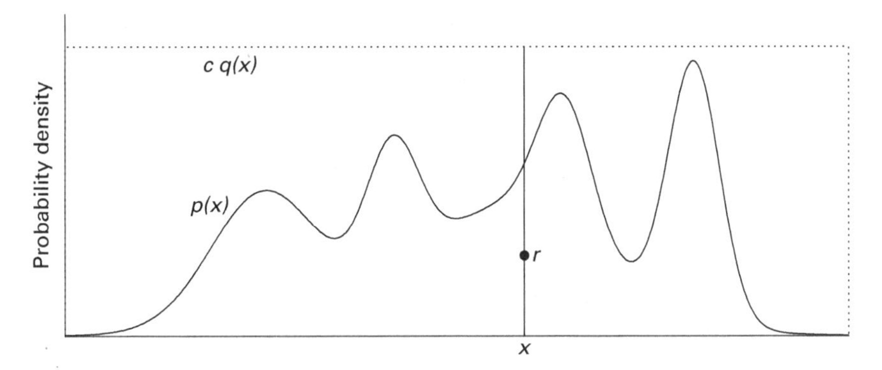

Figure 6.1
Rejection sampling. Here, our target distribution is $p(x)$ and we take the uniform distribution over the range of $p(x)$ as the distribution $q(x)$ from which we generate our initial samples x. We set c such that $p(x) < cq(x)$ for all x. We sample x from $q(x)$ and then r uniformly from $[0, c\,q(x)]$. If $r \leq p(x)$, as in this case, we keep x. Conditioned on keeping x, the probability of sampling a given x is just $p(x)$. The vertical axis here is arbitrary, so this procedure can be followed even if we know $p(x)$ only up to a constant.

Changing the Sampling Problem Say that there is a distribution $q(\mathbf{x})$ that we can sample from easily. We will call this the *proposal distribution*. Further, assume that we can compute the probabilities $p(\mathbf{x})$ up to a constant; that is, we can compute $p^*(\mathbf{x}) \propto p(\mathbf{x})$. Imagine, in addition, that we know a constant c such that $c\,q(\mathbf{x}) \geq p^*(\mathbf{x})$ for all $\mathbf{x}$. We can then generate a sample by performing the following steps:

1. Sample $\mathbf{x} \sim q(\mathbf{x})$,
2. Generate a random real number r uniformly drawn from the interval $[0, c\,q(\mathbf{x})]$,
3. If $r < p^*(\mathbf{x})$, then keep $\mathbf{x}$; otherwise, reject and start again at Step 1.

The result is a sample from $p(\mathbf{x})$.

Why does this work? It is easy to see this intuitively when $p(x)$ is a one-dimensional (1D) distribution (see figure 6.1, where $q(x)$ is taken to be the uniform distribution). The two random values, x and r, give the coordinates of a point in a two-dimensional (2D) plane. This point is drawn uniformly at random from a region that includes the $p^*(x)$ curve (in fact, it is the region under $cq(x)$). If we only keep points under the $p^*(x)$ curve, then we sample each value x with probability proportional to $p^*(x)$. Since $p^*(x) \propto p(x)$, this is equivalent to sampling from the distribution $p(x)$.

Bayesian Inference by Rejection Sampling Rejection sampling can be used to sample from posterior distributions, and hence to approximate Bayesian inference. We will start with the case where the data and hypotheses are discrete, so we want to sample from $P(h|d)$.

The first step is to construct a distribution to sample from. We can write the posterior as a joint distribution on hypothesis h and data d', $P^*(h, d') = P(h)P(d'|h)\delta(d', d)$, where $\delta(\cdot, \cdot)$ is 1 when its arguments match and 0 otherwise. This looks a little odd, but it amounts to the assumption that we are sampling hypothesis h from the prior, then sampling data d' from the likelihood $P(d'|h)$, but only keeping data that match the observed data d. This defines a distribution that puts all its mass on the observed data, while technically being a joint distribution on h and d'.

If we use the proposal distribution $Q(h,d') = P(h)P(d'|h)$, then it is clear that $P^*(h,d') \leq Q(h,d')$ because $\delta(\cdot,\cdot)$ only takes on the values 0 and 1. Consequently, $c = 1$. Furthermore, since $P^*(h,d')$ is equal to $Q^*(h,d')$ when $d' = d$ and is zero otherwise, we will reject a sample if and only if $d' \neq d$.

This provides a simple way to draw from the posterior distribution via rejection sampling: sample hypothesis h from the prior, then sample data d' from the likelihood $P(d'|h)$ associated with this hypothesis and reject the sampled hypothesis if d' does not match the observed data d. The probability of rejecting a hypothesis h is thus $P(d|h)$, and the hypotheses that we keep will be samples from the posterior distribution.

Extensions and Limitations This approach generalizes naturally to continuous hypotheses: we sample from the prior $p(\theta)$ and retain those samples with probability $P(d|\theta)$. However, it cannot be used when the data are continuous because the probability of those data becomes infinitesimally small. In this situation, researchers have used a related method known as *approximate Bayesian computation* (ABC; Beaumont, Zhang, & Balding, 2002), where a second function is used to define a measure of proximity among data sets. Samples are then drawn from the prior, used to simulate data, and retained if the simulated data are close to the observed data.

While rejection sampling is simple, it is often not efficient. Since we only retain each sample with probability $P(d|h)$, the proportion of the time that we retain samples is given by the expectation of $P(d|h)$ with respect to $P(h)$, which is $\sum_h P(d|h)P(h) = P(d)$. This approach will thus only be efficient when the probability of the observed data is high—typically meaning that the amount of observed data is small. However, rejection sampling is a good source of intuitions about approximating Bayesian inference and can be a starting point for defining a strategy for performing Bayesian inference in complex models.

6.4 Importance Sampling

Rejection sampling exactly generates samples from the target distribution $p(\mathbf{x})$, but it can be very inefficient. An alternative approach that starts from the same idea of generating samples from the wrong distribution and then correcting for this is known as *importance sampling.*

Changing the Sampling Problem Assume that we want to evaluate an expectation of a function $f(\mathbf{x})$ with respect to a probability distribution $p(\mathbf{x})$ but have another distribution $q(\mathbf{x})$ that is easier to sample from. Further, assume that $q(\mathbf{x}) > 0$ whenever $p(\mathbf{x}) > 0$.

We can introduce $q(\mathbf{x})$ into our expectation by multiplying and dividing by the same term, giving

$$E_{p(\mathbf{x})}[f(\mathbf{x})] = \int f(\mathbf{x})p(\mathbf{x})\,d\mathbf{x} = \frac{\int f(\mathbf{x})p(\mathbf{x})\,d\mathbf{x}}{\int p(\mathbf{x})\,d\mathbf{x}} = \frac{\int f(\mathbf{x})\frac{p(\mathbf{x})}{q(\mathbf{x})}q(\mathbf{x})\,d\mathbf{x}}{\int \frac{p(\mathbf{x})}{q(\mathbf{x})}q(\mathbf{x})\,d\mathbf{x}}. \tag{6.9}$$

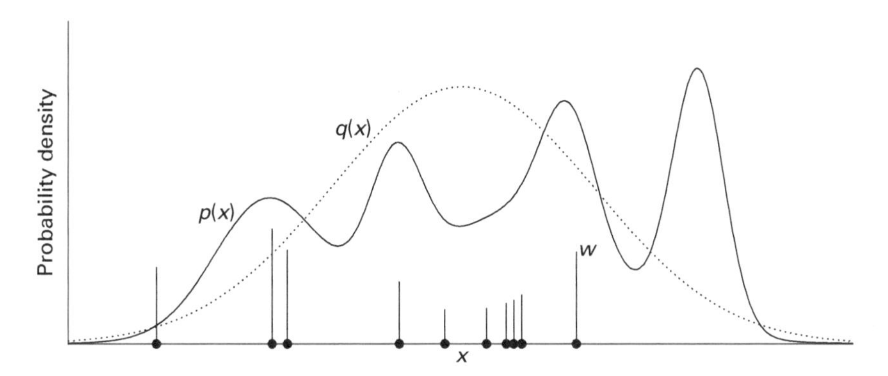

Figure 6.2
Importance sampling. We want to sample values of x from $p(x)$. Instead, we sample values of x from $q(x)$ and correct for the fact that we sampled from the wrong distribution by assigning each sampled x a weight w that is proportional to $\frac{p(x)}{q(x)}$. These weights are higher when $p(x) > q(x)$, corresponding to regions where we undersampled, and lower when $p(x) < q(x)$, corresponding to regions where we oversampled.

The numerator and denominator of the final expression are each expectations with respect to $q(\mathbf{x})$. Applying simple Monte Carlo to approximate these expectations, we obtain

$$E_{p(\mathbf{x})}\left[f(\mathbf{x})\right] \approx \frac{\frac{1}{m}\sum_{j=1}^{m} f(\mathbf{x}_j)\frac{p(\mathbf{x}_j)}{q(\mathbf{x}_j)}}{\frac{1}{m}\sum_{j=1}^{m}\frac{p(\mathbf{x}_j)}{q(\mathbf{x}_j)}}, \tag{6.10}$$

where each sample $\mathbf{x}_j$ is drawn from $q(\mathbf{x})$.

The ratios $\frac{p(\mathbf{x}_j)}{q(\mathbf{x}_j)}$ can be interpreted as "weights" on the sample $\mathbf{x}_j$, correcting for the fact that they were drawn from $q(\mathbf{x})$ rather than $p(\mathbf{x})$ (see figure 6.2). Samples that have higher probability under $p(\mathbf{x})$ than $q(\mathbf{x})$, and thus occur less often than they would if we were sampling from $p(\mathbf{x})$, receive greater weight. More formally, define the weight w_j to be

$$w_j = \frac{\frac{p(\mathbf{x}_j)}{q(\mathbf{x}_j)}}{\sum_{j=1}^{m}\frac{p(\mathbf{x}_j)}{q(\mathbf{x}_j)}}. \tag{6.11}$$

We can then write

$$E_{p(\mathbf{x})}\left[f(\mathbf{x})\right] \approx \sum_{j=1}^{m} f(\mathbf{x}_j)w_j = \hat{\mu}_{IS}. \tag{6.12}$$

Since the only role of $p(\mathbf{x})$ is in defining the weight w_j, importance sampling can still be used even if we know $p(\mathbf{x})$ only up to a constant. Because the weights are normalized, the unknown constant in $p(\mathbf{x})$ appears in both the numerator and the denominator and cancels out.

Analysis of Importance Sampling It is instructive to compare the properties of the the importance sampling estimator $\hat{\mu}_{IS}$ with that of simple Monte Carlo. Under similar assumptions about $f(\mathbf{x})$ and $p(\mathbf{x})$ (with the significant addition that $q(\mathbf{x}) > 0$ for all $\mathbf{x}$ such that $p(\mathbf{x}) > 0$), it is a consistent estimator of μ, with $(\hat{\mu}_{IS} - \mu) \to 0$ almost surely as $m \to \infty$. It is also asymptotically normal, with

$$\sqrt{m}(\hat{\mu}_{IS} - \mu) \rightarrow N(0, \sigma_{IS}^2) \tag{6.13}$$

in distribution, where $\sigma_{IS}^2 = E_{p(\mathbf{x})}\left[(f(\mathbf{x}) - E_{p(\mathbf{x})}[f(\mathbf{x})])^2 \frac{p(\mathbf{x})}{q(\mathbf{x})}\right]$. However, unlike simple Monte Carlo, $\hat{\mu}_{IS}$ is biased, with

$$\hat{\mu}_{IS} - \mu = \frac{1}{m}\left(E_{p(\mathbf{x})}[f(\mathbf{x})]\, E_{p(\mathbf{x})}\left[\frac{p(\mathbf{x})}{q(\mathbf{x})}\right] - E_{p(\mathbf{x})}\left[f(\mathbf{x})\frac{p(\mathbf{x})}{q(\mathbf{x})}\right]\right), \tag{6.14}$$

which goes to zero as $m \rightarrow \infty$, but it can be substantial for smaller values of m.

Despite being biased, there are several reasons why using an importance sampler can make sense even in cases where simple Monte Carlo can be applied. First, it allows a single set of samples to be used to evaluate expectations with respect to a range of distributions, through the use of different weights for each distribution. Second, the estimate produced by the importance sampler can have lower variance than the estimate produced by simple Monte Carlo. If the function $f(\mathbf{x})$ takes on its most extreme values in regions where $p(\mathbf{x})$ is small, the variance of the simple Monte Carlo estimate can be large. An importance sampler can have lower variance than simple Monte Carlo if $q(\mathbf{x})$ is chosen to be complementary to $f(\mathbf{x})$. In particular, the asymptotic variance of the sampler is minimized by specifying $q(\mathbf{x})$ as

$$q(\mathbf{x}) \propto |f(\mathbf{x}) - E_{p(\mathbf{x})}[f(\mathbf{x})]|\, p(\mathbf{x}). \tag{6.15}$$

This is not a practical procedure since finding this distribution requires computing $E_{p(\mathbf{x})}[f(\mathbf{x})]$, but the fact that the minimum variance sampler need not be $p(\mathbf{x})$ illustrates that importance sampling can sometimes provide a lower variance estimate of an expectation than simple Monte Carlo.

Bayesian Inference by Importance Sampling We can use importance sampling to perform Bayesian inference. The distribution that we want to approximate is the posterior, $p(h|d)$. We can do this by choosing another distribution on hypotheses $q(h)$ from which to generate samples, assigning to the resulting samples weights that are proportional to $p(h|d)/q(h)$. One attractive property of this approach is that since the weights are simply proportional to this ratio, we can equivalently assign weights that are proportional to $p(d|h)p(h)/q(h)$. This gives us a method for generating samples from the posterior distribution without having to calculate the normalizing constant—a significant reduction in the cost of computation.

The analysis of the bias and variance of importance sampling presented here makes it clear that this method is most effective when $q(x)$ and $p(x)$ are similar enough that the weights do not vary too much. Bayesian inference using importance sampling will thus be most effective if we can find a proposal distribution $q(h)$ that is close to the posterior distribution $p(h|d)$. In some cases—when the observed data have a small effect on an agent's beliefs, corresponding to a likelihood function $p(d|h)$ that does not take on extreme values—we can simply use the prior distribution $p(h)$ as our proposal distribution. In this case, the weights assigned to each sample are simply $p(d|h)p(h)/p(h) = p(d|h)$, the likelihood. The resulting algorithm for approximating Bayesian inference is extremely simple: generate samples from the prior $p(h)$, and assign those samples weights proportional to the likelihood $p(d|h)$. For obvious reasons, this algorithm is known as *likelihood weighting*.

6.4.1 An Example: Computing Spatial Coincidences

The Bayesian model of spatial coincidences proposed by Griffiths and Tenenbaum (2007a), discussed in chapter 3, requires computing a complicated integral. In general, this is the challenge of applying the Bayesian Occam's razor—calculating the marginal probability of all observed data for a particular model requires integrating over all parameters of that model. In the case of spatial coincidences, this means integrating over the parameters of a mixture model.

The model assumed in the spatial coincidences scenario is relatively simple: either bombs are falling at random (h_0) or some of them are falling around a target (h_1). If n_B bombs fall within some region $\mathcal{R}$ at random, then the probability of any set of bomb locations $\mathbf{x}$ is just $p(\mathbf{x}|h_0) = \frac{1}{|\mathcal{R}|^{n_B}}$. If all the bombs were falling around a target, that would also be relatively easy to assess—we would want to calculate the marginal probability of $\mathbf{x}$ under a Gaussian distribution, integrating over the mean $\boldsymbol{\mu}$ and covariance $\boldsymbol{\Sigma}$ of that Gaussian, $p(\mathbf{x}|h_1) = \int\int p(\mathbf{x}|\boldsymbol{\mu}, \boldsymbol{\Sigma})p(\boldsymbol{\mu})p(\boldsymbol{\Sigma})\, d\boldsymbol{\mu}\, d\boldsymbol{\Sigma}$. Even though this sounds intimidating, it is actually straightforward to do if we use conjugate priors for $\boldsymbol{\mu}$ and $\boldsymbol{\Sigma}$—in this case, a normal distribution on $\boldsymbol{\mu}$ (or even a uniform distribution) and an inverse-Wishart prior on $\boldsymbol{\Sigma}$ (see Minka, 2001, for details).

The tricky part arises when *some* of the bombs are targeted and others fall at random. Then we have a mixture model, where one component is uniform over $\mathcal{R}$ and the other is a Gaussian with unknown mean and variance. If we know which bombs were targeted, then the problem is still easy—we integrate the probability of the bombs that were targeted over $\boldsymbol{\mu}$ and $\boldsymbol{\Sigma}$, and the others follow a uniform distribution over $\mathcal{R}$. Letting $\mathbf{z}$ denote the vector of assignments of bombs to mixture components, we can thus easily compute $p(\mathbf{x}|\mathbf{z})$. The problem is that we don't know $\mathbf{z}$, so we need to sum over all the possible values it could take on.

Writing this out, we have

$$p(\mathbf{x}|h_1) = \sum_{\mathbf{z}} p(\mathbf{x}|\mathbf{z})P(\mathbf{z}), \tag{6.16}$$

where $P(\mathbf{z})$ is the prior distribution on assignments. The problem is that if there are n_B bombs, there are 2^{n_B} possible values for $\mathbf{z}$—a number that becomes large quickly. Using simple Monte Carlo is a tempting solution—we could just sample values of $\mathbf{z}$ from $P(\mathbf{z})$ and then average together the resulting values of $p(\mathbf{x}|\mathbf{z})$. Unfortunately this doesn't work well in cases where some of the bombs really are targeted, because most choices of $\mathbf{z}$ won't pick out the targeted bombs and consequently have very small values for $p(\mathbf{x}|\mathbf{z})$, while others will get close and have very large values. The answer that simple Monte Carlo produces is going to depend more on how lucky the choice of samples was than on what the actual answer is.

Computing marginal probabilities for complex models such as mixture models is one of the places where importance sampling is extremely effective. The trick is to choose a proposal distribution $Q(\mathbf{z})$ that assigns higher probabilities than $P(\mathbf{z})$ to those values of $\mathbf{z}$ for which $p(\mathbf{x}|\mathbf{z})$ is large. One method that has been proposed for doing this is using an inference algorithm such as the Expectation-Maximization (EM) algorithm to estimate the parameters of the mixture model, and then using those parameters to define $Q(\mathbf{z})$. In this case, we could estimate $\boldsymbol{\mu}$ and $\boldsymbol{\Sigma}$ by trying to find values that seemed to correspond to a good location for the target in each bombing scene, and then calculate $P(\mathbf{z}|\mathbf{x}, \boldsymbol{\mu}, \boldsymbol{\Sigma})$ and take this as $Q(\mathbf{z})$.

There are some perils involved in this—it will pick out one possible target, but there could be others that are equally plausible—so the standard practice is to take $Q(\mathbf{z})$ as a mixture of this and the prior $P(\mathbf{z})$. This is the method that was used to compute the probabilities that were compared to human behavior (figuring out how to do this and implementing the resulting algorithm took longer than running the experiment and analyzing the data!).

6.5 Sequential Monte Carlo

A basic problem that arises in probabilistic models of cognition is explaining how learners can deal with the severe computational challenges posed by probabilistic inference. We have discussed how some of these problems can be addressed using Monte Carlo methods in a static setting, where all the data are available and the learner simply has to evaluate a set of hypotheses. However, the challenges seem to become even more severe in contexts where dynamic probabilistic inference is required, either because of intrinsic dynamics of a model, such as the assumption that the state of the world can change over time, or because more data gradually becomes available. These problems can be solved using *sequential Monte Carlo*.

6.5.1 Dealing with a Dynamic World

Sequential Monte Carlo methods are schemes for generating samples from a sequence of probability distributions, typically using the relationship between successive distributions to use samples from one distribution to generate samples from the next. The standard example of such a method is the *particle filter*, which is a sequential version of importance sampling. The particle filter was originally developed for making inferences about variables in a dynamic environment—the problem of "filtering" is inferring the current state of the world given a sequence of observations.

Assume that we have a probabilistic model with a structure similar to a hidden Markov model (HMM), where a sequence of latent variables $z_1, \ldots, z_n$ are responsible for generating a sequence of observed variables $x_1, \ldots, x_n$. In the simplest case, each z_i is generated from a distribution that depends only on z_{i-1}, and each x_i is generated from a distribution that depends only on z_i. Assume that we want to compute $p(z_n|x_1, \ldots, x_n)$. While efficient algorithms exist for computing this probability when the z_i are discrete (such as the forward procedure for HMMs), or when z_i is continuous and $p(z_i|z_{i-1})$ is a Gaussian distribution whose mean is a linear function of z_{i-1} (the Kalman filter discussed in chapter 5), the particle filter provides a way to perform inference that generalizes beyond these cases.

The basic idea behind the particle filter is that we are going to approximate $p(z_n|x_1, \ldots, x_n)$ using importance sampling. In particular, we can perform importance sampling where we use as our proposal the "prior" on z_n, $p(z_n|x_1, \ldots, x_{n-1})$. If we can generate samples from $p(z_n|x_1, \ldots, x_{n-1})$, then we can construct an importance sampler by giving each sample weight proportional to $p(x_n|z_n)$. How can we generate samples from $p(z_n|x_1, \ldots, x_{n-1})$? Well, we know that

$$p(z_n|x_1, \ldots, x_{n-1}) = \sum_{z_{n-1}} p(z_n|z_{n-1})p(z_{n-1}|x_1, \ldots, x_{n-1}), \tag{6.17}$$

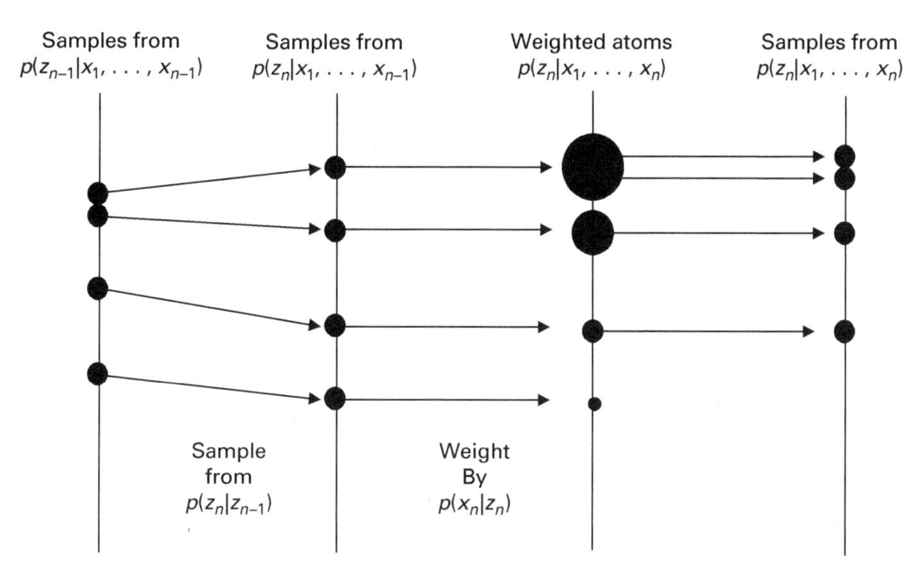

Figure 6.3
The particle filter. We start with samples from $p(z_{n-1}|x_1,\ldots,x_{n-1})$. We then obtain samples from $p(z_n|x_1,\ldots,x_{n-1})$ by sampling from $p(z_n|z_{n-1})$ for each sample of z_{n-1}. We calculate the likelihood $p(x_n|z_n)$ for each of these z_n, using them to assign weights to the samples. If we sample from this weighted distribution, we obtain samples from $p(z_n|x_1,\ldots,x_n)$.

so if we can generate samples $z_{n-1}^{(j)}$ from $p(z_{n-1}|x_1,\ldots,x_{n-1})$ with weights w_j, we can just sample from the distribution

$$p(z_n|x_1,\ldots,x_{n-1}) \approx \sum_j p(z_n|z_{n-1}^{(j)})w_j, \tag{6.18}$$

which is a simple mixture distribution. Alternatively, we could stratify our sampling, making sure that the samples are more diverse, by generating one sample from $p(z_n|z_{n-1}^{(j)})$ for each $z_{n-1}^{(j)}$.

The procedure outlined in the previous paragraph identifies an interesting recursion—it gives us a way to sample from $p(z_n|x_1,\ldots,x_n)$ so long as we can sample from $p(z_{n-1}|x_1,\ldots,x_{n-1})$. This sets us up to introduce the sequential Monte Carlo scheme that underlies the particle filter (see figure 6.3). First, we generate samples from $p(z_1|x_1)$. Then, for each sample $z_1^{(j)}$ we generate $z_2^{(j)}$ from $p(z_2|z_1^{(j)})$ and assign each resulting sample a weight w_j proportional to $p(x_2|z_2^{(j)})$. This step is equivalent to the likelihood weighting procedure introduced previously. We repeat this process for all $i=3,\ldots,n$, multiplying the old weight of each sample $z_i^{(j)}$ (now called a "particle") by $p(x_i|z_i^{(j)})$. There is no need to normalize the particle weights at each step—this can be done at the end of the process.

This simple recursive scheme is known as *sequential importance sampling*. However, it has one big problem: over time, the weights of the particles can diverge significantly since some of the particles are likely to have ended up with a sequence of z_i values that are very unlikely. In some ways, this is a waste of computation, since those particles with very small weights will make little contribution to the final representation of the distribution of z_n.

To address this problem, we can use an alternative approach known as *sequential importance resampling*. Under this approach, we monitor the variance of the normalized particle weights. When this variance exceeds a preset threshold, we sample a new set of particles from a probability distribution corresponding to the normalized weights. This increases the number of particles that are in a good part of the space. The weights are then reset to be uniform over particles, and the process continues. Resampling decreases the diversity in the set of particles, so if the intrinsic dynamics determined by $p(z_i|z_{i-1})$ do not result in a lot of variability, other schemes for introducing diversity (such as perturbing the values of the particles) may need to be used.

While the standard particle-filtering algorithm follows this schema, this only scratches the surface of possible sequential Monte Carlo techniques (for a comprehensive review, see Doucet, de Freitas, & Gordon, 2001; Chopin & Papaspiliopoulos, 2020). Even within particle filtering, many options can be used to improve performance on certain problems. For example, in some cases, it is possible to compute $p(z_i|x_1, \ldots, x_{i-1})$ directly, and it is better to sample from this distribution. In other cases, we can take into account the value of the observation x_i in proposing a value for z_i, increasing the accuracy of our proposals and decreasing the variance of the weights. As with other Monte Carlo methods, tailoring the algorithm to fit a particular probabilistic model is something of an art.

6.5.2 Accumulating Evidence

While working with dynamic models is important, a far more basic context in which we have to perform a challenging probabilistic computation is when we want to update our posterior distribution over hypotheses in light of a new piece of evidence. This requires recomputing the probability of every single hypothesis, which can be extremely demanding when we have a large number of hypotheses. However, sequential Monte Carlo can be useful in this context as well.

Imagine that we have a set of hypotheses $h \in \mathcal{H}$ and a sequence of observations $d_1, \ldots, d_n$. We want to define an efficient scheme for updating our beliefs, allowing us to incrementally compute $p(h|d_1, \ldots, d_n)$ (i.e., giving us the posterior after every observation d_i) without requiring us to sum over all $|\mathcal{H}|$ hypotheses each time we obtain a new observation. Again, imagine performing importance sampling, where we use $p(h|d_1, \ldots, d_{n-1})$ as our proposal. If the d_i are independent given h, then we can just give the resulting samples of h weights proportional to $p(d_n|h)$. So, as before, we have a way to obtain samples from $p(h|d_1, \ldots, d_n)$, provided that we can obtain samples from $p(h|d_1, \ldots, d_{n-1})$. This allows us to set up the kind of recursion used in the particle filter.

The simplest sequential Monte Carlo scheme is as follows. First, generate samples from the prior $p(h)$. Then, as each observation d_i comes in, reweight those samples by $p(d_i|h)$. When we need to compute a posterior distribution, we can normalize the weights. This potentially gives us a set of weighted hypotheses as the approximation to the posterior distribution after each d_i.

While attractive in its simplicity, this scheme has a basic problem: the hypotheses we use in our approximation will be the same regardless of the data we observe, corresponding to the set we sampled from the prior. If the posterior after $d_1, \ldots, d_n$ is very different from the prior, this will result in a poor approximation. As a consequence, we need to introduce techniques such as perturbation of the particles to generate distinct hypotheses at each iteration. When perturbing the particles, our goal is to not change the distribution from which they are drawn—if they are samples from $p(h|d_1, \ldots, d_i)$ before perturbation, they should be samples from that distribution after perturbation. One way to achieve this is to perturb the particles by applying one or more iterations of a Markov chain Monte Carlo (MCMC) algorithm for the distribution $p(h|d_1, \ldots, d_i)$ (for details, see Fearnhead, 2002). We will discuss these algorithms in more detail after considering an example of sequential Monte Carlo in practice.

6.5.3 An Example: Inference for Multiple Object Tracking

In the model of multiple object tracking presented by Vul et al. (2008) and summarized in chapter 5, the presence of multiple objects creates a problem: even though each object follows linear dynamics with Gaussian noise, and can hence be tracked using a particle filter, when objects occupy similar locations they can become confused with one another. Vul et al. (2008) modeled this by introducing an additional variable, γ_t, which is a permutation of the indices of the objects at time t. With eight objects, say, the initial assignment of γ_1 would be the vector $(1, 2, 3, 4, 5, 6, 7, 8)$. However, if objects 1 and 2 crossed paths and became confused with one another at the next time step, γ_2 would become $(2, 1, 3, 4, 5, 6, 7, 8)$. Part of multiple object tracking is thus not just calculating a posterior distribution on the location of the objects, but calculating a posterior distribution on γ_t.

Because of the way that the model is defined, if γ_t is known, then the objects just evolve according to their own linear-Gaussian dynamics. This means that the inference problem reduces to one of keeping track of γ_t. For each value of γ_{t+1}, it is straightforward to compute the probability of the location of the objects at time $t+1$ based on γ_t and their locations at previous times. That means that we have a perfect setup for a particle filter.

Vul et al. (2008) assumed that each γ_t was independent—that the indices of the objects (and hence the potential confusions of different objects) are drawn from the same distribution $P(\gamma_t)$ at each time t and are unrelated to those at the previous time $t-1$. The data consist of a set of observations of the "motion" of each object $\mathbf{m}_t$ containing both the location and velocity for each object at time t. The particle filter takes a set of samples of $\gamma_1, \ldots, \gamma_t$ that capture the posterior distribution on assignments based on $\mathbf{m}_1, \ldots, \mathbf{m}_t$, and updates these samples to reflect the posterior on γ_{t+1} after observing $\mathbf{m}_{t+1}$. This can be done by augmenting each sample with a value of γ_{t+1} sampled from the prior $P(\gamma_{t+1})$, and then assigning each resulting sample a weight proportional to $p(\mathbf{m}_{t+1}|\mathbf{m}_1, \ldots, \mathbf{m}_t, \gamma_1, \ldots, \gamma_{t+1})$, which can be computed by applying the linear dynamics and Gaussian noise assumed by the model to the objects indicated by γ_{t+1}. Conditioned on γ_{t+1}, each object's position evolves as in the Kalman filter discussed in the previous chapter.

6.6 Markov Chain Monte Carlo

One of the most flexible methods for generating samples from a probability distribution is *Markov chain Monte Carlo (MCMC)*, which can be used to construct samplers for arbitrary probability distributions even if the normalizing constants of those distributions are unknown. MCMC algorithms were originally developed to solve problems in statistical physics (Metropolis, Rosenbluth, Rosenbluth, Teller, & Teller, 1953), and are now widely used across physics, statistics, machine learning, and related fields (for more details, see

Newman & Barkema, 1999; Gilks, Richardson, & Spiegelhalter, 1996; Mackay, 2003; Neal, 1993). As with other Monte Carlo methods, MCMC algorithms provide tools for working with probability distributions and hypotheses about processes by which people could approximate Bayesian inference. In chapter 10, we also explore the idea that MCMC algorithms can be used to design behavioral experiments that allow us to sample from people's subjective probability distributions.

6.6.1 Markov Chains

As the name suggests, Markov chain Monte Carlo is based upon the theory of Markov chains—sequences of random variables in which each variable is conditionally independent of all previous variables given its immediate predecessor. The probability that a variable in a Markov chain takes on a particular value conditioned on the value of the preceding variable is determined by the *transition kernel* for that Markov chain. The transition kernel $K(\mathbf{x}^{(i+1)}|\mathbf{x}^{(i)})$ indicates the probability that a chain in state $\mathbf{x}^{(i)}$ at step i changes to state $\mathbf{x}^{(i+1)}$ at step $i+1$. When the states are discrete, this information can be encoded in a matrix where the rows correspond to the values at the next iteration, and the columns correspond to the distributions specified by the kernel. This is called a *transition matrix*.

One well-known property of Markov chains is their tendency to converge to a *stationary distribution*: as the length of a Markov chain increases, the probability that a variable in that chain takes on a particular value converges to a fixed quantity determined by the choice of transition kernel. The stationary distribution is defined to be the distribution $\pi(\mathbf{x})$ that satisfies

$$\pi(\mathbf{x}) = \int K(\mathbf{x}|\mathbf{x}')\pi(\mathbf{x}')d\mathbf{x}, \tag{6.19}$$

meaning that if we sample a state $\mathbf{x}'$ from $\pi(\mathbf{x}')$ and then choose the next state $\mathbf{x}$ by sampling from the distribution $K(\mathbf{x}|\mathbf{x}')$, the probability distribution over states remains $\pi(\mathbf{x})$. This is the sense in which this distribution is "stationary": once the probability distribution over states reaches the stationary distribution, it never changes. Convergence to the stationary distribution can be described formally by defining $p(\mathbf{x}^{(m)}|\mathbf{x}^{(1)})$ to be the probability of the Markov chain being in state $\mathbf{x}^{(m)}$ after m iterations, having started in state $\mathbf{x}^{(1)}$. Provided the Markov chain satisfies conditions of *ergodicity* (for details, see Norris, 1997), this probability will converge to the stationary distribution as m becomes large. That is,

$$\lim_{m\to\infty} p(\mathbf{x}^{(m)}|\mathbf{x}^{(1)}) = \pi(\mathbf{x}^{(m)}) \tag{6.20}$$

for all $\mathbf{x}^{(1)}$ and $\mathbf{x}^{(m)}$. Consequently, if we sample from the Markov chain by picking some initial value and then repeatedly sampling from the distribution specified by the transition kernel, we will ultimately generate samples from the stationary distribution. For the same reason, averaging a function $f(\mathbf{x})$ over the values of $\mathbf{x}^{(m)}$ will approximate the average of that function over the probability distribution $\pi(\mathbf{x})$ as m becomes large.[1]

1. One of the challenges of using MCMC is knowing when a Markov chain has converged to its stationary distribution. It is standard practice to discard the earliest samples as they are likely to be biased toward the values used to initialize the Markov chain (this is referred to as letting the Markov chain "burn in" before saving samples). There are a variety of strategies for checking the convergence of Markov chains, which are discussed in detail in Gilks, Richardson, and Spiegelhalter (1996) and Gelman, Carlin, Stern, and Rubin (1995).

6.6.2 The Metropolis-Hastings Algorithm

In MCMC, a Markov chain is constructed such that its stationary distribution is the distribution from which we want to generate samples. If the target distribution is $p(\mathbf{x})$, then the transition kernel $K(\mathbf{x}^{(i+1)}|\mathbf{x}^{(i)})$ would be chosen to yield $\pi(\mathbf{x}) = p(\mathbf{x})$ as the stationary distribution. Fortunately, there is a simple procedure that can be used to construct a transition kernel that will have that property for any choice of $p(\mathbf{x})$. This procedure is known as the *Metropolis-Hastings algorithm* (Hastings, 1970; Metropolis, Rosenbluth, Rosenbluth, Teller, and Teller, 1953).

The Metropolis-Hastings algorithm defines a kernel $K(\mathbf{x}^{(i+1)}|\mathbf{x}^{(i)})$ as the result of two probabilistic steps. The first uses an arbitrary *proposal distribution*, $q(\mathbf{x}^*|\mathbf{x}^{(i)})$, to generate a proposed value $\mathbf{x}^*$ for $\mathbf{x}^{(i+1)}$. The second is to decide whether to accept this proposal, using information about how the proposal was generated and its probability under the target distribution. The key to making sure that the resulting Markov chain converges to the target distribution is in how the *acceptance function* is defined.

The acceptance function used in the Metropolis-Hastings algorithm is

$$A(\mathbf{x}^*, \mathbf{x}^{(i)}) = \min\left[\frac{p(\mathbf{x}^*)q(\mathbf{x}^{(i)}|\mathbf{x}^*)}{p(\mathbf{x}^{(i)})q(\mathbf{x}^*|\mathbf{x}^{(i)})}, 1\right], \tag{6.21}$$

where $p(\mathbf{x})$ is our target distribution. If a random number generated from a uniform distribution over $[0, 1]$ is less than $A(\mathbf{x}^*, \mathbf{x}^{(i)})$, the proposed value $\mathbf{x}^*$ is accepted as the value of $\mathbf{x}^{(i+1)}$. Otherwise, the Markov chain remains at its previous value, and $\mathbf{x}^{(i+1)} = \mathbf{x}^{(i)}$. Intuitively, equation (6.21) says that we will be more likely to accept proposals if they have a high probability under our target distribution, and less likely to accept them if the proposal distribution strongly favors the proposed value $\mathbf{x}^*$ over the current value $\mathbf{x}^{(i)}$. These two considerations mean that our Markov chain spends more time in regions where $p(\mathbf{x})$ is high and is not unduly affected by the vagaries of the proposal distribution. If you're curious where equation (6.21) comes from, we will explain that in detail in the next section.

So, to summarize the algorithm: Start with an initial value $\mathbf{x}^{(1)}$. Define a proposal distribution $q(\mathbf{x}^*|\mathbf{x}^{(i)})$, and sample $\mathbf{x}^*$ given $\mathbf{x}^{(1)}$. Decide whether to accept $\mathbf{x}^*$ by applying equation (6.21), substituting in the target distribution $p(\mathbf{x})$. If accepted, $\mathbf{x}^{(2)} = \mathbf{x}^*$; if not, $\mathbf{x}^{(2)} = \mathbf{x}^{(1)}$. Repeat many times until the Markov chain has plausibly converged to its stationary distribution, and an average over the resulting $\mathbf{x}^{(m)}$ will approximate the expectation of the same quantity over $p(\mathbf{x})$.

To illustrate the use of the Metropolis-Hastings algorithm, we can use it to generate samples from a Gaussian distribution. For simplicity, we will assume that the Gaussian has zero mean and unit variance, but the same procedure could be used in more interesting cases, such as for sampling from the posterior distribution on the mean of a Gaussian. It is easy to sample from Gaussians in general, so we have no need to use MCMC, but it provides a way to demonstrate the steps that are involved. The state space of our Markov chain is a single dimension that ranges from $-\infty$ to ∞. We need to choose a proposal distribution to take us from state to state. In general, it is good to choose a proposal that is neither too big (meaning that the vast majority of proposals will get rejected) nor too small (meaning that proposals will get accepted but not result in any real movement around the state space). We will use a Gaussian for our proposal, with a mean corresponding to the previous value

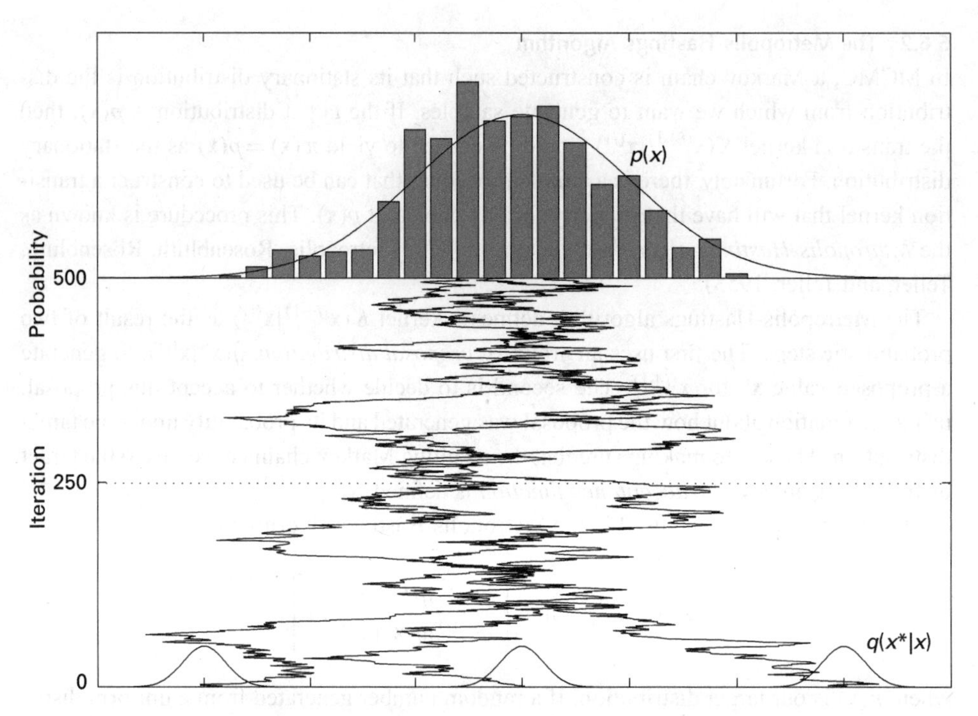

Figure 6.4
The Metropolis-Hastings algorithm. The solid lines shown in the bottom part of the figure are three sequences of values sampled from a Markov chain. Each chain began at a different location in the space, but used the same transition kernel. The transition kernel was constructed using the procedure described in the text for the Metropolis-Hastings algorithm: the proposal distribution, $q(x^*|x)$, was a Gaussian distribution with mean x and standard deviation 0.2 (shown centered on the starting value for each chain at the bottom of the figure), and the acceptance probabilities were computed by taking $p(x)$ to be Gaussian with mean 0 and standard deviation 1 (plotted with a solid line in the top part of the figure). This guarantees that the stationary distribution associated with the transition kernel is $p(x)$. Thus, regardless of the initial value of each chain, the probability that the chain takes on a particular value will converge to $p(x)$ as the number of iterations increases. In this case, all three chains move to explore a similar part of the space after around 100 iterations. The histogram in the top part of the figure shows the proportion of time that the three chains spend visiting each part in the space after 250 iterations (marked with the dotted line), which closely approximates $p(x)$. Samples from the Markov chains can thus be used similarly to samples from $p(x)$. Figure reproduced with permission from Griffiths, Kemp, and Tenenbaum (2008c).

and a variance 0.2 times the variance of the target distribution. Repeating the process of generating a proposal and deciding whether to accept it for 500 steps with a total of three different starting points produces the results shown in figure 6.4.

One advantage of the Metropolis-Hastings algorithm is that it requires only limited knowledge of the probability distribution $p(\mathbf{x})$. Inspection of equation (6.21) reveals that, in fact, the Metropolis-Hastings algorithm can be applied even if we know the value of $p(\mathbf{x})$ only up to a constant, since only the ratio of these quantities affects the algorithm. This can be extremely useful when we are performing Bayesian inference, as we have previously demonstrated that the hard part of applying Bayes' rule is usually computing the normalizing constant. Using Metropolis-Hastings, we do not need to compute this normalizing constant to generate samples from the posterior—we only need to know the likelihood and prior. This makes it far easier to work with probabilistic models since defining the likelihood and prior simply requires thinking about the generative process and inductive biases that are relevant, rather than taking into account concerns about analytic tractability. As a consequence, we can go beyond the restrictive world of conjugate priors and begin to explore richer and more expressive models.

6.6.3 Analyzing the Metropolis-Hastings Algorithm

The Metropolis-Hastings algorithm seems a little like magic—it lets us define a Markov chain that converges to whatever stationary distribution we like. How is this possible? The key is the acceptance function given by equation (6.21). We can derive this acceptance function, and explain why it works, using another property of Markov chains.

Assume that we have a Markov chain that has a transition kernel $K(\mathbf{x}^{(i+1)}|\mathbf{x}^{(i)})$ that satisfies the equation

$$\pi(\mathbf{x})K(\mathbf{x}'|\mathbf{x}) = \pi(\mathbf{x}')K(\mathbf{x}|\mathbf{x}') \tag{6.22}$$

for all $\mathbf{x}$ and $\mathbf{x}'$, where $\pi(\mathbf{x})$ is the stationary distribution of the Markov chain. This property is called *detailed balance*—intuitively, it means that once the Markov chain reaches its stationary distribution, the probability of observing a transition from $\mathbf{x}$ and $\mathbf{x}'$ or $\mathbf{x}'$ and $\mathbf{x}$ is the same (Markov chains that have this property are described as being *reversible*). Detailed balance is a stronger condition than the existence of a stationary distribution—you can check that you can obtain equation (6.19) just by integrating over $\mathbf{x}^{(i)}$ on both sides of equation (6.22).

The acceptance function used in the Metropolis-Hastings algorithm is derived using detailed balance. As before, let $A(\mathbf{x}^*, \mathbf{x}^{(i)})$ be the probability of accepting $\mathbf{x}^*$ in state $\mathbf{x}^{(i)}$. Each time a proposal is rejected, it adds a little bit of probability mass to $\mathbf{x}^{(i)}$, so the probability of no proposal being accepted, which we will denote as $A=0$, is

$$P(A=0|\mathbf{x}^{(i)}) = \int \left[1 - A(\mathbf{x}^*, \mathbf{x}^{(i)})\right] q(\mathbf{x}^*|\mathbf{x}^{(i)})\, d\mathbf{x}^* = 1 - \int A(\mathbf{x}^*, \mathbf{x}^{(i)}) q(\mathbf{x}^*|\mathbf{x}^{(i)})\, d\mathbf{x}^*. \tag{6.23}$$

We can thus write the transition kernel as

$$K(\mathbf{x}^{(i+1)}|\mathbf{x}^{(i)}) = q(\mathbf{x}^{(i+1)}|\mathbf{x}^{(i)})A(\mathbf{x}^{(i+1)}, \mathbf{x}^{(i)}) + \delta(\mathbf{x}^{(i+1)}, \mathbf{x}^{(i)})P(A=0|\mathbf{x}^{(i)}), \tag{6.24}$$

where $\delta(\mathbf{x}^{(i+1)}, \mathbf{x}^{(i)})$ is the delta function picking out $\mathbf{x}^{(i+1)} = \mathbf{x}^{(i)}$. Now we just need to specify $A(\mathbf{x}^*, \mathbf{x}^{(i)})$ so the stationary distribution associated with this kernel is $p(\mathbf{x})$.

We can rewrite detailed balance (equation 6.22) as

$$\frac{\pi(\mathbf{x})}{\pi(\mathbf{x}')} = \frac{K(\mathbf{x}|\mathbf{x}')}{K(\mathbf{x}'|\mathbf{x})}. \tag{6.25}$$

This is trivially satisfied when $\mathbf{x} = \mathbf{x}'$, so we can focus on $\mathbf{x} \neq \mathbf{x}'$, in which case our kernel $K(\mathbf{x}^{(i+1)}|\mathbf{x}^{(i)})$ is just $q(\mathbf{x}^{(i+1)}|\mathbf{x}^{(i)})A(\mathbf{x}^{(i+1)}, \mathbf{x}^{(i)})$. Putting this into the detailed balance equation, we get

$$\frac{\pi(\mathbf{x})}{\pi(\mathbf{x}')} = \frac{q(\mathbf{x}|\mathbf{x}')A(\mathbf{x}, \mathbf{x}')}{q(\mathbf{x}'|\mathbf{x})A(\mathbf{x}', \mathbf{x})}, \tag{6.26}$$

which we can rearrange to

$$\frac{A(\mathbf{x}', \mathbf{x})}{A(\mathbf{x}, \mathbf{x}')} = \frac{q(\mathbf{x}|\mathbf{x}')\pi(\mathbf{x}')}{q(\mathbf{x}'|\mathbf{x})\pi(\mathbf{x})}. \tag{6.27}$$

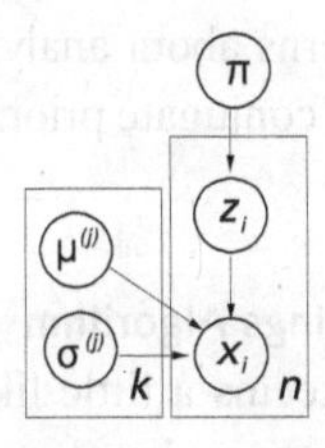

Figure 6.5
Reminder of the graphical model for clustering with a mixture of Gaussians. Here, x_i is the ith data point, z_i is its cluster assignment, π are the parameters of the Discrete distribution on z_i with $P(z_i=k)=\pi_k$, and $\mu^{(k)}$ and $\sigma^{(k)}$ are the parameters of the Gaussian corresponding to the kth cluster. The distribution $p(x_i|z_i=k,\mu,\sigma)$ is Gaussian$(\mu^{(k)},\sigma^{(k)})$. The boxes around the variables are plate indicating that this structure is replicated across n observations and k clusters. In the example considered in the text, $k=2$, so π becomes the parameter of a Bernoulli distribution indicating the probability of choosing the first cluster, $z_i=1$.

If we can define an acceptance function that satisfies this relationship, we will have defined a Markov chain that has $\pi(\mathbf{x})$ as its stationary distribution. Any acceptance function $A(\mathbf{x}^*,\mathbf{x}^{(i)})$ that is proportional to $q(\mathbf{x}^{(i)},\mathbf{x}^*)p(\mathbf{x})$ will thus result in a Markov chain with $p(\mathbf{x})$ as its stationary distribution. It is straightforward to check that this is the case for the acceptance function given in equation (6.21), which is used in the Metropolis-Hastings algorithm because it maximizes the chance of proposals being accepted (and thus minimizes $P(A=0|\mathbf{x}^{(i)})$). However, we will have the opportunity to explore some other valid acceptance functions in chapter 10.

6.6.4 Gibbs Sampling

If we can sample from distributions related to $p(\mathbf{x})$, we can use other MCMC methods. In particular, if we are able to sample from the conditional probability distribution for each variable in a set given the remaining variables, $p(x_j|x_1,\ldots,x_{j-1},x_{j+1},\ldots,x_d)$, we can use another popular algorithm, *Gibbs sampling* (Geman & Geman, 1984; Gilks et al., 1996), which is known in statistical physics as the *heatbath* algorithm (Newman & Barkema, 1999). The Gibbs sampler for a target distribution $p(\mathbf{x})$ is the Markov chain defined by drawing each x_j from the conditional distribution $p(x_j|x_1,\ldots,x_{j-1},x_{j+1},\ldots,x_d)$. You can actually derive the Gibbs sampler from the Metropolis-Hastings algorithm: if you propose to sample just one element of $\mathbf{x}$ from its conditional distribution given the values of the other elements, the proposal is always accepted. This is exactly what is done in Gibbs sampling. You can either cycle through the x_j in turn, or choose one at random to sample at each iteration.

Gibbs sampling has a lot in common with the EM algorithm, but it is typically used to generate samples from a posterior distribution, while the EM algorithm produces a point estimate. The commonalities between the algorithms can be illustrated by considering how to apply Gibbs sampling to the Gaussian mixture model where the components have known variances but unknown means considered in the previous chapter (the graphical model is reproduced in figure 6.5 for convenience). In this case, we want to compute the posterior distribution over the assignments of observations to cluster $\mathbf{z}$ and the component means $\mu^{(1)}$ and $\mu^{(2)}$ and mixing proportions π, given the observations $\mathbf{x}$ (here simplifying things by assuming that our Gaussians are just 1D). Starting from some initial values for $\mathbf{z}$, μ, and π, the Gibbs sampler cycles through these variables, drawing each from its distribution given all other variables. This means that the algorithm effectively alternates between sampling

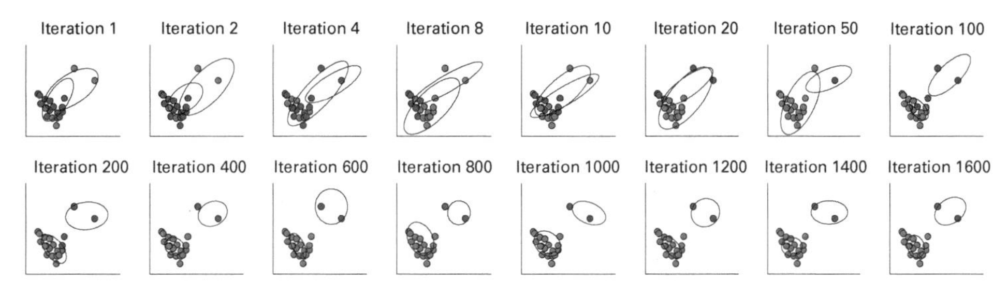

Figure 6.6
Gibbs sampling for a mixture of two Gaussians in two dimensions. The color of the points indicate their assignments, while density ellipses denote the Gaussians. After taking a number of iterations to find a reasonable solution, the algorithm is fairly stable in its assignment of observations to clusters.

the assignments of observations to components and sampling the values of the parameters. These two parts of the algorithm are similar to the E- and M-steps of the EM algorithm, although sampling is used for both latent variables and parameters, while EM uses one process (expectation) to deal with latent variables and another (maximization) to deal with parameters.

Applying the Gibbs sampler in this mixture model is very simple. In each iteration, we cycle through the z_i drawing each from the conditional distribution

$$P(z_i|\mathbf{z}_{-i}, \mathbf{x}, \mu^{(1)}, \mu^{(2)}, \pi) \propto p(x_i|z_i, \mu^{(z_i)})P(z_i|\pi) \tag{6.28}$$

$$= \exp\{-\frac{(x_i - \mu^{(z_i)})^2}{2\sigma^2}\}\pi^{I(z_i=1)}(1-\pi)^{I(z_i=2)}, \tag{6.29}$$

where $\mathbf{z}_{-i}$ indicates $(z_1, \ldots, z_{i-1}, z_{i+1}, \ldots, z_n)$, $I(\cdot)$ is the indicator function taking value 1 when its argument is true and 0 otherwise, and we exploit the conditional independence assumptions incorporated into this model. We then draw $\mu^{(1)}$, $\mu^{(2)}$, and π from their posterior distributions given these values of $\mathbf{z}$. Since conditioning on $\mathbf{z}$ is like observing it, this is extremely easy: we need only compute the posterior distributions of these parameters, assuming that each observation is assigned to the component indicated in $\mathbf{z}$. The posterior distribution on $\mu^{(j)}$ is just the posterior distribution given all x_i such that $z_i = j$ (which we showed is Gaussian in our earlier analyses), and the posterior distribution on π is Beta$(n_1 + 1, n_2 + 1)$, assuming a uniform prior on π. It is easy to generate from both of these distributions. The Gibbs sampler then iterates back and forth between these two steps, sampling assignments, then parameters, then assignments, and so forth. Figure 6.6 shows an example of using this sampler in a 2D Gaussian mixture.

In deriving the Gibbs sampler, we need to compute the conditional distribution for each variable conditioned on all others. Graphical models can be extremely useful in this context. Remember that one of the things that graphical models encode is the pattern of dependencies that exists among a set of variables. In particular, when we condition on all other variables, we know that each variable will be independent of any other variable that is not its parent in the graph, its child, or the parent of a common child. This set of variables is known as the *Markov blanket* of the variable, as it shields the variable from other dependencies. Thus, when we compute the conditional distributions used in Gibbs sampling, we need only consider the variables that appear in the Markov blanket of the target variable.

You can check that this is the case for the mixture of Gaussians given in this discussion by generating the Markov blanket for each variable using the graphical model shown in figure 6.5.

6.6.5 An Example: Gibbs Sampling for Topic Models

Using the topic models introduced in chapter 5 requires solving a difficult probabilistic inference problem—finding the topics that best account for the content of a set of documents. As a reminder, each word w_i is associated with a topic z_i, each topic is a probability distribution over words expressed as a multinomial distribution with parameters ϕ, and the probability distribution over topics within a document is a multinomial distribution with parameters θ. We define a Dirichlet prior (the multivariate generalization of the beta distribution) on both ϕ and θ, with ϕ generated from a symmetric Dirichlet with parameter β and θ generated from a symmetric Dirichlet with parameter α.

We can extract a set of t topics from a collection of documents in a completely unsupervised fashion using Bayesian inference. Since Dirichlet priors are conjugate to multinomial distributions, we can compute the joint distribution $P(\mathbf{w}, \mathbf{z})$ by integrating out ϕ and θ, just as we did in the model selection examples in chapter 3. We can then ask questions about the posterior distribution over $\mathbf{z}$ given $\mathbf{w}$, given by Bayes' rule:

$$P(\mathbf{z}|\mathbf{w}) = \frac{P(\mathbf{w}, \mathbf{z})}{\sum_{\mathbf{z}} P(\mathbf{w}, \mathbf{z})}. \tag{6.30}$$

Since the sum in the denominator is intractable, having t^n terms, we are forced to evaluate this posterior using MCMC. In this case, we use Gibbs sampling to investigate the posterior distribution over assignments of words to topics, $\mathbf{z}$.

The Gibbs sampling algorithm consists of choosing an initial assignment of words to topics (e.g., choosing a topic uniformly at random for each word), and then sampling the assignment of each word z_i from the conditional distribution $P(z_i|\mathbf{z}_{-i}, \mathbf{w})$. Each iteration of the algorithm is thus a probabilistic shuffling of the assignments of words to topics. This procedure is illustrated in figure 6.7, which shows the results of applying the algorithm (using just two topics) to a small portion of the TASA corpus. This portion features 30 documents that use the word MONEY, 30 documents that use the word OIL, and 30 documents that use the word RIVER. The vocabulary is restricted to 18 words, and the entries indicate the frequency with which the 731 tokens (i.e., instances) of those words appeared in the 90 documents. Each word token in the corpus, w_i, has a topic assignment, z_i, at each iteration of the sampling procedure. In the figure, we focus on the tokens of three words: MONEY, BANK, and STREAM. Each word token is initially assigned a topic at random, and each iteration of MCMC results in a new set of assignments of tokens to topics. After a few iterations, the topic assignments begin to reflect the different usage patterns of MONEY and STREAM, with tokens of these words ending up in different topics, and the multiple senses of BANK.

The details behind this particular Gibbs sampling algorithm are given in Griffiths and Steyvers (2004), where the algorithm is used to analyze the topics that appear in a large database of scientific documents. The conditional distribution for z_i that is used in the algorithm can be derived using an argument similar to our derivation of the posterior predictive

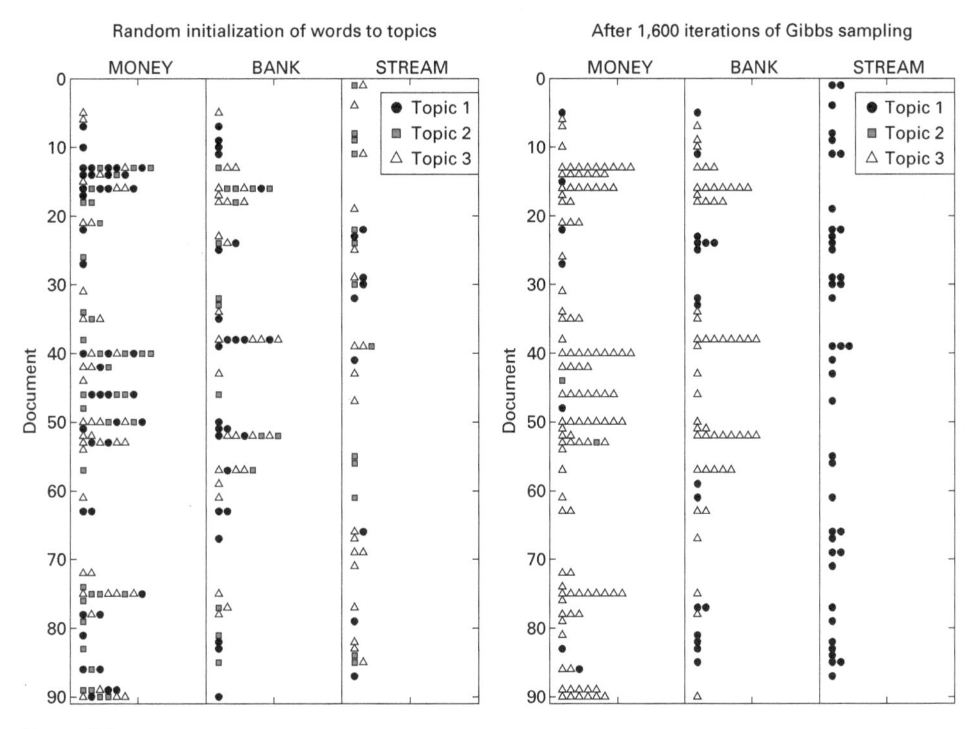

Figure 6.7
Illustration of the Gibbs sampling algorithm for learning topics. Each word token w_i appearing in the corpus has a topic assignment, z_i. The figure shows the assignments of all tokens of three types—MONEY, BANK, and STREAM—before and after running the algorithm. Each marker corresponds to a single token appearing in a particular document, and shape and color indicates assignment: topic 1 is a black circle, topic 2 is a gray square, and topic 3 is a white triangle. Before running the algorithm, assignments are relatively random, as shown in the left panel. After running the algorithm, tokens of MONEY are almost exclusively assigned to topic 3, tokens of STREAM are almost exclusively assigned to topic 1, and tokens of BANK are assigned to whichever of topic 1 and topic 3 seems to dominate a given document. The algorithm consists of iteratively choosing an assignment for each token, using a probability distribution over tokens that guarantees convergence to the posterior distribution over assignments. Figure reproduced with permission from Griffiths, Kemp, and Tenenbaum (2008c).

distribution in coin-flipping, giving

$$P(z_i|\mathbf{z}_{-i}, \mathbf{w}) \propto \frac{n_{-i,z_i}^{(w_i)} + \beta}{n_{-i,z_i}^{(\cdot)} + v\beta} \frac{n_{-i,z_i}^{(d_i)} + \alpha}{n_{-i,\cdot}^{(d_i)} + t\alpha}, \tag{6.31}$$

where $\mathbf{z}_{-i}$ is the assignment of all z_k such that $k \neq i$, and $n_{-i,z_i}^{(w_i)}$ is the number of words assigned to topic z_i that are the same as w_i; $n_{-i,z_i}^{(\cdot)}$ is the total number of words assigned to topic z_i; $n_{-i,z_i}^{(d_i)}$ is the number of words from document d_i assigned to topic z_i; v is the number of words in the vocabulary of the model; and $n_{-i,\cdot}^{(d_i)}$ is the total number of words in document d_i; all of which do not count the assignment of the current word w_i. The two terms in this expression have intuitive interpretations, being the posterior predictive distributions on words within a topic and topics within a document given the current assignments $\mathbf{z}_{-i}$ respectively. The result of the MCMC algorithm is a set of samples from $P(\mathbf{z}|\mathbf{w})$, reflecting the posterior distribution over topic assignments given a collection of documents. A single sample can be used to evaluate the topics that appear in a corpus or to reveal the assignments of words to topics, as shown in figure 6.7. We can also compute quantities such as the strength of association between words using the conditional probability

$$P(w_2|w_1) = \sum_{j=1}^{k} P(w_2|z=j)P(z=j|w_1) \tag{6.32}$$

by averaging over many samples.[2]

While other inference algorithms exist that can be used with this generative model (e.g., Blei, Ng, & Jordan, 2003; Minka & Lafferty, 2002), the Gibbs sampler is an extremely simple (and reasonably efficient) way to investigate the consequences of using topics to represent semantic relationships between words. Griffiths and Steyvers (2002, 2003) suggested that topic models might provide an alternative to traditional approaches to semantic representation, and they showed that they can provide better predictions of human word association data than latent semantic analysis (LSA) (Landauer & Dumais, 1997). Figure 6.8 is a contemporaneous record of an early presentation of this work. Topic models can also be applied to a range of other tasks that draw on semantic association, such as semantic priming and sentence comprehension (Griffiths et al., 2007).

6.7 Variational Inference

Monte Carlo methods are not the only kind of approximation method used for inference—another commonly used method is *variational inference* (Jordan et al., 1999). While Monte Carlo methods deal with computational complexity by approximating the true distribution with stochastic samples, variational methods approximate this distribution by finding the closest member of a simpler family of distributions. Thus, variational methods offer a potentially biased but noise-free approximation, in contrast to the unbiased but noisy Monte Carlo methods discussed earlier in this chapter.

This process of finding the closest family member to the true distribution is what gives variational methods their name: they are based on the calculus of variations, which is concerned with the mapping of functions (i.e., the true and approximating distributions) to scalars (i.e., the deviation between the two distributions). Because the approximating distribution is simpler than the true distribution, it is often possible to calculate the approximation analytically. In cases where this is not possible, the variational approximation is still expressed in a form that can be solved using approximation algorithms such as stochastic gradient descent. For this reason, variational methods have become increasingly popular as advances in training deep neural networks have resulted in new methods and software for solving these optimization problems (see chapter 12).

2. When computing quantities such as $P(w_2|w_1)$, we need a way of finding the parameter ϕ that characterizes the distribution over words associated with each topic. This can be done using ideas similar to those applied in our coin-flip example in chapter 3: for each sample of $\mathbf{z}$, we can estimate ϕ as

$$\hat{\phi}_z^{(w)} = \frac{n_z^{(w)} + \beta}{n_z^{(\cdot)} + v\beta}, \tag{6.33}$$

which is the posterior predictive distribution over new words w for topic z conditioned on $\mathbf{w}$ and $\mathbf{z}$.

Figure 6.8
Record of a lab meeting presentation on the Gibbs sampler for topic models.

We will introduce variational inference via the EM algorithm, which we discussed in chapter 5. Estimating the parameters of complex latent variable models was what motivated the development of variational methods and is a good source of intuition about the underlying formalism. For a more detailed technical introduction, see Blei et al. (2017).

6.7.1 The Variational EM Algorithm

In chapter 5, we introduced the EM algorithm as a method for estimating the parameters θ of a latent variable model without having to evaluate a sum over the latent variable $\mathbf{z}$. In the E-step of the algorithm, we calculate the expected complete log-likelihood

$$E_{P(\mathbf{z}|\mathbf{x},\theta)}[\log p(\mathbf{x},\mathbf{z}|\theta)] = \sum_{\mathbf{z}} P(\mathbf{z}|\mathbf{x},\theta) \log p(\mathbf{x},\mathbf{z}|\theta), \tag{6.34}$$

which is then optimized with respect to θ in the M-step. We showed that this is a lower bound on the log-likelihood, with

$$\log p(\mathbf{x}|\theta) = \sum_{\mathbf{z}} P(\mathbf{z}|\mathbf{x},\theta) \log p(\mathbf{x},\mathbf{z}|\theta) - \sum_{\mathbf{z}} P(\mathbf{z}|\mathbf{x},\theta) \log P(\mathbf{z}|\mathbf{x},\theta) \tag{6.35}$$

$$= E_{P(\mathbf{z}|\mathbf{x},\theta)}[\log p(\mathbf{x},\mathbf{z}|\theta)] + H[P(\mathbf{z}|\mathbf{x},\theta)], \tag{6.36}$$

where $H[P(\mathbf{z}|\mathbf{x},\theta)]$ is the entropy of $P(\mathbf{z}|\mathbf{x},\theta)$, a nonnegative quantity (see Cover & Thomas, 1991).

In the examples we presented, such as mixture models, the individual latent variable z_i are conditionally independent of one another. We were able to exploit that conditional independence to efficiently compute the expected complete log-likelihood. However, in other models, where the latent variables are highly interdependent, computing the expected complete log-likelihood can be intractable, as it still requires summing over all the values of $\mathbf{z}$. The *variational EM algorithm* is one strategy for solving this problem (see Jordan et al., 1999). In this algorithm, the E-step is performed by replacing $P(\mathbf{z}|\mathbf{x},\theta)$ in the expectation with the closest function from a more tractable family of distributions.

Define an arbitrary distribution on $\mathbf{z}$, $Q(\mathbf{z})$, which is nonzero everywhere $p(\mathbf{x},\mathbf{z}|\theta)$ is nonzero. Then, in equation (6.35) take the expectation with respect to $Q(\mathbf{z})$ rather than $P(\mathbf{z}|\mathbf{x},\theta)$. This gives

$$\log p(\mathbf{x}|\theta) = \sum_{\mathbf{z}} Q(\mathbf{z}) \log p(\mathbf{x},\mathbf{z}|\theta) - \sum_{\mathbf{z}} Q(\mathbf{z}) \log P(\mathbf{z}|\mathbf{x},\theta). \tag{6.37}$$

A standard way to measure the relatedness of two probability distributions is the *Kullback-Leibler (KL) divergence*, defined for the distributions $Q(\mathbf{z})$ and $P(\mathbf{z}|\mathbf{x})$ as

$$D_{KL}[Q(\mathbf{z})||P(\mathbf{z}|\mathbf{x})] = \sum_{\mathbf{z}} Q(\mathbf{z}) \log \frac{Q(\mathbf{z})}{P(\mathbf{z}|\mathbf{x})}, \tag{6.38}$$

which is zero when Q and P are the same and increases as they diverge (see Cover & Thomas, 1991). Note that we can rewrite this as

$$D_{KL}[Q(\mathbf{z})||P(\mathbf{z}|\mathbf{x})] = \sum_{\mathbf{z}} Q(\mathbf{z}) \log Q(\mathbf{z}) - \sum_{\mathbf{z}} Q(\mathbf{z}) \log P(\mathbf{z}|\mathbf{x}). \tag{6.39}$$

We can rearrange this formula to obtain

$$-\sum_{\mathbf{z}} Q(\mathbf{z}) \log P(\mathbf{z}|\mathbf{x}) = D_{KL}[Q(\mathbf{z})||P(\mathbf{z}|\mathbf{x})] - \sum_{\mathbf{z}} Q(\mathbf{z}) \log Q(\mathbf{z}). \tag{6.40}$$

We can then use this to replace the last term of equation (6.37), giving

$$\log p(\mathbf{x}|\theta) = \sum_{\mathbf{z}} Q(\mathbf{z}) \log p(\mathbf{x}, \mathbf{z}|\theta) + D_{KL}[Q(\mathbf{z})||P(\mathbf{z}|\mathbf{x}, \theta)] - \sum_{\mathbf{z}} Q(\mathbf{z}) \log Q(\mathbf{z}) \tag{6.41}$$

$$= E_{Q(\mathbf{z})}[\log p(\mathbf{x}, \mathbf{z}|\theta)] + D_{KL}[Q(\mathbf{z})||P(\mathbf{z}|\mathbf{x}, \theta)] + H[Q(\mathbf{z})], \tag{6.42}$$

which implies that the expectation of the complete log-likelihood with respect to $Q(\mathbf{z})$ is also a lower bound on $\log p(\mathbf{x}|\theta)$ since both the KL divergence and the entropy are non-negative.

This result gives us a different way to approach the EM algorithm: we can choose $Q(\mathbf{z})$ from a family that makes it easier to compute the expectation. Taking $Q(\mathbf{z}) = P(\mathbf{z}|\mathbf{x}, \theta)$ potentially results in a tighter lower bound because it minimizes the KL divergence term, but other distributions can be used. In particular, we can choose distributions in which the $\mathbf{z}$ are (conditionally) independent. This approach, in which we define a distribution $Q(\mathbf{z})$ to make inference tractable and then minimize the KL divergence of that distribution to $P(\mathbf{z}|\mathbf{x}, \theta)$, is the foundation of variational methods.

6.7.2 Inference as Optimization

While variational methods were first developed in the context of the EM algorithm, they provide a more general solution to the problem of Bayesian inference. Even if we're not trying to estimate the parameters of a latent variable model, a distribution $Q(\mathbf{z})$ that is close to the posterior $P(\mathbf{z}|\mathbf{x})$ can be used more generally as an approximation to the posterior. Consistent with this emphasis on approximating the posterior, we will suppress the dependence on θ for the remainder of this section.

Unfortunately, while finding $Q(\mathbf{z})$ such that $D_{KL}[Q(\mathbf{z})||P(\mathbf{z}|\mathbf{x})]$ is small makes intuitive sense, it is not a practical strategy. The problem is that computing the KL divergence as defined in equation (6.38) still requires us to evaluate $P(\mathbf{z}|\mathbf{x})$. Using the form given in equation (6.39), we have

$$D_{KL}[Q(\mathbf{z})||P(\mathbf{z}|\mathbf{x})] = \sum_{\mathbf{z}} Q(\mathbf{z}) \log Q(\mathbf{z}) - \sum_{\mathbf{z}} Q(\mathbf{z}) \log P(\mathbf{z}|\mathbf{x}) \tag{6.43}$$

$$= \sum_{\mathbf{z}} Q(\mathbf{z}) \log Q(\mathbf{z}) - \sum_{\mathbf{z}} Q(\mathbf{z}) [\log p(\mathbf{x}, \mathbf{z}) - \log p(\mathbf{x})], \tag{6.44}$$

$$= \sum_{\mathbf{z}} Q(\mathbf{z}) \log Q(\mathbf{z}) - \sum_{\mathbf{z}} Q(\mathbf{z}) \log p(\mathbf{x}, \mathbf{z}) + \log p(\mathbf{x}), \tag{6.45}$$

which contains $\log p(\mathbf{x})$. However, since $\log p(\mathbf{x})$ is constant with respect to Q, we can just focus on the remaining terms. Variational inference thus proceeds by maximizing the *evidence lower bound (ELBO)*:

$$ELBO(Q) = \log p(\mathbf{x}) - D_{KL}[Q(\mathbf{z})||P(\mathbf{z}|\mathbf{x})] \quad (6.46)$$

$$= -\sum_{\mathbf{z}} Q(\mathbf{z}) \log Q(\mathbf{z}) + \sum_{\mathbf{z}} Q(\mathbf{z}) \log p(\mathbf{x}, \mathbf{z}). \quad (6.47)$$

The resulting $Q(\mathbf{z})$ provides the best approximation to $P(\mathbf{z}|\mathbf{x})$. The optimized ELBO is a lower bound on $p(\mathbf{x})$ and can conveniently also be used for model selection.

6.7.3 Choosing the Approximation Family

Mathematical derivations are generally needed to determine how to efficiently select the closest family member to the true posterior distribution. Fortunately, if the prior distributions are exponential family distributions (including the Gaussian, exponential, gamma, and beta distributions, as discussed in chapter 3) and the priors and likelihoods are conjugate (also discussed in chapter 3), then posterior distributions will also be in the exponential family and variational approximations can be found via a simple iterative algorithm (for details, see Blei et al., 2017).

While variational methods often result in computationally simpler inference than Monte Carlo methods, a key choice is selecting the family of approximation distributions. A balance must be struck between an approximation's accuracy (how closely it can mimic the true distribution) and its tractability (how difficult it is to calculate). As much of a distribution's complexity is due to dependencies between the variables, a great deal of complexity can be avoided by assuming complete independence between variables: a *mean-field approximation* (Jordan et al., 1999). However, complete independence is often too strong an assumption to provide a useful approximation, particularly for variables that are expected to be highly dependent. For example, it would not be very accurate to approximate as independent your beliefs about whether a child lives in a blue house and your beliefs about whether her sister lives in a blue house.

A middle ground is to assume independence only where it has the greatest impact on computational tractability, an approach known as a *structured mean-field approximation*. In these cases, the variables are partitioned into set $\mathbf{z}_\ell$ and the approximation is

$$Q(\mathbf{z}) = \prod_{\ell} Q_\ell(\mathbf{z}_\ell). \quad (6.48)$$

Using the calculus of variations, the best member of the family of independent distributions is the one that satisfies

$$Q_\ell(\mathbf{z}_\ell) \propto \exp\left(E_{\mathbf{z}_{-\ell}}[\log p(\mathbf{z}, \mathbf{x})]\right), \quad (6.49)$$

where the expectation is taken over the variable $\mathbf{z}_{-\ell}$ not in the ℓth set.

6.7.4 Alternatives to Variational Inference

A disadvantage of variational inference is that the KL divergence it that minimizes, $D_{KL}[Q(\mathbf{z})||P(\mathbf{z}|\mathbf{x})]$, imposes some strong constraints on $Q(\mathbf{z})$. If you consult equation (6.38), you can see that $Q(\mathbf{z})$ is forced to be zero wherever $P(\mathbf{z}|\mathbf{x})$ is zero; otherwise, the KL divergence goes to ∞. This can make $Q(\mathbf{z})$ too narrow, and thus overconfident compared to $P(\mathbf{z}|\mathbf{x})$ (Minka, 2005).

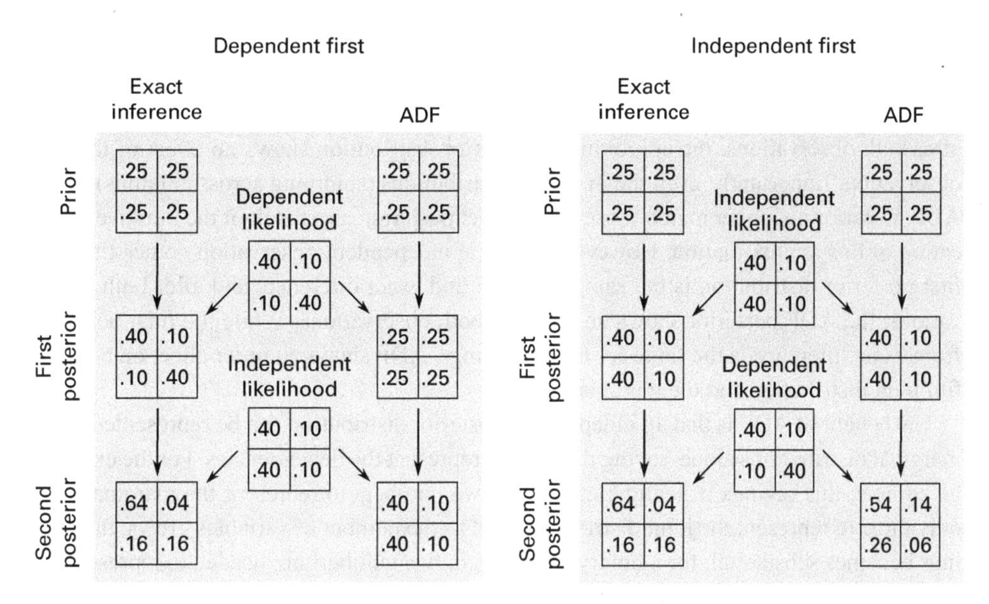

Figure 6.9
Comparison of updating between exact Bayesian inference and assumed density filtering (ADF). Each four-cell box shows the joint probability distribution of two binary variables. The Dependent First and Independent First subplots illustrate how the joint distributions are updated after two observations, with only the order of observations differing between the two subplots. Within each subplot, the likelihoods are in the middle column, while the joint distributions for exact inference are in the left column and the ADF approximation in the right column. The rows illustrate the progressive updating of the joint distribution after each observation.

An alternative is to use an approximation that minimizes the KL divergence in the other direction, $D_{KL}[P(\mathbf{z}|\mathbf{x})||Q(\mathbf{z})]$. This was originally done in *assumed density filtering* (ADF), an algorithm that was designed to deterministically filter sequential data in the same fashion as a particle filter (Boyen & Koller, 1998). Using this version of the KL divergence leads to a different best member of the approximating family in a structured mean field approximation. The best approximation for a partition under this definition is simply the marginal distribution of the posterior:

$$Q_\ell(\mathbf{z}_\ell) = P(\mathbf{z}_\ell|\mathbf{x}), \tag{6.50}$$

and so the approximation of the posterior distribution is just the product of its marginal distributions for each partition. The generalization of ADF to nonsequential data is called *expectation propagation* (Minka, 2001). However, finding the approximating distribution can be more of a challenge using the KL divergence in this direction, and neither of these methods is connected to a global bound like ELBO.

6.7.5 An Example: Locally Bayesian Learning

ADF treats incoming data sequentially, which means that the order in which those data arrive can affect the approximate posterior even in a probabilistic model in which it should have no effect. Such influences of order on inference are known as *order effects*. As a simple example, consider the updating of a joint distribution of a pair of binary variables shown in figure 6.9. The prior is initially independent, and it will be updated with one observation with an independent likelihood and one observation with a dependent likelihood (e.g., a strong correlation between the variables). Here, we demonstrate the effect that the order of the observations has on the final posterior distribution. When the dependent likelihood observation is first, the ADF posterior distribution is unchanged from the prior, so after both observations, the approximate posterior distribution shows no effect of the first observation. Importantly, even the marginal distributions (summing across columns) of the ADF posterior no longer match those of the exact posterior, as a result of the iterative application of this approximation. However, when the independent observation comes first, the first posterior distribution is the same for ADF and exact inference, and after both observations, the ADF posterior shows an effect of both observations. While the final posterior from exact inference is the same for both orderings, ADF shows an order effect on both the full joint distribution and the marginals.

The benefit of ADF is that an independent posterior distribution can be represented by its marginal distributions alone, saving the need to represent the dependencies. For the example given here, this savings is negligible: it takes two numbers to represent the marginals and only three to represent the joint distribution. But as the number of variables grows, this savings becomes substantial: for n binary variables, only n numbers are needed to represent the marginals, while $2^n - 1$ numbers are needed for the joint distribution. However, representing only the marginal distributions is an attractive idea to researchers reverse-engineering the mind using approximate Bayesian cognitive models. Both the approximation's positives and negatives are attractive: the reduction in representational complexity is good for psychological plausibility, and the order effects that are introduced can be compared to those found in human participants.

In associative learning tasks, experimental participants attempt to learn how to map cues to outcomes. This task can be modeled, for example, with the three-layer Bayesian neural network shown in figure 6.10. This network has input cues $\mathbf{x}$, attended cue nodes $\mathbf{y}$, and output nodes $\mathbf{t}$. The activation of the attended cues is a sigmoid function of the hidden weights, $\mathbf{W}_h$, and input cues, $\mathbf{y} = \text{sig}(\mathbf{W}_h\mathbf{x})$; and the activation of the outcomes is a sigmoid function of the outcome weights $\mathbf{W}_o$ and attended cues, $\mathbf{t} = \text{sig}(\mathbf{W}_o\mathbf{y})$. What makes it a Bayesian network is that there is a distribution over the weights, $p(\mathbf{W}_h, \mathbf{W}_o)$. While the two sets of weights may be initially assumed to be independent, $p(\mathbf{W}_h, \mathbf{W}_o) = p(\mathbf{W}_h, \mathbf{W}_o)$, once cue and outcome pairs are observed, the two sets of weights are often no longer independent in the exact posterior distribution. For example, the lack of an outcome in response to a cue can appear when either the relevant hidden weight or the relevant outcome weight is high, but not when both weights are high at once.

While the order of observations does not affect exact inference of the posterior distribution in globally Bayesian learning (GBL), if limited communication is assumed between $\mathbf{W}_o$ and $\mathbf{W}_h$ when updating the posterior, as locally Bayesian learning (LBL) assumes, then order effects are produced. In particular, LBL can produce complex order effects that have been observed in associative learning experiments: retrospective revaluation and highlighting, which combine primacy effects for some learned cue combinations with recency effects with others (Kruschke, 2006). While recency effects could potentially be incorporated into the model under the assumption that the world is changing, and so more recent observations are more valuable and thus should carry more weight (as in the Bayes-Kalman filter introduced in chapter 5), primacy effects are more difficult to ascribe to the model itself, so they

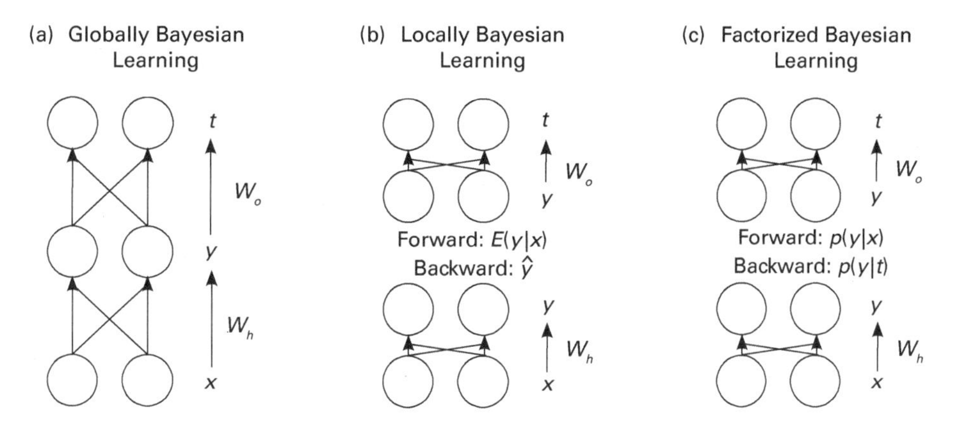

Figure 6.10
Diagrams of three models of associative learning: globally Bayesian learning (GBL), locally Bayesian learning (LBL), and factorized Bayesian learning (FBL). In each model, input cues $\mathbf{x}$ are weighted by hidden weights $\mathbf{W}_h$ and transformed to produce attended cues $\mathbf{y}$. Attended cues $\mathbf{y}$ are weighted by output weights $\mathbf{W}_o$ and transformed to produce outcomes $\mathbf{t}$. In GBL, all the latent variables are inferred together. LBL splits GBL into two modules with copies of the attended cues $\mathbf{y}$ in each module. Messages are passed back and forth between the copies of the attended cues $\mathbf{y}$: the expected value $E(\mathbf{y}|\mathbf{x})$ is passed upward and the single estimate $\hat{\mathbf{y}}$ that maximizes the probability of the outcome is passed backward. FBL uses the same modules as LBL, but with an assumed density filtering approximation instead: the messages passed between modules are distributions over the attended cues rather than a single set of attended cues. Reproduced with permission from Sanborn and Silva (2013).

often are taken as evidence of an approximation. Further work established that these order effects do not depend on using the exact form of the messages used in LBL, and instead an ADF approximation to the posterior distribution, termed factorized Bayesian learning (FBL), produces the same humanlike order effects (Daw, Courville, & Dayan, 2008; Sanborn & Silva, 2013).

6.8 Summary

In many ways, approximate inference algorithms are the hidden secret of Bayesian modeling, as exact inference is impossible in almost all interesting probabilistic models. Understanding different inference algorithms and the circumstances in which they can be used is an important part of the toolkit of researchers working with these models. Often, the ease of inference is a factor that drives the choice of model: it is easy to write down an extremely complex generative model, but there are no guarantees that it will be possible to apply Bayes' rule to compute posterior distributions in that model. Recent innovations such as the development of efficient probabilistic programming languages (see chapter 18) can make working with expressive probabilistic models much easier. One of the benefits of using these languages is that the code used to specify a probabilistic model can be used to automatically apply different inference algorithms to that model. However, having a basic understanding of what these algorithms are and how they work is essential to understanding the situations where they fail.

Approximate inference algorithms also offer an answer to the question of how human minds and brains could possibly perform Bayesian inference. Just as a researcher has a choice of algorithms to use to approximate the posterior probability of different hypotheses in a Bayesian model, our minds and brains have many possible strategies that they could use to approximate Bayesian inference.

We will revisit this theme later in the book. Chapter 11 considers how Monte Carlo methods might account for aspects of human probabilistic reasoning, chapter 13 extends this idea by considering how minds and brains might make the most efficient use of such computational resources, and chapter 12 connects some of the algorithms presented in this chapter to classic and recent work on artificial neural networks.

7

From Probabilities to Actions

Nick Chater, Thomas L. Griffiths, and Mark K. Ho

This book has so far focused primarily on questions of inductive inference: inferring the structure of the environment, a sentence, or a category from samples of data together with background knowledge. But the acquisition of new knowledge is ultimately of practical value to an agent only if it can help guide decisions concerning *actions*. So, for example, an animal may classify possible foodstuffs to decide whether they should or should not be eaten; or interpret a looming shadow to initiate fight or flight. Perceptual inferences about the state of the environment or one's own body may be used to guide reaching, maintain balance, or avoid collision. Causal inferences about the operation of a piece of physical apparatus, or a computer interface, will determine how a user can achieve their goals. And in the social and economic realms, inferences about the motives of another person may determine whether they are friend or foe, what they do or don't know, and other factors, and hence how we should best interact with them.

Understanding decision-making is also important from a methodological point of view because the vast majority of experimental data record behaviors that result from decisions. So, while studying perception using a psychophysical method, we often rely on participants' responses indicating what they saw, or whether a stimulus was visible at all; and such responses are themselves the result of a decision-making process.[1] Moreover, observing people's decisions can be employed to make inferences about the subjective probabilities that underlie those decisions. Indeed, in experimental economics, inference from observed decisions is the primary methodology for inferring a person's subjective probabilities—an approach that is rooted in theoretical results about ideal rational agents, as we shall see. Thus, we can see decision-making both as the ultimate object of most cognitive processes and as a medium through which those processes may be glimpsed.

In this chapter, we summarize the key ideas needed to translate from a probabilistic representation of the world to rational action. This is in itself a topic that could occupy an entire book, and indeed there are several that we recommend (e.g., Robert, 2007; Sutton

1. By contrast, brain imaging, or neural recordings, and typically fine-grained behavioral analysis, such as studying patterns of eye-movements or autocorrelations in reaction time distributions, pick up *incidental* products of cognitive processes that are not the result of deliberate decision. We decide to respond, say, that a stimulus was present on the screen, but we do not decide to respond in 750 ms or with a particular pattern of neural activity.

& Barto, 2018; Russell & Norvig, 2021). Our goal here is to provide an introduction to these ideas at a level of detail that makes it possible to understand the topics presented in the second half of the book. We begin by introducing *statistical decision theory*, which tells us how rational agents should balance probabilities with rewards. We then consider how those rewards should be represented, introducing the idea of *utility functions*, and connect this approach to Bayesian inference via evidence accumulation. Many real-world settings do not involve making a single decision in isolation, but rather a sequence of decisions, which leads us to the topic of *reinforcement learning*. A further extension to Bayesian theories of decision-making is required to account for decisions concerning the learning process itself. Rather than passively processing whatever data comes our way, the brain is engaged in *active learning*, directing its limited information-processing capacities to sample and process information that is likely to prove most valuable or interesting. We conclude the chapter by considering the apparent contrast between Bayesian theories of decision-making (especially of basic cognitive processes, from detecting sensory signals, recalling memories, or planning and executing motor movements) and the vast body of empirical literature in psychology and behavioral economics, which appears to show that people's decisions often radically depart from rational action.

7.1 Minimizing Losses: Statistical Decision Theory

To begin, we consider the question of what an agent should do if it has well-defined subjective probabilities (in line with Bayes' rule and the rest of the laws of probability, of course) and has a well-defined objective that can be quantified in numerical terms. Suppose, for example, that a person is attempting to detect faint targets—such as brief flashes—in an environment where there are also distractors (and also noise in the perceptual system itself). Suppose that we study this detection problem in the lab, running an experiment with a series of discrete trials, where on each trial the experimental participant either presses a button (to signal that the target is present) or does nothing (signalling that the target is absent on that trial). Perhaps the simplest way to evaluate performance is just to count up the number and types of right and wrong answers that the person gives. There are two types of right answer: a "hit," when the target flash was present and the button was pressed; and a "correct rejection," when the target was absent and the button was not pressed. And there are two types of wrong answers: a "miss," when the button was not pressed even though the target was present; and a "false positive," where the button was pressed while there was no target.

This type of setup can be modeled via statistical decision theory (Berger, 1993). For largely historical reasons, in statistical decision theory we normally talk about minimizing a loss rather than maximizing an objective. So, applying the simplest possible loss function, a 0-1 loss function, we can assign a score of -1 to every mistake; and 0 to every correct answer. Then the objective of our hypothetical person can be modeled as minimizing the sum of the losses. When choosing an action, of course, the agent doesn't yet know what these losses will be. Hence, the natural strategy is to choose the action that minimizes *expected* losses, where the expectation is based on one's current subjective probabilities. So, for example, with the 0-1 loss function, on each trial the agent simply needs to press the button if the subjective probability of the target, given the sensory evidence and prior information, is greater than 1/2; if the probability is less than 1/2, then the agent should not press the button. And if it is exactly 1/2, then either pressing or not pressing the button has the same expected loss, so that either option is equally good, and the choice can be made arbitrarily.[2]

7.1.1 Asymmetric Loss Functions

This type of signal-detection task has long been studied by psychologists (Green & Swets, 1966), in applications as diverse as detecting brief flashes of light, radar images indicating approaching enemy aircraft, or medical scans that may indicate cancer. And in general, a 0-1 loss function will be too simple: some errors are much more important than others. So, for example, a "false positive," in which a person is wrongly suspected of having cancer, and then subjective to further testing, is an annoyance; but a "miss," in which a person with cancer is overlooked, and hence does not receive potentially life-saving treatment, is a disaster. To account for this, we need what is called an "asymmetric loss function"—which can apply different penalties for the two losses (and still a loss of zero for "hits" and "correct rejections"). Suppose, for example, that in our cancer detection case, we judge that a "miss" should incur a loss that is 1,000 times greater than a false alarm. For concreteness, let's set these losses to -1 and $-1{,}000$, respectively. Suppose that our priors and sensory information from the scan lead us to a posterior probability p that the person has a cancer, where p is a smallish number, say 0.01.

We have two actions: declaring a "positive" or a "negative" test result. If we were to stick to the original 0-1 loss function, the expected loss of declaring a positive test result is $p.(0)+(1-p)(-1)=p-1=-.99$. The expected loss of declaring a negative test result is $p.(-1)+(1-p).(0)=-p=-.01$. We want to minimize our expected loss, so we should declare the test negative (and presumably send the patient away with no plans for further investigation). But suppose that we switch to our asymmetric loss function, taking account of the fact that missing a cancer is much more serious than a false positive. Now the expected loss of declaring a positive test result is $p.(0)+(1-p)(-1)=p-1=-.99$ as before; but now the expected loss of declaring a negative test result is $p.(-1000)+(1-p).(0)=-p=-10$. Now the expected loss is minimized by declaring a positive test result, even though the actual probability of having cancer, given the positive test, is rather remote. The asymmetric loss function causes the agent to err on the side of caution, and most likely make some further investigations rather than sending the patient home with a clean bill of health.

The type of approach can be generalized in many ways. For example, rather than merely detecting a target, the aim might be to categorize a target and respond appropriately (e.g., eating fruit that is ripe, storing fruit that is unripe, and disposing of fruit that is overripe or moldy). Now there will be a payoff matrix between categories and actions, and the loss function will not be symmetrical, of course. Eating moldy fruit is a much more serious error than occasionally disposing of potentially edible fruit. But the same approach,

2. The prior will typically involve incorporate base rates, concerning how often the target tends to appear, but might capture more complex patterns, as when the present location of the target seems to follow some regularity. Equally, the presence or absence of the target might depend on previous experience concerning what the target looks like, how variable it is, and so on—the calculations concerning the subjective probabilities can be arbitrarily complex.

choosing actions that minimize expected loss given the agent's subjective probabilities, can be applied.

7.1.2 Continuous Actions

In many contexts, the natural measure of performance depends on not just choosing the right category of action, but on the real-valued precision of a continuous-valued action.[3] So, for example, when reaching for an object, it might matter how close we are to the target. In other cases, our output may not be a physical movement but could be a numerical judgment: for example, the market value of an antique, the length of a river, or the population of city. In statistics, two particularly popular loss functions for real-valued outputs are the squared (or quadratic) loss function, where loss is the sum of the squared distance between the estimate y and the target t, $(y-t)^2$; and the absolute value loss function, where loss is the sum of the absolute distance between estimate and the target, $|y-t|$. The squared loss function is, of course, widely used as a default in regression problems in statistics and machine learning (see, e.g., Hastie, Tibshirani, & Friedman, 2009). Both loss functions are minimized if the error is zero: if an action or prediction precisely hits the target value, the quadratic loss function is more sensitive to large errors (because those errors are squared).

From the point of view of choosing actions, these loss functions are very simple special cases. The loss function involved in real-world behavior will usually need to be tailored to the specific behavior being considered. So, for example, in guessing the age of a small child, the "loss" in terms of the irritation of the child might be large for underestimation but small for overestimation. So minimizing expected loss will encourage people to give upwardly biased estimates. Or suppose that our action is making an offer on a second hand car. If an agent has a probabilistic model of the likely minimum price that the seller will accept, given the characteristics of the car, the seller, and so on, how should the agent choose what sum to offer? Here, too, the loss function must be tailored to the situation. If we offer too little, we fail to buy the car and have to continue searching (incurring costs of time and inconvenience); if we offer too much, we lose financially. A Bayesian approach to decision-making requires that we put these different types of loss on a single scale and choose the action (here, our bid), which minimizes the overall expected loss. And in reality, the story is complicated further by the possibility of the initial bid being followed by subsequent bargaining, and so on. So while choosing actions using Bayesian decision theory is conceptually straightforward—we just minimize expected loss—in practice, it is typically very complex. Thus, a realistic cognitive model will typically assume that such computations must be approximated, perhaps drastically.

In simple experimental contexts, this approach to decision-making can provide a good model of behavior. For example, Trommershäuser, Maloney, and Landy (2003) ask participants to rapidly touch a green target on a touch screen (earning points) while avoiding a nearby, or even overlapping, red target (losing points). The experimenters can measure

3. There is also another important generalization, to minimize a function of the entire action trajectory, rather than merely the end point of the action. We do not consider this types of case here. Examples include choosing a trajectory to minimize energy expenditure, imitating another person's actions, or dancing in synchrony to music or in a particular style.

perceptual and (more important) motor noise in this task, roughly correponding to a Gaussian distribution around the true target. Here, then, the participant must choose where to aim, in the light of the gains and losses available (as noted previously, the task could be reframed entirely in terms of losses and remain formally identical).[4] The elegant aspect of this task is that by aggregating over many trials, the experimenters can observe the degree of noise in participants' responses directly, and also infer where they are aiming. It turns out that people's behavior is well predicted by the assumption that they are attempting to minimizing expected loss in this task—that is, they "aim off" from the center of the green target by the appropriate amount, as a function of the degree of loss associated with the red target.

7.1.3 Deviations from Optimality

Results like these seem promising for a Bayesian model of action, and indeed, there is a large body of literature on Bayesian models of motor control that operate within this general framework (e.g., Körding & Wolpert, 2006; McNamee & Wolpert, 2019). On the other hand, though, there are extremely simple tasks in which people's behavior seems to depart dramatically from the optimal Bayesian response. One particular striking example is the phenomenon of probability matching (for a review, see Vulkan, 2000). In a typical task, on each trial a light is either green or red, and the participant has to guess the color on the next trial—the reward is often just the sum of the number of correct answers. Suppose, in reality, that the green and red colors are selected by independent flips of a biased coin, with probability p of green and probability $1-p$ of red. If the participant can figure out this distribution (or, in some variants, is also told about the underlying mechanism), then the Bayesian choice is straightforward. Suppose that the participant's estimate of the probability of the next coin being green is q (which is in general not quite equal to the true p, of course). Sticking with a framing in terms of losses, let us assign a loss of -1 to incorrect guesses and the usual loss of 0 for correct guesses. Then the expected loss for choosing green is just the subjective probability of a red, $1-q$; and then expected loss for choosing red is the subjective probability of a green, q. Given that the aim is to minimize losses, we should consistently choose green when $1-q<q$ (i.e., when the subjective probability of green q is greater than $1/2$), choose red if q is less than $1/2$, and choose arbitrarily if the probability of green and red is equal. This is simply a roundabout way of stating what might appear to be completely obvious: that we should always choose green if we think green is the most likely next item; and always choose red if we think red is most likely.

While this may be the obvious strategy, it is surprisingly rarely observed in experiments. So, for example, Shanks, Tunney, and McCarthy (2002) find that even after hundreds of trials, and where after each block of 50 trials, people are explicitly told how well they are doing, and how much better they would be doing if they used an optimal strategy, only a rather small number of people end up consistently choosing the more probable option. In many experiments, people's choices are better captured by the simple model by which they select green

4. It is well known, of course, that while framing a task in terms purely of losses, of gains, or of a mixture of losses and gains, makes no mathematical difference, it may affect the actions and choices that people actually make, in ways that are typically viewed as inconsistent with statistical decision theory and indeed with the more general rational choice framework, which we shall describe shortly (Kahneman & Tversky, 1979).

with probability q, and select red with probability $1 - q$: that is, their responses *match* the probabilities of each outcome. The precise conditions under which probability matching occurs, and how it is to be explained, have been widely debated. Linking back to the sampling approximations of Bayesian inference that we described in chapter 6, it is interesting to note that one simple explanation is that people are choosing red or green by drawing samples from the underlying distribution, rather than minimizing expected losses (see, e.g., Vul, Alvarez, Tenenbaum, & Black, 2009). But for the moment, the key point is that people seem systematically to fail to solve a Bayesian decision problem despite its extreme simplicity.

7.2 Utilities and Beliefs

So far, we have taken a loss function as given. But the question of what (if anything) human behavior is attempting to optimize in a particular setting (or indeed more broadly) is typically challenging. Only in very restricted circumstances, such as maximizing points in a video game, is there a clear and externally given objective. But our daily lives require us to choose complex courses of action without any externally given, well-defined objective, and where many objectives need somehow to be traded off against each other.

Let us step back a little. In general terms, deciding what to do should, as we have indicated, depends partly on one's beliefs (about the external world, and sometimes also the thoughts and likely actions of other agents that it contains); and the formation of beliefs, and the concepts of which they are constructed, have been the focus of this book so far. But decisions depend not only on what an agent believes, but also on its *desires*, *goals*, or *objectives*, for which we shall use the blanket term *utility*. Most normative theories of decision-making propose a rather strict conception of utility: that each relevant state of the world, S_i (which might arise as a consequence of one's action), is associated with a number, representing the utility $U(S_i)$ of that outcome, for the agent.[5]

A rather stripped-down notion of utility is in play here. So, for example, an agent's utility may not turn purely on its own well-being, or its ability to meet its own objectives, but might depend on the well-being of others, or the attainment of some purely external goal. There is no assumption that utility need be reducible to sensory pleasures and freedom from physical pain (although this viewpoint was popular among early utilitarian economists and political philosophers such as Jeremy Bentham, Francis Edgeworth, and Henry Sidgwick; see, e.g., Cooter & Rappoport, 1984), but might be determined by abstract goals; and there is no assumption that the agent need be aware of their desires, or indeed have awareness of any kind.

In the laboratory, the objective of experimental participants can sometimes be specified externally: to maximize the number of points in a game, such as the experimental game of hitting green (and avoiding red) targets that we discussed earlier. Similarly, our performance might be scored based on our ability to produce the correct answer in an arithmetical calculation, to recall accurately which items on a list have been seen before in an earlier phase of the experiment, or to determine correctly when a signal is seen against a noisy background.

5. We will henceforth use the standard terminology in economics and psychology, in which people are seen as maximizing utilities rather than minimizing losses. Of course, any maximization problem can be recast as a minimization problem, and vice versa, simply by flipping the sign of the function to be optimized.

In this type of case, computational models can straightforwardly be associated with utilities that directly capture the structure of the task—we are in the familiar territory described in the previous section. But we can also make some headway in modeling thought and behavior even when there is no externally given objective.

To start, let us note that, quite generally, when choosing an action, we do not know with certainty what the consequences of that action will be. Indeed, if each action had only one possible outcome,[6] then deciding which action to choose would be fairly straightforward: simply choose the action leading to the outcome with the greatest utility. The standard (although by no means the only) way to choose an action is to aim to maximize *expected* utility—though, crucially, this utility is not an externally given standard but is assumed to reflect the aims of the agent. So consider an agent contemplating an action, a. If the agent takes action a, it believes that each possible state of the world that could be the outcome of this action s has probability $P(s|a)$. Then the expected utility of action a, $EU(a)$, is the sum of the utilities of each possible outcome of the action, weighted by the probability of each outcome:

$$EU(a) = \sum_{s} P(s|a)\, U(s). \tag{7.1}$$

The principle of maximizing expected utility provides a general-purpose criterion that applies, in principle, to a wide variety of decisions and has been applied to foraging, investment, partner choice, shot selections in tennis, and many other decisions. For each action that you might take, just consider the probabilities and utilities of the possible consequences of each action, and hence derive that action's expected utility. Then just choose the action with the greatest expected utility.

This bracingly general and direct formula for deciding what to do is not necessarily easy to follow in practice, however, and we shall explore some of the complexities that arise in building cognitive models of decision-making in the following sections. A first complication is that the very existence of a meaningful utility measure, which the agent can be viewed as maximizing, cannot be taken for granted. It is this problem to which we now turn.

7.3 When Can a Utility Scale Be Defined?

The Bayesian approach to decision-making proposes that behaviors can be understood as maximizing expected utility, perhaps approximately. This approach can get started, of course, only where a notion of utility is well-defined. As mentioned previously, in psychological experiments the participant's objective is often specified directly—for example, to maximize the number of points or to make as few errors as possible. Similarly, in evolutionary arguments in biology (e.g., concerning sex ratios, mating, or child-rearing strategies), some variant of Darwinian "fitness," perhaps defined at the level of the individual gene rather than the whole organism, is a useful externally defined objective (Dawkins, 1978). But, in general, the goals being pursued in human behavior are not prespecified. Indeed, people typically have a myriad of diverse aims that appear to compete for their attention.

6. This special case, where actions have a certain outcome, is known as a deterministic Markov decision process (MDP).

Thus, a driver may want to arrive quickly, drive safely, avoid traffic violations, plan a meeting, and send an urgent message to a colleague. It may be difficult to satisfy these objectives simultaneously: objectives, such as speed and safety, may conflict and need somehow to be traded off against each other.

To apply the expected-utility perspective, we need to be able to combine diverse constraints and aims into a single overall measure (a scale of utility) reflecting the relative importance of each. If such a scale of overall utility can be constructed, then in principle at least, the driver's problem is now clear: the best sequence of actions is that which leads to the maximum expected utility. But when can such a utility scale be defined? That is, what rationality constraints need be imposed on a person's choices, such that it is possible to explain their behavior in expected-utility terms at all (see chapter 2 for a broader discussion of constraints on rational coherence)?

A naive approach to this problem is simply to construct a utility function directly: for example, we might try to measure each objective on a continuous scale and take a weighted sum of these as our overall utility function. But, of course, this approach is unlikely to capture an agent's preferences successfully. For example, it is not clear how objectives concerning properties as diverse as speed, safety, and probability of traffic violations can be measured on comparable scales, how they should be combined, and what weight should be attached to each.[7] Fortunately, though, there are general results that establish the conditions under which a utility scale can be defined merely by looking at the structure of an agent's preferences.

7.3.1 From Preferences to Utilities

Suppose, for a moment, that we ignore any issues of risk and uncertainty and consider an agent's choices between outcomes that are certain, such as between known foods, activities, or consumer goods. An ideally rational decision-maker might be presumed to have preferences that follow a number of natural rules. Suppose, for example, that the decision-maker can compare any two outcomes A and B, either preferring B to A (which we write as $A \preceq B$) or preferring A to B ($B \preceq A$), or being indifferent between them.

Suppose further than the decision-maker's preferences are transitive: if $A \preceq B$, and $B \preceq C$, it seems reasonable that $A \preceq C$. These *completeness* and *transitivity* assumptions are sufficient to ensure the existence of a utility function U, which assigns numbers to outcomes, A, B, so that $A \preceq B$ if and only if $U(A) < U(B)$, and the agent is indifferent between X and Y just when $U(X) = U(Y)$.[8]

This utility function encodes an *ordering* over outcomes, from most to least preferred, but it does not capture the *strength* of preference between those outcomes. Any stretching or squashing of those numerical values, so long as order is preserved, will do equally well as a utility function, if we are choosing, say, whether we would prefer to be given an apple or an orange. All that matters is which items have higher (or lower) utility values. This

7. Indeed, diverse scales may have fundamentally different properties (e.g., being nominal, ordinal, interval, or ratio scales), in the terminology of measurement theory (Narens & Luce, 1986), making combining such scales inherently problematic.

8. This is true if the number of outcomes is finite or countably infinite. If we must choose between outcomes parameterized by one or more real-values, an additional technical continuity assumption is required.

dependence only on order is captured in the term "*ordinal* utility"—and it turns out that the minimal notion of ordinal utility provides a sufficient foundation to construct many parts of microeconomics, such as theory of supply and demand in the formation of market prices (e.g., Kreps, 1990).

From the point of view of cognitive science, however, a richer notion of utility is required. The theme of this book is that cognition involves dealing with an uncertain world, and probability theory provides a framework for understanding how this is possible. Accordingly, we require an account of decision-making that can reflect the fact that our actions may often lead to a variety of possible outcomes. Consider, for example, a simple action such as picking up a cup of coffee. On the one hand, we do not want to expend inordinate amounts of time and care in such a simple action; on the other hand, as our movements become more hurried, the probability of spillage increases; in this, as in many actions, some balance between effort and probability of success must be found. And to make this trade-off sensibly, we need to have to know more about *how much* we prefer different outcomes.

For simplicity, consider the case where actions correspond simply to choosing between monetary gambles (e.g., let us imagine that our decision-maker is at the casino). Suppose, for example, a person has the choice of \$50 for certain, or a .5 probability of either \$0 or \$100. Given only an ordinal scale of utility, we can say only that $U(\$100) > U(\$50) > U(\$0)$, given the minimal assumption that more money is better than less. But to determine whether our decision-maker should gamble or play it safe, we need to know *how much* the utility of \$100 exceeds the utility of \$50, in relation to how much the utility of \$50 exceeds the utility of \$0. Hoping to buy a last-minute ticket to an expensive concert, the decision-maker might very much prefer \$100 to either \$50 or \$0, because only this amount is enough to buy the ticket; such a decision-maker might be expected to take the gamble. Other decision-makers, requiring just \$5 for a pizza, on the other hand, might have the opposite preference, particularly disliking the \$0 outcome, which might leave them hungry. In short, what is required is a *cardinal* utility scale: a scale that assigns a meaningful numerical value to each state so that, in particular, the difference in utility between states can be quantified.

7.3.2 Deriving Real-Valued Utilities

It turns out that our previous assumptions of completeness and transitivity, now applied to *gambles* rather than certain outcomes, imply, along with fairly mild technical assumptions, that outcomes of those gambles can be associated with real-valued utilities so that preferences between lotteries over those outcomes are captured by the *expected* utilities of those lotteries.[9]

9. In addition to a version of the continuity assumption, a more substantive assumption is *independence*: roughly, that the utility assigned to each outcome is not affected by the other outcomes that might alternatively occur. For any three lotteries L, M, and N and any probability $p > 0$, $L \preceq M$ if and only if $pL + (1-p)N \preceq pM + (1-p)N$. Intuitively, if the lottery formed by a probabilistic combination of L and N is not preferable than the lottery formed by a same probabilistic combination of M and N (in the same proportions), then and only then $L \preceq M$. So if I prefer yogurt to ice cream, then I should prefer, say, a 50-50 chance of yogurt or fruit to a 50-50 chance of ice cream or fruit. Note, of course, that independence is a claim about the irrelevance of *alternative* possible outcomes; it is entirely possible that I might prefer yogurt to ice cream when eaten alone, but prefer ice cream to yogurt when eaten with fruit (so that preferences about combinations of items will in general not be independent). Some decision theorists reject the independence axiom (and the closely related "Sure Thing" principle; Savage, 1972), which has

Such a scale can, at least in principle, be constructed from preferences, if we allow preferences to range over gambles rather than just fixed outcomes. Consider the following procedure. First, pick the worst possible outcome under consideration, w, and the best, b, and arbitrarily assign these two to have numerical utilities $U(w)$ and $U(b)$, where, of course, $U(w) < U(b)$. For concreteness, and without loss of generality, let us set $U(w) = 0$ and $U(b) = 1$, so the utilities of all states under consideration will lie on the $[0, 1]$ interval. Then, pick any other outcome, s_i, that is preferred to w but less preferred than b. Using the assumption that any relevant options can be meaningfully compared, just as in the case of ordinal utility, then we can ask whether s_i would be preferred to a gamble having the best outcome, b, with probability p_i and the worst outcome, w, with probability $1 - p_i$. If p_i is high enough, the gamble will be preferred, of course; if p_i is low enough, then it will be rejected. For each s_i, there must be some value p_i at which the balance is struck—the decision-maker is indifferent between the certainty of the outcome s_i, and a gamble with a probability p_i of yielding b and a probability of $1 - p_i$, of yielding w (we shall leave aside a discussion of the assumptions required to make this line of reasoning rigorous, and whether those assumptions are justified (Neumann & Morgenstern, 1944; Edwards, 1954; Kreps, 1988)).

If we follow this procedure for each outcome s_i, the corresponding probabilities, p_i, provide a real-valued measure of the goodness of those outcomes. The best state b has by assumption a value of 1 (b is, of course, trivially equivalent to a gamble that yields b with a probability of 1); the worst state w has a value of zero (because this state is trivially equivalent to the gamble that yields b with probability zero and w with probability 1). Then the higher the value p_i that is associated with the outcome s_i, the higher its utility. Indeed, this value can play the role of the cardinal utility of s; it will allow us to determine the preferences of a decision-maker both over outcomes and gambles over outcomes.

From this standpoint, then, how do we assign a utility to an arbitrary gamble g, which has outcome s_1 with probability p_g and outcome s_2 with probability $1 - p_g$? First, we associate each outcome s_1 and s_2 with the equivalent subgambles involving the best and worst states, b and w, associated with probabilities p_1 and p_2, respectively. Let us call these gambles *best/worst mixtures*. Then the decision-maker should be indifferent between our original gamble g and a gamble with a probability p_g of facing a best/worst mixture parameterized by probability p_1; and a probability $1 - p_g$ of facing a best/worst mixture parameterized by probability p_2 (see figure 7.1). But, assuming that the goodness of a gamble purely depends on its outcomes and their probabilities, then we can collapse this two-stage gamble into a one-stage gamble. Specifically, in the two-stage gamble, there are two independent ways in which the best possible outcome b can arise: with probability p_g, we face the subgamble parameterized by p_1 and win—a sequence with probability $p_g p_1$; and with probability $1 - p_g$, we face the subgamble parameterized by p_2 and win—a sequence with probability $(1 - p_g)p_2$. Thus, the total probability of attaining the best state b is the sum $p_g p_1 + (1 - p_g)p_2$; otherwise, the decision-maker faces the worst outcome w. Thus, we have a new best/worst mixture, parameterized by the probability $p_g p_1 + (1 - p_g)p_2$ of the best option (otherwise, the outcome is, of course, the worst option). This probability can be

---

some counterintuitive implications, the most famous of which is Allais's Paradox (Allais, 1953). Indeed, there are decision theories that involve weakening various principles of standard expected utility theory (for discussion, see, e.g., Bradley, 2017).

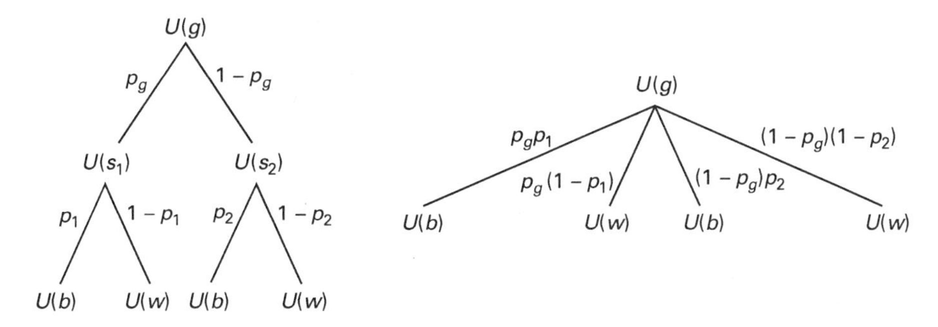

Figure 7.1
Assigning utilities to an arbitrary gamble g using best/worst mixtures of subgambles s_1 and s_2 is equivalent to a single-stage gamble with mixtures of the best and worst gambles b and w.

viewed as a measure of the goodness of a compound gamble—the greater the probability of getting the best rather than the worst outcome, the better.

The generalization to gambles with many outcomes follows the same pattern. A gamble with n possible outcomes $s_1, \ldots, s_n$, where the ith outcome has probability $P(s_i)$, should be equivalent to a best/worst mixture with a probability $\sum_i P(s_i)p_i$, where p_i is a parameterization of the best/worst mixture, which the decision-maker treats as equivalent to the outcome s_i.

To see that the probability of "winning" in the best/worst mixture can be used as a measure of utility, let us blithely rewrite such probabilities as utilities. That is, let us replace p_i with u_i, so our formula for the value of a gamble becomes not $\sum_i P(s_i)p_i$, but $\sum_i P(s_i)u_i$. And this is, of course, just the familiar formula for expected utility: the utility of each possible outcome, weighted by its probability.

So far, we have identified cardinal utility with a particular probability—the probability of winning a best/worst mixture—so these utilities are necessarily defined only on the [0, 1] interval. But this restriction is not necessary. All preferences will be unchanged if all utilities are multiplied by an arbitrary positive number or if an arbitrary constant is added to or subtracted from all utilities. That is, cardinal utility is defined only up to a positive linear transformation—utilities can be defined on any portion of the real line. The absolute size of the numbers used to represent utilities, and whether these numbers are positive or negative, are not important; it is the relative difference between the utilities of different states that matters.

Indeed, more general and sophisticated arguments of this type can be provided. Given surprisingly minimal consistency criteria concerning the preferences of our hypothetical decision-maker (although these criteria may, nonetheless, be violated in reality by human and animal decision-makers), it can be shown that there is a set of utilities *and* subjective probabilities, such that the decision-maker's preferences between options, whether simple states or gambles, perfectly follow the principle of maximum expected utility (e.g., Savage, 1972).[10] The specific utilities and subjective probabilities will vary from

10. The Bayesian approach, here and through out this book, models uncertainty with probabilities; some theorists, particularly in economics, argue that uncertainty can sometimes be so open-ended that probability cannot always meaningfully be applied (see, e.g., Knight, 1921; Binmore, 2008).

person to person, though—even fully rational agents can still have distinct beliefs and preferences, and hence make very different choices. Rationality simply ensures that these choices are coherent within the individual.

7.3.3 Revealed Preference and Cognitive Science

The result presented in the previous section is particularly interesting from a methodological point of view. It suggests that, given sufficient information about a rational agent's preferences, we should be able to infer both the utilities and the probabilities that the agent assigns to different outcomes. In economics, this observation has been the foundation of the *revealed preference* approach to utility (and, by extension, probability; Samuelson, 1938; Savage, 1972)—the idea that probabilities and utilities are revealed by an agent's choices, rather than being directly measurable psychological or neural properties. From this standpoint, choice behavior is seen as primary, and probability and utility are merely convenient theoretical variables for predicting such behavior. The revealed preference standpoint has been viewed as providing a crucial separation between economics (which requires that minimum consistency assumptions are obeyed, so that convenient utility and probability scales can be inferred) and cognitive science. Crudely, from this point of view, economics need only be concerned with what people *choose*, not what they *think*.

This revealed preferences style of argument has been taken to imply that, given fairly minimal consistency and other conditions (which we have skated over in this discussion), there must be a notion of utility such that a rational decision-maker should always prefer actions with the highest expected utility. As we noted here, the conditions required to establish this result may not always apply to real human or animal decision-makers. Nonetheless, the principle that choices should be determined by maximizing expected utility, where a suitable notion of utility is well defined, is a gold standard in rational models of decision-making across a range of disciplines, ranging from economics and the social sciences, to behavioral ecology, artificial intelligence (AI), and cognitive science.

How should we view such explanation, and, in particular, how should we view rational explanation in cognitive science? Taking up the standpoint of traditional economics, one possibility is that we view the type of Bayesian analysis outlined in this book as claiming only that the mind (or brain) behaves as if it makes probabilistic calculations: the probabilities are presumed to be constructions of the theorist rather than descriptions of internal mental or neural states.

While this may indeed be the appropriate interpretation for some Bayesian models, it is also possible that probabilities (and perhaps utilities) *are* mentally represented, and that behavior is not merely patterned as if the brain carries out Bayesian calculations and calculates maximum expected utility, but is the outcome of such calculations. From this perspective, the brain is able to behave as if it were a probabilistic inference and expected utility maximizing engine, precisely because, in some domains at least, it *is* a probabilistic inference and expected utility maximizing engine. And, as we saw in chapter 6, probabilistic inference need not be carried out precise through the symbolic manipulation of the mathematical formulas of probability theory, but through approximate methods, such as sampling. In the next section, we consider how the problem of accumulating the evidence required to make a simple decision might be achieved by simple psychological and neural mechanisms.

7.4 The Accumulation of Evidence

Let us consider a particular illustration of how we might go beyond the "as if" viewpoint. As discussed previously, perhaps the simplest type of decision, and one extensively studied by psychologists, is signal detection. A person is instructed, say, to respond "yes" on every trial in which a brief flash is present and to respond "no" otherwise (Green & Swets, 1966). The optimal strategy is to say "yes" whenever the posterior probability of the light being present is above some threshold determined by the losses incurred in the different outcomes. We can also derive this optimal strategy within the expected-utility framework introduced in section 7.3.

For example, suppose that the participant obtains a reward of $5c$ each time the signal is correctly detected, loses $50c$ in the case of a false alarm, and receives nothing (i.e., $0c$) otherwise. Under this regime, the participant is likely to be extremely tentative. Suppose, on a particular trial, that the participant estimates that the probability that the signal is present is q. Then their expected utility for saying "yes" is $q\,U(5c) + (1-q)\,U(-50c)$. The expected utility for saying "no," by contrast, is $U(0c)$; for convenience, we can set $U(0c) = 0$ (this is possible with no loss of generality because the utility scale is only defined up to a positive linear transformation). Thus, choosing the option "yes" yields strictly greater expected utility than choosing the option "no" when $q\,U(5c) + (1-q)U(-50c) > 0$, assuming that the absolute value of $U(-50c)$ is very much greater than the absolute value of $U(5c)$ (roughly, losing $50c$ is a great deal worse than gaining $5c$), then this will be true only when q is high. In the special case where utility is a linear function of money, losing $50c$ will be exactly 10 times worse than gaining $5c$, and simple algebra shows that the "yes" response will only have strictly greater expected utility than the "no" response when $q > 10/11$. Notice that we have already seen this type of explanation when minimizing a loss function—but here, of course, we are seeing the problem as maximizing a utility. But, as we've seen already, really there is no difference: maximizing utility is just the same as minimizing a loss function which is set equal to negative of that utility.

Signal detection theory has proved to be a highly effective descriptive model across a wide range of psychophysical tasks. And it has traditionally been viewed from the revealed preference standpoint prevalent in economics (i.e., as assuming only that the experimental participants' behavior conforms descriptively with the theory). But it turns out that signal-detection models can also map naturally onto a simple computational mechanism—*diffusion models*—which can capture how the propensity to push for one decision or another can build up over time (Ratcliff, 1978; Usher & McClelland, 2001; Bogacz, Brown, Moehlis, Holmes, & Cohen, 2006; Brown & Heathcote, 2008; Ratcliff, Smith, Brown, & McKoon, 2016; Forstmann, Ratcliff, & Wagenmakers, 2016).

From a Bayesian point of view, these models can be viewed as accumulating a relative strength of evidence that favors making one decision over another (e.g., evidence that a signal is present or that it is not). Or, to pick what has become an important experimental task (Newsome & Pare, 1988; Britten, Shadlen, Newsome, & Movshon, 1992; Mulder, Wagenmakers, Ratcliff, Boekel, & Forstmann, 2012), suppose that we must decide whether a noisy random dot pattern, briefly presented on a computer screen is flowing overall to the left or to the right.

The flow of dots is noisy: that is, while dots are predominantly flowing either left or right, a fraction of the dots is flowing in the opposite direction. Making an overall judgment, therefore, requires accumulating fragments of information from individual dots—a process that will unfold over time. The sum of these fragments traces a random walk in the strength of evidence favoring one decision or the other, sometimes moving toward one decision, sometimes the other. Specifically, suppose that h_L and h_R are the hypotheses concerning whether the dots are drifting left or right; and let us call the different pieces of sensory data $d_1, d_2, \ldots, d_n$. A simple application of Bayes' rule allows us to compare the posterior odds of the two hypotheses, given the data, as follows:

$$\frac{P(h_L|d_1,d_2,\ldots,d_n)}{P(h_R|d_1,d_2,\ldots,d_n)} = \frac{P(h_L)\;P(d_1,d_2,\ldots,d_n|h_L)}{P(h_R)\;P(d_1,d_2,\ldots,d_n|h_R)}. \tag{7.2}$$

Assuming that each piece of data d_i is independent, given the hypotheses h_L and h_R, then the likelihood term for all the data can be decomposed into a product of the likelihoods of each piece of data:

$$\frac{P(h_L|d_1,d_2,\ldots,d_n)}{P(h_R|d_1,d_2,\ldots,d_n)} = \frac{P(h_L)}{P(h_R)} \prod_{i=1}^{n} \frac{P(d_i|h_L)}{P(d_i|h_R)}. \tag{7.3}$$

Taking logarithms of both sides, the likelihood term now becomes a sum of likelihood terms, each reflecting the strength of evidence in favor of one or an other hypothesis, in the light of each new piece of data d_i. Given the noisy nature of the stimulus, some pieces of data will favor one hypothesis and some will favor the other:

$$\log\frac{P(h_L|d_1,d_2,\ldots,d_n)}{P(h_R|d_1,d_2,\ldots,d_n)} = \log\frac{P(h_L)}{P(h_R)} + \sum_{i=1}^{n} \log\frac{P(d_i|h_L)}{P(d_i|h_R)}. \tag{7.4}$$

As more data is processed, the overall sum will gradually drift in the direction of the hypothesis that is best supported by the evidence.[11]

When the random walk hits a predefined boundary—which signifies the strength of evidence required to trigger a decision—a choice is made. The location of these boundaries will reflect the utilities involved in the decision. In a standard signal-detection experiment, these utilities will be shaped by the different numbers of points or monetary payoffs for hits, misses, and false alarms. The same considerations arise, of course, for real-world decisions. For example, if a person or animal foraging for food must decide whether a fungus should be treated as a mushroom or a toadstool, then considerable evidence that it is a mushroom (i.e., is edible) will be required; even a slight suspicion that it may be a toadstool (i.e., poisonous) may be enough for the fungus to be cast aside. This asymmetry in the position of the decision boundaries captures the fact that the utility gain from eating the mushroom

11. The pioneering Bayesian statistician I. J. Good advocated the second term, the log-odds ratio in favor of one hypothesis as against the other, as an general measure of weight of evidence (e.g., Good, 1950). He attributed the idea to Alan Turing during his development of methods to break the Enigma code (Good, 1979). For a broader discussion of other measures of confirmation, see Crupi, Chater, and Tentori (2013). Non-Bayesian approaches to the analysis of sequential data ignore the priors (Wald, 1947; Green & Swets, 1966). For simple decisions such as those considered here, however, this is not a significant restriction, as the role of the prior can still implicitly be captured by shifting the decision bounds—that is, the weight of evidence required before a decision is triggered.

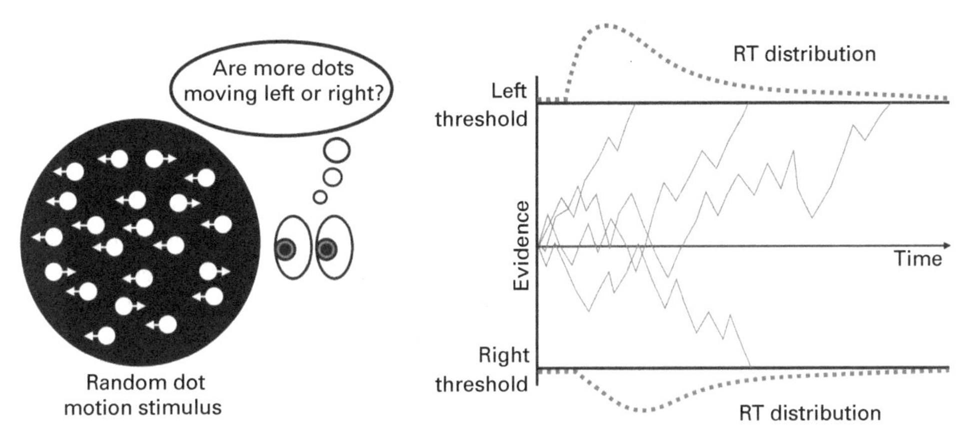

Figure 7.2
The random dot motion paradigm and a visual depiction of the basic drift diffusion model (Ratcliff, 1978). As the perceptual system accumulates evidence for the dots moving to the left or the right, the weight of evidence (depicted with solid lines corresponding to different stimuli) changes in value. When the evidence hits a threshold—either for motion to the left or the right—a decision is made. The response time (RT) distribution results from this stochastic process.

is small compared to the utility loss from being poisoned (though, of course, in extreme circumstances, the utility calculations may be rather different—when near starvation, even hazardous food may have greater expected utility that no food at all).

Diffusion models, whether interpreted in Bayesian terms or not, are widely used and quantitatively successful in many areas of psychology, modeling aspects of perception, categorization, the initiation of movements, and recognition memory, among many other topics (e.g., Hanes & Schall, 1996; Lamberts, 2000; Ratcliff, 1978; Smith & Ratcliff, 2004). One attraction of these models is that they provide fine-grained predictions about distributions of reaction times, speed-accuracy trade-offs, sensitivity of changes of payoffs, and confidence judgments (Pleskac & Busemeyer, 2010; Berg et al., 2016) (see the right panel of figure 7.2 to see how reaction time distributions are generated). Note that, as described, such models are limited to deciding between pairs of options; various generalizations have been proposed (e.g., Usher & McClelland, 2001).

Might the brain be accumulating sensory evidence and making simple decisions by implementing a diffusion model of this kind, thus providing evidence for a neural implementation of Bayesian calculations? An important line of research, involving neural recording in the monkey, suggests that this might be the case (Gold & Shadlen, 2007). In a typical experiment, the monkey is presented with the random dot motion detection task, as described here; and it is rewarded on each trial if it moves its eyes in the same direction as the flow. It turns out that the firing rates of some populations of neurons in the monkey's brain (in the lateral intraparietal cortex) appear closely to track the weight of evidence (e.g., Gold & Shadlen, 2002), rather than, for example, whether a decision is about to be made or which one (although the causal relationship between such accumulation mechanisms and choices has been questioned; Katz, Yates, Pillow, & Huk, 2016). More broadly, a subfield of research mapping neural activity to the mechanism for evidence accumulation and decision-making in a range of perceptual and motor tasks has yielded promising results as part of the general viewpoint that the brain is carrying out approximate Bayesian computations (e.g., De Lafuente, Jazayeri, & Shadlen, 2015; Knill & Pouget, 2004; Pouget, Beck, Ma, & Latham, 2013).

7.5 Sequential Decision-Making

Simple decisions, such as whether a signal is present or absent, or whether a random dot pattern is flowing left or right, are appealing starting points for experimentation and modeling. But, of course, the brain faces decisions of vastly greater complexity on a variety of dimensions. Often the options between which we must decide are themselves highly complex (e.g., when choosing a house, an artwork, a piece of music, or a possible friend), and the process of evaluating sensory and linguistic evidence may also be arbitrarily complex (e.g., in recognizing, making sense of, and evaluating an object, scene, artwork, or person). Here, though, we focus on a specific, and well-studied, aspect of decision complexity: the question of how to choose sequences of actions, or to develop policies for how to act. This is crucial because individual actions typically have no well-defined value except in the light of subsequent actions. Saving rather than spending money now may be advantageous—but not if the decision-maker will squander the accumulated savings in a gambling spree. Similarly, a squirrel caching food for winter will benefit only if it is able to retrieve it later; studying for an exam makes sense only if you intend to take it; reaching for a glass of water makes sense only if you intended to grasp it, and so on. Quite generally, our actions, whether life plans or individual motor actions, make sense only if the individual component actions are part of a larger, coherent framework. This creates particular problems in learning which actions to take because the stream of rewards or punishments received by the agent will depend on combinations of many actions, and it will typically be difficult to determine which individual action should be modified to improve decision-making in the future. Next, we consider some interesting special cases, which have received considerable attention both in machine learning and in the cognitive and brain sciences. We begin by outlining the key mathematical ideas in the abstract—later, we will consider some of the many ways that they can be applied in models of cognition and behavior.

7.5.1 Sequential Decision-Making Problems

Problems related to taking sequences of actions have been studied most extensively in the literature on *planning* and *reinforcement learning*. The basic model of sequential decision-making is the discrete *Markov decision process (MDP)* (Puterman, 1994), which assumes that there is a discrete set of states of the environment $\mathcal{S}$, a discrete set of actions an agent can take in that environment $\mathcal{A}$; a transition function that defines distributions over next states given a previous state and action $T(s, a, s') = p(s_{t+1} = s' \mid s_t = s, a_t = a)$; and a one-step reward function that maps state-action combinations to positive or negative real numbers $R : \mathcal{S} \times \mathcal{A} \to \mathbb{R}$. So in a simple card game, the states of the environment might just be the distribution of cards across players; the actions of the agent might be picking up another card (twist) or doing nothing (stick); the transition function would determine the new state of the cards (which depends on which card is drawn from the pack); and the reward function might be the sum of the points that a person's hand of cards represents (possibly, as in Pontoon, with the crucial complication that if the sum is over some threshold, the person gets zero points). So the challenge of the player is to decide when to stop drawing cards, to maximize the likely points outcome. In the context of MDPs, an agent's behavior is conceptualized as a stimulus-action mapping or *policy*, formalized as a function mapping states to actions, $\pi : \mathcal{S} \rightarrow \mathcal{A}$. Thus, in a card game, a policy specifies, for each distribution of the cards, which action to take (e.g., actions might include picking in up a card from the table, throwing away a card, doing nothing, and others, depending on the game being played).[12]

The first problem that arises in sequential decision-making is that of prediction (also called *policy evaluation*): given a policy and an initial state, how much reward would be obtained in the long term by following the policy? Specifically, suppose that we start with an initial state s_0, repeatedly take the action dictated by a policy $a_t = \pi(s_t)$, calculate the reward $r_t = R(s_t, a_t)$, and sample a new state from the transition function $s_{t+1} \sim T(s_t, a_t, \cdot)$. This generates a trajectory or *rollout* $\langle s_0, a_0, r_0, s_1, a_1, r_1, \ldots \rangle$. What is the long-term reward generated by such a trajectory? While there are different ways of defining what a long-term reward consists of, one standard approach that has nice mathematical properties is to use the *expected cumulative discounted infinite sum of rewards* (i.e., value) associated with a state, given by

$$V^{\pi}(s_0) = E\left[\sum_{t=0}^{\infty} \gamma^t r_t\right], \tag{7.5}$$

where $\gamma \in (0, 1]$ is a discount rate that ensures that the infinite sum converges and the expectation is over the states and rewards generated by following policy π and sampling according to the transition function[13]. The function denoted by V^{π} is typically called the *value function* for policy π (it's worth noting that this is a technical term in reinforcement learning, distinct from intuitive notions of value or uses in economic studies of decision-making). An important property of discounted infinite horizon MDPs is that the value function can be compactly written as a set of stationary recursive equations over states $s \in \mathcal{S}$, known as *Bellman's equations* (Bellman, 1957):

$$V^{\pi}(s) = R(s, a) + \gamma \sum_{s'} T(s, a, s') V^{\pi}(s'), \tag{7.6}$$

where $a = \pi(s)$. Intuitively, Bellman's equation expresses the idea that the value at a state depends on the immediate reward given by $R(s, a)$ and discounted expected value from the next state onward, given by $\gamma \sum_{s'} T(s, a, s') V^{\pi}(s')$.

12. Here, we treat the environment as completely observable—anything that we don't know (such as the order of the cards on the table or the cards of the other players) is simply viewed as a source of randomness, which will affect the results an agent's actions. But this general approach can be extended to deal with *partially observable MDPs (POMDPs)* (Kaelbling, Littman, & Cassandra, 1998), where the agent has to deal with uncertainty about the true state of the environment, as indeed is typically the case in most cognitively realistic scenarios.

13. Note that a discount rate is sufficient but not necessary for the value function to be finite. In cases where the structure of an MDP guarantees that values for a policy are finite, the discount rate can equal 1. In addition, one way to interpret the discount rate is as a constant probability of the task represented by an MDP continuing versus transitioning to a 0-rewarding state forever (i.e., 1 minus the probability of terminating). An alternative interpretation in economics is that *exponentional discounting* is, under certain assumptions, required to avoid what is known as *dynamic inconsistency*, which occurs when a preference between two future options changes purely as a function of time passing. In practice, experiments with humans (and indeed animals) suggest that the psychology of balancing immediate and future rewards is far more complex (Loewenstein & Prelec, 1992), but we put aside such issues here.

It is worth emphasizing that the distinction between reward and value in the MDP model enriches the standard notion of expected utility in an important and cognitively relevant way. Specifically, the reward function models how a decision-maker assigns *intrinsic* utility to states of the world, such as food for a hungry animal.[14] The value function, on the other hand, corresponds to utility that is *derived* from a combination of rewards, the environment, and future behaviors. As we discuss more in section 7.5.4, this makes the MDP model especially useful for modeling learning and computation in sequential decision-making settings.

The second problem that arises in sequential decision-making is *optimal control* (also called *policy optimization*): Given an MDP, what policy would maximize value? Finding a policy with the maximal value function is often what is meant by "solving" an MDP. However, this raises a new question: Since both policies and value functions are functions over states, what exactly do we mean by the maximal function? One that has the highest value at a specific set of states? Any states? All states? Fortunately, one of the attractive mathematical properties of infinite discounted MDPs is that there is a unique optimal value function that has the highest value over all states (there may not be a unique deterministic optimal policy, though, since actions could be tied in value). Furthermore, the optimal value function can also be concisely expressed as a set of recursive *Bellman optimality equations*:

$$V^*(s) = \max_{a \in \mathcal{A}} \left\{ R(s,a) + \gamma \sum_{s'} T(s,a,s') V^*(s') \right\}. \tag{7.7}$$

The intuition here is that the value of a state is determined by the value that we can achieve if we choose the best action—and this action will generate some immediate reward and put us in a new state (according to the probabilistic transition function), which will have its own value. Thus, we can recursively link the value of current and future states.

The Bellman optimality equations express *state values*, but often we are also interested in the closely related value of an action taken at a certain state assuming that we will act optimally from then on. This quantity is often referred to as the *Q-value* (as in "quality") and corresponds to

$$Q^*(s,a) = R(s,a) + \gamma \sum_{s'} T(s,a,s') V^*(s'). \tag{7.8}$$

Once we have Q-values in hand (or some way to quickly compute them from R, T, and V^*), any policy that is greedy with respect to the optimal Q-values (i.e., at each stage chooses the action with the highest, or tied for highest, Q-value) is an optimal policy:

$$\pi^*(s) = \arg\max_{a} Q^*(s,a). \tag{7.9}$$

To summarize, MDPs provide a way to model basic sequential decision-making tasks, and one standard approach to modeling long-term rewards or value is the expected, cumulative, and discounted infinite sum model. This model allows us to concisely define two computational problems: prediction, in which one wants to evaluate a policy at various

14. Whether a notion of intrinsic utility is always applicable is by no means clear, especially for humans.

states, and optimal control, in which one wants to find a policy that maximizes value. Having the Bellman equations is a good start, of course—but we actually need to solve the equations in an efficient way to evaluate policies and determine which policy is optimal. In the following sections, we discuss algorithms from planning and reinforcement learning that can solve these problems under different starting assumptions.

7.5.2 Prediction and Control with a Known Model

Given a known reward $R(s, a)$ and transition model $T(s, a, s')$, several algorithms for prediction and control exist. Control with a known reward and transition model is often referred to as *planning*.

A broad class of sequential decision-making algorithms is based on *dynamic programming*. In dynamic programming, we assume access to the full state space and calculate the value function via backward induction, repeatedly backing up the values of future states onto potential predecessor states until the values of all the states converge. Specifically, starting from an initial value function V_0, we calculate the $k+1$th value at a state from the kth value function (applying the Bellman optimality equation). In the case of policy evaluation, this is

$$V^{\pi}_{k+1}(s) = R(s, a) + \gamma \sum_{s'} T(s, a, s') V^{\pi}_{k}(s'), \tag{7.10}$$

where $a = \pi(s)$. The dynamic programming algorithm for optimal control finds the optimal value function and is known as *value iteration*. It works using backward induction in a similar manner by updating V^*_{k+1} according to

$$V^*_{k+1}(s) = \max_a \left\{ R(s, a) + \gamma \sum_{s'} T(s, a, s') V^*_k(s') \right\}. \tag{7.11}$$

Note that value iteration does not require us to explicitly represent the policy when computing the optimal value function.

Dynamic programming and value iteration form the theoretical foundation for a wide range of other sequential decision-making algorithms, including temporal difference learning algorithms (discussed in section 7.5.3). Moreover, there is also a connection between value iteration and *heuristic search* algorithms, which are used for planning when R and T are known but the state space $\mathcal{S}$ is too large to enumerate completely. In typical heuristic search algorithms, we assume that we are given a set of initial states $\mathcal{S}_0 \subset \mathcal{S}$ and can construct a state transition graph by examining successor states according to T. One approach is to alternate between expanding the transition graph and solving for the optimal solution in that graph, using the solution to guide the next round of expansion. In cases where graph construction is also guided by an admissible heuristic (i.e., one that always underestimates the total cost from a state), this process can be analyzed as a form of *asynchronous value iteration* over a dynamically changing subset of states that is guaranteed to converge to an optimal policy for the initial states. This way of viewing heuristic search provides a unifying perspective on classic deterministic planning algorithms such as A* (Hart, Nilsson, & Raphael, 1968), as well as MDP planning algorithms like LAO* (Hansen & Zilberstein, 2001) and tree search–based algorithms (Kocsis & Szepesvári, 2006). For further details, see Ghallab, Nau, and Traverso (2016).

Aside from dynamic programming, another way to evaluate a policy given a known T and R is to solve the system of linear equations given by equation (7.6). To see why this is the case, we can rewrite the policy evaluation equations in terms of vectors and matrices, where $\mathbf{r}$ denotes a vector of state rewards such that $\mathbf{r}_i = R(s_i, \pi(s_i))$, $\mathbf{T}$ denotes the state transition matrix conditioned on the policy $\mathbf{T}_{i,j} = T(s_i, \pi(s_i), s_j)$, and $\mathbf{v}$ is the vector of state values for which we are solving. In matrix notation, the system of equations specified by equation (7.6) can be written as

$$\mathbf{v} = \mathbf{r} + \gamma \mathbf{T}\mathbf{v}. \tag{7.12}$$

We can then simply rearrange the terms into standard matrix-vector form as follows:

$$[\mathbf{I} - \gamma \mathbf{T}]\,\mathbf{v} = \mathbf{r}, \tag{7.13}$$

where $\mathbf{I}$ is the identity matrix. Once in this form, standard linear algebra algorithms can be used to solve for the value function $V^{\pi} = \mathbf{v}$.

Besides value iteration, an alternative approach to planning in an MDP is to search through the space of policies directly. This idea underlies the *policy iteration* algorithm. Here, we begin with an initial policy π_0 and then repeatedly evaluate it and greedily improve the policy until the policy stops changing. That is, we alternate computing V^{π_k} for a policy π_k (e.g., using dynamic programming) and calculating an updated greedy policy:

$$\pi_{k+1}(s) = \arg\max_{a} R(s,a) + \gamma \sum_{s'} T(s,a,s') V^{\pi_k}(s'). \tag{7.14}$$

This procedure can be iterated until it reaches a fixed point (note, though, that ties between actions must be broken in a consistent manner, or else the algorithm might cycle between equivalent policies and never converge). Reassuringly, and perhaps surprisingly, it can be shown that the resulting policy is globally optimal (Sutton & Barto, 2018).

The algorithms for sequential prediction and control reviewed in this section can be used when the reward and transition functions are known. However, it is often the case that we don't have complete information about the form of a sequential decision problem, so we need to infer at least one of these quantities. We next turn to a family of algorithms for when one function or both functions are unknown.

7.5.3 Prediction and Control with an Unknown Model

How do we evaluate policies or find optimal policies when the model of the environment is not known? This is precisely the type of situation that reinforcement learning algorithms are designed for. Current approaches can be divided into *model-free* approaches, which aim to estimate or optimize the value function without estimating $R(s,a)$ and $T(s,a,s')$ explicitly, and *model-based* approaches, which seek to estimate a model of the environment from which value can be calculated using methods such as those in section 7.5.2 (Sutton & Barto, 2018). Typically, model estimation largely reduces the kind of unsupervised learning problems already discussed in detail in this book—estimating probability densities, inferring latent variables, and building graphical models—so the focus here will be on model-free approaches.

Model-free reinforcement learning algorithms typically define simple learning rules that solve problems of prediction and control without needing to build a model of the environment at all. The aim is simply to learn from experience which actions, and sequences of actions, are the most successful through what can be viewed as a highly sophisticated version of trial-and-error learning. One broad class of model-free algorithms is based on the idea of estimating value functions by *approximate dynamic programming*—that is, by performing a stochastic approximation to true dynamic programming when one can only draw samples from the environment by interacting with it. For example, *temporal difference* (TD) learning methods update a representation of a value function using s, a, s', r samples generated at each time step. The update rule for TD prediction to estimate a value function V^{π} when following the policy π is

$$V^{\pi}(s) \leftarrow V^{\pi}(s) + \alpha(r + \gamma V^{\pi}(s') - V^{\pi}(s)), \tag{7.15}$$

where α is a learning rate. Why does this make sense? The key idea behind the TD update rule is that changes to value estimates are driven by *prediction errors* weighted by the learning rate. More formally, we can observe that at convergence, the value function should match the Bellman equation, so for any state, action, and reward, we expect

$$V^{\pi}(s) = r + \gamma \sum_{s' \in \mathcal{S}} T(s, a, s') V^{\pi}(s'). \tag{7.16}$$

If there is a mismatch between the left and right sides of the equation, then subtracting $V^{\pi}(s)$ from the right side will give us an error signal, indicating whether $V^{\pi}(s)$ is too high or too low based on the values of the other states. Denoting this error signal as δ, we have

$$\delta = r + \gamma \sum_{s' \in \mathcal{S}} T(s, s') V^{\pi}(s') - V(s), \tag{7.17}$$

which justifies the simple learning rule given here, which can now be written as:

$$V^{\pi}(s) \leftarrow V^{\pi}(s) + \alpha\delta. \tag{7.18}$$

However, this still requires us to know $T(s, a, s')$. We obtain the TD prediction rule (equation (7.15)) not be trying to infer this directly, but by treating state s' as a *sample* from the distribution associated with $T(s, a, s')$ – a single sample Monte Carlo approximation to the expectation in equation (7.17); recall the discussion of Monte Carlo sampling from chapter 6. Provided that enough iterations of learning are performed and the learning rate is decreased appropriately over time, the algorithm will converge to a correct estimate of $V^{\pi}(s)$.

In TD control, we want to estimate the optimal action-value function $Q^*(s, a)$ rather than a state-value function since the problem of control is to guide action selection. For example, *Q-learning* uses the following update rule:

$$Q^*(s, a) \leftarrow Q^*(s, a) + \alpha(r + \gamma \max_{a'} Q^*(s', a') - Q^*(s, a)). \tag{7.19}$$

We can understand this update rule using a similar argument to the TD prediction update rule given here. Specifically, the second term on the right side represents a prediction error weighted by the learning rate. Convergence to the true $Q^*(s, a)$ will occur as the number of iterations increase, provided that α is decreased appropriately over time. In addition, a useful property of Q-learning is that it is *off-policy*: that is, the estimate of $Q^*(s, a)$ is independent of the policy followed by the agent and can consequently be based on any sequences of states, actions, and rewards so long as there is sufficient coverage of the state/action space.

7.5.4 Reinforcement Learning and Cognitive Science

Formalisms for planning and reinforcement learning are useful because they provide a unifying, normative framework for understanding adaptation in terms of estimating or maximizing value. In particular, because all correct reinforcement learning algorithms, by design, will converge to a well-defined value function, they inherit some a priori normative justifications as potential models for biological learning. The choice of a *specific* algorithm (e.g., model-based versus model-free learning) reflects different assumptions about computational trade-offs or mechanisms available.

From an historical perspective, the development of the field of reinforcement learning is an excellent example of how attempts to engineer and reverse-engineer intelligent systems can lead to fruitful exchanges of scientific insights across various levels of analysis. The earliest reinforcement learning algorithms were psychological models that formally described behavioral patterns in Pavlovian conditioning (e.g., Bush & Mosteller, 1955; Rescorla & Wagner, 1972). It was later realized that these mechanisms could be recast in the normative framework of dynamic programming (Bellman, 1957) and TD learning (Sutton & Barto, 1987). These basic ideas have served as the foundation for several decades of research on learning in sequential decision-making settings, culminating in a number of breakthroughs in AI over the past decade (e.g., surpassing humans in games such as a wide range of Atari video games, chess, and Go; see Mnih et al., 2015; Silver et al., 2016).

For cognitive scientists, the principles underlying reinforcement learning algorithms provide key insights into adaptation in humans and other species. Here, we review several research threads, starting with cognitively simple models that link TD prediction with Pavlovian conditioning and working our way to more complex models of task hierarchies and model-based planning.

Pavlovian Conditioning and TD Prediction In Pavlovian (or classical) conditioning, an organism learns an association between an unconditioned stimulus that is instrinsically rewarding (e.g., water for a thirsty dog) and a conditioned stimulus (e.g., the sound of a bell that reliably precedes water). In the reinforcement learning framework, the unconditioned stimulus corresponds to a state with a positive reward (s_{UC}, $R(s_{UC}) > 0$) and the conditioned stimulus corresponds to a state without a reward, but that reliably transitions to the unconditioned stimulus (s_C). The estimated value of the conditioned stimulus ($V(s_C)$) then corresponds to the associative strength between the unconditioned and conditioned stimuli, while TD prediction (equation (7.15)) characterizes learning dynamics for building an appropriate association given the organism's experiences. Despite the simplicity of its learning rule, the basic TD algorithm can account for a wide variety of learning phenomena studied in classical conditioning (Sutton & Barto, 1987). Moreover, work in neuroscience paints a compelling picture of how TD learning is implemented in the brain: the reward-prediction error δ described by TD learning has been found to correspond to phasic (i.e., transient) activity of the midbrain dopamine neurons and provides a global signal for synaptic modification (Schultz, Dayan, & Montague, 1997; Glimcher, 2011). These results represent a remarkable convergence of findings at all three of Marr's levels of analysis that are discussed in chapter 1: the problem of value estimation (computational), TD prediction/stochastic approximation (algorithmic), and phasic dopamine (implementation).

Operant Conditioning, Control, and Model-Based versus Model-Free Learning Whereas classical conditioning involves forming value-based associations between states from sequences of observations, operant (or instrumental) conditioning involves forming value-based associations between different states and actions from trial and error (Thorndike, 1898). Specifically, in an operant conditioning experiment, an organism takes an action in a state (e.g., a rat pressing a lever when a light is on), and then an outcome occurs that may be rewarding or punishing (e.g., a food pellet appears). These kinds of scenarios, especially when they involve extended sequences of states, actions, and rewards, are particularly well suited for modeling within the reinforcement learning framework.

As we touched on earlier, one of the most important dichotomies in the space of possible reinforcement learning algorithms is between model-based and model-free learning. Recall that in model-based reinforcement learning, an organism learns a model of the environment (i.e., a transition function and reward function) and then uses this model to compute a value function. Model-based reinforcement learning has been taken to correspond to people engaging in deliberative reasoning about what action makes the most sense in their environment (e.g., Daw et al., 2005). By contrast, in model-free reinforcement learning, the organism learns the value function directly (e.g., using Q-learning). Crucially, from an algorithmic perspective, model-based learning is more flexible but also more cognitively demanding than model-free learning because it involves re-computing the value function as the estimate of the transitions and rewards is updated. In addition, it is interesting to note that the model-based/model-free learning distinction can be mapped onto the familiar psychological distinction between goal-directed and habitual behavior (Wood & Rünger, 2016), although this is not the only way to formalize this distinction (Dezfouli & Balleine, 2013; Miller, Shenhav, & Ludvig, 2019).

In theory, model-based and model-free learning mechanisms are computationally and conceptually distinct, but in real biological systems these two processes are difficult to disentangle completely (Doll, Simon, & Daw, 2012). Over the last two decades, considerable progress has been made in studying the algorithmic interaction between these different forms of learning and control and their neural bases. For example, the *two-step task* (Gläscher, Daw, Dayan, & O'Doherty, 2010) is a simple MDP consisting of two choice stages and an outcome stage with states whose rewards crucially drift over time. The transitions between states in the choice stages and into the outcome stage states are stochastic but predictably above chance, meaning that if participants learn that an outcome stage has the highest reward, then they can engage in model-based planning to reach that outcome state. However, participants could also simply fall back on single-step value estimates provided by a model-free strategy, which would be initially insensitive to new reward information. In critical trials when new reward information is encountered, model-based and model-free learning lead to divergent value updates, thus providing opportunities to distinguish people's algorithmic strategies. Paradigms such as these are often used to study how model-based and model-free learning compete for control of behavior (Gläscher, Daw, Dayan, & O'Doherty, 2010), how they embody different algorithmic and mechanistic trade-offs (Otto, Gershman, Markman,

& Daw, 2013; Daw & Dayan, 2014; Solway & Botvinick, 2015), and how they can interact cooperatively (Kool, Gershman, & Cushman, 2017; Kool, Cushman, & Gershman, 2018).

Distributional RL Standard reinforcement learning algorithms form point estimates of the value function using state, action, and reward samples, but more recent work on *distributional reinforcement learning* explores representing values explicitly as distributions over possible returns (Bellemare, Dabney, & Munos, 2017; Dabney, Rowland, Bellemare, & Munos, 2018; Bellemare, Dabney, & Rowland, 2023). At first blush, it is not obvious why representing a distribution of values would provide any benefit over just representing the expected value—after all, when selecting among actions with different value distributions, we will be computing and comparing expectations. Nonetheless, in practice, value-distributions have been shown to provide a richer target for approximation and thus facilitate representation learning (e.g., with neural networks), mitigate the effects of learning while the policy is changing, and can support a wider variety of downstream behaviors as well as generalization (Bellemare et al., 2017). The success of the distributional approch in deep reinforcement learning has motivated investigations into whether the brain encodes value distributions: Dabney et al. (2020) showed that different dopamine neurons appear to track different levels of value (thus, together, encoding a distribution of values) and thus show a range of positive and negative reward prediction errors during learning. These results enrich the classic picture of how the brain implements scalar reward prediction errors.

Reward Design and Shaping In the standard reinforcement learning problem, we are given a reward function and must find an optimal policy. But we can also go in the opposite direction: given a desired policy, find a reward function that, when maximized, results in an optimal policy that matches the desired policy. This is known as the problem of *reward design* (Singh, Lewis, & Barto, 2009; Sorg, Singh, & Lewis, 2010) and appears in a number of important settings. One example is *reward shaping*, in which we aim to augment an existing reward function in such a way that the optimal policy is preserved, but faster learning is also facilitated. For example, if we want to incentivize a reinforcement learning agent to reach a goal state, we might not just want to provide a single reward for reaching the goal, since that would provide an extremely sparse signal to learn from. Rather, we would want to provide additional shaping rewards for the intermediate steps toward the goal to facilitate faster learning. An important result from reinforcement learning is the *shaping theorem* (Ng, Harada, & Russell, 1999), which provides necessary and sufficient conditions on shaping functions such that they do not change the optimal policy (specifically, that they take the form of "potential functions"). The shaping theorem can be used to design reward functions for people that allow them to achieve a long-term goal but receive more intermediate feedback (Lieder, Chen, Krueger, & Griffiths, 2019). However, it has also been found that when in the role of a teacher, people do not simply provide evaluative feedback consistent with the shaping theorem. For example, people will inadvertently incentivize reinforcement learning algorithms to follow *positive reward cycles*, in which the algorithm systematically deviates from the target behavior to receive a reward for correcting that deviation, followed by further deviation and correction (with further reward), potentially looping indefinitely (Ho, Cushman, Littman, & Austerweil, 2019).

In addition, in the reinforcement learning framework, reward is the driving force behind all adaptation and learning, leading some researchers to propose that maximizing a reward signal is sufficient for explaining all intelligent behavior (Silver, Singh, Precup, & Sutton, 2021). The reward design perspective allows us to pose this thesis as a well-defined question: Given some suitably well specified intelligent behavior, is there a reward function that produces the target behavior when maximized? Abel et al. (2021) analyzed this question for Markov reward functions in MDPs (rewards defined over s, a, s' tuples) and found that for behaviors defined in terms of sets of policies (a generalization of a single optimal policy), such reward functions can fail to exist. For example, in a grid world with a state space corresponding to locations in the grid, the behavioral rule "always go in the same direction" cannot be expressed by a Markov reward function. One important takeaway from these results is that it is not always obvious what the expressivity of certain classes of reward functions are with respect to a given MDP. Such findings motivate ongoing research on learning and optimizing nonMarkov reward functions (Vazquez-Chanlatte, Jha, Tiwari, Ho, & Seshia, 2018; Icarte, Klassen, Valenzano, & McIlraith, 2018).

Representations and Reinforcement Learning Combining latent state inference with reinforcement learning offers one approach to modeling interactions between learning, decision-making, and representations, but it is not the only one. How an organism encodes states or actions has consequences for other processes, such as exploration or internal decision-making algorithms themselves (Ho, Abel, Griffiths, & Littman, 2019). For example, consider that the distinction between model-based and model-free learning is just as much about representation as it is about algorithms: in model-based learning, the value function is computed using a learned representation of the transition function, whereas in model-free learning, the value function is learned directly, without a separate representation of the transition function (Sutton & Barto, 2018).

One prominent alternative to either pure model-based or model-free learning is the *successor representation* (Dayan, 1993; Gershman, 2018; Russek, Momennejad, Botvinick, Gershman, & Daw, 2017), in which states are encoded based on whether they predict visiting other states when following a policy (e.g., an optimal policy or random policy). The simplest version of the successor representation for policy π assigns an expected, discounted, and cumulative visitation count to each future state s^+ from a current state s. This can be expressed as a set of recursive equations much like those for the value function of a fixed policy (equation (7.6)):

$$M^{\pi}(s, s^+) = I(s = s^+) + \gamma \sum_{s'} T(s, \pi(s), s') M^{\pi}(s', s^+), \tag{7.20}$$

where $I(s = s^+)$ is an indicator function that plays a similar role as the reward function in the policy evaluation equations—it counts each visit to state s^+ by evaluating to 1 when the current state s is s^+ and 0 otherwise. The similarity between Equations (7.6) and (7.20) has the important consequence that the value of a state can be expressed as a linear combination of the successor representation and the state reward function:

$$V^{\pi}(s) = \sum_{s^+} M^{\pi}(s, s^+) R^{\pi}(s^+). \tag{7.21}$$

To see why equation (7.21) is equivalent to equation (7.6), we can replace M^π with its definition and do a bit of algebra:

$$V^\pi(s) = \sum_{s^+} M^\pi(s, s^+) R^\pi(s^+)$$

$$V^\pi(s) = \sum_{s^+} \left[I(s = s^+) + \gamma \sum_{s'} T(s, \pi(s), s') M^\pi(s', s^+) \right] R^\pi(s^+)$$

$$V^\pi(s) = \sum_{s^+} I(s = s^+) R^\pi(s^+) + \gamma \sum_{s^+} \sum_{s'} T(s, \pi(s), s') M^\pi(s', s^+) R^\pi(s^+)$$

$$V^\pi(s) = R^\pi(s) + \gamma \sum_{s'} T(s, \pi(s), s') \sum_{s^+} M^\pi(s', s^+) R^\pi(s^+)$$

$$V^\pi(s) = R^\pi(s) + \gamma \sum_{s'} T(s, \pi(s), s') V^\pi(s').$$

Conveniently, the same algorithms used for estimating value (e.g., dynamic programming or TD prediction) can be used to estimate M^π (Dayan, 1993). And as can be seen from equation (7.21), the same successor representation M^π can be used to derive the value of the policy π under *any* state reward function by simple multiplication and summation. This property of the successor representation has motivated its use as a model of policy-based transfer learning in neuroscience (Momennejad, Otto, Daw, & Norman, 2017) as well as in deep reinforcement learning (Barreto et al., 2017). The successor representation also connects to a broader set of ideas based on *predictive maps* in sequential decision-making, which has been used to model grid cells in the hippocampus and other aspects of neural implementations of cognitive maps (Stachenfeld, Botvinick, & Gershman, 2017; Behrens et al., 2018).

Actions can also be represented in different ways. For instance, the same behavior of reaching for a glass of water can be construed at an extremely fine-grained level (e.g., individual muscle contractions) or at a more abstract level (e.g., quenching one's thirst). This intuition motivates *hierarchical reinforcement learning*, in which actions are represented, selected, and learned about at different temporal scales. Although there are a number of formalisms for hierarchical reinforcement learning (Parr & Russell, 1998; Dietterich, 2000), the most widely used is known as the *options framework* (Sutton, Precup, & Singh, 1999). In its simplest form, the options framework distinguishes between actions at two levels of abstraction: ground actions and options. Ground actions are the familiar atomic actions given by the MDP; options assign ground actions to take over multiple ground states before exiting, and so they are essentially policies or partial policies in the ground MDP. Formally, given an MDP and a collection of options $\mathcal{O} = \{o_1, o_2, \ldots, o_n\}$, one can define an *option semi-MDP* that includes options as temporally extended actions (the original ground actions may or may not also be included). Like regular MDPs, semi-MDPs have Bellman equations. For example, the optimal option-level Bellman equation is

$$V^*_{\mathcal{O}}(s) = \max_{o \in \mathcal{O}} \sum_{s', r, t} T^o(s', r, t \mid s) \left[r + \gamma^t V^*_{\mathcal{O}}(s') \right], \tag{7.22}$$

where $T^o(s', r, t \mid s)$ is the distribution over exit states, cumulative intra-option rewards, and exit times induced by initializing option o at state s. With regard to reinforcement learning in humans and animals, we can ask at least two separate but related questions using the options framework: First, how are options used? Second, how are options discovered?

On the question of using options, work on hierarchical reinforcement learning in humans has examined how people learn action values at multiple levels of abstraction (Eckstein & Collins, 2020), how they learn option values via model-free mechanisms (Cushman & Morris, 2015), and how intra-option prediction errors are realized neurally (Botvinick, Niv, & Barto, 2009; Ribas-Fernandes et al., 2011). On the question of discovering options, several proposals have been put forth, including those based on policy compression (Solway et al., 2014), Bayesian inference (Tomov, Yagati, Kumar, Yang, & Gershman, 2020), and resource rationality (Correa, Ho, Callaway, & Griffiths, 2020) (see chapter 13). Nonetheless, the study of how and why people acquire certain hierarchical action representations—and how to best conceptualize their interaction with subgoals, subtasks, and other forms of abstraction—are currently active areas of research.

Attention and Sequential Decision-Making What is represented and how can also be understood as a consequence of the interaction of decision-making with attentional mechanisms (Radulescu, Niv, & Ballard, 2019). Although there is considerable debate about the extent to which attention is a useful construct (James, 1890; Hommel et al., 2019), for our purposes attention can be seen as the process of biasing or filtering information to facilitate efficient learning and computation during decision-making. Thus, if pure inference is about inducing patterns "beyond the data," attention involves "reducing the data" to be more managable. In the context of single-stage decision-making, models that combine selective attention with reinforcement learning can explain how learning is modulated and mapped onto anatomical substrates of attention (Leong, Radulescu, Daniel, DeWoskin, & Niv, 2017; Niv, 2019; Niv et al., 2015).

Recent work has also studied the role that cognitive control—a form of top-down or goal-directed attention (Miller & Cohen, 2001; Shenhav et al., 2017)—plays in planning. Recall that the planning algorithms described in section 7.5.2 all operate on the assumption of a fixed task representation to optimize a policy. For instance, when using heuristic tree search to plan a chess move, one would simulate sequences of moves and countermoves using a model that instantiates the rules for moving the pieces and the conditions for winning the game. However, there are reasons to relax the assumption of a fixed planning model: First, when planning in the real world, there is often no given model, so the cognitive system must regularly face the challenge of constructing models as required. Second, even in domains with a well-defined ground truth model, such as chess, many details will be irrelevant to planning an immediate action. Finally, classic findings in psychology on problem solving, analogical transfer, and insight suggest that people readily switch between different representations of problems to solve them (Duncker, 1945; Ohlsson, 2012; Holyoak, 2012). Motivated by these considerations, Ho et al. (2022) propose and test a normative model of *value-guided task construal* that takes into account interactions between constructing a model (formalized as selecting a simplified MDP) and optimizing a policy in that model (e.g., using one of the algorithms in section 7.5.2). The key idea is to treat model and policy selection as a two-level optimization process: An outer loop selects a simplified model (a *construal*) that is used by an inner loop planning algorithm to compute an optimal policy. In its simplest form, the outer loop seeks to optimize the value of representation (VOR) over task construals:

$$\mathrm{VOR}(c) = U(\pi_c) - C(c), \tag{7.23}$$

where $U(\pi_c) = V^{\pi_c}(s_0)$, the value of a policy computed under construal c when evaluated on the true task from an initial state s_0; and $C(c)$ is a cost associated with the complexity of a construal. Attention comes into play when considering this cost term—specifically, the cost biases selection toward construals that require attending to fewer details (e.g., obstacles in a maze or pieces on a chess board). Across a series of experiments, Ho et al. (2022) found support for the idea that people flexibly form task representations consistent with this account. An important direction for future research is to explore the algorithms that enable people to form simplified representations, as well as to understand the interaction of model construction with other cognitive mechanisms.

7.6 Active Learning

So far, we have discussed cases in which the utility of the consequences of our actions is directly of interest. But in many areas of cognition, consequences may not be primarily of interest in themselves, but rather because they provide further information. This is the domain of *active learning*, where our actions are, at least in part, not in the service of an externally defined goal, but driven by the objective of gathering relevant information as efficiently as possible. In scrutinizing a map, a page of a book, or a face, our eyes do not wander aimlessly; rather, they but search for features of particular relevance or interest. So our eyes will selectively sample the major towns, roads, or harbors rather than uniform areas of sea; they will jump to the paragraphs that seem likely to contain new and interesting information; and alight on those facial features that are mostly likely to give away identity or emotional expression. Similarly, in deciding whom to talk to, what to type into a search engine, or what to read, we are often foraging for information rather than attempting to achieve any concrete external goal. Indeed, large amounts of human activity, especially in the fields of education (when we are learning, say, history or science) and culture (going to movies, reading novels, or listening to music), involve the acquisition and processing of information that is not in the service of any immediate task or goal. Our attention is limited and must be deployed wisely in a complex world with many attractions and distractions. Sometimes, of course, our focus is much narrower—we have a specific decision to make or course of action to pursue, and we want to gather information that will help us solve the challenge of the moment.

In any case, it should be clear that the process of choosing which information to gather (and, similarly, which information to pay attention to and which to ignore, once that information is gathered) is of central importance in the operation of just about every aspect of cognition. There is a continual cycle, in which our current state of knowledge directs our senses and our attention to actively gather new information; and this new information is then used to update our knowledge state; and we then search for still more information guided by this updated knowledge state, and so on. We are relentlessly active learners about our world, searching for useful and interesting information rather than simply passively taking note of whatever data happen to float into view.

To take a mundane example, suppose that we have lost our keys. We do not simply wait for useful evidence about their location to turn up. We actively search for useful clues. We tap our pockets, peer into bags, and look under sofas, hoping to gather sensory information that will give us a (hopefully decisive) lead. And in gathering clues at a crime scene or designing scientific experiments, we are, equally, attempting to choose sets of actions likely to lead to data that are diagnostic between the alternative theories under considerations (e.g., Lindley, 1956; Platt, 1964). In each of these cases, we actively attempt to find information that is as useful, or generally interesting, as possible.

Now what counts as useful or interesting will vary depending on our objectives (finding our keys, catching a criminal, pinning down the best scientfic theory). And in the absence of a specific goal, we find some information interesting and other information dull. Indeed, much of leisure time is spent engaged in searching for and consuming information without any obvious immediate relevance to our lives (watching movies, cheering on sports teams, reading history and fiction, listening to music, and so on). But while the general question of what makes information interesting is hard and open-ended (Chater & Loewenstein, 2016), the same principle of active learning is at work. Our minds are searching for, and attending to, the interesting; and attempting to avoid information that is dull or useless.

There may seem something slightly paradoxical about the very idea that it is possible to actively choose which data we wish to receive. After all, before we move our eyes to a new location or conduct a scientific experiment, we do not know what data we will receive—otherwise, the act of data gathering would be entirely redundant. But if we do not know what data we will receive, how can we assess its potential value?

The answer, as so often in the Bayesian approach, stems from the use of prior knowledge. So, before moving our eyes or conducting the experiment, we can consider our prior probability distribution of possible sets of data (where "prior," here, simply means *prior to the data-gathering act being contemplated*). Suppose that the agent is able to assign a *value* to each possible data outcome; then the informational value of actions that may result in such data can simply be defined as the expected value, where the expectation is relative to the prior distribution over the data. So, for example, our prior knowledge of the human face, alongside current low-fidelity information from the visual periphery, may be enough to narrow down certain locations in visual space as likely to be much more interesting than others, and hence as much more appropriate targets for eye movements. Thus, for example, while scanning images, it is possible for eye movements to jump between informative elements, such as eyes and mouths, and to pay much less attention to patches of the cheek or forehead, or the wall in the background.

Suppose that our Bayesian agent has a range of possible actions $a \in \mathcal{A}$. In the light of each action, and in the light of the current state of knowledge, $\mathcal{K}$, of the agent, each state of the world $s \in \mathcal{S}$, has probability $P(s|a, \mathcal{K})$ and utility (in contexts where this is well defined) $U(s)$. The expected utility, given knowledge $\mathcal{K}$ and action a, is written as follows:

$$EU(a, \mathcal{K}) = \sum_{s \in \mathcal{S}} P(s|a, \mathcal{K}) U(s). \tag{7.24}$$

The rational Bayesian agent, of course, will choose the action that achieves the maximum expected utility: that is, in the light of knowledge $\mathcal{K}$, it will choose an action $a^*(\mathcal{K})$, such that $EU(a^*, \mathcal{K}) \geq EU(a, \mathcal{K})$ for all a. For simplicity, we assume that $a^*(\mathcal{K})$ is the unique action that attains expected utility $EU(a^*, \mathcal{K})$. Now consider that the agent has the possibility of actively finding out some information (e.g., actively carrying out some observation or experiment, yielding some data $d \in \mathcal{D}$). These new data will be added to the agent's background knowledge, $\mathcal{K}$, creating the new background knowledge $\mathcal{K} \cup d$, and will lead the agent to choose a new $a^*(\mathcal{K} \cup d)$ in the light of updated probabilities $P(s|a, \mathcal{K} \cup d)$. How much utility, $U(d)$, should the agent attach to acquiring this data? The agent must evaluate this quantity in the light of its updated probabilities $P(s|a, \mathcal{K} \cup d)$. Specifically, the key question is how much greater is its expected utility, in the light of these probabilities, if it chooses $a^*(\mathcal{K} \cup d)$ over and above its revised expected utility given its original choice of action, $a^*(\mathcal{K})$? The utility to the agent of d is, therefore,

$$U(d) = \sum_{s \in \mathcal{S}} P(s|a^*(\mathcal{K} \cup d), \mathcal{K} \cup d) U(s) - \sum_{s \in \mathcal{S}} P(s|a^*(\mathcal{K}), \mathcal{K} \cup d) U(s). \tag{7.25}$$

Intuitively, we can reason as follows. Really useful data d change our model of the world so we switch from what previously looked like our best choice of action $a^*(\mathcal{K})$ to a new action $a^*(\mathcal{K} \cup d)$, which, according to our updated probabilities, has a much higher expected value. So, for example, data that tell us where to look for treasure, or that the fungus we were about to eat is actually a toadstool, are valuable. Conversely, data that don't change our action (e.g., about the color of the box containing the treasure or the precise weight of the fungus) are from this narrow viewpoint, entirely valueless.[15]

Note that here, and in a wide variety of contexts, the *expected* value of acquiring new information can only be positive or zero (where the expectation comes from the subjective probabilities of the decision-making agent). That is, if a rational agent asks a question, that agent cannot consistently believe that its choice of action in the light of the answer will, in expectation, be worse than if it had received no answer (it will only change its course of action in the light of new information if it believes that the new action has at least as high an expected utility as its previously entertained course of action). This means that, from the point of view of a fully rational agent, in expectation, acquiring new information cannot be harmful. Notice, though, that from an external perspective, an agent learning new information can both be harmful in a specific situation and harmful in expectation. For example, if in a search for buried treasure, I have by pure chance decided to start digging in precisely the right spot, then the expected value (from an outside perspective) of further information (e.g., from old treasure maps or historical reseach) may well be negative because it can only lead me away from my current (lucky) guess.

We have so far been considering the utility of a piece of data. But, of course, the Bayesian agent does not know which data they will encounter *a priori*. Hence, the expected value of

15. The story can be more complex in a variety of ways. For example, new data may cause us to change utilities, $U(s)$, as well as our probabilities; and sequential decision-making may be crucial. I may find that apparently valueless information about the pattern on a fungus becomes crucially valuable in the light of further information (e.g., the information contained in a field mushroom guidebook). We will ignore such complications here for simplicity.

the observation or experiment, d, is the expectation of $U(d)$ in the light of knowledge $\mathcal{K}$ before the experiment has been carried out:

$$EU(d) = \sum_{d \in \mathcal{D}} P(d|\mathcal{K})U(d). \tag{7.26}$$

To get an intuition for how this works, consider a variant of a well-known psychology of reasoning task (Wason, 1966, 1968), in which people must actively select data in the light of a rule, such as *if a person is in the club, they must be at least 21 years old*, which has the form: if p, then q. In the experimental task, the participant is given four cards, each with an age on one side and whether they have entered the club on the other side. But we can only see the cards face up—the task is to say which cards we would like to turn over.

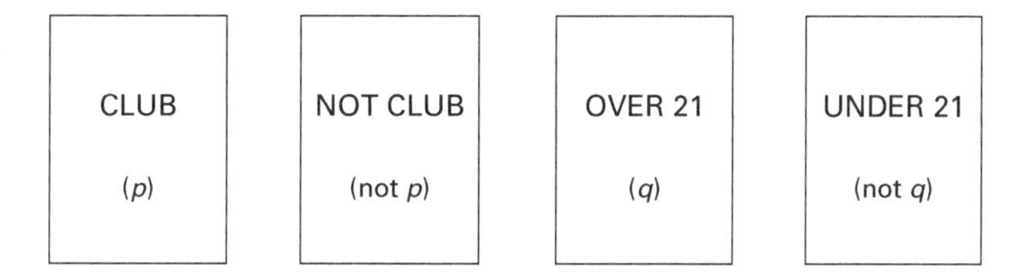

The answer to the question of which cards to turn over (i.e., which information to search for) depends, of course, on our utilities. These are often only vaguely specified in everyday life, and even in many experimental tasks. But these utilities will clearly depend on our objectives. Suppose, for example, that we are a police officer, checking for violations of the rule—and suppose that we get high utility for finding lawbreaking.

Then we can clearly ignore people who are not at the club (we don't turn the not-p card) and who are over 21 (we don't turn the q card). But we do want actively to learn about people who are in the club—the expected payoff for doing so depends on our prior probability, in the light of our background knowledge, that they may be under 21 (and, of course, the utility that we derive from finding any such rule-breakers). And we want to check the under-21's (the not-q card), in case they happen to be in the club. In most realistic scenarios, the expected payoff for turning this card is rather low, though—after all, there is vast numbers of people who are under 21, and the chance that one of them happens to be in the club is low. So turning over the p card will have the greatest expected utility, with some lesser expected utility for the *q card* and zero expected utility for the others (indeed, if we take account of the "effort" of making the inquiry, these options will have a negative expected utility, and hence won't be chosen). This fits with the experimental data (Cheng & Holyoak, 1985; Cosmides, 1989).[16]

But to see how utility is crucial, suppose instead that we are a representative from the student union at a university, checking that people who are over 21 (q card) are not being unfairly turned away (*not-p* card). In this role, we obtain utility not from finding violations of the rule (the p, not-q cases), but from finding exclusions not justified by the rule

16. The Bayesian approach differs, though, from some other theoretical viewpoints. For example, Cosmides (1989) postulates an innate special-purpose "cheater-detection module" to explain this type of experimental data.

(not-p, q cases). To find such cases involves turning over only the not-p and q cards. So which cards we choose—that is, which information we actively choose to investigate—depends not just on the rule, but on our goals; and these changes in card selections depending on the framing of the task are observed experimentally (Gigerenzer & Hug, 1992). Notice that these shifts are not predicted if, as in early discussions of the selection task, the problem of data selection is viewed as purely a matter of "logic," independent of the utilities of the decision-maker.

While this style of analysis may be applicable when we are searching for information to achieve a specific goal (e.g., finding our lost keys or detecting violators of a rule), much active learning has a more disinterested character. The goal of the agent is to find out the state of the world, independent of any immediate implications for action. Indeed, most aspects of perception and cognition seem to fall into this category. Whether browsing through a newspaper, learning a language, or conducting a scientific investigation, we often have little if any sense of whether, or in what way, the information that we are gathering might modify our actions. In such cases, a different approach, based on information theory, can be applied (Lindley, 1956). In this setup, there are no actions, merely mutually exclusive and exhaustive states of the world $s \in \mathcal{S}$, each with probability $P(s)$ (these states might, for example, correspond to alternative categories, scientific hypotheses, or suspects in a murder mystery; note that we omit conditioning on background knowledge K for notational convenience). The amount of uncertainty that we have about which state the world is in is measured by Shannon's entropy:

$$H(s) = \sum_{s \in \mathcal{S}} P(s)) \log 1/P(s). \tag{7.27}$$

In the light of data, d, the so-called conditional entropy, is now

$$H(s|d) = \sum_{s \in \mathcal{S}} P(s|d) \log 1/P(s|d). \tag{7.28}$$

The information gain $IG(s, d)$, the reduction in entropy over s on learning data d, can be written as

$$IG(s, d) = H(s) - H(s|d) \tag{7.29}$$

information gain can be negative or positive. For example, we might be almost certain about some conclusion and then learn information that increases our doubt. In this case, entropy will increase. But following the previous discussion, from the point of view of the agent's own subjective probabilities, the *expectation* of information gain can only be positive or zero (there is something very odd about planning to make an observation that one believes, on average, will leave one more ignorant than before). And this intuition is captured by the information-theoretic approach: therefore, the expected information gain, $EIG(s, d)$, averaged over possible d is

$$EIG(s, d) = \sum_{d \in \mathcal{D}} P(d)[H(s) - H(s|d)] \geq 0. \tag{7.30}$$

The inequality implies that, on average, any new observation or experiment is expected to yield positive or at worst zero information (and this follows from elementary information theory (Cover & Thomas, 1991)). Expected information gain, and closely related notions, have been used as measures of the goodness of experiments, as models of active learning in neural networks (Mackay, 1992b), and to model cognitive phenomena such as how eye movements are directed during reading (e.g., Legge, Klitz, & Tjan, 1997).

Indeed, this approach has also been applied to a variation of the four-card selection task (Wason, 1966, 1968). Suppose that we consider an abstract rule (not one considering clubs, age restrictions, or any other real-world context), such as *If a card has an A on one side, it has a 2 on the other*. And we are presented with the following four cards:

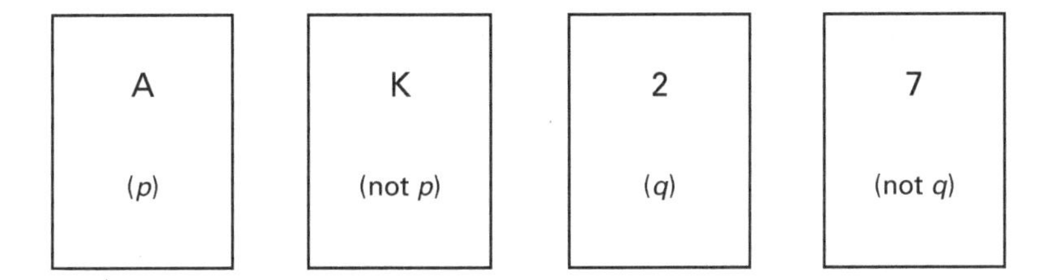

The participant has no particular utilities associated with the rule (they aren't searching for rule violators or cases where people have been treated in a way that the rule does not justify). Instead, the task is simply a matter of gathering information concerning whether the rule is or is not true. In information-theoretic terms, then, we suppose that we begin with some prior assumption (perhaps complete ignorance concerning the truth of the rule), and we wish to turn over the cards that are most likely to reduce our uncertainty, in expectation, as much as possible (Oaksford & Chater, 1994). There are, of course, many ways in which the details of a model of this set-up can be constructed. But to get an intuition about the inferences that people might make, consider a real-world case. Suppose that we are wondering, for example, whether eating tripe makes people sick. The four cards would be as follows:

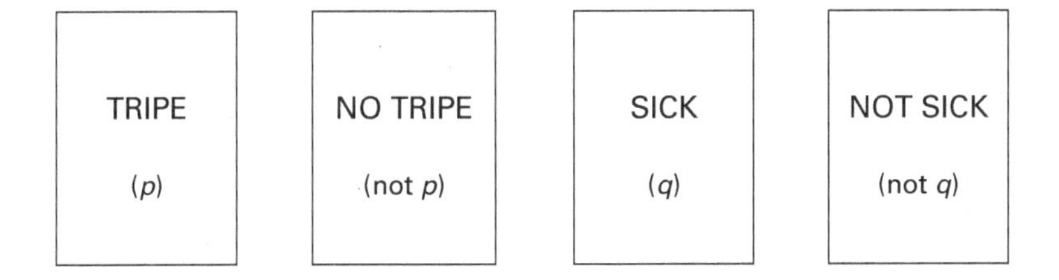

Intuitively, it is clear we should turn over the TRIPE (p) card—it will be very informative to discover whether the person was sick or not. It will also be natural to query the people who were sick (the q card). There are lots of ways of becoming sick—but it might turn out that they were recent tripe-eaters, which would give evidence in favor of the hypothesis. There is also a rather remote possibility of getting useful information from checking people who are not sick (the not-q card), just in case they happen to have eaten tripe—because this would be evidence against the rule. But, as tripe-eating is so unusual, the probability of this outcome is very low, and most likely we will merely sample a healthy nontripe eater, which will tell us little or nothing. So the tendency to actively investigate the cards should have the order p > q > not-q > not-p, which is observed empirically (Oaksford & Chater, 1994), though see Oberauer, Wilhelm IV, and Diaz (1999), which finds that directly manipulating the probabilities of different outcomes can sometimes have at best weak impacts on card choices.

This analysis in terms of expected amount of information gained provides a rational analysis of active data selection in this task, which is particularly intriguing, as it has often been argued that the "logically" correct response to this task is purely to search for falsification of a rule (i.e., thus turning over just the *p* and not-*q* cards) and that turning over the *q* card at all is simply a mistake—a viewpoint that seems to fit with Popper's falsificationist philosophy of science (Popper, 1959/1990), rather than a Bayesian perspective on scientific inference (Howson & Urbach, 1993). The Bayesian active learning framework can also capture a lot of variations of the task, as well as the direction in which changing the probabilities of events *p* and *q* modifies card selection frequencies (Oaksford & Chater, 2003). But people's data selection is not perfectly calibrated with these probabilities—people seem to choose data as if, as is true of the vast majority of real-world rules, *p* and *q* are rare by default, even when this does not hold in a specific experimental context. More generally, this viewpoint helps explain why people frequently adopt a *positive test strategy* (Navarro & Perfors, 2011), searching for instances that confirm a hypothesis of interest in cases where searching for counterexamples has only a small chance of uncovering relevant evidence. Thus, at least in many contexts, the tendency to search for positive instances is not an example of *confirmation bias* but has a rational basis (Klayman & Ha, 1987).

Notice, though, that actively selecting information to gather as much information as possible is defined with respect to a specific set of hypotheses that we wish to test (e.g., whether a specific rule does or does not hold). But often, as noted previously, our aims are much more open-ended—sometimes we may scan a newspaper to see how a particular event turned out, but often we are just wondering if anything interesting has happened. Similarly, in science we are sometimes attempting to design an experiment to test one or more specific hypotheses; but often our inquiries are far more exploratory. The question of how best to capture active learning in these open-ended contexts is important and unresolved—we have only the beginnings of a theory of what makes information interesting (Chater & Loewenstein, 2016).

We have focused here on the problem of determining which data to sample. But at least as important is the parallel question: which computations to carry out, once the data has been sampled. Given presumably severe computational limits on the brain, one of the cognitive system's most important tasks is to carefully direct its computational resources. As with the problem of choosing which information to sample, this very idea has a slightly paradoxical flavor: How is it possible to determine how useful the results of a computation is likely to be before we have carried it out? And, as before, the key is to be able to deploy prior information to determine which computations are likely to be useful, and which not. We explore this question in detail in chapter 13, where we consider how the rational use of limited computational resources might explain some of the ways that human behavior deviates from Bayesian decision theory.

7.7 Forward and Inverse Models

We have seen that minimal consistency conditions on an agent yield the result that the agent can be described as behaving as if it maximizes expected utility, according to some set of probabilities and utilities (which can be inferred from the pattern of decisions that the agent makes). In cognitive science, though, a key goal is to understand the computational processes that support inference and decision-making; hence Bayesian decision theory can be viewed not as a black box model, but rather as a specification of the processes that may underpin decision-making, whether in making simple psychophysics judgments or developing complex sequential policies for interacting with the world. Moreover, neuroscience has begun to provide preliminary evidence that suggests that Bayesian computations may indeed be instantiated in the brain. In this chapter, we have framed our discussion in terms of high-level aspects of decision-making, involving conscious deliberation and choice. But note that Bayesian decision theory applies equally to the analysis of actions over which we exert little or no conscious control (e.g., basic perceptuo-motor processes such as picking up objects, maintaining our balance, or catching a ball). Such tasks involving integrating large amounts of perceptual information and generating plans to control a highly complex motor system. The very fluency of perceptuo-motor control suggests that it may be well approximated by an optimal Bayesian model and indeed, this approach has proved to be very productive (e.g., Körding & Wolpert, 2006)

One key aspect of successful motor control is building a *forward model*, which maps from motor commands to sensory experiences generated by implementing those commands. Such a model is crucial, for example, in distinguishing the sensory consequences of one's own actions (e.g., moving one's head or eyes) from the sensory consequences from changes in the external world (e.g., that the room is being shaken by an earthquake). Bayesian inference can then invert the forward model to create an inverse model, which infers which motor commands will lead either to desired sensory consequences (e.g., in planning an action) or which motor commands did lead to an observed action (e.g., in the interpretation of actions by others).

This logic can be applied not just to one's own actions, but in the interpretation of other people's behavior. Bayesian decision theory provides a forward model, mapping beliefs and desires into actions. But it is also interesting to consider the inverse problem: mapping from observations actions to inferred beliefs and desires. This process of inversion is, of course, the paradigm of Bayesian inference; but it is also of great importance in cognition. The ability to infer people's underlying mental states from their behavior (including, of course, their linguistic behavior) appears to be central to human communication (Baker, Saxe, & Tenenbaum, 2009; Grice, 1957), as well, perhaps as being important for empathy, altruism, and coordinated social behavior (Baron-Cohen, 1997; Houlihan, Kleiman-Weiner, Hewitt, Tenenbaum, & Saxe, 2023). Indeed, a long tradition of research in social psychology suggests that, to some degree, such inverse modeling is involved in the interpretation of our own actions to infer our own beliefs and desires (Bem, 1972; Nisbett & Wilson, 1977). We take up the problems of *inverse decision-making* and *inverse planning* in chapter 14 and draw out links with rational theories of communication in chapter 16.

7.8 The Limits of Reason

The premise of this chapter is that the maximization of expected utility or similar quantities can provide the basis for models of decision-making across a range of domains, from animal foraging to motor control, learning, and high-level decision-making. This viewpoint may appear to clash with the research traditions in the fields of judgment and decision-making and behavioral economics, which appear to show that people frequently and systematically deviate from Bayesian decision theory—and indeed, the basic consistency assumptions that are the foundation of the economic approach to decision-making are routinely and systematically violated (e.g., Kahneman & Tversky, 1984).

Some theorists have argued that the departures from rationality are so widespread that the Bayesian perspective on decision-making, and the rational analysis of behavior more broadly, may is a theoretical blind alley; instead, it has been argued that collections of heuristics or layers of input-output rules might better explain behavior (e.g., Brooks, 1991; Gigerenzer & Todd, 1999; McFarland & Bösser, 1993).

We take the opposite view: that to abandon a rational theory of decision-making is to render human behavior utterly mysterious: indeed, it is to lose the distinction between *behavior* (e.g., picking up a cup, waving to a friend, or typing a message) and mere *movement* (e.g., falling over, having a reflex triggered by the doctor's hammer, or inadvertently leaning on the computer keyboard). Bayesian decision theory helps explain behavior as purposeful activity: our actions are aligned with our preferences and beliefs. Thus, we pick up the coffee cup because we believe that it contains coffee and we want to drink it. The fine details of our motor actions can also be explained in the same terms: our intention to pick up the cup smoothly and efficiently, and to move it without spilling its contents, will help explain the specific way in which we move. More broadly, the Bayesian viewpoint explains how beliefs, preferences, and actions can be linked across many scales (from individual movements, to actions, to momentary plans, to the direction of our entire life), and in a way that is as coherent as possible. If we attempt, by contrast, to see behavior as nothing more than a set of reflexes or a toolbox of special-purpose heuristics, it is difficult to understand the origin of the coherence of human behavior (e.g., Bratman, 1987).[17]

Throughout this book, our use of Bayesian modeling is as a guide to what the ideal solutions to specific inductive problems encountered by humans look like, which can in turn be used as a tool for making sense of human behavior. It would be unrealistic to expect that the approach will always quantitatively model the precise details of human decision-making. We suggest that the Bayesian approach is likely to be particularly effective in domains for which human performance has been shaped by powerful forces of natural selection and learning—motor control, sequencing actions, planning, common-sense reasoning, and so on. It is likely to be much less well-adapted to solving numerical and verbally stated decision problems (e.g., concerning choices between gambles) with which we are unfamiliar.[18] Moreover, the brain cannot follow Bayesian decision theory exactly—in all but the simplest contexts, exact Bayesian calculations are computationally intractable and can only be approximated, for example, by sampling methods (chapter 6 of this book; Chater et al., 2020; Sanborn & Chater, 2016; Vul et al., 2014). But to understand the purposeful nature

17. Of course, our thoughts and behaviors are by no means entirely coherent—but cognition appears to be directed at eliminating incoherence where possible (e.g., Festinger, 1957; Thagard, 2002).

18. We note, though, that understanding how people behave in these unfamiliar contexts, where their behavior may depart substantially from Bayesian decision theory, may be of great practical importance, such as when understanding the tortuous choice processes that people might use to select mortgages, pensions, or courses of medical treatment (Newell, Lagnado, & Shanks, 2022).

of human behavior, it is essential that we view human actions as approximating, a rational model rather than being entirely unconstrained. That is, intelligent decisions can be the product of limited reasoning, but not of no reasoning whatever.

7.9 Summary

The focus of this book is the problem of induction: How is it possible to know the structure of the world from partial and noisy data? But any such learning is useless from the perspective of the survival and reproduction of an organism without the ability to translate knowledge into action (i.e., to combine knowledge with our values to *decide* what to do). Bayesian decision theory provides a solution to this problem, indicating how rational agents should act upon their beliefs. Even simple decisions can involve complex underlying processes of evidence accumulation, and these complexities are magnified when we consider sequences of interdependent decisions. Nonetheless, cognitive scientists have made substantial progress in identifying the mathematical principles underlying human decision-making, building upon and complementing the more general ideas of probabilistic modeling outlined in the previous chapters. As we now transition to the second part of this book, considering more elaborate models and more detailed applications to human cognition, the principles of Bayesian decision theory provide a foundation for linking beliefs and actions.

II ADVANCED TOPICS

Interlude

Our one-word answer to the question of how our minds get so much from so little was simple: Bayes. Part I of this book has expanded that single word into seven chapters, introducing the Bayesian approach and explaining its intricacies. The focus of these chapters has been on articulating the key technical ideas behind Bayesian inference, drawing on research in statistics and machine learning and connecting it to human cognition through examples. Our goal in these chapters was to illustrate how the abstract mathematical principles behind Bayesian reasoning can turn into concrete computational models that capture a variety of aspects of human cognition. The examples that we have provided thus far also hopefully form a blueprint for how a researcher might approach other aspects of human cognition and frame them in terms of Bayesian inference.

Having established these foundations, we are now ready to explore some more advanced topics. Part II of the book introduces some of the ideas that are necessary to turn Bayesian inference into a more complete account of human cognition. What's missing from our story so far? The structure of our knowledge, our languages for representation, are not rich enough. Our understanding of how to perform inference and decision-making with these richer representations, and of how to relate these computations to cognitive processes or neural hardware, is very incomplete. We haven't specified how to measure prior distributions empirically, or explored how this approach applies to core aspects of human cognition, such as language, theory of mind, or reasoning about the physical world.

We are now ready to take up this challenge. The chapters in part II differ in format from those in part I. They are written primarily as reviews rather than tutorials, introducing the broader literature on Bayesian models of cognition and covering a lot of ground rather than going into detail on specific examples of computational models. Each chapter also considers future directions for the topic that it summarizes, pointing to promising paths for further research in these areas.

8

Learning Inductive Bias with Hierarchical Bayesian Models

Charles Kemp and Joshua B. Tenenbaum

To make matters simple, models of learning often focus on a single learning episode. Consider, for example, a child who learns a new concept such as "wombat" after being shown a single labeled example. A Bayesian model might help to explain this outcome using a prior distribution that captures the expectations that the child brings to the episode. This is a useful start, but at least two fundamental questions remain to be addressed. First, where does the prior distribution come from? We have seen that prior distributions play a critical role in Bayesian models, which means that it is important to think carefully about how a learner might end up with an appropriate prior for a given task.

A second, related question asks: How is the current word-learning episode related to other word-learning episodes in the child's life? Children learn the meanings of many words, and it is important to consider how previous learning episodes can help to accelerate future ones. For example, learning previous words may help a child to realize that a novel word such as "wombat" is more likely to refer to an entire object than to a part or a property of an object. Exploiting previous learning episodes is typically possible whenever learners face multiple tasks from the same family—for example, children must learn the causal structure of multiple tools and artifacts, and must discover the patterns of behavior that are appropriate in multiple social contexts. In each case, we would like to understand how children learn to learn—in other words, how learning is accelerated by discovering and exploiting common elements across tasks.

This chapter suggests that hierarchical Bayesian models can help to explain where priors come from and how children learn to learn. The key idea that distinguishes hierarchical models from the simpler models in previous chapters is the notion of *abstraction*. As figure 8.1 suggests, systems of human knowledge are often organized into multiple layers of abstraction. Learning often requires inferences at several of these levels. For example, an infant learning language may recognize that speech sounds are instances of phonemes, strings of phonemes go together to make words, and grammatical rules specify which sequences of words are acceptable sentences (figure 8.1a). Visual experience may allow the infant to recognize that the world contains objects, which are composed of object parts, and that there are higher-level regularities that predict which objects are likely to be found together (figure 8.1d). An infant observing the actions of others may recognize that these actions

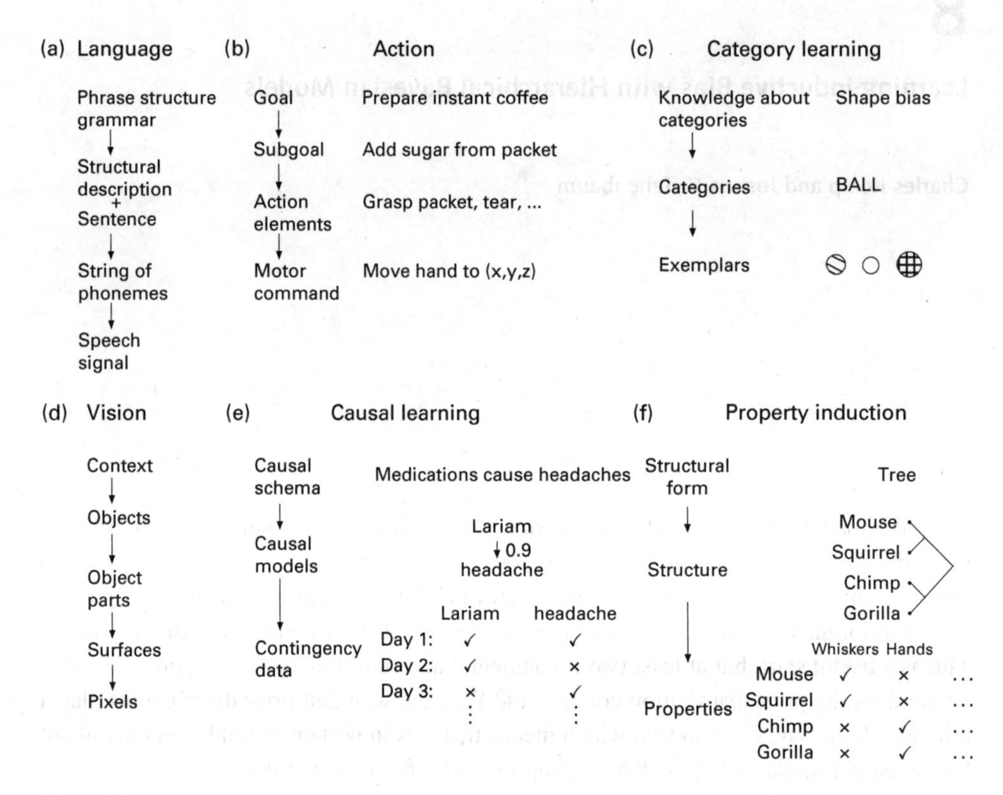

Figure 8.1
Systems of knowledge are often organized into several levels of abstraction. In particular, hierarchies are useful for understanding (a) language, (b) action, (c) categorization, (d) vision, (e) causal learning, and (f) property induction.

are often carried out in the service of goals, and that these goals are achieved by stringing together sequences of lower-level motor commands (figure 8.1b).

This chapter will focus on the remaining three examples in figure 8.1. We consider how children learn multiple categories (figure 8.1c), how children learn multiple causal models (figure 8.1e), and how children learn about multiple properties of a set of objects (figure 8.1f). The hierarchy in figure 8.1c suggests that children may find it easy to learn about individual categories (e.g., balls are round) if they have acquired more abstract knowledge about categories in general (e.g., that objects belonging to any given category tend to have the same shape). The hierarchy in figure 8.1e suggests that learners may find it easy to learn a causal model involving a specific object (e.g., a tablet of Lariam) if they have acquired more abstract knowledge about categories of objects (e.g., medications can cause headaches). Finally, the hierarchy in figure 8.1f suggests that learners may be able to make confident inferences about novel biological features if they have acquired a structured representation (such as a tree) that indicates which animals tend to have features in common.

As we will see, a hierarchical Bayesian model is a probabilistic model defined over an abstraction hierarchy like the examples in figure 8.1. Knowledge at the higher levels sets up prior distributions that are used when reasoning at the lower levels, and probabilistic inference over the hierarchy can explain how the abstract knowledge at the upper levels is

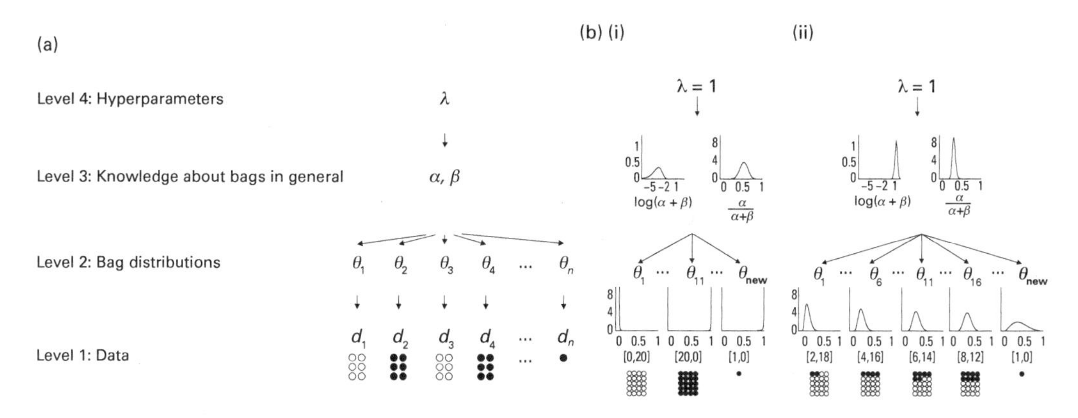

Figure 8.2
A hierarchical model for making inferences about bags of marbles. (a) Graphical model. θ_i captures the color distribution for bag i, and α and β capture more abstract knowledge about the color distributions across the entire population of bags. (b) Inferences made by the model after observing (i) 10 all-white and 10 all-black bags (ii) 20 mixed bags.

acquired. Hierarchical Bayesian models can therefore help to explain how prior distributions are acquired from previous learning episodes, and how these learned priors enable rapid learning in subsequent episodes.

8.1 A Hierarchical Beta-Binomial Model

We will begin by considering a simple hierarchical model that builds on the coin-flipping models discussed in previous chapters. Now, though, we consider bags of marbles instead of coins—bags are analogous to coins, black and white marbles are analogous to heads and tails, and drawing a black marble from a bag is analogous to tossing a coin that comes up heads.

Imagine that you are given a new bag of marbles and you draw a single marble from the bag that turns out to be black. On its own, this single observation does not provide strong evidence about the distribution of colors within the bag. Suppose, however, that you have previously sampled 20 marbles each from 20 previous bags, and observed that each bag contained marbles of only one color—either white or black. Now you might be relatively confident that all the marbles in the new bag are black. Experience with the previous bags has shaped the prior expectations that you bring to the new bag, and the prior that you have acquired allows you to make strong inferences given just a single piece of information about the new bag.

We can capture this inference using the hierarchical model shown in figure 8.2a. The model has four levels. The bottom level records observations that have been made by drawing marbles from one or more bags. Let n_i be the number of balls drawn from bag i, y_i be the number of these balls that are black, and d_i be a pair (y_i, n_i) that summarizes the observations for bag i. The next level up specifies information about the color distribution for each bag. Let θ_i be a variable that specifies the proportion of black marbles in bag i. The next level up includes two variables, α and β, that specify information about bags in general. For example, the information at this level may capture the expectation that each θ_i

is either close to zero or close to 1—in other words, that the marbles in each bag tend to be homogeneous in color.

To convert the hierarchy in figure 8.2a into a hierarchical Bayesian model, we need to define probability distributions that specify how each level is generated given the information at the level immediately above. The prior distribution on each θ_i is Beta(α, β), where Beta() is the beta distribution defined in chapter 3, and the parameters α and β are unknown. We previously considered cases in which the parameters α and β were positive integers, but in general, these parameters can be positive real numbers.[1] As with integer-valued parameters, the mean of the beta distribution is $\frac{\alpha}{\alpha+\beta}$, and the *scale parameter* $\alpha + \beta$ determines the shape of the distribution. The distribution is tightly peaked around its mean when $\alpha + \beta$ is large, flat when $\alpha = \beta = 1$, and U-shaped when $\alpha + \beta$ is small. For example, if α and β are both large and close to each other in size, then most θ_i variables will be close to 0.5, which means that most bags will contain roughly the same proportion of black and white balls. If α and β are both very small and if $\frac{\alpha}{\alpha+\beta} \approx 0.8$, then 80 percent of bags will be almost exclusively black, and the remainder will be almost exclusively white.

The prior distributions on α and β can be defined in terms of one or more hyperparameters. We will consider a model with a uniform distribution on the mean $\frac{\alpha}{\alpha+\beta}$ and an exponential distribution with hyperparameter λ on the scale parameter $\alpha + \beta$. We assume that the hyperparameter λ is fixed, and we set it to 1. In principle, however, we can develop hierarchical models with any number of levels—we can continue adding hyperparameters and priors on these hyperparameters until we reach a level where we are willing to assume that the hyperparameters are fixed in advance.

The resulting model is known by statisticians as a *beta-binomial model*, and can be written as follows:

$$\alpha + \beta \sim \text{Exponential}(\lambda) \tag{8.2}$$

$$\frac{\alpha}{\alpha + \beta} \sim \text{Uniform}([0, 1]) \tag{8.3}$$

$$\theta_i \sim \text{Beta}(\alpha, \beta) \tag{8.4}$$

$$y_i \mid n_i \sim \text{Binomial}(\theta_i). \tag{8.5}$$

Given the hierarchical model in figure 8.2a, inferences about any of the θ_i can be made by integrating out α and β:

$$p(\theta_i|d_1, d_2, \ldots, d_n) = \int p(\theta_i|\alpha, \beta, d_i)p(\alpha, \beta|d_1, d_2, \ldots, d_n)\, d\alpha\, d\beta, \tag{8.6}$$

1. Remember that the general form of the beta distribution is

$$p(\theta) = \frac{\Gamma(\alpha + \beta)}{\Gamma(\alpha)\Gamma(\beta)}\theta^{\alpha-1}(1-\theta)^{\beta-1}, \tag{8.1}$$

where $\Gamma(\alpha) = \int_0^\infty x^{\alpha-1}e^{-x}\, dx$ is the generalized factorial function (also known as the *gamma function*), with $\Gamma(n) = (n-1)!$ for any integer argument n and smoothly interpolating between the factorials for real-valued arguments (e.g., Boas, 1983).

where d_i is the pair (y_i, n_i) that characterizes the observations sampled from bag i. The integral in equation (8.6) can be approximated using the Markov chain Monte Carlo (MCMC) methods described in chapter 6 (see also Kemp, Perfors, & Tenenbaum, 2007).

The upper levels in the model may seem relatively abstract, but they play a critical role—they allow knowledge to be shared across contexts that are related but distinct. In our example, these contexts correspond to different bags. Figures 8.2b.i and 8.2b.ii show two cases in which learning about 20 previous bags influences inferences about a new bag from which a single black marble has been sampled. In figure 8.2b.i, half the samples are entirely white and half are entirely black. The curves at level 2 show inferences about the θ_i variables. For example, the posterior distribution on θ_1 indicates that the model is very certain that θ_1 is near 0. The curves at level 3 show inferences about the α and β variables. The posterior distribution on $\log(\alpha + \beta)$ suggests that α and β are both small, which means that the model has learned that bags tend to be homogeneous in color. The posterior on $\frac{\alpha}{\alpha+\beta}$ indicates that α and β are roughly equal, which means that the model has learned that roughly half the bags contain black marbles and roughly half contain white marbles. Knowing that bags tend to be homogeneous in color allows the model to infer that θ_{new} is close to 1—in other words, that the new bag probably contains a high proportion of black marbles.

Figure 8.2b.ii shows a second example in which 10 percent to 40 percent of the balls in each sample are black. Now the posterior distribution on $\frac{\alpha}{\alpha+\beta}$ is shifted to the left of 0.5, indicating that the model has inferred that there are fewer black balls than white balls across the entire population of bags. The posterior distribution on $\log(\alpha + \beta)$ suggests that $\alpha + \beta$ is relatively large, indicating that each bag is expected to contain a mix of black balls and white balls. As a result, the model infers that θ_{new} is probably not close to 1, even though all balls sampled so far from this new bag have been black.

8.1.1 The Shape Bias in Word Learning

We have focused so far on inferences about bags of marbles, but the hierarchical approach summarized by figure 8.2a can help to explain how children learn words given a single labeled example. Suppose that a mother points at an unfamiliar object lying on the counter and tells her child that it is a "spork." In principle, the child could consider many hypotheses about the meaning of the word: for example, the child might extend the new word to any other object made from the same material, or to any other object that is lying on the counter. By the age of 24 months, however, children are likely to extend the new word to any other artifact that is similar in shape to the original example (Smith, Jones, Landau, Gershkoff-Stowe, & Samuelson, 2002).

The expectation that members of a category tend to be similar in shape is sometimes called the *shape bias*. Extending the hierarchical model in figure 8.2a can help to explain how the shape bias might be learned (see also figures 1.4 and 1.5 in chapter 1). Suppose first that the bags can contain marbles of many different colors instead of just black and white marbles. Now θ_i specifies a distribution over the set of possible colors, and the prior distribution $p(\theta)$ in equation (8.4) is a Dirichlet distribution, which is the n-dimensional generalization of the Beta distribution introduced in chapter 3. Suppose that the model observes marbles drawn from several bags. All marbles drawn from the first bag are red, all marbles drawn from the second bag are green, and so on. Given this evidence, if a single blue marble is drawn from a novel bag, the model will confidently conclude that all marbles in that bag are blue.

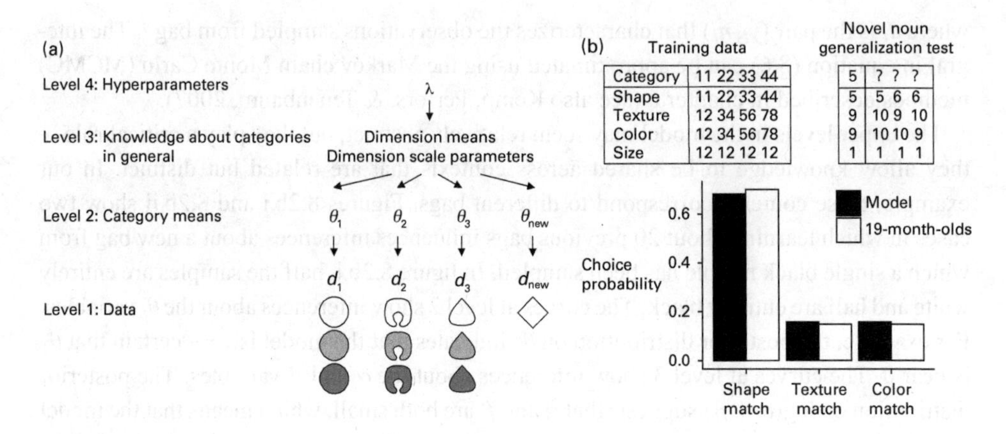

Figure 8.3
An extension of the hierarchical model in figure 8.2 that can be used to learn about categories of objects. After observing three categories that are homogeneous in shape, the model will infer that all objects in the new category are likely to match the single observed exemplar in shape. (b) Model predictions based on a study by Smith et al. (2002), where 19-month-olds were trained on two exemplars of each of four categories. The columns of the first matrix correspond to these exemplars—note that members of the same category have the same shape, but differ in texture, color, and size. The children were then given an exemplar of a novel category (coded here as category 5), and asked to decide which of three choice objects also belonged to this category. Children and the model both prefer the choice object that matches in shape, rather than the choice objects that match in texture or color.

Suppose now that bags of colored marbles are replaced by categories of objects, and the objects vary along multiple dimensions rather a single dimension of color. Figure 8.3 shows a simple example in which a learner observes three objects from each of three categories, and a single object from a new category. Note that the objects vary in both color and shape. We can introduce a copy of the hierarchical model for each dimension: for example, there will be a θ_i^C variable for bag i that captures the color distribution within that bag, and a θ_i^S variable that captures the shape distribution within that bag. Figure 8.3 collects both these variables into a vector labeled $\boldsymbol{\theta}_i$. Just as there is a θ_i variable for each dimension, there are means and scale parameters at level 3 for each dimension.

The observations shown in figure 8.3a suggest that members of any given category can vary in color, and as a result, the posterior distribution on the scale parameter for the color dimension will suggest that this parameter is relatively high. Objects from the same category have the same shape, however, and as a result, the model will infer that the scale parameter for the shape dimension is relatively low. Making these inferences about the scale parameters for the two dimensions means that the model expects that all members of the new category will have the same shape as the single observed example, but members of this new category will vary in color. In other words, learning that the scale parameter for the shape dimension is relatively low has allowed the model to acquire the shape bias, and this bias supports rapid inferences about new categories.

Figure 8.3b shows the outcome when the model is given a novel-noun generalization task inspired by the work of Smith et al. (2002). During training, the model is shown two exemplars of each of four categories. Figure 8.3b shows a matrix of training data where each column represents an exemplar and the rows encode the shapes, textures, colors, and sizes of these exemplars. Each dimension is assumed to take 10 possible values—for example, there are 10 possible shapes, 10 possible colors, and so on. The first two columns represent two objects that both belong to category 1 and both have the same shape, but that vary in texture, color, and size. Note that all pairs of objects belonging to the same category have the same value along the shape dimension.

The model is then tested by presenting it with a "dax"—an exemplar of a novel category that was not presented during training. In figure 8.3, a category label of 5 is used to code "dax," and the first column of the test matrix represents the novel exemplar. The next three columns represent three choice objects with unknown category labels. One choice object matches the dax in shape, the second matches the dax in texture, and the third matches the dax in color. The model is asked to infer which of the three choice objects is most likely to be a dax. The black bars in figure 8.3 show relative choice probabilities for the three objects and indicate that the object that matches in shape is inferred to be the dax. The white bars summarize data reported by Smith et al. (2002), showing that 19-month-olds exposed to the eight training objects also choose the shape match in the novel-noun generalization test.[2] This result is especially interesting because 19-month-olds who have not been exposed to the training objects do not generalize in this way; they choose at random among the three choice objects. The work of Smith et al. (2002), therefore, supports the idea that the shape bias is learned from experience.

The hierarchical model in figure 8.3a has been extended in several ways (Kemp et al., 2007; Perfors & Tenenbaum, 2009), and researchers have proposed other hierarchical models of category learning (Navarro, 2006; Heller, Sanborn, & Chater, 2009; Canini, Shashkov, & Griffiths, 2010), including models that focus on richer, high-dimensional representations of visual categories (Zhu, Chen, Torralba, Freeman, & Yuille, 2010; Salakhutdinov, Tenenbaum, & Torralba, 2013; Li, Fergus, & Perona, 2006; Sudderth, Torralba, Freeman, & Willsky, 2005; Lake, Salakhutdinov, & Tenenbaum, 2015). Although these models make a variety of formal assumptions, a common theme is that knowledge about categories is organized into multiple levels of abstraction, and learning at the more abstract levels helps to explain how humans are able to rapidly learn novel categories from just one or a few examples.

8.2 Causal Learning

Consider now a problem in which a learner must learn about multiple causal systems from the same family. For example, after experience using several previous cell phones, the learner might need to figure out how to use the new phone that she has just bought. We will consider a simple example, where each system can be captured by a simple causal model that includes at most one causal relationship. For example, suppose that a learner is interested in whether blood-pressure medications cause headaches as a side effect. The causal model for each medication is a causal graphical model that indicates whether a cause variable (taking the medication) probabilistically causes an effect variable (experiencing a headache).

Learning causal models for several previous medications can enable rapid learning about a novel medication. Suppose, for example, that a learner has discovered that some blood

2. Smith et al. (2002) indicate that 30 percent of children choose either the texture match or the color match, but they do not indicate the distribution of responses across these two objects. Figure 8.3b therefore assumes that 15 percent of children choose the texture match and 15 percent choose the color match.

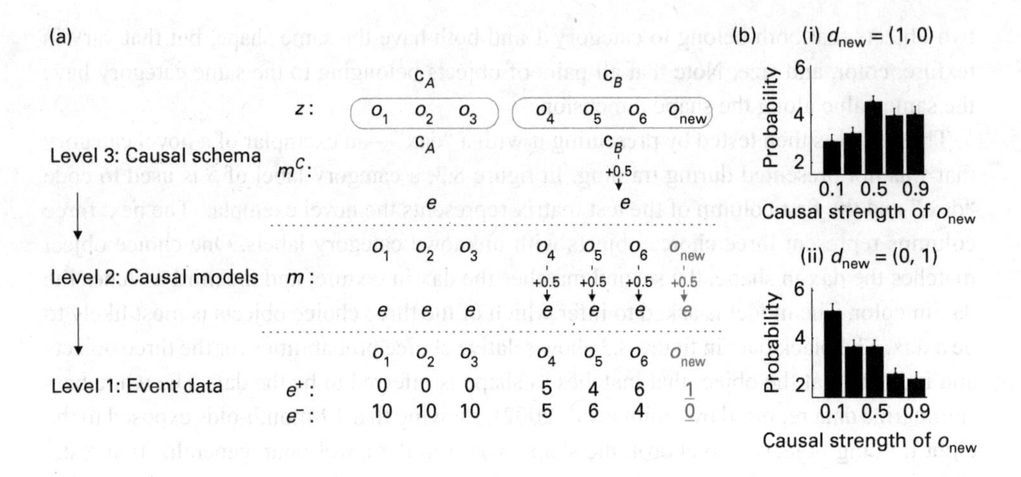

Figure 8.4
A richer hierarchical model. (a) This model that can learn causal models for individual medications (level 2) and causal models for categories of medications (level 3). The information acquired at level 3 allows the model to predict that a sparsely observed new medication o_{new} is likely to produce headaches about half the time. (b) Results from Kemp et al. (2010) showing inferences made by experimental participants who are trained on the event data for o_1 through o_6 shown in (a). (i) Inferences about the causal strength of a new object that is taken once and observed to produce a headache. (ii) Inferences about a new object that is taken once and observed not to produce a headache. All responses were provided on a seven-point scale.

pressure medications never cause headaches, and others cause headaches about half the time. Imagine that the learner now takes a new medication and experiences a headache. Even though she has observed that the new medication has a 100 percent success rate in causing headaches, she might infer that the medication will cause headaches about half the time.

Figure 8.4 shows a hierarchical model that can be used to capture this kind of inference. Level 1 specifies contingency data for each of eight blood pressure medications and shows the number of times that the learner has experienced (e^+) and has failed to experience (e^-) headaches after taking each medication. For example, the learner has taken medication o_1 on 10 occasions, and has experienced a headache on none of these occasions. Note that medications o_1 through o_3 appear not to cause headaches, but medications o_4 through o_6 appear to cause headaches about half the time. For simplicity, we will assume that blood pressure medications are the only possible causes of headaches—in other words, we assume that the learner does not experience a headache on a given day if she does not take medication on that day.

Level 2 shows a causal model m_i for each individual medication. The model for medication o_1 does not include an arrow, indicating that this medication does not cause headaches. The model for medication o_4 does include an arrow, and the numerical label on the arrow indicates that o_4 is a generative cause of headaches with a causal strength of 0.5.

Level 3 specifies a *causal schema* that captures information about medications in general. The schema organizes the medications into categories, where z_i indicates the category assignment for medication i. The schema also includes a set m^C that includes a causal model for each category. Two categories are shown in figure 8.4, and the causal model m^C_A for category A indicates that medications in this category tend not to cause headaches. The causal model m_B^C for category B indicates that medications belonging to this category tend to cause headaches about half the time.

The hierarchy in figure 8.4 can be transformed into a probabilistic model by defining distributions that specify how the variables at each level are generated given the variables at the level immediately above. We provide an informal description of these distributions that is based on the fully specified model presented by Kemp et al. (2010). We assume that all causal events for medication i are independently sampled from the causal model for this medication. At level 2, the causal model m_i for medication i is sampled from a distribution that ensures that the model tends to match the corresponding model for category z_i at level 3. Finally, there are priors on the causal models and category assignments at level 3. The prior on category assignments $P(z)$, is induced by the Chinese restaurant process (CRP; see chapter 9), and captures the expectation that the total number of categories will be small.

Inferences about the causal model m_{new} for a new medication can be made by integrating out the category assignments z and the schema-level causal models m^C at level 3:

$$p(m_{new}|d_1, d_2, \ldots, d_n) = \int p(m_{new}|z, m^C)p(z, m^C|d_1, d_2, \ldots, d_n)\, dz\, dm^C, \tag{8.7}$$

where d_i is the pair (e_i^+, e_i^-) that specifies the contingency data for medication i. As for the model in the previous section, the integral in equation (8.7) can be approximated using MCMC methods.

Figure 8.4a suggests that learning causal models for previous medications can support rapid inferences about a new medication o_7. The event data for this medication indicate that it was taken once, and produced a headache. The hierarchical model infers that the medication probably belongs to category B, and that as a result, the causal strength of the medication is probably around 0.5. Figure 8.4b.i summarizes the inferences that people make in this situation—they infer that the causal strength of the new medication is high, but the peak of the distribution is at 0.5 rather than 1, even though the new medication has a 100 percent success rate in causing headaches. Figure 8.4b.ii shows the corresponding inferences when a new medication is taken once and fails to produce a headache. There is still a possibility that the medication belongs to category A and has a causal strength of around 0.5, but the peak of the distribution is at 0.

The model in figure 8.4a can help to explain how learning the structure of several causal models supports rapid inferences about the structure of a new causal model. Hierarchical Bayesian models have also been applied to several other causal learning problems (Hagmayer & Mayrhofer, 2013). Lucas and Griffiths (2010) have developed a hierarchical model that helps to explain how people learn the functional form of a causal relationship—for example, how people learn whether multiple causes combine according to a conjunctive relationship (an effect occurs only if all causes are present) or a disjunctive relationship (an effect occurs if at least one of the causes is present); Lucas, Bridgers, Griffiths, and Gopnik (2014a) explore the same idea in a developmental setting. Goodman, Ullman, and Tenenbaum (2011) describe a hierarchical model that helps to explain how a learner might come to understand the abstract notion of a causal intervention. In all these cases, hierarchical models help to explain how abstract causal knowledge might be acquired and how this knowledge shapes rapid inferences about new causal systems.

8.3 Property Induction

As our third example of hierarchical Bayesian inference, consider a problem in which a learner finds out that one or more members of a domain have a novel property and must decide how to extend the property to the remaining members of the domain. For instance, given that horses carry enzyme X132, how likely is it that cows also carry this enzyme (Rips, 1975; Osherson, Smith, Wilkie, Lopez, & Shafir, 1990)? Although the learner may know little about enzyme X132, she has previously observed many other features of horses and cows, and noticing that horses and cows share many of these properties suggests that both of them probably carry enzyme X132. This section describes a hierarchical model that captures inferences of this sort, and helps to explain how learning about the properties of a set of objects can support rapid inferences about novel properties of these objects.

The problem that we consider can be formalized as an inference about the extension of a novel property (Kemp & Tenenbaum, 2003). Suppose that we are working with a finite set of animal species. Let e_{new} be a binary vector that represents the true extension of the novel property (figure 8.5). For example, the element in e_{new} that corresponds to cows will be 1 (represented as a black circle in figure 8.5) if cows have the novel property, and 0 otherwise. Let d_{new} be a partially observed version of extension e_{new} (figure 8.5). We are interested in the posterior distribution on e_{new} given the sparse observations in d_{new}. Using Bayes' rule, this distribution can be written as

$$P(e_{\text{new}}|d_{\text{new}}, \mathcal{S}) = \frac{P(d_{\text{new}}|e_{\text{new}})P(e_{\text{new}}|\mathcal{S})}{P(d_{\text{new}}|\mathcal{S})}, \tag{8.8}$$

where $\mathcal{S}$ is a structure that captures the prior knowledge that is relevant to the novel property. The first term in the numerator, $P(d_{\text{new}}|e_{\text{new}})$, depends on the process by which the observations in d_{new} were sampled from the true extension e_{new}. We assume for simplicity that the entries in d_{new} are sampled at random from the vector e_{new}. The denominator can be computed by summing over all possible values of e_{new}:

$$P(d_{\text{new}}|\mathcal{S}) = \sum_{e_{\text{new}}} P(d_{\text{new}}|e_{\text{new}})P(e_{\text{new}}|\mathcal{S}). \tag{8.9}$$

For anatomical and physiological properties (e.g., has enzyme X132), the prior $P(e_{\text{new}}|\mathcal{S})$ will typically capture knowledge about taxonomic relationships between biological species. For instance, it seems plausible a priori that horses and cows are the only familiar animals that carry a certain enzyme, but less probable that this enzyme will be found only in horses and squirrels.

Prior knowledge about taxonomic relationships between living kinds can be captured using a tree-structured representation like the taxonomy shown in figure 8.5. We therefore assume that the structured prior knowledge $\mathcal{S}$ takes the form of a tree, and define a prior distribution $P(e_{\text{new}}|\mathcal{S})$ using a stochastic process over this tree. The stochastic process assigns some prior probability to all possible extensions, but the most likely extensions are those that are smooth with respect to tree $\mathcal{S}$. An extension is smooth if nearby species in the tree tend to have the same status—either both have the novel property or neither does.

For inferences about anatomical properties, the problem of acquiring prior knowledge has now been reduced to the problem of finding an appropriate tree $\mathcal{S}$. Human learners acquire

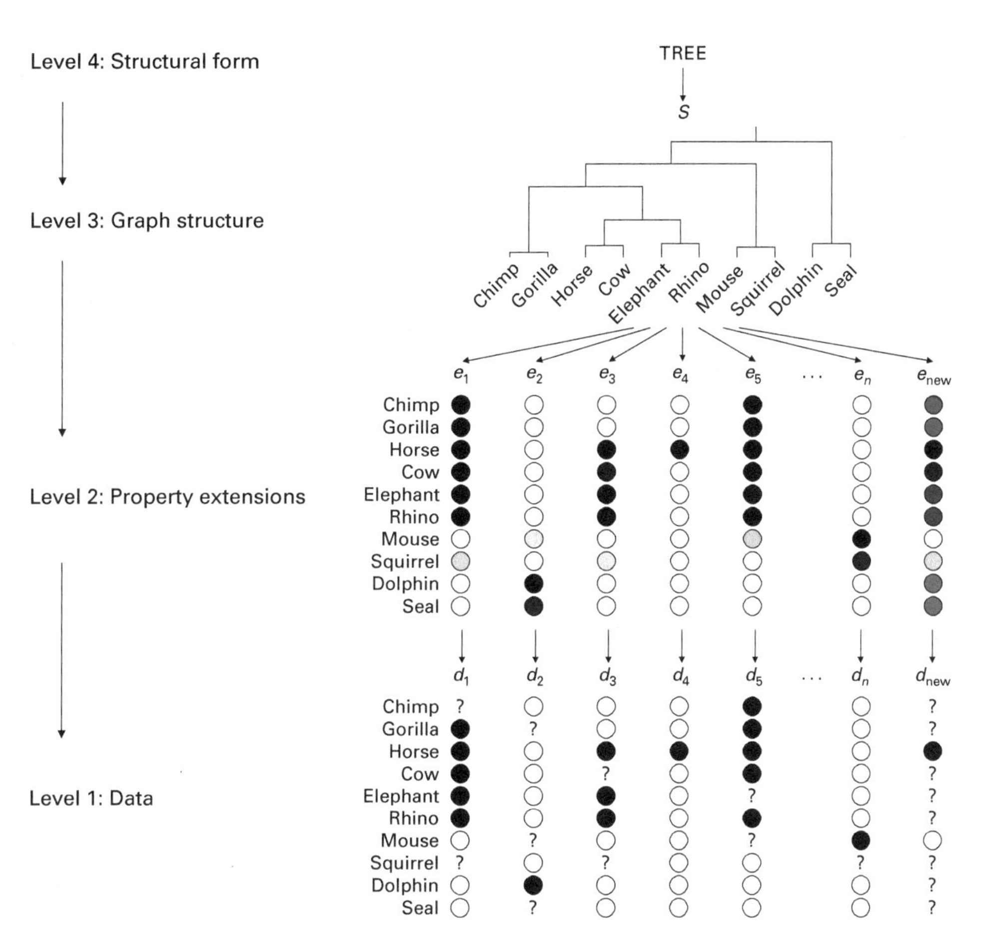

Figure 8.5
Learning a tree-structured prior for property induction. Given a collection of sparsely observed properties d_i (black and white circles indicate that a species is observed to have or lack a given property, and question marks indicate missing observations), we can compute a posterior distribution on structure $\mathcal{S}$ and posterior distributions on each extension e_i. Since each distribution on e_i is difficult to display, we show instead the posterior probability that each species has each property (dark circles indicate probabilities close to 1). Learning the graph structure $\mathcal{S}$ supports confident inferences about the sparsely observed new property—for example, the model is relatively confident that cows have this property, and relatively confident that squirrels do not have this property.

taxonomic representations in part by observing properties of biological species, and noticing, for example, that horses and cows have many properties in common and should probably appear near each other in a taxonomic tree. This learning process can be formalized using the hierarchical Bayesian model in figure 8.5. We assume that a learner has partially observed the extensions of n properties, and these observations are collected in vectors labeled d_1 through d_n. The extensions e_i of these properties are generated from the same tree-based prior that is assumed to generate e_{new}, the extension of the novel property. Again, we see that a hierarchical formulation allows information to be shared across related contexts. Here, information about n partially observed properties is used to influence the prior distribution for inferences about e_{new}. To complete the hierarchical model in figure 8.5, it is necessary to specify a prior distribution on tree $\mathcal{S}$. For simplicity, we will use a uniform distribution over tree topologies, and an exponential distribution with parameter λ over the branch lengths.

Inferences about e_{new} can now be made by integrating the underlying tree $\mathcal{S}$:

$$P(e_{new}|d_1, \ldots, d_n, d_{new}) = \int_{\mathcal{S}} P(e_{new}|d_{new}, \mathcal{S}) p(\mathcal{S}|d_1, \ldots, d_n, d_{new})\, d\mathcal{S}, \tag{8.10}$$

where $P(e_{new}|d_{new}, \mathcal{S})$ is defined in equation (8.8). This integral can be approximated by using MCMC methods to draw a sample of trees from the distribution $p(\mathcal{S}|d_1, \ldots, d_n, d_{new})$ (Huelsenbeck & Ronquist, 2001). Alternatively, a single tree with high posterior probability can be identified, and this tree can be used to make predictions about the extension of the novel property. Kemp, Perfors, and Tenenbaum (2004; see also Kemp & Tenenbaum, 2009) follow this second strategy, showing that a single tree is sufficient to accurately predict human inferences about the extensions of novel biological properties.

The model in figure 8.5 assumes that the extensions e_i are generated over some true but unknown tree $\mathcal{S}$. Tree structures may be useful for capturing taxonomic relationships between biological species, but other kinds of structured representations such as chains, rings, and sets of clusters are useful in other settings. Understanding which kind of representation is best for a given context is sometimes thought to rely on innate knowledge: Atran (1998), for example, argues that the tendency to organize living kinds into tree structures reflects an innately determined cognitive module. The hierarchical Bayesian approach challenges the inevitability of this conclusion by showing how a model might discover which kind of representation is best for a given data set. We can create such a model by adding another level that specifies a prior over level 4 in figure 8.5. Suppose that variable $\mathcal{F}$ indicates whether $\mathcal{S}$ is a tree, a chain, a ring, or an instance of some other structural form. Given a prior distribution over a hypothesis space of possible forms, the model in figure 8.5 can simultaneously discover the form $\mathcal{F}$ and the instance of that form $\mathcal{S}$ that best account for a set of observed properties. Kemp and Tenenbaum (2008) formally define a model of this sort, showing that it chooses appropriate representations for several domains (see illustrations in figure 1.2). For example, the model chooses a tree-structured representation given information about animals and their properties, but it chooses a linear representation (the liberal-conservative spectrum) when supplied with information about the voting patterns of Supreme Court justices.

8.4 Beyond Strict Hierarchies

The key characteristic of a hierarchical model is that it incorporates multiple levels of abstraction. The three models discussed so far are simple examples that have an additional property in common: the variables in each model can be arranged into a tree such that each variable at level k (other than the root) depends on a single variable at level $k+1$. Often, however, it is necessary to work with representations that have multiple levels but are not so strictly organized. For example, figure 8.5 suggests that biological properties are generated over a tree-structured taxonomy at level 3, but some properties (e.g., has high levels of omega-3 acids in its blood) may depend on factors such as habitat and diet that crosscut the taxonomy (Heit & Rubinstein, 1994). A natural extension of the model in figure 8.5 is to introduce multiple structures at level 3, including a tree-structured taxonomy, a food web (Shafto, Kemp, Bonawitz, Coley, & Tenenbaum, 2008), and a classification of the animals by habitat (Shafto, Kemp, Mansinghka, & Tenenbaum, 2011), and to allow a given property at level 2 to depend on one or more of these structures.

The language of graphical models is rich enough to express many kinds of models that have multiple levels of abstraction but are not strictly hierarchical. We have already seen several examples. The topic model introduced in chapter 5 indicates that the words in each document depend on two variables: a topic vector θ, which is specific to the document; and a set of topics ϕ, which is shared across the entire collection of documents. As a result, the graphical model in figure 5.9 does not take the form of a tree. The model in figure 5.14a, combining a topic model with a hidden Markov model (HMM), represents an even greater departure from a strictly hierarchical approach. This model combines a model of semantics (the topic model, latent Dirichlet allocation) and a model of syntax (the HMM), and the ith word in a given document could potentially be generated by either component. Combining the two components produces a graphical model that has multiple levels of abstraction but is very different from a tree.

For our purposes, both these models qualify as hierarchical models in the sense that they incorporate multiple levels of abstraction. In some respects, *multilevel modeling* is the best label for the approach described in this chapter, but we refer to *hierarchical modeling* for compatibility with the cognitive science literature.

8.5 Future Directions

The hierarchical models discussed in this chapter help to explain how experience with previous contexts supports rapid learning about a new context. The most interesting examples of this kind of *learning to learn* may take place relatively early in childhood, when children are assembling the abstract knowledge that will serve as the foundation for much of their subsequent learning. Several research groups have argued that hierarchical Bayesian models can provide insight into cognitive development (Perfors, Tenenbaum, Griffiths, & Xu, 2011; Glassen & Nitsch, 2016; Ullman & Tenenbaum, 2020), and chapter 20 reviews some of the work in this area. Applying the approach to a broader set of developmental phenomena, however, is an important challenge for future work.

Learning does not stop when a child becomes an adult; rather, it continues over the course of a person's lifetime. Machine learning researchers have developed models of *lifelong learning* that aim to match this ability (Thrun & Pratt, 2012), and the hierarchical Bayesian approach provides one promising way to tackle this challenge. In principle, a hierarchical Bayesian framework can keep encountering new learning contexts indefinitely, and future studies should aim to develop models that learn over periods measured in years or decades rather than minutes, hours, or days.

In addition to expanding the time frame over which learning occurs, future work should aim to expand the set of tasks to which a single hierarchical framework can be applied. Inspired by the goal of developing *artificial general intelligence*, recent work in artificial intelligence (AI) has moved from systems that learn one kind of task (e.g., playing Atari games (Mnih et al., 2015)) to systems that learn multiple tasks (e.g., playing Atari games, generating image captions, and stacking blocks (Reed et al., 2022)). Hierarchical Bayesian models (Wilson, Fern, Ray, & Tadepalli, 2007) open up the possibility of a similar trajectory, and future work can explore the extent to which these models account for general-purpose learning in humans.

A final challenge for the hierarchical Bayesian approach is to establish deeper links with models of neural computation. Theories of predictive coding suggest that brain regions are organized into a hierarchy that supports both bottom-up and top-down probabilistic inference (Clark, 2013), and the hierarchical Bayesian approach provides a natural way to formalize these ideas (Lee & Mumford, 2003; Friston, 2009). Ongoing work is using hierarchical Bayesian models to better understand computation in both dysfunctional (Williams, 2018) and normally functioning brains (Rohe, Ehlis, & Noppeney, 2019).

8.6 Conclusion

We began this chapter by suggesting that hierarchical Bayesian models can address two challenges: they help to explain where prior distributions come from, and they help to explain how humans learn to learn. We have only partially addressed the first challenge. The three models in this chapter help to explain how abstract knowledge about categories, causal models, and properties can be acquired, and how this abstract knowledge induces prior distributions that support rapid inferences about novel categories, causal models, and properties. We therefore showed how certain priors could be learned, but in each case, this learning relies on prior assumptions of some sort. For example, the models rely on hyperparameters that are fixed in advance, and we also assumed that the structure of each hierarchical model is known in advance. It is possible to relax these assumptions—for example, hyperparameters can be learned by adding an extra level to the hierarchical model that induces a prior distribution over these hyperparameters. Regardless of how many levels we add, however, a hierarchical Bayesian model will always need to rely on prior knowledge of some sort. The goal of these models is therefore not to explain how probabilistic models can succeed without relying on any kind of prior knowledge. Instead, the goal is to explain how relatively sophisticated knowledge can be acquired by a system that starts out with prior knowledge that is plausibly assumed to be innate.

We addressed the problem of learning to learn by showing how the upper levels in hierarchical Bayesian models can capture knowledge that is relevant across a range of contexts. In cases of this kind, acquiring knowledge at the upper levels supports rapid learning at the lower levels and therefore can be described as "learning to learn." We only touched on the developmental implications of hierarchical models, but these models are appealing in part because they provide a way to understand how learning may change and accelerate over the course of a lifetime.

Psychologists, statisticians, and machine learning researchers have developed many hierarchical models other than the ones discussed here. For example, hierarchical models have been used to explain how people learn about the speech of novel talkers (Pajak, Fine, Kleinschmidt, & Jaeger, 2016), establish communicative conventions with novel partners (Hawkins et al., 2023), predict the outcomes of actions that have never been taken (Gershman & Niv, 2015), and acquire abstract strategies that are relevant across multiple motor tasks (Braun, Waldert, Aertsen, Wolpert, & Mehring, 2010). These models vary in many respects, but all of them rely on probabilistic inference over an abstraction hierarchy. Hierarchical Bayesian models therefore provide yet another example of a central theme developed in previous chapters: probabilistic inference and structured representations—in this case, abstraction hierarchies—can achieve more combined than either approach can achieve in isolation.

9

Capturing the Growth of Knowledge with Nonparametric Bayesian Models

Joseph Austerweil, Adam N. Sanborn, Christopher Lucas, and
Thomas L. Griffiths

Mental representations are often simplifications. When we represent the objects in the world as members of a set of discrete categories, we are imposing a simple structure on the complexity of our experience. Postulating that a set of objects can be described by a small set of features makes thinking about those objects simpler. Likewise, assuming that the causal relationship between two variables can be represented as a smooth function simplifies the underlying reality. Psychologists usually specify the amount of structure in a mental representation—the number of categories, features, or causes—and tend to decide in advance how much complexity the representation can capture. However, we ideally want to simplify the world just enough, imposing no more structure than is needed to make accurate predictions and form reasonable explanations. In learning categories, we don't want to make the unrealistic assumption that there is a fixed number of kinds of things in the world. In identifying features, we don't want to assume that there are only so many features to go around. In learning causal relationships, we don't want to assume that every relationship is linear. The world is boundlessly complex, and we want to be able to emulate that complexity when the available data warrant it.

This need to accommodate potentially unlimited complexity is in tension with the practical challenges of probabilistic inference. Human brains are finite, so if they can form representations of arbitrary complexity, this needs to be done in a way that remains tractable. Likewise, in making probabilistic models of human cognition, we need to be able to perform calculations using those models on a computer. We thus need to be able to define probabilistic models that allow us to work with infinite hypothesis spaces using finite representations and computations.

A natural way for a learner to strike a balance between complexity and tractability is to start with simple representations and add complexity incrementally upon making new observations that require it. This kind of approach is expressed in Jean Piaget's (1954) characterization of cognitive development as a process of *assimilation* and *accommodation*. In Piaget's view, when presented with new information, the child has two options: to assimilate that information to their current understanding of the world, or to change that understanding to accommodate the new information. Despite starting with a simple model of the world, these accommodations accumulate complexity that deepens and enriches the child's representations. Some researchers have tried to capture this process in models that can accumulate complexity in a similar way, such as cascade correlation neural networks (Shultz, Mareschal, & Schmidt, 1994).

This intuitive approach to solving the representational and computational challenges of capturing complexity is made precise in a set of tools developed in *nonparametric Bayesian statistics* (e.g., Muller & Quintana, 2004; Hjort, Holmes, Müller, & Walker, 2010). These tools exploit the methods for approximate inference described in chapters 5 and 6 to work with models that can accommodate unlimited complexity. This approach is "nonparametric" in the sense that it provides a way to work with models that go beyond the simple parametric families that are commonly encountered in statistics. More formally, it covers situations in which the complexity of the models increase with the data—for example, the effective number of parameters that need to be estimated grows as more data are observed. This is in contrast to parametric models, which have a fixed set of parameters, and is more akin to nonparametric frequentist methods such as kernel density estimation, which was briefly mentioned in chapter 5.

To give a concrete example, consider the problem faced by an explorer visiting a new continent. This explorer is familiar with the animals on her own continent and has already organized them into species. Pushing aside a tree branch, she encounters her first animal: Is this a new kind of animal, deserving a new species of its own? Or can its properties be explained by the existing species of animals that she has previously encountered? While her representation at any point in time is finite—she can never postulate more species than individual animals that she has seen—there is no upper limit to the number of species that the world might contain. This is exactly the assumption behind the models considered in this chapter.

In principle, nonparametric Bayesian models have an infinite amount of structure—an infinite number of categories, an infinite number of features, infinite degrees of freedom in a function—but they effectively only instantiate a finite amount of structure in response to any finite observed data set. They postulate no more complexity than is necessary, guided by a version of the Bayesian Occam's razor introduced in chapter 3, and expand only as much as necessary to explain the data. We will consider how this approach can be applied in three settings: categorization, feature learning, and function learning.

9.1 Infinite Models for Categorization

In chapter 5, we saw that psychological models of categorization can be given a probabilistic interpretation. Specifically, these models can be thought of as corresponding to schemes for estimating a probability distribution over objects associated with a category, and how mixture models can be used for this purpose. In this section, we use this formulation of the problem of categorization to introduce one of the most common nonparametric Bayesian models—the *infinite mixture model*. We begin with a more formal treatment of finite mixture models, setting up the mathematical ideas that are then used to generalize this to the infinite mixture model. We then spend a little more time on the key idea that makes it possible to define an infinite model—the *Chinese restaurant process* (CRP)—and discuss how it is possible to perform inference in an infinite model with only finite means.

Assume that we have n objects, with the ith object having d observable properties represented by a row vector $\mathbf{x}_i$. In a mixture model, each object is assumed to belong to a single cluster, z_i, and the properties $\mathbf{x}_i$ are generated from a distribution determined by that cluster. Using the matrix $\mathbf{X} = \left[\mathbf{x}_1^T \, \mathbf{x}_2^T \, \cdots \, \mathbf{x}_n^T\right]^T$ to indicate the properties of all n objects, and the vector $\mathbf{z} = [z_1 \, z_2 \, \cdots \, z_n]^T$ to indicate their cluster assignments, the model is specified by a prior over assignment vectors $P(\mathbf{z})$, and a distribution over property matrices conditioned on those assignments, $p(\mathbf{X}|\mathbf{z})$. These two distributions can be dealt with separately: $P(\mathbf{z})$ specifies the number of clusters and their relative probability, while $p(\mathbf{X}|\mathbf{z})$ determines how these clusters relate to the properties of objects. We will focus on the prior over assignment vectors, $P(\mathbf{z})$, showing how such a prior can be defined without placing an upper bound on the number of clusters.

9.1.1 Finite Mixture Models

Mixture models assume that the assignment of an object to a cluster is independent of the assignments of all other objects. Assume that there are k clusters, θ is a discrete distribution over those clusters, and θ_j is the probability of cluster j under that distribution. Under this assumption, the probability of the properties of all n objects $\mathbf{X}$ can be written as

$$p(\mathbf{X}|\theta) = \prod_{i=1}^{n} \sum_{j=1}^{k} p(\mathbf{x}_i, z_i = j|\theta_j) = \prod_{i=1}^{n} \sum_{j=1}^{k} p(\mathbf{x}_i|z_i = j)\, \theta_j. \tag{9.1}$$

The distribution from which each $\mathbf{x}_i$ is generated is thus a mixture of the k cluster distributions $p(\mathbf{x}_i|z_i = j)$, with θ_j determining the weight of cluster j.

The mixture weights θ can be treated either as a parameter to be estimated or a variable with prior distribution $p(\theta)$. In Bayesian approaches to mixture modeling, a standard choice for $p(\theta)$ is a symmetric Dirichlet distribution, as introduced in chapter 3. The probability of any discrete distribution θ is given by

$$p(\theta) = \frac{\prod_{j=1}^{k} \theta_j^{\alpha_j - 1}}{D(\alpha_1, \alpha_2, \ldots, \alpha_k)}, \tag{9.2}$$

in which $D(\alpha_1, \alpha_2, \ldots, \alpha_k)$ is the Dirichlet normalizing constant

$$D(\alpha_1, \alpha_2, \ldots, \alpha_k) = \int_{\Delta_k} \prod_{j=1}^{k} \theta_j^{\alpha_j - 1}\, d\theta \tag{9.3}$$

$$= \frac{\prod_{j=1}^{k} \Gamma(\alpha_j)}{\Gamma(\sum_{j=1}^{k} \alpha_j)}, \tag{9.4}$$

where Δ_k is the simplex of all possible discrete distributions over k clusters, and $\Gamma(\cdot)$ is the generalized factorial or gamma function, with $\Gamma(m) = (m-1)!$ for any positive integer m. In a *symmetric Dirichlet distribution*, all values of α_j are equal. For example, we could take $\alpha_j = \frac{\alpha}{k}$ for all j. In this case, equation (9.4) becomes

$$D(\tfrac{\alpha}{k}, \tfrac{\alpha}{k}, \ldots, \tfrac{\alpha}{k}) = \frac{\Gamma(\frac{\alpha}{k})^k}{\Gamma(\alpha)} \tag{9.5}$$

and the mean of θ is the distribution that is uniform over all clusters.

The probabilistic model that we have defined is

$$\theta \mid \alpha \sim \text{Dirichlet}(\tfrac{\alpha}{k}, \tfrac{\alpha}{k}, \ldots, \tfrac{\alpha}{k}) \tag{9.6}$$

$$z_i \mid \theta \sim \text{Discrete}(\theta). \tag{9.7}$$

Having defined a prior on θ, we can simplify this model by integrating over all values of θ (i.e., the simplex Δ_k) rather than representing them explicitly. The marginal probability of an assignment vector $\mathbf{z}$, integrating over all values of θ, is

$$P(\mathbf{z}) = \int_{\Delta_k} \prod_{i=1}^{n} P(z_i|\theta)p(\theta)\, d\theta \tag{9.8}$$

$$= \int_{\Delta_k} \frac{\prod_{j=1}^{k} \theta_j^{m_j+\alpha_j-1}}{D(\alpha_1, \alpha_2, \ldots, \alpha_k)}\, d\theta \tag{9.9}$$

$$= \frac{D(m_1 + \frac{\alpha}{k}, m_2 + \frac{\alpha}{k}, \ldots, m_k + \frac{\alpha}{k})}{D(\frac{\alpha}{k}, \frac{\alpha}{k}, \ldots, \frac{\alpha}{k})} \tag{9.10}$$

$$= \frac{\prod_{j=1}^{k} \Gamma(m_j + \frac{\alpha}{k})}{\Gamma(\frac{\alpha}{k})^k} \frac{\Gamma(\alpha)}{\Gamma(n+\alpha)}. \tag{9.11}$$

where m_j is the number of objects assigned to cluster j. The tractability of this integral is a result of the fact that the Dirichlet is conjugate to the multinomial.

Equation (9.11) defines a probability distribution over the cluster assignments $\mathbf{z}$ as an ensemble. Individual cluster assignments are no longer independent. Rather, they are *exchangeable* (Box & Tiao, 1992), with the probability of an assignment vector remaining the same when the indices of the objects are permuted. Exchangeability is a desirable property in a distribution over cluster assignments because the indices labeling objects are typically arbitrary. However, the distribution on assignment vectors defined by equation (9.11) assumes an upper bound on the number of clusters of objects since it allows assignments of objects only to up to k clusters.

9.1.2 Infinite Mixture Models

Intuitively, defining an infinite mixture model means that we want to specify the probability of $\mathbf{X}$ in terms of infinitely many clusters, modifying equation (9.1) to become

$$p(\mathbf{X}|\theta) = \prod_{i=1}^{n} \sum_{j=1}^{\infty} p(\mathbf{x}_i|z_i = j)\, \theta_j, \tag{9.12}$$

where θ is an infinite-dimensional multinomial distribution. To repeat the argument given here, we would need to define a $p(\theta)$ on infinite-dimensional multinomials and compute the probability of $\mathbf{z}$ by integrating over θ. Taking this strategy provides an alternative way to derive infinite mixture models, resulting in something known as a *Dirichlet process mixture model* (Antoniak, 1974; Ferguson, 1983). Instead, we will work directly with the distribution over assignment vectors given in equation (9.11), considering its limit as the number of clusters approaches infinity (Green & Richardson, 2001; Neal, 1998).

Expanding the gamma functions in equation (9.11) using the recursive law $\Gamma(x) = (x-1)\Gamma(x-1)$ and canceling terms produces the following expression for the probability of an assignment vector $\mathbf{z}$:

$$P(\mathbf{z}) = \left(\tfrac{\alpha}{k}\right)^{k_+} \left(\prod_{j=1}^{k_+} \prod_{\ell=1}^{m_j - 1} \ell + \tfrac{\alpha}{k} \right) \frac{\Gamma(\alpha)}{\Gamma(N+\alpha)}, \tag{9.13}$$

where k_+ is the number of clusters for which $m_j > 0$, and we have re-ordered the indices such that $m_j > 0$ for all $k \leq k_+$ (and $m_j = 0$ for all $j > k_+$). There are k^n possible values for $\mathbf{z}$, which diverges as $k \to \infty$. As this happens, the probability of any single set of cluster assignments goes to 0. Since $k_+ \leq n$ and n is finite, it is clear that $P(\mathbf{z}) \to 0$ as $k \to \infty$ since $\frac{1}{k} \to 0$. Consequently, we will define a distribution over *equivalence classes* of assignment vectors—sets of vectors that have the same properties—rather than the vectors themselves.

Specifically, we will define a distribution on *partitions* of objects. In our setting, a partition is a division of the set of n objects into subsets, where each object belongs to a single subset and the ordering of the subsets does not matter. Two assignment vectors that result in the same division of objects correspond to the same partition. For example, if we had three objects, the cluster assignments $\{z_1, z_2, z_3\} = \{1, 1, 2\}$ would correspond to the same partition as $\{2, 2, 1\}$ since all that differs between these two cases is the labels of the clusters. A partition thus defines an equivalence class of assignment vectors, $[\mathbf{z}]$, with two assignment vectors belonging to the same equivalence class if they correspond to the same partition. distribution over partitions is sufficient to allow us to define an infinite mixture model since these equivalence classes of cluster assignments are the same as those induced by identifiability: $p(\mathbf{X}|\mathbf{z})$ is the same for all assignment vectors $\mathbf{z}$ that correspond to the same partition, so we can apply statistical inference at the level of partitions rather than the level of assignment vectors.

Assume that we have a partition of n objects into k_+ subsets, and we have $k \geq k_+$ cluster labels that can be applied to those subsets. Then there are $\frac{k!}{(k-k_+)!}$ assignment vectors $\mathbf{z}$ that belong to the equivalence class defined by that partition, $[\mathbf{z}]$. We can define a probability distribution over partitions by summing over all cluster assignments that belong to the equivalence class defined by each partition. The probability of each of those cluster assignments is equal under the distribution specified by equation (9.13), so we obtain

$$P([\mathbf{z}]) = \sum_{\mathbf{z} \in [\mathbf{z}]} P(\mathbf{z}) \tag{9.14}$$

$$= \frac{k!}{(k-k_+)!} \left(\tfrac{\alpha}{k}\right)^{k_+} \left(\prod_{j=1}^{k_+} \prod_{\ell=1}^{m_j - 1} \ell + \tfrac{\alpha}{k} \right) \frac{\Gamma(\alpha)}{\Gamma(n+\alpha)}. \tag{9.15}$$

Rearranging the first two terms, we can compute the limit of the probability of a partition as $k \to \infty$, which is

$$P([\mathbf{z}]) = \lim_{k \to \infty} \alpha^{k_+} \cdot \frac{\prod_{j=1}^{k_+} (k-j+1)}{k^{k_+}} \cdot \left(\prod_{j=1}^{k_+} \prod_{\ell=1}^{m_j - 1} \ell + \tfrac{\alpha}{k} \right) \cdot \frac{\Gamma(\alpha)}{\Gamma(n+\alpha)} \tag{9.16}$$

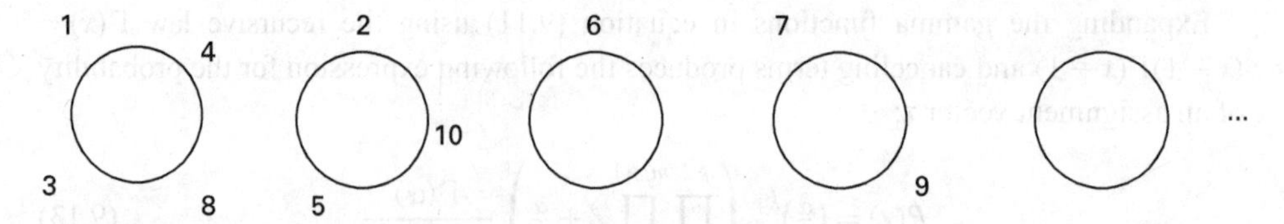

Figure 9.1
A partition induced by the Chinese restaurant process. Numbers indicate customers (objects); circles indicate tables (clusters).

$$= \quad \alpha^{k_+} \cdot \quad 1 \quad \cdot \left(\prod_{j=1}^{k_+} (m_j - 1)! \right) \cdot \frac{\Gamma(\alpha)}{\Gamma(n+\alpha)}. \qquad (9.17)$$

These limiting probabilities define a valid distribution over partitions, and thus over equivalence classes of cluster assignments, providing a prior over cluster assignments for an infinite mixture model. Objects are exchangeable under this distribution, just as in the finite case: the probability of a partition is not affected by the ordering of the objects since it depends only on the counts m_j.

As noted above, the distribution over partitions specified by equation (9.17) can be derived in a variety of ways—by taking limits (Green & Richardson, 2001; Neal, 1998), from the Dirichlet process (Blackwell & MacQueen, 1973), or from other equivalent stochastic processes (Ishwaran & James, 2001; Sethuraman, 1994). Next, we will briefly discuss a simple process that produces the same distribution over partitions: the Chinese restaurant process.

9.1.3 The Chinese Restaurant Process

The Chinese restaurant process (CRP) was named by Jim Pitman and Lester Dubins, based upon a metaphor in which the objects are customers in a restaurant and the clusters are the tables at which they sit (the process first appears in Aldous, 1985, where it is attributed to Pitman). Imagine a restaurant with an infinite number of tables, each with an infinite number of seats.[1] The customers enter the restaurant one after another, and each chooses a table at random. In the CRP with parameter α, each customer chooses an occupied table with probability proportional to the number of occupants and chooses the next vacant table with probability proportional to α. For example, figure 9.1 shows the state of a restaurant after 10 customers have chosen tables using this procedure. The first customer chooses the first table with probability $\frac{\alpha}{\alpha} = 1$. The second customer chooses the first table with probability $\frac{1}{1+\alpha}$, and the second table with probability $\frac{\alpha}{1+\alpha}$. After the second customer chooses the second table, the third customer chooses the first table with probability $\frac{1}{2+\alpha}$, the second table with probability $\frac{1}{2+\alpha}$, and the third table with probabililty $\frac{\alpha}{2+\alpha}$. This process continues until all customers have seats, defining a distribution over allocations of people to tables, and, more generally, objects to clusters. Extensions of the CRP and connections to other stochastic processes are pursued in depth by Pitman (2002).

1. Pitman and Dubins, both probability theorists at the University of California at Berkeley, were inspired by the apparently infinite capacity of Chinese restaurants in San Francisco when they named the process.

The distribution over partitions induced by the CRP is the same as that given in equation (9.17). If we assume an ordering on our n objects, then we can assign them to clusters sequentially using the method specified by the CRP, letting objects play the role of customers and clusters play the role of tables. The ith object would be assigned to the jth cluster with probability

$$P(z_i = k | z_1, z_2, \ldots, z_{i-1}) = \begin{cases} \frac{m_j}{i-1+\alpha} & j \leq k_+ \\ \frac{\alpha}{i-1+\alpha} & \text{otherwise,} \end{cases} \tag{9.18}$$

where m_j is the number of customers currently sitting at table j and k_+ is the number of tables that are currently occupied (i.e., for which $m_j > 0$). If all N objects are assigned to clusters via this process, the probability of a partition of objects $\mathbf{z}$ is that given in equation (9.17). The CRP thus provides an intuitive means of specifying a prior for infinite mixture models, as well as revealing that there is a simple sequential process by which exchangeable cluster assignments can be generated.

9.1.4 Inference by Gibbs Sampling

Inference in an infinite mixture model is only slightly more complicated than inference in a mixture model with a finite, fixed number of clusters. The standard algorithm used for inference in infinite mixture models is Gibbs sampling (Escobar & West, 1995; Neal, 1998). As discussed in chapter 6, Gibbs sampling is a Markov chain Monte Carlo (MCMC) method in which variables are successively sampled from their distributions when conditioned on the current values of all other variables (Geman & Geman, 1984). This process defines a Markov chain, which ultimately converges to the distribution of interest (see Gilks, Richardson, & Spiegelhalter, 1996).

Implementing a Gibbs sampler requires deriving the conditional distribution for each variable conditioned on all other variables. In a mixture model, these variables are the cluster assignments $\mathbf{z}$. The relevant full conditional distribution is $P(z_i | \mathbf{z}_{-i}, \mathbf{X})$, the probability distribution over z_i conditioned on the cluster assignments of all other objects, $\mathbf{z}_{-i}$, and the data, $\mathbf{X}$. By applying Bayes' rule, this distribution can be expressed as

$$P(z_i = j | \mathbf{z}_{-i}, \mathbf{X}) \propto p(\mathbf{X} | \mathbf{z}) P(z_i = j | \mathbf{z}_{-i}), \tag{9.19}$$

where only the second term on the right side depends upon the distribution over cluster assignments, $P(\mathbf{z})$.

In a finite mixture model with $P(\mathbf{z})$ defined as in equation (9.11), we can compute $P(z_i = j | \mathbf{z}_{-i})$ by integrating over θ, obtaining

$$\begin{aligned} P(z_i = j | \mathbf{z}_{-i}) &= \int P(z_i = j | \theta) p(\theta | \mathbf{z}_{-i}) \, d\theta \\ &= \frac{m_{-i,j} + \frac{\alpha}{k}}{n - 1 + \alpha}, \end{aligned} \tag{9.20}$$

where $m_{-i,j}$ is the number of objects assigned to cluster j, not including object i, with $\mathbf{z}_{-i}$ defined analogously. This is the posterior predictive distribution for a multinomial distribution with a Dirichlet prior.

In an infinite mixture model with a distribution over cluster assignments defined as in equation (9.17), we can use exchangeability to find the full conditional distribution. Since it is exchangeable, $P(\mathbf{z})$ is unaffected by the ordering of objects. Thus, we can choose an ordering in which the ith object is the last to be assigned to a cluster. It follows directly from the definition of the CRP; namely, that

$$P(z_i = j | \mathbf{z}_{-i}) = \begin{cases} \frac{m_{-i,j}}{n-1+\alpha} & m_{-i,j} > 0 \\ \frac{\alpha}{n-1+\alpha} & j = k_{-i,+} + 1 \\ 0 & \text{otherwise} \end{cases}, \tag{9.21}$$

where $k_{-i,+}$ is the number of clusters for which $m_j > 0$, not including the assignment of object i. The same result can be found by taking the limit of the full conditional distribution in the finite model, given by equation (9.20) (Neal, 1998).

When combined with some choice of $p(\mathbf{X}|\mathbf{z})$, equations (9.20) and (9.21) are sufficient to define Gibbs samplers for finite and infinite mixture models respectively. Demonstrations of Gibbs sampling in infinite mixture models are provided by Neal (1998) and Rasmussen (2000). Similar MCMC algorithms are presented in Bush and MacEachern (1996), West, Muller, and Escobar (1994), Escobar and West (1995), and Ishwaran and James (2001). Algorithms that go beyond the local changes in cluster assignment allowed by a Gibbs sampler are given by Jain and Neal (2004) and Dahl (2003).

9.1.5 Modeling Human Category Learning

Infinite mixture models provide a way to solve one of the problems that came up when we first considered mixture models as a tool for understanding human category learning: they indicate how a learner could select what kind of representation to use for a given set of objects. If the objects are well characterized as belonging to a single cluster, then the model might form a representation dominated by a single cluster—a prototype model. If the objects are so dispersed as to have nothing in common, the number of clusters might end up being closer to the number of objects—an exemplar model. Normally, the infinite mixture model will produce a result somewhere between these two extremes. Consequently, it seems that exploring this class of nonparametric Bayesian models might provide some insight into the flexibility of human category learning.

Interestingly, infinite mixture models were proposed as an account of human category learning before they became widespread in statistics and machine learning. Anderson (1990) proposed a model of categorization in which people assigned objects to clusters, with the possibility of increasing the number of clusters if there were a poor match between the object and all existing clusters. The probabilistic model at the heart of this account was an infinite mixture model, as pointed out by Neal (1998). Recognizing this relationship, Sanborn, Griffiths, and Navarro (2006, 2010a) showed that the use of the more sophisticated inference algorithms subsequently developed in statistics and machine learning could improve the predictions made by this model.

Anderson's categorization model, also known as the rational model of categorization, has been applied to understand how the category representations that people learn change as they age. Figure 9.2a shows a set of classic category structures introduced by Shepard, Hovland, and Jenkins (1961). These category structures are defined eight stimuli that differed

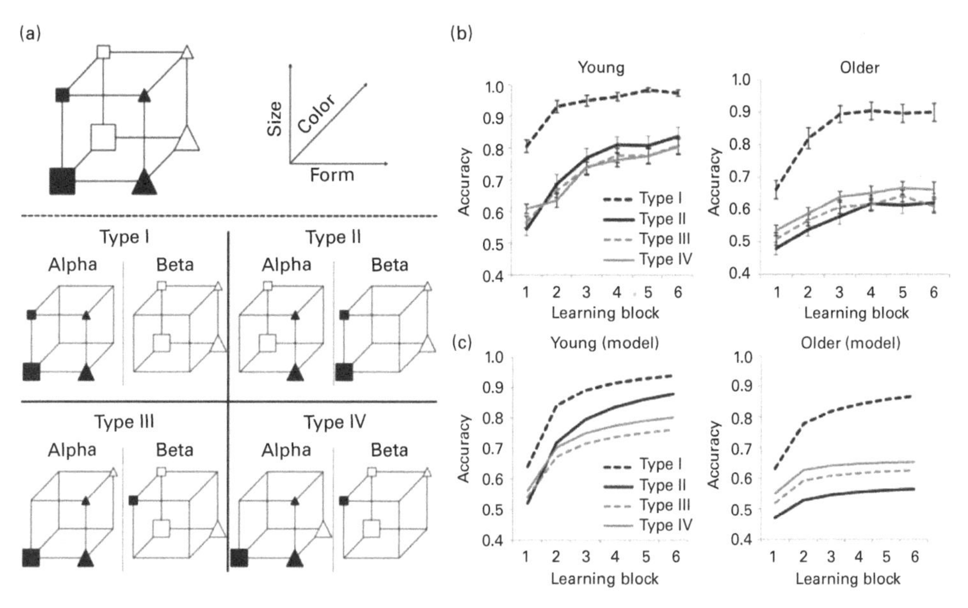

Figure 9.2
Category learning in young and older adults. (a) Illustration of four different category types from Shepard et al. (1961). For each type participants had to learn which stimuli had category label Alpha and which had category label Beta. The stimuli with each label are illustrated on each cube. While the Type IV structure can be captured by one cluster per category, the Type II structure needs two clusters per category. (b) Accuracy over learning for young and older adults for each type. (c) Anderson's categorization model fit to each group of participants. Adapted from figures 1, 2, and 6 of Badham et al. (2017).

on three binary dimensions (e.g., shape, color, and size). The stimuli are divided into two categories, with four stimuli per category. This results in six distinct category structures, labeled type I to type VI. Shepard, Hovland, and Jenkins (1961) found that these structures varied in how easy they were to learn, with type I being the easiest, type II harder, types III, IV, and V harder still, and type VI the hardest. These structures are also represented differently by mixture models: type IV problems can be accurately represented by a single cluster per category, while type II problems need two clusters per category to be accurately represented.

While Shepard et al. (1961) found that type II problems are easier to learn than type IV problems, this effect was found with the usual experimental population of young adults. When older adults are faced with this task, they show the opposite pattern: type IV is easier than type II (see figure 9.2b; Badham, Sanborn, & Maylor, 2017; Rabi & Minda, 2016). This pattern is reproduced by Anderson's categorization model, assuming that older adults have a substantially lower α parameter than young adults, so they produce fewer clusters (see figure 9.2c). Further support comes from Davis, Love, and Maddox (2012), who found a similar result for Anderson's model for different categorization problems. Interestingly, the difference in α may be related to lower cognitive capacity in older adults: Dasgupta and Griffiths (2022) showed that higher values of α are consistent with a higher cognitive cost for representing a probability distribution. It may be that older adults have fewer representational resources than young adults, explaining why they appear to use fewer clusters in categorization tasks.

Infinite mixture models can be extended in a variety of ways to capture different aspects of categorization. For example, Anderson's categorization model assumes that the category label is just another feature of an object. Hence, clusters are shared across categories, with the distribution over clusters associated with each category being obtained by conditioning on the feature corresponding to the category label. An alternative approach is to explicitly associate each category with a distinct distribution over clusters, but allow for the possibility that some of those clusters are shared. This assumption can be captured via the *hierarchical Dirichlet process* (Griffiths, Canini, Sanborn, & Navarro, 2007).

Another generalization of the infinite mixture model allows the possibility that different clustering schemes might be appropriate for explaining the distribution of distinct subsets of observed features. For example, a piece of furniture could have features that describe its shape—having legs, a large flat surface—or the materials used to construct it—maple wood. The first set of features supports clusters based on function—tables, chairs, and so on—while the second set of features supports clusters based on material. This kind of distinction can be captured in a generative model that clusters the features themselves, and then groups objects for each cluster of features (the CrossCat model, discussed in chapter 1; Shafto, Kemp, Mansinghka, Gordon, & Tenenbaum, 2006; Mansinghka et al., 2016).

A different generalization of the infinite mixture model can help explain the strong effects that separable dimensions—those that are easily identifiable from the stimuli—have on the category representations. Categories that are aligned with separable dimensions tend to be easy to learn, while those that are not are more difficult. This can be explained as a prior over the shape of the clusters in the mixture: clusters are expected to be aligned with the separable dimensions, and category structures that match this prior are easier to learn (e.g., Shepard, 1987; Austerweil, Sanborn, & Griffiths, 2019). But how can this prior over cluster alignments itself be learned? This can be done using another infinite mixture: if the prior on the shape of each cluster is itself an infinite mixture, then different types of clusters that correspond to the separable dimensions can be learned across a lifetime of experience. A two-level infinite mixture model can thus explain a wide range of dimensional biases (Sanborn, Heller, Austerweil, & Chater, 2021).

Finally, infinite mixture models can also be extended to settings where objects are described not just by the features they possess, but also by the relations in which they participate with other objects. For example, when trying to make sense of a new social environment, you may pay attention to which pairs of people seem to be friends. Based on these relations, you could try to infer an underlying cluster structure, where the probability that any two people are friends depends only on the clusters they belong to. More formally, people a and b belong to clusters z_a and z_b, with the probability that they are friends being given by $\eta_{z_a z_b}$. In statistics, this kind of model is known as a *stochastic blockmodel*, and it is the relational equivalent of a mixture model. Defining a prior distribution over cluster memberships using the CRP results in the *infinite relational model*, which has been used to explain aspects of how humans learn relational theories (Kemp, Tenenbaum, Griffiths, Yamada, & Ueda, 2006).

9.1.6 Beyond Categorization

Chinese restaurant processes and related distributions have applications in cognitive science that go well beyond categorization. They can be used as prior distributions in any setting where inferences are being made about a latent variable that has a discrete but potentially infinite set of values.

One setting where the CRP has been used successfully is in making inferences about the latent causes that might explain observed events. For example, imagine that you went to a café, ordered a drink, and enjoyed it. A week later, you returned to the same café and order the same drink, but this time it is terrible. One way that you could make sense of this experience is by inferring that something changed at the café—perhaps the coffee beans were different on the two occasions that you visited. In doing so, you're postulating a latent cause for the phenomenon. As you have more experiences, you might infer more latent causes as you hypothesize that the café uses several kinds of beans with different flavors.

This "latent cause" perspective has been used to explain patterns of results and animal conditioning, where suddenly removing a reward is less effective at reducing a behavior than gradually reducing the rate at which the reward is provided (Gershman, Blei, & Niv, 2010). Intuitively, the sudden change suggests a different latent cause should be inferred, and the animal learns that while that cause is present the action no longer produces the reward. The original relationship between the action and reward is thus preserved and can manifest again if the animal thinks that the environment has reverted to the original latent cause. Gradually reducing the rate of reward doesn't result in a change in the inferred latent cause, and the relationship is eliminated.

Another setting where unknown numbers of discrete latent variables appear is in language. Phonemes, words, and syntactic categories are all discrete sets that need to be inferred from the environment. In these cases, the CRP can be useful in defining prior distributions. For example, Goldwater, Griffiths, and Johnson (2006a) defined a probabilistic model of word segmentation—explaining how a child may go from hearing a continuous stream of phonemes to recognizing discrete words within that stream—in which the CRP was used to define the prior distribution on words.

Variants of the CRP can be used to more precisely capture the probability distributions that appear in language. In the original CRP, as the number of customers in the restaurant increases the number who sit at each table follows a power-law distribution, with $P(m_j) \propto m_j^{-1}$. This is a "heavy-tailed" distribution, where a small number of tables end up with very large numbers of customers. Power-law distributions arise often in language—for example, the frequencies with which different words are used follow a power-law distribution (Zipf, 1932). However, in the CRP, the exponent of the power law (the negative power to which m_j is raised) is 1, while linguistic power laws often have exponents closer to 2.

By introducing additional parameters into the CRP, it is possible to define models that produce power laws with a range of exponents. In particular, in the *two-parameter Pitman-Yor process*, the ith customer would be assigned to the jth table with probability

$$P(z_i = j | z_1, z_2, \ldots, z_{i-1}) = \begin{cases} \frac{m_j - a}{i-1+b+ak_+} & j \le k_+ \\ \frac{b+ak_+}{i-1+b+ak_+} & \text{otherwise,} \end{cases} \tag{9.22}$$

where a and b are parameters of the process. The resulting distribution in the number of customers per table is a power law with an exponent of $1 + a$. This model can thus be used to better capture the distributions that arise in language, and models based on the Pitman-Yor process have been shown to have deep connections with sophisticated smoothing schemes used in estimating probability distributions over words (Goldwater, Griffiths, & Johnson, 2006b; Teh, 2006). In fact, the distribution induced by the Pitman-Yor process is the most general distribution over exchangeable partitions (Pitman, 2002).

Applications of ideas from nonparametric Bayesian statistics to language don't stop at the level of words. A standard problem that arises in natural language processing is estimating the probability distributions associated with the rules of probabilistic grammars (see chapter 16). In these grammars, a rule identifies a discrete set of possible ways in which a symbol can be rewritten, each associated with a probability. Using distributions based on the CRP to represent these probabilities has the consequence of "caching" the outcomes of previous applications of a rule: at each point where the rule is applied, you can choose to use a previously generated outcome or create a new one (Johnson, Häubl, & Keinan, 2007a). This property makes it possible to capture some of the rich dependencies in language that are otherwise missing from simple grammars. A similar approach has been used as a form of *stochastic memorization* in probabilistic programming languages (Goodman et al., 2008a), which we discuss in more detail in chapter 18.

9.2 Infinite Models for Feature Representations

Section 9.1 showed how methods from nonparametric Bayesian statistics could be used to define models of category learning that do not require assuming that there is a fixed set of kinds of things in the world. The same strategy can be applied to feature learning. In feature learning, the goal is to identify the latent features that explain the observed properties of a set of objects. In the simplest case—which we will focus on here—the assignment of features to objects is binary, with a latent variable indicating whether an object possesses each feature. The challenge is in simultaneously deciding what features the objects have and how many features should be used to represent the set of objects, just as in the category learning case, the challenge is inferring both the category assignments and their number.

Feature learning can be thought of as a similar problem to category learning. If we imagine each object being associated with a binary vector of features, category learning corresponds to the case where there is a constraint that each row can contain only one nonzero entry. Feature learning is the general case, in which there can be multiple nonzero entries per object. In other words, one discrete unit is associated with each data point in category learning, whereas zero or more discrete units are associated with each data point in feature learning. In this section, we introduce a distribution that is similar to the CRP, but covers this more general case. This distribution can then be used as a prior in models of feature learning, or in other cases where we seek to infer a binary vector but do not wish to limit its length. As in the previous section, we derive this infinite binary prior as the limit of a finite distribution.

9.2.1 A Finite Feature Model

We have n objects and k features, and the possession of feature j by object i is indicated by a binary variable z_{ij}. Each object can possess multiple features. Thus, the z_{ij} form a binary $n \times k$ feature matrix, $\mathbf{Z}$. We will assume that the entries in this matrix are generated by the model

$$z_{ij} \sim \text{Bernoulli}(\theta_j)$$
$$\theta_j \sim \text{Beta}(\frac{\alpha}{k}, 1),$$

where θ_j is the probability that any object has feature j. Each z_{ij} is independent of all other assignments, conditioned on θ_j. Here, the θ_j are independent, so each z_{ij} depends only upon whether other objects possess feature k.

We can compute the joint probability of all assignments of features to objects using this model, defining a probability distribution over matrices $\mathbf{Z}$:

$$\begin{aligned} P(\mathbf{Z}) &= \prod_k \int \left(\prod_i P(z_{ij}|\theta_j) \right) p(\theta_j)\, d\theta_j \\ &= \prod_k \frac{B(m_j + \frac{\alpha}{k}, n - m_j + 1)}{B(\frac{\alpha}{k}, 1)} \\ &= \prod_j \frac{\Gamma(m_j + \frac{\alpha}{k})\Gamma(n - m_j + 1)}{\Gamma(\frac{\alpha}{k})} \frac{\Gamma(1 + \frac{\alpha}{k})}{\Gamma(n + 1 + \frac{\alpha}{k})}, \end{aligned} \tag{9.23}$$

where $B(r, s)$ is the standard beta function and m_j is the number of objects in cluster j. Again, the result follows from conjugacy, this time between the binomial and beta distributions. This distribution is exchangeable, depending only on the count m_j.

It is straightforward to compute the full conditional distribution for any z_{ij} as follows:

$$\begin{aligned} P(z_{ij} = 1 | \mathbf{z}_{-i,j}) &= \int_0^1 P(z_{ij}|\theta_j) p(\theta_j|\mathbf{z}_{-i,j})\, d\theta_j \\ &= \frac{m_{-i,j} + \frac{\alpha}{k}}{n + \frac{\alpha}{k}}, \end{aligned} \tag{9.24}$$

where $\mathbf{z}_{-i,j}$ is the set of assignments of other objects, not including i, for feature j, and $m_{-i,j}$ is the number of objects possessing feature j, not including i.

9.2.2 Taking the Infinite Limit

We can now examine the consequences of taking $k \to \infty$. The use of $\frac{\alpha}{k}$ in defining this model guarantees that the resulting infinite matrices remain sparse. As with the CRP, we need to define a distribution on equivalence classes of matrices, since the probability of any particular matrix will go to zero as $k \to \infty$. In this case, we calculate the probability of the equivalence class of matrices that are the same up to the order of their columns (for details, see Griffiths & Ghahramani, 2005). Taking the limit of equation (9.23) gives

$$\lim_{k \to \infty} P([\mathbf{Z}]) = \exp\{-\alpha H_n\} \frac{\alpha^{k_+}}{\prod_{h>0} k_h!} \prod_{j \le k_+} \frac{(n - m_j)!(m_j - 1)!}{n!}. \tag{9.25}$$

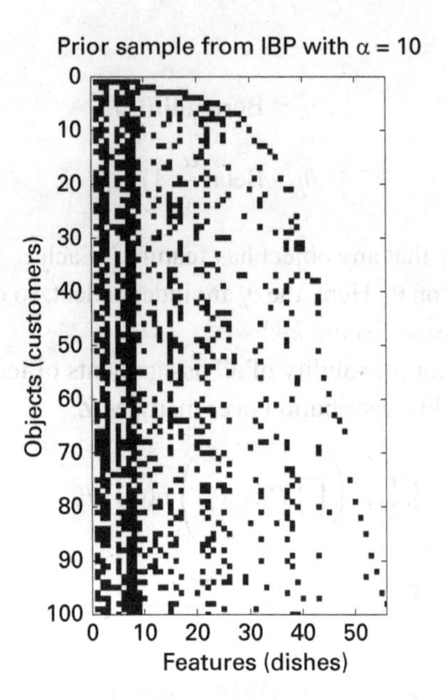

Figure 9.3
A binary matrix sampled from the Indian buffet process with $\alpha = 10$.

Again, this distribution is exchangeable: neither the number of identical columns nor the column sums are affected by the ordering on objects.

9.2.3 The Indian Buffet Process

The joint probability given in equation (9.25) is not immediately intuitive, but can be produced by a simple generative process known as the *Indian buffet process (IBP)* (Griffiths & Ghahramani, 2005). As with the CRP, this process assumes an ordering on the objects, generating the matrix sequentially using this ordering, and objects correspond to customers in a restaurant. Many Indian restaurants in London offer lunchtime buffets with an apparently infinite number of dishes. The first customer starts at the left of the buffet and takes a serving from each dish, stopping after a number of dishes drawn from a Poisson(α) distribution as her plate becomes overburdened. The ith customer moves along the buffet, sampling dishes in proportion to their popularity, serving herself with probability $\frac{m_j}{i}$, and trying a Poisson($\frac{\alpha}{i}$) number of new dishes. The customer-dish matrix $\mathbf{Z}$ is our feature matrix, with customers along the rows, dishes along the columns, and entries indicating which dishes were sampled by which customers. The probability of producing a member of each equivalence class is just the probability given in equation (9.25). An example of a matrix sampled from this process is shown in figure 9.3.

Inference in a model that uses the IBP as a prior may require full conditional distributions. If we care about the identities of the columns, as when they are associated with different parameters in a hierarchical model, the best way to derive these distributions is in terms of the second generative process outlined previously, drawing each z_{ij} as a Bernoulli trial for j such that $m_j > 0$, and then re-sorting the columns. Using the fact that the distribution is exchangeable, we treat the ith object as the nth in the generative process outlined here, to give

$$P(z_{ij} = 1 | \mathbf{z}_{-i,j}) = \frac{m_{j,-i}}{n}. \tag{9.26}$$

The same result can be obtained by taking the limit of equation (9.24) as $k \to \infty$. By the same set of assumptions, the number of new clusters should be drawn from a Poisson($\frac{\alpha}{n}$) distribution. This can also be derived from equation (9.24), using the same kind of limiting argument as that presented here to obtain the terms of the Poisson.

9.2.4 Modeling Human Feature Learning

The IBP provides a simple way to define probabilistic models that can identify the features that should be used to represent a set of objects. This potentially provides an account of how people form feature representations and how those representations depend on context (in particular, the other objects that a person is familiar with). Austerweil and Griffiths (2011) explored the predictions of this kind of account, showing that people seemed to form different representations of objects depending on the distributional properties of the set of objects in which they appear. Figure 9.4 shows how two sets of objects were generated from the same set of parts. Each object had three of the six parts, creating 20 possible combinations of parts. One set of objects, the *correlated* set, repeated the same 4 combinations of three parts four times each. The other set of objects, the *independent* set contained 16 of the 20 unique combinations. When shown the *correlated* set, a probabilistic model based on the IBP forms a representation in which each repeated combination is a single feature. When shown the *independent* set, the six parts from which the objects were actually constructed are identified as features. People seem to form different representations of the objects when shown these two sets as well: when asked whether the four unobserved combinations of objects are likely to appear with the others, participants shown the *correlated* set were far less willing to generalize to these new objects than participants shown the *independent* set.

Analogous to the CRP and the associated Dirichlet process, there have been many extensions and generalizations of the IBP and its associated continuous stochastic process, the *beta process* (Hjort, 1990). One technique for doing so is to elaborate the culinary metaphor of the IBP. For example, transformation-invariant feature learning models have been produced by having customers take a "spice" that is applied to each dish, which transforms the taste of the dish (Austerweil & Griffiths, 2013). Recent work has also explored ways to combine the IBP with neural networks, defining a prior that can be used to help neural networks learn distinct representations of related tasks over time (Kessler, Nguyen, Zohren, & Roberts, 2021).

9.3 Infinite Models for Function Learning

So far, we have focused on cases where the latent structure to be inferred is discrete—either a category or a set of features. However, a similar problem of wanting to accommodate infinite complexity while maintaining simplicity arises in other settings. One of the most prominent examples is *function learning*—learning a relationship between two or more continuous variables. This is a problem that people often solve without even thinking about it, as when learning how hard to press the pedal to yield a certain amount of acceleration when driving a rental car. Nonparametric Bayesian methods also provide a way to solve this problem that

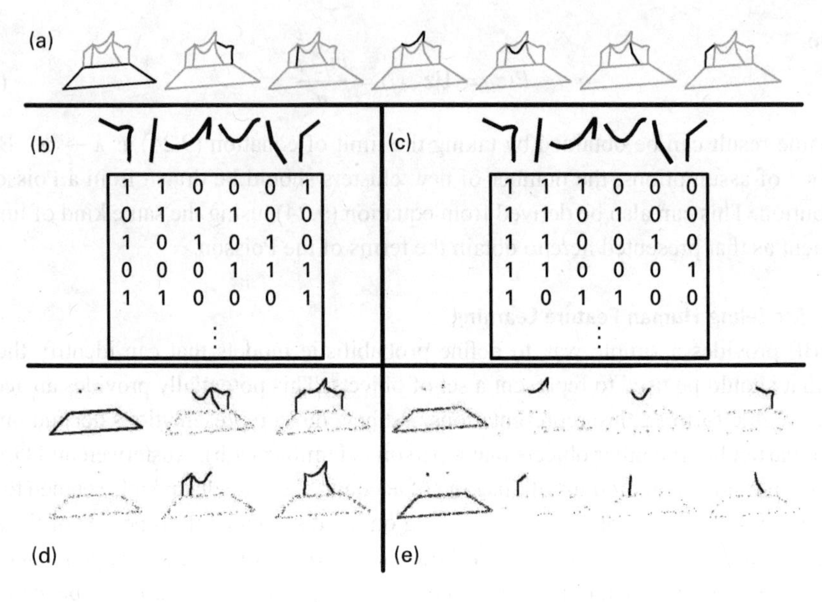

Figure 9.4
Inferring different feature representations depending on the distributional information. (a) The bias (on left) and the six features used to generate both object sets. (b)–(c) The feature membership matrices for (b) *correlated* and (c) *independent* sets respectively. (d)–(e) The feature representations inferred by model for (d) *correlated* and (e) *independent* sets respectively, here represented by single samples drawn from the posterior distribution. Reproduced with permission from Austerweil and Griffiths (2011).

makes it possible to learn complex functions in a way that remains tractable and favors simple solutions.

Viewed abstractly, the computational problem behind function learning is to learn a function f mapping from x to y from a set of real-valued observations $\mathbf{x}_n = (x_1, \ldots, x_n)$ and $\mathbf{t}_n = (t_1, \ldots, t_n)$, where t_i is assumed to be the true value $y_i = f(x_i)$ obscured by some kind of additive noise. In machine learning and statistics, this is referred to as a *regression* problem. In this section, we discuss how this problem can be solved using Bayesian statistics, and how the result of this approach is related to a class of nonparametric Bayesian models known as Gaussian processes. Our presentation follows that in Lucas, Griffiths, Williams, and Kalish (2015).

9.3.1 Bayesian Linear Regression

Ideally, we would seek to solve our regression problem by combining some prior beliefs about the probability of encountering different kinds of functions in the world with the information provided by $\mathbf{x}_n$ and $\mathbf{t}_n$. We can do this by applying Bayes' rule, with

$$p(f|\mathbf{x}_n, \mathbf{t}_n) = \frac{p(\mathbf{t}_n|f, \mathbf{x}_n)p(\mathbf{f})}{\int_{\mathcal{F}} p(\mathbf{t}_n|f, \mathbf{x}_n)p(f)\, df}, \tag{9.27}$$

where $p(f)$ is the prior distribution over functions in the hypothesis space $\mathcal{F}$, $p(\mathbf{t}_n|f, \mathbf{x}_n)$ is the probability of observing the values of $\mathbf{t}_n$ if f were the true function (the likelihood), and $p(f|\mathbf{x}_n, \mathbf{t}_n)$ is the posterior distribution over functions given the observations $\mathbf{x}_n$ and $\mathbf{t}_n$. In many cases, the likelihood is defined by assuming that the values of t_i are independent given f and x_i, and each follow a Gaussian distribution with mean $y_i = f(x_i)$ and variance σ_t^2. Predictions about the value of the function f for a new input x_{n+1} can be made by integrating over this posterior distribution:

$$p(y_{n+1}|x_{n+1}, \mathbf{t}_n, \mathbf{x}_n) = \int_f p(y_{n+1}|f, x_{n+1}) p(f|\mathbf{x}_n, \mathbf{t}_n)\, df, \quad (9.28)$$

where $p(y_{n+1}|f, x_{n+1})$ is a delta function placing all its mass on $y_{n+1} = f(x_{n+1})$.

Performing the calculations outlined in the previous paragraph for a general hypothesis space $\mathcal{F}$ is challenging, but it becomes straightforward if we limit the hypothesis space to certain specific clusters of functions. If we take $\mathcal{F}$ to be all linear functions of the form $y = b_0 + xb_1$, then our problem takes the familiar form of linear regression. To perform Bayesian linear regression, we need to define prior $p(f)$ over all linear functions. Since these functions can be expressed in terms of parameters b_0 and b_1, it is sufficient to define a prior over the vector $\mathbf{b} = (b_0, b_1)$, which we can do by assuming that $\mathbf{b}$ follows a multivariate Gaussian distribution with mean zero and covariance matrix $\mathbf{\Sigma}_b$. Applying equation (9.27) then results in a multivariate Gaussian posterior distribution on $\mathbf{b}$ (see Bernardo & Smith, 1994 for details) with

$$E[\mathbf{b}|\mathbf{x}_n, \mathbf{t}_n] = \left(\sigma_t^2 \mathbf{\Sigma}_b^{-1} + \mathbf{X}_n^T \mathbf{X}_n\right)^{-1} \mathbf{X}_n^T \mathbf{t}_n, \quad (9.29)$$

$$\mathrm{cov}[\mathbf{b}|\mathbf{x}_n, \mathbf{y}_n] = \left(\mathbf{\Sigma}_b^{-1} + \frac{1}{\sigma_t^2} \mathbf{X}_n^T \mathbf{X}_n\right)^{-1} \quad (9.30)$$

where $\mathbf{X}_n = [\mathbf{1}_n\ \mathbf{x}_n]$ (i.e., a matrix with a vector of ones horizontally concatenated with $\mathbf{x}_{n+1}$) Since y_{n+1} is simply a linear function of $\mathbf{b}$, applying equation (9.28) yields a Gaussian predictive distribution, with y_{n+1} having mean $[1\ x_{n+1}]E[\mathbf{b}|\mathbf{x}_n, \mathbf{t}_n]$ and variance $[1\ x_{n+1}]\mathrm{cov}[\mathbf{b}|\mathbf{x}_n, \mathbf{t}_n][1\ x_{n+1}]^T$. The predictive distribution for t_{n+1} is similar, but with the addition of σ_t^2 to the variance.

While considering only linear functions might seem overly restrictive, linear regression actually gives us the basic tools that we need to solve this problem for more general clusters of functions. Many clusters of functions can be described as linear combinations of a small set of basis functions. For example, all k-th degree polynomials are linear combinations of functions of the form 1 (the constant function), x, x^2, ..., x^k. Letting $\phi^{(1)}, \ldots, \phi^{(k)}$ denote a set of functions, we can define a prior on the class of functions that are linear combinations of this basis by expressing such functions in the form $f(x) = b_0 + \phi^{(1)}(x)b_1 + \ldots + \phi^{(k)}(x)b_k$ and defining a prior on the vector of weight $\mathbf{b}$. If we take the prior to be Gaussian, we reach the same solution as outlined in the previous paragraph, substituting $\mathbf{\Phi} = [\mathbf{1}_n\ \boldsymbol{\phi}^{(1)}(\mathbf{x}_n)\ \ldots\ \boldsymbol{\phi}^{(k)}(\mathbf{x}_n)]$ for $\mathbf{X}$ and $[1\ \phi^{(1)}(x_{n+1})\ \ldots\ \phi^{(k)}(x_{n+1})$ for $[1\ x_{n+1}]$, where $\phi(\mathbf{x}_n) = [\phi(x_1)\ \ldots\ \phi(x_n)]^T$.

9.3.2 Gaussian Processes

If our goal were merely to predict y_{n+1} from x_{n+1}, $\mathbf{y}_n$, and $\mathbf{x}_n$, we might consider a different approach, simply defining a joint distribution on $\mathbf{y}_{n+1}$ given $\mathbf{x}_{n+1}$ and conditioning on $\mathbf{y}_n$. One surprisingly general and powerful way to do this is to take the $\mathbf{y}_{n+1}$ to be jointly

Gaussian, with the covariance matrix

$$\mathbf{K}_{n+1} = \begin{pmatrix} \mathbf{K}_n & \mathbf{k}_{n,n+1} \\ \mathbf{k}_{n,n+1}^T & k_{n+1} \end{pmatrix}, \tag{9.31}$$

where $\mathbf{K}_n$ depends on the values of $\mathbf{x}_n$, $\mathbf{k}_{n,n+1}$ depends on $\mathbf{x}_n$ and x_{n+1}, and k_{n+1} depends only on x_{n+1}. If we condition on $\mathbf{y}_n$, the distribution of y_{n+1} is Gaussian with mean $\mathbf{k}_{n,n+1}^T \mathbf{K}_n^{-1} \mathbf{y}$ and variance $k_{n,n+1} - \mathbf{k}_{n,n+1}^T \mathbf{K}_n^{-1} \mathbf{k}_{n,n+1}$. This approach to prediction is often referred to as using a *Gaussian process*, since it assumes a stochastic process that induces a Gaussian distribution on $\mathbf{y}$ based on the values of $\mathbf{x}$. This approach can also be extended to allow us to predict t_{n+1} from x_{n+1}, $\mathbf{t}_n$, and $\mathbf{x}_n$ by adding $\sigma_t^2 \mathbf{I}_n$ to $\mathbf{K}_n$, where $\mathbf{I}_n$ is the $n \times n$ identity matrix, to take into account the additional variance associated with the observations $\mathbf{t}_n$.

The covariance matrix $\mathbf{K}_{n+1}$ is specified using a two-place function in x known as a *kernel*, with $K_{ij} = K(x_i, x_j)$. Any kernel that results in an appropriate (symmetric, positive-definite) covariance matrix for all $\mathbf{x}$ can be used. Common kernels include a radial basis function, with

$$K(x_i, x_j) = \theta_1^2 \exp(-\frac{1}{\theta_2^2}(x_i - x_j)^2), \tag{9.32}$$

indicating that values of y for which values of x are close are likely to be highly correlated, where θ_1 and θ_2 are free parameters of the kernel controlling the overall level of covariation and the speed with which it falls off as a function of the distance between points. Gaussian processes thus provide an extremely flexible approach to regression, with the kernel being used to define which values of x are likely to have similar values of y.

9.3.3 Taking the Infinite Limit

Bayesian linear regression and Gaussian processes appear to present two quite different approaches to the problem of regression. In Bayesian linear regression, an explicit hypothesis space of functions is identified, a prior on that space is defined, and predictions are formed by computing the posterior distribution over functions and then averaging over that distribution. In contrast, Gaussian processes simply use the similarity between different values of x, as expressed through a kernel, to predict correlations in the corresponding values of y. It might thus come as a surprise to know that these two approaches are equivalent: continuing the theme of this chapter, we can derive standard Gaussian process models as the infinite limit of Bayesian linear regression.

Showing that a Bayesian linear regression model of the kind outlined here is a form of Gaussian process prediction is straightforward. The assumption of linearity means that vector $\mathbf{y}_{n+1}$ is equal to $\mathbf{X}_{n+1}\mathbf{b}$. It follows that $p(\mathbf{y}_{n+1}|bfx_{n+1})$ is a multivariate Gaussian distribution with mean zero and covariance matrix $\mathbf{X}_{n+1}\mathbf{\Sigma}_b\mathbf{X}_{n+1}^T$. Bayesian linear regression thus corresponds to prediction using Gaussian processes, with this covariance matrix playing the role of $\mathbf{K}_{n+1}$. Using a richer set of basis functions corresponds to taking $\mathbf{K}_{n+1} = \mathbf{\Phi}_{n+1}\mathbf{\Sigma}_b\mathbf{\Phi}_{n+1}^T$. Thus, Bayesian linear regression corresponds to using the kernel function $K(x_i, x_j) = [1\ x_i][1\ x_j]^T$, and richer basis functions simply make this $K(x_i, \acute{x}_j) = [1\ \phi^{(1)}(x_i)\ \ldots\ \phi^{(k)}(x_i)][1\ \phi^{(1)}(x_i)\ \ldots\ \phi^{(k)}(x_i)]^T$.

It is also possible to show that Gaussian process prediction can always be interpreted as Bayesian linear regression, albeit with potentially infinitely many basis functions. Just as we can express a covariance matrix in terms of its eigenvectors and eigenvalues, we can express a given positive definite kernel $K(x_i, x_j)$ in terms of its eigenfunctions $\boldsymbol{\phi}$ and eigenvalues $\boldsymbol{\lambda}$, with

$$K(x_i, x_j) = \sum_{k=1}^{\infty} \lambda_k \phi^{(k)}(x_i)\phi^{(k)}(x_j) \tag{9.33}$$

for any x_i and x_j. Using the results from the previous paragraph, any kernel can be viewed as the result of performing Bayesian linear regression with a set of basis functions corresponding to its eigenfunctions, and a prior with covariance matrix $\boldsymbol{\Sigma}_b = \text{diag}(\boldsymbol{\lambda})$.

These equivalence results establish an important duality between Bayesian linear regression and Gaussian processes: for every prior on functions, there is a kernel that defines the similarity between values of x, and for every positive-definite kernel, there is a corresponding prior on functions that yields the same predictions. Bayesian linear regression and prediction with Gaussian processes are thus just two views of the same class of solutions to regression problems.

9.3.4 Modeling Human Function Learning

The duality between Bayesian linear regression and Gaussian processes provides a novel perspective on human function learning. Previously, theories of function learning had focused on the roles of different psychological mechanisms. One class of theories (e.g., Carroll, 1963; Brehmer, 1974; Koh & Meyer, 1991) suggests that people are learning an explicit function from a given cluster, such as the polynomials of degree k. This approach attributes rich representations to human learners, but it has traditionally given limited treatment to the question of how such representations could be acquired. A second approach (e.g., DeLosh, Busemeyer, & McDaniel, 1997.; Busemeyer, Byun, DeLosh, & McDaniel, 1997) emphasizes the possibility that people could simply be forming associations between similar values of variables. This approach has a clear account of the underlying learning mechanisms, but it faces challenges in explaining how people generalize beyond their experience. More recently, hybrids of these two approaches have been proposed (e.g., McDaniel & Busemeyer, 2005; Kalish, Lewandowsky, & Kruschke, 2004), with explicit functions being represented, but associative learning.

Bayesian linear regression resembles explicit rule learning, estimating the parameters of a function, while the idea of making predictions based on the similarity between predictors (as defined by a kernel) that underlies Gaussian processes is more in line with associative accounts. The fact that, at the computational level, these two ways of viewing regression are equivalent suggests that these competing mechanistic accounts may not be as far apart as they once seemed. Just as viewing category learning as density estimation helps us to understand the common statistical basis of prototype and exemplar models, viewing function learning as regression reveals the shared assumptions behind rule learning and associative learning.

Gaussian process models also provide a good account of human performance in function learning tasks. Griffiths et al. (2008b) compared a Gaussian process model with a mixture of kernels (linear, quadratic, and radial basis) to human performance (see also Lucas, Griffiths, Williams, & Kalish, 2015). Figure 9.5 shows mean human predictions when trained on a linear, exponential, and quadratic function (from DeLosh et al., 1997), together with

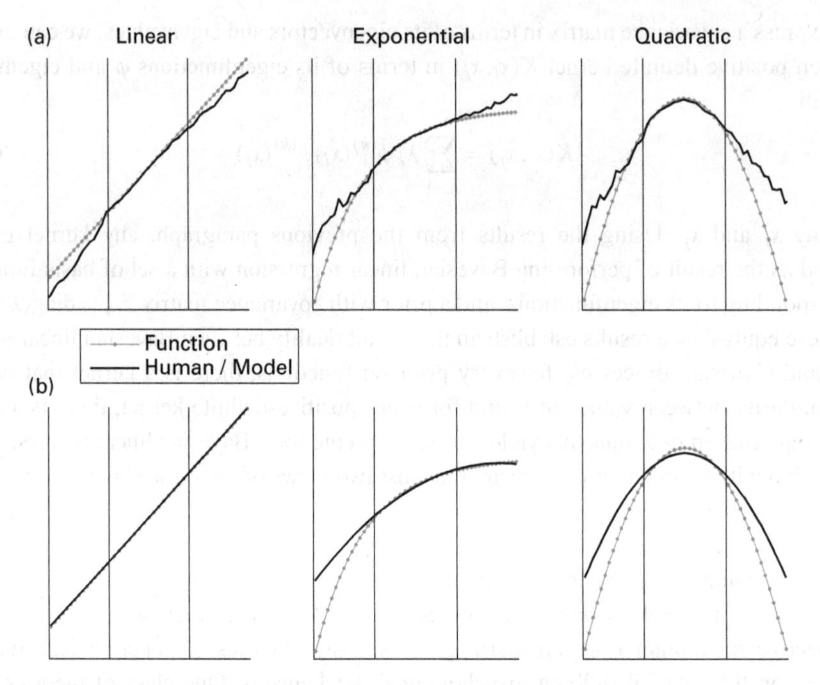

Figure 9.5
Extrapolation performance. (a)–(b) Mean predictions on linear, exponential, and quadratic functions for (a) human participants (from Delosh et al., 1997) and (b) a Gaussian process model with Linear, Quadratic, and Nonlinear kernels. Training data were presented in the region between the vertical lines, and extrapolation performance was evaluated outside this region. Reproduced with permission Griffiths et al. (2008).

the predictions of the Gaussian process model. The regions to the left and right of the vertical lines represent extrapolation regions, being input values for which neither people nor the model were trained. Both people and the model extrapolate nearly optimally on the linear function, and reasonably accurate extrapolation also occurs for the exponential and quadratic function. However, there is a bias toward a linear slope in the extrapolation of the exponential and quadratic functions, with extreme values of the quadratic and exponential function being overestimated.

Subsequent work using Gaussian processes to model human function learning has dug more deeply into the kinds of kernel functions needed to capture human expectations about functions. Wilson, Dann, Lucas, and Xing (2015) directly estimated kernels from human function learning data and found that humans tend to prefer smoother functions than those assumed in typical machine learning approaches. Schulz, Tenenbaum, Duvenaud, Speekenbrink, and Gershman (2017) explored how different kernels could be combined to capture the compositional structure of functions, using a simple grammar to define a distribution over kernels that allow different properties of functions (such as being linear or periodic) to be combined.

9.4 Future Directions

While nonparametric Bayesian methods have been used to study a variety of topics in cognitive science, most of these applications have used the small family of tools presented in this chapter—the CRP, the IBP, and Gaussian processes. The body of literature on nonparametric Bayesian statistics covers a much wider range of topics and continues to expand, creating other opportunities for cognitive science. For example, methods similar to those used to define the CRP and IBP can be used to define probability distributions on infinite ranked sequences (Caron & Teh, 2012) and graphs (Caron, 2012).

While our focus in this chapter (and in the book more broadly) has been on Monte Carlo methods, variational inference can also be used for inference in nonparametric Bayesian models (e.g., Blei & Jordan, 2006). While Monte Carlo methods emphasize the discrete structure of the CRP and IBP, variational methods turn inference into a continuous optimization problem. As a consequence, these methods potentially have a different interpretation in terms of the underlying cognitive processes and have the potential for establishing stronger links to methods based on artificial neural networks.

In general, integrating nonparametric Bayesian models with deep learning potentially offers a new way to think about the trade-off between structure and flexibility that is intrinsic to human cognition. For example, artificial neural networks are known to suffer from *catastrophic forgetting*, where training on one task replaces the knowledge acquired when performing a previous task (McCloskey & Cohen, 1989). The discrete structure offered by the CRP is potentially a way to prevent this: if the system is capable of recognizing that a task is different from what it was previously doing, it can perform that task without modifying the representation of previous tasks (Jerfel, Grant, Griffiths, & Heller, 2019). Likewise, the IBP has been used to define a structured prior to support continual learning in neural networks (Kessler, Nguyen, Zohren, & Roberts, 2021). The integration of a capacity to recognize discrete distinctions in the environment with continuous learning suggests a path toward systems that appropriately balance structure and flexibility.

9.5 Conclusion

Human minds have to grapple with a world that contains unknown numbers of clusters, features, and causes, as well as unknown forms of relationships between variables. Nonparametric Bayesian models provide a way to define meaningful prior distributions for such a world, allowing us to model how people assimilate information into existing representations and modify those representations to accomodate inconsistent results. This capacity can be used as a component of more complex Bayesian models, with the prior distributions discussed in this chapter being useful in any situation in which there is uncertainty over the dimensionality or complexity of latent variables.

We opened this chapter with the example of an explorer encountering a new kind of animal—a case that is readily addressed by the models that we have described. But being able to postulate that something is of a kind that we haven't seen before isn't the sole province of explorers. It's a problem faced by scientists who push the boundaries of their knowledge, and by every human child too. Piaget highlighted assimilation and accommodation as essential forces of cognitive development because so much of our early experience requires expanding what we know in different ways. Nonparametric Bayesian models give us a way to understand these forces—a precise account of when to assimilate and when to accommodate. Using these models, we can capture a part of what it means to grow up in a world of boundless complexity.

10

Estimating Subjective Probability Distributions

Thomas L. Griffiths, Adam N. Sanborn, Raja Marjieh, Thomas Langlois, Jing Xu, and Nori Jacoby

It should be clear from the previous chapters that the predictions of Bayesian models of cognition depend intimately upon the choice of particular probability distributions—how people learn will reflect the prior probabilities of different hypotheses, and how people categorize objects will be determined by the distributions that represent different categories. Estimating these distributions is thus an important part of defining Bayesian models.

Sometimes we can measure these distributions from the world. For example, in the "predicting the future" experiments of Griffiths and Tenenbaum (2006), discussed in chapter 3, prior distributions for various everyday quantities could be estimated from online data sets. Likewise, the explanation of the perceptual magnet effect offered by Feldman, Griffiths, and Morgan (2009), highlighted in chapter 5, required representing phonetic categories using a mixture model, but clues about the parameters of the mixture components could be taken from the human speech signal. Using distributions that are derived from the world is attractive because it minimizes the assumptions that we have to make about how our subjective probabilities—our internal degrees of belief—might differ from the objective, measurable probabilities of the world around us. For this reason, Anderson (1990) recommended this approach in his definition of rational analysis and demonstrated how it could be used to explain phenomena such as power-law curves for forgetting in terms of mimicking the statistical structure of our environment.

In other cases, we are interested in seeing what the consequences are of assuming different distributions, comparing the resulting models with human behavior. Griffiths and Tenenbaum (2006) used this approach to infer the form of people's prior distributions for phenomena such as waiting on the telephone when trying to buy tickets, a phenomenon for which it is hard to obtain objective data. From people's judgments—in this case, the fact that the longer people waited, the longer they expected their additional wait to be—they were able to diagnose that people assume that such wait times follow a power-law distribution. Under this approach, we might define a model and then try to find the distribution that results in the best fit between that model and human behavior. This is an effective strategy, but it faces two kinds of risks. One risk is *underfitting* human behavior, because there are relatively few parametric families of distributions and the particular distribution that best captures human behavior may not be in one of these families. The other risk is *overfitting* human behavior, ending up with a distribution that captures performance on the particular task being modeled but doesn't generalize to other closely related tasks.

In this chapter, we consider a different approach to estimating subjective probability distributions, focused on designing novel experimental methods for measuring those distributions directly. The key idea is to design experiments that let us *sample* from subjective probability distributions. In this way, we can form an estimate of the distribution from the samples. We begin by summarizing standard methods for eliciting subjective probability distributions that have been used in statistics, and then turn to a set of experimental methods based on the sampling algorithms introduced in chapter 6.

10.1 Elicitation of Probabilities

Statisticians, social scientists, and computer scientists often need to capture people's beliefs about a continuous quantity in the form of a probability distribution. To solve this problem, they have developed a variety of *elicitation methods* that use a combination of asking people quantitative questions and then inferring a distribution that corresponds as closely as possible to the answers (for reviews, see Garthwaite, Kadane, & O'Hagan, 2005 and O'Hagan et al., 2006).

A standard approach to elicitation is to ask people to provide quantiles for quantities, or quantities for quantiles. For example, if the goal were to estimate the probability distribution that somebody assigned to the grosses of movies, this could be done by asking people to name a dollar amount corresponding to the lowest 5 percent of grosses, the lowest 10 percent, etc. Alternatively, people could be asked to assign a percentile rank to various dollar amounts, indicating where they think those dollar amounts fall in the overall distribution. Either set of questions will provide a set of numbers that can be used to approximate a cumulative density function, from which an estimate of a probability density function can be recovered.

These traditional elicitation methods can be effective in settings in which the goal is to estimate a distribution over a single quantity, and they do not have any limitations in terms of the form of the resulting distribution (although specific schemes for analyzing the data, such as finding the Gaussian distribution that best corresponds to people's estimates, can introduce additional constraints). However, they have two weaknesses as a general method for estimating probability distributions to be used in Bayesian models of cognition.

First, traditional elicitation methods are feasible to use only for simple, low-dimensional quantities. The grosses of movies all fall along a single dimension—dollar amounts—and the corresponding distribution can be captured by a univariate probability density function. Even generalizing to two dimensions creates challenges in terms of assessing appropriate quantiles and quantities, although it's possible to navigate these. Higher-dimensional distributions over more complex spaces with no natural ordering or representational format, such as people's prior distributions over categories, functions, or causal relationships, lie outside the scope of these methods.

Second, these methods rely on people having veridical access to their subjective probabilities. For a one-dimensional (1D) quantity, this might be a reasonable assumption, although there is plenty of evidence that asking people for explicit probability judgments can be problematic (e.g., Tversky & Kahneman, 1974), which is one reason why the experiments

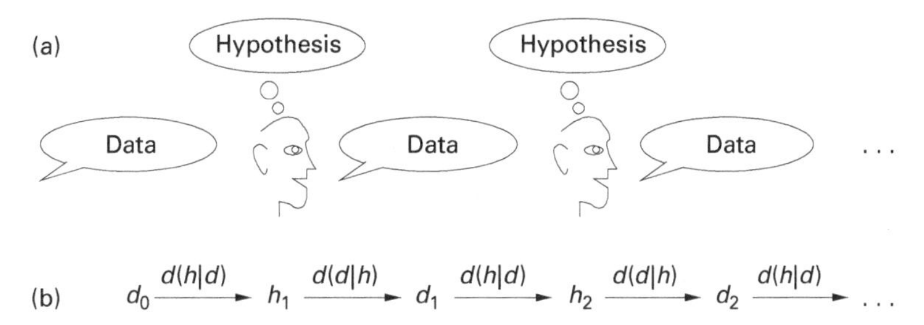

Figure 10.1
Iterated learning. (a) Information is passed down a chain of learners, with each learner forming a hypothesis based on data generated by the previous learner and then generating data in turn. (b) This process defines a Markov chain on data d and hypotheses h. With Bayesian learners, we sample h from $p(h|d)$, and then d from $p(d|h)$. Figure reproduced with permission from Griffiths and Kalish (2005).

presented in this book typically try to avoid asking people to state probabilities. However, people might not have the same kind of access to the distributions that Bayesian models use to characterize prior probabilities or category representations. Measuring people's distribution over the physiognomy of things that fit into the category of dogs, or the prior probability that they assign to deterministic causal relationships, might be a challenge.

For this reason, the methods presented in the remainder of the chapter are designed to be effective for estimating subjective probability distributions over arbitrarily complex objects, using naturalistic judgments that do not require people to state subjective probabilities. They also make no assumption about the form of the underlying distribution. To do so, they employ techniques that computer scientists and statisticians have developed for sampling from complex probability distributions. However, the inspiration for these methods came from neither of these disciplines—it came from linguistics.

10.2 Iterated Learning

When a child learns a language, she learns it from speakers who in turn learned it from other speakers. Languages are transmitted via a process that has been called *iterated learning* (Kirby, 2001), being passed from learner to learner. Figure 10.1a provides a schematic illustration of the simplest version of iterated learning, in which a language is passed along a single chain of learners. Each learner observes linguistic data generated by the previous learner, forms a hypothesis, and then generates data provided to the next learner based on that hypothesis.

A natural question to ask is how the process of transmission by iterated learning should be expected to influence the structure of languages. Figure 10.1b shows that we can analyze this simple form of iterated learning as a Markov chain on data d and hypotheses h. If we assume that the learners are applying Bayesian inference, then the transition probabilities in this Markov chain result from sampling h from the posterior $p(h|d)$, and then d from the corresponding likelihood function $p(d|h)$.

Formulating iterated learning as a Markov chain allows us to ask the question of what the stationary distribution of this Markov chain might be. Recall that a Markov chain will converge to its stationary distribution, provided that it satisfies the conditions for ergodicity (see chapter 6). Griffiths and Kalish (2005; 2007) proved that if all the learners have the

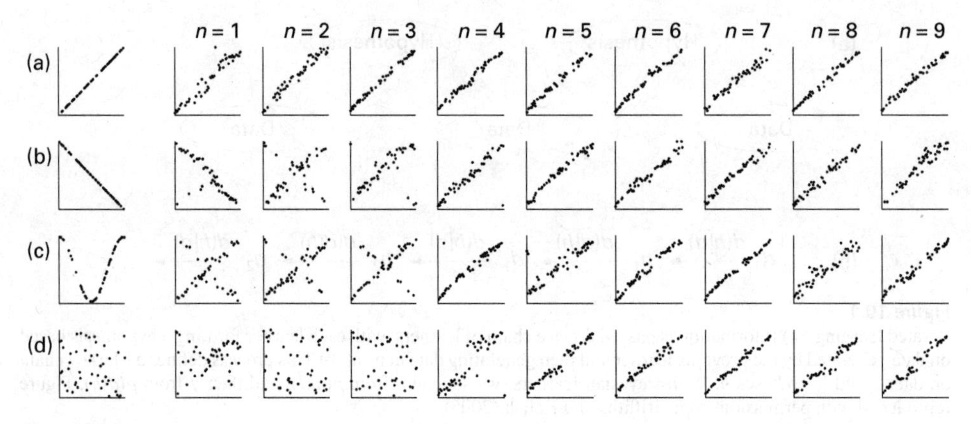

Figure 10.2
Iterated function learning. Each row shows the results produced by one chain of learners, initialized with different functions (shown in the first column). Each column is the predictions of a single learner, which become the training data for the next learner in the row. Regardless of initialization, after just nine iterations the functions have transformed into positive linear functions. Kalish et al. (2007) ran 32 such chains, of which 28 finished as positive linear functions and 4 finished as negative linear functions. Panels (a)–(d) show chains initialized with positive linear, negative linear, nonlinear, and random relationships, respectively. Figure adapted from Kalish et al. (2007).

same prior, the stationary distribution on h is the prior distribution $p(h)$. In the context of language learning, this implies that over time, we should expect languages to change to become easier to learn, conforming more closely with human inductive biases as reflected in this prior distribution.

This theoretical analysis potentially has interesting implications for understanding cultural transmission, but needs to be validated empirically. Kalish, Griffiths, and Lewandowsky (2007) conducted an experiment that provides a good test of the theory, using the function-learning task discussed in chapter 9. People were taught a relationship between two variables, represented by colored bars on a computer screen. They saw 50 pairs of values for these variables, and then they were asked to generate 50 predictions of the value of one variable given the other. These 50 predictions were taken as the data for the next participant, creating an iterated learning chain.

Function learning provides a good test of the theory because it is a case where people's inductive biases are well understood. Decades of research on human function learning has shown that people find it easiest to learn positive linear functions (i.e., functions that are linear with positive slope), followed by negative linear functions, followed by nonlinear functions. We can translate this information into a prior distribution. If a hypothesis has higher prior probability, then it should require less data consistent with that hypothesis to end up with a high posterior probability—that is, it should be easier to learn. Thus, we should expect positive linear functions to have high probability under people's prior on functions. Consequently, the analysis of iterated learning given by Griffiths and Kalish (2007) predicts that we should expect positive linear functions to emerge with high probability from iterated function learning.

Figure 10.2 shows the results from Kalish et al. (2007). Regardless of how chains were initialized, they were dominated by positive linear functions after just nine iterations of transmission. These results provide strong support for the idea that iterated learning produces outcomes that are consistent with people's inductive biases. The positive feedback process that it establishes—where the initial data is repeatedly passed through a biased learning system—magnifies those biases significantly. It's possible to detect that people find it easier to learn positive linear functions by looking at the first iteration—there are fewer errors for the positive linear function, and the errors that people make on other functions tend toward positive linear—but it is much more obvious in the final iteration.

These results raise another possibility: that we could use iterated learning as an experimental paradigm for measuring people's prior distributions. There is no need to actually have information transmitted between participants—we can still construct a Markov chain within participants, with each person seeing a sequence of trials in which the stimuli seen on subsequent trials are determined by their responses on previous trials.

Lewandowsky, Griffiths, and Kalish (2009) explored this possibility in an experiment designed to measure people's priors for everyday quantities using iterated learning. The task was the "predicting the future" problem introduced by Griffiths and Tenenbaum (2006) and discussed in chapter 3: given an observed quantity so far, t, such as the amount of money that a movie has made, predict the total, t_{total}, such as the total gross for the movie. Those values of t_{total} could be used to generate stimuli on the next trial by using assumption of uniform sampling in the likelihood of the corresponding Bayesian model, $p(t|t_{\text{total}})$. In this case, that means sampling the next value of t uniformly from between 0 and t_{total}. If people's responses are samples from $p(t_{\text{total}}|t)$, then over time, the resulting Markov chain will converge to the distribution $p(t_{\text{total}})$.

Figure 10.3 shows the estimated stationary distributions produced by applying iterated learning to the "predicting the future" task. The stationary distributions were estimated by aggregating the last half of all chains across participants. There is a close correspondence between the true distributions of these quantities and the estimated stationary distributions, consistent with the hypothesis that iterated learning can be used to estimate human prior distributions. Subsequent work has used the same approach to estimate prior distributions on concepts (Griffiths, Christian, & Kalish, 2008a; Canini, Griffiths, Vanpaemel, & Kalish, 2014) and causal relationships (Yeung & Griffiths, 2015).

The priors inferred by iterated learning can be quite revealing and can improve the predictions of Bayesian models. Figure 10.4 shows inferred priors on the parameters of the noisy-OR and noisy-AND-NOT functions used in causal models for elemental causal induction (see chapter 4 for more details on the noisy-OR). The noisy-OR is used to capture people's assumptions about generative causes, where the cause increases the probability of the effect, and the noisy-AND-NOT corresponds to preventive causes, where the cause decreases the probability of the effect. In both functions, w_0 is the background rate of the effect and w_1 is the strength of the cause. The joint distribution over these two parameters tells us what people's expectations are about the rate at which effects occur and the assumed strength of causes. Yeung and Griffiths (2015) identified this distribution using an iterated causal learning task, in which people saw contingency data, estimated w_0 and w_1, and then subsequently saw new contingency data generated using the probability distribution that resulted from their estimates.

Figure 10.4 shows the stationary distribution of this process—our best estimate of people's prior on w_0 and w_1—for generative and preventive causes. The results suggest that people are relatively indifferent about the background rates at which effects occur (the distribution on w_0 is roughly uniform), but expect causal relationships to be relatively strong

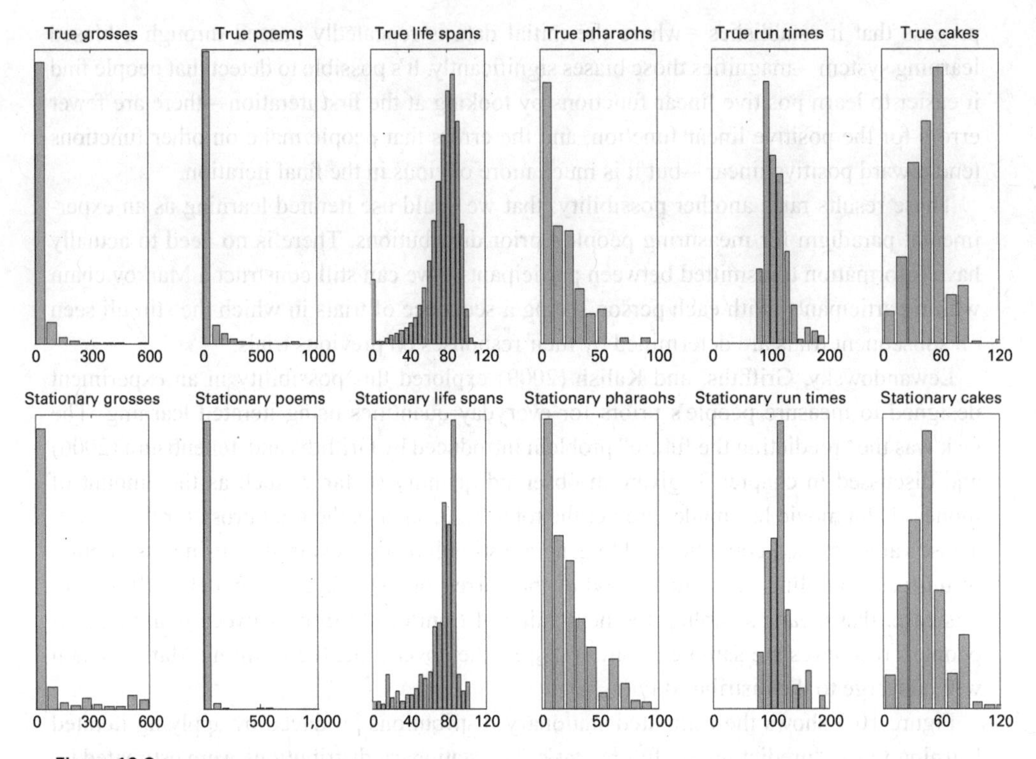

Figure 10.3
Predicting the future. The top row shows the actual distributions of a set of everyday quantities. The bottom row shows the stationary distributions over those quantities resulting from applying the iterated learning paradigm to the predicting the future task of Griffiths and Tenenbaum (2006). Figure reproduced with permission from Lewandowsky et al. (2009).

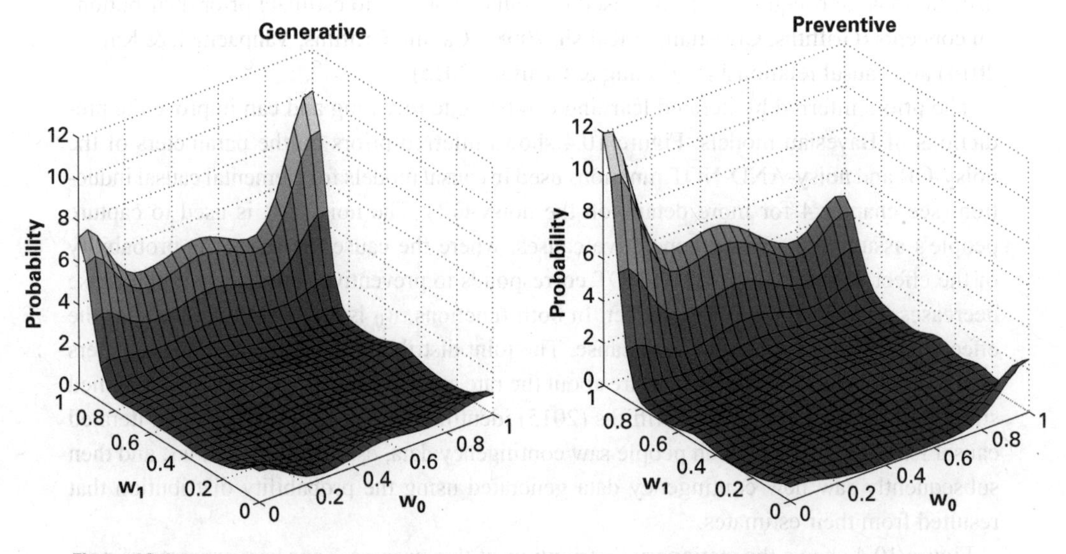

Figure 10.4
Smoothed empirical estimates of human priors on the parameters of Bayesian networks used for elementary causal induction—the background rate of the effect occurring, w_0, and strength of the cause, w_1—produced by iterated learning. Figure adapted from Yeung and Griffiths (2015).

(the distribution assigns higher probabilities to higher values of w_1). Taking this into account results in models that fit human judgments about causal relationships better than those that use other priors (Yeung & Griffiths, 2015).

While we have focused on inferring priors here, it is worth noting that iterated learning can be used to study the expected effects of human perception, learning, and memory on the cultural artifacts that are transmitted across generations. Xu, Dowman, and Griffiths (2013) studied the effect of cultural transmission on systems of color terms (see figure 10.5) using an array of Munsell color chips (see figure 10.5a) that were originally used to collect a large cross-cultural sample of systems of color terms from nonindustrial societies in the World Color Survey (WCS; (Kay, Berlin, Maffi, Merrifield, & Cook, 2009). Participants were initially presented with a random subset of colors (figure 10.5b, top), which were classified into arbitrary categories labeled with pseudowords, and were then asked to generalize what they had learned for the remaining colors (i.e., categorize new colors). The results of one generation of learners became the input for the next generation. Importantly, the number of color terms varied from one condition to another, simulating how the number of "basic" color terms varies across languages. Remarkably, a striking similarity emerged between the resulting artificial color systems and those found in different cultures around the world after as few as 13 iterations (Compare figures 10.5b and 10.5c).

One way to understand why iterated learning converges to the prior is to recognize that it is a form of Gibbs sampling (see chapter 6). In Gibbs sampling, we construct a Markov chain that converges to a particular stationary distribution on a set of variables by iteratively sampling a value for each variable from its conditional distribution, given the current values of all other variables. If the Bayesian learners have a prior $p(h)$ and a likelihood function $p(d|h)$, we can define the joint distribution $p(d,h)=p(d|h)p(h)$. It is then easy to recognize that sampling from the posterior $p(h|d)$ and then the likelihood $p(d|h)$ corresponds to iteratively sampling from the two conditional distributions of this joint distribution. As a result, the distribution on d and h will converge to $p(d,h)$ over time, and the marginal distribution on h will converge to $p(h)$.

This gives us another way to think about iterated learning as an experimental method: it's an implementation of a Gibbs sampling algorithm in which samples are generated by people rather than the computer. This establishes a link between sampling algorithms and experimental paradigms that can potentially be used to convert methods that computer scientists use to generate samples from distributions represented by computers into methods that cognitive scientists can use to generate samples from subjective distributions inside people's heads.

10.3 Serial Reproduction

Methods closely related to iterated learning have previously been used in psychology to study the effects of human cognition on the cultural transmission of information. The most famous of these methods is the *serial reproduction* paradigm introduced by Bartlett (1932). In this paradigm, a participant is shown a stimulus, such as a story or image, and reproduces it from memory after a delay. The reproduction produced by that first participant is then shown to a second participant, who produces another reproduction. As this process is repeated, the stimulus changes significantly.

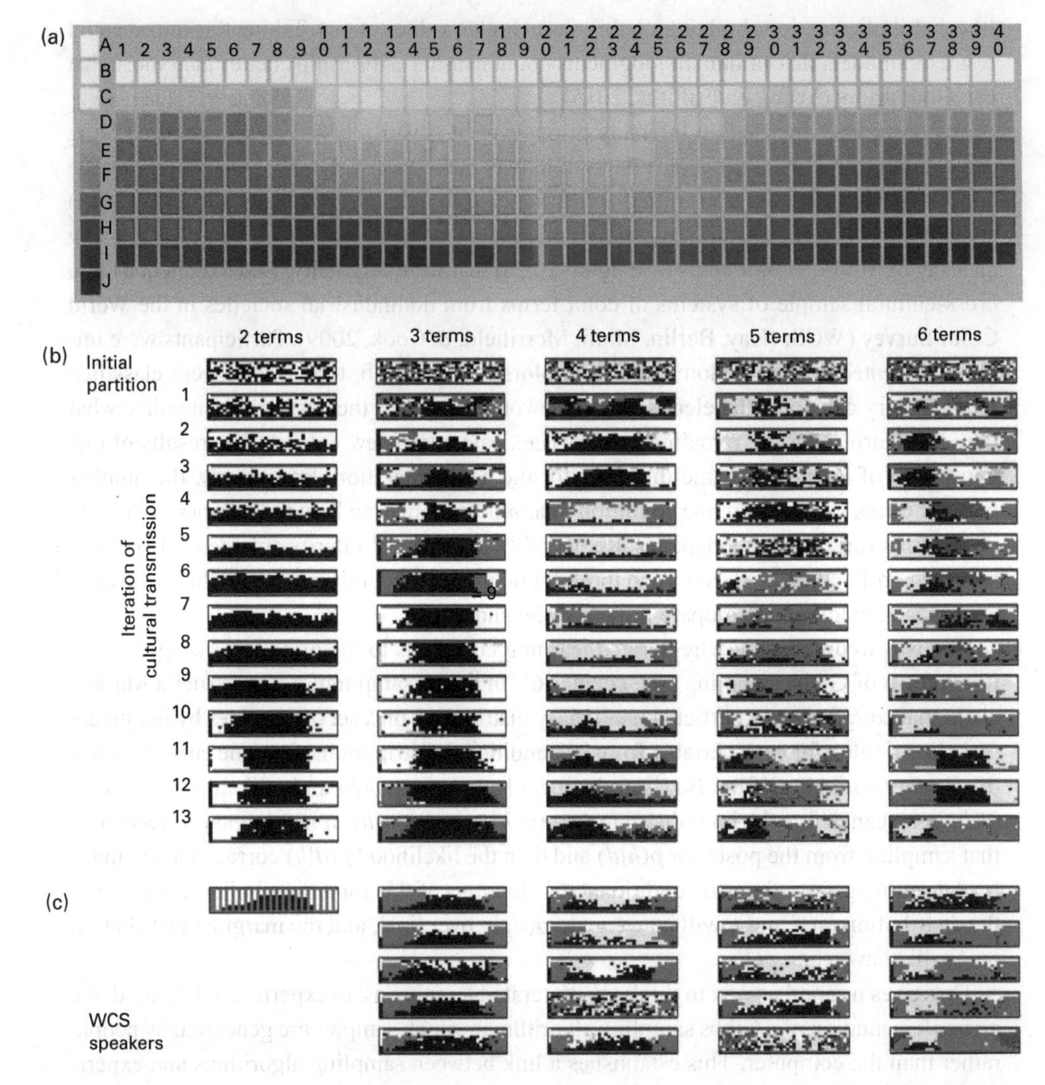

Figure 10.5
Simulating the cultural transmission of color term systems. (a) The stimulus set is based on that used in the WCS, which collected systems of color terms from the languages of 110 nonindustrial societies (Kay et al., 2009). (b) Iterated learning chains for the different term divisions. (c) Sample speakers from WCS languages that are similar to the experimental results. The results for two terms show an estimated partition for the Dani language, as no two-term languages appear in the WCS. Figure reproduced with permission from Xu et al. (2013).

Perception and memory are noisy processes; images and sounds are rarely transmitted without distortion, and first impressions of stories and images are far from permanent. To deal with such uncertainties, humans often rely on prior information to fill the gaps. The reliance on prior information will often lead to biases, where the average participant's response deviates from the real response. Such prior information may reflect the distribution of stimuli in the world, x (Jacoby & McDermott, 2017; Langlois, Jacoby, Suchow, & Griffiths, 2021), or the states that people infer from those stimuli, μ (Xu & Griffiths, 2010). Serial reproduction capitalizes on this observation to construct a process that allows the precise characterization of human prior information. By repeatedly observing and reproducing stimuli from memory, systematic biases due to internalized priors can build up and become manifest. In other words, serial reproduction amplifies biases of perception and memory in a way that reveals the shared priors that generate them.

Formally, serial reproduction implements a Gibbs sampler in the form of a Markov chain $\ldots \rightarrow x_t \rightarrow \mu_t \rightarrow x_{t+1} \rightarrow \ldots$ over the space of (x, μ) pairs. Our goal is then to characterize the observed stationary distribution of this process $p(x)$ as a function of the prior probabilities. Two different analyses of the inference process exist in the literature, leading to subtle differences in the interpretation of $p(x)$. We shall describe both of them, for completeness and to avoid future confusion.

First, the model of Xu and Griffiths (2010) posits that when presented with a noisy stimulus, x_t, participants attempt to infer the true state of the world, μ_t. This can be modeled as a Bayesian inference of the form $p(\mu_t|x_t) \propto p(x_t|\mu_t)\pi(\mu_t)$ where $\pi(\mu_t)$ is the prior over the possible states of the world μ_t and $p(x_t|\mu_t)$ is the likelihood of observing x if the true state of the world is μ_t. If predictions are then generated by simply sampling from the posterior (see Griffiths & Kalish, 2007, for a discussion of other possibilities), the resulting stationary distribution over stimuli x is the posterior predictive distribution $p(x) = \int p(x|\mu)\pi(\mu)d\mu$ (Xu & Griffiths, 2010). Second, Jacoby and McDermott (2017) and Langlois, Jacoby, Suchow, and Griffiths (2021) interpreted serial reproduction as follows: given a *true* incoming stimulus, x_t, a noisy percept, μ_t, is generated through the likelihood $p(\mu_t|x_t)$. At the reconstruction stage (and assuming no production noise for simplicity), participants attempt to infer the true underlying stimulus by incorporating prior information regarding the distribution of stimuli in the world $\pi(x)$. This can be modeled using the posterior $p(x_{t+1}|\mu_t) \propto p(\mu_t|x_{t+1})\pi(x_{t+1})$. Assuming, as before, that participants generate inferences by sampling from the posterior, it is possible to show that the stationary distribution over stimuli converges to the prior itself $p(x) = \pi(x)$ (Jacoby & McDermott, 2017; Langlois et al., 2021).

Serial reproduction has been used to study priors in a variety of domains. Xu and Griffiths (2010) demonstrated the practical soundness of the paradigm by applying it to simple 1D domains. For example, in one task, participants were trained to distinguish between two types of fish (namely, fish-farm fish and ocean fish). Fish stimuli were generated schematically and varied only in terms of their width. The width of fish-farm fish was normally distributed with a certain mean and variance, whereas that of ocean fish was uniformly distributed. By training participants on different farm-fish distributions, and then running a serial reproduction task in which participants saw a fish and had to reproduce it knowing that it came from a farm (initial fish were not necessarily from the fish-farm distribution), the authors showed that the process gradually recovers the training distributions.

In a more complex application of serial reproduction, Langlois et al. (2021) revealed shared priors in spatial memory by iterating a task in which participants reproduced precise point locations within images (figure 10.6). In the task, participants viewed a red point positioned at random on top of an image, such as a gray circle or triangle. Following a delay, the image reappeared on the screen without the red point, and participants were instructed to indicate the exact location of the red point shown during the stimulus phase (see figure 10.6a for illustration of the task and serial reproduction procedure). Past work (Huttenlocher, Hedges, & Duncan, 1991; Wedell, Fitting, & Allen, 2007) highlighted consistent biases

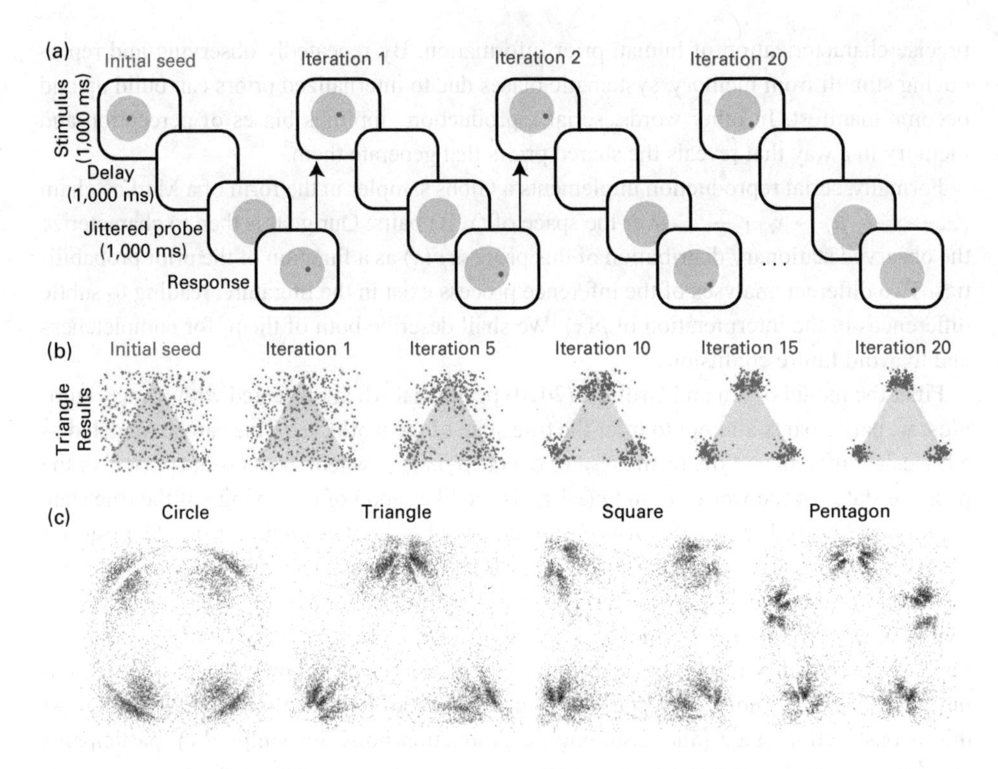

Figure 10.6
Estimating priors for reconstruction from visuospatial working memory. (a) Illustration of the serial reproduction method for an image of a circle. The first participant views an image (a circle) with a point overlaid in a random position and is then asked to reproduce its location from memory. The next participant views the same image, but with the point located at the position reconstructed by the previous participant. The process is repeated for a total of 20 iterations. (b) Serial reproduction results for the remembered position of point locations overlaid on an image of a triangle. The initial uniform distributions of 500 points are shown (far left) becomes increasingly structured with more iterations of serial reproduction. (c) Experimental results showing superposition of responses across all iterations of the chains for images of simple geometric shapes: a circle, triangle, square, and pentagon, highlighting which regions of these shapes are estimated to have high prior probability. Figure adapted from Langlois et al. (2021).

in spatial memory, such as a clear tendency for point reconstructions within an image of a triangle to be biased toward the triangle vertices. Serial reproduction revealed details that had eluded past experimental approaches (figure 10.6c). In particular, it revealed that spatial memory for point locations over a circle image are biased toward quadrant edges, not the quadrant centers, in a departure from previous work (Huttenlocher et al., 1991; Wedell et al., 2007) (figures 10.6a and c).

Priors can originate from short-term interactions with the stimuli, possibly within the duration of the experiment (Xu & Griffiths, 2010; Jazayeri & Shadlen, 2010) but they can also correspond to lifelong culturally dependent learning, such as in the case of language and music. For example, Jacoby and McDermott (2017) used serial reproduction to reveal culturally dependent priors in rhythm perception. In their task, participants were presented with a simple random rhythm and were asked to reproduce it using finger tapping. Western participants were shown to have very different rhythm representations compared with participants recruited from the Bolivian Amazon (see figure 10.7). In an extension of that

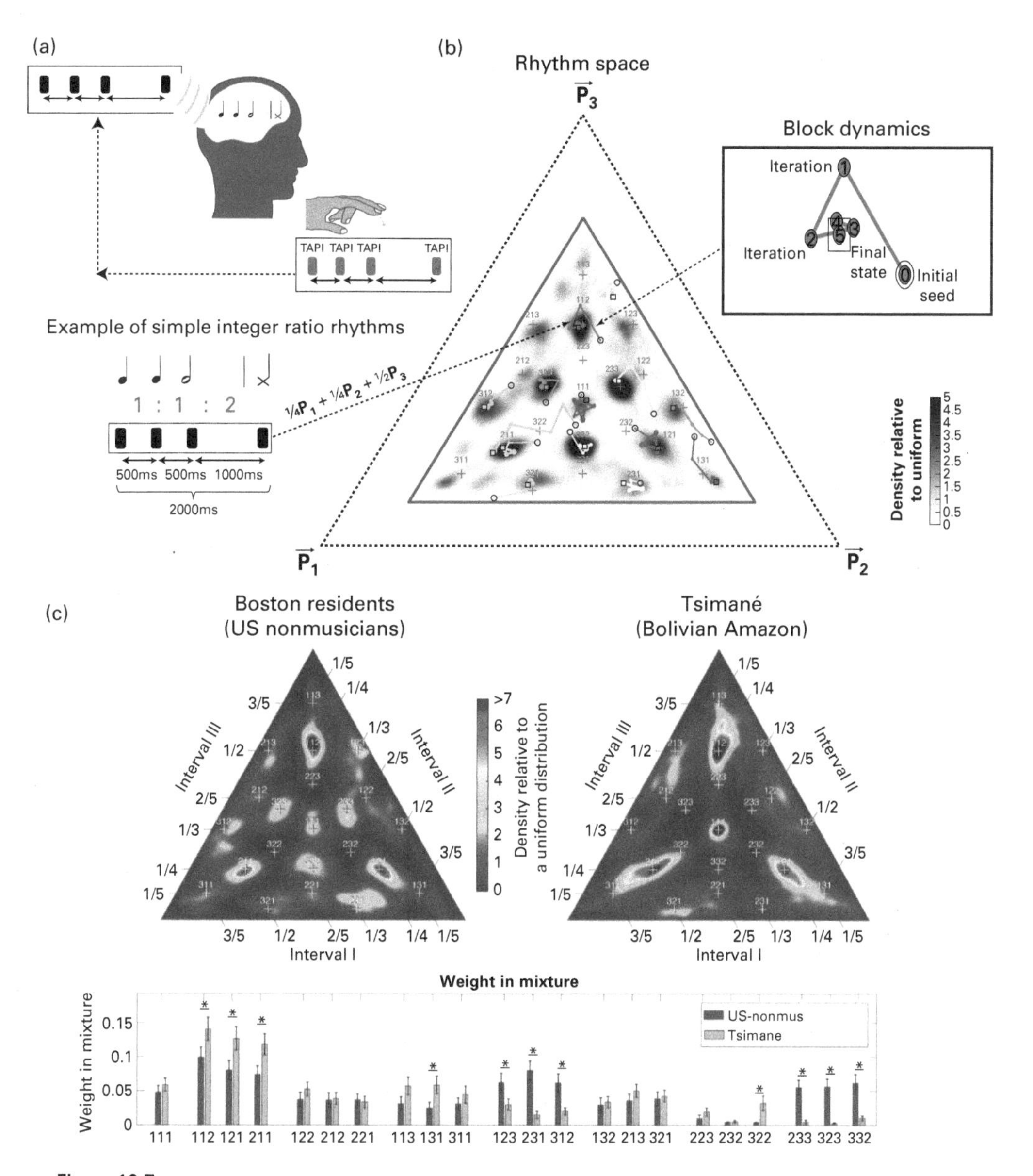

Figure 10.7
Serial reproduction experiment with musical rhythm from Jacoby and McDermott (2017). (a) Schematic of experiment. Participants are presented with random rhythms (a repeating cycle of three clicks, defined by three interbeat time intervals) and reproduce them by tapping, which becomes the stimulus on the following trial. This is iterated five times. (b) The rhythm space, where each axis represents one of the three intervals in a rhythm. Integer ratio rhythms occupy a subset of points in the rhythm space. An example rhythm (1:1:2) is displayed on the left. The colored dots connected by lines show trajectories from example experimental trials. Inset shows one example trial in more detail, converging to the 1:1:2 rhythm. (c) Experimental results showing cross-cultural differences between Tsimane' participants from the Amazon and US nonmusicians. Despite the marked difference, there is overlap between the empirical results and integer-ratios. Results shows kernel density estimates of the responses in the last iterations. The panel below shows the relative importance ("weight") of different integer-ratio categories in a mixture model fitted to the data. The results show categories with significant differences (such as 1:1:2/1:2:1/2:1:1 and 2:3:3/3:2:3/3:3:2) possibly reflecting different lifelong exposure to music. Figure adapted from Jacoby and McDermott (2017).

project, Jacoby et al. (2024) studied participants from 39 groups from 15 countries. They found that priors depend on the nature of musical practices in each culture, but they also share universal features such as the existence of discrete rhythm categories at small integer ratios. Viewed together, these studies highlight the prospect of serial reproduction as a modern tool for studying perceptual priors in a wide range of contexts and for creating meaningful comparisons of these priors across groups.

10.4 Markov Chain Monte Carlo with People

Iterated learning and serial reproduction are effective methods for studying subjective probability distributions of specific kinds: iterated learning can reveal the prior distributions that inform learning, and serial reproduction can reveal the prior distributions that inform perception and memory. However, the Bayesian models presented in this book assume subjective probability distributions of many kinds that do not fall into these two classes. For example, models of categorization assume that categories are associated with probability distributions over stimuli. How could those distributions be estimated?

One way to engage with the broader problem of estimating subjective probability distributions is to take the key insight behind iterated learning and serial reproduction—that Markov chain Monte Carlo (MCMC) algorithms (such as Gibbs sampling) can be implemented with people—and generalize it. Fortunately, there are other types of MCMC algorithms that we can use with people, including the most famous such algorithm: the Metropolis-Hastings algorithm (Metropolis et al., 1953; Hastings 1970).

Sanborn and colleagues (Sanborn & Griffiths, 2008; Sanborn, Griffiths, & Shiffrin, 2010b) explored the potential of Metropolis-Hastings as a scheme for sampling from subjective probability distributions. This algorithm does not require participants to generate new examples as done in iterated learning. Instead, they are given a choice between two items, where those items are selected in such a way that they implement an MCMC algorithm. In the Metropolis-Hastings algorithm, a Markov chain that converges to a particular stationary distribution is constructed by using a proposal distribution to propose a variation on the current state and an acceptance rule that depends on the target distribution to decide whether to accept that variation. Sanborn and colleagues realized that this kind of structure could naturally be translated into an experimental paradigm.

The experimental paradigm that Sanborn et al. developed makes it possible to sample from a probability distribution $p(x)$ proportional to any nonnegative subjective quantity $f(x)$, such as the probability or utility assigned to outcome x. The key is to design a task where people choose between two alternatives x^* and x such that the probability that they choose x^* is

$$P(\text{choose } x^* | x, x^*) = \frac{f(x^*)}{f(x) + f(x^*)}. \tag{10.1}$$

This is potentially straightforward to do, as equation (10.1) is simply the Luce choice rule (Luce, 1959), which is widely used to model people's choices. If a task of this kind can be identified, a Markov chain can be constructed in exactly the same way as in the Metropolis-Hastings algorithm, presenting the current value of the chain and a proposed alternative to people, asking them to choose between these options, and taking the result of the choice

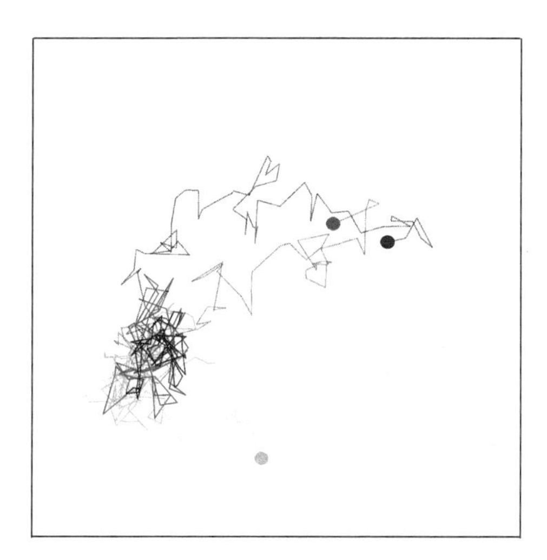

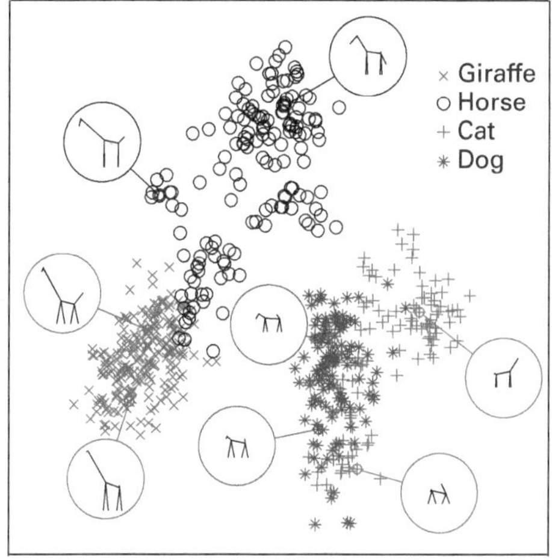

Figure 10.8
Markov chains and samples for one participant in an experiment exploring categories of animals. Both panels are a two-dimensional (2D) projection of a nine-dimensional space, where the original dimensions corresponded to the lengths and angles of lines in a stick-figure quadruped. The left panel shows three Markov chains for the "Giraffe" category, started at disparate points in the space but converging on a fixed region relatively quickly. The right panel shows samples for all four categories, taken from the corresponding Markov chains after an appropriate burn-in. The bubbles show specific examples of stick figures from these categories, illustrating that they seem to do a good job of capturing both the content and the variation associated with each category. Figure reproduced with permission from Sanborn and Griffiths (2008).

as the new current value. Equation (10.1) thus becomes the acceptance probability in this algorithm, resulting in a Markov chain that has $p(x) \propto f(x)$ as its stationary distribution. While equation (10.1) is not identical to the acceptance rule used in the Metropolis-Hastings algorithm, it corresponds to another valid acceptance rule known as the "Barker rule" (Barker, 1965; Neal, 1993), and it is easy to check that it satisfies the detailed balance condition discussed in chapter 6.

One sample application of *Markov chain Monte Carlo with People (MCMCP)* is estimating the structure of natural categories. If category c containing object x is represented by a probability distribution $p(x|c)$, then we can construct a task that satisfies equation (10.1) with $f(x) = p(x|c)$ by presenting people with two objects and asking them to indicate which is most likely to belong to the category (for details, see Sanborn et al., 2010). This provides a way to explore the structure of categories that people have learned through experience ("natural categories"). Sanborn et al. (2010) conducted an experiment where participants were shown stick-figure animals that varied along nine dimensions, such as the angle of the head and the length of the neck, and asked to make judgments about the membership of particular stick figures in four categories—cats, dogs, horses, and giraffes. The proposal distribution was a Gaussian distribution in this nine-dimensional space, with a small probability of a big jump to another more distant point. After a few hundred choices, the Markov chains produced by people's responses tended to converge on specific regions of the space, picking out different distributions for the different categories. The results from one participant are shown in figure 10.8. Other examples include investigating mental representations in intuitive physics (Cohen & Ross, 2009), expressions in cartoon faces (McDuff, 2010),

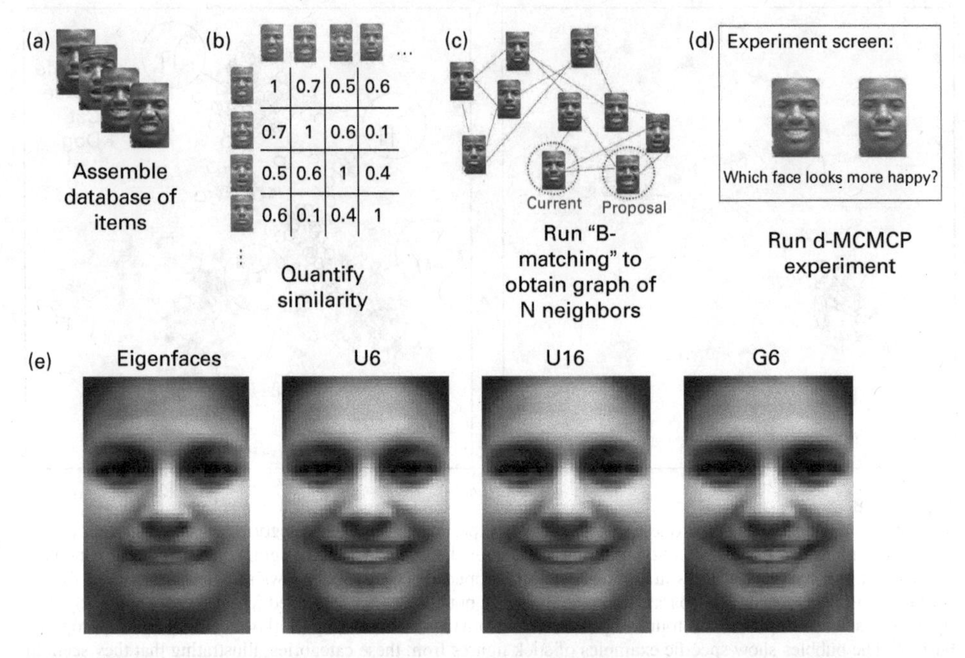

Figure 10.9
Using Markov chain Monte Carlo with people for a discrete set of stimuli. (a) Stimulus images. (b) A computer vision algorithm is used to build a similarity matrix. (c) The *b*-matching algorithm constructs a graph based on the similarity matrix, ensuring that each node has the same number of neighbors. (d) The MCMC algorithm consists of people making decisions about which of a pair of neighboring images in the graph best captures the category. (e) This approach results in better estimates of the average happy face than running an MCMC algorithm in a continuous space where faces are represented by the eigenvectors of the set of images in the database. U6 is a uniform random walk on a graph with 6 neighbors per stimulus, U16 a uniform random walk with 16 neighbors, and G6 a geometric proposal distribution in which the number of steps is first chosen from a geometric distribution and those steps are then taken via a uniform random walk on a graph with 6 neighbors. Figure adapted from Hsu et al. (2019).

and even what people mean when they say they have had "a good night's sleep" (Ramlee, Sanborn, & Tang, 2017).

This method can be scaled up to spaces with many more dimensions than the nine used in the stick figure experiment. Martin, Griffiths, and Sanborn (2012) applied it to investigating facial expressions in a 175-dimensional space defined by the eigenvectors of images of faces ("eigenfaces"). Participants in this experiment saw pairs of faces and were asked to judge which face of the pair was happier or sadder. The Markov chains in this experiment were linked over participants: the face last chosen by one participant was the starting point for the next participant. The result of this procedure were realistic-looking facial expressions of the type that would be difficult to describe in words, as shown in the eigenface result in figure 10.9e.

Of course, even the Metropolis-Hastings algorithm will have trouble sampling from complex probability distributions in high dimensions. While the sampler will converge to the correct distribution eventually, it may take too many trials for the approach to be feasible for use with human participants. Fortunately, Metropolis-Hastings is an extremely flexible algorithm, and researchers have developed many clever ways in which to increase its efficiency. One way is to introduce the idea of "momentum" to the sampler, so if the Markov chain is traveling along a ridge of high probability, then it will tend to stay on that ridge and not waste time probing the low-probability valleys. Interleaving trials in which participants can select the future direction that the sampler will travel in can increase its efficiency (Blundell, Sanborn, & Griffiths, 2012).

A problem with trying to construct a feature space for images is that it is very difficult, and often, as in the case of eigenfaces, points in a feature space do not correspond with any sensible image. An alternative to constructing a feature space is to use only real images. Hsu, Martin, Sanborn, and Griffiths (2019) extended this approach to use in a setting where the goal is to estimate a distribution over a discrete set of stimuli. In this case, the proposal distribution is constructed by using a similarity measure on the stimuli to define a b-matching, which is a graph in which every node is connected to b other nodes. Each node in the graph corresponds to one stimulus, and the connections are made so that stimuli are linked to other stimuli to which they are similar. The proposal distribution for the Markov chain Monte Carlo with People (MCMCP) algorithm, then, is a random walk on this graph—picking one of the b edges at each node uniformly at random. Since each node has the same number of edges, this distribution is symmetric.

Figure 10.9 shows how this works for a set of stimuli corresponding to faces displaying different emotions. A computer vision algorithm is used to construct a similarity matrix, and the b-matching algorithm constructs the graph. People see pairs of faces that are neighbors in this graph and are asked to choose which face of the pair is a better match for a category—in this case, happy faces. Hsu et al. (2019) found that this approach outperformed an MCMC algorithm in which the faces are first transformed into a continuous space and a Gaussian proposal distribution is applied in that space. This approach has subsequently been applied to better understand how surgeons mentally represent fractures of the humerus (Jabbar et al., 2013).

10.5 Gibbs Sampling with People

Despite the flexibility of MCMCP, there are certain aspects of the paradigm that can make it hard to apply in certain domains. Specifically, the two-alternative forced choice interface of MCMCP provides a single bit of information per trial. This, in turn, can be quite time consuming in highly multimodal domains that require a fair bit of exploration. Similarly, the performance of MCMC algorithms critically depends on the choice of a proposal distribution; a distribution that is too narrow may not converge in practice and a distribution that is too broad may miss important details in the subjective distribution. When computer simulations are involved, this may not be such a big problem, as it is often relatively cheap to experiment with a variety of proposal widths and pick the best parameter (e.g., through cross-validation). However, this is not the case when the sampler involves human recruitment in the loop, which can be quite expensive.

To remedy this, Harrison et al. (2020) proposed an alternative paradigm called *Gibbs Sampling with People (GSP)*, which instantiates another variation on the Gibbs sampling process discussed earlier. Unlike serial reproduction and iterated learning, where the conditional sampling alternates between hypotheses and stimuli, here the process alternates directly over the stimulus dimensions. Specifically, given a d-dimensional stimulus space (e.g., colors) and a parameterization $(x_1, \ldots, x_d)$ (e.g., RGB channels), in a GSP

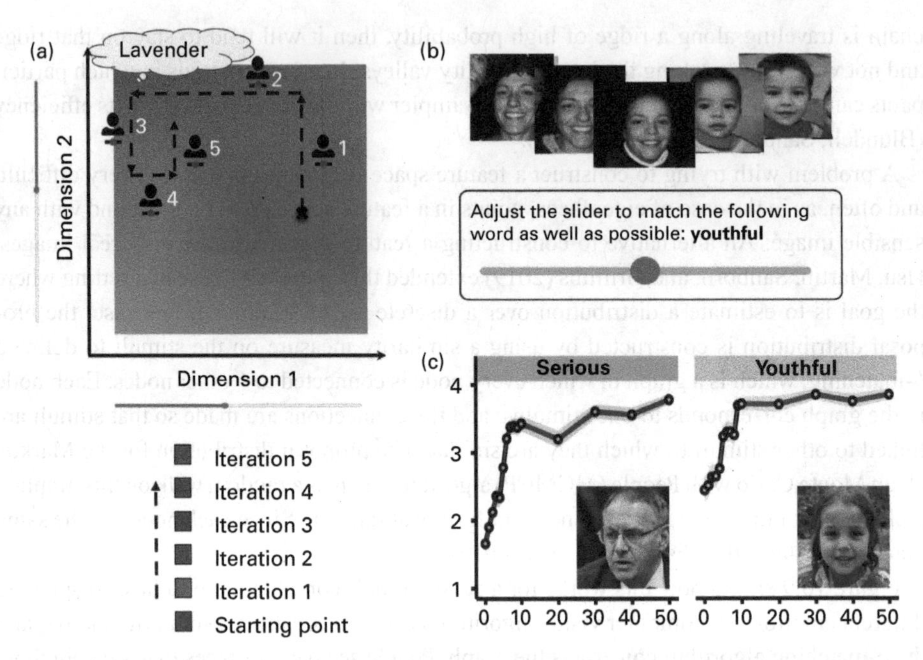

Figure 10.10
Gibbs Sampling with People (GSP). (a) Schematic procedure. At each iteration, participants control one stimulus dimension by moving a slider to optimize for a certain category (e.g., lavender). The result of one generation of participants becomes the input for a new generation. (b) GSP applied to faces. Participants interact with a slider and seek faces that best match a target categories. (c) Example faces sampled from GSP chains for the categories "serious" and "youthful." Curves denote average sample-quality ratings elicited from a separate group of raters as a function of chain iterations. Samples quickly converge on high-quality images with respect to the target categories. Figure adapted from Harrison et al. (2020).

trial, we fix $d-1$ parameters and have a participant explore and select the value of one free dimension (e.g., the R channel) using a slider (see figure 10.10a), so that the generated stimulus best matches a certain target category (e.g., strawberry). The resulting stimulus then gets passed to a new participant, where they get to explore a different dimension and so on. Each such iteration constitutes a sample from the conditional probability $p(x_i|x_1,\ldots,x_{i-1},x_{i+1},\ldots,x_d)$, and by circulating over the dimensions, this process instantiates a Gibbs sampler from the subjective distribution $p(x_1,\ldots,x_d)$. A related paradigm with a similar interpretation was applied to the study of subjective randomness of discrete coin-flip sequences in Griffiths, Daniels, Austerweil, and Tenenbaum (2018).

Harrison et al. (2020) applied GSP to the study of subjective categories over a variety of domains ranging from simple colors and up to fully naturalistic faces. For example, by coupling GSP with the first 10 principal dimensions of the latent space of StyleGAN (Härkönen, Hertzmann, Lehtinen, & Paris, 2020), a modern neural architecture for naturalistic image synthesis, the authors showed that GSP can be effectively used for the study of human bias with respect to face categories (such as what constitutes a "serious" or a "youthful" face; figure 10.10). GSP has also been applied for the study of emotional prototypes in prosody (Van Rijn et al., 2021), the study of the acoustic pleasantness of chords (Marjieh, Harrison, Lee, Deligiannaki, & Jacoby, 2022), and for the generation of structured task distributions in reinforcement learning (Kumar et al., 2022).

10.6 Future Directions

Both MCMC and GSP illustrate how algorithms that were originally designed for computers can be reconstrued as new methods for studying human cognition. The critical step is designing a task for which it is reasonable to interpret human behavior as a sample from a particular probability distribution. Many other algorithms use random samples to solve particular problems, raising the possibility that there are other algorithms that can be equally useful as experimental methods.

For example, so far, we have discussed MCMCP chains as separate entities; however, modern samplers often use multiple simultaneous chains to adaptively optimize their proposal function (Goodman & Weare, 2010). It is conceivable, therefore, to come up with new paradigms in which information is shared across multiple sampler chains to achieve better convergence behavior. Likewise, MCMC and deterministic optimization can be viewed as two limits of a continuum of algorithms in which one manipulates the stochasticity (or *temperature*) of the process: MCMC is the "high-temperature" case, where the goal is to bounce around the distribution and it is OK to visit regions with lower probability, while optimization is the "low-temperature" case, in which changes to the state should increase its probability. As shown by Harrison et al. (2020), one way to control this level of stochasticity in MCMC algorithms with people is to aggregate over multiple human judgments per iteration, driving the sampler toward more deterministic behavior. This can be particularly useful if we're interested in characterizing the modes of distributions rather than their general shape, or optimizing over subjective losses defined by a population of participants. There are probably as many optimizers out there as samplers, and bringing them into the toolbox of the modern psychologist can be of great value.

More generally, the methods that we use in cognitive science are beginning to undergo an important change. Studying the mind became a science in the twentieth century, and for most of that century, it used a specific methodology: people would come into a lab and do a task for around an hour, perhaps being assigned to one of a few different conditions. This methodology was in part a consequence of the constraints of running a physical laboratory: people had to travel to get there, so it made sense to have them stay for a while, and experiments would be administered by research assistants and consequently couldn't manipulate too many variables.

Twenty-first-century cognitive science is in a very different situation. Experiments are increasingly run using online crowdsourcing services. Often, these experiments are just scaled-up versions of lab experiments. However, crowdsourcing offers a completely different profile from a physical lab: people can be paid to make as little as a single decision, and the tasks that people are presented with are selected by a computer that has access to all previous decisions. This sets up an environment where there is far greater freedom to explore innovative experimental designs, and where experiments look much more like algorithms that are run with people. Thinking intelligently about how to design those algorithms, drawing on ideas from computer science and statistics, has the potential to shed a lot more light on the nature of human cognition (for further discussion of this point, see Suchow & Griffiths, 2016b).

10.7 Conclusion

Bayesian models of cognition assume probability distributions over complex, high-dimensional objects that would be difficult to estimate using traditional elicitation methods. However, a variety of approaches exist for estimating these distributions, drawing on algorithms for estimating probability distributions that are used in computer science and statistics. Iterated learning can be used to estimate prior distributions that inform learning. Serial reproduction can be used to infer prior distributions that result in biases in perception and memory. MCMC and GSP offer more general algorithms that can be used to reveal the structure of psychological representations in a remarkable variety of settings. Just as computer scientists and statisticians constantly innovate on the methods that they used to estimate probability distributions, we see these methods as providing a foundation on which further tools for gaining insight into human cognition can be built.

11

Sampling as a Bridge Across Levels of Analysis

Thomas L. Griffiths, Edward Vul, Adam N. Sanborn, and Nick Chater

As noted in chapter 2, probabilistic models of cognition are typically defined at what Marr (1982) called the computational level, characterizing an abstract problem that people need to solve and the logic of its solution. Research proceeds by comparing these ideal solutions to human behavior, using the results to try to refine the assumptions behind our models. This approach is quite different from traditional methods used to study the mind. Historically, the models defined by cognitive psychologists have engaged with what Marr termed the "algorithmic level," focusing on identifying the cognitive processes that underlie behavior. This involves spelling out the calculations that the brain makes to solve the problem that has been specified at the computational level. Neuroscience adds analyses at the implementational level, suggesting how these cognitive processes might be implemented in the brain. This raises a basic question: How are insights at these levels of analysis connected?

Understanding the relationship between models at different levels of analysis is central to evaluating the contribution that probabilistic models of cognition can make to psychology. Marr (1982) introduced the notion of the computational level 40 years ago, and the work of Shepard (1987) and Anderson (1990) provided early examples of the promise of this kind of approach. However, the use of probabilistic models of cognition to explain the various phenomena discussed in this book makes understanding the implications of computational-level analyses for algorithmic- and implementational-level accounts a pressing concern (e.g., Anderson, 1991a). Elucidating this relationship is relevant to both the interpretation of empirical data and the compatibility of different theoretical approaches.

On the empirical side, we need to understand when deviations of human behavior from the ideal solutions to the presumed computational problems reflect mistaken assumptions in the formulation of those problems (or the mistaken assumption that people behave in an ideal fashion) and when they provide clues about the cognitive and neural processes that allow people to approximate those ideal solutions. The identification of cognitive processes by looking for deviations from ideal theories is a strategy that has proved extremely effective in the past, with the heuristics and biases research program (Tversky & Kahneman, 1974) providing one of the best examples. By extending the scope of the problems for which we can identify ideal solutions, probabilistic models of cognition could potentially make this strategy applicable in a wider range of domains.

On the theoretical side, we need to know whether showing that people behave in a way that is consistent with a particular probabilistic model provides evidence against a particular theory at the algorithmic or implementational level, or whether these accounts have no bearing on one another simply because they lie at different levels of analysis. For example, one of the key issues in the debate between proponents of probabilistic and connectionist models (Griffiths, Chater, Kemp, Perfors, & Tenenbaum, 2010; McClelland et al., 2010) is that many probabilistic models are defined in terms of structured, discrete representations, while connectionist models use continuous, graded representations that can mimic discrete structure when appropriate. However, since probabilistic models are typically defined at the computational level and connectionist models are typically defined at the algorithmic or implementational level, it is not clear whether this reflects a fundamental incompatibility. Without identifying when probabilistic inference with discrete representations can be approximated by neural networks using continuous representations and general-purpose learning algorithms, we cannot know whether a particular probabilistic model is inconsistent with a particular connectionist model.

In this chapter, we consider how it may be possible to build bridges between different levels of analysis within the framework of probabilistic models of cognition. In particular, we consider how the algorithms that are used for approximating probabilistic inference can also provide a source of hypotheses about the algorithms used by human minds and brains. The chapter is based upon and expands Griffiths, Vul, and Sanborn (2012b).

11.1 A Strategy for Bridging Levels of Analysis

In proposing the idea of different levels of analysis for information-processing systems, Marr (1982) expected that there would be constraints that hold between levels. Computational-level analyses impose a strong constraint on analyses at the algorithmic and implementational levels: whatever form those cognitive and neural processes take, they need to approximate the ideal solution to the computational problem. In turn, considerations at the algorithmic and implementational levels impose constraints on theorizing at the computational level: we know that the ideal solutions to computational problems need to be approximated using specific cognitive and neural processes. Just as physical, biochemical, and physiological factors shape the solutions that evolution can find to the problems posed by the environment (ruling out, presumably, wheels, jet engines, and silicon chips), the structure of the human mind and brain should constrain the solutions that we consider to computational problems.

Following this logic, showing that people behave in a way that is consistent with the predictions of a probabilistic model implies that the cognitive and neural processes that people are using in solving that problem somehow approximate probabilistic inference. This suggests a strategy for bridging levels of analysis: focus on algorithms that are used in computer science and statistics for approximating probabilistic inference, and explore those algorithms as candidate models of cognitive and neural processes. The resulting models are *rational process models* (Sanborn et al., 2010a; Shi, Griffiths, Feldman, & Sanborn, 2010), accounts that are intended to bring us closer to an understanding of the processes operating at the algorithmic and implementational levels but that are motivated by the principle of rationality that guides models developed at the computational level.

Rational process models take a different approach from traditional strategies for making computational models in cognitive psychology, which start with a hypothesized set of psychological mechanisms and examine how those mechanisms can be combined to model behavior. In a rational process model, we begin with an algorithm for approximating probabilistic inference, ask whether the components of the algorithm are consistent with what we know about cognitive processes, and then examine how well the model fits behavior. The result is a class of models that are guaranteed to approximate probabilistic inference but deviate from ideal solutions in ways that can be instructive about the processes that underlie human judgments. Indeed, the hope is that the ways in which a rational process model deviates from perfect rationality will turn out to be the very ways in which human behavior deviates from the ideal rational solution.

11.2 Monte Carlo as a Psychological Mechanism

What cognitive and neural processes are candidates for approximating the ideal solutions identified by probabilistic models of cognition? As discussed in chapter 6, one of the main strategies that is used for approximating probabilistic inference in computer science and statistics is the Monte Carlo principle: performing computations with samples from a probability distribution rather than the distribution itself. A variety of sophisticated Monte Carlo algorithms have been defined, which can be used to approximate probabilistic inference across a range of circumstances. These algorithms provide a rich source of hypotheses about possible rational process models.

The idea that people might approximate probabilistic inference by sampling has links to a long-standing literature in cognitive psychology. Sampling appears as a basic component of a variety of psychological theories of choice and decision-making (Luce, 1959; Busemeyer, 1985; Stewart, Chater, & Brown, 2006). Sampling processes can explain why the probability of responses in simple estimation tasks seems to track the posterior probability of the corresponding hypotheses (Vul, Goodman, Griffiths, & Tenenbaum, 2014), and even how children might choose among hypotheses when making causal inferences (Denison, Bonawitz, Gopnik, & Griffiths, 2013). Different forms of sampling can show close correspondence to the availability, representativeness, and anchoring and adjustment heuristics (Chater et al., 2020).

Modern Monte Carlo methods (of the kind discussed in chapter 6) provide innovative ways of drawing samples from probability distributions, which provide an opportunity to define more complex rational process models. In the remainder of this chapter, we explore how three of these Monte Carlo methods—importance sampling, particle filters, and Markov chain Monte Carlo (MCMC)—can be used to define rational process models for different aspects of cognition, and how they can help us bridge levels of analysis.

11.3 Exemplar Models as Importance Samplers

One of the simplest Monte Carlo methods introduced in chapter 6 is importance sampling, in which samples are drawn from a distribution other than the target distribution and then reweighted to approximate a sample from the target distribution. A simple importance-sampling algorithm for Bayesian inference is to sample hypotheses h from the prior distribution $p(h)$ and then weight those by the likelihood function $p(d|h)$ to obtain an approximation to the posterior distribution $p(h|d)$. This algorithm—likelihood weighting—translates into a natural process model for approximating simple probabilistic inferences, in which a person remembers events from the past and then retrieves them from memory according to their similarity to a current event. The events from the past act as samples from the prior, and the similarity function corresponds to the likelihood. This is a kind of exemplar model (as discussed in chapter 5), and importance sampling can be shown to be equivalent to a formal definition of an exemplar model used as a process model in the past. The result is a simple rational process model that approximates human behavior in a variety of tasks that have been analyzed using probabilistic models of cognition (Shi et al., 2010).

For example, for the "predicting the future" problem of Griffiths and Tenenbaum (2006) discussed in chapter 3, in which people are asked to predict the total of a quantity (e.g., the time it takes to bake a cake) given the value so far (e.g., how long the cake has been in the oven), it seems unlikely that people explicitly represent probability distributions of everyday events. Importance sampling provides a more psychologically plausible solution: recalling events, such as experienced cake-baking times, that exceed the given value and weighting each according to the likelihood that it will be seen (e.g., the chance that you would see the cake at a particular time given its total time in the oven) to compute the expected total cake-baking time.

An importance sampling method for approximating Bayesian inference—likelihood weighting, introduced in chapter 6—can also be shown to be formally equivalent to the exemplar models that are widely used in cognitive psychology (e.g., Nosofsky, 1986). An exemplar model consists of a set of stored exemplars $\mathcal{X} = \{x_1, x_2, \ldots, x_{n-1}\}$, a similarity function η_{ni} relating a new observation x_n to an exemplar x_i. On observing x_n, all exemplars are activated in proportion to η_{ni}. The use of the exemplars depends on the task. In an identification task, where the goal is to identify an x_i that matches x_n, the probability that x_i will be selected is

$$P_e(x_i|x) = \frac{\eta_{ni}}{\sum_{j=1}^{n-1} \eta_{nj}}, \tag{11.1}$$

where $P_e(\cdot)$ denotes the probability distribution resulting from the exemplar model. In a categorization task, where each exemplar x_i is associated with category c_i, the probability that the new object x will be assigned to category c is given by

$$P_e(c_n = c|x_n) = \frac{\sum_{i|c_i=c} \eta_{ni}}{\sum_{i=1}^{n-1} \eta_{ni}}, \tag{11.2}$$

where we assume no biases β_c toward particular categories. The general form of the response predicted by an exemplar model is

$$\text{response}(x) = \frac{\sum_{i=1}^{n-1} f(x_i)\eta_{ni}}{\sum_{i=1}^{n-1} \eta_{ni}}, \tag{11.3}$$

where $f(\cdot)$ is a function of the stored exemplar x and possibly some information stored with it, such as a category label. For identification, $f(x_i)$ is just the delta function $\delta(x_i, x_n)$, and for categorization, it is the category membership function $I(x_i = c)$.

Now we will compare the predictions of this generalized exemplar model to a Bayesian model approximated using likelihood weighting. Assume that we observe stimulus x, which we believe to be corrupted by noise and potentially missing some accompanying information, such as a category label. Let x^* denote the uncorrupted stimulus. Our goal is simply to reconstruct x, finding the value of x^* to which it corresponds. Using the probability distribution $p(x|x^*)$ to characterize the consequences of the noise process, and $p(x^*)$ to encode our a priori beliefs about the probability of seeing a given stimulus, we can apply Bayes' rule to compute the posterior distribution $p(x^*|x)$. Specifically, we have

$$p(x^*|x) = \frac{p(x|x^*)p(x^*)}{\int p(x|x^*)p(x^*)\,dx^*}, \tag{11.4}$$

where $p(x|x^*)$ is the likelihood and $p(x^*)$ is the prior. The resulting posterior distribution to can be used to answer questions about the properties of x^*.

Both simple Monte Carlo and importance sampling can be applied to the problem of evaluating the expectation of a function $f(x^*)$ over a posterior distribution on x^*. Simple Monte Carlo would draw values of x^* from the posterior distribution $p(x^*|x)$ directly. Importance sampling would generate from another distribution, $q(x^*)$, and then reweight those samples. One simple choice of $q(x^*)$ is the prior, $p(x^*)$. If we take m samples from the prior, then the weight assigned to each sample is proportional to the ratio of the posterior to the prior:

$$w_i \propto \frac{p(x_i^*|x)}{p(x_i^*)} = \frac{p(x|x_i^*)}{\int p(x|x^*)p(x^*)\,dx^*}, \tag{11.5}$$

where $p(x^*)$ has now been cancelled from the numerator of equation (11.4). Since the weights sum to 1, we can rewrite this as

$$w_i = \frac{p(x|x_i^*)}{\sum_{i=1}^{m} p(x|x_i^*)} \tag{11.6}$$

because we can cancel $p(x^*) = \int p(x|x^*)p(x^*)\,dx^*$ from the numerator and denominator when we normalize. Substituting these weights into the importance sampling estimate,

$$E\left[f(x^*)|x\right] \approx \sum_{i=1}^{m} f(x_i^*)w_i, \tag{11.7}$$

we obtain

$$E\left[f(x^*)|x\right] \approx \frac{\sum_{i=1}^{m} f(x_i^*)p(x|x_i^*)}{\sum_{i=1}^{m} p(x|x_i^*)}. \tag{11.8}$$

This is simply the likelihood-weighting procedure introduced in chapter 6.

We can now observe a formal equivalence between the generalized exemplar model and Bayes approximated by likelihood weighting. If our set of exemplars $\mathcal{X}$ is drawn from the probability distribution $p(x^*)$, then the expectation of any function $f(\cdot)$ over the distribution $P_e(x_i|x)$ defined in equation (11.1) is an importance sampler for the expectation of $f(x^*)$

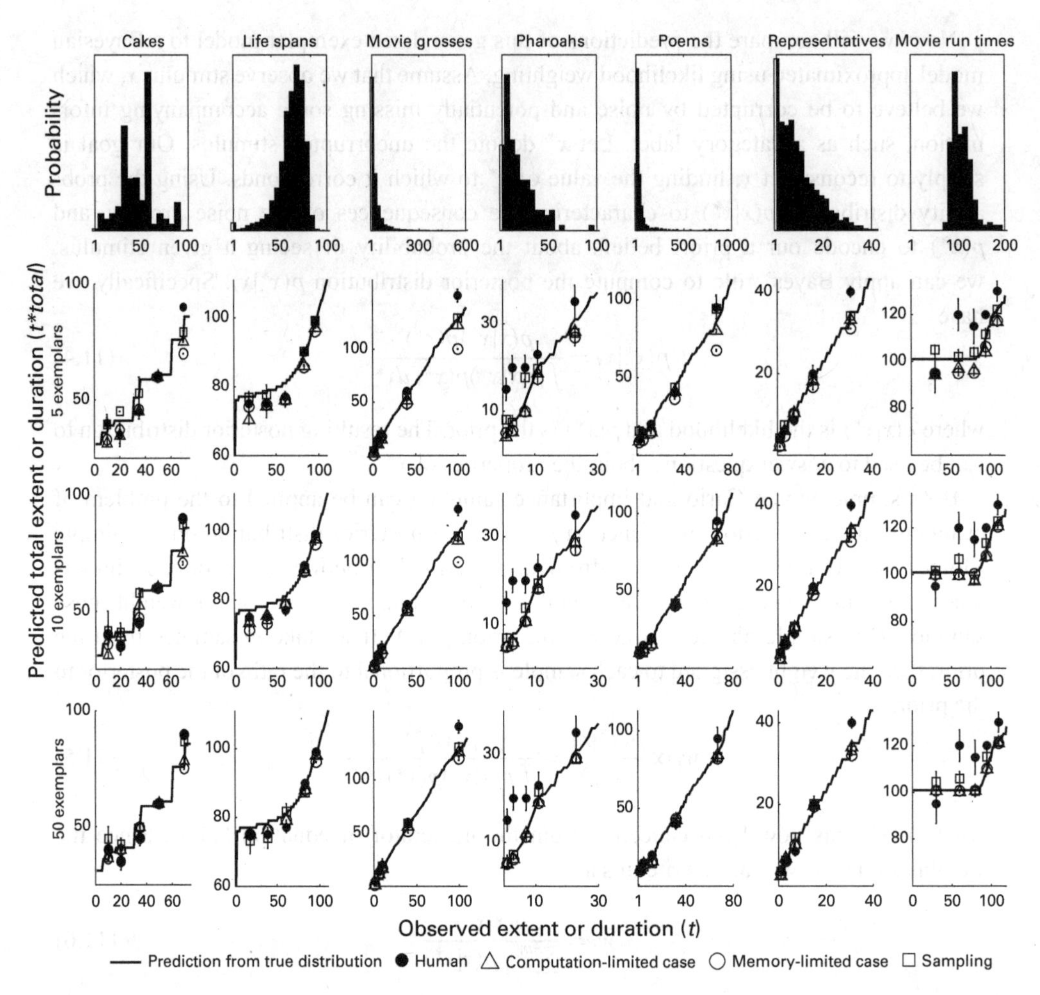

Figure 11.1
Results of using an exemplar model to approximate Bayesian inference in the "predicting the future" task of Griffiths and Tenenbaum (2006) introduced in chapter 3. Two different methods for limiting the number of exemplars—computation-limited, where a fixed number of exemplars are given nonzero weight, and memory-limited, where a fixed number of exemplars are available in memory—are used to approximate the median of the posterior, and are compared to directly sampling from the posterior distribution. Reproduced with permission from Shi et al. (2010).

with respect to the posterior distribution over x^*, as given in equation (11.4), for the Bayesian model with prior $p(x^*)$ and likelihood $p(x = x_n | x^* = x_i) \propto \eta_{ni}$.

Figure 11.1 shows the results of using this kind of exemplar-based approach to approximate Bayesian inference in the "predicting the future" problem. In this case, two kinds of exemplar models were evaluated. The first kind were "computation-limited" models in which the amount of computation was treated as a bottleneck, limiting the number of exemplars with nonzero weight. The second kind was "memory-limited," and the total number of exemplars was restricted regardless of their weight. Both kinds of exemplar models resulted in a good approximation of the median of the posterior even with a small number of samples, comparable to directly sampling from the posterior.

11.4 Particle Filters and Order Effects

The particle filter is a particularly attractive candidate for explaining how people might be able to produce behavior that is consistent with Bayesian inference despite limits on their memory or computational capacity. This is because it provides a general recipe for sequentially updating beliefs (i.e., using importance sampling to weight samples by the likelihood of new observations; see chapter 6) that can be applied in a wide range of models.

As an example, we will show how particle filters can be used for inference in an infinite mixture model applied to inferring how stimuli cluster. Specifically, we will illustrate this algorithm when applied to Anderson's rational model of categorization (Anderson, 1991a), introduced in chapter 9, which uses a Chinese restaurant process (CRP) prior on cluster size. A Gibbs sampling algorithm for performing inference with a CRP was also introduced in chapter 9, but categorization judgments often need to be made while observations are incoming, so here we apply a particle filter instead. Figure 11.2 illustrates how the particle filter with two particles works for three sequentially presented stimuli. As there is only one possible clustering of the stimulus after the first observation, both particles represent this clustering. Next, following observation of the second stimulus, the likelihood of each "descendant" of each particle is calculated—that is, the probability of assigning the new observation to each existing cluster in addition to assigning it to its own unique cluster.[1] Because the proportion of clusterings of past observations acts as the prior, the posterior distribution over clusters of all observed stimuli is then the relative likelihood for each descendant, constrained so that the posteriors associated with each ancestor particle are equal. To keep the representational complexity constant, the two new particles are sampled from this posterior, and then the process repeats.

Particle filters can exhibit a characteristic bias that is often seen in human behavior, being more sensitive to data presented earlier in the sequence, known as *primacy effects*. These kinds of order effects have been observed in categorization tasks. For example, in an experiment by Anderson and Matessa (reported in Anderson, 1990), participants were given the same set of stimuli, and after observing all the stimuli, they were asked to place them into two equal-sized groups. The order of the stimuli was manipulated between participants to emphasize one or another set of features, and participants tended to group the stimuli according to the features that their ordering emphasized first. This order effect is produced by the particle filter because past clusterings are not revisited. Every particle offers a deterministic interpretation of the clustering of past observations, and uncertainty is reflected by the distribution of clusterings over particles. With an infinite number of particles, every possible past clustering will be represented, but with a finite number, many low-probability clusterings will be missing. If later observations greatly increase the probability of these hypotheses, then the particle filter may not be able to quickly adapt. As shown in figure 11.2, by chance the first two stimuli always appear in the same cluster in the final particles. If later stimuli

1. The likelihoods are computed based on the iterative CRP prior and the probability of features belonging to the same cluster, which for binary features is determined by independent beta-Binomial distributions for each feature. See chapter 9 for more details on this model.

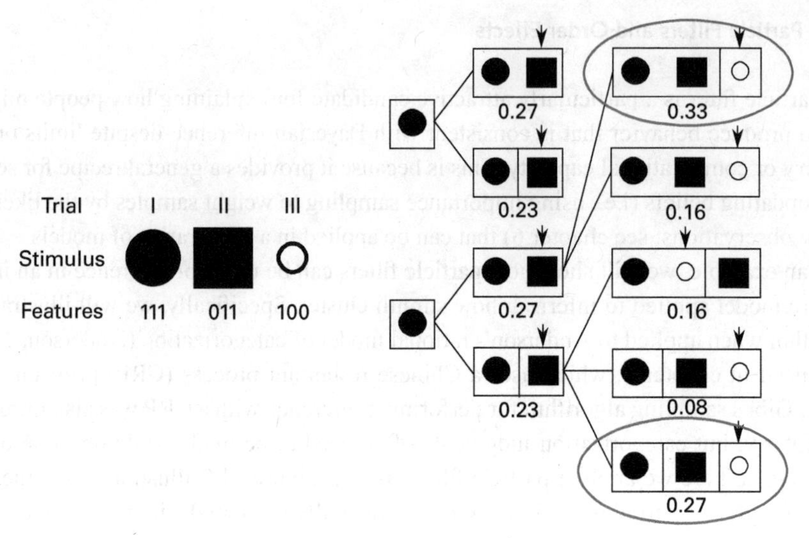

Figure 11.2
Three sample stimuli with three binary features each, along with an illustration of how a particle filter approximation to nonparametric Bayesian model infers how to cluster the stimuli. The first feature codes as a circle or a square, the second solid or empty, and the third big or small. The right side of the figure illustrates how a particle filter sequentially updates its representation of how the stimuli cluster as the stimuli are observed. Adapted from Griffiths et al. (2011).

begin to demonstrate that the best way to cluster the items is by whether they are circles or squares, the particle filter adapt more slowly than it ideally should.

In addition to these primacy effects, which results from resampling the particles, perturbation of particles can produce a *recency effect*, in which more recent data have a stronger influence (Abbott & Griffiths, 2011). The presence of primacy and recency effects in human behavior has been used to argue against probabilistic models of cognition, since many such models assume that observations are independent and are thus insensitive to order (e.g., Kruschke, 2006). Particle filters illustrate how such order effects can be produced as a consequence of approximating Bayesian inference. Models of behavior based on particle filters have been applied to a range of problems, including category learning (Lloyd, Sanborn, Leslie, & Lewandowsky, 2019; Sanborn et al., 2006, 2010a), associative learning (Daw & Courville, 2008), causal learning (Abbott & Griffiths, 2011), visual tracking of objects (Vul et al., 2009), and comprehension of sentences (Levy, Reali, & Griffiths, 2008).

11.5 Markov Chain Monte Carlo and Stochastic Search

In chapter 6, we discussed MCMC algorithms as a method for approximating a posterior distribution by constructing a Markov chain that converges to that distribution over time. MCMC algorithms have a very different flavor from other sampling schemes, typically maintaining a single sample that incrementally approaches the posterior distribution and then randomly explores within it. This kind of algorithm is suggestive of cognitive processes that involve searching through a space of hypotheses with the goal of finding one that provides a good explanation for the available evidence—engaging in a kind of *stochastic search*. Two contexts in which processes of this kind are prominent are memory and theory development.

When trying to retrieve information from memory, it's not unusual to have the sense of searching through a set of candidate memories to find something that fits what you're looking for. One task that makes this explicit is the *Remote Associates Test* (Mednick, 1962), in which people are provided with three words that are each related to a target word and are asked to identify the target. This task can be formulated as a problem of probabilistic inference: how likely is it that these three words would be generated if a particular word were the target? Accordingly, it can be solved via a process like the Metropolis-Hastings algorithm: start with an initial hypothesis as to the target word, generate a proposed variation on it using an appropriate semantic representation, and compare the relative posterior probabilities of the original hypothesis and the variation. Smith, Huber, and Vul (2013) showed that people's responses on the Remote Associates Test demonstrate patterns of sequential dependency that are consistent with stochastically searching through a set of possible answers, and Bourgin, Abbott, Griffiths, Smith, and Vul (2014) demonstrated that this behavior could be explained by a variant on the Metropolis-Hastings algorithm.

Searching through our memory is not the only time when we struggle to come up with good answers. In many ways, this is the fundamental problem faced by scientists: seeking good hypotheses—theories—to account for observed data. However, scientists are not the only ones who need theories. An influential approach to understanding human cognitive development has argued that children develop intuitive theories of the world by a process that is very similar to that of scientists (Carey, 1985; Gopnik & Meltzoff, 1997; Wellman & Gelman, 1992). Understanding how people explore the space of theories is thus potentially relevant to understanding how we come to build causal models of the world around us.

Ullman, Goodman, and Tenenbaum (2012) argued that this process of theory formation could be understood as a result of *stochastic search algorithms* such as MCMC. They showed that this approach could be used to discover meaningful theories expressed in a simple logical language, capturing notions from intuitive biology and physics such as taxonomies and magnetism. In particular, the stochastic search algorithm explored hypotheses in a way that recapitulated aspects of human cognitive development, considering hypotheses more like those entertained by younger children earlier in the search process.

MCMC algorithms future and explanations of other aspects of cognition. For example, Gershman, Vul, and Tenenbaum (2012) argued that *perceptual multistability*—where a stimulus moves back and forth among different possible percepts—could be accounted for by the inference problem underlying perception being solved via something like the Metropolis-Hastings algorithm. The iterative, incremental nature of the updates to beliefs that result from these algorithms also provides a way to think about the sequence in time because of the hypotheses that people generate when solving judgment problems (Lieder, Griffiths, Huys, & Goodman, 2018b, 2018c; Dasgupta, Schulz, & Gershman, 2017)—a topic that we will return to in chapter 13.

11.6 A More Bayesian Approach to Sampling

In Bayesian models, the probability distributions, whether or not they are accurate, are known precisely. However, for sampling algorithms such as importance sampling, particle filters, and MCMC this is not the case except in the limit of an infinite number of samples. For finite sample sizes, the relative frequencies in the set of samples provide only a noisy estimate of the underlying probabilities. In statistics and machine learning, this noise is generally suppressed by generating a very large number of samples. But such a computationally intensive solution is not plausible for people who seem to be able to consider only a handful of samples when making a judgment or decision (Juslin, Winman, & Hansson, 2007; Sundh, Zhu, Chater, & Sanborn, 2023; Weber et al., 2007).

Fortunately, there are ways to improve judgments and decisions based on a small number of samples. Instead of relying on the relative frequencies in the samples, the Bayesian approach can be taken a step further, placing a prior on the underlying probabilities themselves so that the estimated probability depends on both this prior and the samples that are generated. This approach has been termed *Bayesian Monte Carlo* in statistics and machine learning (Rasmussen & Ghahramani, 2003), and in effect it involves two stages of Bayesian inference: first, a sampling approximation to Bayesian inference is performed on information gleaned from the external world, and second, Bayesian inference is performed on those internally-generated samples to produce a judgment.

The priors for Bayesian inference on samples from a probability distribution will necessarily be fairly generic because there is no detailed knowledge about that probability distribution—that information generally has to come from the samples themselves. A simple example is estimating the probability of a binary variable from a set of samples, such as estimating the probability that it will snow tomorrow. If k out of n samples indicate snow, then the frequentist estimate is simply k/n. However, prior knowledge about the sample probabilities can be incorporated by placing a prior on these probabilities, with a common choice being the symmetric beta distribution with parameter β. If $\beta = 1$, this is a uniform distribution in which every probability is equally likely, if $\beta < 1$, then the extreme probabilities of 0 and 1 are more likely than 0.5, and if $\beta > 1$, then 0.5 is more likely than 0 and 1. This prior does not incorporate information about the specific question of snow tomorrow; instead, it has to be applicable to the range of questions that could be asked. Incorporating this prior information changes the mean probability estimate to $\frac{k+\beta}{n+2\beta}$, so it is shifted toward the middle of the range whenever $\beta > 0$. This moderation of the estimate decreases its squared error when the prior correctly reflects the range of situations in which it is used, and it often also works well when β is incorrect (Zhu, Sanborn, & Chater, 2020).

Using this scheme, which has been termed the *Bayesian sampler*, allows rational process models to explain a wide range of biases in human probability judgment that previously were out of reach (Zhu et al., 2020). For example, people's probability estimates are famously incoherent as demonstrated by the conjunction fallacy: when asked to judge the probability that it will snow tomorrow, $P(s)$, and the probability that it will snow and be cold tomorrow, $P(s, c)$, a greater-than-chance proportion of participants judge $P(s, c) > P(s)$ (Tversky & Kahneman, 1983). For simple setups such as this one, where it is unlikely that the conjunction causes participants to think of examples they would not otherwise (cf. Sanborn & Chater, 2016), this cannot be explained by judgments based on the relative frequencies of samples no matter the sampling algorithm—these judgments will be coherent on average. But if we assume that people incorporate fewer samples n', where $n' < n$, of conjunctions into their judgments, then the Bayesian sampler can produce conjunction fallacies. This is because the prior will have a larger moderating effect on the conjunction judgments, so it can pull the average conjunction judgment above that of the average judgment of a simple event (Zhu et al., 2020). Moreover, the assumption that smaller samples will be drawn for conjunctions of events, rather than simple events, is independently plausible (e.g., it is plausibly more difficult to generate memories or imagine instances of exemplars that have two properties X and Y than to generate or imagine exemplars that are merely X, or merely Y). This claim can also be tested empirically, of course.

A second example are the probabilistic identities first explored in Costello and Watts (2014). These identities are the sums and differences of the average judgments of different events, constructed so that probability theory predicts that they exactly equal zero. In addition, while a response based on the relative frequency from a set of samples will be noisy, responses will on average equal zero for these identities. However, while people's average judgments are close to zero for some identities, others deviate strongly from zero (see figure 11.3). Following the noise model of Costello and Watts (2014), the Bayesian sampler can explain both the matches and deviations from the predictions of probability theory (or counting the relative frequency of samples) by the moderation of estimates by the prior (see figure 11.3).

An important question for the Bayesian sampler is the extent to which it is real or simply an "as-if" model. Pertinently, the same kind of influence on judgments that the Bayesian sampler produces can also result from a noise process in which there is a fixed probability of incorrectly counting each sample when tallying the relative frequencies (e.g., there is a probability of 0.2 of reading a "snow" sample as "not snow" and vice versa). Indeed, on average, this noise process can imitate the Bayesian sampler perfectly (Costello & Watts, 2017, 2019; Zhu et al., 2020). It requires considering the relationship between the mean and variance of judgments to show differences between these approaches, and there is evidence that people use a prior that matches the distribution of probabilities outside the laboratory (Sundh, Zhu, Chater, & Sanborn, 2023).

11.7 Making Connections to the Implementational Level

While probabilistic models of cognition typically focus on the computational level, a complete understanding of human cognition will require answering questions at all of Marr's levels of analysis. Rational process models provide one way to begin to systematically seek answers to those questions, linking the computational and algorithmic levels. It may be the case that this strategy can also be used to develop hypotheses about the implementational level. For example, exemplar models are easy to implement using artificial neural networks (e.g., Kruschke, 1992), so importance sampling is similarly straightforward to implement in a simple neural circuit (Shi & Griffiths, 2009). A variety of neural implementations have been proposed for particle filters as well (Huang & Rao, 2014; Legenstein & Maass, 2014; Kutschireiter, Surace, Sprekeler, & Pfister, 2015).

Exploring these connections between levels of analysis is likely to be a major project for advocates of Bayesian models of cognition in the future. There has been a significant

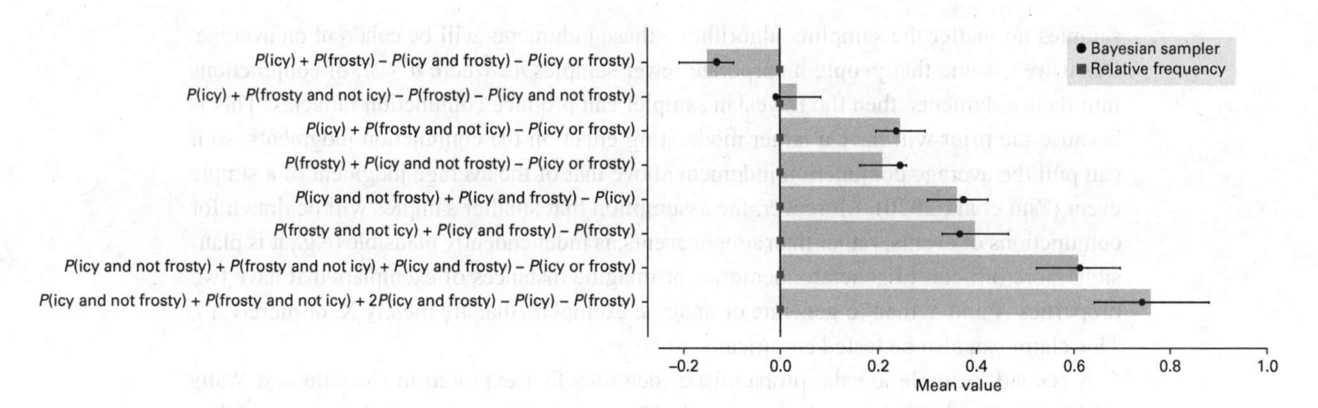

Figure 11.3
Empirical results and model predictions for a series of probabilistic identities. Participants were asked about the probability that a random day in England would be X, where X was a weather event. Mean responses were collected for each event and these means were added or subtracted as indicated on the vertical axis. These probablistic identities, first explored in Costello and Watts (2014), are constructed so that both probability theory and the relative frequency of samples (green squares) predict zero on average for each identity. The Bayesian sampler model makes predictions that can deviate from zero (red dots). Figure adapted from Chater, Zhu, Spicer, Sundh, León-Villagrá, and Sanborn (2020).

amount of work exploring schemes by which brains might implement Bayesian inference (see Doya, Ishii, Pouget, & Rao, 2007). However, these schemes typically focus on a single neural mechanism that can be used for all problems throughout the brain. Thinking in terms of rational process models suggests a different approach. Since different sampling methods are better for solving different problems and seem to provide good models of different psychological phenomena, we might expect that the brain would employ not just one mechanism for probabilistic inference, but many. Different algorithms can be implemented or approximated in different ways. Considering the algorithmic level before jumping to the implementational level thus supports a kind of mechanistic pluralism rather than a search for a single solution.

11.8 Future Directions

Rational process models have successfully explained a variety of empirical effects, but challenges remain to be addressed. A first challenge is to characterize which algorithm is used in which task, as the data collected thus far have not proved diagnostic: many different sampling algorithms could produce the qualitative empirical effects that have been discussed in this chapter. More diagnostic effects need to be found, with initial work along these lines showing that order effects in causal learning differentiate between kinds of particle filters (Abbott & Griffiths, 2011) and dependencies between responses can differentiate between MCMC algorithms (Castillo, León Villagrá, Chater, & Sanborn, 2021; Zhu, Sanborn, & Chater, 2018). Future work should compare a wider variety of algorithms both within and between tasks.

A second challenge is to quantitively relate the outputs of sampling algorithms to standard behavioral measures such as judgments, decisions, and response times. While there is broad agreement on how to map samples to decisions, estimates, and responses (Hamrick,

Smith, Griffiths, & Vul, 2015; Vul et al., 2014; Zhu, Sundh, Chater, & Sanborn, 2023), there remains much to work out about which variables are sampled and how samples carry over and evolve from trial to trial. Addressing these challenges is well worth the effort: rational process models potentially promise explanations of human behavior that combine the powerful generality of Bayesian models with the descriptive accuracy of process models.

Finally, recent advances in machine learning offer new opportunities for defining models that approximate Bayesian inference. In this chapter, our focus has primarily being on Monte Carlo methods, but recent work has shown how optimization-based methods, including those based on artificial neural networks, can be used to perform probabilistic inference at scale. We discuss some of the possibilities raised by this work in chapter 12, but see this as a path toward exciting new kinds of rational process models.

11.9 Conclusion

Algorithms for approximating Bayesian inference are useful not just when working with probabilistic models, but also when thinking about the cognitive processes that humans may engage in when facing inductive problems. Many of the algorithms that are used by computer scientists and statisticians to approximate Bayesian inference have natural psychological interpretations. Exploring these models offers a way to bridge the computational and the algorithmic levels of analysis, allowing us to generate hypotheses about how human minds and brains might take on the computational challenges of probabilistic inference. Further ideas for bridging these levels of analysis are provided by recent work on artificial neural networks, which we explore in chapter 12.

12

Bayesian Models and Neural Networks

Thomas L. Griffiths, Ishita Dasgupta, and Erin Grant

Artificial neural networks have a long history as models of human cognition (McClelland & Rumelhart, 1986) and have undergone a recent resurgence to become a dominant approach to machine learning (LeCun, Bengio, & Hinton, 2015). Given their history and current popularity, understanding how neural networks relate to probabilistic models of cognition is important both for contextualizing this work and developing an understanding of human cognition that draws upon both theoretical traditions.

There are two paths toward establishing connections between Bayesian inference and neural networks: using neural networks as systems for performing Bayesian inference and thinking about neural networks as just another probabilistic model to which Bayesian inference can be applied. Each path offers distinctive insights that are relevant to understanding human cognition, helping us imagine how human brains could approximate Bayesian inference and how human learning could be guided by the equivalent of prior distributions without anything that looks explicitly like Bayesian inference taking place.

If we follow the first path, the way to integrate Bayesian models of cognition with neural networks is to think about them as operating at different levels of analysis: Bayesian models capture Marr's (1982) computational level, while neural networks address the algorithmic. Neural networks thus complement—and potentially implement—algorithms for approximate Bayesian inference, such as the Monte Carlo and variational methods discussed in chapters 6 and 11, and provide a source of hypotheses about some of the systematic deviations between rational action and human behavior mentioned in chapter 7.

If we follow the second path, creating what are sometimes referred to as *Bayesian neural networks* (e.g., MacKay, 1995), the concepts that make Bayesian inference a powerful tool for understanding human cognition, such as priors that capture specific inductive biases, are used to understand and constrain neural networks. In doing so, we have the opportunity to explore methods for making neural networks that instantiate inductive biases that are better aligned with those of human learners.

Our goal in this chapter is to chart these two paths, summarizing some of the key ideas that serve as landmarks along the way. The body of literature on these topics is vast, with many contributions having been made in the last few years. Rather than providing a comprehensive review, we focus on the conceptual connections that are valuable for linking these theoretical perspectives in the context of understanding human cognition. However, we also provide pointers to resources that can be used to explore this literature more deeply.

12.1 What Is a Neural Network?

There are many types of artificial neural networks, unified by the idea of defining a system of simple computational units that interact with one another via weighted connections. These various formalisms are all, to greater or lesser extent, inspired by brains—systems of neurons that interact with one another via synaptic connections. Some artificial neural networks are especially formulated as probabilistic models. For example, Boltzmann machines can be interpreted as a kind of undirected graphical model (see chapter 4).

For simplicity, in this chapter we will primarily focus on *multilayer perceptrons* (for more detailed treatments of these and other neural network architectures, see McClelland & Rumelhart, 1986; Goodfellow, Bengio, & Courville, 2016). A multilayer perception is a feed-forward neural network in which computational units—"nodes"—are organized into layers and information flows from one layer to the next via weighted connections. The activation of the ith node in a given layer of the network, y_i, is determined by its input and an *activation function*. The input is the sum of the activations of all nodes in the previous layer of the network, weighted by the strength of their connections:

$$\text{input}_i = \sum_j w_{ji} z_j, \tag{12.1}$$

where z_j is the activation of the jth node in the previous layer and w_{ji} is the weight from that node to the ith node in the next layer. These weights can be collected into a *weight matrix* $\mathbf{W}$. Nodes can also have a bias term, w_{0i}, that determines their default activation.

The activation function transforms the input into the activation of the node. This transformation is nonlinear, inspired by the way that biological neurons accumulate input until it exceeds a threshold and then fire. This nonlinearity also serves a practical purpose, as without it, a multilayer neural network could be expressed as a single linear function. A classic choice for the activation function is the sigmoid

$$g(\text{input}) = \frac{1}{1 + \exp\{-\text{input}\}}, \tag{12.2}$$

but contemporary applications of neural networks use other activation functions that better support learning in neural networks with many layers (e.g., rectified linear units; Nair & Hinton, 2010).

Learning is typically performed by stochastic gradient descent. Each output node y_i in a network has some target value that it should produce. We can define a *loss function* $\mathcal{L}(\mathbf{W})$ that captures the difference between the outputs and the targets as a function of the weights $\mathbf{W}$. A standard loss function is the squared-error loss, which we can write for a single observation as

$$\mathcal{L}(\mathbf{W}) = \sum_i (t_i - y_i)^2 \tag{12.3}$$

$$= \sum_i (t_i - g(\sum_j w_{ji} z_j))^2. \tag{12.4}$$

The gradient descent algorithm finds weights that reduce this loss by calculating the gradient of the loss—its derivative with respect to the weights—and then moving in the direction that reduces that loss.

If we differentiate this loss function with respect to weight w_{ji}, we obtain

$$\frac{d\mathcal{L}}{dw_{ji}} = -2(t_i - g(\sum_j w_{ji}z_j))g'(\sum_j w_{ji}z_j)z_j \tag{12.5}$$

via the chain rule for derivatives. Since we want to minimize the loss, we update the weights by moving in the opposite direction of the gradient (i.e., going in the direction where the loss goes down). In this case, that means setting w_{ji} to $w_{ji} - \eta\frac{d\mathcal{L}}{dw_{ji}}$, where η is a *learning rate*. Typically, we want to minimize the loss over an entire data set, which would require computing the gradient of the loss function across all the observations in that data set, but stochastic gradient descent approximates this by taking the gradient for single observation or a small number of observations at a time. If the learning rates are gradually decreased over time, this algorithm is guaranteed to converge to a local minimum of the loss.

In a multilayer perceptron, we have many layers of nodes and weights. The network receives input x and produces output y, with the intermediate layers z being referred to as *hidden layers* as they do not correspond to variables that are observed in the data set. Stochastic gradient descent can be used to update all the weights in the network, using the chain rule to calculate the derivative of the last with respect to each weight. This derivative has terms that capture the contribution of each node to the overall loss, which can be interpreted as propagating the loss back through the network, resulting in this algorithm being called *backpropagation* (Rumelhart et al., 1986a). Contemporary software for training neural networks automatically computes the required derivatives, making it easy to define and train neural networks with arbitrarily complex architectures (e.g., Abadi et al., 2015; Paszke et al., 2019).

12.2 Bayesian Inference *by* Neural Networks

We will begin by considering how neural networks can be used to perform Bayesian inference. Our starting point is a classic observation of equivalence between a simple neural network and a simple Bayesian model. We then consider how other approximation schemes—such as the Monte Carlo and variational methods discussed in chapter 6—can be implemented in neural networks.

12.2.1 A Simple Neural Network That Performs Bayesian Inference

Consider a simple classification problem: we have a set of objects that are drawn from two classes, with each object having d observed binary features $x_1, \ldots, x_d$. We want to define a Bayesian model that captures classification in this setting.

To simplify things, we assume that the features are independent.[1] Letting y denote the class, we have

$$P(x_1, \ldots, x_d|y) = \prod_j P(x_j|y) \tag{12.6}$$

1. This is known as a *naive Bayes* model, since it makes a naive assumption about independence, but often performs well in classification settings, as it has few parameters to estimate and is hence fairly robust (e.g., Rish, 2001).

for each class. Since there are only two hypotheses (call them $y=1$ and $y=0$), we can write the posterior in odds form:

$$\frac{P(y=1|x_1,\ldots,x_d)}{P(y=0|x_1,\ldots,x_d)} = \frac{P(x_1,\ldots,x_d|y=1)\,P(y=1)}{P(x_1,\ldots,x_d|y=0)\,P(y=0)} \tag{12.7}$$

$$= \frac{P(y=1)}{P(y=0)} \prod_j \frac{P(x_j|y=1)}{P(x_j|y=0)} \tag{12.8}$$

and take logarithms to obtain

$$\log\frac{P(y=1|x_1,\ldots,x_d)}{P(y=0|x_1,\ldots,x_d)} = \log\frac{P(y=1)}{P(y=0)} + \sum_j \log\frac{P(x_j|y=1)}{P(x_j|y=0)}. \tag{12.9}$$

If we want to convert the log posterior odds back to a posterior probability, we can do so by exploiting the property of the sigmoid:

$$p = \frac{1}{1+\exp\{-\log\frac{p}{1-p}\}} \tag{12.10}$$

so

$$P(y=1|x_1,\ldots,x_d) = \frac{1}{1+\exp\{-\log\frac{P(y=1|x_1,\ldots,x_d)}{P(y=0|x_1,\ldots,x_d)}\}} \tag{12.11}$$

or, equivalently,

$$P(y=1|x_1,\ldots,x_d) = \frac{1}{1+\exp\{-\log\frac{P(y=1)}{P(y=0)} - \sum_j \log\frac{P(x_j|y=1)}{P(x_j|y=0)}\}}. \tag{12.12}$$

Now imagine solving the same problem using a simple neural network. In fact, take the simplest such network, with a single output y and no hidden layers—just a set of weights w_j mapping directly from x_j to y. Using a sigmoid activation function, we have

$$y = \frac{1}{1+\exp\{-w_0 - \sum_j w_{ji}x_j\}}, \tag{12.13}$$

where w_0 is an additional weight—known as a *bias*—that is included to modify the value of y when all values of x_j are 0.

Comparing equations (12.12) and (12.13) suggests that there might be a relationship between these two models: both are a sigmoid of a linear function. Here, $\log\frac{p(y=1)}{p(y=0)}$ is a single fixed value that can potentially be captured by w_0. The catch is that $\log\frac{P(x_j|y=1)}{P(x_j|y=0)}$ takes different values for $x_j=1$ and $x_j=0$, while w_jx_j takes the value w_j when $x_j=1$ and 0 when $x_j=0$. We can accommodate this by defining

$$w_j = \log\frac{P(x_j=1|y=1)}{P(x_j=1|y=0)} - \log\frac{P(x_j=0|y=1)}{P(x_j=0|y=0)} \tag{12.14}$$

$$w_0 = \log\frac{P(y=1)}{P(y=0)} + \sum_j \log\frac{P(x_j=0|y=1)}{P(x_j=0|y=0)}, \tag{12.15}$$

and then the two models are directly equivalent: the value of output y in the neural network is $P(y=1|x_1,\ldots,x_d)$ in the Bayesian model.

This simple example illustrates how two models that start in quite different places can end up being formally equivalent, and how it is possible for neural networks to directly implement Bayesian inference. One interesting difference between these approaches is in how they formulate the problem. In the Bayesian approach, we start with a *generative model* specifying how features are generated based on the class, and then use Bayes' rule to work backwards from features to class labels. The neural network starts with the inverse problem, learning a function that maps from features to class labels directly. This is called a *discriminative model*. These approaches make quite different assumptions about the properties of the data—the generative approach explicitly models the distribution of the features, while the discriminative approach only models the distribution of the class labels. These various assumptions can have implications for the way that the observed data are interpreted, and in particular how missing data are handled (Hsu & Griffiths, 2009). However, generative-discriminative pairs like the relationship between naive Bayes and a one-layer neural network (also known as *logistic regression*) show that there is significant potential for bridging these two perspectives (for more discussion of this point, see Efron, 1975; Ng & Jordan, 2001)

12.2.2 A Neural Implementation of Importance Sampling

The structure of a neural network can also be used to implement approximate algorithms for Bayesian inference. In this section, we illustrate this by showing how importance sampling—one of the Monte Carlo algorithms introduced in chapter 6—can be implemented in a simple neural network (Shi & Griffiths, 2009).

Assume a generative model in which observation x are generated from a Gaussian distribution centered on an unknown value z. We have a prior distribution over these unknown values of $p(z)$, and want to compute the expectation of a function $f(z)$ over the posterior distribution $p(z|x)$, $E[f(z)|x]=\int f(z)p(z|x)\,dz$. For example, taking $f(z)=z$ would allow us to compute the posterior mean.

Evaluating expectations over the posterior distribution can be challenging: it requires computing the posterior, and potentially also a multidimensional integration. The expectation $E[f(z)|x]$ can be approximated using importance sampling. Recall from chapter 6 that importance sampling approximates an expectation over the posterior by using a set of samples from a surrogate distribution $q(z)$ and assigning those samples weights proportional to the ratio $p(z|x)/q(z)$. We then have

$$E[f(z)|x]=\int f(z)\frac{p(z|x)}{q(z)}q(z)dz \simeq \sum_j f(z_j)\frac{p(z_j|x)}{q(z_j)} \qquad z_j \sim q(z). \tag{12.16}$$

If we choose $q(z)$ to be the prior $p(z)$, the weights reduce to the likelihood $p(x|z)$, giving

$$E[f(z)|x] \simeq \frac{\sum_{z_j} f(z_j)p(x|z_j)}{\sum_{z_j} p(x|z_j)} \qquad z_j \sim p(z), \tag{12.17}$$

which is the likelihood weighting algorithm discussed in chapter 6, in which we approximate the posterior distribution with samples from the prior weighted by the likelihood.

We will construct an analog of this algorithm using a particular kind of neural network—a *radial basis function* (RBF) network. This is a multilayer neural network architecture in which the hidden units are parameterized by locations in a latent space z_j. Upon presentation of stimulus x, these hidden units are activated according to a function that depends only on the distance $||x - z_j||$, such as $\exp(-|x - z_j|^2/2\sigma^2)$. RBF networks are popular because they have a simple structure with a clear interpretation and are easy to train, and they have been used as models of pattern recognition in neuroscience (Kouh & Poggio, 2008) and category learning in psychology (Kruschke, 1992).

Implementing importance sampling with RBF networks is straightforward. We create a network where the inputs correspond to x and the single output node approximates $E[f(z)|x]$. Each hidden unit represents a stored value, z_j, which is sampled from the prior. The activation function is taken to be proportional to $p(x|z_j)$. After the activations are computed for all the hidden units, they are normalized. The weight from hidden unit j to the output unit is set to $f(z_j)$. Such a network produces output exactly in the form of equation (12.17). The same set of samples from the prior can be used to perform inference for any x, so this network instantiates a simple neural circuit for approximating the posterior mean.

This importance sampler is a neural network implementation of the exemplar model for Bayesian inference discussed in chapter 11 (Shi, Griffiths, Feldman, & Sanborn, 2010). The exemplars correspond to the z_j represented by the hidden units. Given this insight, it is straightforward to define other neural network architectures that can approximate Bayesian inference by memorizing and generalizing from exemplars. For example, Abbott, Hamrick, and Griffiths (2013) showed how the sparse distributed memory architecture of Kanerva (1988) can be used to perform Bayesian inference.

12.2.3 Learning to Perform Bayesian Inference

In chapters 5 and 6, we saw that Bayesian inference in latent variable models can be very computationally expensive, and often intractable as well. More specifically, the posterior distribution $P(z|x)$ over a latent variable z given data x often cannot be computed exactly and requires us to make approximations using methods like Monte Carlo or variational inference. These approximations are expensive to compute, requiring drawing many samples for Monte Carlo or many optimization steps for variational inference.

One way to reduce the cost of probabilistic inference is to try to store and re-use computations wherever possible. When we compute a distribution $Q_x(z)$ that approximates the posterior $P(z|x)$ for some x, we can simply store and re-use this previous estimate when we encounter x again. However, this quickly becomes infeasible. For example, when we have a large number of possible observations x, the chance of seeing the same x twice is small. We can re-use an inference only when we encounter the exact same x, but we have to start from scratch when we encounter observations that are *similar* but not exactly the same.

We can formulate learning how to approximate $P(z|x)$ as a problem to be solved by neural networks. In particular, we can learn a function that takes in x and produces (an estimate of) $Q_x(z)$, by learning from a database of $(x, Q_x(z))$ pairs generated by exact Bayesian inference. This is called *amortization* since the costs of doing a new inference are spread out (or amortized) over several previous inferences (for a review of amortization in the context of cognitive science, see Dasgupta & Gershman, 2021). The network that implements this function is called a *recognition network* or an *inference network*. We refer to the estimate of

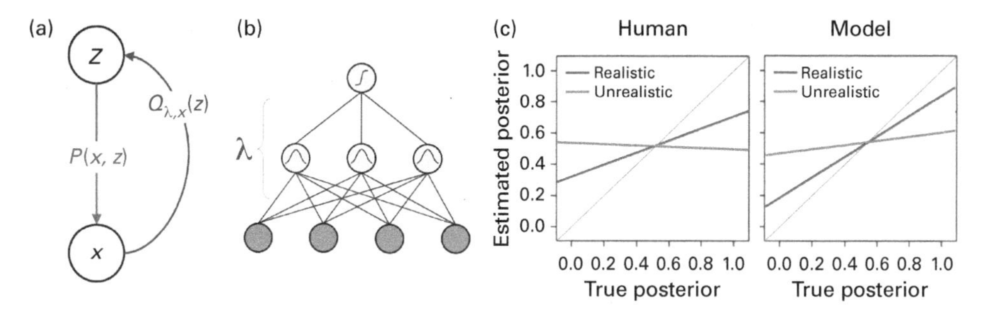

Figure 12.1
Amortizing posterior inference. (a) $P(x, z)$ is a generative model that produces latent variables z, and data x given z. $Q_{\lambda,x}(z)$ is an inference network, parameterized by λ, that maps observable data to the posterior distribution over underlying latent variables. (b) The inference network. (c) Humans are better at Bayesian inference when the provided probabilities are believable (replotted from Cohen et al., 2017). The same pattern arises in a recognition network with a small number of hidden units (replotted from Dasgupta, Smith, Schulz, Tenenbaum, & Gershman, 2018). Figure adapted from Dasgupta and Gershman (2021).

the function representing the approximate posterior as $\tilde{Q}$. Note that this estimate contains two sources of error from the true P—the error inherent in the initial approximation of the posterior ($P \rightarrow Q$, called the *approximation error*) and the error from an imperfect inference network ($Q \rightarrow \tilde{Q}$, the *amortization error*).

To make this more concrete, we discuss amortization in the context of a variational approximation. As discussed in chapter 6, a variational approximation $Q_\lambda(z)$ (where λ are the parameters of the distribution) to a true posterior $P(z|x)$ can be derived by maximizing the evidence lower bound (ELBO). Maximizing the ELBO is equivalent to minimizing the Kullback-Leibler (KL) divergence between $Q_\lambda(z)$ and $P(z|x)$. For standard variational inference, this is computed for a given x, that is, for a given set of observed data. When we observe new data x', we have to recompute Q from scratch. The key conceptual leap for amortizing this computation is that we can instead learn a $\tilde{Q}_{\lambda,x}(z)$ that is a function of x. We can do this by learning a function mapping from x to some parameters that uniquely identify $\tilde{Q}$ (e.g., the mean and variance of a Gaussian distribution; see figure 12.1). The parameters λ are optimized to minimize the KL divergence (or in actual practice, maximize the ELBO) in expectation over the distribution of x:

$$E_{P(x)}\left[D_{KL}(\tilde{Q}_{\lambda,x}(z)||P(z|x))\right], \tag{12.18}$$

where x comes from some query distribution $p(x)$. This optimization is easily done with gradient descent (for mathematical details, see Ranganath, Gerrish, & Blei, 2014).

Evidence that humans actually amortize inferences comes from investigating a core prediction of amortization—that past inferences influence future inferences by imprinting themselves onto the parameters of the recognition network. Since we are minimizing the KL divergence in expectation over the query distribution, frequent queries will be prioritized over less frequent ones and $\tilde{Q}$ will more closely approximate the true posterior for these frequent queries. In Dasgupta, Schulz, Tenenbaum, and Gershman (2020), this is used to model findings from human probabilistic inference showing that people are much better at Bayesian inferences when the probabilities are realistic and the problems are embedded in believable real-world scenarios (Evans, Handley, Over, & Perham, 2002; Cohen, Sidlowski,

& Staub, 2017). Concretely, a neural network with two hidden units in a single hidden layer with a radial basis activation function is trained as the inference network on a distribution of queries. A query consists of a prior and likelihood (both Bernoulli parameters in the example) and data (a sample from a Bernoulli distribution) that are sampled such that they result in a posterior sampled from a fixed distribution. The model is then tested on either the same distribution of posteriors or a different distribution of posteriors, emulating the difference between believable (previously encountered) or unbelievable (unfamiliar) posteriors. This model performs significantly better on the believable than the unbelievable distribution, reflecting human behavior (figure 12.1c). This finding refutes the view that the brain relies on a general-purpose inference engine that operates equally well on arbitrary probabilities, whereas a model where humans amortize and reuse past inferences captures these effects. A series of other findings in human behavior that can be explained by amortized inference are detailed in Dasgupta and Gershman (2021).

Another important use case for inference networks in cognitive science is as components of models that implement Bayesian inference in complex domains where inference would otherwise be intractably expensive. A notable example is in the probabilistic programming methods discussed in chapter 18.

12.3 Bayesian Inference *for* Neural Networks

The second path for developing correspondences between neural networks and Bayesian methods is based on treating neural networks as probabilistic models and using Bayesian methods to estimate the parameters of these models. Since neural networks are typically large and complex, performing Bayesian inference in this setting can be challenging. However, some of the algorithms that are used to train neural networks already admit a Bayesian interpretation. In this section, we first outline the Bayesian perspective on neural networks and then discuss how this relates to neural network learning algorithms, ultimately highlighting a surprising connection to hierarchical Bayesian inference.

12.3.1 Bayesian Neural Networks

A neural network can be viewed as a probabilistic model. For example, a multilayer perceptron specifies a function that can be interpreted as a mapping from its input to a probability distribution over outputs. The parameters that define this mapping are the weights and biases of the network. Using θ to denote these parameters, we can define $p(t|\mathbf{x},\theta)$ as the probability given to the target value t by the network given input $\mathbf{x}$. A training set d consists of many $(t,\mathbf{x})$ pairs, and training the neural network yields parameters θ that minimize a loss function over this data set. This training process also has a probabilistic interpretation.

Exactly how we specify $p(t|\mathbf{x},\theta)$ depends on the nature of t. If t is continuous, then we can assume that $p(t|\mathbf{x},\theta)$ is a Gaussian distribution with variance σ_t^2 centered on the output of the network y:

$$p(t|\mathbf{x},\theta)=\frac{1}{2\sigma_t^2}\exp\{-(t-y)^2/2\sigma_t^2\}. \tag{12.19}$$

The part of this expression that is affected by θ is $-(t-y)^2$, so maximizing $p(d|\theta)$ is equivalent to minimizing the squared-error loss between t and y (equation (12.3)). If t is discrete, then it makes sense to use the cross-entropy loss, where we score the network's predictions for each output by $-\log p(t|\mathbf{x}, \theta)$. Minimizing this loss is equivalent to maximizing the likelihood $p(d|\theta)$. Given the correspondence between traditional methods for training neural networks and maximum likelihood estimation, it is natural to consider Bayesian approaches to neural network learning.

In this context, the Bayesian approach requires defining a prior distribution on the parameters of the neural network, $p(\theta)$. This implicitly defines a prior distribution over the functions that the neural network represents. Bayesian inference involves calculating the posterior distribution $p(\theta|d)$ or finding the maximum a posteriori (MAP) value of θ rather than the maximum likelihood estimate. There are many ways that we can imagine defining such a prior, but the simplest is to assume that the weights and biases of the neural network are drawn from a Gaussian distribution with zero mean and variance σ_w^2.

Actually computing posterior distributions over the weights of a neural network is potentially extremely computationally costly, as neural networks typically have very large numbers of weights and we are not able to rely upon conjugate priors or the other tricks that are used to make Bayesian inference tractable. However, there is one interesting case where increasing the size of neural networks turns out to be beneficial: Neal (1993) showed that in the limit of infinitely many hidden units, a Bayesian multilayer perceptron becomes a Gaussian process with a specific kernel defined by the activation function of the hidden units (see chapter 9 for a more detailed discussion of Gaussian processes). This deep connection between neural networks and nonparametric Bayesian statistics makes it possible to use models inspired by neural networks but retain desirable characteristics of probabilistic models, such as being able to express various degrees of uncertainty in their predictions.

12.3.2 Implicit Priors and Learning Algorithms

While full Bayesian inference is typically intractable for neural networks, finding approximations to the MAP estimate of θ can be relatively straightforward. In fact, existing algorithms for training neural networks have been shown to correspond to MAP inference under specific prior distributions.

Consider the Gaussian prior on the weights of a neural network mentioned in section 12.3.1. Using this prior, we can revisit the simple one-layer network that we used to introduce the gradient descent algorithm. In this case, the parameters of the network are just the weight matrix $\mathbf{W}$. To perform Bayesian inference, we need to define the likelihood $p(d|\mathbf{W})$ and the prior $p(\mathbf{W})$. If we focus on a single observation with a set of target outputs t_i, our likelihood is

$$p(d|\mathbf{W}) = \prod_i \frac{1}{2\sigma_t^2} \exp\{-(t_i - y_i)^2/2\sigma_t^2\} \tag{12.20}$$

$$\propto \exp\left\{-\frac{1}{2\sigma_t^2}\sum_i (t_i - y_i)^2\right\}. \tag{12.21}$$

Assuming a Gaussian prior with zero mean on each w_{ji} we have

$$p(\mathbf{W}) = \prod_{ij} \frac{1}{2\sigma_w^2} \exp\{-w_{ji}^2/2\sigma_w^2\}$$

$$\propto \exp\left\{-\frac{1}{2\sigma_w^2}\sum_{ij} w_{ji}^2\right\}. \tag{12.22}$$

The posterior probability $p(\mathbf{W}|d)$ is proportional to $p(d|\mathbf{W})p(\mathbf{W})$. Since we just care about maximizing $p(\mathbf{W}|d)$, we can focus on the log posterior probability, which is $\log p(\mathbf{W}|d) = \log p(d|\mathbf{W}) + \log p(\mathbf{W})$. Taking the logarithms of the expressions (12.21 and 12.22) and summing, we have

$$\log p(\mathbf{W}|d) = -\frac{1}{2\sigma_t^2}\sum_i -(t_i - y_i)^2 - \frac{1}{2\sigma_w^2}\sum_{ij} w_{ji}^2 + C, \tag{12.23}$$

where C is a constant that does not depend on $\mathbf{W}$. Multiplying this by a scalar doesn't change the optimal value of $\mathbf{W}$ so we can multiply by $2\sigma_t^2$ and change the sign to obtain

$$\arg\max_{\mathbf{W}} \log p(\mathbf{W}|d) = \arg\min_{\mathbf{W}} \left[\sum_i (t_i - y_i)^2 + \frac{\sigma_t^2}{\sigma_w^2}\sum_{ij} w_{ji}^2\right], \tag{12.24}$$

where the first term on the right side is easily recognized as $\mathcal{L}(\mathbf{W})$ from equation (12.3). Consistent with results in the previous chapters, the MAP solution can thus be interpreted as adding another regularization term to the function optimized for the maximum-likelihood estimate.

Differentiating equation (12.24) with respect to w_{ji} yields $\frac{d\mathcal{L}}{dw_{ji}} + 2\frac{\sigma_t^2}{\sigma_w^2} w_{ji}$. If we apply the gradient descent algorithm, the weight update rule becomes

$$w_{ji} = w_{ji} - \eta\left(\frac{d\mathcal{L}}{dw_{ji}} + 2\frac{\sigma_t^2}{\sigma_w^2} w_{ji}\right) \tag{12.25}$$

$$= w_{ji}\left(1 - 2\eta\frac{\sigma_t^2}{\sigma_w^2}\right) - \eta\frac{d\mathcal{L}}{dw_{ji}}. \tag{12.26}$$

The second term on the right side is just the standard weight update from gradient descent. The Bayesian version of this algorithm introduces the first term, which shrinks w_{ji} toward zero with each weight update.

This idea of reducing w_{ji} with each weight update was independently developed in the neural network research community, where it goes by the name *weight decay* (Hanson & Pratt, 1988). It helps to stop weights from becoming overly large during training, and implicitly it has the same effect as assuming a Gaussian prior on those weights. Weight decay is easy to implement and converges to a local maximum of the posterior under the same conditions that allow gradient descent to converge to a local minimum of the loss.

The choice of a regularizer such as weight decay is one of many choices that must be made when setting up a neural network model—we must also decide on a neural network architecture (such as how many layers, how many hidden units, and what activation functions to use), the hyperparameters of the learning algorithm (such as the learning rate and the schedule on which it is modified), and schemes to initialize the parameters of the model. Surprisingly, we can show that many of these choices can be interpreted in terms of implicitly defining different priors. One such choice is the number of gradient descent steps used to optimize the parameters of the neural network model. It can be shown that stopping the optimization early, at t steps, is equivalent in special cases to fully optimizing a regularized loss in which the regularization penalty scales as $1/t$ (Santos, 1996; Ali, Kolter, & Tibshirani, 2019); in other words, the regularization penalty diminishes as the number of iterations of gradient descent increases. As in weight decay, the regularizer can be seen as an implicit Gaussian prior on the weights of the neural network, here with a variance that scales with the number of iterations of training t.

Contemporary algorithms used for training deep neural networks have also been suggested as having connections to Bayesian inference. For example, the *dropout* algorithm (Srivastava, Hinton, Krizhevsky, Sutskever, & Salakhutdinov, 2014), in which a subset of weights are not included in each weight update, has been connected to the idea of having a distribution over parameter θ, and in some cases, this distribution can be shown to align with the Bayesian posterior (Gal & Ghahramani, 2016). Even the stochastic gradient descent algorithm itself has been characterized as performing approximate Bayesian inference (Mandt, Hoffman, & Blei, 2017).

12.3.3 Meta-learning and Hierarchical Bayes

Section 12.3.2 introduced the idea of implicit priors that result from an algorithmic choice such as the form of a regularizer or the early stopping iteration. These priors express relatively simple preferences for models that have weights that are close to zero or close to their initial state. However, there are settings in which we would like the prior distribution over a model itself to depend on data. One such is the setting of meta-learning, where a learner is not presented with a single task, such as learning a particular concept, but with many tasks that all have a similar character (Schmidhuber, 1987; Thrun & Pratt, 2012). The ideal learner in this setting uses of commonalities across these tasks not only to become better at solving each individual task, but also to solve future tasks better and more quickly, effectively "learning to learn."

We can consider a straightforward way to implement a meta-learner in the context of the one-layer neural network from before. Recall that this model has a matrix of weights $\mathbf{W}$, which we tune by taking steps to solve the problem $\min_{\mathbf{W}} \mathcal{L}(\mathbf{W})$, where $\mathcal{L}(\mathbf{W})$ is a loss function such as the one from equation (12.3). The meta-learning setting captures the case in which we have multiple losses, $\mathcal{L}_1, \mathcal{L}_2, \ldots$, to be simultaneously minimized:

$$\min_{\mathbf{W}} \mathcal{L}(\mathbf{W}) = \min_{\mathbf{W}} \sum_i \mathcal{L}_i(\mathbf{W}). \tag{12.27}$$

These individual losses might represent, for example, losses on subgroups of data that are more similar within subgroups than across subgroups, or losses corresponding to different types of tasks. Rather than using a single set of weights $\mathbf{W}$ for all losses—which would correspond, in our example, to a single network for all the losses—a meta-learning algorithm allows a separate set of weights $\mathbf{W}_i$ for each loss $\mathcal{L}_i$, and relates the weights $\mathbf{W}_i$ by the global parameters θ. With this parameterization, the objective becomes

$$\min_{\theta} \mathcal{L}(\theta) = \min_{\theta} \sum_{i} \mathcal{L}_i(\mathbf{W}_i(\theta)). \tag{12.28}$$

Each set of weights $\mathbf{W}_i$ in equation (12.28) is not individually learned, as in section 12.1, but are somehow derived from the global parameter θ; this setup allows the weights of the individual models $\mathbf{W}_i$ to be adapted for each loss while capturing information that is redundant across the losses via their dependence on the global parameters θ.

There are various ways in which the individual model weights $\mathbf{W}_i$ could relate to the global parameters θ. One simple way to set up a meta-learning algorithm is by taking θ to be the weight initialization for optimizing $\mathbf{W}_i$ with gradient descent on each of the individual losses (Finn, Abbeel, & Levine, 2017). Since the gradient descent algorithm is differentiable with respect to its parameter initialization, we can treat the whole procedure as a computational graph in which we backpropagate the error corresponding to the *final* value of each loss $\mathcal{L}_i$ all the way to the parameter initialization. If we truncate each optimization of $\mathbf{W}_i$ to a fixed number of gradient descent steps, this meta-learning objective takes a simple form; in the case of one step ($t=1$), we can write it as

$$\min_{\theta} \mathcal{L}(\theta) = \min_{\theta} \sum_{i} \mathcal{L}_i \left(\theta - \eta \frac{d\mathcal{L}_i}{d\theta} \right), \tag{12.29}$$

which means that for each loss $\mathcal{L}_i$, we evaluate the loss of the weight $\mathbf{W}_i = \theta - \eta \frac{d\mathcal{L}_i}{d\theta}$, and use this to tune θ using backpropagation and gradient descent just as in section 12.1.

Under certain conditions, the effect of early stopping—using a fixed and small number of gradient descent steps in the weights $\mathbf{W}_i$—in this meta-learning algorithm is the same as in section 12.3.2, in that it corresponds to a Gaussian prior with variance proportional to t over the weights $\mathbf{W}_i$. However, in contrast to the previous section, the mean of this prior is at the initialization parameter, θ. The meta-learning objective given here thus learns the mean of a Gaussian prior over the weights $\mathbf{W}_i$ that can be used for MAP inference on a new data set (Grant, Finn, Levine, Darrell, & Griffiths, 2018); figure 12.2 visualizes this perspective.

Estimating the parameters of a prior via meta-learning in this way is an instance of hierarchical Bayes (see chapter 8). While Bayesian inference indicates how a learner should integrate data with a prior distribution over hypotheses, a hierarchical Bayesian model learns that prior distribution. This idea has been widely used in Bayesian models of cognition and in that examples we have considered throughout this book. For example, hierarchical Bayes can be used to learn the properties of objects that words tend to label (such as shape) while learning the meaning of individual words (Kemp, Perfors, & Tenenbaum, 2007), and to identify different kinds of causal relationships while learning those relationships (Mansinghka, Kemp, Tenenbaum, & Griffiths, 2006).

Hierarchical Bayesian models have been used extensively in cognitive science, but the computational costs involved can make them difficult to use for models outside specific classes (e.g., where conjugacy applies; see chapter 3). Consequently, establishing a link between hierarchical Bayes and meta-learning—which can be implemented efficiently for a wide range of models with continuous parameterizations, such as neural networks—potentially

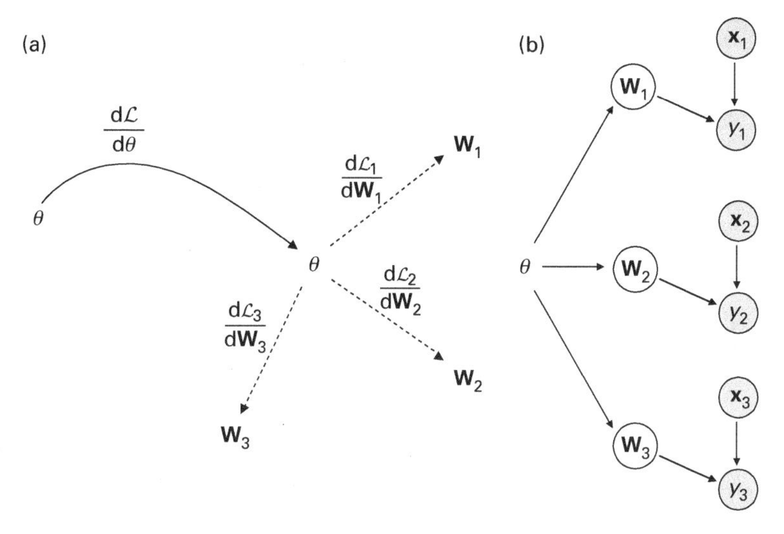

Figure 12.2
Meta-learning and hierarchical Bayes. (**a**) The gradient-based meta-learning algorithm of Finn et al. (2017) optimizes via $\mathcal{L}$ the parameter θ of a set of models so that when one or a few gradient descent steps are taken from the initialization at θ using the loss $\mathcal{L}_i$, each model obtains new weights $\mathbf{W}_i$ that result in good generalization performance on examples $\mathbf{x}_i, y_i$ associated with that loss. (**b**) The probabilistic graphical model for which the algorithm in (**a**) provides a parameter estimation procedure (Grant et al., 2018). Each task-specific set of weights $\mathbf{W}_i$ is distinct from but influences the estimation of the others through the parameters of a prior θ shared across all $\mathbf{W}_i$. Figure adapted from Griffiths et al. (2019).

expands the scope of Bayesian modeling. For example, McCoy, Grant, Smolensky, Griffiths, and Linzen (2020) demonstrated that the meta-learning algorithm described in this chapter can be used to create neural networks with an implicit prior distribution that makes it easy to learn languages from a simplified language typology; this can be viewed as a step toward a neural network instantiation of a "universal grammar" that supports language learning (see figure 12.3).

12.4 Future Directions

Research on deep learning continues to develop rapidly, and there are many further innovations in amortized inference, Bayesian neural networks, and meta-learning that have yet to be absorbed into cognitive science. All these topics provide fertile ground for innovation in developing probabilistic models of cognition and for understanding how human minds and brains might deal with the computational challenges of Bayesian inference.

The development of novel neural network methods also offers the opportunity to push the limits of the kinds of inferences that these models are able to capture. Memory-based meta-learning is a recent approach where an algorithm for sequentially updating the state of a neural network is learned directly from data. Using this approach, meta-learned agents can solve problems that have traditionally been addressed using structured probabilistic models, including Bayesian inference (Mikulik et al., 2020), model-based reinforcement learning (Wang et al., 2016) and causal learning (Dasgupta et al., 2019). These models are fully amortized, and therefore very efficient at run time, and adapt very well to structure in the

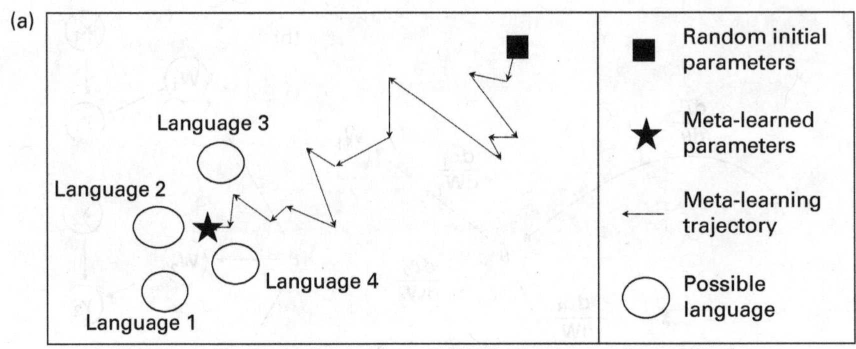

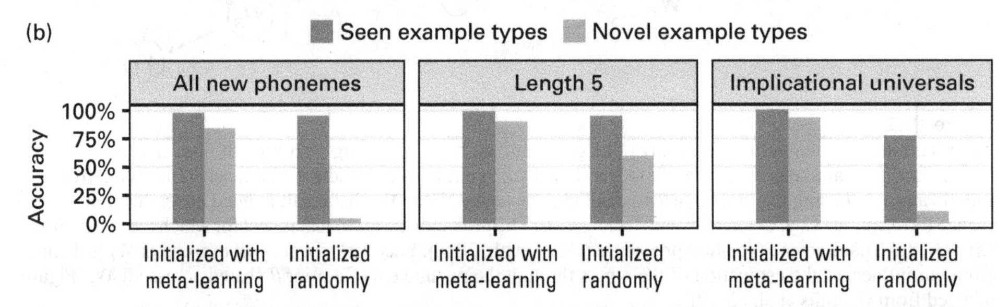

Figure 12.3
Identifying linguistic inductive biases with metalearning. (a) The size and shape of syllables across the world's languages follow strong cross-linguistic tendencies that suggest universal constraints on human language learning. McCoy et al. (2020) used meta-learning to investigate such inductive biases by analyzing the initial state of a language model trained with meta-learning on a data set that reflects the typology of natural language syllables. (b) Analysis of the initial state revealed, for example, a prior for implicational universals—that certain input-output mappings in a language imply other input-output mappings—which are widely attested in the syllable structure of natural languages. Figure adapted from McCoy et al. (2020).

environment that is hard to express in explicit structured models. However, these models also inherit the same issues as neural networks in general—they require large amounts of data to learn from and can generalize poorly. Evaluating the capacities and limits of these models is an important topic for future work (for preliminary steps in this direction, see Kumar, Dasgupta, Cohen, Daw, & Griffiths, 2021).

In addition to Bayesian inference *by* neural networks and Bayesian inference *for* neural networks, another productive direction to explore is Bayesian inference *as a model of* neural networks. With the ever-increasing complexity of deep learning models, it becomes harder and harder to understand the implicit assumptions that underlie these models. Our analysis of the inductive biases of neural networks in section 12.3 relied on simplifying assumptions about the structure of the model and the form of the loss function due to the complexity of analyzing even simple neural networks with one hidden layer. This raises addressing the following possibility: Are artificial neural networks themselves complex enough that we can build cognitive models of them that can help us understand them? Li, Grant, and Griffiths

(2021) explored this perspective by using a Bayesian model that was previously used to study inductive biases in human function learning (Wilson, Dann, Lucas, & Xing, 2015) to interpret and make predictions about the inductive biases of neural network models.

12.5 Conclusion

Rather than being competing frameworks for understanding human cognition, we view probabilistic models and neural networks as providing complementary insights that can be used for reverse-engineering the mind. These two approaches are at different levels of analysis and have different strengths and weaknesses: probabilistic models provide a powerful set of tools for abstractly characterizing human inductive biases, particularly in cases where those inductive bases are concisely expressed in terms of structured representations; neural networks are a flexible framework for understanding how efficient approximations to Bayesian inference can be learned from data, often making it possible to engage with problems at scales that go beyond the current limits of probabilistic models. Together, these two approaches provide a toolkit for building models that engage with a wide range of questions about human cognition.

13

Resource-Rational Analysis

Falk Lieder, Fred Callaway, and Thomas L. Griffiths

As outlined in chapter 2, rational analysis (Anderson, 1990) derives predictions of human behavior from the assumption that the mind is optimally adapted to people's goals and the structure of the environment. In deriving these predictions, rational analysis makes minimal assumptions about the mind's computational limits. However, beginning with the foundational work of Herbert Simon (1956), a substantial body of research on *bounded rationality* has demonstrated that people's computational limitations are far from minimal and impose substantial constraints on human reasoning and decision-making. This limits the applicability of rational analysis to phenomena where people's cognitive resources are sufficient to accurately approximate optimal behavior. When these assumptions are not met, rational analysis runs the risk of rationalizing suboptimal inferences and decisions as the optimal pursuit of hypothetical goals under hypothetical assumptions about the structure of the environment when those errors are, in fact, a result of people's cognitive limitations.

Resource-rational analysis is an extension of rational analysis that takes people's limited cognitive resources more seriously (Griffiths, Lieder, & Goodman, 2015; Lieder & Griffiths, 2020). While rational analysis strives to predict human behavior from people's goals and the structure of the environment alone, resource-rational analysis strives to uncover cognitive mechanisms and representations by taking into account the cognitive architecture that people have available to pursue their goals. Rational analysis derives its predictions from classical theories of rationality that effectively assume that rational agents have infinite computational power, such as expected utility theory (Neumann & Morgenstern, 1944) and probability theory. In contrast, resource-rational analysis relies on a more realistic notion of rationality that takes into account that real people have to make efficient use of their limited time and bounded cognitive resources (Russell & Subramanian, 1994; Horvitz, 1987).

Resource-rational analysis evolved from the rational process models introduced in chapter 11. Rational process models acknowledge that the mind has to approximate optimal inference and optimal decision-making with a limited amount of time and computation. Rational process models assume that this is accomplished by an algorithm that converges to the optimal inference or decision as the amount of time and computation increases. This assumption provides some useful guidance in modeling human cognition, but there are still infinitely many ways to approximate optimal inference and decision-making. By contrast, resource-rational analysis leads to the single most effective approximation strategy that is possible under the given constraints on time and computation.

In this chapter, we summarize the key ideas behind resource-rational analysis and provide some examples of how it can be used to reverse-engineer the mind. We begin by considering how a rational agent should make best use of its limited cognitive resources, relating this approach to other concepts of rational action. We then use this formalism to define resource-rational analysis and present two examples where this approach can be used to make sense of behaviors that otherwise seem irrational: anchoring effects and the over representation of extreme events. These examples illustrate how heuristics that have previously been discovered by psychologists can be understood from the perspective of resource-rational analysis. We then turn to the more profound question of how effective heuristics can be derived within this framework. By recognizing that the construction of a cognitive strategy can be expressed as a sequential decision problem, we draw on tools from the literature on planning and reinforcement learning introduced in chapter 7 to solve this problem. We illustrate this approach in the context of simple choice and planning.

13.1 The Rational Use of Cognitive Resources

Research on human judgment and decision-making has established that people do not conform to the norms of logic, probability theory, and expected utility theory that are the basis of rational analysis (e.g., Tversky & Kahneman, 1974; Wason, 1968; Kahneman & Tversky, 1979). The brain's finite computational power limits people's ability to achieve these normative standards. As a result of this bounded rationality, the ideals of maximizing expected utility, reasoning according to the laws of logic, and handling uncertainty according to the laws of probability are out of reach for people.

So how should the mind process information, given that its computational resources are limited? This question is pertinent not only to understanding the human mind, but also to the creation of artificial intelligence (AI). We can therefore employ the general insights that AI researchers have gained into the design of intelligent bounded agents to make sense of the design of a particular class of intelligent agents called "humans." In particular, we can use the theory of *bounded optimality* (Russell & Subramanian, 1994; Horvitz, 1987). The mathematical framework of bounded optimality was developed as the theoretical foundation for designing optimal agents that run on performance-limited hardware and have to interact with their environment in real time. A program is bounded optimal if running it on the agent's performance-limited hardware generates decisions that lead to world states whose expected utility is at least as high as for any other program that the agent's hardware can execute. The bounded optimal program, therefore, is given by

$$\text{program}^{\star} = \underset{\text{program} \in \mathcal{P}_{\text{HW}}}{\arg\max} \; E_{s_{1:t}|e, a_t = \text{program}(o_{1:t}, a_{1:t-1})} \left[u\left(s_{1:t} \right) \right], \tag{13.1}$$

where $s_{1:t} = (s_1, s_2, \cdots, s_t)$ is the sequence of states that the environment e was in throughout the agent's life, and u denotes the utility function that the agent is designed to optimize, indicating how good a sequence of states is. The subscript of the expectation operator E indicates that the sequence of states depends on both environment e and the agent's actions, a_t. The agent chooses its actions using a program whose output depends on the agent's full life experience (to account for learning), formalized as a sequence of noisy observations $o_{1:t}$ of the state, as well as the agent's past actions, $a_{1:t-1}$. Finally, $\mathcal{P}_{\mathrm{HW}}$ is the set of programs that the agent's hardware can execute. This restriction is what differentiates bounded optimality from perfect optimality.

By solving the optimal program problem defined in equation (13.1), it is sometimes possible to derive optimal algorithms. For instance, Russell and Subramanian (1994) derived an optimal mail-sorting program. This suggests the intriguing possibility that it might also be possible to derive *optimal cognitive strategies* for people. Applied to the special case of human rationality, the principle of bounded optimality suggests that being rational is to reason and decide according to cognitive strategies that perform as well as or better than any other strategies that people could use. This new normative standard for human reasoning and decision-making is the one that we adopt in resource-rational analysis (Griffiths et al., 2015; Lieder & Griffiths, 2020).

Resource-rational analysis identifies the best biologically feasible minds out of the infinite set of bounded-rational minds. We can formalize this idea by applying the definition of bounded optimality (equation (13.1)) to the human mind. By analogy to the definition of bounded optimal programs, we can define the resource-rational mind $m^\star$ for a brain with respect to the utility function u as

$$m^\star = \underset{m \in \mathcal{M}_{\mathrm{B}rain}}{\arg\max}\ E_{s_{1:t}|e, a_t = m(o_{1:t}, a_{1:t-1})}\left[u(s_{1:t})\right], \tag{13.2}$$

where $a_{1:t}$ are the choices that a person with mind m would live in the environment e, u measures how well they did, and $\mathcal{M}_{\mathrm{B}rain}$ is the set of minds that are biologically feasible, given the biophysical constraints of their brain. This equation is exactly the same as equation (13.1), except that the hardware is a brain rather than a computer and the program is the mind.

The cognitive limitations inherent in the biologically feasible minds $\mathcal{M}_{\mathrm{B}rain}$ include a limited set of elementary operations (e.g., counting and memory recall are available, but exact Bayesian inference is not), limited processing speed (each operation takes a certain amount of time), and potentially other constraints, such as limited working memory. Critically, the world state s_t is constantly changing while the mind m deliberates. Thus, to perform well, the bounded optimal mind $m^\star$ not only has to generate good decisions, but it also has to generate them quickly. Since each cognitive operation takes a certain amount of time, this entails that bounded optimality often requires computational frugality.

Unfortunately, it might be intractable to compute the resource-rational mind defined by equation (13.2) because it requires optimizing over an entire lifetime. To provide a more tractable definition that can be used to derive predictions about which heuristic h a person should use to make a particular decision or inference, it will be assumed that life can be partitioned into a sequence of episodes, each of which starts with a state $s_0 = (w_0, b_0)$ that comprises the unknown state of the external world w_0 and the person's internal belief state b_0. Furthermore, let $\mathrm{result}(s_0, h)$ denote the judgment, decision, or belief update that results from applying heuristic h in the initial state s_0. In this setting, we can decompose the value of having applied a particular strategy into the utility of its termination state $u(s_\perp)$ and the computational cost of its execution. The latter is critical because the time and cognitive resources that a person expends on any one decision or inference (current episode) take away from their budget for other decisions and inferences (future episodes). To capture this, let the random variable $\text{cost}(t_h, \rho, \lambda)$ denote the total opportunity cost of investing the cognitive resources ρ used or blocked by heuristic h for the duration t_h of its execution, when the agent's cognitive opportunity cost per unit of cognitive resources and unit of time is λ. In this setting, we can define the resource-rational heuristic $h^\star$ for a particular brain to use in the belief state b_0 as

$$h^\star(b_0, Brain, e) = \arg\max_{h \in \mathcal{H}_{\text{Brain}}} E_{\text{result}|s_0,h,e}\left[u\left(\text{result}\right)\right] - E_{t_h,\rho,\lambda|h,s_0,Brain,e}\left[\text{cost}(t_h, \rho, \lambda)\right], \tag{13.3}$$

where $\mathcal{H}_{Brain}$ is the set of heuristics that the brain can execute. Which heuristics the brain can execute is constrained by the available elementary operations and memory constraints on the representation of the heuristic. Moreover, the representation of the heuristic has to be a sequence of elementary operations.

The cost of thinking is an opportunity cost. Concretely, it is the total utility that the agent could have obtained by investing the time and cognitive resources that it took to execute the heuristic into other pursuits. Formally, this opportunity cost can be defined as

$$\text{cost}(d, \rho, \lambda) = \int_0^{t_h} \rho(t) \cdot \lambda(t)\, \text{dt}. \tag{13.4}$$

For simplicity, we can assume that the heuristic's cognitive demands ρ and the agent's opportunity cost λ are roughly constant, while the heuristic h is being executed. In this case, the cost of thinking can be approximated by $\text{cost}(t_h, \rho, \lambda) = t_h \cdot \rho \cdot \lambda$. To further simplify this analysis, $\rho \cdot \lambda$ can be approximated by the agent's reward rate in the environment e; this corresponds to the assumption that (1) the agent cannot multitask and (2) the current reward rate is an accurate estimate of the value of the agent's time. In brief, the key assumption is that people's cognitive mechanisms should trade accuracy against opportunity cost in an adaptive, near-optimal manner.

Notions related to bounded optimality have previously been proposed in psychology (for a review, see Gershman, Horvitz, & Tenenbaum, 2015). Most prominently, Lewis, Howes, and colleagues have argued for the importance of taking constraints into account in rational analysis (Howes, Lewis, & Vera, 2009) and connected this idea to bounded optimality (Lewis, Howes, & Singh, 2014). They introduced a framework called *computational rationality*, which focuses on identifying optimal programs for bounded agents to execute and highlighted the potential for this approach to produce *ecological-bounded-optimal* explanations of behavior, in which both the environment and computational constraints are taken into account. The definition of rationality that we use in resource-rational analysis differs from this framework in three significant ways: First, it explicitly captures the opportunity cost of the time and computation that applying strategy h to the current problem incurs at the expense of the agent's ability to solve other problems concurrently or in the future. Second, it weighs the states that the environment might be in according to the person's belief state (b_0) rather than their overall frequency in the environment. This accounts for people's ability to adapt their cognitive strategy to individual problems based on their imperfect knowledge about the state of their environment (Payne, Bettman, & Johnson, 1993). Third, the utility function is allowed to depend on the belief state $b_\perp$ that results from reasoning according to

h. This captures the potential benefits of belief updates resulting from computations in the current episode for decisions made in future episodes.

Resource-rational analysis differs from the classic notion of rationality, which stipulates that people should reason according to the laws of logic and probability theory and choose their actions according to expected utility theory, in three major ways: First, it evaluates reasoning by its utility for subsequent decisions rather than by its formal correctness—a form of pragmatism. Second, it takes into account the cost of time and the boundedness of people's cognitive resources. And third, rational action is defined with respect to the distribution of problems in the environment rather than a set of arbitrary laboratory tasks. Arguably, all three changes are necessary to obtain a normative, yet realistic, theory of human rationality. Unlike the decision theoretic and Bayesian accounts, resource-rational solutions to problems are not defined by the quality of the people's actions or the truthfulness or coherence of their beliefs, but rather, in terms of the underlying cognitive strategies. Unlike logic and probability theory, it measures the quality of these strategies not by their adherence to rules that preserve truth or coherence, but rather by their practical effects on people's actions and their consequences.

To cast the distinction between classical rationality and resource-rational analysis in psychological terms, classical rationality is essentially a behaviorist characterization of rational action. That is, it stipulates the optimal action that the agent should take based on the problem posed by that agent's environment: it says nothing about the mental states of the agent. By contrast, resource-rational analysis takes a cognitive approach to defining rational action: it emphasizes the cognitive operations that the rational agent should perform, leading to beliefs that then guide actions. For researchers interested in analyzing the actions of agents with finite time and computational resources, and particularly researchers who want to answer questions about the mental states of those agents, resource-rational analysis provides a more productive way to define rational action.

13.2 The Process of Resource-Rational Analysis

Herbert Simon famously argued that to understand people's cognitive strategies, one has to simultaneously consider their cognitive constraints and the structure of their environment (1956, 1982). By focusing on the structure of the environment while de emphasizing the role of computational limitations, Anderson's (1990) rational analysis realized only half of Simon's vision for understanding bounded rationality. This was a reasonable compromise because there were formal tools for deriving optimal behavior for unbounded agents, but no equivalent tools for deriving optimal cognitive strategies for bounded agents. But the ideas introduced in section 13.1 allow us to go beyond the simplifying assumptions of unbounded optimality and derive more realistic models of human reasoning and decision-making.

Resource-rational analysis achieves this by modifying the optimality assumption of rational analysis to take into account that people can perform certain costly cognitive operations at only a limited speed (Griffiths et al., 2015). Resource-rational analysis is a four-step methodology (see figure 13.1) that uses the theoretical ideas introduced here to derive process models of cognitive abilities from formal definitions of their function and assumptions about the mind's computational architecture. This function-first approach starts at the computational level of analysis (Marr, 1982). When the problem solved by the cognitive capacity

1. Start with a computational-level (i.e., functional) description of an aspect of cognition, formulated as a problem and its solution.
2. Posit a class of algorithms for approximately solving this problem, a cost to computational resources used by these algorithms, and a utility of more accurately approximating the correct solution.
3. Find the algorithm in this class that optimally trades off resources and approximation accuracy (equation 13.3).
4. Refine by revising the model, algorithms, or costs (Steps 1, 2, or 3), or by proceeding to the next level down: approximating the algorithms in Step 2 to capture further resource constraints.

Figure 13.1
The four steps of resource-rational analysis.

under study has been formalized, resource-rational analysis postulates an abstract computational architecture, which is a set of elementary operations and their costs with which the mind might solve this problem. Next, resource-rational analysis derives the algorithm that is optimal for solving the problem identified at the computational level with the abstract computational architecture (equation (13.3)). The resulting process model is used to predict people's responses and reaction times in a given experiment, and those predictions are then tested against empirical data. Based on this evaluation, the assumptions about the computational architecture and the problems to be solved are revised and the analysis cycle is repeated. The iterative refinements of the assumed cognitive architecture proceeds from abstract, minimal assumptions to an increasingly more realistic model of the underlying neurocognitive architecture (see figure 13.2). In this way, resource-rational analysis can be used to connect Marr's (1982) computational and algorithmic levels.

By explicitly positing a class of possible algorithms and a cost to the resources used by these algorithms, we can invoke an optimality principle to derive the algorithm that the mind should be using. This makes resource-rational analysis a methodology for analyzing information-processing systems at an intermediate level defined by an idealized family of computational mechanisms that corresponds to a particular computational architecture. This method enables us to reverse-engineer not only the problem that a system solves (computational level of analysis), but also the system's computational architecture.

To identify a family of potential cognitive strategies and the corresponding cognitive architecture (Step 2), resource-rational analysis draws on previous research in the computational sciences. Concretely, having formulated the problem to be solved in precise mathematical terms allows us to mine the literature of AI, machine learning, operations research, and other areas of computer science and statistics for classes of algorithms that have been developed to efficiently solve such problems. Such a literature search generally yields one or more parametric families of algorithms. Different settings of an algorithm's parameters often produce qualitatively different behaviors and different speed-accuracy trade-offs. For instance, particle filtering is a general approach that leads to specific algorithms varying in the number of particles, the re-sampling criteria, and other elements (Abbott & Griffiths, 2011). This results in an infinite collection of algorithms, some of which have qualitatively different properties (e.g., one particle versus millions of particles). Steps 2 and 3 allow us to find reasonable points within this space of algorithms, which can then

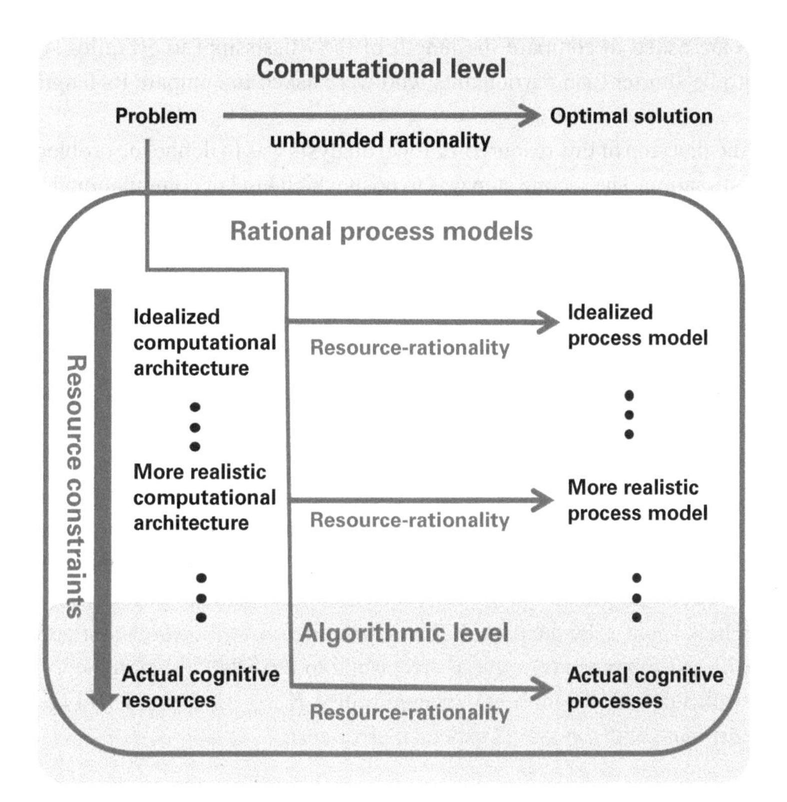

Figure 13.2
Illustration of how resource-rational analysis connects levels of analysis. Figure reproduced with permission from Griffiths et al. (2015).

be compared to human behavior. To the degree that evolution, development, and learning have adapted the system to make optimal use of its finite computational resources, resource-rational analysis can be used to derive the system's algorithm from assumptions about its computational architecture.

Resource-rational analysis is a new cognitive modeling paradigm that can be applied to all aspects of human cognition (Lieder & Griffiths, 2020), including decision-making, reasoning, memory, perception, judgment, planning, learning, and problem-solving. In the next two sections, we present two case studies that illustrate the methodology of resource-rational analysis in domains of numerical estimation and decision-making.

13.2.1 An Example: Numerical Estimation

Our first case study highlights the four steps of resource-rational analysis, shown in figure 13.1, in an investigation of the computational mechanisms that give rise to the anchoring bias in numerical estimation (Lieder, Griffiths, Huys, & Goodman, 2018b). This is the phenomenon where, when people are first asked whether a quantity is above or below some arbitrary value, their subsequent estimates are systematically biased toward the arbitrary value that they previously compared it to. For example, people might be asked to estimate the length of the Mississippi River, first comparing it to either 50 miles or 10,000 miles.

People who are asked to compare the length of the Mississippi to 50 miles subsequently estimate it to be shorter than participants who were asked to compare its length to 10,000 miles.

In brief, the first step of this resource-rational analysis was to define the problem solved by numerical estimation. The second step was to posit which kind of computational architecture the mind might employ to solve this problem. The third step was to derive the optimal solution to the numerical estimation problem afforded by the computational architecture. And the fourth step was to evaluate the resulting predictions against people's estimates of numerical quantities under various experimental conditions.

Step 1: Computational-Level Analysis In numerical estimation, people have to make an informed guess about an unknown quantity X based on their knowledge $\mathcal{K}$. In general, people's relevant knowledge $\mathcal{K}$ is incomplete and insufficient to determine the quantity X with certainty. For instance, people asked to estimate the boiling point of water on Mount Everest typically do not know its exact value, but they do know related information, such as the boiling point of water at normal altitude, the freezing point of water, the qualitative relationship between altitude, air pressure, and boiling point, and other statistics. We formalize people's uncertain beliefs about X by the probability distribution $p(X|\mathcal{K})$, which assigns a plausibility $p(X=x|\mathcal{K})$ to each potential value x. According to Bayesian decision theory, the goal is to report estimate $\hat{x}$ with the highest expected utility, $E_{p(x|\mathcal{K})}[u(\hat{x},x)]$. This is equivalent to finding the estimate with the lowest expected error cost:

$$x^{\star} = \arg\min_{\hat{x}} E_{p(x|\mathcal{K})}[\text{cost}(\hat{x},x)], \tag{13.5}$$

where $x^{\star}$ is the optimal estimate, and $\text{cost}(\hat{x},x)$ is the error cost of the estimate $\hat{x}$ when the true value is x.

Step 2: Posit a Class of Possible Algorithms How the mind should solve the problem of numerical estimation (see equation (13.5)) depends on its computational architecture. Thus, to derive predictions via resource-rational analysis, one has to specify the mind's elementary operations and their cost. To do so, Lieder et al. (2018b) built on the models reviewed in chapter 11, which assume that the mind's elementary computation is *sampling*.

Sampling stochastically simulates the outcome of an event or the value of a quantity such that, on average, the relative frequency with which each value occurs is equal to its probability. According to Vul, Goodman, Griffiths, and Tenenbaum (2014), people may estimate the value of an unknown quantity X using only a single sample from the subjective probability distribution $P(X|K)$ that expresses their beliefs. However, for the complex inference problems that people face in everyday life, generating even a single perfect sample can be computationally intractable. Thus, while sampling is a first step from computational-level theories based on probabilistic inference toward cognitive mechanisms, a more detailed process model is needed to explain how simple cognitive mechanisms can solve the complex inference problems of everyday cognition. Lieder et al. (2018b) explored a more fine-grained model of mental computation whose elementary operations serve to approximate sampling from the posterior distribution. In statistics, machine learning, and AI, sampling is often approximated by *Markov chain Monte Carlo* (MCMC) methods (Gilks, Richardson, & Spiegelhalter, 1996). As discussed in chapter 6, MCMC algorithms allow the drawing of

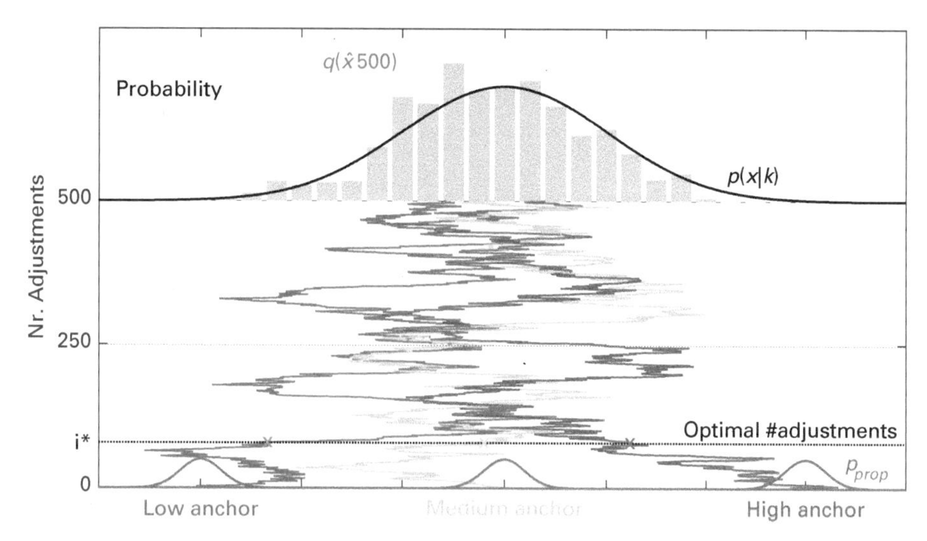

Figure 13.3
Resource-rational anchoring and adjustment via the Metropolis-Hastings algorithm. The three jagged lines are examples of the stochastic sequences of estimates that the adjustment process might generate starting from a low, medium, and high anchor, respectively. In each iteration, a potential adjustment is sampled from a proposal distribution p_{prop} illustrated by the bell curves. Each proposed adjustment is stochastically accepted or rejected, such that over time, the relative frequency with which different estimates are considered $q(\hat{x}_t)$ becomes the target distribution $p(x|\mathcal{K})$. The top of the figure compares the empirical distribution of the samples collected over the second half of the adjustments with the target distribution $p(x|\mathcal{K})$. Importantly, this distribution is the same for each of the three sequences. In fact, it is independent of the anchor because the influence of the anchor vanishes as the number of adjustments increases. Yet when the number of adjustments (iterations) is low (e.g., 25), the estimates are still biased toward their initial values. The optimal number of iterations i^* is very low, as illustrated by the dotted line. Consequently, the resulting estimates indicated by the red, yellow, and red crosses are still biased toward their respective anchors. Figure reproduced with permission from Lieder et al. (2018b).

samples from arbitrarily complex distributions using a stochastic sequence of approximate samples, each of which depends only on the previous one.

Lieder et al. (2018b) assumed that the mind's computational architecture supports MCMC by two basic operations. The first operation takes in the current estimate and stochastically modifies it to generate a new one. The second operation compares the posterior probability of the new estimate to that of the old one and accepts or rejects the modification stochastically. The cost of computation was taken to be proportional to how many such operations have been performed. These two basic operations are sufficient to execute an effective MCMC strategy for probabilistic inference, known as the Metropolis-Hastings algorithm (Hastings, 1970). This algorithm (described in more detail in chapter 6) is the basis for our anchoring-and-adjustment models, as illustrated in figure 13.3.

To be concrete, given an initial guess $\hat{x}_0$, which can be interpreted as the anchor a ($\hat{x}_0 = a$), this algorithm performs a series of adjustments. In each step, a potential adjustment δ is proposed by sampling from a symmetric probability distribution. The adjustment will either be accepted (i.e., $\hat{x}_{t+1} = \hat{x}_t + \delta$) or rejected (i.e., $\hat{x}_{t+1} = \hat{x}_t$). If a proposed adjustment makes the estimate more probable ($p(X = \hat{x}_t + \delta|\mathcal{K}) > p(X = \hat{x}_t|\mathcal{K})$), then it will always be accepted. Otherwise, the adjustment will be made with probability $\alpha = \frac{p(X=\hat{x}_t+\delta|\mathcal{K})}{p(X=\hat{x}_t|\mathcal{K})}$; that is, according to the posterior probability of the adjusted relative to the unadjusted estimate. This strategy ensures that regardless of which initial value you start from, the frequency with which each

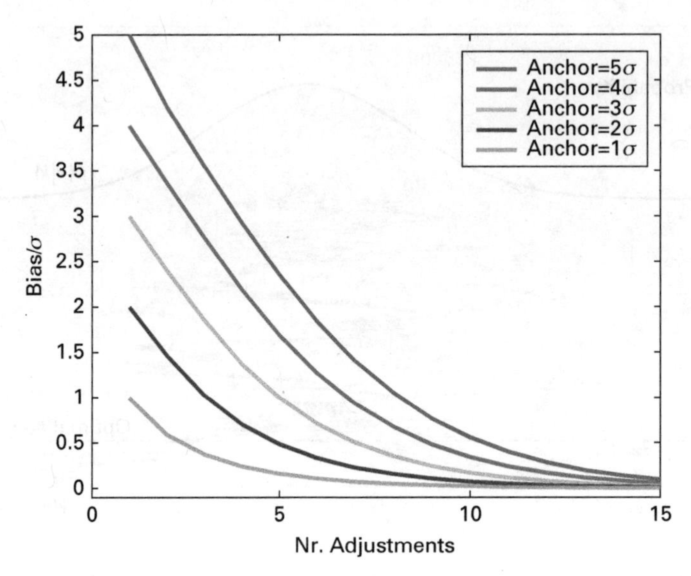

Figure 13.4
In resource-rational anchoring and adjustment, the bias of the estimate is bounded by a geometrically decaying function of the number of adjustments. The plot shows the bias of resource-rational anchoring and adjustment as a function of the number of adjustments for five initial values located $1, \cdots, 5$ posterior standard deviations (i.e., σ) away from the posterior mean. The standard normal distribution was used as both the posterior $p(X|\mathcal{K})$ and the proposal distribution. Figure reproduced with permission from Lieder et al. (2018b).

value x has been considered will eventually equal its subjective probability of being correct (i.e., $p(x|\mathcal{K})$). This is necessary to capture the finding that the distribution of people's estimates is very similar to the posterior distribution $p(x|\mathcal{K})$ (Vul et al., 2014; Griffiths & Tenenbaum, 2006). More formally, as the number of adjustments t increases, the distribution of the estimates $p(\hat{x}_t)$ converges to the posterior distribution $p(x|\mathcal{K})$. This model of computation has the property that each adjustment decreases an upper bound on the expected error by a constant multiple (Mengersen & Tweedie, 1996). This property is known as geometric convergence and illustrated in figure 13.4.

There are several good reasons to consider this computational architecture as a model of mental computation in the domain of numerical estimation. First, the success of MCMC methods in statistics, machine learning, and AI suggests that they are well suited for the complex inference problems that people face in everyday life. Second, MCMC can explain important aspects of cognitive phenomena ranging from category learning (Sanborn et al., 2010a) to the temporal dynamics of multistable perception (Moreno-Bote, Knill, & Pouget, 2011; Gershman, Vul, & Tenenbaum, 2012), causal reasoning in children (Bonawitz, Denison, Gopnik, & Griffiths, 2014a), and developmental changes in cognition (Bonawitz, Denison, Griffiths, & Gopnik, 2014b). Third, MCMC is biologically plausible, in that it can be efficiently implemented in recurrent networks of biologically plausible spiking neurons (Buesing, Bill, Nessler, & Maass, 2011). Last but not least, process models based on MCMC might be able to explain why people's estimates are both highly variable (Vul et al., 2014) and systematically biased (Tversky & Kahneman, 1974).

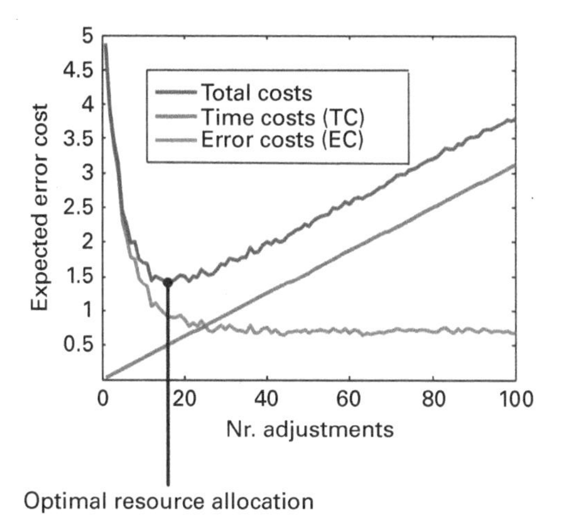

Figure 13.5
The expected value of the error cost $\text{cost}(x, \hat{x}_n)$, shown in green, decays nearly geometrically with the number of adjustments n. While the decrease in the error cost diminishes with the number of adjustments, the time cost $\gamma \cdot t$ shown in red continues to increase at the same rate. Consequently, there is a point when further decreasing the expected error cost by additional adjustments no longer offsets their time cost so that the total cost, shown in blue, starts to increase. That point is the optimal number of adjustments $t^{\star}$. Figure reproduced with permission from Lieder et al. (2018b).

Step 3: Find the Resource-Rational Strategy Resource-rational anchoring and adjustment makes three critical assumptions. First, the estimation process is a sequence of adjustments such that after a sufficient number of steps, the estimate will be a representative sample from the distribution of beliefs about the unknown quantity X given the knowledge $\mathcal{K}$, $p(X|\mathcal{K})$. Second, each adjustment costs a fixed amount of time. Third, the number of adjustments is chosen to achieve an optimal speed-accuracy trade-off. It follows that people should perform the optimal number of adjustments; that is,

$$t^{\star} = \arg\min_{t} E_{p(\hat{x}_t)} \left[\text{cost}(x, \hat{x}_t) + \gamma \cdot t\right], \tag{13.6}$$

where $p(\hat{x}_t)$ is the distribution of the estimate after t adjustments, x is its unknown true value, $\hat{x}_t$ is the estimate after performing t adjustments, $\text{cost}(x, \hat{x}_t)$ is its error cost, and γ is the time cost per adjustment.

Figure 13.5 illustrates this point, showing how the expected error cost—which decays geometrically with the number of adjustments–and the time cost—which increases linearly—determine the optimal speed-accuracy trade-off. We inspected the solution to equation (13.6) when the belief and the proposal distribution are standard normal distributions (i.e., Gaussians with mean 0 and variance 1) for different anchors. Lieder et al. (2018b) found that for a wide range of realistic time costs, the optimal number of adjustments (see figure 13.6a) is much smaller than the number of adjustments that would be required to eliminate the bias toward the anchor. Consequently, the estimate obtained after the optimal number of adjustments is still biased toward the anchor, as shown in figure 13.6b. This is a consequence of the geometric convergence of the error (see figure 13.4), which leads to quickly diminishing returns for additional adjustments. This is a general property

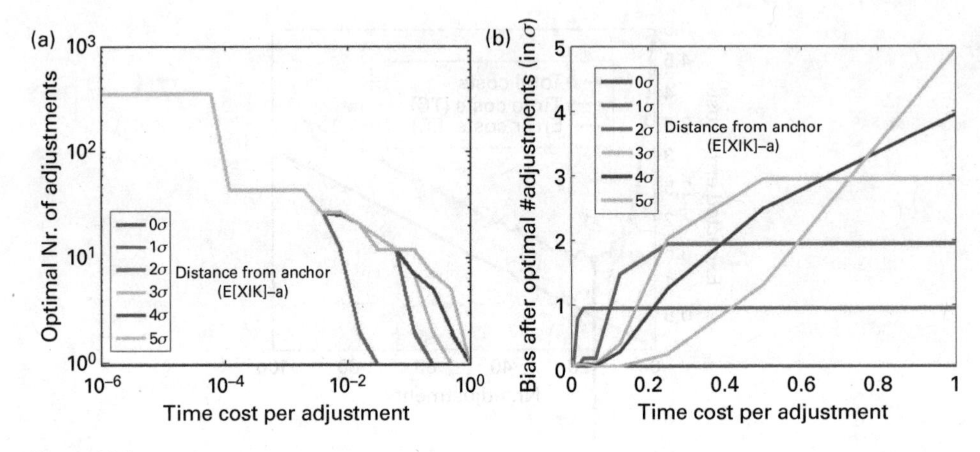

Figure 13.6
Optimal number of adjustments (a) and the bias after optimal number of adjustments (b) as a function of relative time cost and distance from the anchor. Figure reproduced with permission from Lieder et al. (2018b).

of this rational model of adjustment that can be derived mathematically (Lieder, Griffiths, & Goodman, 2012).

Step 4: Evaluate the Model and Refine It The predictions of the resource-rational anchoring-and-adjustment model were evaluated both against existing data and through a new experiment. First, Lieder et al. (2018b) applied this model to simulate people's judgments in previously conducted anchoring experiment and found that it captured a wide range of empirical phenomena, including insufficient adjustment from anchors, an increase in anchoring bias with the extremity of anchors, and the effects of uncertainty and incentives on the magnitude of the bias. Second, Lieder et al. (2018c) designed two experiments specifically to test the model's prediction that the anchoring bias should increase with time pressure but decrease with error cost. The first experiment confirmed this prediction in a task where people generated their own anchors, and the second experiment confirmed it in a task where people's anchors were provided by leading questions.

13.2.2 An Example: Decision-Making Under Uncertainty

In our second case study, Lieder, Griffiths, Huys, and Goodman (2018a) applied resource-rational analysis to elucidate how people make decisions under uncertainty. To illustrate the methodology, we summarize their work in terms of the four steps of resource-rational analysis shown in figure 13.1.

Step 1: Computational-Level Analysis In the first step of their resource-rational analysis, Lieder et al. (2018a) specified the function of decision-making as choosing actions to maximize their *expected utility* (Neumann & Morgenstern, 1944; see also chapter 7). Formally, computing the expected utility $E_{p(s|a)}\,[u(s)]$ of action a requires integrating the probabilities $p(s|a)$ of its possible outcomes s with their utilities $u(s)$. Unlike simple laboratory tasks where each choice can yield only a small number of possible payoffs, many real-life decisions have infinitely many possible outcomes. As a consequence, the expected utility of action a becomes an integral:

$$E_{p(s|a)}[u(s)] = \int p(s|a) \cdot u(s)\, ds. \tag{13.7}$$

In the general case, it is intractable to compute the exact value of this integral. Thus, decision-makers have to approximate it in one way or another.

Step 2: Posit a Class of Possible Algorithms In the second step of resource-rational analysis, Lieder et al. (2018c) explored the implications of resource constraints on decision-making under uncertainty. To do so, they modeled the cognitive resources available for decision-making based on the assumption that people can generate samples from probability distributions. Sampling methods can provide an efficient approximation to integrals, such as the expected utility in equation (13.7) (Hammersley & Handscomb, 1964).

Lieder et al. (2018c) then expressed people's time and resource constraints as a limit on the number of samples, where each sample is a simulated outcome. Thus, the decision-maker's primary cognitive resource is a probabilistic simulator of the environment. The decision-maker can use this resource to anticipate some of the many potential futures that could result from taking one action versus another, but each simulation takes a nonnegligible amount of time. Since time is valuable and the simulator can perform only one simulation at a time, the cost of using this cognitive resource is thus proportional to the number of simulations (i.e., samples).

Importance sampling is a popular sampling algorithm in computer science and statistics (Hammersley & Handscomb 1964; Geweke, 1989; see also chapter 6) with connections to both psychological process models (Shi, Griffiths, Feldman, & Sanborn, 2010; see also chapter 11) and neural networks (Shi & Griffiths, 2009; see also chapter 12). It estimates a function's expected value with respect to a probability distribution p by sampling from an importance distribution q and correcting for the difference between p and q by down-weighting samples that are less likely under p than under q and up-weighting samples that are more likely under p than under q. Concretely, *self-normalized importance sampling* (Robert & Casella, 2009) draws m samples $x_1, \cdots, x_m$ from distribution q, weights the function's value $f(x_j)$ at each point x_j by $w_j = \frac{p(x_j)}{q(x_j)}$, and then normalizes its estimate by the sum of the weights:

$$x_1, \cdots, x_m \sim q, \quad w_j = \frac{p(x_j)}{q(x_j)} \tag{13.8}$$

$$E_p[f(x)] \approx \hat{E}^{\text{IS}}_{q,m} = \frac{1}{\sum_{j=1}^{m} w_j} \cdot \sum_{j=1}^{m} w_j \cdot f(x_j). \tag{13.9}$$

With a finite number of samples, this estimate is generally biased. Following Zabaras (2010), Lieder et al. (2018c) approximated its bias and variance by

$$\text{Bias}[\hat{E}^{\text{IS}}_{q,m}] \approx \frac{1}{m} \cdot \int \frac{p(x)^2}{q(x)} \cdot (E_p[f(x)] - f(x))\, dx \tag{13.10}$$

$$\text{Var}[\hat{E}^{\text{IS}}_{q,m}] \approx \frac{1}{m} \cdot \int \frac{p(x)^2}{q(x)} \cdot \left(f(x) - E_p[x]\right)^2 dx. \tag{13.11}$$

Lieder et al. (2018c) hypothesized that the brain uses a strategy similar to importance sampling to approximate the expected utility gain $E_{p(s|a)}[\Delta u(s)]$ of taking action a and approximate the optimal decision $a^\star = \arg\max_a E_{p(s|a)}[\Delta u(s)]$ by

$$\hat{a}^\star = \arg\max_a \overline{\Delta U}^{\text{IS}}_{q,m}(a), \quad \overline{\Delta U}^{\text{IS}}_{q,m}(a) \approx E_{p(s|a)}[\Delta u(s)] \tag{13.12}$$

$$\overline{\Delta U}^{\text{IS}}_{q,m}(a) = \frac{1}{\sum_{j=1}^{m} w_j} \sum_{j=1}^{m} w_j \cdot \Delta u(s_j), \quad s_1, \cdots, s_m \sim q. \tag{13.13}$$

Note that importance sampling is a family of algorithms: each importance distribution q yields a different estimator, and two estimators may recommend opposite decisions. Thus, in the third step of their resource-rational analysis, Lieder et al. (2018c) investigated which distribution q yields the best decisions.

Step 3: Find the Resource-Rational Algorithm If a decision has to be based on only a very few simulated outcomes, then what is the optimal way to generate them? Formally, the agent's goal is to maximize the expected utility gain of a decision made from only m samples. The utility foregone by choosing a suboptimal action can be upper-bounded by the error in a rational agent's utility estimate. Therefore, the agent should minimize the expected squared error of its estimate of the expected utility gain $E[\Delta U]$, which is the sum of its squared bias and variance; that is, $E\left[(\overline{\Delta U}^{\text{IS}}_{q,m} - E[\Delta U])^2\right] = \text{Bias}\left[\overline{\Delta U}^{\text{IS}}_{q,m}\right]^2 + \text{Var}\left[\hat{\Delta U}^{\text{IS}}_{q,m}\right]$ (Hastie, Tibshirani, & Friedman, 2009). As the number of samples m increases, the estimate's squared bias decays much faster than its variance; see equations (13.10)–(13.11). Therefore, as the number of samples m increases, minimizing the estimator's variance becomes a good approximation to minimizing its expected squared error.

According to variational calculus, the importance distribution

$$q^{\text{var}}(s) \propto p(s) \cdot |\Delta u(s) - E_p[\Delta U]| \tag{13.14}$$

minimizes the variance (equation (13.11)) of the utility estimate in equation (13.13) (Geweke, 1998; Zabaras, 2010). This means that the optimal way to simulate outcomes in the service of estimating an action's expected utility gain is to over represent outcomes whose utility is much smaller or much larger than the action's expected utility gain. Each outcome's probability is weighted by how disappointing ($E_p[\Delta U] - \Delta u(s)$) or elating ($\Delta u(s) - E_p[\Delta U]$) it would be to a decision-maker who anticipated receiving the gamble's expected utility gain ($E_p[\Delta U]$). But unlike in *disappointment theory* (Bell, 1985; Loomes & Sugden, 1984, 1986), the disappointment or elation is not added to the decision-maker's utility function; rather, it increases the event's subjective probability by prompting the decision-maker to simulate that event more frequently. Unlike in previous theories, this distortion was *not* introduced to describe human behavior but instead was derived from first principles of resource-rational information processing: importance sampling over simulates extreme outcomes to minimize the mean-squared error of its estimate of the action's expected utility gain. It tolerates the resulting bias because it is more important to shrink the estimate's variance.

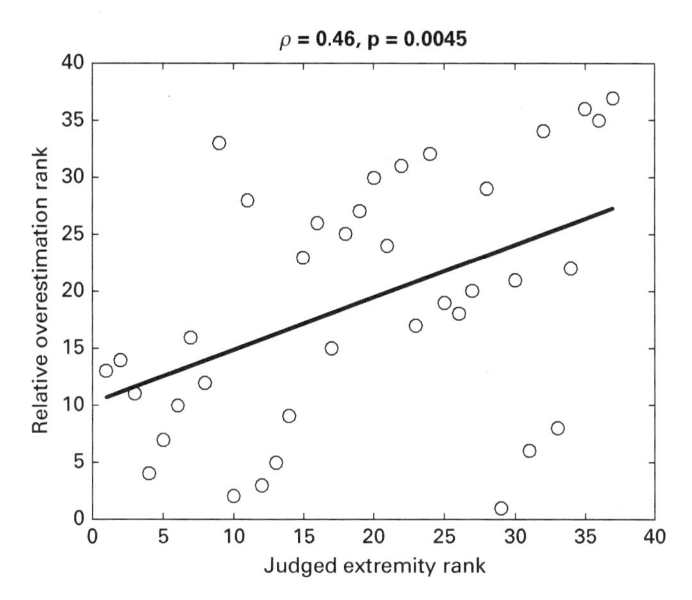

Figure 13.7

Relative overestimation ($\frac{(\hat{f}_k - f_k)}{f_k}$) increases with perceived extremity ($|u(o_k)|$). Each circle represents one event's average ratings. Figure reproduced with permission from Lieder et al. (2018c).

Unfortunately, importance sampling with q^{var} is intractable because it presupposes the expected utility gain $E_p[\Delta U]$ that importance sampling is supposed to approximate. However, the average utility $\overline{\Delta u}$ of the outcomes of previous decisions made in a similar context could be used as a proxy for the expected utility gain, $E_p[\Delta U]$. That quantity has been shown to be automatically estimated by model-free reinforcement learning in the midbrain (Schultz, Dayan, & Montague, 1997). Therefore, people should be able to sample from the approximate importance distribution:

$$\tilde{q}(s) \propto p(s) \cdot \left|\Delta u(s) - \overline{\Delta u}\right|. \tag{13.15}$$

This distribution weights each outcome's probability by the extremity of its utility. Thus, on average, extreme events will be simulated more often than other equally probable outcomes of moderate utility. We therefore refer to simulating potential outcomes by sampling from this distribution as *utility-weighted sampling* (UWS).

Step 4: Evaluate the Model and Refine It In the fourth and final step of their analysis, Lieder et al. (2018c) evaluated their resource-rational model against empirical data. The UWS model predicts that people should overestimate the frequency of extreme events. Lieder et al. (2018c) experimentally tested this prediction by asking people to judge the extremity and the relative frequency of mundane events, stressful life events, and lethal events. A significant rank-order correlation between participants' extremity judgments and the extent to which they overestimated each event's frequency confirmed this prediction (see figure 13.7). UWS also predicts that people should overweight extreme events in decision-making. A wide range of previously reported biases in decisions from experience and decision from

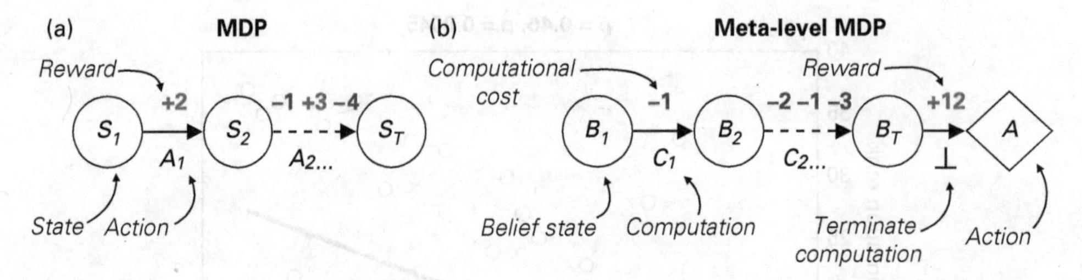

Figure 13.8
Meta-level Markov decision processes. (a) An MDP formalizes the problem of acting adaptively in a dynamic environment. The agent executes actions that change the state of the world and generate rewards, which the agent seeks to maximize. (b) A meta-level MDP formalizes the problem of *deciding how to act* when computational resources are limited. The agent executes computations that update their belief state and incur computational costs. When the agent executes the termination operation $\perp$, they take an external action based on their current belief state.

description are consistent with this prediction (Lieder et al., 2018c). Finally, the recent discovery of biases in human memory that favor extreme events (Madan, Ludvig, & Spetch, 2014) further corroborated UWS as a psychologically plausible mechanism.

13.3 Cognition as a Sequential Decision Problem

In the previous sections, we saw examples of how resource-rational models can predict what people will think about and how long they'll think before making a judgment or decision. However, both of these models implicitly assume that the decision of what to think about must be made in advance—before one has done any thinking. In contrast, intuition suggests that our decisions about what to think about are dynamic, constantly changing based on what we have thought about before. For example, when picking a destination for a vacation, we might briefly consider many countries to identify a few top contenders, which we then consider in more detail.

This observation—namely, that what we think about now can inform what we think about next—suggests that thinking efficiently requires solving a sequential decision problem. The idea that cognition (or more generally, computation) can be modeled as a sequential decision problem was pioneered by Russell and Wefald (1991) in their work on *rational metareasoning* for AI systems. They were interested in building computer programs that could decide which computation to perform next based on the current state of the program. As cognitive scientists, we face a similar problem: identifying cognitive processes that decide which cognitive operation to perform next based on the current mental state. This suggests that we can use the formal tools developed in rational metareasoning to identify resource-rational cognitive processes.

One such tool is the *meta-level Markov decision process*, illustrated in figure 13.8. Recall from chapter 7 that a Markov decision process (MDP) is the standard way to model sequential decision problems in which an agent engages in an extended interaction with an environment. An MDP is defined by a set of possible environment states, $\mathcal{S}$; a set of actions the agent can execute, $\mathcal{A}$; a reward function, R, which the agent seeks to maximize; and a transition function T, which specifies how actions change the state. A meta-level MDP (or meta-MDP) applies this same framework to the case where an agent is interacting with their own internal environment, that is, their computational architecture (Hay, Russell, Tolpin, & Shimony, 2012).

In a meta-MDP, the states correspond to the agent's *beliefs* about the world. Formally, a belief state $b \in \mathcal{B}$ is a distribution over world states. The actions correspond to *computations*. A computation $c \in \mathcal{C}$ is a primitive operation afforded by the agent's computational architecture; it updates the agent's belief in much the same way that an external (or "object-level") action updates a state. The meta-level transition function, T_{meta}, describes precisely how computation updates beliefs. Typically, the transition function is derived by assuming that computations generate information about the world state that is integrated into the new belief by Bayesian inference. The meta-level reward function R_{meta} describes both the costs and benefits of computation. For the former, R_{meta} assigns a strictly negative reward for all computational operations except for one special operation, denoted as $\perp$. This operation is the *termination* operation; when it is executed, the agent takes the object-level (external) action that produces the maximal expected reward:

$$a^*(b) = \arg\max_{a \in \mathcal{A}} \mathop{E}_{s \sim b} \left[r_{\text{object}}(s, a) \right]. \tag{13.16}$$

The meta-level reward for executing $\perp$, then, is the expected object-level reward for the chosen action. Because this is the quantity that the chosen action maximizes, we can simply replace the argmax with a max:

$$R_{\text{meta}}(b, \perp) = \max_{a \in \mathcal{A}} \mathop{E}_{s \sim b} \left[r_{\text{object}}(s, a) \right]. \tag{13.17}$$

Intuitively, it is rewarding to stop assessing options when you have identified an action that you think will yield a high reward.

In the following sections, we show how the framework can be applied to understand the dynamics of people's decision-making strategies, beginning with simple one-step choices and then moving on to more complex planning problems.

13.3.1 An Example: Attention in Preferential Choice

Consider the problem faced by a diner at a buffet table and a shopper looking at a supermarket shelf. They are presented with a number of possible options and must choose the one that they like most; that is, the one that will provide maximal utility. However, the utility of each option is generally not immediately apparent. Instead, the decision-maker must spend some time evaluating each option to determine which one they prefer. Given that this time comes at a cost, which options should a resource-rational agent evaluate, and for how long, before making a choice? Here, we frame this problem as a meta-MDP and show that the optimal policy for that meta-MDP captures patterns in what people look at when choosing among snack foods.

We consider simple choice problems in which an agent is presented with a set of items (e.g., snacks) and must choose one. Each item i is associated with some true but unknown value, $u^{(i)}$, the utility that the agent would gain by choosing it. Following previous work in psychology and neuroscience (Krajbich, Armel, & Rangel, 2010), we assume that the agent informs the choice by collecting noisy samples of the items' true values, each providing a small amount of information, but incurring a small cost.

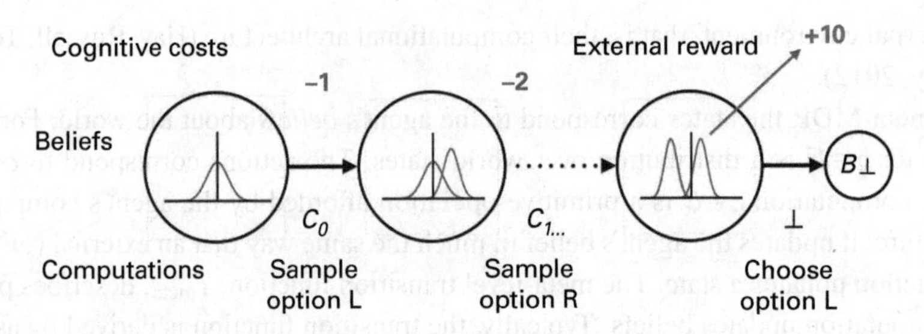

Figure 13.9
Attention allocation as a meta-MDP. A distribution over the reward associated with each option is updated by sampling from the attended option at each timestep.

We model attention by assuming that the agent can sample from one item only at each time point. This sets up a fundamental problem: How should one allocate one's attention (samples) to make good decisions without incurring too much cost? Importantly, the agent cannot simply attend to the item with the highest true value because they do not know the true values. Rather, they must decide which item to pay attention to based on their current value estimates and their uncertainty.

As illustrated in figure 13.9, this problem is naturally modeled as a meta-MDP where the beliefs correspond to estimated value distributions and the computations correspond to attending to an item and updating its estimate. Formally, a belief, $b \in \mathcal{B}$, corresponds to a set of posterior distributions over each item's value. Because the distributions are Gaussian, the belief can be represented by two vectors, $\boldsymbol{\mu}$ and $\boldsymbol{\lambda}$, that specify the mean and precision of each distribution (the precision is the inverse of the variance). That is,

$$p(u^{(i)} \mid b) = \text{Normal}(u^{(i)}; \mu^{(i)}, 1/\lambda^{(i)}).$$

The initial belief state, b_0, captures the agent's prior beliefs about the distribution of values in the environment. For simplicity, we assume a standard Gaussian prior; thus, $\mu_0^{(i)} = 0$ and $\lambda_0^{(i)} = 1$ for all i.

A computation, $c \in \mathcal{C}$, corresponds to drawing a noisy sample of one item's value and updating the corresponding estimated value distribution by Bayesian inference. The transition function, T_{meta}, describes this process; it can be represented by the following generative process:

$$\begin{aligned} x_t &\sim \text{Normal}(u^{(c)}, \sigma_x^2) \\ \lambda_{t+1}^{(c)} &= \lambda_t^{(c)} + \sigma_x^{-2} \\ \mu_{t+1}^{(c)} &= \frac{\sigma_x^{-2} x_t + \lambda_t^{(c)} \mu_t^{(c)}}{\lambda_{t+1}^{(c)}} \\ \lambda_{t+1}^{(i)} &= \lambda_t^{(i)} \text{ and } \mu_{t+1}^{(i)} = \mu_t^{(i)} \text{ for } i \neq c. \end{aligned} \tag{13.18}$$

The first line of equation (13.18) defines the noisy value sample, the next two lines specify the Bayesian belief update given the sample, and the final line states that the beliefs about nonattended items do not change. The rules for updating the means and precisions of the

Gaussians are just those that follow from Bayesian inference, introduced for means and variances in chapter 3.

Finally, the meta-level reward function incorporates both the cost of sampling and the utility of the chosen item. The meta-level reward for sampling is

$$R_{\text{meta}}(b_t, c_t) = -(\gamma_{\text{sample}} + I(c_t \neq c_{t-1})\,\gamma_{\text{switch}}).$$

where $I(\cdot)$ is the indicator function. This includes a fixed cost, γ_{sample}, as well as an additional switching cost, γ_{switch}, that is paid when sampling from a different item than that sampled on the last time step. The reward for terminating is the expected value of the chosen item (the one with maximal expected value):

$$R_{\text{meta}}(b_t, \perp) = \max_i \mu_t^{(i)}. \tag{13.19}$$

Approximating the optimal policy to the meta-MDP defined here yields an optimal strategy for allocating attention when making choices. To provide an intuitive understanding of this strategy, we focus on two key properties of belief states: (1) uncertainty about the true values and (2) differences in the value estimates. Figure 13.10a shows the probability of the optimal policy sampling an item as a function of these two dimensions. We see that the optimal policy tends to fixate on items that are uncertain and have estimated values similar to the other items. In the case of trinary—but not binary—choice, we additionally see a stark asymmetry in the effect of relative estimated value. While the policy is likely to sample from an item whose value is substantially higher than the competitors, it is unlikely to sample from an item with a value well below them. In particular, the policy has a strong preference to sample from the items with best or second-best value estimates. Intuitively, this is because sampling those items is most likely to change the choice that one makes by switching the order of the top two contenders.

Do people use a similar strategy to allocate attention when making choices? To address this question, we use two data sets collected by Krajbich, Armel, and Rangel, in which participants chose between junk food snacks (either two or three per trial) while their gaze was recorded with an eye tracker (Krajbich et al., 2010; Krajbich & Rangel, 2011). We can simulate this kind of data from the model by assuming that the attended (sampled) item is fixated and that each sample takes 100 ms. Then we can compare the simulated data to the actual recordings. Callaway, Rangel, and Griffiths (2021) conducted just such a comparison, showing that the optimal policy captures many patterns in the human fixation data, sometimes quite closely.

Here, we focus on just one key predictions of the model—namely, that attention should be allocated to the two items with the highest estimated value. Although we cannot directly measure participants' evolving value estimates, we can use the ratings that participants provided for each snack as a proxy. Specifically, we can ask how people's tendency to look at the worst-rated item in the current choice set changes over time. In the two-item case, both items are necessarily in the top two, so we should see no effect (a flat line at 50 percent). In the three-item case, however, we should see that people become increasingly less likely to look at the worst-rated item. This is because, as the value estimates become increasingly precise, the worst-rated item is increasingly likely to have the worst estimate. As shown in figure 13.10b, this is just what we see.

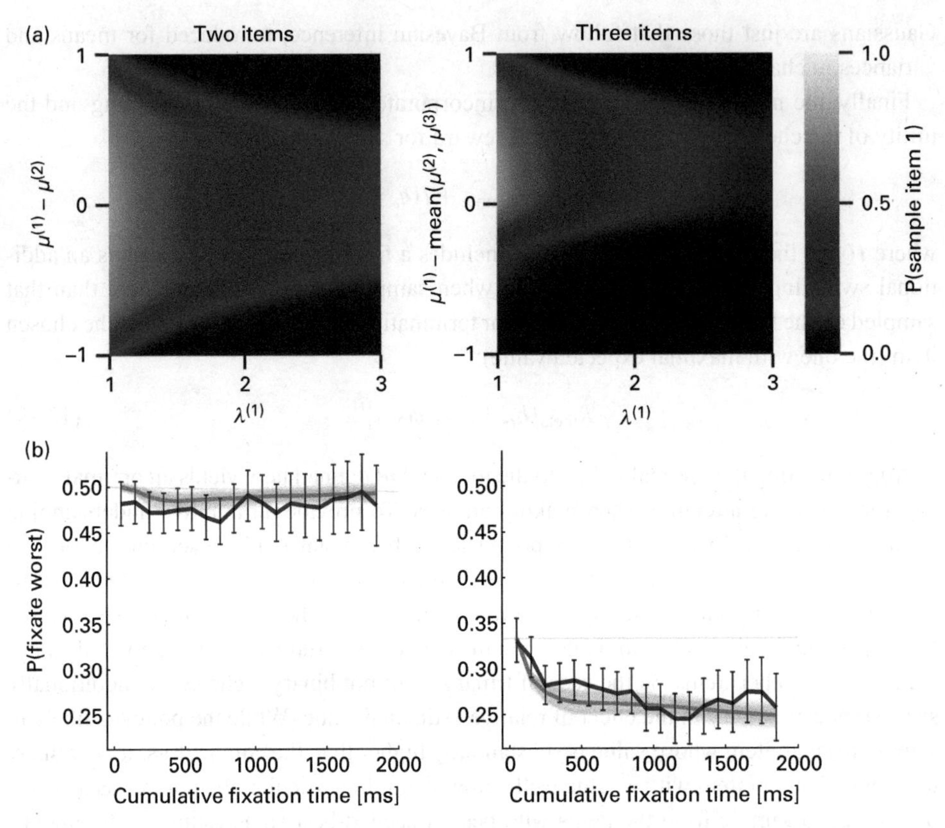

Figure 13.10
Resource-rational attention allocation. (a) Optimal policy. The heat maps show the probability of attending to item 1 (as opposed to attending to item 2 or terminating the sampling process) as a function of the precision of its value estimate, $\lambda^{(1)}$, and the mean of its relative value estimate, $\mu^{(1)} - \text{mean}(\mu^{(2)}, \mu^{(3)})$. (b) Comparison with human fixations. The lines show the probability of fixating the lowest-rated item as a function of the cumulative fixation time to any of the items. Each panel compares human data (black) and model predictions (purple). Error bars indicate 95 percent confidence intervals computed by 10,000 bootstrap samples. The light-purple region indicates uncertainty in the model prediction due to noise in the policy optimization.

13.3.2 An Example: Planning

In the example from section 13.3.1, we applied the meta-MDP framework to a simple, one-step decision. Unfortunately, many of the problems that people face in the world aren't so simple; they require taking multiple actions in sequence. Returning to our vacation example, after choosing a country, we would need to decide which specific cities and sights to visit. To minimize time spent in transit, we would want to pick destinations that are close to each other. This means that we have to think about where we eventually want to end up when planning each leg of the journey. If a particularly beautiful beach town would take us far from other desirable destinations, we might be better off skipping it. This process of making a sequence of interdependent choices is called *planning* (see chapter 7). More generally, planning involves using a model of the world to simulate, evaluate, and select among possible courses of action

One classic way of formalizing planning is *decision-tree search*. A decision tree, illustrated in figure 13.11, represents a set of hypothetical future states and actions. Every

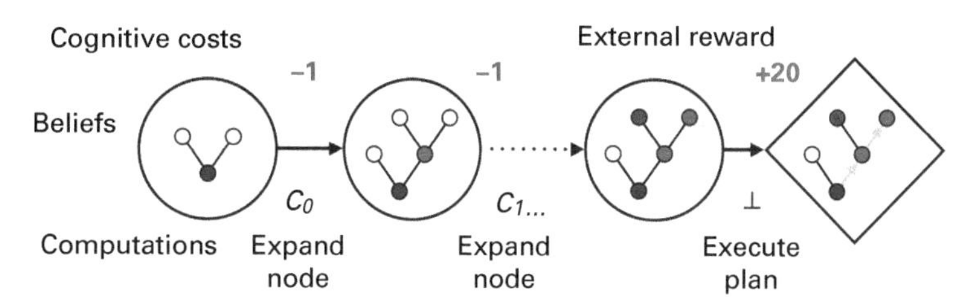

Figure 13.11
Decision-tree search as a meta-MDP. The rewards associated with different nodes are updated by choosing which nodes to expand.

branching point corresponds to a decision that one might have to make. In principle, one can identify the best plan by considering every possible decision point. However, traversing the full decision tree is infeasible because the size of the tree grows exponentially with the number of steps ahead that one looks at. In the earliest days of AI research, Newell and Simon (1956) recognized that the success of human planners (and any hope for success of artificial planners) depended critically on the use of heuristics to circumvent this exponential growth. Recent work on human planning has largely followed a similar vein, proposing and testing possible heuristics that people could use to reduce the cost of planning (Huys et al., 2015).

While useful, the approach of postulating and testing specific heuristics faces two major challenges. First, it is limited by the creativity of the researchers who must generate hypotheses about different possible heuristics that people use. Second, it does not provide a straightforward way to predict which heuristics will be employed in new situations. To address these challenges, we can model planning as a meta-MDP. By solving this meta-MDP with different assumptions about the structure of the environment, we can see which planning heuristic is optimal in each case and ask whether people adapt their planning strategies in a similar way.

As illustrated in figure 13.11, planning can be modeled as a meta-MDP where the beliefs correspond to partially constructed decision trees and the computations correspond to operations that expand the trees. A decision tree represents a set of possible action sequences as a tree-structured directed graph, in which nodes correspond to hypothetical future states and edges correspond to actions that bring the agent from one state to another. The internal nodes are labeled with the reward that the agent would receive if they visited that state. The leaf nodes of the tree are called the *search frontier*; these are states that the agent has not yet considered but could consider next. A belief, $b \in \mathcal{B}$, defines a possible configuration of this tree. If we assume a constant transition structure, the belief can be represented as a vector where b_i is either the reward at state i (if that state has been considered) or a null value otherwise.

A computation, $c \in \mathcal{C}$, corresponds to *node expansion*. This operation determines the cost or reward for visiting a state, integrates that value into the total value of the path leading to that state, and adds the immediate successors of the target state to the search frontier; that is, the set of nodes that can be expanded on the next iteration. These dynamics (including the distribution of rewards that could be revealed at each node) are encoded in the metalevel transition function, T_{meta}. In addition to expanding a node, the agent can execute the $\perp$

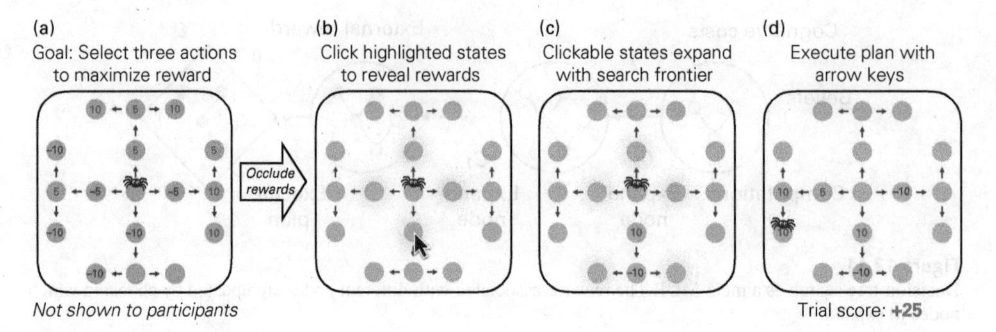

Figure 13.12
Planning task. (a) Participants are presented with a sequential decision problem displayed as a graph. Gray circles indicate states, arrows indicate actions, and green and red numbers indicate rewards and punishments. (b) Rewards are initially occluded, but can be revealed by clicking on the corresponding state. Only highlighted states can be clicked. (c) The clickable states expand with the search frontier, which includes all states adjacent to either the initial state or an already-clicked state. (d) At any point, participants can execute a plan by pressing a sequence of three arrow keys.

operation. The agent then stops planning and executes an action sequence that has maximal expected value according to the decision tree that it has built up to that point.

Finally, the metalevel reward function incorporates the cost of node expansion and the quality of the plan that is ultimately executed. For the former, we assume a fixed cost for each expansion. For the latter, the expected value of a plan is the sum of rewards up to and including the associated node, plus (for an incomplete plan) the expectation of the unknown future rewards. The chosen plan is the one that maximizes this expected value. Thus, the reward for the termination action is equal to the maximal expected value of any plan:

$$r_{\text{meta}}(b, \perp) = \max_{p} V(b, p), \tag{13.20}$$

where p is a complete plan (i.e., a sequence of states beginning with the current state and ending with a terminal state) and $V(b, p)$ is the expected value of executing a plan given the current belief:

$$V(b, p) = \sum_{i \in p} \begin{cases} s_i & \text{if } b_i \text{ has been expanded} \\ \mathrm{E}[R_i] & \text{otherwise.} \end{cases} \tag{13.21}$$

Solving the meta-level MDP defined in this discussion yields an optimal algorithm for planning in the environment described by the meta-MDP (in particular, the distribution of rewards and the object-level transition structure). Do these optimal planning algorithms resemble human planning? Answering this question is challenging because planning typically occurs entirely in one's head. To circumvent this challenge, Callaway et al. (2022b) designed a task that makes people's planning directly observable (figure 13.12). In particular, the task requires participants to click future states to see what reward they would gain if they visited that state. The sequence of clicks thus reveals the order in which the participant considered each state. In the model, this corresponds to a sequence of node-expansion computations.

A planning algorithm can be described in terms of two major components: a *selection rule* that decides which node to expand next and a *stopping rule* that decides when to stop

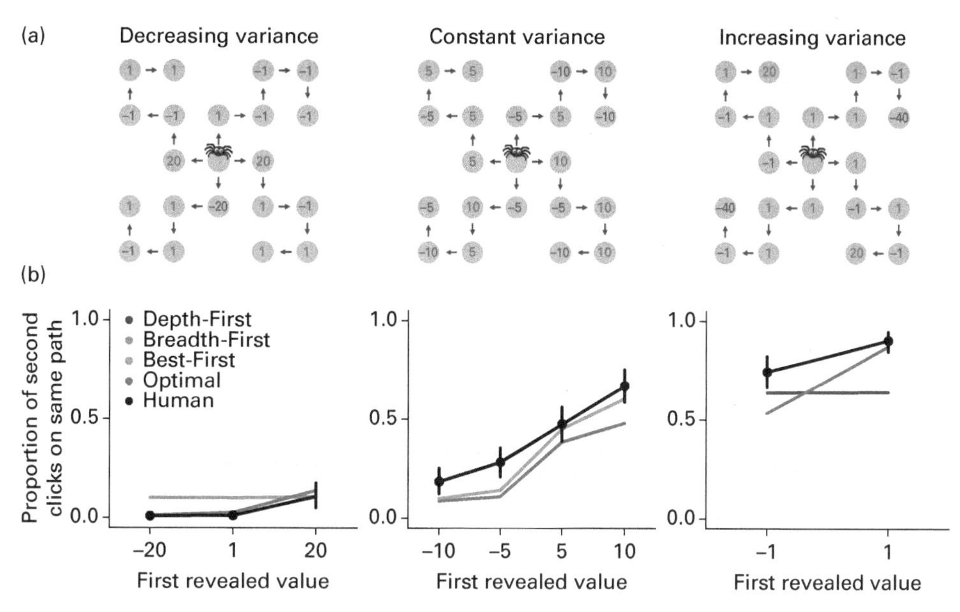

Figure 13.13
Resource-rational adaptation of planning strategies. (a) Environment manipulation. Each environment is characterized by a different location-dependent reward distribution in which large values are found at the beginning of each path (left), at any location (center), or at the end of each path (right). (b) Behavioral indicator of planning strategy. Each panel shows the probability of making a second click on the same path as the first, depending on the value revealed by that first click. Models that capture the qualitative trend are highlighted.

planning and take action. Callaway et al. considered both in detail, finding qualitative consistency between the optimal model and human planning on both dimensions. Here, we focus on the selection rule. In the environment illustrated in figure 13.12 (where each state can take the value −10, −5, 5, or 10 with equal probability), the optimal selection rule closely resembles an algorithm commonly used in an AI, *best-first search*. This algorithm expands nodes that lie on the frontier of plans with maximal expected value; that is, it focuses its attention on the plan that currently looks best. Consistent with the optimal planning algorithm, participants expanded a path with maximal expected value on average 81.5 percent of the time.

However, best-first search is not always the optimal planning algorithm. Indeed, a key assumption of resource-rational analysis is that people's mental strategies adapt to the structure of the environment. To investigate the effect of environment structure on human planning strategies, we constructed three new experimental environments that had different reward distributions (see figure 13.13a), each designed to benefit a different planning algorithm. In the "constant variance" environment, all states had the same reward distribution, as in the previous experiment. Best-first search performs well in this environment. In the "decreasing variance" environment, most states had small rewards (−1 or 1); only states on the first step of the tree had large rewards. Breadth-first search, a strategy that expands every node at each level before continuing to the next level, performs well in this environment. Finally, in the "increasing variance" environment, the large rewards could be found only at the *last* step of the tree. Depth-first search, a strategy that expands the tree as far as possible in one direction before considering other directions, performs well here.

To see if people adapt their planning strategy to the structure of the environment, we can look at the second click on each trial, which provides a simple diagnostic of their overall strategy. Specifically, we can ask how often people use their second click to continue down the path that they began with their first, depending on the value revealed by that first click. An overall tendency to continue down the same path is consistent with a depth-first strategy, the reverse tendency is consistent with a breadth-first strategy, and high sensitivity to the revealed value is consistent with a best-first strategy. As shown in figure 13.13b, people's second clicks were consistent with the environment-appropriate search order. However, while each of these hand-specified search algorithms captured participant behavior well in one of the environments, only the optimal model captures it well in every environment.

13.4 Future Directions

By providing a way to derive the optimal sequence of cognitive operations from the specification of a problem posed by the environment and a description of the computations available to an agent, resource-rational analysis offers a new perspective on many of the classic questions of cognitive psychology. In this chapter, we focused on problems related to decision-making, in part because in this area, the ways in which people deviate from classical rationality and the kind of heuristics and strategies they follow have already been studied extensively. However, perhaps the greatest potential of resource-rational analysis is in applications to other phenomena at the heart of cognitive psychology, with the potential to provide insight into how people manage the limited resources of memory (Dasgupta & Gershman, 2021; Berg & Ma, 2018; Yoo, Klyszejko, Curtis, & Ma, 2018; Gershman, 2021; Sims, Jacobs, & Knill, 2012; Suchow & Griffiths, 2016a), attention (Gabaix, 2014; Callaway et al., 2021; Wiederholt et al., 2010), and cognitive control (Lieder, Shenhav, Musslick, & Griffiths, 2018; Lieder & Griffiths, 2017; Lieder & Iwama, 2021; Shenhav et al., 2017), and how this shapes people's mental representations (Ho et al., 2022), reasoning (Dasgupta, Schulz, & Gershman, 2017; Icard & Goodman, 2015; Dasgupta, Schulz, Tenenbaum, & Gershman, 2020), learning (Bramley, Dayan, Griffiths, & Lagnado, 2017), goal-setting (Correa, Ho, Callaway, & Griffiths, 2020), and goal-pursuit (Prystawski, Mohnert, Tošić, & Lieder, 2020).

Another important direction for future work is understanding how people develop effective resource-rational strategies (He & Lieder, 2022; Jain et al., 2022; Rule, Tenenbaum, & Piantadosi, 2020). In the examples presented in this chapter, we showed that people follow strategies that are consistent with the rational use of limited cognitive resources. But how did people come to these strategies? As with rational analysis, resource-rational analysis appeals to the various adaptive mechanisms that could lead people to approximate ideal solutions, such as evolutionary pressures, learning over the course of the life span, or reasoning about an effective strategy in the context of a specific task. However, by framing these cognitive strategies as the result of solving a sequential decision problem, resource-rational analysis makes a connection to an extensive body of literature on human reinforcement learning (Niv, 2009) that offers a variety of potential learning mechanisms to explore. These learning mechanisms also have well-established neural correlates, leading to the tantalizing possibility that we might be able to understand meta-level reinforcement learning in terms of existing neural mechanisms (Krueger, Lieder, & Griffiths, 2017; He, Jain, & Lieder, 2021).

Finally, resource-rational analysis provides a particular explanation for why people systematically deviate from classical rational action: to the extent that people act in ways that are consistent with resource-rational models, we can understand those actions as being a consequence of them making intelligent use of limited cognitive resources. This suggests that interventions that focus on teaching people the "correct" way to think and make decisions—without taking into account the associated computational costs—are likely to be ineffective. Rather, we should focus on identifying resource-rational strategies that people could execute that will improve their performance (Becker et al., 2022; Callaway et al., 2022a; Consul, Heindrich, Stojcheski, & Lieder, 2022; Mehta et al., 2022; Skirzyński, Becker, & Lieder, 2021; Becker, Skirzynski, van Opheusden, & Lieder, 2022) or on modifying the environments in which people make decisions in ways that make the relevant computations easier (Callaway, Hardy, & Griffiths, 2020, 2023). We have taken preliminary steps in both of these directions, but there is substantial work to be done on supporting decision-making by resource-rational agents.

13.5 Conclusion

Probabilistic models of cognition are typically framed at the computational level, leading to a variety of critiques. Why should psychologists who want to understand the cognitive mechanisms underlying human behavior care whether that behavior is rational? And how can these models account for the substantial body of literature showing that people deviate systematically from the prescriptions of probability theory and maximization of expected utility? We view resource-rational analysis as providing a path toward addressing these critiques. By defining rational action in terms that emphasize the internal computations of cognitively limited agents, this approach makes it possible to define theories of cognitive processes that have the same kind of optimality assumptions that make probabilistic models attractive. Agents following this prescription will deviate from classical rationality, and through the examples we have presented in this chapter, we have shown that in numerous instances those deviations are consistent with human behavior. We look forward—to the full extent of our limited planning horizons—to seeing where this approach takes us next.

14

Theory of Mind and Inverse Planning

Julian Jara-Ettinger, Chris Baker, Tomer Ullman, and Joshua B. Tenenbaum

To effectively interact with other people, we must continually infer and monitor their mental states: what they think, what they want, and what they know about the world—and what they think, want and know about our own mental states and those of others. Even as passive observers, the ability to understand other people's behavior in terms of mental states provides a powerful tool for social learning. Watching a more knowledgeable person's behavior can reveal to us how the world works. Watching a more experienced person can teach us when persistence or practice is helpful and when it's not. Attending to how people act toward others can let us determine who is nice, opportunistic, or mean, and how we ourselves should act to be (and be perceived as) positive social partners.

Whenever we explain, predict, or judge each other's behavior, we do so by thinking about their minds. Yet other people's minds are unobservable, making the ability to infer mental states from observable actions a prerequisite for human like social intelligence. This capacity is known as a *Theory of Mind*. The hypothesis of this chapter is that these abilities in humans can be understood as approximate Bayesian inferences over a mental model of how people think and act. We show how the same Bayesian framework used to model rational action planning (chapter 7) can also be used to model other people's mental processes and infer their latent mental states, as a form of *inverse planning*. In contrast to chapter 7, which focuses on generating high-value actions given a world model and a utility function, here we focus on attributing a world model and utility function to other agents, with the goal of explaining their behavior under the assumption that they are planning rationally: choosing actions that they expect to have high value, given their world models and utility functions that we seek to infer.

14.1 Representing and Inferring Desires

We begin by considering one of the simplest social situations: watching someone with perfect knowledge choose an option from a finite set of possibilities. (We'll consider more complex cases later in the chapter, where knowledge is less than perfect, or choices unfold over a sequence of actions and a range of other real-world complications come into play.) Imagine, for instance, watching a friend choose between chocolate cake and ice cream for dessert. Intuitively, your friend's choice (an observable action) reveals their preference (a mental state). In this setting, inverse planning reduces to a simpler form of inverse decision-making.

To formalize this intuition, we can begin by defining an event as a set of possible world states and a set of possible actions that an agent can take to change the state of the world. In the example given here, the state space is $\mathcal{S} = \{\emptyset, \text{Cake}, \text{Ice cream}\}$, which consists of the state where your friend doesn't have dessert ($\emptyset$), the state where your friend has cake (Cake), and the state where your friend has ice cream (Ice cream). The action space is $\mathcal{A} = \{$"order cake," "order ice cream"$\}$. Your friend's preferences then can be represented by a reward function $R : \mathcal{S} \rightarrow \mathbb{R}$ that associates each state of the world with a scalar that can be positive (meaning that the agent *likes* the state) or negative (meaning that the agent *dislikes* the state). In this situation, we can make four assumptions. First, you and your friend know the current state of the world, which is initially $\emptyset$, as your friend hasn't ordered dessert yet. Second, you and your friend also know the state space and the action space (i.e., the dessert options and the fact that these can be ordered are both known). Third, actions are always successful: taking the actions $a =$ "order cake" and $a =$ "order ice cream" always result in having cake ($s =$"cake") and ice cream ($s =$"ice cream"), respectively. Finally, we'll assume that only your friend knows their own rewards (i.e., how much they like ice cream and cake).

After your friend takes observable action $a \in A$, we can infer their underlying reward function by computing the posterior probability through Bayesian inference:

$$p(R|a) \propto p(a|R)p(R). \tag{14.1}$$

Here, $p(R)$ is the observer's prior distribution over possible reward functions, capturing the observer's expectations about what other people generally like; and $p(a|R)$ is the probability that your friend would take action a if their preferences were correctly represented by reward function R. If we assume that our friend's attitude toward cake is independent of their attitude toward ice cream, we can treat each component of the reward function as independent. That is, the prior probability over any combination of preferences for cake and ice cream is given by the prior probability over ice cream rewards times the prior probability over cake rewards. Formally:

$$p(R) = \prod_{s \in \mathcal{S}} p(R(s)). \tag{14.2}$$

To compute the probability $p(a|R)$ of an observed action given a reward function, we need a model of how people act. Empirical data with children and adults suggests that in a situation like this one, people expect agents to take the action leading to the highest possible reward (Lucas et al. 2014b; Lucas, & Kemp 2017). This can be captured through a simple decision model where

$$a = \begin{cases} \text{order cake,} & \text{if } R(\text{cake}) > R(\text{ice cream}) \\ \text{order ice cream,} & \text{if } R(\text{cake}) < R(\text{ice cream}) \\ \text{Bernoulli}(0.5), & \text{otherwise.} \end{cases} \tag{14.3}$$

Under this model, $p(a|R)$ is 1 whenever a selects the state with the highest reward, 0 when it does not, and 0.5 when the two rewards are identical. Figure 14.1a shows the posterior distribution over your friend's preferences after they take action $a =$ "order ice cream" using this simple decision model, and a uniform prior on rewards over the range [0, 1].

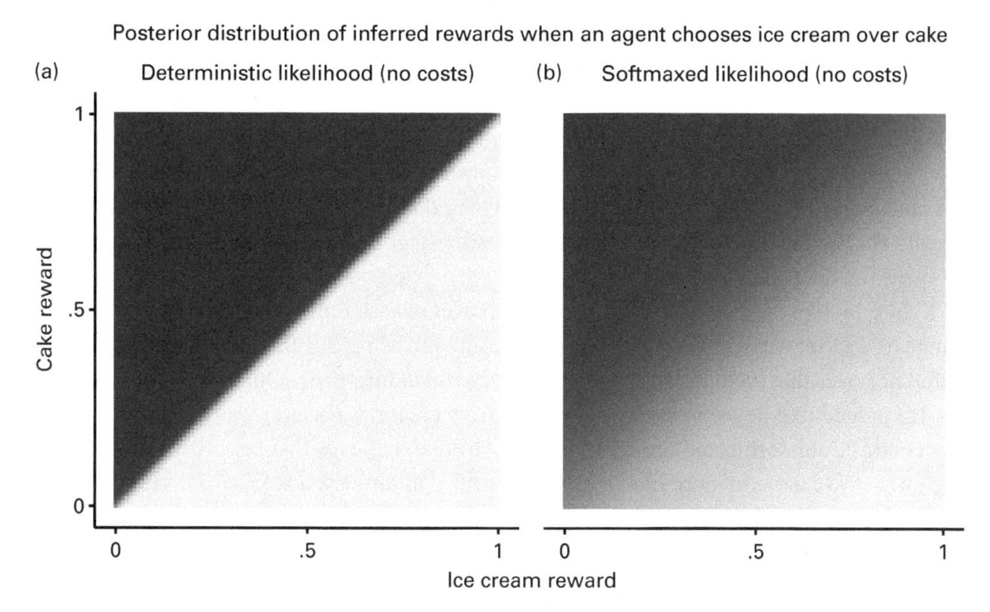

Figure 14.1
Posterior distribution over the reward of ice cream and cake after watching someone choose ice cream. Probabilities are represented with a graded color range from dark blue (lower probabilities) to bright yelow (high probabilities). (a) Posterior distribution using the deterministic likelihood function (equation (14.3)). (b) Posterior distribution using the softmaxed likelihood function (equation (14.4) with $\beta = 0.25$).

In practice, models that expect agents to always maximize rewards are unrealistically strict. For instance, suppose that your friend likes both ice cream and cake but has a very small preference for ice cream. According to a strict reward-maximizing model, your friend should order ice cream every single time they face this choice, but intuitively, we would expect your friend to sometimes order cake instead. This can be accounted for by relaxing the expectation that agents strictly maximize rewards to an expectation that agents probabilistically maximize rewards. We can achieve this by building a likelihood function that applies a softmax function to the agent's rewards:

$$p(a|R) \propto \exp(\beta R(s_a)). \tag{14.4}$$

Here, a is the agent's observable action, R is the unobservable reward function guiding their action, and s_a is the resulting state (e.g., if $a =$"order cake," then s_a="cake"). This element of randomization is also known as adopting a *Boltzmann policy* for selecting actions (Sutton & Barto, 1998).

As equation (14.4) shows, the softmax is a simple transformation where each value (in this case, the reward $R(s_a)$) is multiplied by a scalar (the parameter β) and exponentiated. After applying this transformation to all values, the full set is normalized (by dividing the right term by a constant $\sum_{a' \in A} \exp(\beta R(s_{a'}))$, so the terms for choosing the different action sum to 1). This process transforms a set of scalars (in this case, rewards) into a probability distribution over actions. The rationality behind this transformation is modulated by temperature parameter $\beta \in [0, \infty)$. The higher β is, the more the resulting distribution concentrates probability on the options with the highest possible rewards. As $\beta \to \infty$, the probability in

$p(a|R)$ increasingly concentrates on the action associated with the highest possible reward, converging to the deterministic reward-maximizing model from equation (14.3). In contrast, the lower that β is, the resulting distribution spreads the probability across all the options, while still assigning a higher probability to options with higher values. At the limit, when $\beta = 0$, $p(a|R)$ becomes a uniform distribution over actions, expressing the idea that agents do not act in response to their rewards at all. Thus, β allows us to relax the expectation that agents strictly maximize rewards to an expectation that agents probabilistically maximize rewards (by decreasing the value of β).

Figure 14.1b shows the posterior distribution over reward functions that take values in the range [0, 1] after we watch our friend choose ice cream, using $\beta = 0.25$. As this figure shows, inference over this probabilistic model shows a graded inference. Our initial model (figure 14.1a) judged that any set of rewards where R(ice cream)> R(cake) was equally probable. By contrast, our softmaxed model (figure 14.1b) now believes that reward sets with R(ice cream)>> R(cake) are more likely. This is because, intuitively, a weaker preference for ice cream would give the agent a higher chance of choosing to eat cake (which we did not observe).

Jern et al. (2017) showed how this approach produces human like preference inferences. In one of their tasks, participants watched an agent choose among different meals, each consisting of multiple food items. In figure 14.2a, for instance, the agent could take an eggplant dish and a cookie, a chicken dish and a slice of cake, or a fish dish and an apple. After watching the agent choose the eggplant dish with the cookie, participants were asked to infer the agent's preference for different food items. As figure 14.2b shows, people's inferences correlated highly with the inferred rewards using the Bayesian framework (where the reward associated with each food option was given by the sum of the rewards of each food item; see the schematic of the full space of stimuli in figure 14.2c). Further, this same model unifies a range of inferences that young children make (Lucas et al. 2014b) suggesting that these inferences are at work from early in childhood.

People's actions usually incur a cost (in terms of time and physical effort), and people's mental-state inferences account for how these costs affect agents' choices (Jara-Ettinger, Gweon, Schulz, & Tenenbaum, 2016). We can capture this by extending our framework to include cost functions and utility functions. A cost function $C : \mathcal{A} \rightarrow \mathbb{R}^+$ is a mapping from actions to positive scalar values that represent negative consequences associated with taking different actions. As we will see in section 14.4, this term can capture highly abstract aspects of cost, but we can begin by thinking of costs in terms of money (in economic contexts) or energy (in biological contexts).

A utility function $U : \mathcal{A} \times \mathcal{S} \rightarrow \mathbb{R}$ is a mapping that associates every possible combination of states and actions with the difference between the attained rewards and the incurred costs:

$$U(a, s) = R(s) - C(a). \tag{14.5}$$

This utility function captures the expectation that agents value action plans that yield high rewards while incurring the lowest possible costs (Jara-Ettinger et al., 2016; Liu, Ullman, Tenenbaum, & Spelke, 2017; Csibra, Bíró, Koós, & Gergely, 2003).

If the action costs are known, we can infer agents' unobservable rewards by now assuming that agents act to probabilistically maximize their utilities (rather than just their rewards),

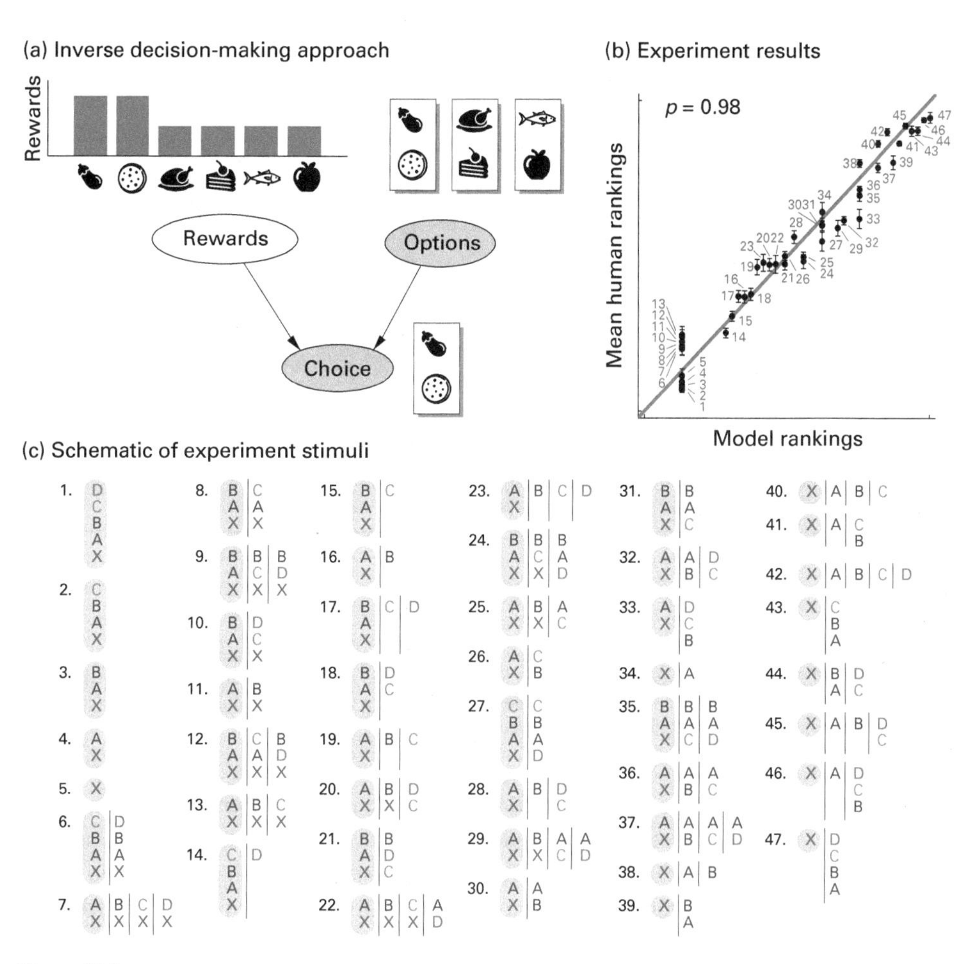

Figure 14.2
Reward inferences from Jern et al. (2017); positive attributes condition of experiment 1. (a) Schematic of task and model setup. The space of available choices, and the choice that is selected, are observable, and the participant's task is to infer the reward associated with different features (in this case, the food items) of the choice. (b) Experiment results. Each point represents a trial with the model prediction on the horizontal axis and participant judgments on the vertical axis. (c) Experimental stimuli. Trials are numbered 1–47. In each trial, each column represents a potential option. Within each choice, letters represent different features. The shaded columns show the agent's selected choice. Figure reproduced with permission from Jern et al. (2017).

using the likelihood function

$$p(a|R; C) \propto \exp(\beta U(a, s_a)), \tag{14.6}$$

where the left side of the equation is the probability of choosing an action, given a specific reward and cost function, and the right side is the softmax of the utility function (which is in turn just the reward minus the cost).

Figures 14.3a–b show the posterior distribution over reward functions when your friend chooses ice cream in a context where each choice incurs a different cost. When cake is more costly, seeing your friend choose ice cream no longer implies that the reward for ice cream is higher than the reward for cake (figure 14.3a); your friend might have preferred

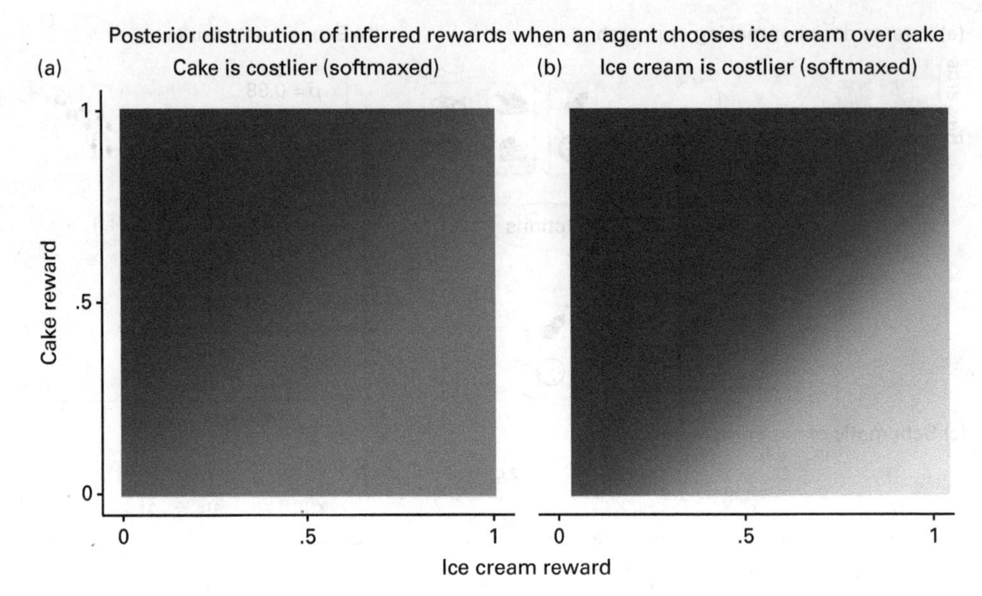

Figure 14.3
Posterior distribution over the reward of ice cream and cake after watching someone choose ice cream in a context where costs are in play. (a) The posterior distribution using a softmaxed likelihood function ($\beta = 0.25$) when cake incurs a cost of 0.25 and ice cream has cost 0. (b) The posterior using a softmaxed likelihood function ($\beta = 0.25$) when ice cream incurs a cost of 0.25 and cake has a cost of 0.

it just because the cost was lower (therefore making the utility higher). Conversely, when ice cream is more costly, seeing your friend order it provides even stronger evidence that they prefer ice cream over cake. After all, the reward must have been high enough to justify the additional cost. This inference is shown in figure 14.3b, where the posterior distribution now concentrates the probability on regions where the reward for ice cream is much higher than the reward for cake. As we've seen, cost need not be money—the concept represents any factors that an agent might find aversive.

In many situations, however, we do not know other people's costs. In these cases, it is possible to jointly infer an agent's costs and rewards from their choices, via bayesian inference:

$$p(R, C|a) \propto p(a|R, C)p(R, C), \tag{14.7}$$

where the priors over cost and reward can be assumed to be independent, such that $p(R, C) = p(R)p(C)$, and the likelihood function is given by equation (14.6). In principle, any sequence of actions can be explained by many (an infinite number, in fact) different combinations of costs and rewards. We revisit this in section 14.3, where we show how spatial information, priors over costs and rewards, and access to multiple observations constrain these inferences and make the problem tractable.

14.2 Representing and Inferring Beliefs

Many situations involve reasoning about agents acting under incomplete or incorrect knowledge, and interpreting their behavior involves inferring what they know or believe.

14.2.1 Beliefs About Costs and Rewards

Continuing with the dessert-selection example (whether to order cake or ice cream), suppose that your friend isn't sure about how much they will like each option. In this case, we cannot represent an agent as having a single reward associated with each dessert. Instead, we can capture their uncertainty about their rewards using probability distributions.

To illustrate how, we'll begin by assuming that the range of possible rewards is finite, falling in the $[0, 1]$ range (although the framework can trivially be extended to infinite reward ranges). In this range of rewards, a wide range of possible beliefs can be represented through beta distributions (see chapter 3; the logic presented here can be applied to any parameterized probability distribution). Because the shape of beta distributions are entirely determined by two parameters—α and β—inferring your friend's beliefs becomes equivalent to inferring these two parameters. Formally, if $b_{ic} = \{\alpha_{ic}, \beta_{ic}\}$ and $b_c = \{\alpha_c, \beta_c\}$ represent your friend's beliefs about how much they will enjoy ice cream (b_{ic}) and cake (b_c), respectively, the posterior over their beliefs about their rewards is given by Bayes' rule:

$$p(b_{ic}, b_c|a) \propto p(a|b_{ic}, b_c)p(b_{ic}, b_c). \tag{14.8}$$

To set the prior distribution $p(b_{ic}, b_c)$, we can assume that your friend's belief about how much they'll enjoy ice cream is independent of their belief about how much they'll enjoy cake, so $p(b_{ic}, b_c) = p(b_{ic})p(b_c)$. Note that because b_{ic} and b_c represent probability distributions (although each distribution technically consists of two parameters, α and β), their priors consist of a mapping that assigns a probability to each possible probability distribution. That is, these priors capture our belief that your friend might have different kinds of beliefs about their rewards (e.g., we might assign a low prior probability to distributions that reflect a belief that the desserts have a low reward, and a higher prior probability to distributions that reflect a belief that the desserts have a high reward). Each of these prior distributions can be set by assigning a prior distribution to parameters α and β. Because these two parameters can take any value in range $(0, \infty)$, the prior can be represented through any probability distribution defined over positive real numbers, such as an exponential or a gamma distribution (see chapter 3).

The likelihood can then be computed by integrating over the possible rewards that your friend expects to obtain:

$$p(a|b_{ic}, b_c) = \int_{R_{ic}=0}^{1} \int_{R_c=0}^{1} p(a|R_{ic}, R_c)p(R_{ic}, R_c|b_{ic}, b_c)\, dR_{ic}\, dR_c, \tag{14.9}$$

where $p(a|R_{ic}, R_c)$ is the probability that your friend would take observed action a if the rewards for ice cream and cake were R_{ic} and R_c (computed using a softmax choice model from equation (14.4)), and $p(R_{ic}, R_c|b_{ic}, b_c)$ is your friend's belief that each dessert will yield these rewards:

$$p(R_{ic}, R_c|b_{ic}, b_c) = \text{Beta}(R_{ic}; b_{ic})\text{Beta}(R_c; b_c). \tag{14.10}$$

In the generalized case with m options, computing the posterior distribution becomes prohibitively expensive but can be solved through sampling-based methods (chapter 6). This framework can be easily extended to include inferences over beliefs about costs (see Jara-Ettinger, Floyd, Tenenbaum, & Schulz, 2017, for an applied model that shows how young children's mental-state inferences about others can be explained by a model that accounts for the possibility that agents can be uncertain about their own costs and rewards).

14.2.2 Uncertainty over States of the World and World Dynamics

So far, we have assumed that each action is deterministically associated with a corresponding state (e.g., the action "Order cake" leads to the state "cake"). This is rarely the case, as people's actions can fail to have their intended consequence. If people consider the chance that their actions will be successful when deciding what to do, then our inferences about their behavior must account for this. To achieve this, we can introduce an uncertainty model through a transition function $T : \mathcal{S} \times \mathcal{A} \times \mathcal{S} \rightarrow [0, 1]$, where $T(s, a, s')$ is the agent's belief that taking action a in state s will change the state to s'. Notice that this transition function can express either the true probabilistic structure of the environment or simply the agent's uncertainty about the structure of the environment.

Returning to our dessert example, suppose that your friend notices that the waiter is extremely busy and might forget their order. Because the cake takes more time to prepare, choosing it increases the chance that the waiter will forget their order. If the waiter has a 20 percent chance of forgetting to bring ice cream and a 40 percent chance of forgetting to bring cake, we can represent your friend's expectations through the following transition function T:

$$a = \text{ice cream} \qquad\qquad a = \text{cake}$$

$$\begin{array}{r} \\ \emptyset \\ \text{cake} \\ \text{ice cream} \end{array} \begin{array}{c} \begin{array}{ccc} \emptyset & \text{cake} & \text{ice cream} \end{array} \\ \begin{pmatrix} 0.2 & 0 & 0.8 \\ 0 & 0.2 & 0.8 \\ 0 & 0 & 1 \end{pmatrix} \end{array} \qquad \begin{array}{r} \\ \emptyset \\ \text{cake} \\ \text{ice cream} \end{array} \begin{array}{c} \begin{array}{ccc} \emptyset & \text{cake} & \text{ice cream} \end{array} \\ \begin{pmatrix} 0.4 & 0.6 & 0 \\ 0 & 1 & 0 \\ 0 & 0.6 & 0.4 \end{pmatrix} \end{array}$$

In each matrix, entry (i,j) shows the probability of switching from state i to state j when taking the action "order ice cream" (left matrix) and "order cake" (right matrix). For instance, the first row of the first matrix indicates that if your friend doesn't have dessert yet ($s = \emptyset$) and orders ice cream, there is a 20 percent chance that they will not get any dessert, a 0 percent chance that they will get cake, and an 80 percent chance that they will get ice cream. Similarly, the second row indicates that your friend has cake and asks to exchange it for ice cream, there is a 0 percent chance that they will be left without dessert (as they already have cake), a 20 percent chance that the waiter will forget and they will be left with cake, and an 80 percent chance that the waiter will replace the cake with ice cream.

Under this world model, your friend might order cake because they do not want to risk being left without any dessert. To formalize this intuition, we can assume that other people act under an *expected* utility function, given by

$$U(a, s) = \left(\sum_{s' \in S} R(s')T(s, a, s') \right) - C(a). \tag{14.11}$$

Equation (14.11) is a simple extension of equation (14.5), replacing the determinstic reward for an expected reward and calculated by integrating the uncertainty about the outcome that

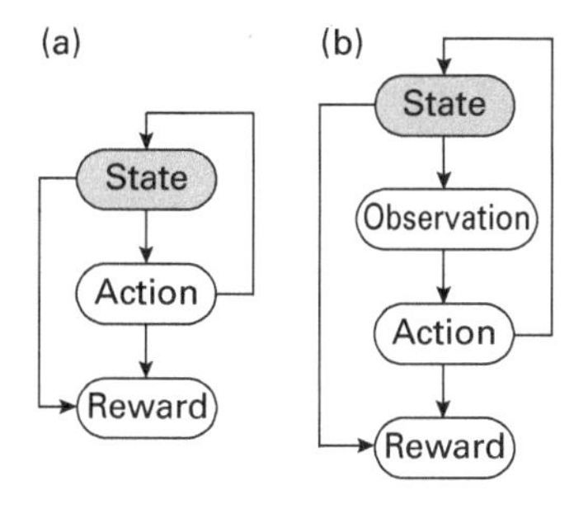

Figure 14.4
Formal models of sequential decision-making. (a) Simple schematic of an MDP. (b) Schematic of a partially observable MDP. States of the world are unobservable, but produce an observation, which guides how the agent acts (by updating their model of the world).

the chosen action might produce. Using this new expected utility function, we can use the same likelihood function expressed in equation (14.6) to infer agents' preference under a probabilistic environment.

14.3 Action Understanding in Space and Time

So far, we have focused on simple situations where agents make a single choice to reach a single outcome. In more realistic situations, action understanding involves interpreting agents that navigate in space over extended periods of time. To infer other people's beliefs and preferences in this situation, we need a model of decision-making that captures sequential planning rather than one-shot decision-making. The framework of *Markov decision processes* (MDPs), introduced in chapter 7 (and illustrated in figure 14.4a), achieves this goal (Sutton & Barto, 2018). For simplicity, we focus on domains where the agents' task is to navigate and explore in a two-dimensional (2D) grid world (much like watching an agent from a bird's-eye view), as actions in these domains are sufficiently rich for people to infer beliefs, desires, emotions, and even social relations (Heider & Simmel, 1944).

MDPs represent the world as a state s from the set $\mathcal{S}$ of all possible world states. For instance, in the map shown in figure 14.5, the world has 20 possible world states, where each state captures the agent's position in space (the world is a 5×5 grid world, but 5 positions are occupied by walls). In each state, the agent can take an action a from a set of possible actions $\mathcal{A}$, such as $\mathcal{A} = \{$move north, move south, move east, move west, eat$\}$.

When the agent takes an action, the state of the world changes as determined by the transition function $T : \mathcal{S} \times \mathcal{A} \times \mathcal{S} \rightarrow [0, 1]$. For simplicity, suppose that agents' actions change the state of the world deterministically (although this assumption can be easily relaxed; see section 14.2.2): The agent successfully moves in the intended direction unless they attempt to cross a map border or a wall (in which case they remain in the same state), and the action "eat" does not change the state of the world (but can produce a reward, depending on the state in which it's executed).

The state space, action space, and transition function specify how an agent can act in the world. Next, to capture how an agent chooses to behave, we can use a utility function $U(s, a) = R(s, a) - C(s, a)$. Notice that costs and rewards now depend both on the state and the action, which increases the expressiveness relative to the simpler utility formulation from

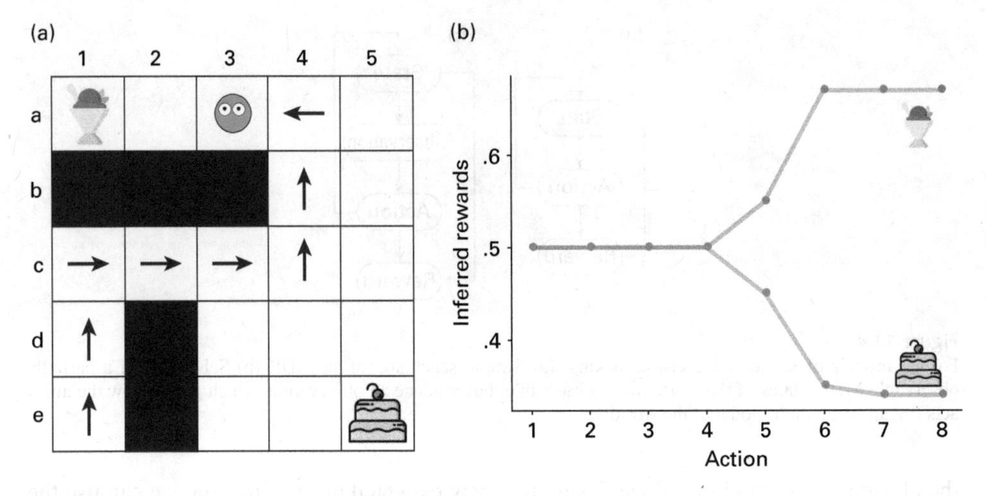

Figure 14.5
A simple example of how action sequences can be used to infer reward functions. (a) Example grid-world where an agent can move in the four cardinal directions and consume the food items. (b) Inferred rewards as a function of the observed actions shown in panel a.

equation (14.5). In the context of spatial navigation, for instance, the reward can depend on taking the right action ("eat") in the right state (one where the agent has direct access to food), and costs might be different as a function of the chosen action ("eat" might have a different cost than the four physical movements) and the state (e.g., attempting to cross a wall could be set to incur a cost of 0 because the agent fails to move in any direction).

In chapter 7, we saw how to compute an MDP's optimal policy—a function that associates each state of the world with the action that will guarantee that the agent maximizes its utilities in the long run. This is achieved by choosing actions with the highest optimal value $V^*_U(s, a)$, given by

$$V^*_U(s,a) = U(s,a) + \max_{a' \in A} \lambda \sum_{s' \in S} T(s,a,s') V^*_U(s',a'), \tag{14.12}$$

where $V^*_U(s, a)$ expresses the immediate utility $U(s, a)$ that the agent obtains by taking action a in state s, plus the expected value obtained if the agent continues to take actions that maximize this value function, weighted by a future-discount parameter $\lambda \in (0, 1)$. This parameter intuitively captures the idea that agents are guaranteed to obtain immediate rewards, but future expected rewards may never materialize due to unexpected events or unaccounted changes in the world.

In classical MDPs, the optimal policy is built by maximizing the value function. Thus, the optimal policy $\pi : S \to \mathcal{A}$ associates each state with the action that maximizes the value function $V^*_U(s, a)$. From an action understanding standpoint, however, we need to account for planning errors. Agents can occasionally make suboptimal choices due to mistakes or accidents. This can be accounted for by softmaxing an MDP's value function to obtain a probabilistic policy, such that

$$\pi_U(a|s) \propto \exp(\beta V^*_U(s,a)). \tag{14.13}$$

Note that in MDPs, agents' actions depend only on the current state. Therefore, given an observed trajectory $t = (\mathbf{s}, \mathbf{a})$—an ordered sequence of $|t|$ pairs of states $\mathbf{s}$ and actions $\mathbf{a}$—the probability that the action would take each action in the corresponding state is given by

$$p(\mathbf{a}|U, \mathbf{s}) = \prod_{i=1}^{|t|} \pi_U(a_i|s_i). \tag{14.14}$$

Thus, given an observed trajectory $t = (\mathbf{s}, \mathbf{a})$, the posterior distribution over utility functions is given by

$$p(U|t) \propto \left[\prod_{i=1}^{|t|} \pi_U(a_i|s_i) \right] p(U). \tag{14.15}$$

Figure 14.5b shows an example inference of this model. Here, an agent begins on the bottom-left part of a grid world and can move in any cardinal direction. The map contains two walls—one spanning $(b, 1-3)$ and a second one spanning $(d-e, 2)$—and two sources of rewards, ice cream on state $(a, 1)$ and cake on state $(e, 5)$. For simplicity, we can assume that all actions, regardless of the state they're executed in, incur a cost of 1 (formally, $C(a, s) = 1, \forall(a, s) \in A \times S$), and rewards are always 0, except when the action "eat" is taken in positions $(a, 1)$ or $(e, 5)$.

Figure 14.5 shows the inferred reward associated with eating cake and ice cream as a function of how many actions have been observed. The first four actions do not reveal the agent's rewards because the agent would behave the same way regardless of their preference. The fifth action—moving east—is still consistent with pursuing the cake or the ice cream. Yet the model begins to infer that the agent prefers ice cream. This is because in state $(c, 3)$, an agent with a reward function where $R(\text{ice cream}) > R(\text{cake})$ will always take the action "move east." By contrast, an agent with a reward function where $R(\text{ice cream}) < R(\text{cake})$ should be equally likely to "move east" or "move south." Finally, when the agent takes their sixth action—move north—the action is probable only when $R(\text{ice cream}) > R(\text{cake})$, allowing the model to infer a preference for ice cream.

This approach to modeling human goal inference has been validated experimentally by several behavioral studies (Baker, Saxe, & Tenenbaum, 2009), which presented scenarios like those in figure 14.6. Participants were asked to make goal inferences at several points along the paths of agents navigating around obstacles toward one of several marked locations (see caption for details). The conditions in figures 14.6(a), (b) and (c) show the same agent path, but they differ in the presence of a gap in the obstacle (a), or the location of one goal object (c). These slight differences in the environment have large effects on people's goal inferences: In figure 14.6(a), after just three steps, people immediately infer that goal A is much more likely than B or C. Figures 14.6(b) and (c) increase the ambiguity, with goals A and B assigned similar probability in (b), and goals A, B, and C rated similarly in (c) until the agent approaches goal A after 11 steps.

The same framework be easily extended to break down inferred utility functions into the underlying costs and rewards. To achieve this, it is only necessary to treat the cost function as unobservable and variable across agents and terrain types, using equation (14.7) with an MDP as the generative model. Consider the event shown in figure 14.7. If we assume that actions in the same terrain must have the same cost, then cost inferences reduce to

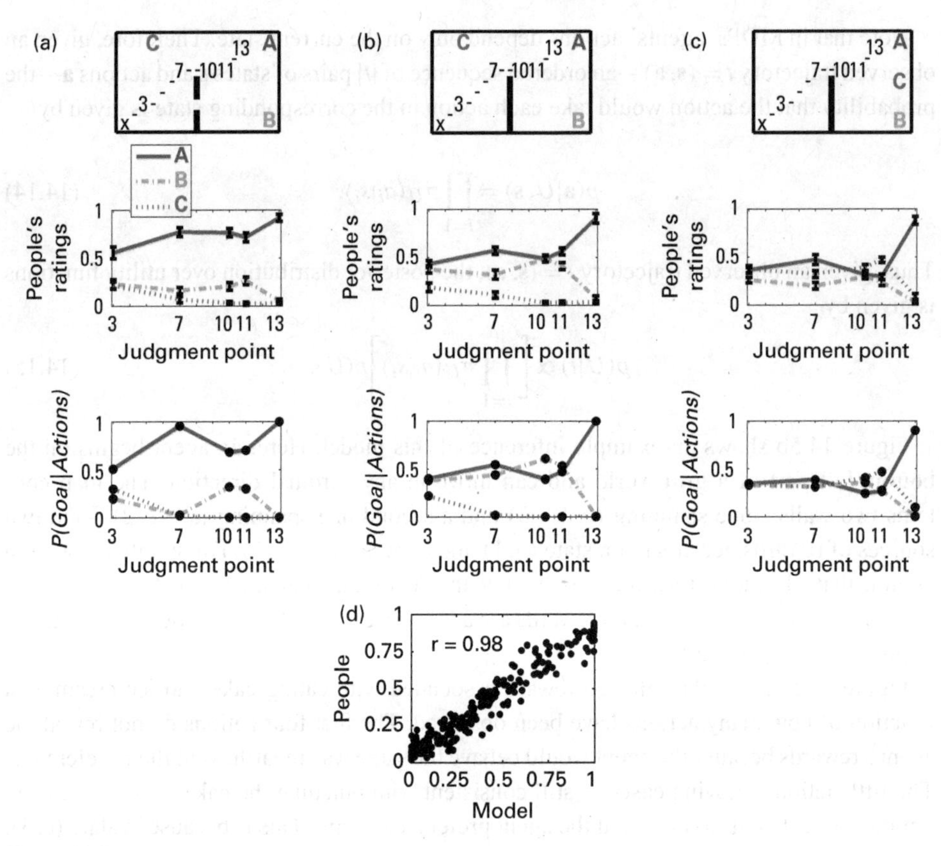

Figure 14.6
Behavioral experiment comparing human goal inferences with the predictions of a Bayesian model based on the principle of rationality (Baker et al., 2009). (a)–(c) Comparing human and model inferences in three conditions of the experiment. The top row shows a map with walls (in black), three goals (labeled A, B, and C), and the agent's trajectory. The agent's starting point is marked with an X and the numbers indicated time points where participants were asked to rate the probability that the agent was pursuing each of the three goals. The second row shows people's ratings for each goal as a function of time point in the path, and the third row shows the model predictions. (d) Quantitative comparison of human and model inferences across all conditions of the experiment.

inferring two cost values and two reward values using Bayesian inference, with the likelihood term computed through equation (14.14). Figure 14.7 shows the inferred expected costs and rewards as a function of the observed actions.

14.3.1 Distinguishing Between Decision-Making and Action Planning

MDPs provide a normative solution when multiple sources of costs and rewards are in play. But this formulation implicitly blends decision-making (which rewards will the agent attempt to collect?) with action planning (what actions does the agent need to take to collect them?). Yet action planning is a hierarchical process, where agents must first choose a goal (decision-making) and then take the actions to pursue it (action planning). Blending these two processes into a single computation (equation (14.13)) also limits us in our ability to distinguish between suboptimal choices and suboptimal action planning: Softmax models with high noise (when β is low) assume that agents make poor choices and take poor actions, while softmax models with low noise (when β is high) assume

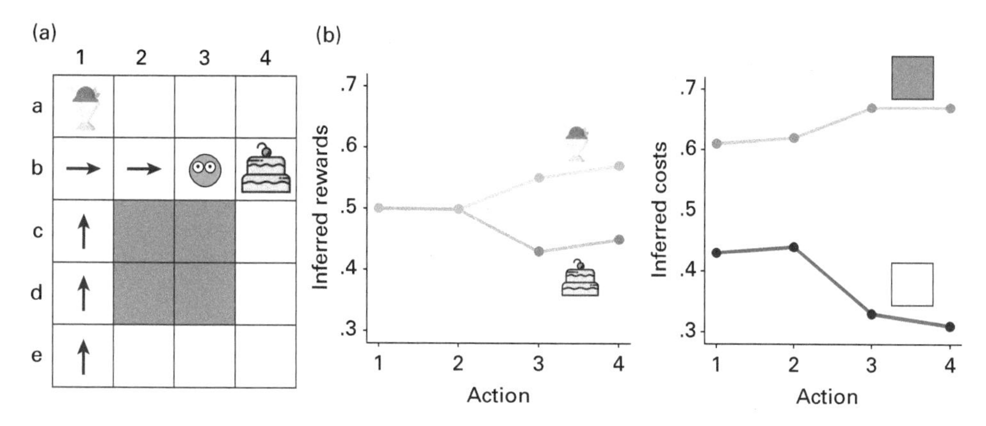

Figure 14.7
Inferring costs as well as rewards. (a) Example grid-world where it is possible to jointly infer the agent's costs associated with walking through each terrain and the agent's reward associated with each food item. (b) Joint inferences over costs and rewards as a function of time.

that agents make optimal choices and take optimal actions. Yet, intuitively, these processes are dissociable and might be subject to different degrees of suboptimality (see figure 14.8).

To distinguish between decision-making and action planning, we can build a hierarchical model where a utility-based model identifies which goals to pursue (described in equations (14.5)–(14.7); section 14.1), and an MDP computes the action plan that will fulfill each goal. More formally, consider an event with n sources of reward, each in a different position in space (i.e., a physical environment with n objects scattered around). Let $\mathcal{G}$ be the set of possible goals—defined as every state where at least one action can yield a positive reward (i.e., state s is a possible goal if $R(s,a) > 0$ for some action a)—and U_g the utility that the agent obtains when pursuing goal g (as determined by equation (14.5)). The agent's probability of selecting goal g, then, is given by

$$p(g|U) \propto \exp(\beta_D U_g), \tag{14.16}$$

where β_D is the softmax parameter that regulates the agent's ability to select the highest-utility goal. The utility U_g associated with each goal is given by the reward associated with the final state minus the expected cost for reaching it. Note that this implies that each goal's cost will depend on the agent's initial position and the goal's location. We can calculate this cost by solving an MDP to maximize the goal-specific reward function R_g, defined as

$$R_g(s,a) = \begin{cases} R(s,a), & \text{if } s = g \\ 0, & \text{otherwise.} \end{cases} \tag{14.17}$$

That is, this goal-specific reward function sets all rewards to 0 for any possible state in the environment, with the exception of the state identified in goal g. This reward function R_g, combined with the model's general cost function that determines the cost of different actions, allows the MDP to generate an action plan that can be used to estimate the cost of fulfilling goal g. Critically, building the probabilistic action policy π_g also involved softmaxing the

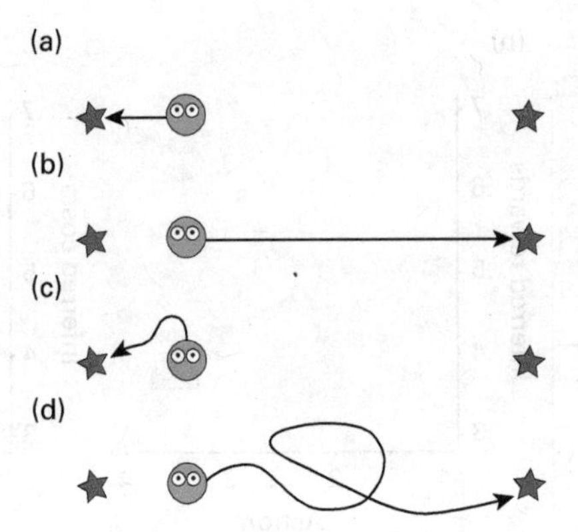

Figure 14.8
Four types of actions that distinguish between rational choice and rational action. (a) Rational choice and rational action; (b) irrational choice and rational action; (c) rational choice and irrational action; (d) irrational choice and irrational action.

action plan (equation (14.13)), which can be done using a separate softmax parameter β_A that captures the agent's ability to navigate efficiently toward its goals.

Note that because the MDP policies are probabilistic, the exact cost will depend on whether the agent makes any errors during planning. Therefore, the utility associated with each goal U_g uses the expected cost

$$C_g = \sum_{t \in \mathcal{T}} C(t)p(t), \tag{14.18}$$

where $\mathcal{T}$ is the set of all possible trajectories $t = (\mathbf{a}, \mathbf{s})$ that reach goal g from the agent's starting point. $C(t)$ is the trajectory's cost, given by

$$C(t) = \sum_{i=1}^{|t|} C(a_i, s_i), \tag{14.19}$$

and $p(t)$ is the probability that trajectory t happens, given by

$$p(t) = \prod_{i=1}^{|t|} \pi_g(a_i|s_i)T(s_i, a_i, s_{i+1}), \tag{14.20}$$

where $\pi_g(a_i|s_i)$ is the probability that the agent takes action a_i in state s_i and $T(s_i, a_i, s_{i+1})$ is the probability that this action in that state transitions the world to the next state in the trajectory. Naturally, considering every possible trajectory is intractable, but equation (14.20) can be approximated through sampling-based methods.

From an action understanding standpoint, inference over this model requires integrating over the unobservable goal that the agent selected and is acting to achieve. Thus, given an observed trajectory $t = (\mathbf{a}, \mathbf{s})$, the likelihood is given by

$$p(t|U) = \sum_{g \in G} p(t|g)p(g|U), \tag{14.21}$$

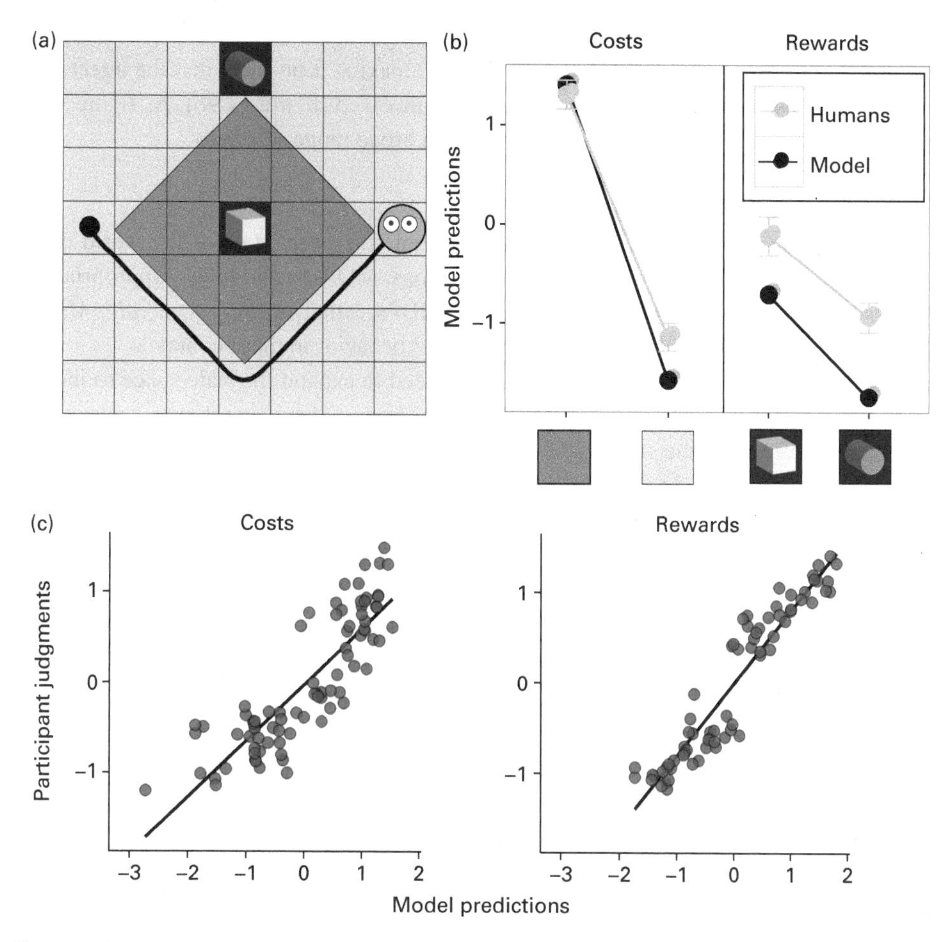

Figure 14.9
Results from joint inferences over costs and rewards. (a) Example scenario: An agent navigates from a starting location to a target location. The map contains different types of terrains and objects that the agent can collect if they desire. (b) People's inferences about the underlying costs and rewards, and model predictions for the example in (a). (c) Overall comparison between human joint cost-reward inferences and model inferences.

where $p(t|g)$ is the action-planning model given a target goal (equation (14.13) with reward function (14.17)), and $p(g|U)$ is the decision-making model given the set of rewards (equation (14.16)). Finally, this likelihood, weighted by a prior over utility functions $p(U)$, yields the posterior distribution over agents' underlying utilities.

Jara-Ettinger, Schulz, and Tenenbaum (2020) showed that this model captures people's capacity to jointly infer other people's costs and rewards based on how they act, using scenarios like the one shown in figure 14.9a. Here, an agent must travel from a starting point (middle left) to a target location (middle right), but it has the option of collecting one of two objects on the way (a white cube or an orange cylinder). The agent's path immediately reveals that navigating through the blue terrain is less costly than navigating through the purple terrain (otherwise, why take the longer path?). Both the model and the people infer that the agent does not like the orange container because the agent could have gotten it by taking an equally costly path. By contrast, both the model and the people are more uncertain about the white box. Although the agent chose not to get that box, it's also in the middle of a region that we inferred was costly to go through, making it possible that the agent liked the box but chose not to pursue it due to the costs involved (figure 14.9b). As figure 14.9c shows, this model captures people's inferences in a broad range of events.

14.3.2 Planning Under Uncertainty

Agents often face situations where they do not know the exact state of the world or the position of the various target goals. In these situations, MDPs are no longer appropriate, as they assume perfect information. An extension of MDPs, called *partially observable Markov decision processes* (POMDPs), helps model agents' behavior in these contexts.

To model planning under uncertainty, we first need to expand the state space to include world states that the agent may consider plausible, even if they cannot ever occur. In the example in figure 14.5, the state space consisted of 20 possible states, each capturing the agent's position in space. To model an agent who may not know whether the ice cream is in the top-left position or in the bottom-right position, we would need to expand the state space to 40 states, with each state capturing the agent's position in space and the position of the ice cream and the cake. In this extended state space, the world is always in one of the original 20 states (where the ice cream is always on the top left), but the agent may believe that they are in a world state that does not match reality (e.g., believing that the cake is in the top-left position). Under this expanded state space, we can define an agent's beliefs $B: \mathcal{S} \to [0, 1]$ as a probability distribution over states of the world.

When agents act under their beliefs about the state of the world (instead of acting based on the true state of the world), their policy must now map beliefs onto actions (rather than states onto actions). Formally, we require a policy $\pi_U: \mathcal{B} \times \mathcal{A} \to [0, 1]$, such that $\int_{a \in A} \pi_U(b, a) = 1$ for any belief b and any utility function U (i.e., for any belief, the probabilities over all actions must add up to 1, ensuring that they express a proper probability distribution).

In addition, we need to specify how agents' beliefs change as they interact with the world. To achieve this, POMDPs assume that each combination of states and actions produces observations about the world that the agent can use to make inferences about the true world state. For instance, an agent walking into a room may receive an observation that reveals what is inside.

Formally, let Ω be the set of possible observations that the agent can receive (i.e., the full set of all possible information that the agent could get as they interact with the world). In a POMDP, the agent receives one observation $o \in \Omega$ at each time step, determined by an observation function $O: \mathcal{A} \times \mathcal{S} \times \Omega \to [0, 1]$, where $O(a, s, o)$ is the probability of getting observation o in state s after taking action a. As the agent receives observations, they update their beliefs through

$$p(s|a, o; b) = \sum_{s_o \in S} T(s_o, a, s) O(a, s, o) b(s_o). \tag{14.22}$$

The left side represents the agent's belief that they're in state s after taking action a and receiving observation o, given that they previously had beliefs b. The right side computes this term by considering all the possible states that the agent might have previously been in ($s_o \in S$). For each potential previous state, $b(s_o)$ is the agent's belief that they were in such a state, $T(s_o, a, s)$ is the probability that action a in state s_o would transition to state

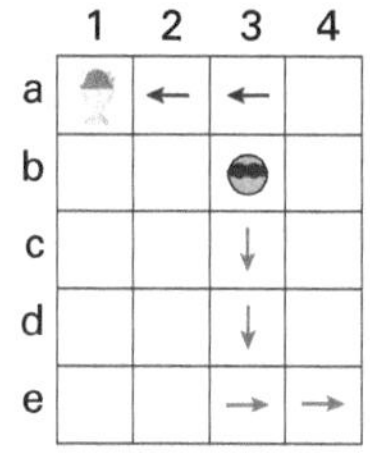

Figure 14.10
Example of how an agent with partial knowledge about the world can take actions to infer its location and obtain rewards. Here, an agent wearing a blindfold is initially unsure about their position in space. If they can detect walls, then the agent can infer their location in space by reaching the northwest (black arrows) or southeast (red arrows) corner. Reaching the northwest corner, however, is a better strategy because it also guarantees that the agent will reach the goal state.

s, and $O(a, s, o)$ is the probability of receiving observation o after taking action a to reach state s.

To illustrate the dynamics in a POMDP, consider an agent with a blindfold moving in a 5×4 grid world (figure 14.10). The state space S consists of 20 states, each indicating the agent's position in space. Suppose that the agent cannot see anything, but they can feel the walls whenever they try to cross one (and hence hit it). The space of observations is then $\Omega = \{\emptyset, \text{wall}\}$. Any action-state pair within the map would produce observation $\emptyset$, and any action in a state where the agent hits a wall produces an observation "wall." An agent with complete uncertainty ($b(s) = 1/16$ for all $s \in S$) can determine their position in space by moving south until hitting the observation "wall" (at which point they will know that they must be in one of the bottom four states) and then moving east until hitting the observation "wall" again (the red path in figure 14.10). At this point, the agent will know that they must be in the bottom-right corner. If actions change the state of the world in a deterministic way, the agent will then know the exact state of the world at every time point by simply tracking which actions they took, enabling them to navigate toward the reward. A better strategy, however, would be to move north until hitting the wall and then move west until hitting a wall again (black path in figure 14.10). This is because the top-left corner not only reveals the state of the world, but also leaves the agent in the state that has a reward. Solutions to POMDPs naturally produce policies that combine actions in the service of reducing uncertainty with actions in the service of obtaining rewards, making them a natural framework for understanding agents whose behavior is a combination of exploration and exploitation.

Under partial knowledge, we can define the utility (from the subjective point of view of the agent) of taking action a under beliefs b as

$$U(b, a) = \sum_{s \in S} b(s) U(s, a) \tag{14.23}$$

and the optimal value of belief b as

$$V_U^*(b) = \max_{a \in A} \left(U(b, a) + \lambda \sum_{o \in \Omega} O(s, a, o) V_U^*(b') \right). \tag{14.24}$$

Equation (14.24) is equivalent to equation (14.12), extended to account for the agent's uncertainty. Here, the optimal value of belief b is calculated by considering the action that yields the highest utility, plus the expected future value (discounted in time by parameter λ) by integrating over the possible information that the agent might receive, and the agent's updated beliefs $b'(a, o, b)$ after taking action a, receiving observation o with initial beliefs b. These updated beliefs are calculated through equation (14.22).

As in MDPs, we can build a probabilistic policy by softmaxing this value function (equation (14.24)). Using the resulting policy as the generative model, joint belief (the probability distribution over states), desire (the latent reward function), and competence (the underlying cost function) can be inferred through Bayesian inference, where given an observed trajectory t,

$$p(B, R, C|t) \propto p(t|B, R, C)p(B)p(R)p(C). \tag{14.25}$$

Baker et al. (2017) developed and tested the model that we have just presented, asking adult participants to make joint inferences about agents' beliefs and desires based on how they navigated an environment. Figure 14.11 shows several scenarios from this experiment, in which a hungry graduate student leaves their office to walk to lunch at one of three food trucks: Korean (K), Lebanese (L), or Mexican (M). There are two parking spots for the trucks (marked in yellow), and trucks can park in different spots on different days, or not show up at all, so the student may not know where each truck is parked and must plan carefully where to walk to get lunch from the best truck available as quickly as possible. Using a POMDP for a generative model, the agent's desires can be captured using a reward function that represents their preferences over trucks, and the agent's initial beliefs can be represented as a probability distribution over each of three partially observable world states: the Northeast parking spot being occupied by (1) Lebanese (L) or (2) Mexican (M), or (3) being empty (N for none). Finally, observations of the trucks are determined by line of sight, with a small probability of observation failure.

Consider figure 14.11c, in which the student can initially see the Korean truck in the Southwest parking spot but cannot see the Lebanese truck parked in the Northeast parking spot due to the building blocking the view. The student walks past the Korean truck, continues walking around the building, sees that the Lebanese truck is in the Northeast spot, and then turns and walks back to eat at the Korean truck. Based on this information, which truck did they want the most? And which truck did they believe was in the Northeast spot? Participants here infer that the student wanted Mexican food most and Lebanese food least. Participants also attribute an optimistic initial belief that the Mexican truck was in the Northeast spot.

The model captures people's inferences that Mexican is most preferred, and Lebanese least, and people's attribution of a false belief that the Mexican truck was present. The POMDP model naturally explains the action of going around the building to check which truck is in the Northeast spot as rational exploration: seeking information here is rational if either Mexican or Lebanese have greater reward than Korean, and if the prior belief that Lebanese or Mexican could be in the Northeast spot is nonzero. Figure 14.11b shows a case where the student paused after going around the building. Here, both people and the model attribute greater belief and desire for Lebanese and Mexican (the slightly greater values for Mexican over Lebanese are due to perceiving the "pause" in the trajectory

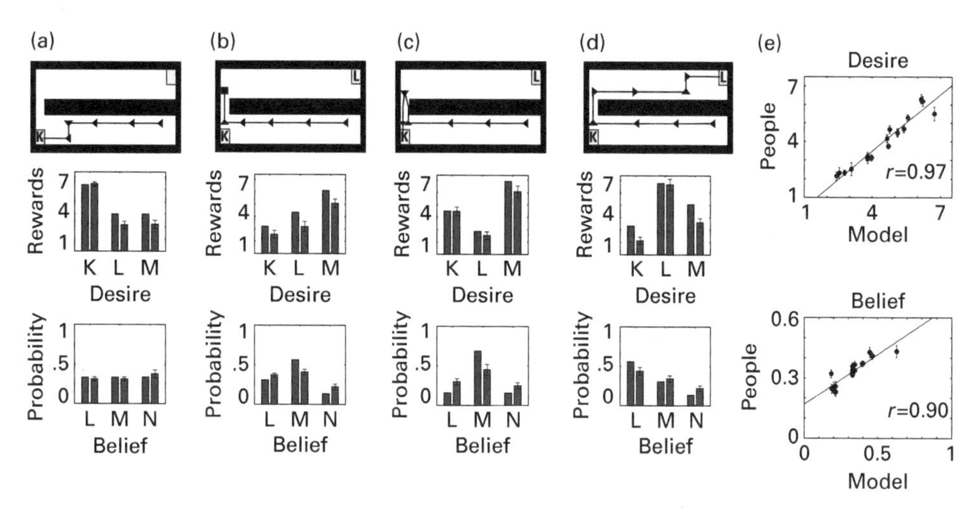

Figure 14.11
Experimental validation of Bayesian inference over beliefs and desires using a POMDP as a generative model. The experiment presented scenarios in which a hungry student seeks lunch from one of several food trucks: Korean (K), Lebanese (L), or Mexican (M). People made different belief and desire attributions in each scenario, which the model captures with high accuracy. (a)–(d) Comparing participant and model inferences across a range of experimental scenarios. In each panel, the top row shows the stimuli, the second row shows inferred rewards for each of the three food trucks, and the final row shows inferences about what the student expected would be on the top-right spot (with N corresponding to the belief that there was nothing there). (e) Correlation between participant judgments and model inferences.

animation as hesitation on seeing Lebanese, which is interpreted as a preference for Mexican).

Once the agent turns back after observing Lebanese in the Northeast spot (figure 14.11c), only reward functions where Mexican is preferred to Korean, and both Korean and Mexican are preferred to Lebanese, can explain the entire trajectory, and these are inferred by the model. The model's false belief inference stems from the fact that hypotheses that assign low prior probability to Mexican being in the Northeast spot are not consistent with the observed behavior; for these hypotheses, the rational action is to go straight to the Korean truck without seeking to observe the Northeast spot first. Figure 14.11 shows more scenarios in which the model captures people's judgments, and figure 14.11e shows correlations between average human desire and belief judgments and model predictions across all scenario types.

14.4 Minds Thinking About Themselves and Other Minds

So far, we have considered how to infer beliefs, costs, and rewards given someone's behavior. However, mental-state inferences are often in the service of other tasks such as learning from others, deciding if someone is nice or mean, or learning what goals are worth pursuing. In this section, we describe how to use Bayesian Theory of Mind models to solve these problems.

14.4.1 Tracking Knowledge and Learning from Experts

While we have now considered cases where agents have incomplete or incorrect knowledge about the world, all our examples always used situations where the observer (i.e., the reader,

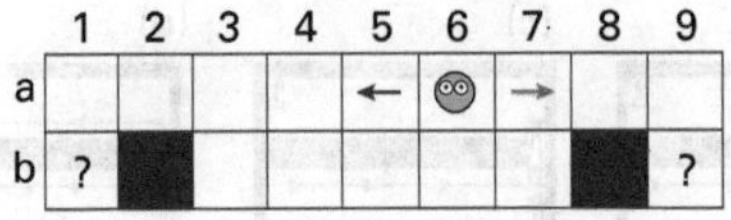

Figure 14.12
Example of how it is possible to learn about the world by watching how other agents act. A reward is located in $(b, 1)$ or $(b, 9)$. If the agent navigates left (gray arrow), the agent must believe that the reward is in state $(b, 1)$. If, instead, the agent navigates right (red arrow), the agent may know that the reward is located in state $(b, 9)$ or may be uncertain and hence check the spot closest to the starting location.

or the participant in the experiment) had perfect information about the world. In everyday social situations, we often have to interpret the actions of agents who may know more about the world than we do.

Consider figure 14.12, which shows a grid world with an object that could be located in state $(b, 1)$ or in state $(b, 9)$. If we assume that the agent knows the object's location, then their actions will reveal this: if the agent moves left, then the reward must be in $(b, 1)$, and if the agent moves right, the reward must be in $(b, 9)$. We can formalize these intuitions by inferring the state of the world based on the agent's behavior, given by $p(s|a) \propto p(a|s)p(s)$. Here, $p(s)$ is the prior probability that the world is in state s (which is known to the agent), and $p(a|s)$ is the likelihood, given by the softmaxed policy that an agent with perfect knowledge (hence modeled as an MDP) would take the observed action a if the true state of the world were s.

In more complex cases, both the observer and the actor may have incomplete knowledge about the world. Continuing with the previous example, suppose that you do not know whether the agent knows the object's location. If you watched the agent navigate to the right, reach state $(a, 9)$ and then retrace their steps, you could immediately make two inferences: (1) the agent did not know where the object was located, and (2) the object must be in position $(b, 1)$. By contrast, if the agent began by taking a step to the left, a single action would already give us some confidence that the object is in position $(b, 1)$ and the agent knew this (otherwise, why would the agent begin searching by incurring a high cost?).

We can model these intuitions as a joint inference over agents' initial beliefs b and the true state of the world s (using a POMDP as the generative model). To achieve this, we must compute the probability that an agent with belief b in world state s would take action a, $p(a|b, s)$, which can be obtained by considering how a rational agent would act under different beliefs in different world states.

Figure 14.13 shows the results from a study testing this capacity in humans (Jara-Ettinger, Baker, & Tenenbaum, 2012). Participants watched an agent navigate a simple environment (figure 14.13a) and they had to infer the location of the different food carts. The world contained three food carts (see figure 14.13b for all arrangements), and the agent preferred A over B and B over C. However, each food truck could be either open or closed, such that the agent couldn't always eat at their favorite truck. In the path in figure 14.13a, for instance, we can infer that there is an open cart in the north position because the agent ultimately went to it. However, this cart cannot be cart A. Otherwise, the agent would have walked straight there rather than checking other pathways. The fact that the agent never checked the west aisle suggests that cart A is also not there; otherwise, the agent would have gone there to check if it was open. Together, this suggests that the agent first saw cart B open in the north

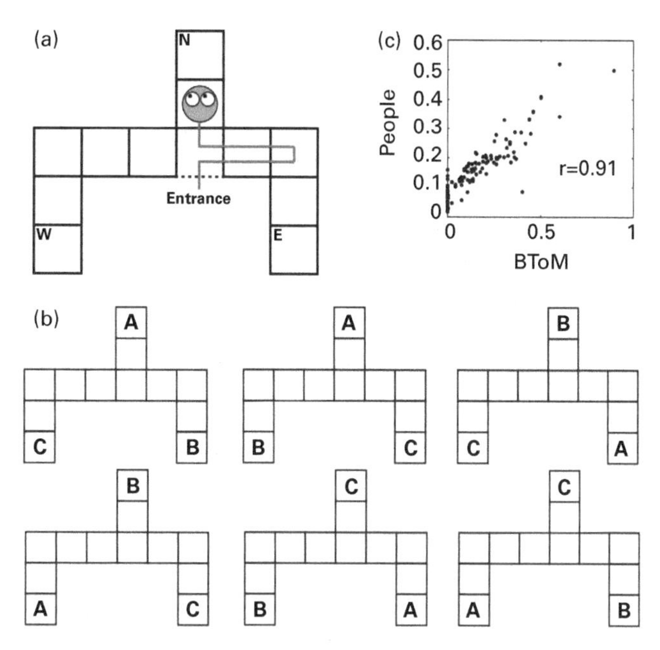

Figure 14.13
A more complex example of learning about the world via another agent's actions. (a) Example trial of experiment 2 in Baker et al. (2017). An agent navigates a set of pathways with three food carts (positioned in N, W, and E) to decide where to eat. The agent always prefers A over B and B over C, but these food carts are sometimes closed. (b) Possible world states. Participants had to rate the probability of each world given how the agent behaved. (c) Overall correlation between our model and participant judgments.

spot, found cart A closed in the east spot, inferred that cart C must be in the west spot, and therefore went back to eat at cart B. Figure 14.13c shows that this computational model tracked participant inferences with quantitative accuracy.

14.4.2 Helping and Hindering

The rewards and costs described so far were associated with concrete world states and actions: Does my friend want cake or ice cream? How much do they cost? But people also care about other people. People's goals, costs, rewards, and intentions can take into account and be directly related to the goals, costs, rewards, and intentions of others. This is true both for our own planning and decision-making, and in our inference over mental states. To capture this kind of reasoning, we need to extend our framework to include multiagent planning and allow beliefs and utilities to operate over other mental variables.

To continue with our running example, if my friend forgot their wallet and wants to get some ice cream, I may order one for them. My preference is not for them to have ice cream, per se, but for my friend to have what they want. That's part of what being a friend is. And if I see a person paying for a person's ice cream, I might reasonably conclude that they are friends. Even young children seem capable of such reasoning (Hamlin, Wynn, & Bloom, 2007; Hamlin & Wynn, 2011; Hamlin, Ullman, Tenenbaum, Goodman, & Baker, 2013). How can we formalize this common-sense understanding in our framework?

Let us consider a simple case in which one agent, A, has a utility function U^A that maps every combination of states and actions into the difference between the costs and the rewards.

Meanwhile a second agent, B, has a utility function U^B that is a function of agent's A utility:

$$U^B(s, a^B) = \rho[U^A(s, a^B)] - c(a^B). \tag{14.26}$$

Here, agent B's utility for taking action a^B in state s is given by a constant $\rho < 0$ multiplied by the expected utility that agent A will receive when that action is taken, minus the cost that agent B incurs to take this action. Note that in many cases, agent B might not know agent A's exact utility, in which case the first term of the right side can be replaced with the expected utility rather than an exact one. Here, ρ is a parameter that controls the direction and degree of the social preference—how much B cares about A. When $\rho > 0$, B will help A by taking actions to achieve states that are favorable to A. When $\rho < 0$, B will hinder A by doing the opposite. Both of these are balanced by the cost: if I'm your friend, I'll buy you ice cream, but I'm not giving you my car. Given an agent with such a utility function, social reasoning about goals can be turned into a straightforward question of inference over the ρ parameter.

As before with joint reasoning about costs and rewards, if we do not know another agent's ρ nor their cost, their specific actions could be explained in different ways. An agent's lack of willingness to help could be explained by the cost being too high or the motivation (ρ) not being high enough. If an agent refuses to help when the cost is known to be low (e.g., the agent is adjacent to the known reward and easily can hand it to the other agent), the model will infer a lower value of ρ than when an agent's costs is high (e.g., if the social agent is more distant to the inferred reward than the first agent, such that even if they tried to help, they would get to the reward long after the first agent did).

The inference problem remains largely the same as that described for most of this chapter: given a set of actions, infer the underlying utility (rewards, costs), beliefs, intentions, and now we add the parameter weighting the utility of others. As before, the likelihood for actions is given by a model of planning. In a social setting, this means a multiagent planning model, with possible uncertainties and partial observability depending on the assumptions about the beliefs of the agents involved.

It is worth pausing here to consider an alternative account of social goal inference, one that does not rely on utility functions that operate over utility functions. Social goal inference has also been cast as a problem of relating perceptual cues (in particular, visual cues) directly to inferences over mental states. Certainly, some actions seem to relate perception directly to mental inference. Slapping a person means that you're not friends with them; so why go through the computationally expensive calculation of inverting a planning procedure to figure that out? If someone asked for ice cream and you gave it to them, you're their friend. Simple. Various accounts have tried to formalize such cue-based social inferences. For example, Barrett, Todd, Miller, and Blythe (2005) related a host of motion patterns directly to the inference of social intentions such as "guarding" or "courting."

On the face of it, cue-based accounts are simpler than reasoning about probabilistic planning. However, they also require learning and training many cues for many situations, and they have trouble with generalizing. Consider that the ρ parameter is not tied to a particular state or action. In a specific situation, helping another person may require getting close to them. Other times, helping may require stepping away. The same action could be interpreted as harmful or helpful depending on the mental states of the agents involved.

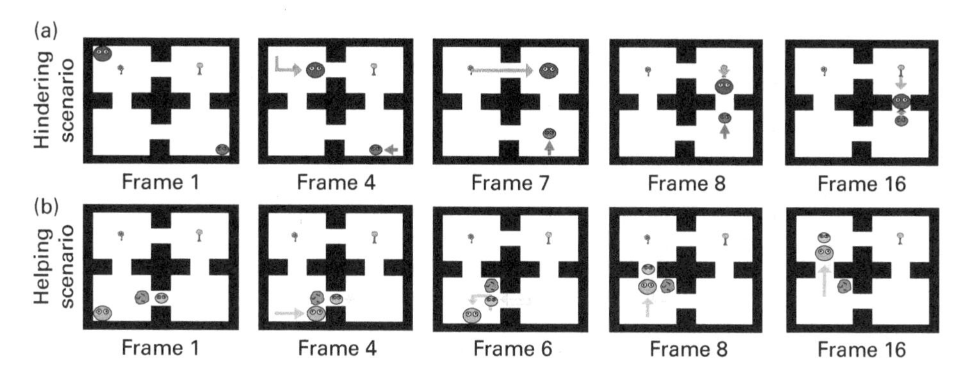

Figure 14.14
Still frames of example behavior sequences used in the helping and hindering domain used in Ullman et al. (2009). Both agents begin in frame 1 and progress in discrete steps, as shown by the colored arrows. The sequences were paused at different probe points, and participants were asked to infer the agents' goals.

To test this idea, Ullman et al. (2009) contrasted an inverse-planning model of social goal inference to a cue-based one, in a domain similar to that used by Baker et al. (2009). In this domain, two agents pursued selfish and social goals in a small maze environment (see figure 14.14). The small agent was always selfish, in that it was trying to get to one of two goal points (a flower or tree). The large agent was either selfish or social. Social agents try to help or hinder the little agent. The small agent occasionally failed in its action, mimicking the behavior of agents in Hamlin et al. (2007). Large agents never failed in their actions and could push small agents around. The environment occasionally included a boulder that only large agents could move. Different scenarios used different initial locations of the goals, agents, and boulder, as well as different goals for the agents.

Participants were asked to judge the goals of the large agent at different time points, and their judgments were compared to a Bayesian Theory of Mind model that included multiagent planning, as well as to a cue-based model that included 10 visual cues from previous applicable cue-based models, including cues such as geodesic distance, changes in distance, and relative movement. As shown in figure 14.15, people's judgments about the large agent's goal included moments of certainty and confusion, rapid switches, and growth in confidence. The cue-based model was able to recover the correct goal when the goal was selfish (using cues such as shrinking distance to the tree or flower), but it was far less accurate about social goals. The cue-based model was also poor at generalizing: when trained on scenarios that involved boulders and tested on nonboulder scenarios or vice versa, its performance suffered greatly. By contrast, the planning-based model was able to recover human judgments with high precision, its accuracy was unaffected by whether the goal was selfish or social, and it performed equally well on boulder and nonboulder scenarios.

While the simple toy-domain shown in figure 14.14 is a useful arena for pitting different accounts against one another, it is also far simpler than real-world situations. As the most basic extension, the situation can be made more complicated when we consider that agents could be balancing social rewards with their own rewards (I have other things going on besides your ice cream), but this is a straightforward extension of equation (14.26). Of course, situations in which multiple agents are reasoning and acting with regard to the social

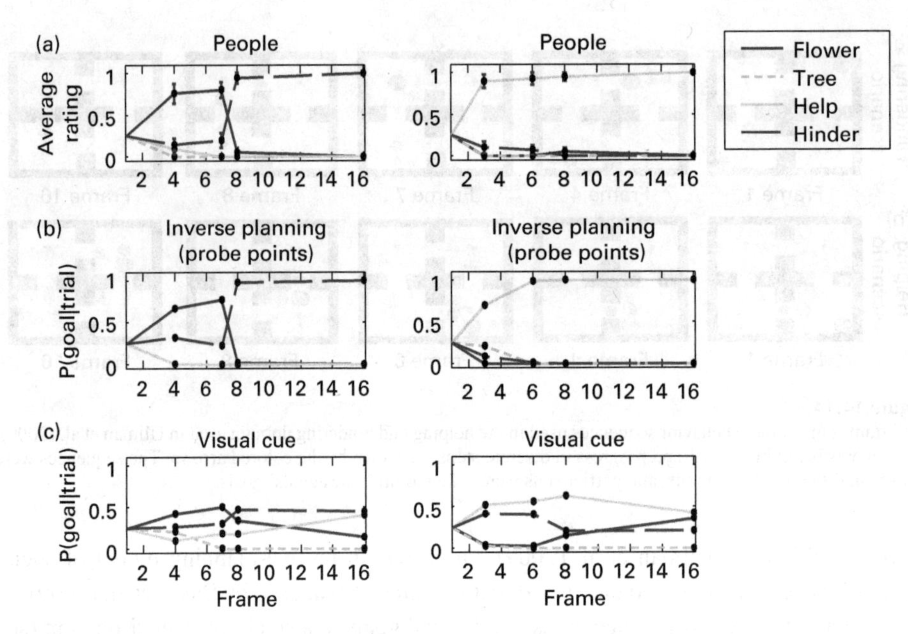

Figure 14.15
Example human judgments compared to model predictions for the task used in Ullman et al. (2009). The first column corresponds to the hindering scenario shown in figure 14.14a, and the second column corresponds to the helping scenario shown in figure 14.14b. (a) Average participant judgements about the large agent's goal at different probe points. The horizontal black lines show standard error bars. (b) Predictions of Bayesian Theory of Mind model at probe points. (c) Predictions of cue-based model used in Ullman et al. (2009).

goals of others (while being uncertain about the rewards and costs of others) are complicated. One could quickly end up with situations in which the model would reason that A might be trying to help B even though A knows that B doesn't want A's help, but A thinks that's only because B believes that A doesn't understand what B's goal is, and if A could convince B, and so on, and so forth.

Another major over-simplification is that this model required fully specifying the small space of possible goals. This led on occasion to odd situations in which people's understanding of social behavior and the planning-based model diverged. For instance, consider a situation in which the small agents struggles and struggles and eventually reaches a flower, while the large agent simply hangs out in a far-off corner. What is its goal? The model used in Ullman et al. (2009) quickly identifies this as "helping."

The logic is the following: If the large agent were trying to get to the flower or tree, it would move toward them. If it were trying to hinder, it would move to hinder. Since it did neither of these, it must be that the large agent knew that by the time it reached the small agent, the small agent would already have reached the goal. Since movement is costly, why bother budging? People clearly did not share this intuition. This does not suggest that the planning-based modeling approach is wrong, but rather that this simple instantiation is too simple. A fuller model would include a richer space of goals, beliefs, and communication (e.g., moving slightly toward the other agent to indicate willingness to help while not committing). But these are all additional cogs in the basic machinery, not a new machine altogether.

14.4.3 Honesty and Reputation

In social situations like the ones we presented here, inferences about helping or hindering were achieved by allowing agents to have desires over other people's desires. The same approach can be used to formalize the idea that agents can also have desires over other people's beliefs.

People's desires over other people's beliefs capture a range of types of social behavior that are common, like inferring that an agent is motivated to share their knowledge or that they want the agent to have an incorrect belief (i.e., lying). When we consider that agents have beliefs about each other, this approach can help model people's reputational concerns. We can achieve this by building a meta reward function over beliefs $\text{MR} : \mathcal{B} \rightarrow \mathbb{R}$, where $\text{MR}(b^B)$ is the reward that an agent gets when agent B's beliefs are b^B, and $\mathcal{B}$ is the space of possible beliefs. In many cases, agents do not care about the full set of beliefs that an agent has, and they only care about a particular dimension. For instance, A might want to ensure that B believes that A is helpful, without caring about B's remaining beliefs about the world or about others. These types of situations can be captured by associating a reward with different critical features of others' beliefs and computing MR as the sum of these features.

Given a personal reward function R_p and a meta reward function MR, we can define the final reward function R_f as a function that maps states of the world and observer beliefs onto rewards through

$$R_f(s, b^A) = R_p(s) + \theta \int_{b^B \in \mathbb{B}} \text{MR}(b) b^A(b^B)\, db^B, \tag{14.27}$$

where ρ is once again a parameter indicating how much the agent cares about their reputation, and $b^A(b^B)$ is agent A's belief that B has belief b^B.

14.4.4 Effort and Motivation

Finally, the models developed so far assumed that the relation between states and actions is given by a transition function where the probability of switching from state s to state s' after taking action a is constant, determined by $T(s, a, s')$. More realistically, agents can exert different degrees of effort that affect the probability that their actions will be successful. We can expand past frameworks by assuming that agents' behavior is a combination of an action and an effort $e \in [0, 1]$ that affects the action's cost and its probability of success. Given an action space $\mathcal{A}$, we can define the effort-based action space $\text{CA} = (\mathcal{A} \times [0, 1])$ and an effort-based cost function $C : \mathcal{A} \times [0, 1] \rightarrow \mathbb{R}$ can be expanded to take effort into account, such that

$$C(a, e) = C(a)^{(c+e)}. \tag{14.28}$$

Here, the cost of an action is expontentiated by a constant value c and the amount of effort e put into the action.

Similarly, the transition function now must account for the amount of effort the agent placed, which can be achieved by building a transition function that takes effort as a parameter and concentrates the distribution of expected outcomes accordingly. The exact nature of this function, however, depends on the event. In some cases, low effort may lead to failure and cause the agent to remain in their previous state; in other cases, low effort may cause unintended negative consequences and leave the agent in a less favorable state than their original one. Note that because the action space and the transition function are continuous, these problems must be solved in a non-discrete framework, such as continuous MDPs (Sutton & Barto, 2018).

14.5 Future Directions

The framework presented in this chapter captures the core computations behind our ability to make sense of other people's behavior. In the remainder of this chapter, we discuss what's missing and how this framework provides a starting point for tackling questions that any complete theory of human social cognition must explain: How can we capture human reasoning where goals are richer or more abstract than simple navigation toward objects? How do we represent and reason about different types of minds? How do these mental-state inferences work in more realistic situations where we see bodies moving in three dimensions rather than points in a 2D world? And how are these computations implemented in the brain?

14.5.1 Inferences over Types of Goals and Types of Minds

Theory of Mind models and tasks have historically focused on a limited class of goals: reaching for objects or navigating toward different positions in space (and helping or preventing others from reaching certain objects or navigating toward a position in space). The space of goals that people pursue, however, is much broader, including purely epistemic goals (learning something for the sake of learning it), communicative goals (moving with the goal of sharing a message, such as when we gesture), and even creative goals like improvising tools for ad hoc tasks.

Intuitively, these complex goals can be expressed as compositions of simpler ones that involve moving in space to elicit rewards that can be obtained from a relatively abstract space (e.g., rewards for obtaining information or rewards for revealing communicative intent). One way to expand these models is by building hierarchical goals where more abstract goals are fulfilled by performing sequences of basic types of actions. Under this view, recognizing abstract goals requires parsing other people's behavior in terms of simple actions, and then having access to a space of abstract goals that can be fulfilled through those actions. Velez-Ginorio, Siegel, Tenenbaum, and Jara-Ettinger (2017) showed how a space of unbounded goals can be formalized within a probabilistic context-free grammar that combines primitive goals to represent richer ones. In this approach, desires are no longer represented as rewards. Instead, they are represented as propositions which, when true, elicit a reward. Inferring an agent's desire is then a process of inferring what type of proposition an agent was attempting to make true.

More broadly, people are not only able to reason about different types of goals, but also about different types of minds. Intuitively, the behavior of a human adult is driven by more complex psychological machinery than the behavior of a toddler, and the psychological machinery behind the behavior of a toddler is incommensurable relative to the psychological machinery of nonhuman mammals. These intuitive differences suggest that the beliefs, desires, and cognitive processes that we attribute to others have different representational scopes (e.g., while we think of humans and hamsters as both capable of having beliefs about the existence of an object, we might intuitively believe that only humans can have beliefs about its historical significance). Even in simple situations where we can assume that all agents have similar beliefs and desires, we nonetheless expect different types of agents to create plans with different degrees of complexity. Recent work has found initial evidence that inferences about types of minds can be expressed as Bayesian inference of a space of planning and decision models of varying complexity (Burger & Jara-Ettinger, 2020). More research is needed about characterizing the variability in the different types of minds that we can represent and formalizing this space to explain how people can jointly infer another agent's type of mind, as well as their mental states.

14.5.2 Mental-State Reasoning from Visual Input

The models presented here operate over simple spaces of actions in 2D displays. A key challenge lies in extending these models to full-body control. Modeling moving bodies rather than dots moving in simple 2D space not only expands the space of situations that these models can reason about, but also provides a more natural framework for understanding additional inferences that people make in social contexts. For instance, as observers, when watching another agent pursue a goal, we may want to estimate the activity's difficulty in deciding whether we should engage in it ourselves. Recent work shows that these problems can be solved by reasoning about motion planning. Given a set of arranged blocks (e.g., blocks sorted by colors or arranged into a tower), Yildirim et al. (2019) showed that difficulty can be estimated by calculating the energy that an agent would have to expend to build the tower (through an ability to simulate full-body control), as well as the risk of the construction, given by the chance that the arrangement might collapse.

A further challenge is to model different degrees of granularity depending on the task at hand. When we reason about agents moving around in a city, we intuitively conceptualize them in a similar way to how the models here handle 2D motion. When reasoning about events in smaller spatio-temporal scales, like an agent moving in a room and interacting with objects, however, we need to model their arm movements to interpret reaching and grasping events. And when reasoning about more fine-grained situations, like an agent manipulating a tool, we require an even more precise model of hand control. Thus, a critical challenge lies in building models of action-understanding that can flexibly switch between levels of granularity depending on the goal that the agent is pursuing.

14.5.3 Neural Substrate of Inverse Planning

Finally, our Bayesian models are best conceptualized at a computational level of analysis, and we know little about their underlying algorithmic implementation. One outstanding challenge is to explain how these models may be implemented in the brain. Broadly, two types of solutions that have been used in similar problems can serve as a starting point.

A first possibility is that complex computations are implemented as a collection of simple heuristic that, combined, approximate normative inferences. If so, cheap rules that exploit contingencies between low-level visual features and the corresponding mental-state inferences may allow people to avoid doing any kind of Bayesian inference, while still obtaining some approximate solution (although note that previous attempts at this have been unsuccessful; see section 14.4.2). One challenge to this idea, however, is that people's mental-state inferences are not only qualitatively but also quantitatively consistent with Bayesian models, and a solution of this type would have to explain how a collection of rules may approximate these models with such quantitative accuracy.

A second possibility is that Bayesian inference is compiled into subsymbolic computations that allow these inferences to be performed in more of a feed-forward fashion than is typical for Bayesian inference algorithms work, similar to how face recognition implements Bayesian computations (Yildirim, Belledonne, Freiwald, & Tenenbaum, 2020). It is also possible that only some aspects of action understanding can be implemented in a feed-forward fashion, perhaps explaining the boundary between the perceptual component of how we detect agents and goals and the more cognitive component of inferring beliefs and desires (Scholl & Tremoulet, 2000).

14.6 Conclusion

In 1944, the psychologists Fritz Heider and Marianne Simmel published "An Experimental Study of Apparent Behavior" (Heider & Simmel, 1944), in which they showed how a simple animation of geometrical shapes moving in a 2D space can elicit rich social inferences. People watching this animation for the first time often report seeing these shapes spontaneously come to life, and inferences about the shapes' goals, beliefs, desires, and intentions are so strong that they almost feel *visible*. The framework presented here—expressing actions in terms of abstract mental states in a hierarchical model that captures how minds relate to behavior—provides a theoretical foundation for explaining how social reasoning often goes beyond the data, allowing us to determine what others think and want, what they think they want, how they feel toward other people, and even what they think about themselves.

At the same time, watching Heider and Simmel's classical animation also reveals the long road ahead. A few seconds of this video not only reveal each individual agent's mental states; they also reveal the nature of their social relations—who is brave, who is a bully, and which agents are friends and which are not, to name only a few. The work presented here, we hope, will be a foundation for building richer and more powerful models that capture inferences not only about individual minds, but also about inferences about the complex social actions that compose social relationships.

15

Intuitive Physics as Probabilistic Inference

Kevin A. Smith, Jessica B. Hamrick, Adam N. Sanborn, Peter W. Battaglia, Tobias Gerstenberg, Tomer D. Ullman, and Joshua B. Tenenbaum

To reason about and interact with the world around us, we must understand how it changes over time. Crucially, we consider not just one possible future, but a range of possible outcomes: we can tell when a ball *almost* knocks another into a goal (Gerstenberg, Peterson, Goodman, Lagnado, & Tenenbaum, 2017; Gerstenberg & Tenenbaum, 2016), when a tower of blocks is precariously stacked and might fall down (Battaglia, Hamrick, & Tenenbaum, 2013), or that an object moving behind an occluder will reappear even while we are unsure exactly where or when it will appear (Smith & Vul, 2015). This suggests that our internal models of the physical world are probabilistic, translating uncertainty about the world's state or dynamics into a distribution of beliefs over possible future outcomes or latent object properties.

In this chapter, we demonstrate how techniques from probabilistic modeling can be used to explain the predictions and inferences that people make when reasoning about physical systems. We first describe why physical reasoning is an interesting problem and why a probabilistic framing is important for tackling it. We then lay out one theory of probabilistic physical reasoning—the *Intuitive Physics Engine* (IPE; Battaglia, Hamrick, & Tenenbaum, 2013). We discuss how probabilistic modeling with the IPE can explain a wide range of ways that people reason about physics. Next, we describe how the mind might perform this reasoning efficiently, through approximations to both probabilistic reasoning and the IPE. We end by discussing current and future directions for probabilistic models of physical reasoning.

15.1 The Ecological Nature of Physical Reasoning

Physical reasoning is an attractive domain for studying how cognition uses complex, probabilistic generative models for three reasons. First, people have extensive experience with the physical world. Starting from infancy, we grow our understanding of physics from the building blocks of "core knowledge" (Spelke & Kinzler, 2007) to mature physical intuitions according to systematic developmental trajectories (Spelke, Breinlinger, Macomber, & Jacobson, 1992), driven by consistent changes in the way that infants interact with the world (e.g., developing motor skills to grasp objects; Baillargeon, 2002). Thus, by adulthood, we would expect that interactions with the world should be guided by consistent physical intuitions that are compatible with accurate, Newtonian principles.[1]

Second, as researchers, we have access to normative computational models that can determine what the future state of a scene will be. This is in contrast to other instances of probabilistic cognition that rely on rich generative models (e.g., social cognition) for which it is challenging or impossible to determine normative accounts of how the world behaves. Access to this ground truth allows us to study when human inferences might deviate from the true future state of the world, and whether these errors might be the result of a rational inference process (e.g., Sanborn, Mansinghka, & Griffiths, 2013).

Finally, there is a set of computational models that serve as proxies for understanding how people simulate physics. At the core of any probabilistic model of cognition is the forward causal model, which predicts how causes give rise to effects. This forward model allows us to calculate likelihoods and posterior distributions (see chapters 3 and 4). If researchers want to model human physical reasoning in a probabilistic framework, they need a causal model that approximates the way that the world works. Fortunately, there is a suite of models that are designed to approximate realistic physical interactions: computer physics engines, such as those in games and graphics software. Using these game engines to approximate the cognitive systems underlying physical reasoning has led to successful modeling of human physical predictions (Battaglia et al., 2013; Smith & Vul, 2013; Smith, Dechter, Tenenbaum, & Vul, 2013; Gerstenberg et al., 2017), and the shortcuts that game engine designers have taken to model physics both realistically and quickly have provided ideas about how the mind performs efficient approximations of physics (Ullman, Spelke, Battaglia, & Tenenbaum, 2017).

15.2 The Psychological Nature of Physical Reasoning

It is, of course, sometimes possible to exactly calculate the posterior probability distributions that are necessary for probabilistic reasoning. Indeed, classic work demonstrating that human judgments match Bayesian inference often uses analytic probabilistic models. For example, when modeling how people integrate two sources of uncertain perceptual information, researchers have used priors, likelihoods, and loss functions that result in a Bayesian solution that is simply the weighted average of the observable information (Ernst & Banks, 2002; Körding & Wolpert, 2004).

However, because of the inherent complexity of many physical processes and aspects of physical reasoning, analytic solutions cannot be used to solve many real-world problems. Instead, we must consider what sorts of approximations the mind makes—how knowledge is represented, accessed, and used in behavior—to efficiently solve these problems. Consider the simple, analytically tractable case of determining the relative mass of two rigid objects from their velocities before and after they have collided with each other. While there are simple algebraic expressions for calculating the relative mass if the velocities

1. While there are many instances of human physical reasoning that rely on incorrect principles (e.g., McCloskey, Caramazza, & Green, 1980; Caramazza, McCloskey, & Green, 1981; Gilden & Proffitt, 1989; Vasta & Liben, 1996), these errors may be based on a separate cognitive system that is used for more abstract problems. For further discussion, see Smith, Battaglia, and Vul (2018) and the section on "Errors in Physical Reasoning" later in this chapter.

are known, accounting for perceptual uncertainty in these situations greatly complicates the problem and makes pure analytic solutions intractable (Sanborn et al., 2013). These equations become more difficult to solve as the complexity of the system increases. For example, it's impossible to analytically predict the state of a system with three objects colliding (Diacu, 1996), much less precisely characterize systems with complex dynamics, like fluids. Yet people have no problems stacking multiple dishes on top of each other and regularly pouring liquids from one container to another.

How, then, can people do probabilistic physical reasoning? Approximations of some sort seem mandatory. Following from the work in chapter 11, on rational process models, it is useful to look at the approximations to probabilistic inference from computer science and statistics that have been used as algorithmic models of human behavior in tasks such as categorization, decision-making, and causal inference. These algorithms provide a tractable way of performing probabilistic inference, and also make systematic errors that often match the errors that people make.

Perhaps the simplest rational process model for probabilistic physical reasoning is the exemplar model (Shi, Griffiths, Feldman, & Sanborn, 2010). Instead of maintaining an internal physical model of the world, probabilistic physical reasoning could instead be performed by remembering previous experiences and weighing them according to their similarity to the current situation. In simple tasks, such as inferring which of a pair of colliding objects is heavier by observing their movement, a weighted average of only 50 prior experiences captured human-level performance across various settings of the underlying physical variables (Sanborn et al., 2013). But in more complex domains (e.g., predicting whether and in what direction a stack of blocks will fall), the number of possible object configurations is very large. Yet even in such domains, we can still predict what will happen for configurations of objects that we have never seen before, suggesting that the exemplar model cannot explain much complex physical reasoning. As we outline next, people seem to represent the external physical world with an internal physical model that supports Bayesian inference. To make this Bayesian inference tractable, the mind might use a number of approximations, including model-based sampling, learning a recognition model for rapid inference, or using an approximate form of the physical model itself.

15.3 A Mental Model of Physics

A key component of probabilistic cognition is the forward causal model, which allows us to make inferences by understanding how the world works. For instance, when two objects collide, we can reason about unobserved variables (the masses) based on observed variables (the trajectories; see Sanborn et al., 2013). This can be considered a simple instantiation of Bayes' rule, where we reason about the causes (c) based on the observed effects (e):

$$P(c|e) \propto P(e|c)P(c). \tag{15.1}$$

A crucial part of this equation is the likelihood model $P(e|c)$, which requires understanding how effects come from causes—for example, how likely is it that we would observe the objects' trajectories for a given specification of the objects' masses? This likelihood can be instantiated by mental models of the world that provide us with information on how causes translate into effects (Craik, 1943), potentially using a mechanism of approximate probabilistic simulation. But how are these mental models for physical reasoning structured?

Extending prior research into spatial reasoning via continuous simulation, recent work has suggested a method for performing this model-based physical reasoning—namely, that people have an IPE that can simulate the world in ways analogous to the game physics engines that underlie many modern video games. According to this theory, the IPE takes a mental representation of the world and iteratively steps it forward in time using approximately correct physical principles. However, while game physics engines are deterministic, the IPE is probabilistic to account for uncertainty in both initial world conditions and physical dynamics: people can never be perfectly certain of exactly how heavy an object is or how collisions will resolve. The IPE, therefore, provides us with a belief distribution over possible futures, such as where a thrown ball will end up or the range of ways that a stack of blocks might topple. We define the IPE as Φ, which can transform a world state s at a given time into a distribution of future world states:

$$s^{t+1} \sim \Phi(s^t). \tag{15.2}$$

This belief distribution can be used as an input to other probabilistic cognitive models, forming a bridge between perception and other cognitive systems.

15.3.1 Mental Simulation and Spatial Reasoning

Many theories suggest that spatial reasoning relies on representations that are analogous in both form and content to the three-dimensional spatial structure of the real world. (Kosslyn, Ball, & Reiser, 1978, but cf. Pylyshyn, 2002 for alternative theories on the nature of spatial representations). These representations can be transformed via simulation: transforming the mental representations in a way similar to how their real-world counterparts would change through time. For instance, if we are asked to determine if two shapes are the same, the time that it takes to make this judgment is related to the time that it would take to rotate the shapes into alignment, suggesting that we are mentally performing this rotation (Shepard & Metzler, 1971). If we are asked whether two edges of an unfolded paper cube will touch when refolded, our reaction times are related to the time that it would take to fold the cube enough to check those edges (Shepard & Feng, 1972).

This view of mental simulation has two crucial components. First, the object representations and transformations that underlie mental simulation must reflect the objects and transformations that exist in the world (Fisher, 2006). If we wish to use simulation to understand how the world will unfold, this correspondence is necessary to ensure that the results of our simulations approximate reality. Second, simulation acts in a step wise fashion: one cannot predict a future state of the world without predicting intermediate states (Moulton & Kosslyn, 2009).

The same cognitive systems that let us mentally traverse through space or rotate objects might also include the capability of understanding how objects interact. Indeed, mental simulation underlies reasoning about mechanical events: our speed of reasoning about the kinematics of pulley systems depends on the number of components that must be set in motion (Hegarty, 1992), and the time that it takes to judge how turning a gear in a chain will affect gears further in the chain depends on the number of intervening gears (until people discover rules that can shortcut this process; Schwartz & Black, 1996b). However, while these tasks do involve reasoning about physical events, they could be accomplished either by piecewise simulation or by sequential reasoning about the components (e.g., using a causal logic to assess the interaction between gear A and gear B, then gear B and gear C, etc.). We therefore turn to instances of physics where the continuous dynamics of the scene are important—understanding how objects collide, fluids pour, or things fly through the air—and discuss how a simulator that includes physical principles accounts well for human judgments about these scenarios.

15.3.2 The Intuitive Physics Engine

The IPE proposed by Battaglia et al. (2013), was motivated by previous theories of mental models underlying spatial and mechanical reasoning. While this mental model is theorized to reproduce the dynamics of the world well enough to make useful predictions (Sanborn et al. 2013; Smith, Battaglia, & Vul, 2018), it is not supposed to perform these calculations analytically according to idealized physics; instead, the IPE is suggested to "favor speed and generality over the degree of precision needed in engineering problems" (Battaglia et al., 2013, p. 18328). These constraints are also found in a similar class of problems: modeling physics for video games, which require dynamics that are good enough to be acceptable to the game players, but also fast enough to run in real time. These game physics engines function by eschewing analytic solutions, and instead simulating physics in a step wise fashion with state transition functions that are locally consistent without explicitly modeling fundamental physical properties (e.g., conservation of energy; Gregory, 2018). The IPE is theorized to function in a similar fashion, using step wise, approximate physical principles to model the world (Ullman et al., 2017).

As with game physics engines, the IPE takes as input a description of the state of the world, and it yields as output simulations of hypothesized future world states. These state representations are comprised of a set of object descriptions, as objects are a basic mental building block (Spelke et al., 1992). Each object representation describes not just the shape, position, or motion of the object, but also latent properties such as mass and friction. Put together, the full state representation is similar to those used by computer-aided design programs to represent scenes, but includes additional information needed to understand the causal mechanisms that describe how the scene should unfold. However, unlike computer representations of scenes, mental representations have various memory limitations and will not include all items in a scene; instead, the mind may represent only a limited set of objects that are in motion and relevant to the judgments that we must make (Ullman et al., 2017).

These object representations are multimodal, drawing on information from vision, audition, and touch. There is ample evidence that we can integrate information from vision and audition (Battaglia, Jacobs, & Aslin, 2003; Alais & Burr, 2004) or haptics (Ernst & Banks, 2002; Yildirim & Jacobs, 2013) to make nonphysical judgments, which suggests that information from each of these modalities is integrated into a coherent representation in the brain (Taylor, Moss, Stamatakis, & Tyler, 2006; Erdogan, Chen, Garcea, Mahon, & Jacobs, 2016). Because the IPE relies on these integrated representations, it can also make predictions not just about how physics will transform the visual location of objects over time, but also what will be heard or felt. Conditioning on auditory information allows us to reason about material properties based on the sound of a collision (Traer & McDermott, 2016), infer the

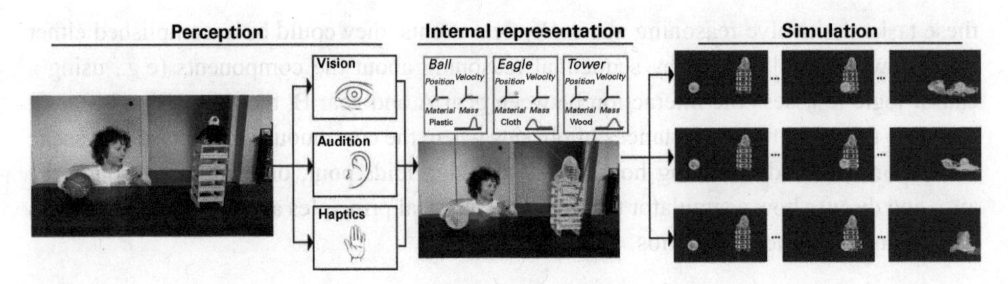

Figure 15.1
People perceive a scene through multiple sensory modalities (*left*) to form an internal representation of the world. This is an object-centric representation, containing probabilistic information about the locations, extents, and properties of objects (*center*). The IPE uses this representation to stochastically simulate ways that the world might unfold using approximately accurate dynamics (*right*). These simulations give rise to a range of possible future states of the world that feed into other cognitive systems to make predictions, decisions, and other behavioral choices.

number and type of objects in an opaque box that is shaken (Siegel, Magid, Tenenbaum, & Schulz, 2014), or figure out in which hole a ball was dropped in a "plinko" box (in which a ball dropped at the top of the box bounces off a series of typically hidden objects before emerging at the bottom of the box) by integrating the sequence of sounds with information about where obstacles are positioned in the box (Gerstenberg, Siegel, & Tenenbaum, 2018).

One crucial difference between game physics engines and the IPE is that while game physics engines are deterministic, both the inputs and outputs of the IPE are belief distributions over states of the world. This distribution of beliefs over world state s comes from two sources. First, there is perceptual uncertainty in constructing mental models of the world: we are unable to exactly perceive the properties of objects given our sensory input (such as their location and velocity). In addition, the state transitions within the IPE are themselves stochastic, especially for physical events that are intrinsically hard to predict such as collisions (Smith & Vul, 2013).

Thus, the IPE can be thought of as a stochastic transition function over hypothetical world states. Because both the input and the output of this model take the same form, the same queries on the current belief state of the world can be applied to hypothetical belief states—for example, "Where is the ball now?" is the same function applied to current beliefs as "Where will the ball go after it is tossed?" is to predictions of future world states. Thus, we can define world state queries Q such that the query on the current world state ($Q(s)$) and the query on the output of the IPE ($Q(\Phi(s))$) produce similar types of output. This provides a key link between perception and higher-level cognition, providing generalized output about hypothetical futures that we can use for prediction, inference, planning, reasoning, and learning.

The Physics Engine in the Brain Within the brain, there are specialized neural regions dedicated to performing ecologically important tasks like recognizing faces (Kanwisher, McDermott, & Chun, 1997) and judging the mental states of others (Saxe & Kanwisher, 2003). Understanding and interacting with the physical world constitute another task important for our survival, so it might be expected that the brain dedicates cortical areas to the IPE. Indeed, Fischer, Mikhael, Tenenbaum, and Kanwisher (2016) found that there are areas of the brain that respond preferentially to making predictions about, or just watching, physical events. Furthermore, these brain regions encode information about physically relevant properties such as weight (Schwettmann, Tenenbaum, & Kanwisher, 2019) and stability (Pramod, Cohen, Tenenbaum, & Kanwisher, 2022) and are in fact the only parts of the brain from which this information can be decoded.

These "physics areas of the brain" are located in premotor/supplementary motor cortex and parietal somatosensory association cortex, which are brain regions that have been previously implicated in spatiotemporal prediction (Schubotz, 2007), motor action planning (Chouinard, Leonard, & Paus, 2005), and tool use (Goldenberg & Spatt, 2009). This further suggests that the IPE acts as an interface between perception and other cognitive modules that can be used, for instance, to plan our actions.

Errors in Physical Reasoning To produce reasonably accurate predictions, the IPE is believed to transform mental representations of the world using principles that are approximate but generally capture how the world itself unfolds (Battaglia et al., 2013; Sanborn et al., 2013; Smith et al., 2018). This claim is distinct from a separate body of literature that finds significant errors in human reasoning about physical principles: that we display errors when reasoning about ballistic motion (Caramazza et al., 1981; Hecht & Bertamini, 2000), inappropriately believe that objects exiting curved tubes retain curvature in their motion (McCloskey et al., 1980) or fail to understand how water acts in a tipped container (Kalichman, 1988).

However, these studies that find errors in physical reasoning typically use abstract diagrams or ask for explanations of physical principles, both of which are thought to require more abstract, rule-based reasoning than more realistic, predictive tasks (Schwartz & Black, 1996a). Furthermore, tasks that rely on explicit reasoning about physical concepts activate a wider range of brain areas (Jack et al., 2013) than tasks that use more perceptual or action-oriented information (Fischer, Mikhael, Tenenbaum, & Kanwisher, 2016). Thus, cases where people behave according to incorrect physical principles may be instances of reasoning with a different cognitive system than the IPE discussed here (for further discussion, see Hegarty, 2004; Smith et al., 2018; Zago & Lacquaniti, 2005).

This is not to say that the IPE always produces accurate predictions. As described later in this chapter, certain physical approximations produce biases and errors in predictions. Furthermore, it is possible that there are physical principles that are encountered rarely or have little impact on our predictions, and so are not accurately modeled by the IPE. However, for many scenarios with relatively simple shapes and dynamics that are presented in a realistic fashion, models that assume unbiased, accurate physical principles do a good job of explaining human physical reasoning.

15.4 Human Physical Reasoning

As a probabilistic generative model, the IPE supports many different ways of reasoning about the world. The simplest way is through prediction: running the model forwards on the current state of the world to form a belief about how the world will turn out. But principles of probabilistic cognition suggest how the IPE can support various ways of reasoning about physics: inverting a generative model to form inferences about the world, reasoning about counterfactual models of the world to determine causality, conditioning on outcomes to plan

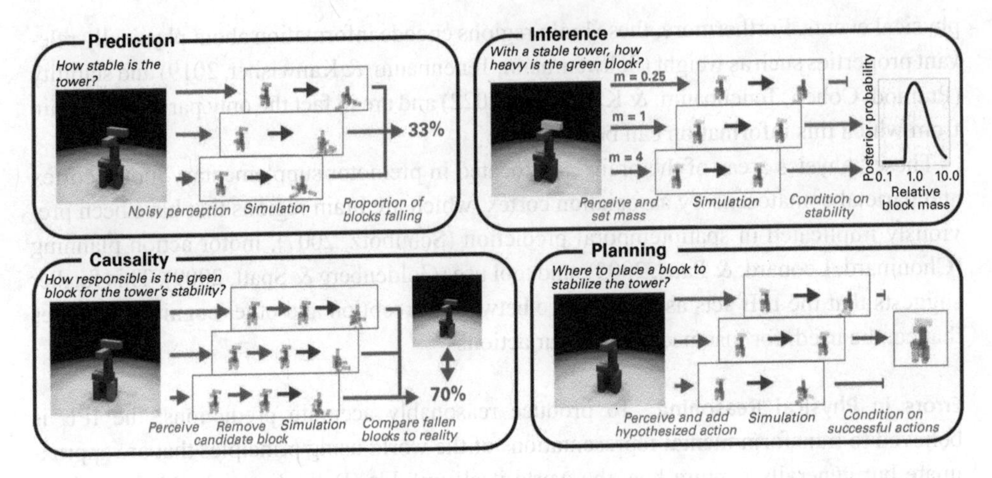

Figure 15.2
The IPE as a generative model can support a variety of ways of reasoning about physics via probabilistic cognition. *Prediction* is running the IPE forwards and querying the results. *Inference* requires conditioning belief based on how well a world with the relevant parameters would match observations. *Causal reasoning* requires comparing the expected result of hypothetical worlds without the causal agent to actual observations. *Planning* involves selecting actions that are expected to produce the desired outcome.

our actions, and so on. In the following sections, we provide evidence for and explain how the IPE supports these various facets of cognition.

15.4.1 Prediction

Prediction is the simplest use of the generative models of physics: running the IPE forwards and querying the simulated outcomes to make judgments about possible future states of the world. Here, probabilistic reasoning allows us to make graded predictions across a wide variety of scenarios. For example, we may predict how towers fall (Battaglia et al., 2013), balls bounce around (Deeb, Cesanek, & Domini, 2021; Smith & Vul, 2013; Smith et al., 2013; Smith & Vul, 2015; Gerstenberg et al., 2017) or roll down slopes (Ceccarelli et al., 2018; Ahuja & Sheinberg, 2019), how objects fly under ballistic motion (Smith et al., 2018), and how fluids pour (Bates, Yildirim, Tenenbaum, & Battaglia, 2019; Kubricht et al., 2016, 2017).

This is equivalent to developing a posterior belief over future world states (s^t) given the current belief over the world state (s^0) and the physics engine (Φ):

$$p(s^t) = p(s^t | s^0, \Phi) p(s^0). \tag{15.3}$$

Because these equations are often analytically intractable, in most cases the prior and posterior beliefs are approximated using Monte Carlo methods: treating a belief distribution as a collection of samples from a probability distribution ($\mathbf{s} = [s_0, s_1, \ldots s_n]$). In this way, each sampled state can be iteratively updated with the physics engine until a final state is reached (where Φ^* indicates iteratively applying the physics engine):

$$s_i^t = \Phi^*(s_i^0). \tag{15.4}$$

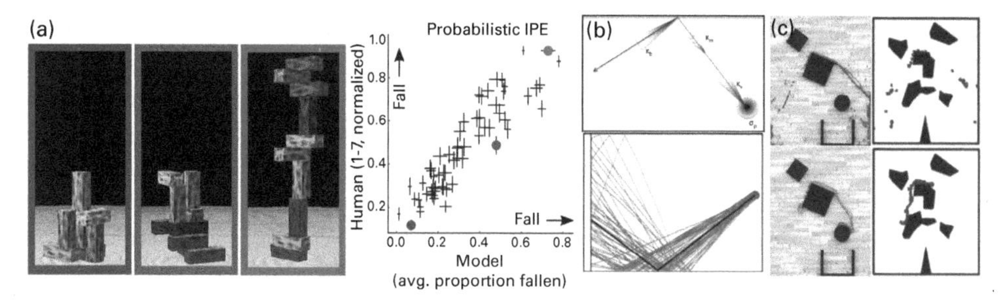

Figure 15.3
Instances of probabilistic prediction in intuitive physics. (a) The IPE captures predictions about general events such as stability. For instance, even though the tower with a red outline is stable, both people and an IPE model treat it as unstable (*right*) because small changes in the location or pose of almost any block will cause it to come crashing down (Battaglia et al., 2013). (b) Uncertainty in the IPE is driven both by noise in perception and accumulating stochasticity throughout prediction (*top*), which gives rise to a distribution over possible paths that objects might take (*bottom*; Smith & Vul, 2013). (c) Probabilistic prediction can also explain judgments of how fluids pour by approximating the fluid with a set of interacting particles. This can differentiate between water (*top*) and honey (*bottom*) by modeling more viscous liquids as having stronger interparticle forces (Bates et al., 2018).

Battaglia et al. (2013) applied this approach to understanding physical prediction. In this work, participants viewed images of block towers like those in figure 15.3a, and were asked to predict whether the tower will fall or remain stable under the effects of gravity. They found that participants' stability judgments could be better captured by a probabilistic simulation model than alternative, feature-based heuristics (such as the height of the tower). This model assumes that an observer has perceptual uncertainty about the exact location of the different blocks in the tower and uses a deterministic IPE to simulate how the world will unfold under these different initial conditions. Since each initial scene will have a slightly different block configuration, the output of the IPE is a distribution over possible future scenes. Participants' judgments are then explained by aggregating the IPE's predictions across these scenes, such as the average proportion of blocks that fall. The same model also explained participants' physical intuitions across a variety of other tasks that included judging in which direction the tower will fall, or which objects would be more likely to fall off a table if it were bumped; conversely, no single feature-based heuristic could capture performance across all these tasks.

Smith and Vul (2013) explored the extent to which noise in physical dynamics themselves affected participants' physical predictions, using a task in which participants were asked to view a ball bouncing around a computerized table and predict where that ball would travel while occluded. Like Battaglia et al. (2013), Smith and Vul assumed that participants may have perceptual uncertainty about the exact position and velocity of the ball when it disappeared behind the occluder, but they also investigated dynamic sources of uncertainty: that the ball's trajectory would be perturbed in each time step and additionally perturbed whenever the ball collided with a wall (figure 15.3b). They found that assuming uncertainty about how the physical dynamics will unfold over time was critical for explaining participants' predictions in this task, which implies that the physical transition function Φ is itself stochastic. In other experiments, this uncertainty in dynamics was also required to explain participants' judgments about their overall uncertainty about their own predictions

(Smith & Vul, 2015), and how people update their predictions as a scene unfolds (Smith et al., 2013).

The proposal of the IPE has been extended beyond rigid bodies to soft bodies, cloths, and fluids. As early as five months of age, infants demonstrate rich expectations about the dynamics of fluids and other nonsolid substances, distinct from their expectations about solids (e.g., Hespos, Ferry, Anderson, Hollenbeck, & Rips, 2016). Van Assen, Barla, and Fleming (2018) found that the human visual system supports accurate inferences about fluid viscosity, which can be modeled as hierarchical estimation over midlevel visual features, such as "compactness," "elongation," "pulsing," and "clumping."

Recent work has suggested that people understand these fluid dynamics using simulation (Bates et al., 2019; Kubricht et al., 2016, 2017). Bates et al. (2019) asked participants to predict how liquids with different viscosities (water and honey) would flow down a set of obstacles, and to judge what proportion of that liquid would fall into a bucket on the ground. They found that participants' predictions were well approximated by a model that captures the complex dynamics underlying fluid motion through representing the liquid by a number of interacting particles. The results showed that participants' predictions were sensitive to the liquid's viscosity, making different predictions for how honey will flow, or how water will spill (figure 15.3e). Relatedly, a model of fluid dynamics with uncertain viscosity was used to explain people's intuitions about the angle at which a filled container would start to pour out a liquid (Kubricht et al., 2016) or sand (Kubricht et al., 2017, but see also Schwartz & Black, 1999).

15.4.2 Inference

It is clear how to use a forward model like the IPE to make predictions about the world: start with a set of initial conditions and run the IPE forwards. However, people can make inferences about the hidden states of a physical system just by observing how it unfolds, which requires using the IPE to make judgments in the opposite direction.

These inferences are naturally captured by Bayesian models of cognition. Here, we define a set of latent properties (l) that may not be directly observable (e.g., object weight or elasticity), and observable properties (o) that may or may not change over time, such as object shape, position, or velocity. Thus, a scene is a collection of latent and observed properties ($s^t = [o^t, l]$). After watching a scene unfold, posterior beliefs over the latent properties can be calculated by a simple application of Bayes' rule, conditioned on how the observed scene properties have unfolded:

$$p(l|o^t) \propto p(o^t|l, o^0, \Phi)p(o^0|l)p(l). \tag{15.5}$$

Sanborn et al. (2013) demonstrates how this approach can explain biases in judgment arising from human physical inferences. When observing two rigid objects colliding on a computer screen, people can infer the relative masses of the objects from observing their velocities, which requires reasoning backwards from these observed velocities to the masses that would have caused that collision. These mass judgments have been found to depend on the elasticity of the collision: when the collision is especially "bouncy," people are more likely to correctly judge the heavier object to be heavier than when they observe a less elastic collision between two objects of the same masses. But according to the laws of mechanics,

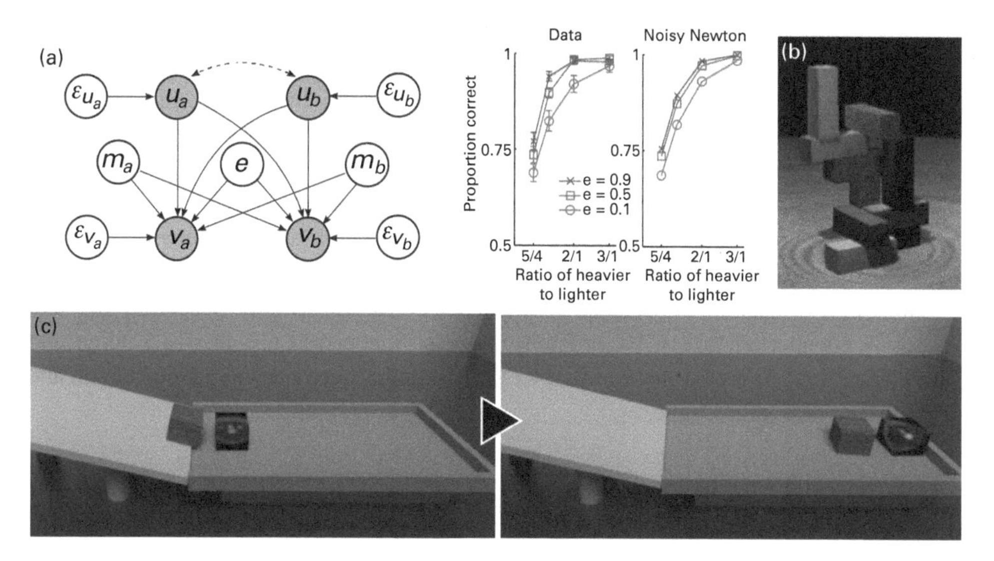

Figure 15.4
Probabilistic models support our ability to make judgments about latent physical properties such as mass. (a) A graphical model used to infer the masses (m) and elasticity (e) of two blocks colliding based on the initial and final velocities (u and v), which are perturbed by perceptual noise (ϵ; *left*). This noisy inference model explains why people's mass judgments are biased by the elasticity of the collision (*right;* Sanborn et al., 2013). (b) Since the tower is stable, we judge that the purple blocks must be heavier than the green blocks because if they were not, we would expect some of the blocks to fall (Hamrick, Battaglia, Griffiths, & Tenenbaum, 2016). (c) We might expect wooden blocks to be lighter than iron ones; however, if we see a block with a wood texture launching a block with an iron texture, we quickly update our beliefs about their relative weights (Yildirim et al., 2018).

the relative masses of two objects can be calculated just from observations of the starting and ending velocities and should not depend on the elasticity of the collision. This dependence on an "irrelevant" variable has in the past been taken as evidence that we do not use accurate physical principles in these situations (Todd & Warren, 1982; Gilden & Proffitt, 1989). However, viewing this mass judgment through the lens of probabilistic reasoning shows that the sensitivity to elasticity is not necessarily due to a simple heuristic or errors in understanding Newton's laws of motion. Instead, because of perceptual uncertainty, collisions with slower speeds (which result from inelastic collisions) are simply harder to distinguish than collisions with faster speeds. Similarly, people seem to be biased toward assuming that objects in motion are heavier than stationary objects (Stocker & Simoncelli, 2006); adding a prior expectation that objects move slowly results in an interaction with Newtonian mechanics that captures this bias. In this way, human judgments are consistent with Bayesian inference using an accurate model of collision dynamics (figure 15.4a; Sanborn et al., 2013; Sanborn, 2014).

People are also able to infer relative masses from scenes with more complex arrangements of objects—for example, towers of blocks (figure 15.4b)—and even update their beliefs about these relative masses across trials (Hamrick et al., 2016). These impressive feats of physical inference are not limited to vision, or to adults. Before they are one year old, infants understand that objects that compress a pillow are heavier than those that don't (Hauf, Paulus, & Baillargeon, 2012). And by shaking a box, children can infer what objects are inside and how many of them there are. Children can even use information about what they would *expect* to hear to determine how difficult a discrimination task would be without having to physically shake the box. For instance, children know that two different pencils will make similar noises when shaken in a box, so that this is a difficult discrimination task, but a pencil and a cotton ball will make distinct noises and so present an easier choice (Siegel et al., 2014).

Inferences about physical properties in turn can recalibrate the simple perceptual judgments on which they seem to be based. For example, if people see a slope with a shallow slant, but observe a ball bouncing off of the slope as if it were steep, they will adjust their perception of the orientation of the slope to be steeper, consistent with the behavior of the ball (Scarfe & Glennerster, 2014). This inference suggests that people use physical inference to build internal world representations that are consistent between their direct perception and their observations of dynamics.

15.4.3 Causal Reasoning

Two billiard balls, ball A and ball B, collide with one another, and ball B goes into the pocket of the billiard table. Did ball A cause ball B to go into the pocket? Is it sufficient to notice that the two balls collided to answer this question about causation, or is more required? In philosophy, there are two broad families of theories that try to analyze what causation is. According to *process theories* of causation, causes bring about effects via a spatio temporal contiguous process, such as the transmission of physical force (Dowe, 2000). According to *dependence theories* of causation, causes and effects are related via probabilistic or counterfactual dependence, such that for c to qualify as a cause of event e, e would not have happened if c hadn't happened (see Gerstenberg & Tenenbaum, 2017; Waldmann, 2017).

Both of these approaches have had a strong influence on psychological accounts of causal reasoning. The *force dynamics model* developed by Wolff (2007) is a process theory that suggests that people judge an event to be causal based on the force transferred between the agent and the patient. For example, to decide whether ball A was the cause of ball B going into the pocket, this theory suggests that we look at the configuration of forces associated with the patient and the agent at the time of the collision. This theory has been used to map various force configurations onto descriptions like "caused" or "helped" (Wolff, 2007; Wolff, Barbey, & Hausknecht, 2010). Crucially, the force dynamics model suggests that people consider only what actually happened to judge whether an event was causal.

Dependence theories, on the other hand, predict that people's judgments about causality are based on what might have happened in a counterfactual situation in which the causal event had been absent or different. The belief in what would have happened is often represented as a distribution over possible alternative outcomes, and many variants of probabilistic theories of causation exist that aim to capture people's inferences about the strength of the relationship between putative cause and effect (Griffiths & Tenenbaum, 2005; Cheng, 1997; Jenkins & Ward, 1965).

Counterfactual theories of causation naturally capture causal relationships between particular sets of events, such as whether the bump of the table caused the tower to fall or whether the gust of wind that happened at the same time would have been sufficient to bring about the same result. These theories posit that c is a cause of event e to the extent that a counterfactual outcome e' would be different if c were removed from scene s:

$$\text{CAUSE}(c \rightarrow e) \propto P(e' \neq e | s, \text{remove}(c)). \tag{15.6}$$

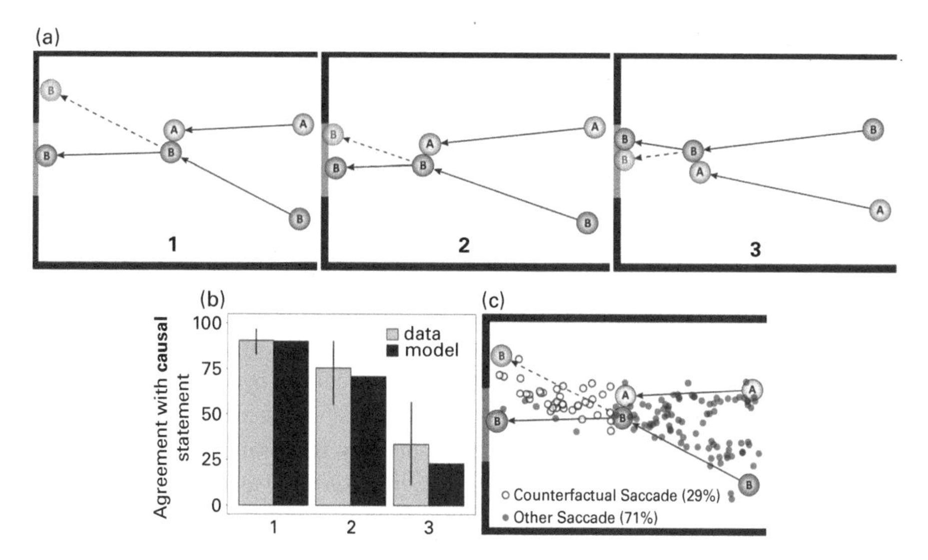

Figure 15.5
Probabilistic models allow us to determine what *might* have happened in the absence of a possible cause. (a) Instances where ball A certainly (1), maybe (2), or did not (3) cause ball B to go through the gate. (b) The counterfactual simulation model suggests that we make judgments about ball A's causal relevance by simulating what would have happened to ball B if ball A had not been there and comparing the outcome in this counterfactual situation to what actually happened. This model predicts human causal judgments well. (c) Supporting this theory, when making causal judgments, people spontaneously look toward where ball B would have gone without ball A (Gerstenberg et al., 2017).

Gerstenberg, Goodman, Lagnado, and Tenenbaum (2021) developed the *counterfactual simulation model* of causal judgment to quantitatively capture dependence theories. According to this model, people make causal judgments by comparing what actually happened with what would have happened in relevant counterfactual situations. For example, when asked to judge whether ball A caused ball B to go into the pocket, the model not only considers that the two balls actually collided and ball B went into the pocket, it also considers would have happened if ball A hadn't been in the scene. These counterfactual considerations are in general probabilistic because the model—like people—cannot always be sure what would have happened. The model predicts that an observer's causal judgments will increase the more certain that she is that the outcome would have been different if the cause hadn't been in the scene (see figures 15.5a and 15.5b).

It's worth making explicit what's assumed to be involved in this process. To make causal judgments under a counterfactual theory, people first observe what actually happened. They then go back in time (mentally) and make a change to the scene to undo the causal event of interest (e.g., by mentally "removing" the candidate cause ball from the scene). Finally, they predict what the outcome in this counterfactual situation would have been through simulating the counterfactual course of events (see figure 15.2). A probability distribution over different counterfactuals arises naturally from the probabilistic nature of simulations in the IPE.

In some of the counterfactual scenarios, the simulated outcome might often be the same as what actually happened (i.e., ball B would likely still have gone into the pocket even if ball A hadn't been there, in example 3 in figure 15.5a), whereas in other scenarios, the outcome might well have been different. People's causal judgments were well explained by the counterfactual simulation model's uncertainty about whether the cause made a difference as to whether the outcome happened. The more certain participants were that the outcome would have been different, the more they said that the candidate caused the outcome to happen (Gerstenberg et al., 2021).

Gerstenberg et al. (2017) tested a key algorithmic-level prediction of the counterfactual simulation model: that people reach their causal judgment by spontaneously imagining what would have happened if the cause hadn't been present in the scene. They asked people to watch video clips of two balls colliding (A and B) and judge whether ball A caused or prevented ball B from going into a goal while using eye-tracking to determine where people looked as they made these judgments under the hypothesis that eye movements would (at least to some extent) follow imagined trajectories. As predicted by the model, participants looked not only at the balls and their actual motions, but also at locations where ball B *would have gone* if ball A had not been in the scene. Importantly, these same eye movements were not observed in a condition in which participants were asked only to make a judgment about what actually happened. So counterfactual simulations were specifically recruited in the service of making causal judgments, but without any explicit instruction in the experiment to consider counterfactual contrasts. Furthermore, participants whose explicit judgments were better predicted by the counterfactual simulation model made more counterfactual eye movements, and participants on average made more counterfactual looks in cases where they were less extreme (more uncertain) about their causal judgment, as an efficient sampler would do as well.

Extensions of the counterfactual simulation model have been shown to capture many other aspects of causal reasoning, including people's judgments about whether an event "almost happened" as a function of how much a causally relevant variable would have needed to change to produce the event (e.g., the force with which a ball is kicked; Gerstenberg & Tenenbaum, 2016), judgments about causation by omission (e.g., a ball aimed to knock another ball off its course might have caused it to go in a hole by virtue of having missed it; Gerstenberg & Stephan, 2021), and judgments of the causal factors responsible for maintaining a situation as stable (e.g., to what extent a single block in a tower is responsible for the tower's not falling over, by simulating what would happen if the block was removed from the tower; Gerstenberg, Zhou, Smith, & Tenenbaum, 2017; Zhou, Smith, Tenenbaum, & Gerstenberg, 2023).

15.4.4 Planning and Action Selection

Being able to make predictions and inferences about the physical world is about more than reasoning: it also supports rich interaction with physical systems. Specifically, a model of the physical world like the IPE can also be used to choose the best sequence of actions to take in a given scenario. This process of action selection on the basis of a model is referred to as *planning* (Sutton & Barto, 2018) and has been found to occur in the context of physical reasoning at multiple levels of abstraction, ranging from low-level motor control to high-level problem solving.

A large body of work has shown that the motor system represents forward models of how motor commands affect the motion of our bodies, the dynamics of external objects, and how our bodies might interact with those objects (Miall & Wolpert, 1996; Wolpert & Kawato, 1998; Wolpert, Miall, & Kawato, 1998; Kawato, 1999; Flanagan & Wing, 1997;

Davidson & Wolpert, 2005). These forward models are used by the motor system to compute optimal actions or trajectories as follows. First, the forward models estimate possible current states in the world, either through Bayesian inference (Wolpert, 2007) or through a filtering procedure like a Kalman filter (Grush, 2004). After an action is taken, these forward models compute expectations about what the next state of the world will be, and combine these expectations with actual sensory data to compute a posterior distribution over states.

This distribution over states can be used to compute the expected cost or reward of possible actions, marginalizing over all possible states. action selection consists of computing this expected cost (L) for all possible actions ($a \in \mathcal{A}$) and then choosing the action (a^*) with the lowest cost (including action costs and costs of not accomplishing our goals) based on the IPE ($\Phi^*(s)$) across all plausible states ($s \in \mathcal{S}$; Wolpert, 2007):

$$a^* = \arg\min_{a \in \mathcal{A}} \sum_{s \in \mathcal{S}} L(a, \Phi^*(s))p(s). \quad (15.7)$$

In addition to cases where the current state of the world is uncertain, forward models aid in computing the costs of actions when *future* states are uncertain. For example, Dasgupta, Smith, Schulz, Tenenbaum, and Gershman (2018) showed that when trying to launch a ball into a goal, people make predictions about where the ball will end up given a particular action. Where the ball ends up determines the utility of the action: if the ball makes it into the goal, there is net positive utility for accomplishing the objective, while if the ball misses the goal, a cost is incurred. To actually choose which actions to evaluate, Dasgupta et al. (2018) used a model of decision-making known as "Bayesian optimization" (Hernández-Lobato, Hoffman, & Ghahramani, 2014) and showed that this model not only predicted people's action evaluations, but also captured how they combined information from both mental simulations and real physical experiments. Li et al. (2019) also showed how physical simulations can help compute an intrinsic reward that encourages the exploratory behaviors that are needed to uncover the causal properties of a physical system and that are similar to the exploratory behaviors produced by human participants.

However, even with a model that provides an estimate of the utility of an action, it is not always clear which actions should be considered in the first place: there are always many things we *could* do, but the vast majority of those actions will not be useful. While in theory, action selection can be accomplished by exploring the space of possible actions and conditioning on those that are successful (see figure 15.2, lower right), in reality, it is impossible to consider the outcome of every possible action that could be taken. Allen, Smith, and Tenenbaum (2020) studied how people choose and use tools to accomplish goals in physical problem solving with a large space of possible actions. They find characteristics of rapid trial-and-error problem solving: people search stochastically but in a structured way at first, then exploit promising solutions to quickly solve these problems. Allen et al. (2020) propose that this rapid search requires not just a model to assess actions, but also prior expectations about what general sorts of actions are likely to be successful to avoid considering useless actions, as well as generalization mechanisms that take into account both simulated expectations and real-world observations to update posterior beliefs about what actions might be useful.

In physical problem-solving tasks that require multiple steps—such as stacking blocks into a tower—a model of physical dynamics can be used to score plans depending on physical constraints. For example, Yildirim, Gerstenberg, Saeed, Toussaint, and Tenenbaum (2017) examine a block-stacking task in which a set of blocks must be assembled into a given target configuration. To model this task, they first search for a symbolic plan specifying which blocks should be stacked with one or both hands, and in what order, and then score plans according to the physical stability of the tower in each step (along with other geometric and spatial constraints). Yildirim et al. (2017) showed that this model captures how likely human participants are to use one or two hands to solve the task, suggesting that this choice in humans may also be informed by estimates of physical stability. Yildirim et al. (2019) demonstrated how an extension of the model that takes into account physical effort and physical risk accurately captures people's intuitions about how difficult it would be to build certain block towers.

Finally, it is worth noting that planning is not necessarily limited to scenarios involving physical reasoning, and an exciting direction for future work is to combine insights from the literature on nonphysical planning and learning with forward physical models like the IPE. For example, hippocampal replay and preplay during spatial navigation tasks in rats strongly resemble rollouts of a forward model (Pfeiffer & Foster, 2013; Ólafsdóttir, Barry, Saleem, Hassabis, & Spiers, 2015), and various theories have suggested that this replay occurs during a consolidation process of model-based experience into model-free action policies (Momennejad, Otto, Daw, & Norman, 2017; Mattar & Daw, 2018). Related work has explored how people trade off between model-based and model-free accounts of learning in environments with nonstationary rewards (Gläscher, Daw, Dayan, & O'Doherty, 2010; Daw, Gershman, Seymour, Dayan, & Dolan, 2011; Dolan & Dayan, 2013; Kool, Cushman, & Gershman, 2016; Keramati, Smittenaar, Dolan, & Dayan, 2016). While such work examines the use of models during learning processes, other work has explored the use of planning at decision time, looking at how people construct and traverse trees of possible future states (Huys et al., 2012; Solway & Botvinick, 2015; Opheusden, Galbiati, Bnaya, Li, & Ma, 2017). A number of recent advances in artificial intelligence (AI) suggest other possible mechanisms for model-based planning (Hamrick, 2019), which could be integrated with models like the IPE to build process-level models of physical planning and action selection. Indeed, recent work combining model-based planning with physical models has demonstrated how to build AI systems that can reason about complex physical scenes, such as deciding how to stack blocks into a tower (e.g., Janner et al., 2019; Fazeli et al., 2019; Bapst et al., 2019). Such methods, when combined specifically with an IPE model, may also prove useful in explaining how people interact with everyday physical scenes.

15.5 Efficient Physical Reasoning

While section 15.4 demonstrates the various ways in which the IPE can be used within the framework of probabilistic cognition to explain different facets of human physical reasoning, features of physics and the IPE make it such that applying generalized probabilistic algorithms to these problems is computationally intractable. First, because there are no analytic equations to describe how physics unfolds except in the most trivial scenarios (Diacu, 1996), general probabilistic prediction requires running the IPE forward a limited number of times to *approximate* the posterior belief about the future state of the world. Second, generalized probabilistic inference algorithms require applying the likelihood function—here, the IPE—hundreds or thousands of times to produce a well-formed posterior distribution, and even more if the algorithm is initialized poorly. Yet we use the IPE to make predictions and inferences about physics in real time. In this section, we describe possible shortcuts that the mind might take to more efficiently approximate probabilistic physical reasoning.

15.5.1 Sampling Simulations

A *sample* is a random value drawn from a probability distribution (see chapter 6). Because the IPE is a probabilistic system (Battaglia et al., 2013; Smith & Vul, 2013), every simulation from the IPE is a sample from the probability distribution over future states of the world, conditioned on current observations. The most straightforward way to use samples from the IPE is through a brute-force Monte Carlo approximation, in which a large number of samples is drawn from the IPE to give a reasonable expectation of the future (see chapter 6 for further details). For example, Battaglia et al. (2013) and Smith and Vul (2013) used large numbers of simulations (48 and 500, respectively) to form predictions and explain how people behave in aggregate. However, using such a large number of samples from the IPE seems rather at odds with limits on an individual's working memory and attention. Do people really sample tens or hundreds of mental simulations before making a decision?

There is an priori reason to think that people may not require a large number of simulations to make a decision. Vul, Goodman, Griffiths, and Tenenbaum (2014) performed a theoretical analysis asking what an optimal decision-making agent ought to do under time constraints. Specifically, if an agent has a limited amount of time to make as many decisions as possible, how many samples should be taken per decision? The answer is a trade-off among the utility of each correct decision, the amount of time it takes to draw a sample, and the reliability of each sample. Intuitively, if it takes a lot of time to take a sample, then fewer decisions can be made, thus resulting in lower utility. However, if each sample is very noisy, then decisions are more likely to be wrong, and therefore it might be advantageous to take more samples. Through formal analyses of this trade-off, Vul et al. (2014) found that in plausible scenarios, it can actually be optimal for an agent to take only a single sample to support a decision. Making a decision based on a single sample also naturally explains the classic cognitive bias of probability matching: in experiments in which people are asked to predict whether a high-probability or low-probability outcome will occur, they tend to predict the outcomes according to their probabilities rather than always predicting the highest-probability outcome as they should (Vulkan, 2000).

To determine the number of samples that people require from the IPE to support physical judgments, Hamrick, Smith, Griffiths, and Vul (2015) ran an experiment in which participants had to predict whether a ball would go through a hole. Crucially, they varied the difficulty of each trial by changing the size of the hole (i.e., either small or large) or the margin by which the ball would go through or miss the hole. On some trials, the ball would go through or miss the hole with high probability according to the IPE (e.g., there might be a 90 percent or a 10 percent chance of going through the hole), while on others, it was very unclear whether it would go through (e.g., closer to a 50 percent chance). Participants in this experiment took longer to make judgments when the IPE predictions were very uncertain, suggesting perhaps that they were taking more samples in these cases.

Through a model of response time based on an optimal model of decision making known as the sequential probability ratio test, Hamrick et al. (2015) showed that differences in their participants' response times could be due to a process in which samples are accumulated until a particular level of confidence is reached. Through this model, they showed that while the number of samples varied across stimuli depending on their difficulty, on average the number of samples ranged from two to four per decision. These results corroborate other more informal analyses by Battaglia et al. (2013) and Hamrick et al. (2016), suggesting that their participants relied on one to six simulations from the IPE to make decisions about towers of blocks. Thus, although each individual simulation from the IPE might be expensive, these results suggest that people can—and do—rely on only a few simulations to still achieve reasonable levels of accuracy in their judgments.

A number of questions remain regarding the computational efficiency of sampling from the IPE as well. If each sample taken from the IPE is actually a noisy physical simulation, then there are additional parameters that can be set that affect the amount of time it takes to run that simulation. For example, there is a choice of how long each simulation should be run for (e.g., how many time steps should occur). Another simulation parameter that can be adjusted is the level of detail the simulation should be run at (e.g., the length of each time step). Similar analyses of the speed-accuracy trade-off can be performed to answer these questions, and these are exciting directions for future research.

15.5.2 Rapid Inferences

A long-standing tradition in psychology has been to treat perception as inference: if we have a generative model of optics, we can condition on our retinal inputs to understand how objects are segmented in the world and where they are located (Von Helmholtz, 1867). This tradition has been carried forwards to suggest that people perceive latent physical properties (e.g., mass) from dynamic scenes by conditioning those variables based on how well their observations match what they should expect to see based on their IPE with different settings of those parameters (Hamrick et al., 2016; Sanborn et al., 2013). In practice, this inference is often carried out by "analysis-by-synthesis" (Yuille & Kersten, 2006): setting the initial conditions of the scene (e.g., the masses and densities of objects), running the IPE forwards, and then perturbing those initial conditions via a process like Markov chain Monte Carlo (MCMC) until the predictions of the IPE match the observations. However, this approach has been criticized for being computationally infeasible for cognition, as in the general case, it requires running the generative model hundreds or thousands of times to form a good posterior estimate over those latent physical variables. This approach would also seem to be at odds with the findings that people use only a handful of physical simulations in most scenarios (as described in section 15.5.1).

If the mind is to produce these inferences as rapidly as it does, it must therefore have ways of speeding up this inference process. One method for doing so is to initialize the inference process with an intelligent guess from bottom-up features (Yuille & Kersten, 2006). Poor initializations require running the generative model to assess model parameterizations that are unlikely to explain the world; conversely, a good initialization can speed up inference by ensuring that each sample from the generative model is informative. Models that implement this rapid initialization via pattern recognition (using deep networks; Wu, Yildirim, Lim, Freeman, & Tenenbaum, 2015) or trained features (Ullman et al., 2018) have been found to describe human inferences better than either pattern recognition or full reasoning over the space of hypotheses.

The analysis-by-synthesis approach to inference traditionally is applied to problems with a fixed amount of information—for example, judging relative masses after observing a video of two objects colliding (Sanborn et al., 2013; Wu et al., 2015). But physical events by their nature are dynamic, unfolding over time. Yildirim et al. (2018) demonstrated that human inferences about weight change along with the unfolding observations from the world. Furthermore, they suggest that additional approximations to the inference process are required to explain how these judgments change over time. Following the theory of rational process models (chapter 11; see also Griffiths et al., 2012b), they suggest that these inference dynamics can be explained by a model based on particle filters, in which a belief about masses is formed as a limited set of hypotheses that are tracked and updated over time. But they also propose another possible explanation: that people might have an approximate inverse IPE that can go directly from observations to latent scene causes. This is similar to other proposals for amortized inference over generative models (Stuhlmüller, Taylor, & Goodman, 2013; Le, Baydin, & Wood, 2016), which suggest that we can use our IPE to imagine scenes that can be used to train an approximate inverse model. This inverse model will be less flexible than analysis-by-synthesis, but it will also be much more efficient and therefore might be useful to have for inference tasks that we must do often or quickly (e.g., judging masses in common scenarios). Determining what approximations the mind uses for online physical inferences, therefore, remains an open area of research.

15.5.3 Physics Hacks and Game Engine Approximations

Many of the approximations that are relevant for IPEs are also relevant for general efficient inference schemes, including sampling and the heuristic use of bottom-up features. However, an IPE may also contain domain-specific *conceptual* approximations, useful for physical reasoning. Engineers that develop physics-engines for video games work under the constraint of generating "good enough" simulations in real time and at everyday scales. Such engineers are not working to create a high-fidelity model of fluid dynamics, cloud mechanics, or molecular interactions, but rather to make a splash of water look reasonable enough. To achieve this, engineers use principled workarounds and shortcuts to overcome limitations of time, memory, and computation. Such workarounds are useful regardless of the specific implementation language or environment of the physics engine (for general game engine concepts, see Gregory, 2018). As the human mind is under similar constraints of simulating physically plausible objects at everyday scales with a limited computational budget, we may find a convergent conceptual evolution between the workarounds and notions used in physics engines and those used by the IPE. Next, we focus on three examples of major short cuts and approximations, but see Ullman et al. (2017) for more detail.

Consider first the notion of *shape* as opposed to *body* in physics-engine software. The shape of an object is what is eventually rendered on the screen, while the body of an object is what is used for actual dynamic calculations and collision detection and resolution. The body is often an approximation of the shape, using bounding boxes and convex hulls (see figure 15.6). As a simplifying example, consider a character in a video game hurling an ornate vase at a wall. While the player may see rendered on the screen an embellished object flying toward the wall (the shape), from the point of view from a physics engine, it would be a waste of resources to exactly and accurately simulate every ridge and dip in the vase

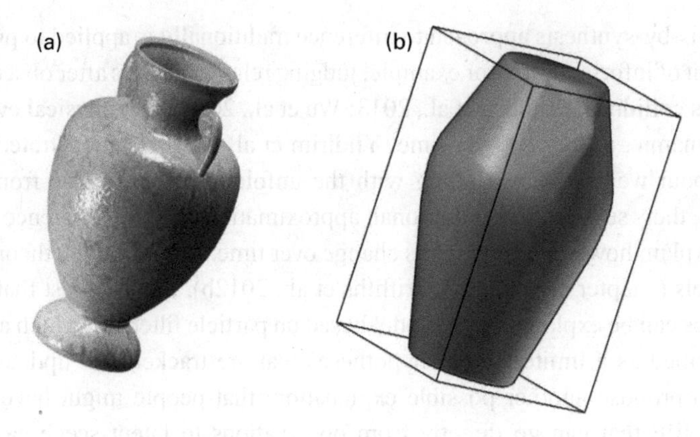

Figure 15.6
Difference between (a) visual shape and (b) physical body representations of an entity in a game engine. The body is an approximation of the meshes used to represent the shape, such as with a convex hull or bounding box. Images are based on an object file created by user kc8qzo on BlendSwap (https://www.blendswap.com/blend/4906).

as it flies and makes contact with the wall. The complex shape of the vase is represented instead by a simple convex hull (the body), or even a box. Such a hull is much easier to store in memory, and it is easier to check when this hull overlaps with another hull or surface to trigger a collision event.

Physical reality does not make such a distinction, of course, but the shape/body split is a useful conceptual scheme in a game engine that runs on hardware with finite memory and computational power, and it may be advantageous for the mind to have such a split as well. Such an approximation may also help to explain why young infants do not use detailed shape representations to track an object's identity as it moves in space, even though they can distinguish them perceptually (Xu & Carey, 1996; Xu, 2005; Ullman et al., 2017; Smith et al., 2019). While game engines do not often set object bodies in a dynamic way, one can imagine the mind making different body approximations depending on the computational budget and task at hand. Figuring out when a vase will strike a surface with limited time to spare, a person may approximate the vase using only a coarse bounding box. By contrast, attempting to grasp a vase by the handle would require a more fine-grained body approximation that takes into account the "hole" that the handle makes in the convex hull.

Another major way that game physics engines save on memory and computation is by assigning entities to the *static* or *dynamic* category. Static items are those that are immobile—objects like the ground or walls—whereas dynamic objects can move and be affected by forces. Crucially, static objects are not treated as large dynamic masses, but instead have undefined mass and so are unaffected by collisions and other forces. As with body and shape, this distinction between static and dynamic entities obviously does not exist in real physics. But it is an extremely useful approximation from an engineering perspective (e.g., it would be wasteful to calculate the infinitesimal effect that dropping an object on the ground has on the motion of the Earth), and thus one that the IPE might use. Such a distinction can help explain why extended surfaces are used earlier in development for navigation compared to everyday objects (Hermer & Spelke, 1994; Lee & Spelke, 2008), why shifting a wall causes changing posture and loss of balance in children and adults (Lee & Aronson, 1974), and why even very young infants expect an object made of disparate parts to move together when it is lifted, but not to take the floor with it (Spelke, Breinlinger, Jacobson, & Phillips, 1993).

Finally, the IPE may reduce memory and computational requirements by judicious choices of what to include in its representations. To some extent, this is necessary: the world around us contains a great number of objects that exceed the limits of our memory. But the representations in the IPE might be simplified beyond these representations to include only the objects believed to be relevant to the simulations. This oversimplification can explain a curious error in physical reasoning: people rate the conjunction of two events as more likely than just one (e.g., that "Ball A hits ball B and ball B ends up in a goal zone" is more likely than just "Ball B ends up in a goal zone"; Ludwin-Peery, Bramley, Davis, & Gureckis, 2020). This error—known as the *conjunction fallacy*—is an impossibility according to probabilistic reasoning. Yet by assuming that people are failing to consider ball A to be relevant if they are not directly cued with the conjunction question, the presence and magnitude of this conjunction fallacy can be quantitiatively predicted across a range of scenarios (Bass, Smith, Bonawitz, & Ullman, 2021).

There are other such concepts and shortcuts that help to organize a simulation and simplify computation, and some of them seem to explain otherwise puzzling psychological phenomena (see Ullman et al., 2017). And the inspiration can flow in the other direction—by studying the principled concepts and workarounds the IPE uses, cognitive scientists can help to develop useful tools for engineers who develop game engine simulations. Of course, it is possible that many of the concepts and workarounds in game physics engines are only the result of explicit development by engineers, with no correlate in the IPE. But given the similar need to create approximate physical representations, the possibility that there may be a connection is well worth exploring.

15.6 Future Directions

Despite the significant progress that has been made in developing probabilistic models of intuive physics, there are still a number of exciting directions that remain open for future research.

15.6.1 Learning Models of Physics

Much of the work described in this chapter has focused on mature models of physics, where the rules governing dynamics are known and all learning is about the objects in the world, including their configuration and properties (e.g., learning about an object's mass). But there are cases where people must learn about entirely new dynamics. Infants develop an understanding of the physical world over the first few years (see chapter 20), and adults can learn to handle novel physics, such as when they encounter a new video game world.

Learning physical dynamics can naturally be formulated in a probabilistic framework, by extending inference to include the structure of the physics engine. In equation (15.5), inference was defined as forming a posterior over latent object properties (l) such as mass, conditioned on a set of observations ($o^{1..t}$), and given a fixed physics engine (Φ). If we now consider a particular instantiation of a physics engine as $\Phi = \phi$, then learning new physics becomes updating a posterior over a space of possible physics engines:

$$p(\phi | o^t, l) \propto p(o^t | l, o^0, \phi) p(o^0 | l) p(l) p(\phi). \tag{15.8}$$

This formulation allows the physics engine to be any function that transforms one scene state into another, and thus presents an intractable space to search through. It is unlikely that people are reasoning about all aspects of the physics engine in a given situation, so a hierarchical scheme is useful here in which the top most level of the hierarchy assumes only the existence of objects, properties, and laws of dynamics, with more specific properties or dynamics learned as one moves down the hierarchy (e.g., elasticity or a repulsive force; Ullman et al., 2018). These top-level concepts might map onto the representations that enable "core knowledge" of physics even in early infants (see Spelke et al., 1992; Baillargeon, 1994, 2004; Kinzler & Spelke, 2007, and chapter 20): allowing object permanence by storing things in the world as discrete items that persist over time, or representing expectations that constrain motion to require smooth spatio temporal paths as part of the most basic rules of dynamics. This framework suggests a structured representation for expressible models of dynamics that limit the space of physics engines that we might search through.

The machine learning community has also made progress on studying how physics might be learned by introducing approaches for training models from observed experience to predict the physical dynamics of objects and materials over time. While earlier approaches simply trained neural networks to mimic the dynamics of articulated physical systems (e.g., how a simulated robot arm's limbs move; Grzeszczuk, Terzopoulos, & Hinton, 1998), more recent approaches often use structured architectures that assume that physical systems can be represented as graphs, with objects representing the nodes and the relations (e.g., the possibility that two objects could interact) as the edges (Battaglia, Pascanu, Lai, Jimenez Rezende, & Kavukcuoglu, 2016; Chang, Ullman, Torralba, & Tenenbaum, 2016). This approach builds aspects of core knowledge (e.g., that objects exist) into the architecture, as well as critical dynamical invariances (e.g., collisions between different objects are all governed by the same laws of physics). Beyond the architecture, however, learning is assumed to be driven by the training data, without the need for structured knowledge representations.

The machine learning approach has both benefits and drawbacks compared to the hierarchical Bayesian modeling approach. The hierarchical approaches rely on prespecified primitive bits of knowledge that are composed into a physics engine, while machine learning algorithms can function in environments where these primitives are not readily available. Using similar architectures as for rigid-body physics, they can learn to simulate or make inferences about nonrigid materials and fluids (Li, Wu, Tedrake, Tenenbaum, & Torralba, 2018; Guevara et al., 2018; Bouman, Xiao, Battaglia, & Freeman, 2013; Sanchez-Gonzalez et al., 2020; Pfaff, Fortunato, Sanchez-Gonzalez, & Battaglia, 2021), as well as nonphysical dynamics such the movements and interactions among intentional agents (Sukhbaatar, Szlam, & Fergus, 2016; Hoshen, 2017; Tacchetti et al., 2018; Sun, Karlsson, Wu, Tenenbaum, & Murphy, 2019), suggesting a method for joint physical and social prediction without specifically engineering how these systems are conjoined. However, despite the flexibility of these learned models, they are not easily interpretable, which makes it difficult to understand how they might represent and use physical constants that are required by the IPE (e.g., gravity or mass), and therefore how they might generalize to novel scenarios.

While there are preliminary studies of what physical knowledge might be captured by these models (e.g., Piloto et al., 2018; Riochet et al., 2018), further work is required to understand how learned models of dynamics capture human physical concepts. In addition, while graph-based machine learning models can in theory learn dynamics from raw image input, robust methods to do so are still being developed and this work has so far largely focused on rigid-body physical dynamics (e.g., Watters et al., 2017; Veerapaneni et al., 2020; Kipf et al., 2021). In addition, simulation models that make predictions from images do not always predict distributions over future states; even when they do, this stochasticity is learned from training data (Babaeizadeh, Finn, Erhan, Campbell, & Levine, 2018) and is not a tunable parameter that could be used to explain human performance. This limits their utility as a probabilistic cognitive model. More research is thus needed to understand how a general, probabilistic physics simulator might be learned.

15.6.2 Combining Simulation with Rule-Based Reasoning

This chapter has focused on the IPE as the forward model of physics that people use, as its stochastic nature makes it easily interpretable within the framework of probabilistic cognition. But there are also theories of physical reasoning suggesting that people do not use simulation for physical reasoning, and this reasoning is instead based on a set of axioms and logical rules (Hayes, 1979; DiSessa, 1993). These logic-based theories have been used to explain how people reason about containment relationships (Davis, Marcus, & Frazier-Logue, 2017), use biased rules to judge whether objects will balance on a beam (Siegler, 1976), or use heuristics to predict how water will settle in a tipped container (Vasta & Liben, 1996).

Although simulation theory and rule-based reasoning make very different assumptions about the underlying representations and mental processes that support physical reasoning, they describe separate capabilities that we are able to bring to bear to understand the world depending on the situation. For instance, Smith et al. (2018) found that when people are asked to catch an object in ballistic motion, their predictions are consistent with simulations from an IPE, but when those same people are asked to draw the motion of objects in identical situations, their drawings demonstrate idiosyncratic biases. This result supports theories suggesting that we typically use simulation in scenarios that are more dynamic and realistic, but they use rules and heuristics when encountering more abstract diagrams or explicit problems (see also Schwartz & Black, 1996a; Hegarty, 2004; Zago & Lacquaniti, 2005).

The cognitive systems that underlie logical reasoning are often posed as mostly deterministic, which makes them difficult to reconcile with probabilistic cognition. It is, therefore, important to understand how simulation and rules can be combined into a probabilistic framework. Prior work has focused on how people can learn these rules from simulation and feedback, where it is easy if the rule is physically relevant (Schwartz & Black, 1996b), but more difficult with unrelated cues (Callaway, Hamrick, & Griffiths, 2017). These findings often assume that once a good rule is learned, it will supplant the use of simulation (Schwartz & Black, 1996b).

But there are many scenarios where we do not use just simulation or just rules. For instance, there are cases for which logical analysis of a scene provides a clear answer but people still rely in part on simulation. When predicting the motion of a ball that is contained within a box, it should be easy to judge that the ball will never reach an area outside the box based on the containment relationships alone (Davis et al., 2017), but people will at least sometimes use simulation to make those judgments (Smith et al., 2013; Smith, Peres, Vul, & Tenenbaum, 2017). Similarly, there are situations where people use rules that are biased and less accurate than physical simulation: the rules that people use for balance judgments produce biases that privilege weight over leverage when comparing torques around a center point (Siegler, 1976), but these biases cannot be derived from an IPE (Marcus & Davis, 2013). In these cases, people must choose between inaccurate but cheaper heuristics versus more accurate but more cognitively expensive simulation. While there have been initial proposals for how this trade-off is performed (e.g., based on an implicit cost/benefit comparison; Smith et al., 2018), it is an open question how people choose among and combine these various systems for physical reasoning, and how this combination of systems fits within the general framework of probabilistic cognition.

15.6.3 Joint Physical and Social Reasoning

Intuitive physics and intuitive psychology deal with seemingly different domains—objects and agents, things and people. Even infants have diverging expectations when an entity is seen as a physical body compared to a perceiving agent, and some cognitive development researchers propose that different reasoning systems form two separate modules for handling these separate domains (Kinzler & Spelke, 2007), with a classification scheme that triggers different expectations depending on the type of entity that is being considered. Ongoing work in cognitive neuroscience has also identified dissociations in brain region activity when processing physical and social scenes (Isik, Koldewyn, Beeler, & Kanwisher, 2017; Fischer et al., 2016). However, even if these two domains are handled by two different computational modules, they must work in concert to produce reasonable interpretations of common scenes. Agents are physical beings that are subject to physical constraints, and these constraints help make sense of their goals, beliefs, and intentions. Consider, for example, a simple scene in which 10-month-olds see an agent jump over a barrier to get to a goal (Gergely, Nádasdy, Csibra, & Bíró, 1995). When the goal is removed, both adults and infants expect the agent to make a bee line for the goal, rather than repeat the spatio temporal trajectory that it took previously (jumping over a now nonexistent barrier). Such an expectation is obvious and intuitive, but only if we take agents to have goals, to act efficiently to achieve their goals, and—crucially for the current point—not to be able to pass through solid barriers.

Using the framework of Bayesian Theory of Mind to capture intuitive psychology (Baker, Saxe, & Tenenbaum, 2009; Baker, Jara-Ettinger, Saxe, & Tenenbaum, 2017; see also chapter 14), the link between psychological and physical reasoning can be made in several ways. First, physics provides the baseline *transition function* for the world, which is needed for planning actions. That is, in order to plan, an agent needs to know $P(s'|s, a)$, the probability of moving to a new state s' conditioned on being in a specific state s and taking a specific action a (which may be not to act at all). In general, such a transition function is arbitrary and can apply to any planning context, (e.g., it can describe the possible legal moves in an abstract game of tic-tac-toe), but in a real-world dynamic context this transition function is provided in part by the IPE ('"If I throw this apple, what will happen?"). By inverting such a planning procedure, people can work backwards to reason about the goal that generated that plan (see Holtzen, Zhao, Gao, Tenenbaum, & Zhu, 2016, for an implementation that infers people's hierarchical goals from videos of them moving in an everyday environment).

Second, physics provides a natural notion of cost, which can be used to estimate the reward of the agent. A great deal of psychological reasoning can be reduced to the Naive

Utility Calculus (Jara-Ettinger, Gweon, Schulz, & Tenenbaum, 2016):

$$U(a, s) = R(s) - C(a). \qquad (15.9)$$

Here, the utility U of an agent is determined by the reward of state s and the cost of action a. If we have a good estimate for C, we can reason about the likely rewards that drove an agent to pay that cost (see chapter 14 for further details). There can be different types of cost, coming from mental effort, opportunity cost, temporal discounting, and others. But a basic, natural type of cost is physical effort. The more that an agent is willing to physically exert itself to get to a particular state s, the more that s must be worth. Even young infants can infer value from cost in this way, reasoning that if an agent was willing to climb a steep hill to get to goal A, but only a shallow hill to get to goal B, then A must be worth more to the agent than B (Liu, Ullman, Tenenbaum, & Spelke, 2017), although more work is needed to establish whether the physical effort here is related to force or distance. In a social situation, young children can use a similar calculus to reason that if person A is unwilling to spend some small amount of physical effort to help B, then person A must not really like B (Jara-Ettinger et al., 2016). In this way, the IPE and Naive Utility Calculus can jointly provide a unified computational framework for explaining the everyday inferences we make about the plans of others given their physical constraints (see also Sosa, Ullman, Tenenbaum, Gershman, & Gerstenberg, 2021).

15.7 Conclusion

We regularly reason about our physical world by making predictions about what will happen next, updating our beliefs about the properties of objects, and planning how we will act. While these tasks often intuitively seem effortless, performing them requires both rich generative models of the world and the ability to deal with the underlying uncertainty in perception and dynamics. Probabilistic models of cognition can help us explain how we can simulate physics under uncertainty, as well as how those simulations support a range of ways of reasoning about the world. Conversely, studying physical reasoning can help develop an understanding of how the mind approximates Bayesian principles in complex domains, as many of the problems that we solve easily are in principle computationally intractable. Thus, physical reasoning is a quintessential domain in which to use and extend probabilistic models of cognition.

16

Language Processing and Language Learning

Nick Chater, Andy Perfors, and Steven T. Piantadosi

A main objective of this book has been to illustrate how probabilistic inference over rich structured representations provides a powerful machinery for modeling human intelligence. We have seen that structured representations from graphical models to logic can help represent knowledge and categories, encode the perceptual world, or reason about naive physics or other minds. Yet the domain in which structured representations are most transparently relevant to cognition is, of course, human language. In this chapter, we apply some of the foundational ideas developed in this book to understand the cognitive processes underpinning language processing.

Probabilistic ideas have often been overlooked or even actively pushed aside in the study of language (e.g., Chomsky, 1969; see also Norvig, 2012). One reason for this is that it has sometimes been assumed that a probabilistic approach to language can work only if language has a very simple structure, corresponding to statistics over pairs or triples of phonemes or words (for a discussion, see Jurafsky & Martin, 2008), or through learning associations between words or distributional patterns linking words and their contexts (Landauer & Dumais, 1997; Redington, Chater, & Finch, 1998). Indeed, probabilistic approaches to language, and by extension connectionist approaches, have sometimes been viewed as something close to a covert return to behaviorism (Fodor & Pylyshyn, 1988). But we have seen that this reaction against probabilistic ideas as incompatible with structured symbolic representations is out of date (Chater & Manning, 2006). Indeed, understanding how language is processed and learned requires the integration of both structured representations and probabilistic methods.

Probability enters into the cognitive science of language in two crucial ways. First, we have the problem of interpreting language—that is, creating a rich representation of the phonemes, words, syntactic structure, and, critically, the meaning, of linguistic input from a noisy and highly ambiguous stream of speech (or the similar ambiguous stream of visual input in the context of sign language). The problem of inferring the most likely structure from a noisy input is, of course, a paradigm example of probabilistic inference—and as we have seen throughout this book, the standard Bayesian approach to this type of problem is to attempt to invert a generative model of the language. Thus, to work out the most likely analysis of the speech input requires inverting a model for generating, or synthesizing, this speech input. This general line of thinking has a long history in the psychology of language, tracing back to the analysis-by-synthesis models of speech perception developed at the Haskins laboratory in the 1950s and 1960s (e.g., Halle & Stevens, 1962), and is, of course, in line with the Bayesian viewpoint on cognition explored throughout this book. We shall see that many of the computational problems associated with symbolic approaches to language processing, such as the spectacular ambiguity of natural language, are greatly eased when multiple levels of probabilistic constraints can be applied to prune the vast number of possible readings of a sentence, or indeed, an acoustic wave form. We note that there is an increasing body of experimental evidence across a range of linguistic levels that fits well with the probabilistic framework.

The second way in which probability enters the story concerns how the generative model of language is *learned*. Here, the objective is not merely to infer the structure of speech or other linguistic input in real time. It is instead to infer a model of the entire language itself, and to thus be able to use this model in using language—to correctly produce language, to understand novel sentences that may never have been heard before, and to distinguish between sentences that are grammatically acceptable and those that are not. This model is learned through experience, particularly crucially in the first years of life. Learning language is, then, yet another problem of probabilistic inference, where now the aim is not to infer the structure of a particular speech signal, given a generative model of the language, but to infer the generative model of the language itself, given a history of linguistic (and potentially other) input. Bayesian models of learning natural language grammars also throw a radically new perspective on nativist arguments by Chomsky (1980), Pinker (1994), and many others. These authors have argued for the necessity of an innate universal grammar due to the supposed impossibility of learning an infinite language from the finite, and indeed somewhat noisy, sample of utterances available to children. The claim that learning a language without a great deal of innate information is impossible is backed up by *poverty of the stimulus* arguments: that the child has too little, and too poor-quality, data for learning to be possible (Chomsky, 1986).

As in the case of understanding individual utterances, purely symbolic models of language acquisition have no principled way of prioritizing between the vast range of possible models of the language that will fit with a particular body of linguistic data (and indeed, they also tend to fare badly when dealing with noisy data). The Bayesian approach addresses this problem by focusing on grammars that are a priori more plausible,[1] but that also fit well with the observed data.

Mathematical arguments show that this approach can work in principle, to overcome the apparent "logical" problems of language acquisition, without building in strong language-specific prior information (e.g., Chater, Clark, Goldsmith, & Perfors, 2015; Chater & Vitányi, 2007). Here, we will describe recent computational work that has demonstrated that Bayesian learning models can acquire a wide range of aspects of natural language from language corpora, without needing to build in language-specific prior information. Later in this chapter, we review work that focuses on the acquisition of the key grammatical patterns hypothesized by Chomsky and others as central to language, using probabilistic inference over programs. This work uses *adapter grammars* and similar approaches, which allow the

1. One natural a priori bias is in favor of simpler grammars, as we will discuss further in chapter 21.

possibility of gradually constructing an increasingly complex grammar as more linguistic data is encountered, using the principles of nonparametric Bayesian modeling introduced in chapter 9).

This approach is readily compatible with item- or construction-based models of language that are currently prevalent in linguistics and language acquisition research. We also consider how Bayesian learning over symbolic structures relates to the astonishingly high-levels of performance in natural language tasks by very large deep neural networks trained on vast linguistic corpora. Nonetheless, we argue that such *large language models* do not yet provide a plausible cognitive model of human language processing, for a number of reasons, including the lack of a natural interface with meaning and the pragmatic use of language (Chater, 2023).[2]

16.1 Language Processing

According to conventional approaches in linguistics and psycholinguistics, language is governed by many layers of representations, which can be ordered in increasing levels of abstraction (see figure 16.1). When interpreting speech, for example, we might begin with an acoustic representation of the input arriving at the ear (which might be something akin to Fourier analysis, picking out the spectral power at each frequency). The acoustic input arriving at the ear, of course, will contain not merely the speech that we are attempting to understand, but background noises of all kind—ranging from the chatter of other speakers to background music and the rumble of traffic. One immediate challenge is, then, to separate out the acoustic signals associated with speech. Another challenge is to infer the phonemes in the speech signal regardless of the enormous variations between accents, individual speakers, acoustic environments (down a phone line, in an echoey swimming pool, etc), patterns of intonation, and many more variations. Then there is the task of inferring how the stream of phonemes splits into words (and morphemes, such as verb endings, case markings, and other elements), and how these words cohere to convey meaningful phrases and sentences and more abstract levels of meaning concerning how what is being said fits into the rest of the conversation, or how it relates to current perceptual input or to background knowledge. Each of these steps is formidably challenging and potentially interdependent (so that, for example, information about meaning may help us decode noisy or corrupted speech (Mattys, Davis, Bradlow, & Scott, 2012)).

In view of our focus on probabilistic models of structured representations, we focus here on language interpretation at or above the level of the word. Language scientists typically assume the patterns in language above the level of the word (or more strictly, the morpheme, to include meaningful sound units such as tenses and cases) is internally represented as a grammar: that is, a system of symbolic rules that can generate the sentences in a language and provide an analysis of sentence structure that provides a framework for semantic analysis (i.e., the analysis of the meaning of a sentence).

2. Nonetheless, Contreras Kallens, Kristensen-McLachlan, and Christiansen (2023) convincingly argues that the spectacular success of large language models does severely undermine the credibility of in-principle arguments that language learning is impossible without substantial innate grammatical knowledge.

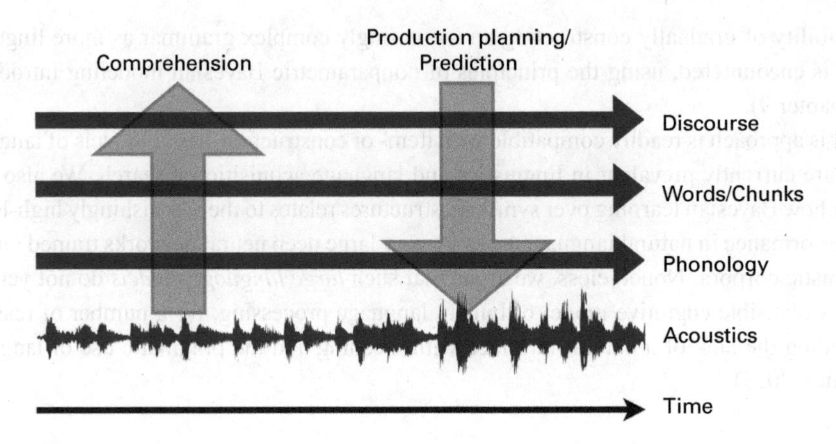

Figure 16.1
Computing levels of linguistic representation in real-time language understanding and production. In language understanding, the brain is assumed to begin with acoustic representations and to successively compute more abstract and temporally extended representations (from phonemes, to words, all the way up to representations of the entire discourse)—note that theorists differ concerning the number and nature of the specific representations. In language production, the process is reversed, beginning with abstract representations of meaning and generating an acoustic signal. Yet the computational mechanisms underpinning understanding and production may be tightly coupled (Pickering & Garrod, 2013): when we are listening to speech, we are engaging in top-down processing to reconstruct what a person is saying to us in real time. This "analysis-by-synthesis" perspective on the perception of speech is naturally aligned with the Bayesian approach to interpreting sensory input more broadly (Yuille & Kersten, 2006). Figure redrawn and adapted from Christiansen and Chater (2016a).

16.1.1 The Challenge of Ambiguity

The process of grammatical analysis is fraught with difficulties. One set of problems concerns the spectacular ambiguity of individual words. You most likely interpreted the word "set" in the last sentence without pause or difficulty. But, taken in isolation, it has at least 13 possible meanings, assuming that "set" is functioning as a noun, and even more verb and adjective meanings, according to a psycholinguistic database of meanings like WordNet (Miller, 1995). We have a set lived in by a badger; a set of tennis; we get set; read set texts; we can set our heart on something; set off on a journey; wonder whether our ice has set; be set up; and many, many more—where the same word has many syntactic functions as well as different meanings. This ambiguity likely helps to make language efficient by allowing us to leave out information that is redundant with the context (Piantadosi, Tily, & Gibson, 2012b), but in turn, it makes language processing and understanding more complex because comprehenders need to sort out which meaning was intended. If we created distinct words for every possible meaning of "set," for example, these words would necessarily be, on average, longer and more complex—there are, after all, only a limited number of short words to go around. But the same point applies to resolving ambiguity in any way—it takes extra linguistic material to remove ambiguity, which slows communication. The cognitive system must find a balance between using snappy but highly ambiguous language, which needs to be disambiguated by inferences from context, and less ambiguous but more ponderous language. It turns out that humans prefer a surprisingly high level of ambiguity, with which our brain copes remarkably well.

The challenges increase when we turn to syntactic ambiguity, where a single sequence of words can have multiple possible sentence structures, and often these alternative structures have distinct meanings. For example, "I tripped the clown with the skateboard" could mean that I *used* the skateboard to trip the clown, or it could mean that the clown possessed the skateboard.

Psycholinguists—who use primarily experimental methods to study how the brain represents and processes language—have studied even more diabolical examples, such as *garden path* sentences. To choose a famous example from Bever (1970), consider the sentence fragment "The horse raced . . ." This is naturally (and, one might think, inevitably) viewed as a sentence in which "the horse" is the subject of the verb "raced"; but if the sentence turns out to be "The horse raced past the barn fell," this assumption is revealed to be incorrect, and the language processor will be confounded. To understand the sentence, we have to reanalyze the beginning and resolve the ambiguity of "The horse raced . . ." another way. The only way to make sense of the structure of this sentence is to see it as a contraction of "The horse that was raced past the barn fell"—a structure analogous to "The picture painted by the artist fell." The horse turns out not to be the subject of the verb "raced" after all, but of the verb "fell"—and we are picking out the particular horse "that was raced past the barn" by person or persons unknown. This example shows that even a simple string of three words "The horse raced . . ." can turn out to be unexpectedly locally syntactically ambiguous—until we see the rest of the sentence, the language processor often cannot know for sure that it has inferred the right structure.

The language processor does tend to jump to conclusions, though. Indeed, it turns out that the brain typically uses all the information it can to resolve syntactic and other ambiguities as quickly as possible. But this approach to ambiguity resolution will sometimes lead to trouble if the brain's first guess is incorrect. Thus, once the language processor has jumped to the conclusion that "The horse" is the subject of the past-tense verb "raced," the arrival of the next word, "fell," leads the system to run aground: a recalculation is required.

This type of garden path phenomenon is no mere curiosity. Indeed, one of the remarkable discoveries of symbolic computational linguistics has been that lexical ambiguity, as well as local syntactic ambiguity, is everywhere—so the number of possible readings of the parts of a sentence that the brain has to choose among typically grows exponentially with sentence length. It is easy to imagine that we can dismiss such phenomena by the riposte that *in context*, and with the right intonation (if the sentence is spoken, not written), it will almost always be "obvious" which structure is the right one. This is quite right (Piantadosi et al., 2012b; Miller, 1951)—but it raises the scientific questions targeted by linguistics and psycholinguistics of *how* the language-processing system is able to succeed in using context to resolve such ambiguities, and to do so in real time.

One simple strategy is for the language processor to have a bias in favor of common syntactic strutures. And indeed, people have been shown to use cues like the baseline frequency of different readings to make the best guess about the appropriate structure. But lots of other factors matter too, such as the specific words involved, prior context, stress, intonation, and many more. The integration of different types of probabilistic cues to provide the best overall interpretation is precisely where a probabilistic approach to inference is especially helpful. Jurafsky (1996) outlines a pioneering model of how to frame and solve this type of problem in probabilistic terms, accounting for a wide range of psycholinguistic phenomena (see also Jurafsky, 2003).[3] This model ranks possible ambiguous meanings and syntactic constructions by their conditional probability, pruning low-ranked options using a popular heuristic breadth-first search algorithm called "beam-search." The pruning of unpromising interpretations in parsing explains why, for example, the reading of "The horse raced past the barn": as meaning "The horse that had been raced by the barn" has been rejected as highly unlikely, before the arrival of "fell." So when "fell" is encountered, the language processor becomes stuck and finds it difficult to recover to make sense of the sentence.

16.1.2 Probabilistic Parsing

Let's look at the problem of assigning syntactic structures to sentences—the problem known as "parsing"—more formally from a probabilistic point of view. Probabilistic parsing involves estimating the probability of different parse trees t (or whatever grammatical formalism we favor, which might take the form of dependency diagrams or attribute value matrices, among many others), given a sequence of words s.[4] Suppose that we have a probabilistic model P_m of the language. Then, using the normal Bayesian formula, we have

$$P_m(t|s) = \frac{P_m(s|t)P_m(t)}{\sum_{t'} P_m(s|t')P_m(t')}. \tag{16.1}$$

So we need a prior $P_m(t)$ over tree structure t; and we also need a way of working out the conditional probability $P_m(s|t)$ of a sentence s, conditional on a particular t. However, to work out the denominator, we face the usual problem of summing over a potentially very large set of possible trees—which can be computationally costly.

The prior $P_m(t)$ can perhaps most naturally be determined by the complexity of the parse tree.[5] Specifically, if the shortest code that could express the parse tree has length(t), then one natural prior is proportional to $2^{-\text{length}(t)}$. But the task of finding the *shortest possible* code for a parse tree (or almost any other representation) will not in general be computable—the space of possible codes is too difficult to search. But a crude heuristic is to use instead the length of the parse tree when expressed in a standard form, and a simple coding language (or even more crudely, simply to count the number of grammatical rules invoked). The more crucial question concerns working out the conditional probability of the sentence, given the tree. If the language model takes a simple form, such as a stochastic phrase structure grammar, then this may be fairly straightforward.

Figure 16.2a shows a simple grammar fragment, with *phrase structure* rules and rules for converting *syntactic categories* into specific words. Each of these rules can be associated with a probability. So, for example, generating any sentence begins with an S symbol, which

3. Many influential early psycholinguistic models deliberately ignored probabilistic cues and focus instead on attempting to prune syntactic ambiguities based on principles based on syntactic structure alone, with principles such as "minimal attachment" and "late closure" (e.g., Frazier & Fodor, 1978). This perspective was partly driven by the idea that syntactic processes in language should be independent of other linguistic levels (e.g., Fodor, Bever & Garrett, 1974).

4. We'll abstract away from details of the speech input henceforth, and consider s to be a string of words—but this is an oversimplification. Intonation, in particular, provides useful guidance about the syntactic structure of a sentence.

5. We will return to this issue in chapter 21, on algorithmic probability and related ideas.

(a)
S → NP VP (1)
VP → V NP (.75)
VP → V NP PP (.25)
NP → Det N (.7)
NP → NP PP (.3)
PP → P NP (1)
V → saw (.8)
V → prodded (.2)
N → telescope (.2)
N → stick (.3)
N → girl (.3)
N → boy (.1)
N → cat (.1)
Det → the (1)
P → with (1)

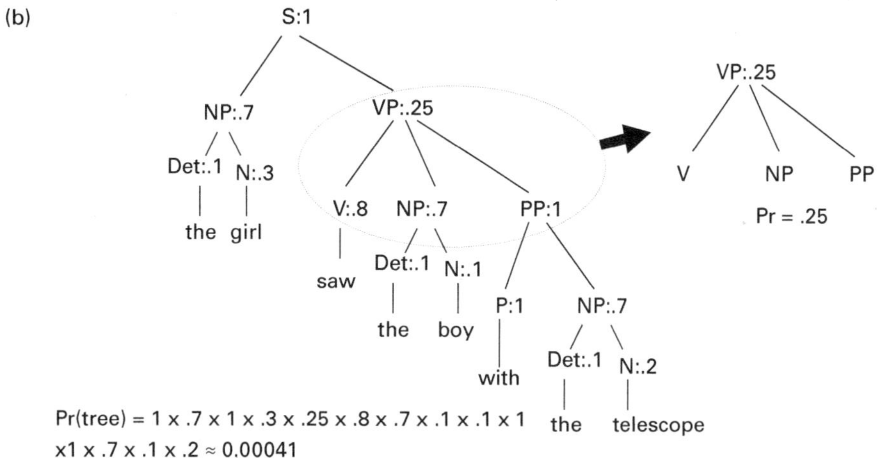

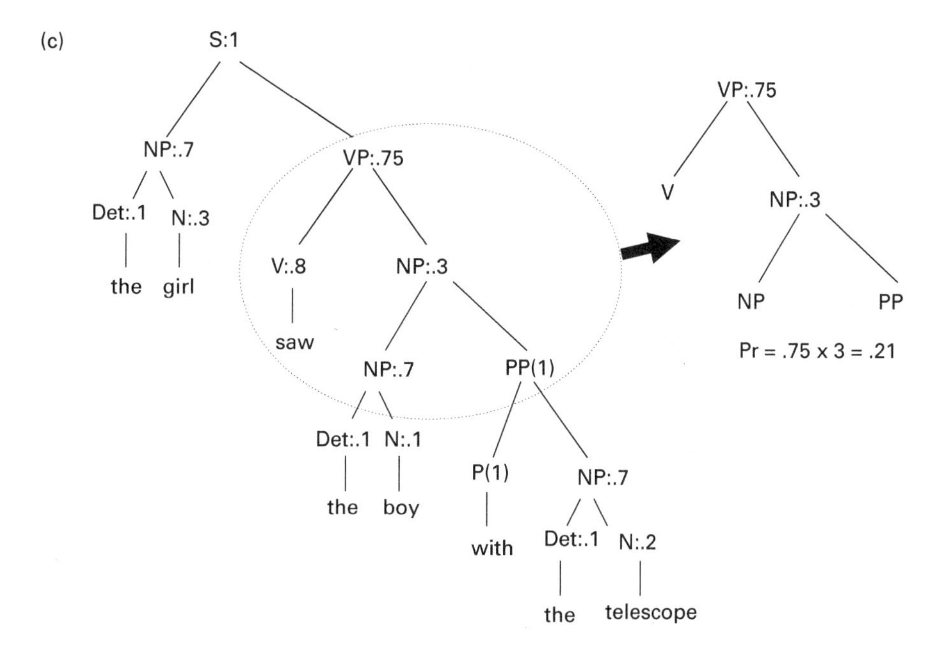

Figure 16.2
Ambiguity and phrase structure. (a) A fragment of a stochastic, context-free phrase structure grammar, which can generate a simple sentence in two ways (b), (c). A Bayesian parser should prefer trees with higher probability. Focusing on the conditional probability of the sentence, given the tree, note that the two options (b) and (c) differ only with regard to whether the prepositional phrase "with a telescope" attaches to the verb (modifying how the seeing is done) or the object noun phrase "the boy" (i.e., it is the boy with the telescope who is seen). The parts of the tree that differ are highlighted to the right. Here, the flatter tree structure invokes one less grammatical rule and assigns the word string a higher conditional probability, and hence, it should be preferred. Figure adapted from Chater and Manning (2006).

in this simple grammar generates NP VP with probability 1. The NP then generates either V NP with probability .75 or a V NP PP (i.e., adding a prepositional phrase) with probability .25. A crucial simplifying assumption here is that the rules operate independently—each NP has the same probability of converting to a V NP or a V NP PP structure, for example, whatever its role in the rest of the sentence. The process of applying rules continues until we have a string of words, such as "The girl saw the boy with the telescope." This whole process provides a generative probabilistic model P_m for syntactic structures, and thereby over strings of words (the parse tree t specifies a string of words s, such that $P_m(s|t) = 1$). Thus, the probability $P_m(t, s)$ of a particular string of words s with a parse tree t is simply the product of the probabilities of the different rules in the parse tree. Then the total probability of a string of words, $P_m(s)$, is just the sum of these probabilities for the different trees, t', that yield that string of words: $\sum_{t'} P_m(t', s)$ (this is just a rearrangement of the denominator in the Bayesian equation given here).

Figures 16.2b–c illustrate how two different syntactic trees can generate the same string of words—generating a syntactic (and here also semantic) ambiguity. The structures differ regarding whether "with the telescope" is a prepositional phrase modifying how the action of seeing was achieved: "the girl [saw] [the boy] [with the telescope]," or whether this phrase picks out a particular boy who was seen: the girl [saw] [the boy with the telescope].

When faced with a string of words which is ambiguous, the usual Bayesian procedure will prefer the parse tree with the highest probability. For example, if Bayesian inference is approximated by sampling, it may be biased toward choosing high-probability trees in proportion to their probability. As can be seen in figure 16.2, these preferences can sometimes be determined locally by considering only the relevant parts of the parse tree that differ. From a probabilistic viewpoint, which structure is preferred depends on the specific probabilities in play, in contrast with previous structural models of parsing where the shape of the syntactic tree is decisive. Experiments have indicated that parsing preferences do seem to follow probabilistic, rather than a purely structural, principles across a number of languages (Levy, 2008; Desmet & Gibson, 2003; Desmet, De Baecke, Drieghe, Brysbaert, & Vonk, 2006).

Note, though, that capturing such psychological data requires a much richer probabilistic model than specified here, in which the resolution of syntactic ambiguities can be influenced by specific lexical items, the hearer's recent experience and general background knowledge (Traxler, 2014). Thus, for example, if the hearer is listening to a report of evidence that boys have been stealing equipment from the observatory, then the much more likely interpretation is that the boy, not the girl, has the telescope. Similarly, the parallel sentence "The girl saw the boy with the microscope," while equally syntactically ambiguous, will not typically be interpreted as implying that the girl is looking through the microscope; on the other hand, as part of a story in which some children have been magically shrunken to a minuscule size, this interpretation will suddenly become more probable. This last example illustrates that the probabilistic approach to language focuses on the probability that particular strings of words will be said (and the underlying meanings and syntactic structures that might underlie people's generation of these strings), not the probability that these sentences are true. After all, the language processor has no difficulty understanding fairy tales by assuming that each new sentence is a plausible continuation of the story, even though the probability of the events being true may be close to zero.

What counts as a plausible sentence in a particular communicative context given a complex and noisy speech input can depend in principle on information of any and every sort—what is plausible depends on the speech signal itself, the immediate environment, the prior linguistic context, the hearer's model of the mind of the other person, their general knowledge about the world (and hence, indirectly, what it is reasonable to say about that world), and many other factors.

This full complexity is surely too great to be mentally represented in a probabilistic model of the language; and, in any case, even moderately complex probabilistic models are too complex for exact calculation, a problem exacerbated by the time pressure under which language processing operates. And, indeed, the language processor does often settle for what seem to be "good enough" parses, using incomplete probabilistic analysis, although these analyses may turn out to be incorrect. Thus, for example, people typically, but wrongly, interpret "While Anna dressed the baby spit up on the bed" as implying that Anna dressed the baby (Christianson, Hollingworth, Halliwell, & Ferreira, 2001; Ferreira, Christianson, & Hollingworth, 2001). Indeed, people's interpretations of anomalous sentences can be well explained by Bayesian decoding that takes into account both plausibility and the probability of different kinds of mistakes or mishearings of the sequence of words (Gibson, Bergen, & Piantadosi, 2013).

The calculations in figure 16.2 may suggest that the language processor considers the probability of different readings of a sentence in its entirety—but this would require waiting for the end of the sentence before beginning a probabilistic analysis. This would make communication painfully slow; but more importantly, it would run into fundamental limitations of human memory, which requires linguistic information to be chunked and recoded as soon as it is received, because otherwise it will immediately be overwritten by the onrushing torrent of speech (Christiansen & Chater, 2016b). So the language processor has to make probabilistic guesses in the moment, in the light of whatever information is available; and, as fresh words arrive, these guesses will sometimes prove to be incorrect (as we saw with the "The horse raced past the barn fell" example).

From this point of view, we can think of language processing as requiring continual anticipation of what is likely to come next, suggesting a tight coupling between language understanding and language production (Pickering & Garrod, 2013). This viewpoint captures the fact that we are often able to finish one another's sentences, and that turn-taking handovers in fluent dialogue are astonishingly fast, implying that we often know what people will say before they have finished saying it and hence can already begin to prepare our reply (Levinson, 2016). Moreover, a wide range of psycholinguist experiments and models have indicated a powerful role for predictive processes (Levy, 2008; Lowder, Choi, Ferreira, & Henderson, 2018). Indeed, it is natural to think of language understanding as analogous to word-by-word language production. This viewpoint fits, of course, with the broader analysis-by-synthesis perspective on language understanding, mentioned earlier, and aligns accounts of language processing with Bayesian accounts of perception (Yuille & Kersten, 2006).

16.1.3 Rational Speech Acts: Inferring What People Are Doing with Language

Language processing, of course, must go beyond the analysis of individual lexical items, and syntactic and semantic structure, to work out what message the speaker is attempting to convey—and, again, to do this in real time. This "pragmatic" interpretation of linguistic utterances is extremely complex and will involve knowledge of language, social conventions, the nature of the conversational interaction, and arbitrary background knowledge about the world. While a full discussion is far beyond the scope of this chapter, note that one approach is to view the problem of pragmatic inference as one of inferring people's intentions from what they say, just as we attempt to infer people's actions from what they do: using an inverse planning approach, as we discussed in chapter 15. The idea is that in communication, both parties assume that communicative signals are chosen to convey the meaning of interest as efficiently as possible. So, when a person says, "Some of my fish are black," we tend to infer that they aren't all black—because otherwise the person could communicated more precisely (and at no extra communicative cost) that "All my fish are black." Equally, if we see an escaped cow running down the street, we remark, "Look at that!" without having to give any further details, because it is clear that the escaped cow is the most unexpected aspect of the scene (indeed, saying, "Look at the escaped cow running down the street!" would seem bizarrely prolix). But if you happened to want to draw attention to some other aspect of the scene, you would clearly have to be much more precise. The pragmatic principles underlying these types of inference can be formulated in Bayesian terms, in the *Rational Speech Act* framework (Frank & Goodman, 2012; Goodman & Frank, 2016; Goodman & Stuhlmüller, 2013).

The Rational Speech Act approach assumes that both speaker and hearer have common knowledge of literal meanings of their linguistic terms.[6] In particular, we can begin the analysis with the notion of a *literal listener*, who is presumed to use possible utterances from a speaker, combined with background knowledge, to make inferences about the state of the world. The literal listener, by virtue of being literal, makes no assumptions about the speaker being, for example, as helpful or informative as possible. But, in reality, of course, the speaker will choose messages in a thoughtful way. Thus, an utterance such as "I have a dog" means, to the literal listener, that the speaker has at least one dog. But in most contexts, if the speaker had two dogs, it would be more informative and helpful to say, "I have two dogs"—this would provide the literal listener with more precise information. More generally, the speaker is assumed to choose what to say to maximize some utility function.[7] Knowing that the speaker will do this, the real listener (rather than the hypothetical literal listener) will assume that if the speaker had more than one dog, she would probably have said so; and hence that it is likely, though not certain, that the speaker has exactly one dog. This pattern of reasoning is not specific to language or communication—the listener is inverting a model of the speaker's actions, to infer the speaker's intentions.

6. The very idea of literal meanings is somewhat controversial in philosophy and the language sciences. Some theorists, for example, argue that context-specific meanings are primary, and abstract literal meaning that applies across contexts is at best a useful approximation (see, e.g., Christiansen & Chater, 2022). But even so, assuming literal meanings may provide a useful assumption.

7. The speaker is assumed to choose probabilistically, using a softmax function, rather than deterministically choosing to maximizing utility—otherwise the probabilities in the calculations all become jammed at 0 or 1.

Now we can take a further step: knowing that the listener will make such inferences, the speaker can deliberately choose what she says by inverting the model of the nonliteral listener's inferences. So, for example, the speaker might not choose to say, "I have one dog," precisely because the speaker knows that the more informal and slightly easier-to-say phrase "a dog" will be interpreted as implying a single dog in any case. Thus, the listener is inverting the speaker's model, which will itself involve inverting the model of the literal listener. In principle, this hierarchy might continue further, though many iterations may be both unnecessary and cognitively infeasible.

The Rational Speech Act approach has been applied to capture a variety of what are known as the conversational implicatures of language—that is, inferences that go beyond the literal meaning of what is said in highly predictable ways (Grice, 1975). Thus, following the logic described here, the approach can explain why "Some of the people enjoyed the party" seems to preclude the speaker knowing that "All the people enjoyed the party" (or the speaker would have chosen to provide this more informative message). Related, though distinct, reasoning can help explain other nonliteral uses of language, such as hyperbole, as when a person describes the weather as "boiling" or objects that there are "millions of reasons" why a project won't work (Kao, Wu, Bergen, & Goodman, 2014); or why unusual ways of expressing a meaning tend to suggest that this meaning applies in some unusual way, so that "Maria has the ability to finish the homework" seems to have very different import than the plain statement "Maria can finish the homework" (Bergen, Levy, & Goodman, 2016). The model captures the following line of reasoning: if there is nothing unusual about the situation being described, then the simplest way of expressing it would have been chosen. But it wasn't—so the speaker will intend, and the hearer will infer, that the case is not usual. Here of course, the unusual aspect of the claim is that while Maria has the ability to finish her homework later, the usual implication that she is likely to do so is blocked.

There are, of course, other unusual situations that might be relevant instead. For example, suppose that the homework consists of a very hard set of math problems, and Maria is the class's star math student. So one teacher might remark to the another: "Maria has the ability to finish the homework, but none of the other students have the slightest chance." Here, the "can" is avoided to stress Maria's specific aptitude for math. The open-ended nature of pragmatic reasoning, as well as its dependence on background knowledge of all kinds, make it challenging to model. This also, of course, suggests that such inference will need to be seen as continuous with common-sense reasoning about the physical and social worlds, which we have argued should be formulated in Bayesian terms in this book.

We have so far assumed the existence of a literal listener as a "base case" from which successive layers of recursive inference can arise. If this assumption is removed, then the Bayesian reasoning strategy based on inverse planning described here runs the risk of circularity. Indeed, this type of case arises when people use communicative signals that have no conventional meaning (whether or not those conventions are linguistic, or concern, say, facial expressions or gestures). For example, suppose that a professor ostentatiously uses a rival's book to stabilize a wobbly table (where other books and papers would work equally well; and perhaps the wobbly table is not a problem in any case). This action may be intended, and interpreted, as expressing contempt for the book (or perhaps even its author)—and might successfully convey this dismaying message to the audience (particularly if its members know about the rivalry). All may agree that such an action conveys a sense that the book is most useful as a physical object rather than deserving to be read. But there is, of course, no such prior convention (and, of course, no literal meaning) behind it. The general problem of inferring meanings from signals in cases like this raises new and perhaps unexpected challenges. As Clark (1996) points out, communication is a joint activity—the parties have to have a common understanding of what signals convey which messages.[8] And finding this understanding requires that each can align with the mind of the other.

Thus, a traditional Bayesian mind-reading approach to this problem can appear to lead to regress (Chater, Zeitoun, & Melkonyan, 2022). The receiver tries to guess what the sender intends the signal to convey. But in choosing her signal, the sender should therefore try to second-guess what the receiver will infer (so that she infers whatever the sender intended). But now the receiver has to "third-guess" what that second-guess might be, and so on, indefinitely. Following Clark (1996), one way to proceed is to assume that both parties should aim not to read the mind of the other, but rather to jointly infer the most appropriate signal-meaning mappings based on their common ground (Chater & Misyak, 2021). This type of problem is particularly pressing if we suspect that viewing communication as involving joint action and joint reasoning applies even to linguistic communication, perhaps because of doubts that literal meaning is well defined (Clark, 1996; Christiansen & Chater, 2022). Capturing this type of reasoning in a Bayesian framework is an interesting challenge for future research (for related work, see Stacy et al., 2021; Wang et al., 2021)).

16.2 Language Acquisition

Children's astonishing learning abilities are nowhere better exemplified than in language: in just a few years, babies transform themselves from helpless, nonverbal blobs to linguistic whizzes with a vast vocabulary, mastery of complex abstract syntax, and even the ability to indulge in wordplay and sarcasm. How do they do this? This is a vast question, impossible to do justice to in one small part of one (relatively) small chapter. Here, we address three topics in language acquisition (learning how to recognize phonemes, how to segment speech into words, and learning grammar), providing a quick glimpse at how Bayesian approaches have been useful for shedding light on enduring questions within these areas.[9] The goal is to highlight the scientific insight that can result from the ability of Bayesian models to clarify *what can be learned*, and *from what input*, as well as (most importantly) *why*.

16.2.1 Phoneme Learning

Phoneme learning refers to the process of acquiring the speech sounds specific to the language that one speaks. One of the most interesting issues in this area is the *perceptual magnet effect*, which occurs after successful learning: discriminability between vowels is reduced for near-prototypical vowel sounds. The pattern underlying this shrunken perceptual space is marked by decreased distance between items within a phonetic category and increased distance between items across categories: as an example, all /i/ exemplars sound

8. For experiments illustrating the sophistication of human joint reasoning with simple communicative setups, see, for example Galantucci (2005) and Misyak, Noguchi, and Chater (2016).

9. One obvious omission here is learning the meanings of individual words, which we omit because many of the relevant issues are discussed in chapters 3 and 8.

more similar than their raw acoustic representations suggest, while /a/ and /i/ exemplars sound more dissimilar.

The perceptual magnet effect has been thoroughly empirically studied. For a long time, most computational models either implicitly assumed that it was a categorization effect parallel to categorical perception (Iverson & Kuhl, 1995) or focused on purely process-level accounts of how the effect might be implemented (Vallabha & McClelland, 2007). But key questions remained unanswered: Why should prototypes exert a pull on nearby speech sounds? And why should a learner shrink phonetic space in particular directions as they learn phonetic categories? Feldman, Griffiths, and Morgan (2009) presented a Bayesian analysis that answered these questions by approaching speech perception as a problem of Bayesian statistical inference. They asked to what extent learners perform this task optimally, given certain assumptions about their hypotheses, likelihoods, and priors. Their model assumes that vowel sounds are generated from phonetic categories by sampling them from a target and adding noise; the learner must then work backwards from the exemplars of sounds that they hear to infer the nature of the most probable target production. The model predicts that, with experience, learners will realize that sounds near the center of categories are more frequent; they will then compensate for the noisy speech signal by biasing perception to the center of the category. The model captures this intuition mathematically and also explains a range of empirical effects while also making novel predictions. For instance, it predicts that category variance and degree of noise in the environment should affect the strength of the perceptual magnet. For more details on the model, see chapter 4.

Extensions of this approach maintain the idea of speech perception as optimal statistical inference and use Bayesian models to investigate how additional knowledge or assumptions can help. They indicate that the *categorical perception of consonants* can indeed be accounted for within the same framework by assigning consonants less variability (compared with the variation due to noise) than vowels (Kronrod, Coppess, & Feldman, 2016). Other work indicates that learning to segment words at the same time as phonetic category learning can make both tasks more tractable since word-level information is useful for disambiguating English vowel categories (Feldman, Goldwater, Griffiths, & Morgan, 2013). This may help to explain how tasks that individually might seem too difficult for an infant can jointly constrain and simplify each other (rather like the interweaving of answers in a crossword makes solving the individual clues easier rather than more difficult, as the answers mutually constrain each other).

Other work, similar in approach, can help clarify at what level the perceptual reorganization underlying the perceptual magnet effect occurs (Kuhl, 2004). Are people more sensitive overall to dimensions that encompass many potential phonetic contrasts, such as the general voice-onset time distinction that helps identify many phonemes? Or do they become differentially sensitive only to specific contrasts, like the /b/-/p/ distinction? Results suggest that perceptual reorganization involves making inferences about general dimensions, and this constitutes a hierarchical learning of the sort captured by the hierarchical Bayesian models discussed in detail in chapter 8 (Pajak & Levy, 2014). One consequence of this is that second-language sound categories may be filtered through the prior learning shaped by a person's native-language phonetic inventory (e.g., Strange, 2011). This might, of course, lead to a systematic distortion of the phonetic categories in the second language, as is evident in the distinctive accents of second-language speakers with different first languages.

Still other work has adapted this approach to the question of how speakers adapt to the phonetic variability in the world. Within a language, dialects vary systematically; and even within a speech community, individual speakers vary markedly from each other, such that one person's /b/ might sound like another person's /v/. Kleinschmidt and Jaeger (2015) developed a Bayesian model that views adaptation and learning as parts of the same process—inferring the correct generative model for the current speaker—but operating over different time scales. This model accounts for phenomena as disparate as perceptual recalibration, selective adaptation, and generalization across groups of speakers. It has also been extended to capture aspects of the patterns by which phonology changes over successive generations of language users, whereby the phonetic cues that are more likely to be reduced over time are those that carry less information (Hall, Hume, Jaeger, & Wedel, 2018).

16.2.2 Word Segmentation

The problem of *word segmentation* involves identifying a lexicon by segmenting words out of the continuous streams of speech that learners hear. There has been a long tradition of research exploring the possibility that people solve this problem at least in part, by learning the transitional probabilities (TPs) between phonemes or syllables. Specifically, it is assumed that low TPs are an indication of a word boundary (Saffran et al., 1996). Learning based on TPs is well studied experimentally, but this work has mostly been limited to exploring how well people use TPs, sometimes in combination with other information, to segment artificial languages, typically learned in less than an hour. By contrast, modeling is valuable for exploring how different assumptions, as well as the utility of different kinds of information, scale with extremely large amounts of data—an amount comparable to what people hear over multiple years of life.

Consider one of the most influential Bayesian models of word segmentation (Goldwater, Griffiths, & Johnson, 2009), built on an earlier model developed by Brent (1999). This model, given continuous speech input, was able to infer a vocabulary whose size did not need to be prespecified (thanks to a prior that assigned a positive probability to all vocabulary sizes, though with a strong bias toward small vocabularies, based on the ideas from nonparametric Bayesian statistics introduced in chapter 9), and was originally used to explore the impact of different assumptions that learners might make about how words are generated. One question was whether words are presumed to be generated independently or are thought to be predictive of other words in the sentence. The model showed that assuming that each word is independent from all others means that undersegmentation errors are more likely (e.g., thinking that "thedoggie" is one word rather than two), while assuming that each word constrains its neighbors reduces such errors considerably. Intuitively, this follows because the model can allow that the frequent co-occurrence of "the" and "doggie" can be explained as a result of a statistical connection between these words, rather than the assumption that they must form a single word—the independence model only has the latter option.

Frank, Goldwater, Griffiths, and Tenenbaum (2010) compared a variety of segmentation models and found that a Bayesian model was better able to capture human performance on an artificial segmentation task with more varying word and sentence lengths, exposure times, and total vocabulary size than simpler models. Interestingly, however, all models did poorly unless they were modified to take human memory limitations into account. Other related models have been used to explore different ways to implement and test such limitations (Borschinger & Johnson, 2011; Phillips & Pearl, 2015). Still other questions have been addressed with Bayesian models of word segmentation. One is how (and whether) different kinds of information besides TPs are useful in learning word segmentation—information such as stress, phonotactics, or referential information (Doyle & Levy, 2013; Borschinger & Johnson, 2014). These models have also been used as a basis for explanation or comparison while investigating how segmentation performance is affected by the distribution of words (Kurumada, Meylan, & Frank, 2013) or the size of the input (Borschinger, Demuth, & Johnson, 2012). Overall, while it is possible to defined moderately successful algorithms of segmentation by looking directly at TPs, viewing the problem of segmentation in the framework of Bayesian inference allows a range of productive extensions to be defined and explored.

16.2.3 Abstract Linguistic Structure

One of the most important topics in linguistics is the abstract structure of language, including questions about *morphology* (roughly, the system that composes the form and meaning of a word from its parts, including markers for tense, case, or plurals and, of course, word stems themselves) and *syntax* (roughly, the principles that determines allowable arrangements of words and morphemes to compose phrases and whole sentences). Developing full Bayesian models for learning these patterns is difficult because the space of possible patterns is very large. Nonetheless, such models have proved useful for investigating the learning of specific aspects of linguistic structure, or casting new light on the apparently severe problems inherent in learning abstract linguistic patterns from partial and noisy linguistic data.

Consider the *no-negative-evidence problem* (Baker, 1979), which centers around the issue of how language learners can correctly pick up on the right linguistic generalizations (when there are often so many possibilities) in the absence of evidence about which ones are *incorrect*. Thus, the child hears positive examples of what can be said (e.g., by hearing the speakers around her)—but does not seem to have access to negative examples of what cannot be said. This problem arises, for example, in the problem of learning *verb-argument constructions*. Such constructions correspond to the set of possible arguments each verb can take, and they are highly specific to individual verbs. Crucially, they vary considerably and are hard to predict based from underlying features like phonology or meaning. In English, a verb like "give" can occur in two constructions, one taking a direct object dative (as in "He gave her the package") and one indicating the recipient using a prepositional phrase (as in "He gave the package to her"). Other verbs, like "donate," occur in only one of those constructions: in most dialects, it is ungrammatical to say "He donated her the package." This is particularly puzzling, given that the meanings of "give" and "donate" are otherwise extremely similar. The puzzle is that the child doesn't make the apparently reasonable assumption that "he donated her the package" is likely to be acceptable. Of course, she never hears it—but this will be true of an almost limitless number of perfectly viable sentences. One might suspect that part of the story is that the child might get negative feedback from caregivers, either through direct correction or indirectly, through incomprehension. But many language acquisition researchers, following Brown and Hanlon (1970), have assumed that such feedback is rarely available and/or is not sufficient even when it is provided (although see Chouinard & Clark, 2003; Hirsh-Pasek, Treiman, & Schneiderman, 1984).

In any case, let us suppose that negative evidence is at least not required for successful language acquistion. Without such evidence, how might a learner figure out which constructions go with which verbs without being told when she incorrectly overgeneralizes? Bayesian modeling suggests an answer, at least in principle. Just as hypotheses in any domain that more tightly predict the data should be preferred, so should a learner (given enough data) be able to learn that some verbs take different constructions than others. Indeed, take the parallel with science: all we can ever observe are things that *can occur*. But we develop principles (including, for example, the conservation of energy or the second law of thermodynamics) which are powerful precisely because they specify what *can't* occur (e.g., no energy mysteriously coming into being out of nowhere; no spontaneous heat flow from cool bodies to hot bodies, and so on). So we can't decisively rule out the possibilities that these principles don't really hold, and merely appear to hold by coincidence. If so, we might expect to see violations at any moment. But for the Bayesian, while the coincidence story is not impossible, it is spectacularly unlikely. This general perspective is part of the motivation for the Bayesian approaches in the philosophy of science (e.g., Earman, 1992; Horwich, 1982). Can it work in the more confined domain of learning verb argument structure and other puzzling aspects of language learning?

This approach is embodied by multiple specific computational models that differ dramatically in their representational assumptions, ranging from the language-specific and intricate (Alishahi & Stevenson, 2008) to the domain-general (Hsu & Griffiths, 2009; Perfors, Tenenbaum, & Wonnacott, 2010). Overall, such models demonstrate that it is the abstract behavior of the probabilistic reasoning instantiated within the likelihood that drives the effect, while a prior preference for simplicity prevents overfitting.[10] Bayesian models have also been applied to extensions of the purely syntactic problem of learning verb constructions, such as the question of how the learner links the syntactic argument positions of a verb with the thematic roles specified by its semantics (Pearl & Sprouse, 2019).

The trade-off between likelihood (which favors models that tightly fit the input data) and prior (which favors simpler models with shorter representations) is evident in many Bayesian models of morphology and grammar learning. Work by Perfors, Tenenbaum, and Regier (2011) demonstrates that a Bayesian learner given typical child-directed language can infer that grammars with hierarchical phrase structure provide a better representation for that data than grammars without it. This emerges from a general prior favoring simplicity, since grammars with hierarchical phrase structure can capture typical English input more parsimoniously while still fitting the data well. Other research shows that human learning of artificial grammars can be captured by models that implement this kind of likelihood-prior trade-off (Frank & Tenenbaum, 2011), and children's early utterances can be captured by item-based grammars (Bannard, Lieven, & Tomasello, 2009). Indeed, Bayesian models have been applied to many classic learnability problems, from syntactic island effects (Pearl & Sprouse, 2013) to interpreting the anaphoric "one" in English (Foraker, Regier, Khetarpal, Perfors, & Tenenbaum, 2009) to linking logical and syntactic forms (Abend, Kwiatkowski,

10. See also Chater et al. (2015) and Pearl (2021) for more general overviews of how Bayesian models can be applied to syntactic questions in language acquisition.

Smith, Goldwater, & Steedman, 2017). In some cases, the question is framed purely in terms of simplicity (we'll explore the deep connection between Bayesian and simplicity-based ideas in chapter 21). Specifically, where specific hypotheses about the abstract struture of language are being compared, we can estimate whether the additional complexity in formulating the hypothesis (e.g., specifying that "give" and "donate" have different argument structures) pays off, by providing a more precise encoding of the observing linguistic data (i.e., a higher likelihood and a shorter code), and hence a shorter, simpler description overall (Hsu & Chater, 2010; Hsu, Chater, & Vitányi, 2011 see chapter 21.5.3 for further discussion). These various types of analysis allow us to quantify the amount of linguistic data that must be encountered by a language learner (typically, a child) to overcome the problem of lack of negative evidence. Each case of the problem can be dealt with by a separate analysis—giving different verdicts for different linguistic phenomena. It turns out, for example, that having a few years of language input easily suffices to distinguish the argument structures for "give" and "donate." Other aspects of language, such as Noam Chomsky's observation that, in many dialects at least, there are linguistically subtle restrictions on when we can, and cannot, contract "want to" as "wanna" in casual speech, seem to require infeasibly large amounts of data.[11]

Finally, as in other areas, Bayesian models of abstract linguistic structure have been useful in investigating to what extent synergies between different kinds of knowledge might lead to improved learning. For instance, Johnson, Demuth, and Frank (2012) show that a learner who acquires collocational structure (roughly which words go next to each other) at the same time as attempting to link words to their referents and following social cues does better than a learner without these cues. This is because the tasks mutually constrain each other, so that a learner that is sensitive to all types of information can make more headway than a learner who tries to learn only one type of information at a time.

Many of the models that investigate the role of multiple sources of information use an approach called the "adaptor grammar framework" (Johnson, Griffiths, & Goldwater, 2007b). This posits multiple layers of representation, some capturing the frequency of observed items (e.g., sentences, rules, or constructions), others capturing which linguistic patterns are permissible. Such models have been extended to incorporate contextual information (Synnaeve, Dautriche, Börschinger, Johnson, & Dupoux, 2014) and function word learning (Johnson, Griffiths, & Goldwater, 2007b) into models of word segmentation. Adaptor grammars have also been applied to problems like native language identification (Wong, Dras, & Johnson, 2012) and discourse structure (Luong, Frank, & Johnson, 2013). A similar framework, "fragment grammars" (O'Donnell, 2015), has been used to capture aspects of phonotactic learning (Futrelle, Albright, Graff, & O'Donnell, 2013), sentence processing

11. So, for example, we can contract "Who do you wanna take to the party?" to "Who do you wanna take to the party?"; but it seems distinctly odd to contract "Who do you want to take Aishah to the party?" with "*Who do you wanna take Aishah to the party?" using the conventional linguists' asterisk to denote ungrammaticality. A standard linguistic analysis focuses on the restriction that there is a "hidden" gap *want _ to* in this case, where the name of a person, say, "Sarah," might be inserted. And the restriction appears to be that contractions can't occur across gaps. This type of constraint could arise from innate aspects of language processing, or constraints required to construct the best model of the systematic patterns across the rest of the language (i.e., just as we can figure out the answer to one crossword clue more easily by cross-checking with the answers to other clues).

(Luong, O'Donnell, & Goodman, 2015), and argument structure (Bergen, Gibson, & O'Donnell, 2013).

Overall, then, from the most basic of learning tasks to the most complex—from learning phoneme categories to making abstract grammatical generalizations—it is evident that children accomplish a great deal of linguistic learning in the first years of life. Bayesian models of language acquisition have been especially useful in explaining how that learning is possible: and clarifying how different assumptions about the knowledge and representational capacities of the learner lead to different predictions. We now turn to another, still more abstract, aspect of the challenge of learning a language, which has been highly influential in shaping nativist thinking about language acquisition.

16.3 Ascending the Chomsky Hierarchy

One of the most important developments in the history of both linguistics and computer science was the discovery of the *Chomsky hierarchy* of formal languages. In this way of thinking, a *formal language* is a technical term referring to a set of strings (Chomsky, 1957): for example, the set of strings $\{a, baa, aba\}$ is an example of a language; so is the set of English words containing the sequence *ing*; so is the set of all binary sequences that never contain 11 (e.g., $\{0, 1, 01, 00, 10, 001, \ldots\}$); and so is the set of sequences of words that are considered grammatical sentences in a natural language such as Tagalog, Swahili, or Basque. Any such set may be finite or infinite, but primarily sets are distinguished by the computational resources that are required to generate or recognize them. This is why this area is considered to be foundational not only in linguistics, but also in the theory of computation (Hopcroft, Motwani, & Ullman, 2001), which seeks to characterize what kinds of problems computers are able to solve using different resources.

16.3.1 Finite and Regular Languages

Perhaps the simplest kinds of languages are *finite languages*. These are sets that contain a finite number of elements and often are described just by listing these elements. The set of English words containing "ing" is one such example because there are a finite number of English words, so that there can only be finitely many containing "ing." What is interesting, and perhaps not obvious at first, though, is that we are able to define *infinite* sets of strings by using finite resources. *Regular languages* (also called *finite-state languages*) can be generated by a computational device that contains a finite set of states. Strings in the language correspond to "walks" between these states. A simple example is

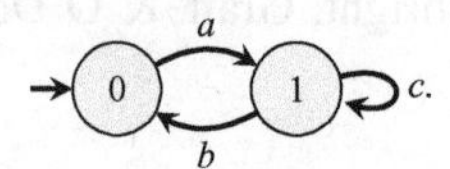

This machine has two states, 0 and 1. When this device generates strings, one must begin at the start state 0 (the starting state has an incoming arrow on the left) and then follow edges while emitting the symbols ("a", "b", and "c") that label those edges. For example, one could emit the string "accb," starting in state 0, looping three times in state 1, and then transitioning back to state 0. As should be clear from this example, there is no upper bound on the length of the strings that this machine can produce. Moreover, all strings follow a specific pattern—and that pattern is determined by the nodes, edges, and labels of the machine. For example, this machine could never emit a string with two consecutive "a"s or consecutive "b"s because each "a" must be followed by either a "b" or a "c" according to the edges shown; and each "b" must be immediately followed by an "a".

We note that formal presentations of finite state machines include several additional technical components, such as labeling of some states as "accepting" states (which must be the last state visited for a string to be valid), transitions between states that don't emit characters, as well as other variations like labeling states rather than edges with symbols (Hopcroft et al., 2001). Regular languages—the sets of strings generated by these machines—are often written in a notation known as *regular expressions*, which describe the strings generated. For example, if we require that strings end with the machine in the 0 state, we could write the expression $(ac^*b)^*$, where x^* means that x (either a symbol or a sequence of symbols) can be repeated zero or more times. Thus, we always emit an "a," any number of "c"s," and a "b"; and then we can repeat that whole thing any number of times.

Finite state machines can have any finite number of states. Here is an example of the English system of auxiliary verbs, adapted from Berwick and Pilato (1987):

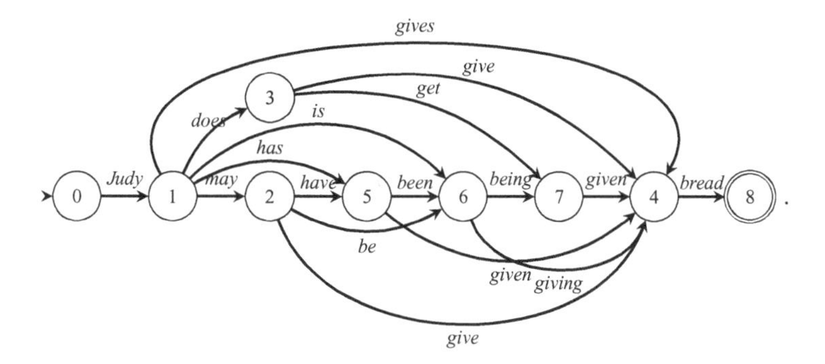

Walks on this graph provide acceptable strings of English; for example, we can say, "Judy does get given bread" but not "Judy does have give bread."

In addition, a common variant of finite state machines introduces *probabilities* on transitions (e.g., all outgoing edges must sum to 1), so the machine generates a distribution on strings rather than a set. This is useful because it allows machines to generate probabilistic predictions about upcoming symbols:

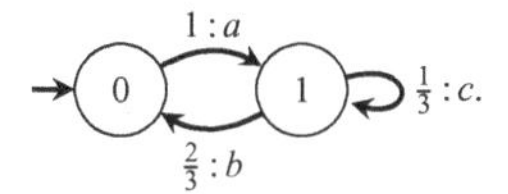

Here, we have annotated probabilities on each transition. Now, when we see an "a," it will be followed by "b" two-thirds of the time, and "c" otherwise. Moreover, the number of "c"s we see in a row will be geometrically distributed since state 1 will have a one-third chance of generating "c" and returning to state 1. Such probabilistic machines—which we discussed in chapter 5 as *hidden Markov models* (HMMs)—are amenable to a wide range of efficient algorithms for learning and inferring states from strings (Manning & Schütze, 1999). In turn, these devices that specify particular kinds of probability distributions on strings are widely applied throughout language technology, from speech recognition to statistical language analysis.

It's interesting to consider the range of possible uses for finite state machines: can *any* pattern on strings be captured with such a device? You might suspect the answer is yes because a finite state machine can have any finite number of states, so if we need more, we can always add them. However, Chomsky showed that there are intuitively simple languages that generate sentences that cannot be captured by *any* finite state machine (Chomsky, 1956, 1957). One example is the language {*ab*, *aabb*, *aaabbb*, *aaaabbbb*, . . .}, consisting of n "a" symbols followed by exactly n "b" symbols, sometimes written as $a^n b^n$.

The key insight is that a finite machine with, say, k states can only "remember" k different values. The machine's knowledge about what characters it has produced, or can produce next, is entirely determined by the state that it is in, and there are only finite many states. However, $a^n b^n$ requires the machine to "remember" an unbounded number of values because n can be arbitrarily high. To put this a bit more formally, note that a path of length $n > k$ on a finite state machine *must* visit the same state twice since each character requires us to visit a state. But if a path is in a state twice, that shows that there must be a loop (a state where we can follow a sequence of states to get back to where we started). So, we could, in principle, follow that loop as many times as we want, generating acceptable strings with as many repetitions of that subsequence loop as we would like. Thus, if $n > k$, accepting the string $a^n b^n$ *also* means that the machine accepts strings with *other* numbers of *a*s (since that sequence of *a*s must be in some state twice, so we can follow that loop again if we like, as many times as we want). This shows that no finite state machine can accept exactly the language $a^n b^n$. This style of argument is called "pumping" and can be found in studies such as Hopcroft et al. (2001).

Chomsky (1956) argued that some structures in language, like *if-then* structures, follow this pattern, where in English each "if" (think "a") must be followed by exactly one "then" (think "b"). Accordingly, for example, we can say "*If if* coffee is good for you, *then* you'll drink it, *then* you are making sensible choices" (though, as the reader will be well aware, such sentences can be extremely difficult to understand and are, needless to say, vanishingly rare). Perhaps, though, it makes sense to imagine that *but for memory limitations*, we could embed sentences indefinitely in English, yielding something like an $(\text{if})^n(\text{then})^n$ language, and thus English cannot be captured by a finite state machine. On the other hand, people *do* reliably break down after a few of these embeddings (e.g., $n > 2$) (Gibson & Thomas, 1999), which would indicate that these structures may be processed with a finite number of states by the human language processor—for example,

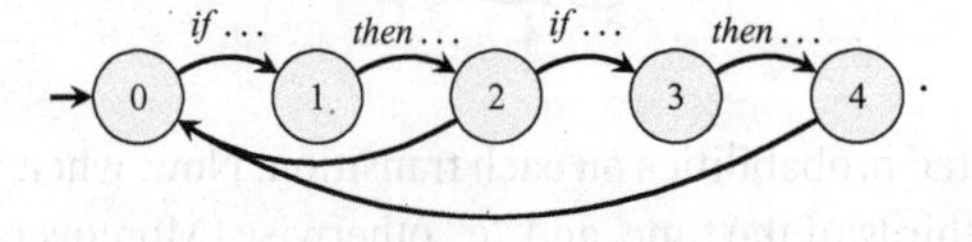

There is much discord between psycholinguistic and linguistic theories about whether we should formulate scientific accounts of language that capture what people actually do (break down for $n > 2$) or what people could "in principle" do (e.g., have arbitrarily many embeddings). Advocates of the latter viewpoint typically believe that an abstract knowledge of the language can be separated from the limited and error-prone mental-processing operations that deploy that knowledge to help understand and produce sentences—and it is this abstract knowledge that should be the focus of interest. By contrast, others doubt that such a separation is possible (or perhaps that the abstract structure of a language is a theoretical convenience for linguists but has no corresponding mental representation)—and hence prioritize modeling what people actually say and understand.

16.3.2 Context-Free Grammars and Beyond

One alternative to finite-state machines, outlined by Chomsky, was to consider systems of *rewrite rules*, constituting a *grammar*, that work in a fundamentally different way from finite-state machines. Rewrite rules start with a symbol, typically S, and then specify rules for how S may be replaced by other symbols; and then rules for how those symbols may be replaced, as we saw in figure 16.2. For example,

$$S \rightarrow aSb$$

$$S \rightarrow ab$$

says that any symbol S on the left can be replaced by either aSb (creating a new S) or ab (yielding no additional Ses). If we follow this, we might get

$$S \rightarrow aSb \rightarrow aaSbb \rightarrow aaabbb,$$

where we followed the first rule twice and then the second rule. You can see that this example generates exactly the language $a^n b^n$, which shows immediately that this system of rewrite rules is more powerful than finite state machines: it can model a language that no finite state machine can. This kind of grammar is called a *context-free grammar*: the context-free here means that when you replace a symbol like S, it doesn't matter what other characters (context) are before and after it.

In fact, context-free grammars are *strictly* more powerful than finite state machines, meaning that any language that you can describe with a finite state machine, you are also able to describe with a system of rules. The proof of this idea is simple: you imagine that the symbols that get rewritten are the states of the machine, and there is one rewrite rule for each edge, which gives the next state. For example, the finite state machine that we started with might be

$$0 \rightarrow a1$$

$$1 \rightarrow c1$$

$$1 \rightarrow b0,$$

where following the rules of the grammar is identical to walking around states of the machine, with the current state represented as the last character (we may also need a rule like $0 \rightarrow \epsilon$, saying that the computation can end by yielding an empty string ϵ in state 0). Thus, finite-state languages are a subset of context-free languages (although the containment changes somewhat when probabilities are introduced).

Importantly, we may also form a *probabilistic context-free grammar* which assigns probabilities to each expansion, like

$$S \rightarrow aSb \quad p = 0.4$$

$$S \rightarrow ab \quad p = 0.6.$$

This in turn specifies a distribution on strings that have a context-free structure (similar to those in figure 16.2). Systems with dozens or even hundreds of rules have often been used in natural language processing tasks, where the probabilities and the rules are well suited to natural language usage.

Because such grammars can do everything that a finite state machine can do, people talk about the Chomsky hierarchy of different kinds of computational devices: finite languages are a subset of finite-state languages, which are a subset of context-free languages. Further mathematical development uncovers other classes of languages. For example, $a^n b^n c^n$ can be shown not to be generated by any context-free grammar, but it can be generated by a *context-sensitive grammar* where the allowed rewrite rules depend on neighboring characters. Other formal languages are not describable with even context-sensitive grammar, and eventually one ends up with computational systems that posses the full power of Turing machines. One such formalism focuses on allowing *transformations* of what has been produced by a phrase structure grammar by moving and manipulating branches of the trees in systematic ways.

A considerable amount of work has tried to determine where natural language falls in the hierarchy and document structures that require different kinds of computational processes (Jäger & Rogers, 2012). Transformations were initially widely used in formal models of natural languages in linguistics (Chomsky, 1957), but their theoretical role has gradually reduced either to a single transformation (Chomsky, 1995) or none (Pollard & Sag, 1994; Steedman, 2000). It now is typically thought that language requires more than context-free power but somewhat less than context-sensitive power (e.g., Joshi, 1985)), and the reasons why our communication system should occupy that place in the hierarchy remain unclear. However, the characterization of human language in computational terms may depend on fairly philosophical questions of how to handle people's finite memories. Should a theory of language capture what humans might do with unbounded memory (e.g., process $(\text{if})^n(\text{then})^n$ for any n?). Or should our theory attempt to explain what people actually can do with their limited cognitive systems? After all, people's memories are finite, which implies that the set of strings that we can process is also limited. However, at the same time, the regularities that we can process seem well approximated by computational devices like grammars that seem very naturally unbounded, including hierarchical patterns in music, planning, problem-solving, and other areas.

16.3.3 Learnability and the Chomsky Hierarchy

This brief overview of language structures and computation has highlighted some of the key conceptual tools that mathematicians, computer scientists, and linguists have developed for thinking about patterns in sets of strings. Given these patterns, the natural question about human nature is whether the computations underlying natural language must be "built in" in some sense, or whether they could be learned. Can a probabilistic approach really meet the challenge of learning the types of grammatical structures found in natural language, from reasonably small amounts of data, without prior assumptions that are so strong that they amount to a built-in universal grammar? Many theorists working on language acquisition have assumed not.

Gold (1967) mathematically studies a learning situation where a learner sees strings from a formal language and must use these examples to infer the complete (usually, infinite) language (see Johnson, 2004). For example, a learner might see the strings $\{aaaabbb, ab\}$ and infer $a^n b^n$. But Gold shows that even among regular languages, not all string sets could be discovered by learners who observe examples from the set of strings. That is, no matter *what* the learner does with data, there will be languages that they cannot learn. This has often been taken to indicate that children's learning spaces must be severely constrained by, for example, an innate grammar (Carnie, 2013). As a result, Gold's proof spurred the development of complex theories of language learning under Gold-style assumptions (Wexler & Culicover, 1983), which often required learners to have a highly constrained set of hypotheses and transition between them in a particular order when data was observed.

However, for reasons of mathematical tractability, Gold's setup required assuming that parents (teachers) could provide maximally unhelpful examples to learners—that is, it studies the learning situation in the *worst* case. But the case where parents or teachers actively try to mislead children about the rules of language is unlikely to be relevant to actual language acquisition. In a milder formal setting, where learners observe strings sampled probabilistically from a target grammar, it can be shown that learners can in theory learn the correct languages out of the space of all computations (Chater & Vitányi, 2007). This work draws on a prior idea in general inductive inference (Solomonoff, 1964 and see chapter 21), where learners try to find concise programs to describe the data that they see. It can be shown that learners who do this will make optimal inferences about the structures in the world (Hutter, 2005). This work also helps address the "no negative evidence" problem outlined earlier in the chapter—with sufficient data and computational resources, it turns out that it is possible to learn language from positive evidence alone.

Yang and Piantadosi (2022) implemented these ideas in a Bayesian inference model that observed strings and inferred programs that generate strings. They showed, for example, that most of the simple formal languages used in linguistics and experimental learning studies could be discovered with a domain-general inference scheme that does Markov chain Monte Carlo (MCMC) over programs (Goodman, Tenenbaum, Feldman, & Griffiths, 2008b). Earlier work by studies such as Elman (1990) and Christiansen and Chater (1999) using neural networks indicated that this is possible in principle, but generalization was very limited, and no explicit representation of the grammatical regularities was created.

In Yang and Piantadosi (2022), learners were assumed to have access to a family of simple, domain-general computational primitives, like the ability to pair tokens in a list, call probabilistic coin flips, recurse, and other simple computational operations. The model took some observed strings and inferred what program was likely to have generated the strings. For example, when given a few strings from $a^n b^n$, the model learns the program

$$F(x) := append(a, pair(if(flip(1/3), x, F(\epsilon)), b)),$$

which will combine an "a" at the beginning of a string with a "b" at the end and flip a coin to decide whether to recurse in the middle. Thus, it implements the probabilistic $S \rightarrow aSb$ grammar described here. Note that the probabilistic operation *flip* is very important for this model because it allows the program to generate a distribution of outputs. In this probabilistic setting, straightforward Bayesian inference can compute the probability of a program h given data d as $P(h \mid d) \propto P(h)P(d \mid h)$, where $P(d \mid h)$ is the program H's probability of generating the data. As Yang and Piantadosi (2022) show, the model is able to construct programs that implement finite-state machines, context-free grammars, context-sensitive grammars, and others. It constructs these programs as ways of explaining the data that it observes, much as a scientist would formulate a computational theory to explain relevant experimental or observational data.

The question of whether the learned grammars are innate to the model is somewhat subtle. There is a sense in which the learned computational devices are innate for this model because they are constructed from built-in primitives and built-in rules of combination; but there is also an important sense in which the computational devices are constructed. In this model, *every possible* computation can be represented, so that the model is in some sense maximally unconstrained in what it can learn—as Yang and Piantadosi (2022) argue, this builds in the *least* amount of information into the learning model. Perhaps analogously, when one opens a word processor, it is possible to write any book; and it takes real work to construct the best one.

Critically, the model does require children to have the ability to form program-like representations and evaluate them as probabilistic theories of data. Given the range of domains in which children learn new algorithmic structures (Rule, Tenenbaum, & Piantadosi, 2020)—from social rules to arithmetic to games—it is plausible that children deploy these general-purpose capabilities in language acquisition to represent the computational structures required for language.

16.4 Have Deep Neural Networks Solved the Problem of Processing and Learning Language?

The Bayesian approach to the acquisition of language aims to learn a model of the language from linguistic experience, and, as we have seen, a model of the language has typically been viewed as specified by abstract formal rules. We have seen that Yang and Piantadosi's work shows that patterns governed by these highly abstract rules can be learned from surprisingly small amounts of data using a general-purpose probabilistic programming language—and we will see other examples of learning using probabilistic programs in chapter 19. However, the high level of abstraction embodied in the rules of grammar outlined by Chomsky and learned successfully for a range of artificial languages without any built-in language-specific knowledge does not represent the only type of regularity in human language. Indeed, linguists and language acquisition researchers have increasingly been focusing on *construction grammars* (e.g., Goldberg, 2006; Tomasello, 2009), which capture the observation that language consists of a mix of regularity, subregularity, and outright exceptions, which seem better captured by a more flexible linguistic formalism (e.g., Dunn, 2017) consisting of constructions: linguistic patterns (at a range of level of abstraction) paired with their meanings. A long-term challenge for Bayesian models of grammar learning is to learn the patterns of both the highly abstract and the very specific constructions that comprise a language from naturally occurring corpora from that language. The promising results obtained so far indicate that this may be possible without requiring the strong, language-specific, innate constraints that many authors have assumed to be required—that is, without recourse to any innate universal grammar (as proposed by, e.g., Chomsky, 1980; Pinker, 1994; Crain & Lillo-Martin, 1999).

Recent developments in neural network models of language processing, though, might appear to suggest that the very attempt to build a specific model of language and language acquisition is unnecessary. In 2020, the results from the Large Language Model (LLM) GPT-3, a giant neural network with an extraordinary 175 billion trainable parameters (Brown et al., 2020), astonished many in both the academic community and the media (at this time, GPT-4 is fast evolving, alongside many competing LLMs, with some remarkable results, as we'll discuss next—but let us focus on GPT-3 for the moment). GPT-3 was trained to predict the next items in a corpus based on previous items—and the corpus was almost the entire contents of the World Wide Web, totalling around a trillion words. The computer power required was correspondingly vast.

GPT-3, of course, has no symbolic representational language (whether a conventional grammar or a representation using logical formulas). Nonetheless, it is able to generate surprisingly natural-looking language in an apparently highly flexible way. For example, seeded with the author Jerome K. Jerome, the title "The Importance of Being on Twitter," and the first word "It," GPT-3 provided a number of remarkably convincing opening paragraphs before drifting into incoherence (Klingeman, 2020). Similarly, seeded with the identity of the philosopher David Chalmers and a series of philosophical questions on consciousness (Shevlin, 2020), it provided brief responses giving a tolerable account of Chalmers's views.

While impressive, a closer analysis of GPT-3's performance highlights that it is better viewed as building a model of the language to which it has been exposed rather than a model of the world that the language describes. GPT-3 has been trained to map text onto text; and it does this in an astonishingly sophisticated way. But is it really learning about the world, or the nature of the mapping between language and the world, or indeed the nature of communication itself? With GPT-3, caution seems appropriate. For example, it might seem that GPT-3 may provide insights into how children are able to learn the complex mixture of rules, subrules, and exceptions that make up a human language. But this appearance may be somewhat misleading. From the point of view of construction grammar, learning a language involves acquiring a collection of pairs of linguistic regularities and their meanings. But GPT-3 appears to have no representation of meaning and no representation of syntactic structure onto which meaning may naturally be mapped (though see Pavlick, 2023).

So despite its remarkable ability to write stories and answer questions, GPT-3 may have no real understanding of any of the language that it produces. Lacker (2020) nicely illustrates that while GPT-3 can answer an impressive range of general knowledge queries (it has almost the entire contents of the web at its disposal, after all), it generates complete bizarre answers when asked bizarre questions: *Q: How many eyes does my foot have? A: Your foot has two eyes*. And it happily gives nonsense answers when asked nonsense questions: *Q: How many rainbows does it take to jump from Hawaii to seventeen? A: It takes two rainbows to jump from Hawaii to seventeen.*

GPT-3 is flummoxed here, presumably, because these bizarre strings of words don't occur on the web. This suggests that the vastness of the corpus used by GPT-3 is probably crucial to its success—it generalizes successfully where its data are rich. Human children learn language from exposure to tens of millions of words, along with the pragmatic and physical contexts in which those words are spoken—rich symbolic representations of the structure of language and the structure of the world may be critical to generalizing so effectively.[12]

But what about successors to GTP-3? How will this picture change as LLMs become increasingly sophisticated? At this time, later iterations of GPT, including the rapidly evolving GPT-4, have been trained (with the assistance of human users) to block many nonsensical responses and, more generally, to tune out unacceptable outputs, which are otherwise difficult to avoid given that the training corpus is the largely unfiltered contents of the internet (Ouyang et al., 2022). Moreover, the performance of LLMs continues to develop rapidly, including building surprisingly good links betwen language and visual images and showing what some have described as "sparks" of general intelligence (Bubeck et al., 2023), including high-end human-level performance on IQ tests and academic and professional exams.

How far such models will progress over the coming years is difficult to predict, of course—but the kinds or errors that GPT-3 makes should at least give us pause. They suggest that LLMs may have a strategy for mimicking intelligent behavior by mining vast quantities of data that may rely on a shallower analysis of language and the world that we, as humans, might imagine.[13] On the other hand, it is possible that further iterations of LLMs are leading those models to a fundamentally different, and less shallow, mode of operation, which does indeed involve building rich and flexible representations of the world, at least to some degree. Indeed, it is also possible that matching human intelligence requires some combination of two types of process: general but fairly shallow representations of language and the world (through the analysis of vast quantities of linguistic and sensory input), alongside limited-capacity mental operations over explicit symbolic representations, which may be required in deliberative reasoning and planning, especially in highly novel contexts.

In the context of language research, we can see Yang and Piantadosi's work, and LLMs such as GPT-3 and GPT-4, as representing the opposite ends of a continuum of approaches to learning language. Yang and Piantadosi use Bayesian inference over rich, general-purpose, logical representations, and they obtain high levels of generalizations from relatively small training sets, albeit in simple artificial languages. LLMs involve training very large neural networks using nonsymbolic representations of linguistic input, generalizing much less from the inputs that it is given, but working successfully with a training corpus that is large enough that only modest generalization is often sufficient. It is likely that future cognitive models of grammar learning may require combining insights from both approaches (McCoy & Griffiths, 2023). Moreover, we suspect that successful models of language acquisition will need to capture the fact that using language effectively involves the skilful use of communicative signals, including the repertoire of complex symbols comprising human languages, in social interactions (e.g., Christiansen & Chater, 2022). In short, models of language acquisition will need to see the learning of grammar as part of a wider process

12. For further discussion, see the epilogue in Christiansen and Chater (2022), Pavlick (2023), and Schultz and Frank (2023).

13. Although, of course, some aspects of human cognition may themselves operate on shallower representations of the world that we imagine (Chater, 2018).

of learning to communicate and socially interact (e.g., Chater, McCauley, & Christiansen, 2016; Clark, 2009, 1996).

16.5 Future Directions

The last few decades have seen remarkable strides in reverse-engineering the human language acquisition and processing system, in parallel with a rapidly changing understanding of the nature of language itself. But, of course, as in other areas of cognition, the astonishing performance of the human brain far outstrips any computational model that has yet been devised, whether from the perspective of Bayesian models, deep neural networks, classical symbolic conversational models, or any combination of these. Artificial intelligence (AI) models of various kinds may perhaps be acquiring some sparks of general intelligence—but they still appear far from the rich, creative, and flexible performance achieved by the human mind.

One theme that we believe is likely to play an increasing role in future work is understanding the rich pragmatic inferences that allow even the simplest communicative signals (including pointing, gestures, and facial expressions, as well as full-blown language) successfully to convey information in a highly flexible way, depending on recent discourse context, past interactions of the speaker and listener, the current goals and environment, and background knowledge of every kind (Galantucci, 2005; Sperber, 1985). More broadly, seeing language not just as a complex system of interlocking patterns (principles of phonology, morphology, prosody, syntax and so on), but as a set of tools for guiding human interaction, may be productive. This will require, of course, a much better understanding of the interface between language, social interaction, and thought.

A second, related theme is the interactive, conversational nature of most language (Clark, 1996; Pickering & Garrod, 2021). Researchers have often treated language as monologue first and dialogue second; but recent developments in the language sciences suggest that the reverse may be the case. Indeed, this perspective may also be crucial for acquisition: children appear primarily to learn language through interactions with caregivers and other children rather than mere exposure to linguistic input (e.g., from the television or radio). Language acquisition is primarily the ability to acquire a skill that allows children to join in with the many and varied conversational games played by those around them. Reverse-engineering the nature of this skill and how it can be learned from realistic amounts of linguistic interaction (measured in the tens of millions of words rather than the trillion or so used by typical LLMs) is clearly a huge challenge. As we saw in chapter 1, a crucial question for cognitive science is how the human brain is able to create so much from so little.

A third area of future development, we suggest, may be to understand how language itself has been shaped by its function in social interactions and by the underlying computational machinery recruited by the brain to achieve that function (Christiansen & Chater, 2016a; Kirby & Tamariz, 2021). Like any cultural product, human languages will be strongly shaped by the biases of our cognitive systems (and, of course, the perceptual and motor machinery through which speech and signs are generated and perceived)—a Bayesian analysis of mechanisms of cultural transmission that could have this kind of effect (Griffiths & Kalish, 2007) appears in chapter 11. Reverse-engineering the language system successfully, therefore, should shed light on how a child's language changes through development, how languages gradually change over time (through, e.g., the process of grammaticalization, Hopper & Traugott, 2003), and how language has evolved over the long term. Language is shaped to fit with human processing and learning biases (rather than being a purely abstract mathematical system), and this may greatly ease the problem of acquisition. The guesses that the child makes about structured language will tend to be correct precisely because the language has been shaped by similar guesses by past learners (Chater & Christiansen, 2010; Zuidema, 2002).

Finally, reverse-engineering the language system requires explaining how the computational processes underlying language are rooted in neural hardware. A Bayesian perspective on language acquisition, as well as the success of computational models that learn linguistic structure from experience, have suggested that poverty of the stimulus arguments for the necessity of an innate universal grammar may be unpersuasive. But the problem of explaining how neural machinery is recruited to the challenge of acquiring the skill of conversational interaction with others from experience, and how language is processed with such nuance and subtlety in real time, remains formidable. More broadly, though, while the challenges of understanding human language remain substantial, progress in computational modeling, together with important developments in linguistics and cognitive neuroscience, suggest that real and rapid advances may be possible.

16.6 Conclusion

Language is perhaps humanity's most remarkable, and far-reaching, collective invention. It underpins our ability to formulate, transmit, and record knowledge; to work together over long periods on complex joint projects; to create and enforce legal and moral codes; to invent ideologies, religions, and scientific theories; and to create social, economic, and technological worlds of astonishing flexibility and richness. Yet the fact that language is a structured symbolic system has sometimes led researchers to conclude that probabilistic ideas can be of no more than marginal interest in understanding language. But throughout this book, we've seen that symbolic and probabilistic ideas are better viewed as complementary rather than competing. The study of language illustrates how productive this complementarity can be: showing that the processing of language at every scale, from recognizing a word, to parsing a sentence, to acquiring a language, requires rich probabilistic inference over sophisticated, structured linguistic representations.

17

Bayesian Inference over Logical Representations

Charles Kemp, Noah D. Goodman, and Thomas L. Griffiths

Our world contains many things, such as carrots, children, and rabbits, and all of us know a great deal about these things. We know about properties of objects—for example, we know that carrots are orange and rabbits are furry. We also know about relationships between objects—for example, we know that rabbits eat carrots and that each child has exactly two biological parents. Understanding how this kind of *commonsense knowledge* is organized and acquired is a central challenge for cognitive science, and this chapter discusses probabilistic approaches that can help to address this challenge.

Any successful account of commonsense knowledge must satisfy two basic criteria. First, it must work with representations that are rich enough to capture the content of commonsense knowledge. Previous chapters of this book have described representations that can express some aspects of this knowledge: for example, the causal Bayesian networks introduced in chapter 4 can represent relationships between properties such as "having wings" and "flying." It is difficult, however, to see how a Bayesian network would represent the fact that each child has exactly two biological parents. Cognitive scientists have discussed many kinds of representations that are more expressive than simple Bayesian networks, including logical theories, schemas, frames, scripts, and semantic networks. In this chapter, we use *predicate logic* to capture knowledge about objects, properties, and relations. Predicate logic includes symbols that refer to objects, features, and relations, and these symbols can be combined to construct theories that capture laws and principles of many kinds.

Working with expressive representations is a useful starting point, but the second basic requirement is to provide an account of how these representations are acquired and used to make inductive inferences. Chapter 16 showed how probabilistic models can be defined over grammars and how probabilistic inference can help to explain how these grammars are used to parse sentences and how these grammars are acquired. Along the same lines, this chapter will discuss how probabilistic models can be defined over logical representations and how probabilistic inference can explain how logical theories are used and acquired.

Logic has been part of cognitive science from the very beginning, but often it is proposed as a normative account of deductive reasoning—the goal, for example, is to explain how humans infer what must follow from a given set of statements. This chapter takes a different approach and treats logic as an account of knowledge representation rather than reasoning. Instead of using logic for inference, we propose that probabilistic inference can explain how logical representations are learned and used for reasoning. Previous approaches in the AI and machine learning literature have combined logic and probability in exactly this way and have shown that logical representations can be usefully combined with probabilistic inference (Muggleton, 1996; Richardson & Domingos, 2006; DeRaedt, Frasconi, & Muggleton, 2008). This chapter builds on this body of work to suggest that combining logic and probability can help to explain how human knowledge is acquired and used.

17.1 Logical Theories

We will introduce predicate logic by explaining how it can capture knowledge about a social system shown in figure 17.1a. The system captures mating relationships among a group of red-winged blackbirds. The nodes in the network represent birds, and there is a link between node x and node y if x mates with y.

The right side of figure 17.1b shows some logical statements that capture information about the system in figure 17.1a. The logical language includes symbols that stand for specific objects (e.g., a and b are specific birds) and variables that range over the set of objects (e.g., x and y). The language also includes symbols that refer to features and relations. For example, $\texttt{C}(\cdot,\cdot)$ (which means "couples with") denotes the mating relation, and $\texttt{M}(\cdot)$ is a binary feature that indicates whether a given object is male. The language includes one special relation $= (\texttt{x}, \texttt{y})$ that is true when x and y refer to the same object. Following traditional mathematical notation, we will write statements of this kind as $\texttt{x} = \texttt{y}$ rather than $= (\texttt{x}, \texttt{y})$.

The simplest logical statements specify a single fact. For example, $\texttt{M(a)}$ indicates that bird a is male, and $\texttt{C(a, e)}$ indicates that bird a mates with bird e. The not symbol $\neg$ can be used to negate a statement: for example, $\neg\texttt{M(e)}$ indicates that bird e is not male. Simple statements can be combined using four binary connectives: and ($\wedge$), or ($\vee$), if ($\rightarrow$), and if and only if ($\leftrightarrow$). For example, $\texttt{M(a)} \wedge \neg\texttt{M(e)}$ indicates that a is male and that e is not, and $\texttt{M(a)} \vee \neg\texttt{M(a)}$ indicates that either a is male or a is not male.

In addition to the four binary connectives, the logical language used in figure 17.1b includes quantifiers that range over the set of objects. The "for all" quantifier ($\forall$) indicates

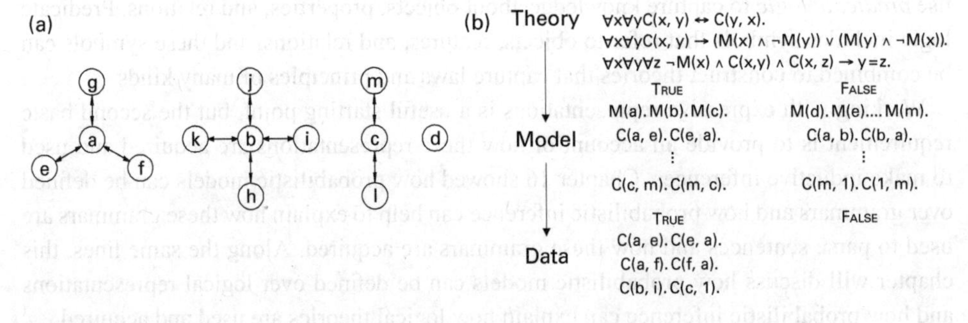

Figure 17.1
Representing theories. (a) A network that indicates which red-winged blackbirds mate with each other. A link from x to y indicates that x mates with y (i.e., that $\texttt{C(x, y)}$ is true). (b) A hierarchical framework that can be used for theory learning. The top level specifies a theory expressed in predicate logic. The middle level is a model that specifies the extension of each predicate mentioned in the theory. For example, the model here indicates that birds a through c are male and birds d through m are not. The bottom level specifies the observations available to the learner.

that a statement is true for all objects. For example, $\forall x[M(x) \vee \neg M(x)]$ indicates that for all birds x, either x is a male or x is not male. The "there exists" quantifier ($\exists$) indicates that a statement is true for an object. For example, $\exists x M(x)$ indicates that some bird is a male.

The theory in figure 17.1b uses the representational resources just described to capture some abstract properties of the social network in figure 17.1a. The first law in the theory indicates that the mating relation $C(\cdot, \cdot)$ is symmetric: in other words, for all pairs of birds x and y, x mates with y if and only if y mates with x. The second law indicates that if x mates with y, then either x is male and y is not, or y is male and x is not. The final law indicates that females have at most one mate.

The representation language just introduced is a standard variety of first-order predicate logic, but many other logical languages have been developed. These languages can be grouped into families with different expressive powers. For example, propositional logics do not allow quantification, first-order languages allow quantification over objects but not relations, and second-order languages allow quantification over both objects and relations. There are also logics that include modal connectives that go beyond the example in figure 17.1b and can be used for multiple purposes, including formulating statements about knowledge, ignorance, and belief (Verbrugge, 2009). Modal logic is beyond the scope of this chapter, however, and we will mostly look at first-order predicate logic.

17.2 A Hierarchical Bayesian Account of Theory Learning

To explain how theories are learned and used, we use a hierarchical Bayesian framework (see chapter 8) with three levels. An example of the framework is shown in figure 17.1b. The top level specifies a theory expressed in predicate logic. The middle level specifies a model of the theory. Note that *model* is a technical term in this context, which we have inherited from standard treatments of logic (Chang & Keisler, 1973). The model specifies the extension of each feature and relation in a way that is consistent with the theory. For example, the extension of the male feature $M(\cdot)$ specifies whether each bird is male. The extension of the mating relation $C(\cdot, \cdot)$ specifies for each pair (x, y) whether x mates with y. These extensions are specified in full and leave no room for uncertainty about whether a given bird is male or whether the mating relation holds between a certain pair. The bottom level of the model specifies the incomplete and possibly noisy data that a learner might observe. Figure 17.1b specifies a case in which a learner observes several positive instances of the mating relation $C(\cdot, \cdot)$, but negative instances of $C(\cdot, \cdot)$ and the feature $M(\cdot)$ are not observed.

To turn the hierarchy in figure 17.1b into a probabilistic model, we must specify a prior distribution $P(t)$ on theory t, a distribution $P(m|t)$ that specifies how model m is generated from an underlying theory, and a distribution $P(d|m)$ that specifies how the observed data d are generated from an underlying model. The applications described in the following sections make different assumptions about these distributions, but one common theme is that the prior $P(t)$ should favor simpler theories over complex theories.

After defining probability distributions over the hierarchy in figure 17.1b, the resulting hierarchical model can be used for both bottom-up and top-down reasoning. Here, we focus on cases in which a learner observes incomplete information at the bottom level and must infer both the model at the middle level and the theory at the top level. The learning problem is especially interesting when the model and theory include features and relations that are not directly observed. For example, feature M(·) is not observed at the bottom level in figure 17.1, but based solely on the observed network, a learner might notice that there are two kinds of birds—those with multiple mates and those with at most one mate. Discovering unobserved features and relations is sometimes called "predicate invention" (Kok & Domingos, 2007).

Although we have focused so far on a relatively standard version of predicate logic, the hierarchical framework in figure 17.1b is compatible with many representation languages. A key step when choosing a representation language is to decide how expressive the language should be. For example, the language used in figure 17.1 is more expressive than the propositional language that is similar in many respects but does not include the quantifiers ∀ and ∃. In general, there is a trade-off between the expressiveness of a language and the tractability of learning a theory expressed in that language. The more expressive a language, the greater the number of possible theories and the more difficult it will be to identify the theory with maximum posterior probability given the observed data.

17.3 Learning a Kinship Theory

Figure 17.1 focuses on a single relation, but we now turn to a more complex setting involving multiple kinship relations (e.g., mother, aunt, and grandmother) that hold between blood relatives. We consider a problem in which a learner observes the relationships that hold between certain pairs of individuals and must learn a theory that specifies, for example, that the sister of a parent is an aunt. Several researchers have developed models that learn logical kinship theories (Quinlan, 1990; Pericliev & Valdés-Peréz, 1998; Katz, Goodman, Kersting, Kemp, & Tenenbaum, 2008; Mollica & Piantadosi, 2021), and the approach described here is based on the work of Kemp and Regier (2012).

Figure 17.2a shows a family tree that includes 32 relatives of an individual labeled as Ego. We consider kinship systems that partition the tree into categories such that Ego refers to individuals in the same category using the same term. The colors in figure 17.2a show a system with 14 categories. For example, Ego's mother's mother (MM) and father's mother (FM) have the same color because Ego uses the same term ("grandmother") to refer to them both. Different languages have different kinship systems, but we will focus on the English kinship system shown in figure 17.2a.

Figure 17.2b shows a logical theory that captures the English kinship system. For example, the first rule indicates that for all individuals x and y, x is the mother of y if and only if x is a parent of y and x is a female. The theory is formulated using four primitive concepts: PARENT(·, ·), CHILD(·, ·), MALE(·) and FEMALE(·). If desired, the set of primitives could be reduced to two since CHILD could be defined in terms of PARENT (or vice versa), and FEMALE could be defined as the negation of MALE (or vice versa). All of the relations in the kinship system must ultimately be defined in terms of the primitives: for example, sister(·, ·) is defined using the concept daughter(·, ·), which in turn is defined using the primitives CHILD(·, ·) and FEMALE(·). Note that the logical language used requires that the right side of each rule is a conjunction with exactly two components.

The theory in figure 17.2 requires 15 rules to define the 14 relations in the English kinship system. One of the rules is used to define the relation sibling, which does not appear in

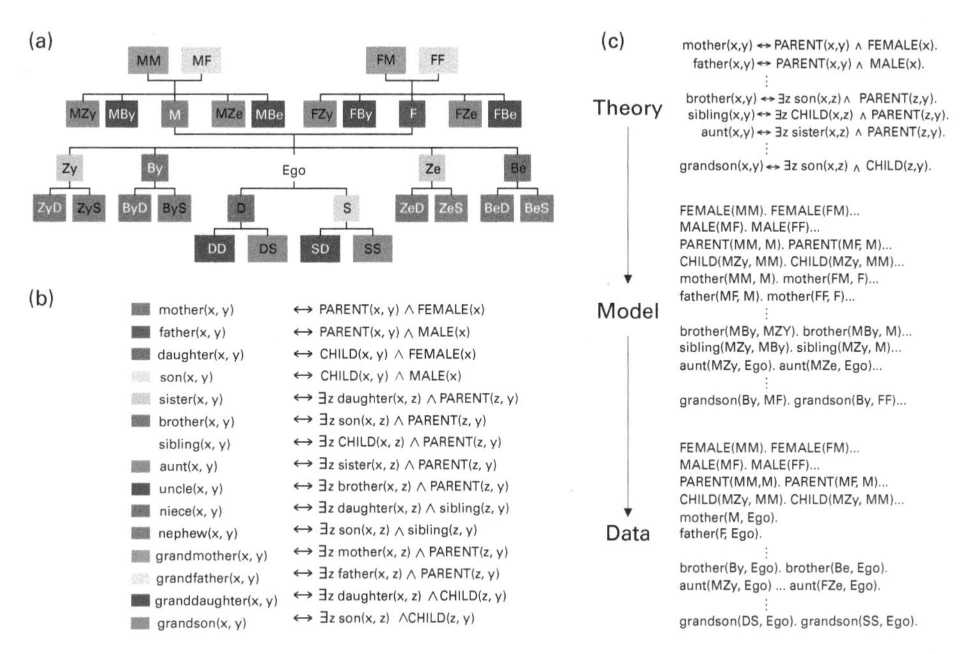

Figure 17.2
Modeling kinship relations. (a) The kinship system of English. The individual labeled Ego uses the same term to refer to individuals with the same color: for example, Ego's mother's mother (MM) and father's mother (FM) are both called "grandmother." The symbols Z, B, D, S, y, and e stand for "sister," "brother," "daughter," "son," "younger," and "elder," respectively. (b) A kinship theory that captures the English kinship system. (c) A hierarchical framework that can be used for learning kinship theories.

the system shown in figure 17.2a but is used to define the relations `niece` and `nephew`. It turns out that at least 15 rules are needed to describe the English kinship system, which means that the theory in figure 17.2b is one of the simplest possible representations of the English kinship system. Note, however, that other 15-rule representations are possible, including a theory that defines `sister` as a female sibling.

The hierarchical Bayesian approach introduced in section 17.2 can help to explain how theories like the example in figure 17.2b can be learned. Figure 17.2c shows a hierarchical model that specifies theory t at the top level. Learning at this level requires a prior $P(t)$ over theories, and here, we use a description length prior such that $P(t)$ is inversely proportional to the number of rules in t. The model m at the middle level specifies the extension of each concept mentioned in the theory. The model in figure 17.2b lists all and only the positive examples of each concept. For simplicity, we consider only theories t that specify a complete kinship system—in other words, we consider only theories that specify a complete partition of the family tree in figure 17.2a. The distribution $P(m|t)$, therefore, takes the value 1 for the unique model m that is consistent with t, and the value 0 for all other models. Finally, the bottom level specifies the data that are available for learning. We assume that the extensions of all primitive concepts are observed, and the relations that Ego uses to refer to all 32 relatives are also observed. For example, the data set shown in figure 17.2c specifies that M is the mother of Ego, F is the father of Ego, and so on. Note, however, that some of the relations are unobserved: for example, the data do not specify that MM is the mother of M, and the sibling relation is completely unobserved.

Given the assumptions that we have made, the posterior distribution on theories $P(t|d)$ takes the following simple form:

$$P(t|d) \propto P(d|t)P(t) \propto \begin{cases} P(t), & \text{if } d \text{ is consistent with } t \\ 0, & \text{otherwise.} \end{cases} \tag{17.1}$$

Identifying the theory t that maximizes $P(t|d)$, therefore, is equivalent to finding the shortest possible theory t that accounts for the data. As is often the case, the set of possible theories forms a large combinatorial space, and identifying the best theory in this space is computationally challenging. The theory in figure 17.2b was identified using a depth-first search strategy described by Kemp and Regier (2012).

We assumed that the data d specify the kinship terms that Ego uses for every relative in the family tree, but if some of these terms were unobserved, they could be predicted using distribution $P(t|d)$. AI and machine learning researchers have focused on the problem of predicting held-out kinship relations (Hinton, 1986; Paccanaro & Hinton, 2002), and Quinlan (1990) describes an approach to this problem that relies on logical representations.

Quinlan's approach to predicting kinship relations belongs to a broader body of literature on *inductive logic programming*, and a key idea in this literature is that logical theories can be acquired by finding the simplest theory that accounts for some data. Simplicity can be formalized in different ways (see chapter 21), but counting the number of symbols in a theory is one standard approach. Models that formalize the complexity of a theory as the length of its shortest description can be described as *minimum description length* (MDL) approaches (Chater & Vitányi, 2003b; Mackay, 2003; Grünwald, 2007), and the hierarchical framework in figure 17.2c can be viewed as one example. As this framework suggests, the idea that the prior probability of a theory should be inversely proportional to its length is the link that connects probabilistic approaches and MDL approaches to learning and inference.

17.4 Learning Relational Categories

Logical representations are appealing in part because they can capture relatively complex systems, including kinship systems that have many interlocking relations. Kinship theories, however, can take a relatively long time for humans to learn, and we therefore turn to a simpler setting that supports experimental studies of theory learning. The theories that we consider are logical representations that specify a single category. The members of each category are groups of figures, and several examples of these figures are shown in figure 17.3. The figures vary along three dimensions—size, shading, and circle position—and there are five possible values along each dimension.

The middle level of figure 17.3 shows one example of a group with five figures. The group is one member of a category that includes all groups where the three dimensions are aligned. As the circle position moves from left to right, note that the size of the figures increases and the shading becomes darker. Other groups also belong to the same category—for example, the group that includes only the first three figures shown in figure 17.3. Other possible categories include the category such that all figures in a group must be the same size, and the category such that all figures in a group must be different along some dimension.

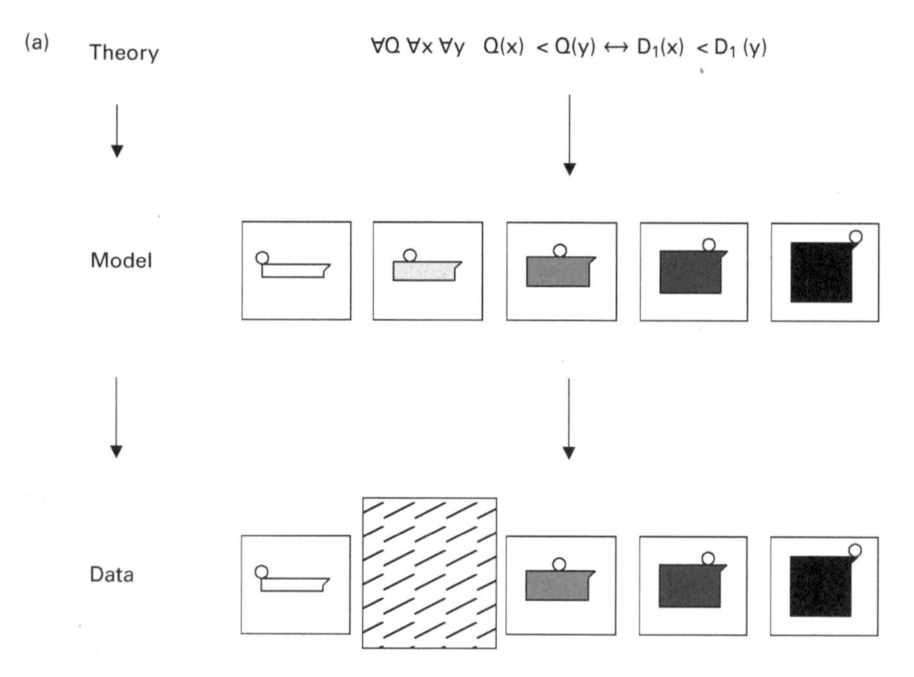

Figure 17.3
A hierarchical framework for learning relational categories. The theory at the top level indicates that all dimensions Q must be aligned. The middle level shows one five-figure group that satisfies this regularity. The bottom level shows this group with one of the figures concealed.

The same hierarchical framework used in previous sections can help to explain how these relational categories are learned. The top level is a logical theory that specifies which groups are valid instances of the category. The theory is formulated in a language that supports quantification over dimensions, as well as quantification over objects. For example, the theory expressed in figure 17.3 specifies that for all dimensions Q and all figures x and y, x is less than y along dimension Q if and only if x is less than y along dimension D_1. If D_1 is the dimension of size, then the theory states that all dimensions must be aligned with the size dimension, which means that all three dimensions must be aligned. As for the kinship model, we use a description length prior $P(t)$ over the logical theories at the top level:

$$P(t) \propto \lambda^{|t|}, \tag{17.2}$$

where $|t|$ is the number of symbols in t, and λ is a constant between 0 and 1. Longer theories t have larger values of $|t|$, and therefore smaller prior probabilities $P(t)$.

The model m at the middle level is a group that must be consistent with the theory at the top level. In general, multiple groups will be consistent with the theory, and we assume that m is sampled uniformly at random from all groups consistent with t:

$$P(m|t) \propto \begin{cases} 1, & \text{if } m \text{ is consistent with } t \\ 0, & \text{otherwise.} \end{cases} \tag{17.3}$$

Finally, data d at the bottom level specify an observation of the model m. Figure 17.3 shows a case in which data d are generated by concealing one figure uniformly at random. The distribution $P(d|m)$, therefore, is

$$P(d|m) \propto \begin{cases} 1, & \text{if } d \text{ is consistent with } m \\ 0, & \text{otherwise.} \end{cases} \tag{17.4}$$

To infer the figure that is missing from D, we can compute the posterior distribution $P(m|d)$:

$$P(m|d) \propto P(d|m)P(m) = P(d|m) \sum_t P(m|t)P(t). \tag{17.5}$$

The final term on the right side specifies a sum over the space of theories. Kemp and Jern (2009) describe how equation (17.5) can be approximated by enumerating a large but finite space of theories and summing over this space.

Figure 17.4 summarizes the results of an experiment in which participants were shown four of the five figures in a group and asked to infer the unobserved figure. The four observed figures are shown as triples that represent the value of each figure along the three

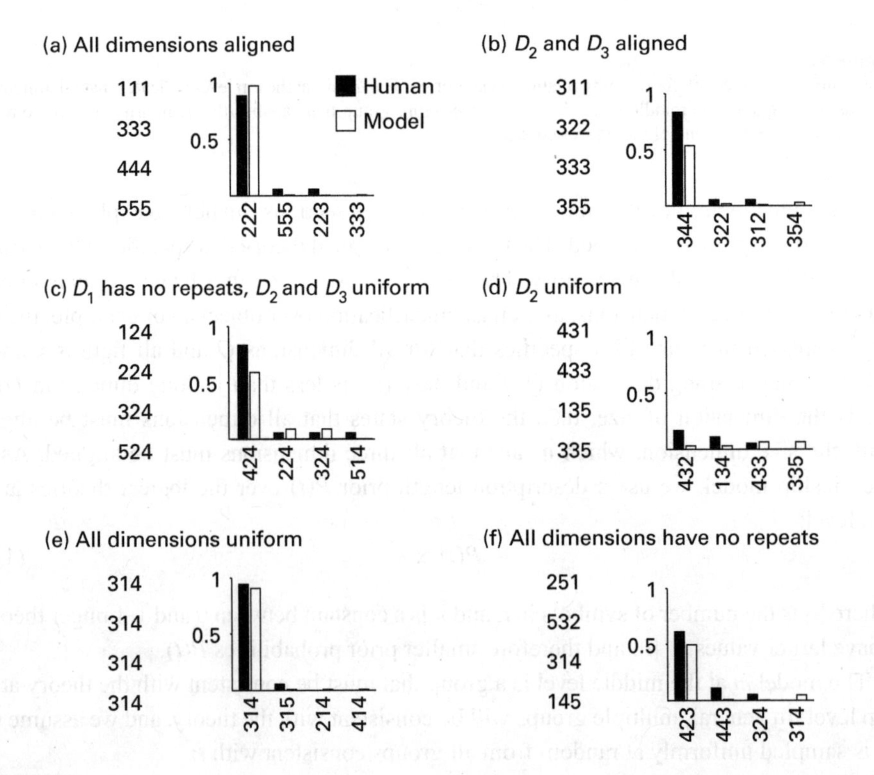

Figure 17.4
Human responses and model predictions about a task in which learners observe four of the five figures in a group and must infer which figure is missing. Each figure is represented as a triple that specifies its value along the three dimensions of size, shading, and color position. The four triples at the left of each panel represent the four observed figures, and the bar charts show inferences about the unobserved figure in each group. Each chart includes the top two responses according to humans and the two top responses according to the model in figure 17.3.

dimensions. For example, the four observed figures in figure 17.4a correspond exactly to the four observed figures at the bottom of figure 17.3. Given these four figures, humans and the model both overwhelmingly predict that the missing figure is 222, a prediction consistent with the inference that all dimensions are aligned. figure 17.4b shows a second example in which dimensions 2 and 3 are aligned, and the observed figures all have value 3 on dimension 1. In this case, humans and the model both agree that the missing figure is most likely to be 344. In five of the six problems in figure 17.4, the best response according to the model is the same as the most common human response. In the remaining problem (figure 17.4d), the model generates a diffuse distribution over all cards with value 3 on dimension 2, and all human responses satisfy this regularity.

This section described a category-learning model that is based on the work of Kemp and Jern (2009), but other researchers have developed accounts of category learning that combine logical representations and probabilistic inference (Kemp, Han, & Jern, 2011; Piantadosi, Tenenbaum, & Goodman, 2016). The next section describes one such model that uses a grammar to specify a prior distribution over theories.

17.5 Specifying Priors over Logical Theories Using Grammars

So far, we have given several examples of representation languages that rely on predicate logic, but our treatment of each language was relatively informal, and we did not give a formal specification of the theories that can be formulated in each language. One way to provide this specification is to add a new level to the hierarchical framework in figure 17.1b that defines a grammar for generating logical theories at the level immediately below it. In addition to providing a precise characterization of the space of the theories, the generative grammar can be used to formulate a prior definition over the space of theories. In this section, we illustrate both ideas using a probabilistic account of concept learning known as the *rational rules* model (Goodman, Mansinghka, Roy, Bonawitz, & Tenenbaum, 2008a). Many earlier accounts of concept learning also rely on logical representations (e.g., Bruner, Goodnow, & Austin, 1956; Nosofsky, Palmeri, & McKinley, 1994; Feldman, 2000), but we focus on the rational rules model because of its emphasis on both grammars and probabilistic inference.

Following classic work (Bruner, Goodnow, & Austin, 1956; Shepard, Hovland, & Jenkins, 1961), we will focus on concepts that are defined over objects that have binary features and will consider theories that correspond to a single logical rule. For example, suppose that objects vary in their color (red or blue), shape (circle or square), and size (small or large). Consider the concept corresponding to all blue circles. This concept can be represented as the rule $\forall \mathtt{x}\ \mathtt{l}(\mathtt{x}) \Leftrightarrow \mathtt{f}_1(\mathtt{x}) = \mathtt{1} \wedge \mathtt{f}_2(\mathtt{x}) = \mathtt{0}$, where $\mathtt{l}(\mathtt{x})$ is a label that indicates whether $\mathtt{x}$ is an instance of the concept, and the rule indicates that for all objects $\mathtt{x}$, $\mathtt{x}$ is an instance of the concept if and only if the first feature of $\mathtt{x}$ takes the value $\mathtt{1}$ and the second feature takes the value $\mathtt{0}$.

Even though this setting is simple, there are potentially infinitely many logical rules that could be considered. For example, $\forall \mathtt{x}\ \mathtt{l}(\mathtt{x}) \Leftrightarrow (\mathtt{f}_1(\mathtt{x}) = \mathtt{1} \wedge \mathtt{f}_2(\mathtt{x}) = \mathtt{0} \wedge \mathtt{f}_2(\mathtt{x}) = \mathtt{0})$ picks out exactly the same concept as our simpler rule above, and we can continue to add further redundant clauses indefinitely. Defining a prior distribution over this infinite space may seem

$$
\begin{array}{ll}
(S1) & \mathtt{S} \rightarrow \forall \mathtt{x}\, \mathtt{l(x)} \Leftrightarrow (\mathtt{D}) \\
(D1) & \mathtt{D} \rightarrow (\mathtt{C}) \vee \mathtt{D} \\
(D2) & \mathtt{D} \rightarrow \mathtt{False} \\
(C1) & \mathtt{C} \rightarrow \mathtt{P} \wedge \mathtt{C} \\
(C2) & \mathtt{C} \rightarrow \mathtt{True} \\
(P1) & \mathtt{P} \rightarrow \mathtt{F}_1 \\
 & \vdots \\
(PN) & \mathtt{P} \rightarrow \mathtt{F}_\mathtt{N} \\
(F_1 1) & \mathtt{F}_1 \rightarrow \mathtt{f}_1(\mathtt{x}) = 1 \\
(F_1 2) & \mathtt{F}_1 \rightarrow \mathtt{f}_1(\mathtt{x}) = 0 \\
 & \vdots \\
(F_N 1) & \mathtt{F}_\mathtt{N} \rightarrow \mathtt{f}_\mathtt{N}(\mathtt{x}) = 1 \\
(F_N 2) & \mathtt{F}_\mathtt{N} \rightarrow \mathtt{f}_\mathtt{N}(\mathtt{x}) = 0
\end{array}
$$

Figure 17.5
Production rules for the disjunctive normal form grammar for logical concepts used by Goodman et al. (2008a). S is the start symbol, D, C, P, $\mathtt{F_i}$ are the other nonterminals, and $\mathtt{f_i}$ represents the features of the objects that the hypotheses are defined over.

challenging at first, but grammars provide an elegant solution, and we can define a prior distribution over rules by defining a probabilistic grammar that produces these rules.

Figure 17.5 shows one such grammar. We start with the symbol S, which is rewritten to $\forall \mathtt{x}\,\mathtt{l(x)} \Leftrightarrow (\mathtt{D})$. Next we rewrite D, say to $(\mathtt{C}) \vee \mathtt{D}$. Rewriting D to False and C to $\mathtt{P} \wedge \mathtt{C}$ yields the formula $((\mathtt{P} \wedge \mathtt{C}) \vee \mathtt{False}$. If we replace P with $\mathtt{F}_1$ and C with $\mathtt{P} \wedge \mathtt{C}$, then P with $\mathtt{F}_2$ and C with True, we obtain

$$\forall \mathtt{x}\, \mathtt{l(x)} \Leftrightarrow ((\mathtt{F}_1 \wedge \mathtt{F}_2 \wedge \mathtt{True}) \vee \mathtt{False}).$$

Replacing $\mathtt{F}_1$ with $\mathtt{f(x)} = 1$ and $\mathtt{F}_2$ with $\mathtt{f}_2(\mathtt{x}) = 0$ yields the formula

$$\forall \mathtt{x}\, \mathtt{l(x)} \Leftrightarrow ((\mathtt{f}_1(\mathtt{x}) = 1 \wedge \mathtt{f}_2(\mathtt{x}) = 0 \wedge \mathtt{True}) \vee \mathtt{False}),$$

which is equivalent to the rule $\forall \mathtt{x}\, \mathtt{l(x)} \Leftrightarrow \mathtt{f}_1(\mathtt{x}) = 1 \wedge \mathtt{f}_2(\mathtt{x}) = 0$ introduced previously. Of course, many other logical formulas can be generated by applying different sequences of productions.

This grammar uses a finite and compact set of production rules to specify an infinite set of possible rules. To obtain a very natural prior, we need only add probabilities to the productions, creating a probabilistic, context-free grammar (see chapter 16). The resulting prior over formulas has an important property: formulas that take fewer steps to build will have higher prior probability. Thus, there is a natural preference for *syntactically simpler* formulas, which connects with the MDL approach described earlier in this chapter.

A drawback to simply assigning probabilities to the production rules in this grammar is that we do not usually have a priori reasons to assign particular probabilities, resulting in a large set of free parameters. A solution is to include uncertainty about these production probabilities τ. We assume a uniform prior over the possible values of τ. The probability of a particular derivation t becomes

$$\begin{aligned} P(t|\mathcal{G}) &= \int P(\tau) \prod_{s \in t} \tau(s) d\tau \\ &= \int \prod_{s \in t} \tau(s) d\tau \\ &= \prod_{y} \beta(|\{y \in t\}| + 1), \end{aligned} \tag{17.6}$$

where s denotes a rule of the grammar; $\tau(s)$ is its probability; $s \in t$ includes those rules used in deriving t; $\beta(\mathbf{v})$ is the multinomial beta function (i.e., the normalizing constant of the Dirichlet distribution with vector of parameters $\mathbf{v}$, see chapter 3); and $|\{y \in t\}|$ is the vector of counts of the productions for the nonterminal symbol y in derivation t. Allowing uncertainty about the probabilities τ thus introduces long-range statistical dependencies into the prior. These dependencies turn out to be useful in accounting for selective attention to more frequently used parts of hypotheses (see Goodman et al., 2008a).

To compute the likelihood, we need to specify a function that computes the probability that an object is in the concept that is being learned, given its feature values. In this case, the logical rule can be used to compute whether a given object is in the concept. We also need to be able to accommodate errors—cases where the observed features differ from those predicted by the logical formula. Goodman et al. (2008a) assumed that these errors occur independently with some fixed probability, resulting in a likelihood function proportional to $\exp\{-bk\}$, where b is a parameter of the model and k is the number of mismatches between the logical rule and the observed features.

Using a grammar to define a prior distribution on logical concepts also helps to solve the problem of posterior inference. Each formula can be represented by a tree structure showing how it was generated by the grammar (see chapter 16 for more details on probabilistic grammars). Goodman et al. (2008a) defined a Markov chain Monte Carlo (MCMC) algorithm for sampling from the posterior distribution on logical formulas, which was based on stochastically modifying the structure of these trees, making it possible to efficiently search the infinite space of formulas. Because shorter formulas are favored by the prior, this algorithm spends most of its time exploring short descriptions of a given concept, consistent with patterns in human concept learning previously documented by Feldman (2000).

This model of concept learning has only a single free parameter, b, which determines how well a formula needs to capture observations. However, its predictions align remarkably well with human concept learning. Table 17.1 shows a classic concept learning problem studied by Medin and Schaffer (1978). In this problem, people observed labeled examples of two concepts and then generalized to a new set of objects. As shown in the table, the proportions of people choosing a particular label for each new object closely match the predictions of the rational rules model.

The rules used in this version of the model are relatively simple, and they can in fact be replaced by propositional rules that do not involve quantification. A similar grammar-based approach, however, can be used to develop models that rely on richer logical languages. Piantadosi et al. (2016) considered languages that can express both first-order

Table 17.1
The category structure of Medin and Schaffer (1978) with human categorization probabilities from Nosofsky, Palmeri, and McKinley (1994) and the predictions of the Rational rules model (b = 1)

Object	Feature Values	Human Judgments	Rational rules
A1	0001	0.77	0.82
A2	0101	0.78	0.81
A3	0100	0.83	0.92
A4	0010	0.64	0.61
A5	1000	0.61	0.61
B1	0011	0.39	0.47
B2	1001	0.41	0.47
B3	1110	0.21	0.21
B4	1111	0.15	0.07
T1	0110	0.56	0.57
T2	0111	0.41	0.44
T3	0000	0.82	0.95
T4	1101	0.40	0.44
T5	1010	0.32	0.28
T6	1100	0.53	0.57
T7	1011	0.20	0.13

Note: Participants were trained to categorize objects A1–A5 into one category and B1–B4 into another, and then asked to generalize to objects T1–T7.

and second-order quantification, and found that first-order (but not second-order) quantification seemed critical for capturing people's judgments. Grammar-based approaches can also be used to capture inferences about real-world concepts such as numerical concepts (Piantadosi, Tenenbaum, & Goodman, 2012) and kinship concepts (figure 17.2). For example, Mollica and Piantadosi (2021) describe a grammar-based approach that helps to explain how learners across cultures acquire logical representations of kinship terms, including the English kin terms shown in figure 17.2.

17.6 Future Directions

The models presented in this chapter rely on various representation languages, and a key challenge for ongoing work is to identify the representation language that provides the best account of human behavior (Piantadosi et al., 2016; Denić & Szymanik, 2022). A natural working hypothesis is that there is a single language of thought and characterizing this language will lead to accounts of learning and reasoning across multiple contexts. Any unifying account of this kind, however, will need to explain why some aspects of the language are exploited in certain contexts but not others. For example, people clearly have the conceptual machinery to think about quantification over predicates (Kemp & Jern, 2009; Kemp, 2012), but this capacity appears not to be used in some contexts (Piantadosi et al., 2016).

If there is a single language of thought, it can almost certainly express both declarative descriptions (a square has four equal sides and four equal angles) and generative descriptions (a square can be traced by repeating an operation that includes a forward step and a 90-degree turn). The logical approaches reviewed in this chapter are best suited for formulating declarative descriptions, while generative descriptions are naturally formulated using probabilistic programs (Piantadosi, 2021; Ellis et al., 2021; Dehaene, Al Roumi, Lakretz, Planton, & Sablé-Meyer, 2022). Additional work is needed to clarify the relative merits of declarative and generative descriptions (cf. Pullum, 2020) and how they might be combined. Although probabilistic programs provide a unifying language for representing many kinds of knowledge (see chapter 18), it seems likely that some aspects of cognition are best captured by declarative descriptions. For example, a learner may grasp the concept of a magic square (a grid of numbers from 1 to n such that the sums of every row and every column are identical) without being able to generate examples of the concept.

Although this chapter has focused on compositional representation languages, human learners almost certainly rely on other kinds of mental representations, including some that are qualitatively different from those considered here. Sensory information, for example, may often be encoded using mental images that are qualitatively different from logical theories (Paivio, 1991; Pearson & Kosslyn, 2015). Although approaches that focus on compositional representation languages are sometimes pitted against those that focus on mental imagery, the more promising direction is to understand how people integrate both kinds of representations. Researchers have begun to develop probabilistic models that address this challenge (Erdogan, Yildirim, & Jacobs, 2015), but further work in this direction is needed.

17.7 Conclusion

This chapter has argued that probabilistic models can help to explain how rich systems of knowledge are acquired and used. We suggested that logical theories provide a natural way to capture rich systems of knowledge, and we demonstrated that a logical approach to knowledge representation can be combined with a probabilistic approach to reasoning and inference. To illustrate these ideas, we showed how logic and probability can combine to support inferences about social networks, kinship systems, and concepts and categories.

The models discussed in this chapter help to explain how concepts and theories can be learned in some relatively simple settings, but accounting for theory-learning in the real world is a much greater challenge. In particular, computational accounts of theory learning should eventually aim to explain how children acquire intuitive theories over the course of development (chapter 20), and how scientists discover new theories (Langley, Simon, Bradshaw, & Zytkow, 1987; Schmidt & Lipson, 2009; Ellis et al., 2021). Psychologists and AI researchers have begun to explore both of these questions, and probabilistic models suggest ways to make additional progress in both areas.

18

Probabilistic Programs as a Unifying Language of Thought

Noah D. Goodman, Tobias Gerstenberg, and Joshua B. Tenenbaum

The previous chapters of this book have successively opened up a landscape of probabilistic models with increasing power and generality. At each step, we introduced a new class of models and argued that it was able to capture important aspects of cognition that models in earlier chapters could not. Some of these models stricly generalized the simpler ones that preceded them, but many of the new model classes were disjoint from each other: grammars cannot be expressed as graphical models (nor vice versa), nonparametric models cannot be expressed as logical models (nor vice versa), and so on. This proliferation is worrying if our goal is to develop a simple set of principles for cognition, and it leaves us wondering if there is a formalism for probabilistic models that can unify all of these model classes.

At the same time, the model classes that we have discussed are increasingly general and representationally powerful, and we may wonder how far this process can go. From the point of view of cognitive science, a natural upper limit on complexity is given by the constraint of *computability*. If cognition is computation, then the cognitive system must be limited to employing probabilistic models that are computable; that is, given some random input, the process of drawing a sample from the probability distribution can be carried out by a computer program.

In this chapter, we consider how far it is possible to develop the Bayesian approach to cognitive science given only this constraint of computability. We will develop a formalism—*universal probabilistic programs* based on the *stochastic λ-calculus*—that unifies and extends all of the model classes described in previous chapters. We introduce a particular *probabilistic programming language* known as *Church* (Goodman, Mansinghka, Roy, Bonawitz, & Tenenbaum, 2012), which allows us to express any computable probabilistic model and to write a simple and effective (if not generally practical) procedure for performing any computable Bayesian inference in such a model. First, we outline the theoretical foundations of this approach, introducing stochastic λ-calculus, then Church, and the fact that such a language is universal. We then go on to illustrate how Church can express a range of familiar and novel modeling concepts from other chapters of the book. Indeed, we have suggested that Church (or any computationally universal probabilistic programming language) provides a framework for modeling all human concepts, as functions in a *Probabilistic Language of Thought* (PLoT). We review the PLoT hypothesis (Goodman,

Tenenbaum, & Gerstenberg, 2014) and the unifying perspective it provides on concepts as the basic units of human cognition.

18.1 Probabilistic Programs and the Stochastic Lambda Calculus

We begin by defining the stochastic λ-calculus, a mathematical system that is able to represent complex generative models using only a few basic constructs. It is based on the λ-calculus, which is one of the principal models of deterministic computation. After introducing this mathematical basis, we will add a number of features to make modeling easier, making this a more practical *probabilistic programming language* (PPL) for cognitive science. In particular, we will adopt the PPL Church (Goodman et al., 2012), which elegantly extends the sparse mathematical system of stochastic λ-calculus based on the Scheme dialect of Lisp.

How should we begin to build a compositional formal system—a language—for expressing probability distributions over complex world states? Intuitively, a first clue comes from the idea of representing distributions as *generative processes*: the series of random steps by which the world comes to be as it is. But while generative processes are a useful way to represent probabilistic knowledge, adopting such a representation only transforms our problem into one of finding a compositional language for generative processes. The solution to this version of the problem comes from a simple idea: if we have described a deterministic process compositionally in terms of the computation steps taken from start to end, but then inject noise at some point along the way, we get a stochastic process that unfolds in the original steps except where a random choice is made. In this way, a distribution over outputs is generated, not a single deterministic output, and this distribution inherits all the compositionality of the original deterministic process. The stochastic λ-calculus realizes this idea formally by extending a universal computational system (λ-calculus) with points of primitive randomness.

The essence of λ-calculus is a simple but general way to construct new functions and to apply functions to arguments. To this, we will add a choice operation: a way to randomly choose which of two subcomputations to use. The resulting system, *stochastic* λ-calculus, describes random computation as in the following definition.

Definition *Let* V *be a countable set, called the* variables. *The* lambda terms *are specified inductively by*

- $v \in V$
- $(\lambda(x)\ M)$ *for terms* M, $x \in V$
- $(M\ N)$ *for terms* M *and* N
- $(M \oplus N)$ *for terms* M *and* N

Intuitively, $(\lambda(x)\ M)$ specifies a new function whose inputs will be called x and which performs the computation specified by the *body* M; the term $(M\ N)$ specifies an application of the *operator* function M to the *operand* N; the term $(M \oplus N)$ specifies a random choice between M and N. To make this precise, we will need to describe how we substitute the operand for the input variable when it can be done (and some additional bookkeeping).

Definition *The free variables of a lambda term are specified inductively by*

- $FV(v) = \{v\}$ *for* $v \in V$
- $FV((\lambda(x)\ M)) = FV(M) \setminus \{x\}$
- $FV((M\ N)) = FV(M) \cup FV(N)$
- $FV((M \oplus N)) = FV(M) \cup FV(N)$

Substitution replaces all free occurrences of a variable by a specified term, written as M[x:=N] *when* x *is to be replaced by* N *in* M. *Substitution is also given inductively:*

- $x[x := N]$ *is* N
- $y[x := N]$ *is* y *if* $x \neq y$
- $(M1\ M2)[x := N]$ *is* $(M1[x := N]\ M2[x := N])$
- $(M1 \oplus M2)[x := N]$ *is* $(M1[x := N] \oplus M2[x := N])$
- $(\lambda(y)\ M)[x := N]$ *is* $(\lambda(y)\ M[x := N])$ *if* $x \neq y$ *and* $y \notin FV(N)$

In λ-calculus, we model function application by substitution. However, notice that substitution is not always defined: there can be "name collisions" between a lambda term and a substitution, which require us to rename variables before trying to substitute. Stochastic λ-calculus is specified by the following *reduction rules*, which change a lambda term into a simpler one by renaming, substitution, and random choice as follows:

- α-conversion: if $x \notin FV(N)$ and y does not appear in N, then N is replaced by $N[x := y]$ (i.e., bound variables can be renamed).
- β-reduction: $((\lambda(v)\ E)\ E')$ is replaced by $E[v := E']$.
- η-conversion: if $x \notin FV(F)$, $(\lambda(x)\ (F\ x))$ is replaced by F.
- Choice-reduction: $(N \oplus M)$ is replaced uniformly at random by either N or M.

This set of reduction rules gives a model of computation. To make the connection with more typical ideas of computation, such as arithmetic, we would need to specify how standard values (such as true, false, $0, 1, 2, \ldots$) and standard operators (and, or, $+, *, \ldots$) should be encoded by lambda terms. As an example, Boolean logic can be encoded by `True` = (λ(x) (λ(y) x)), `False` = (λ(x) (λ(y) y)), `and` = (λ(x) (λ(y) ((x y) x))), and so on. (It is a good exercise to show that these definitions behave as they should: for instance, the function `and` composed with `True` and `False` reduces to `False`, and so on.) Rather than describing all these encodings directly as lambda terms, which is possible, we will use them as primitives without specifying their encoding—a more convenient way to deal with them, and also standard for programming languages.

Notice that for a given lambda term, there are often many different sequences of reductions that could be applied. For the ordinary lambda calculus (the subset of stochastic λ-calculus lambda terms with no choice operator), the Church-Rosser theorem guarantees that the order of reductions does not matter.[1] For stochastic λ-calculus, this no longer holds. For instance, the term $((\lambda(x)(= x\ x))\ (0 \oplus 1))$ will always be true if we first reduce the operand (using choice-reduction), but it will sometimes be false if we first substitute into the operator (using β-reduction). We therefore must choose a convention; we assume that at

1. More precisely, it says that any two ways to reduce a lambda term until no further reductions are possible must result in the same final term.

an application, the operand is always reduced as far as possible before applying β-reduction to substitute it into the operator (this is called "eager" or "call by value" order). Under this convention, the term $((\lambda(x)(=x\,x))\ (0 \oplus 1))$ will always reduce to true (though the random choice may result in 0 or 1).

We can think of lambda terms as specifying generative processes: they give us instructions for the computational steps (including elementary random choices) that we should go through to compute a final value (which will be a lambda term that cannot be reduced further). We expect generative models to specify distributions, and indeed we can talk about the distribution on final values that result from reducing a lambda term until no reduction is possible. The probability of a particular sequence of reductions will be related to the number of times choice-reduction is used, and the probability of a final value is the sum over the possible reductions.

Definition *For lambda terms* M,N, *let* $\mu_M(N) = \sum_{r \in R_{M,N}} 2^{-C(r)}$, *where* $R_{M,N}$ *is the set of finite reduction sequences from* M *resulting in* N *and* C(r) *is the number of times that choice-reduction is used in* r. *If* $\sum_N \mu_M(N) = 1$, *we call* μ_M *the distribution induced by* M, *and we say that the distribution* μ_M *is represented by* M.

Not every lambda term induces a proper distribution: some have an uncountable number of infinite reduction sequences ("non-halting computations"), which means the probability of the finite reductions will be less than 1.

As we will describe in more detail later in this chapter, stochastic λ-calculus is general enough to capture all the distributions that we might plausibly want in cognitive science. It might seem natural to simply use the stochastic computations that we can now describe directly to express cognitive processes—as input-to-output functions intended to describe the algorithmic nature of the mind. A key move in Bayesian cognitive science is instead to use distributions to represent generative knowledge about the world, which are then put to work for reasoning, decision-making, and other cognitive processes via conditional inference. Using stochastic λ-calculus in this way allows us to describe a very large set of potential input-output functions as different conditional inferences for the same set of beliefs about the world. Surprisingly, we can make this move without further increasing the computational expressivity of our modeling language because stochastic λ-calculus itself can already express the computations underlying conditional inference. Before we can explain this further, however, we will first make stochastic λ-calculus into a more convenient language for expressing mental representations and processes by adding some syntactic features borrowed from computer programming languages.

18.2 A Probabilistic Programming Language: Church

We now move from an essentially mathematical object, stochastic λ-calculus, to the PPL Church (Goodman et al., 2012), named after Alonzo Church, who invented λ-calculus. This is not a change in fundamental expressivity, but rather a set of added features and libraries that make it more straightforward to write useful models. Many of these additions will be treated very informally, especially when they are standard features of the Scheme programming language (see, e.g., Abelson & Sussman, 1996), which Church builds on.

We will give only a brief introduction here to the syntax and ideas of probabilistic programming using Church. Further details and many examples can be found in the *Probabilistic Models of Cognition* online book, available at http://v1.probmods.org. We encourage readers to explore this resource for a complementary perspective on many of the ideas presented in this book, and also to explore the many runnable coding examples in the browser-based *WebChurch* dialect of Church, which is also accessible at http://v1.probmods.org. We also recommend looking at the second-edition online book, http://probmods.org, which uses a newer probabilistic programming language called *WebPPL* (http://WebPPL.org/), based on JavaScript.

Probabilistic programming languages are similar to directed graphical models in that they provide both a strictly more general mathematical formalism for describing complex probabilistic models, as well as practical implementation tools for performing automated or semi-automated scalable inference in those models. But probabilistic programs are both finer-grained—including more of the model details within the specification language—and wider in coverage, capturing models with dynamic, unbounded, and recursive structure. A universal language such as Church naturally expresses Bayesian inference algorithms in the same language that we write our generative models and lets us work with classes of meta-generative models or generative models whose outputs are any of our more familiar generative models—so Bayesian learning for arbitrary models can also be expressed as inference in the same framework that we use to reason within those models.

Church uses a syntax inherited from the Lisp family of languages (McCarthy, 1960). Thus, operators precede their arguments and are written inside grouping parentheses: for instance, `(+ 1 2)` encodes the operation "add 1 and 2." Two additions to λ-calculus that make it easier to specify functions are allowing multiple arguments and attaching values to names by `define`. For instance, we will write `(define add2 (lambda (x y) (+ x y)))` for the function `add2` that sums its two inputs. We can now reuse this function to define a new function that performs an analogous sum on four inputs: `(define add4 (lambda (w x y z) (add2 (add2 x y) (add2 w z))))`. We also allow the shortcut of dropping the lambda when we define a function: `(define (add2 x y) (+ x y))`. When we define a function in this way, we allow it to refer to its own name—that is, to be defined *recursively*.[2]

Perhaps the biggest difference between stochastic λ-calculus and Church is the addition of *elementary random procedures* (ERPs) to replace the uniform choice operation. For instance, the ERP `flip` represents the Bernoulli distribution (it "flips a coin" with "faces" `true` and `false`, with a given probability): `(flip 0.7)` results in `true` 70 percent of the time. Every time the function is called, a new random value is generated—the coin is flipped anew each time.

These random primitives can be combined just as we do for ordinary functions. For instance, `(and (flip 0.5) (flip 0.5))` is the more complex process of taking the conjunction of two random Booleans. This returns `true` if and only if both flips come up `true`. We can use ERPs plus recursion to define more interestingly structured

2. It is always possible to convert a recursive function definition into pure stochastic λ-calculus by using a *fixed point combinator*.

generative processess. For instance, we can use `flip` and the conditional `if` function[3] to define the geometric distribution recursively: `(define (geom p) (if (flip p) (+ 1 (geom p)) 0))`.

Notice that `flip` is effectively a replacement for the choice-reduction operator because we can define $(M \oplus N)$ as `(if (flip 0.5) M N)`; below we discuss how to write `flip` using $\oplus$.

A Church program specifies not a single computation, but a distribution over computations. This *sampling semantics* for probabilistic programs (see Goodman et al., 2012, for more details) means that composition of probabilities is achieved by ordinary composition of functions, and it allows us to specify probabilistic models using all the tools of representational abstraction in a modern programming language. We will not provide a primer on the power of function abstraction and other such tools here, but we will use them in ways that we hope are intuitive and illustrative.

Several other useful additions to stochastic λ-calculus that will be broadly useful are constructing arbitrary symbolic expressions, equality operations, and memoization. These language features in Church parallel core proposals for human conceptual representations that have long been studied and debated in cognitive science. Perhaps the most familiar (and most controversial) for cognitive modeling is the use of arbitrary *symbols*. In Church (as in Lisp) a symbol is a basic value that has only the property that it is equal to itself, not to any other symbol: `(equal? 'bob 'bob)` is true, while `(equal? 'bob 'jim)` is false. (The single quote syntax simply indicates that what follows is a symbol.) Critically, symbols can be used as unique identifiers on which to hang some aspect of conceptual knowledge. For instance, they can be used to refer to functions, as before when we used `define` to create the `add2` function and then reused this function by name to create `add4`. Symbols can also be used together with functions to represent knowledge about an unbounded set of objects. For instance, the function

```
(define eyecolor (lambda (x) (if (flip) 'blue 'brown)))
```

takes a person `x` and randomly returns an eye color (e.g., `(eyecolor 'bob)` might return `'blue`). That is, the function definition wraps up the knowledge about how eye color is generated *independent* of which person is being asked about—a person is simply represented by a symbol (`'bob`) that is associated with another symbol (`'blue`) via the `eyecolor` function.

Of course, this representation of an object's property has a flaw: if we ask about the eye color of Bob twice, we may get different answers! Church includes an operator `mem`, which takes a function and returns a *memoized* version: one that makes its random choices only once for each distinct value of the function's arguments, and thereafter, when called, returns the answer stored from that first evaluation. For instance, a memoized version of the `eyecolor` function,

```
(define eyecolor (mem (lambda (x) (if (flip) 'blue 'brown)))),
```

3. In Church, the conditional has a traditional but possibly cryptic syntax: `(if a b c)` returns `b` if `a` is true, and `c` otherwise. Thus `(if (flip) b c)` randomly returns `b` or `c`.

could output either `'blue` or `'brown` for Bob's eye color, but only one of these possibilities, to be determined the first time that the function is called. This ensures that `(equal? (eyecolor 'bob) (eyecolor 'bob))` is always true.

Thus symbols can be used as "indices" to recover random properties, or as labels that allow us to recover stored information about objects and their relations. These uses are conceptually very similar, though they have different syntax and can be combined. For instance, we can access one function inside the definition of another by its name, passing along the current objects of interest:

```
(define eyecolor
  (mem (lambda (x)
    (if (flip 0.1)
      (if (flip) 'blue 'brown)
      (if (flip) (eyecolor (father x)) (eyecolor (mother x))))))).
```

This (false, but perhaps intuitive) model of eye color asserts that the color is sometimes simply random, but most of the time, it depends on the eye color of one of a person's parents—which is accessed by calling the `father` or `mother` function from inside the `eyecolor` function, and so on. Symbols and symbolic reference are thus key language constructs for forming representations of complex concepts and situations out of simpler ones.

The rest of this chapter will provide more examples and motivation for how and why we should think about Church, and probabilistic programming more generally, as a unifying computational framework for cognitive modeling. In particular, we will focus on its potential for unifying the representation of diverse human concepts and conceptually driven reasoning in what has been called a "probabilistic language of thought." But first, we will use Church and its capactiy for recursive definition to explain how we can think of the underlying mathematical formalism, stochastic λ-calculus, as a universal theoretical formalism for describing generative models.

18.3 Universality

We began this chapter with the goal of exploring the space of cognitive models limited only by computability, but what exactly is a computable model? One reasonable notion of computable probabilistic model is a model defined by a *computable distribution.*

Definition *A computable probability distribution is specified by a computable function (a deterministic lambda term)* $\mathrm{p}:\mathrm{D}\times\mathbb{N}\rightarrow\mathrm{Bool}$*:* $\mathrm{p}(\mathrm{x},\mathrm{n})$ *gives the* nth *(binary) digit of the probability of* x.

It is straightforward to see that the distribution induced by a term of stochastic λ-calculus is computable (the definition of this distribution given here, plus a strategy for enumerating reduction sequences, results in an algorithm to compute the probability to a given precision). Is it also the case that stochastic λ-calculus can represent any computable distribution? We first examine a simple case and then state the more general theorem.

Lemma 18.1 *Every computable binomial distribution is represented by some stochastic λ-calculus term.*

Proof: Let $p : \mathbb{N} \rightarrow Bool$ be the computable function that gives the nth binary digit of the probability p of `True` (i.e., the coin weight). We will construct a stochastic λ-calculus function to sample from the binomial distribution. We start by imagining that we've already sampled a (computable) uniform random number in [0, 1]—call it $a : \mathbb{N} \rightarrow Bool$—and we need compare this number to p: only if $a < p$ we return `True`; otherwise, we return `False`. To make this comparison, we check the digits one by one, looking for the first difference between a and p (at which point we can decide which is bigger):

```
(define (L p nth)
  (if (a nth)
      (if (p nth) (L p (+1 nth)) False)
      (if (not (p nth)) (L p (+1 nth)) True))).
```

However, since a is uniform, the nth bit of a is a uniform random choice between `True` and `False`, and we may make this choice within the comparison function instead of choosing all the digits of a in advance:

```
(define (L p nth)
  (if (flip)
      (if (p nth) (L p (+1 nth)) False)
      (if (not (p nth)) (L p (+1 nth)) True))).
```

(Note that `(flip)` is simply a short form for (`True` ⊕ `False`).)

To finish, we will show that the binomial distribution `(flip p)` is represented by `(L p 0)`. Each recursion of `L` leads to one choice, and it is possible to stop after n recursions only if all previous choices have matched p; thus, there is a unique reduction sequence that halts after n choices. Further, only if the nth digit of p is false can `L` result in `True` on the nth recursion, so there is a reduction sequence with n choices if and only if the nth digit is false. This leads to $\mu_{(\texttt{L p 0})}(\texttt{True}) = \sum_n 2^{-n\delta_{p(n)}}$. By using the definition of the binary representation, $p = \sum_n 2^{-n\delta_{p(n)}}$. QED.

We now state the more general theorem without proof—it is parallel in spirit to this, but more complex.

Theorem 18.1 *Any computable distribution is represented by some stochastic λ-calculus expression.*

There is thus an equivalence in expressivity between stochastic λ-calculus and models which can be described as a computable distribution. Why would we choose to use one representation or the other in practice? Probabilistic programs have two related advantages: they make the generative structure of models explicit, and they are naturally compositional (whereas the laws for composing distributions are complex, requiring integration). The equivalence between these formalisms gives us confidence that we have a sufficiently broad class of models—limited only by computability—so we hazard a thesis.

Thesis (Universality) *Any computable probabilistic model can be represented as an expression in stochastic λ-calculus.*

This is a thesis, rather than a theorem or hypothesis, because we mean for it to apply to any reasonable notion of computable probabilistic model, rather than specifying a particular one. (Cf. the *Church-Turing thesis* that ordinary λ-calculus exhaustively captures computable functions.)

18.4 Conditional Inference

How does inference enter into this system? In Bayesian cognitive science, we are concerned not only with specifying generative models, which capture an agent's beliefs about the world, but with conditional inferences over these models to describe learning, reasoning and other proccesses of belief updating. It would be natural to think that we needed to introduce a new operator into stochastic λ-calculus to express conditioning. Remarkably, it turns out that this is not necessary. Conditioning can be directly defined within Church via a recursive function that corresponds to the notion of rejection sampling (see chapter 6)—a stochastic recursion that embodies the naive (but always reliable) "guess-and-check" approach to inference.

Assume that we have distribution `dist` (a stochastic function with no input arguments) that represents a generative model of interest by drawing samples from it, and predicate `pred` (a deterministic function taking as input a value of the type output by `dist`, and returning `True` or `False`) that represents some piece of information on which we would like to condition belief updates. Then we can define a higher-order stochastic function `condition`, which takes as input the model `dist` and the information `pred`, and returns a sample from a new distribution equal to `dist` conditioned on the predicate being `True`:

```
(define (condition dist pred)
    (define sample (dist))
    (if (pred sample) sample (condition dist pred))).
```

That is, we keep sampling from `dist` until we get a sample that satisfies `pred`, and then we return this sample. This directly implements Bayesian belief updating via the sampling semantics: `condition` takes a prior model specified by `dist` and effectively constructs a new model (specified via a process for generating a sample from it) that represents the posterior, corresponding to `dist` conditioned on the evidence that `pred` is true. In Church, we usually use a more convenient syntax for conditionals in the form of the `query` function, which specifies a particular random variable of interest or "query-expression" and which can be converted automatically into the following form:

```
(query
  ...definitions...
  query-expression
  condition-expression).
```

Our initial distribution is now specified by the `query-expression` evaluated in the context of the generative world model given by `...definitions...`, and our predicate for conditioning is the `condition-expression` evaluated in the same context.

For instance, if we wanted to compute the conditional distribution of two samples from the geometric distribution, conditioned on their sum being less than 4, we could write

```
(query
  ;; ...definitions...
 (define (geom p) (if (flip p) (+ 1 (geom p)) 0))
 (define a (geom 0.5))
 (define b (geom 0.5))
```

```
;; query-expression
(list a b)

;; condition-expression
(< (+ a b) 4)).
```

Or referring again to the eye-color example, if we wanted to ask about Bob's mother's likely eye color, given that Bob has blue eyes, we could write

```
(query
  (define eyecolor
    (mem (lambda (x)
      (if (flip 0.1)
        (if (flip) 'blue 'brown)
        (if (flip) (eyecolor (father x)) (eyecolor (mother x)))))))
        ;; ... and other definitions

  ;; query-expression
  (eyecolor (mother 'bob))

  ;; condition-expression
  (equal? (eyecolor 'bob) 'blue)).
```

In these examples, the ";;..." syntax indicates code comments. Notice that there is a distinction here between the definitions, which represent probabilistic knowledge reusable across many queries, and the query and condition expressions, which represent the particular question of interest at the moment. In this example, the particular people need to be introduced only in the question of interest because the conceptual knowledge is defined over arbitrary symbols. But also note that to make this a well-defined model, the definitions need to include some specifications for the functions `(mother ...)` and `(father ...)` that return valid values. This could be something sophisticated, like a generative process for family trees, or as simple as `(define (mother x) gensym)`, which returns a randomly generated, arbitrary "generic symbol" (effecitvely a placeholder name for an individual's mother—but someone who can have an eye color, as far as the `eyecolor` function is concerned).

Although we may *define* conditional inference by using the rejection sampler, it is not the case that `query` must be *implemented* by rejection sampling. Indeed, PPLs routinely provide implementations of `query` based on other algorithms borrowed from computational Bayesian statistics (see chapter 6), such as Metropolis-Hastings (Hastings, 1970) and sequential Monte Carlo (Doucet, de Freitas, & Gordon, 2001), and, of course, exact enumeration (see chapter 3). Since the development of Church and other early probabilistic programming approaches (e.g., BLOG, Milch & Russell, 2006; or Anglican, Wood, Meent, & Mansinghka, 2014), work on the engineering front has focused on developing increasingly powerful inference algorithms and more scalable inference frameworks, important innovations include Hamiltonian Monte Carlo (HMC) (Carpenter et al., 2017); variational inference using neural network encoders to learn a variational approximate posterior, as in Pyro (Bingham et al., 2019) and Turing (Ge, Xu, & Ghahramani, 2018); and meta-programming and programmable inference, for flexibly combining different modeling and inference motifs, as in Venture (Mansinghka, Selsam, & Perov, 2014) and Gen (Cusumano-Towner, Saad, Lew, & Mansinghka, 2019).

These engineering innovations—and many more that are still under development—will be crucial to realizing the promise of probabilistic programming. But it must be emphasized that even the PPLs that we already have represent remarkable artifacts and intellectual achievements, in their universality and their simplicity. With Church, for example, we have a programming language that is close enough to the grain of conscious thought that anyone can learn it, and that can be used to describe any probabilistic generative model and any probabilistic inference algorithm that any computer could possibly implement. The simple inference algorithms implemented in the online books http://v1.probmods.org and http://probmods.org, for Church and its descendant WebPPL are far from the most efficient and not designed for practical deployment on large-scale problems. But they are available to you right now to run in your web browser, and they are *universal* inference algorithms: in principle, if you are willing to wait long enough, they can compute any conditional Bayesian inference that can be computed. As the engineering side of probabilistic programming moves forward, the field should move from these in-principle possibilities to increasingly practical tools: programming languages and platforms that provide the easiest, most robust way to implement all the models in this book, as well as an effective route to scaling up the Bayesian approach to much more integrative and large-scale accounts of human cognition.

18.5 From Probabilistic Programs to a Probabilistic Language of Thought

So far in this chapter, we have defined a unifying theoretical formalism, stochastic λ-calculus, spanning the different probabilistic modeling formalisms presented in this book (and indeed any computable model), embedded this formalism in an elegant and conceptually natural programming language for modeling cognition, Church, and described a very general and useful class of Bayesian inference algorithms that implements the `query` function, which you can access in your web browser via the probmods online books. The remainder of the chapter presents a deeper look at how probabilistic programs can be used to model human thought, with two specific goals. We want to show more concretely how a PPL like Church can express a diverse range of cognitive models, as well as how it provides a way to think in reverse-engineering terms about fundamental aspects of thinking that are harder to approach with the more familiar tools of Bayesian modeling presented in earlier chapters.

The core message of this book has been that robust, resource-rational probabilistic inference is at the heart of human cognition. Reverse-engineering the mind is enabled by the right toolkit of theory, representations, and algorithms for learning and reasoning with uncertain knowledge—including probabilistic versions of hierarchies, relational systems, logic, grammars, and many other forms of symbolic structure and abstraction. But we have yet to bring the rational probabilistic inference approach into contact with perhaps the most fundamental property of human minds: their productivity. The sets of thoughts we can think, and the situations we can think about, are open-ended. They are not unconstrained, but they are unbounded. Our experience in the world, however, is finite and limited in many ways that our thinking is not. This is why generalization must also be at the heart of cognition. The ability to generalize both rationally and productively, to reason robustly under uncertainty in an endless range of novel situations, remains a central open challenge for computational cognitive science. And addressing *this* challenge is the most important reason why the Bayesian approach to cognitive modeling needs the tools of probabilistic programming.

The oldest and arguably best proposal for how productive generalization in cognition can be possible is that our thinking is compositional: that is, our thoughts are built from *concepts*—elements of knowledge—that can be combined and recombined in something like a *language of thought* (LoT) to reason flexibly about an unbounded set of novel situations (Fodor, 1975, 1998; Carey, 2009; Spelke, 2022). Probabilistic programming gives us a way to extend this LoT view, to think about concepts as a combinatorial system for probabilistic reasoning about an open-ended world of situations; each such situation specifies a probability distribution over what we expect to see in the world, given what we have seen. Indeed, we can recast the stochastic λ-calculus and Church as formal instantiations of a *probabilistic* LoT in which mental representations are built from language like composition of concepts, but the semantic content of those representations is not a logical specification of possible worlds as in traditional LoT proposals; rather it specifies a distribution on more or less probable worlds, conditioned on everything else that we know or could imagine being true.

Goodman, Tenenbaum, and Gerstenberg (2014) originally introduced the *Probabilistic Language of Thought (PLoT) hypothesis* as a way to unify different views of human concepts and their role in mental life, which have traditionally been approached from distinct and incommmensurate computational paradigms. At the risk of oversimplifying, we could call these the *statistical* and *symbolic* views. The statistical view emphasizes rational probabilsitic inference, and concepts as the locus of our predictive generalizations—the kinds of inferences that have been the focus of the early chapters of this book. Concepts give ways of picking out structure in the world that can be learned and generalized from small numbers of positive examples via Bayesian inference (Tenenbaum, 2000). They summarize stable regularities in the world, such as typical distributions of objects, properties, and events as encoded by prototype and exemplar models (Murphy, 2002) and their generalizations in nonparametric Bayesian models (Sanborn et al., 2010a). The symbolic view emphasizes the ways concepts provide the building blocks of compositional thought, as just discussed: how they can be flexibly combined to form an infinite array of thoughts to reason productively about an infinity of situations, and how they can be composed to make new concepts, which are building blocks of yet more complex thoughts. Related is the notion that concepts get much of their meaning and their function from the role that they play in explanatory structures: both larger-scale systems of abstract thought, and fine-grained mechanistic accounts of how causes give rise to effects. These explanatory aspects of concepts are traditionally addressed by the "theory theory" (Gopnik, 2003) and other accounts based on inferential or conceptual roles (Block, 1998), which have been formalized using causal graphical models, hierarchical Bayesian models, and relational nonparametric models (see examples in chapter 1, and more detail in chapters 4, 8, and 9). For our purposes, we group all these accounts under the heading of *symbolic* views because some kind of logical, relational, or other expressive symbolic representational machinery has traditionally been seen as essential to their implementation, as opposed to or in addition to statistical inference machinery.

We can phrase the PLoT hypothesis, informally, as follows:

Probabilistic language of thought hypothesis (informal version):
Concepts encode probabilistic generative knowledge about possible entities, relations, events, situations, and types, through the roles they play in a language-like (compositional and recursive) system for world modeling and inference.

The PLoT gives us a way to think about concepts that integrates their statistical, compositional, and explanatory functions, unifying across these views and accounting for more of the richness of human reasoning than could be captured using any of the traditional computational approaches alone. It can be seen as a way of making the statistical view of concepts more flexible and systematic by enriching it with a fine-grained notion of composition coming from symbolic approaches. It can also be seen as making symbolic approaches to concepts more useful for reasoning in an uncertain world by embedding them in a probabilistic framework for inference and decision.

Let's unpack these claims. Because concepts are mental representations that are simultanesouly probabilistic and composable with language like means, they can support graded inferences under uncertainty and inductive learning in ways that extend productively to an unbounded range of new situations. Because they are both generative and recursive, they can represent world knowledge that spans from high levels of abstraction to highly specific causal mechanisms and rich spaces of possibility; thus, they support probabilistic inference in the service of causal reasoning, explanation, planning, and imagination—not just statistical prediction in the actual world.

Stochastic λ-calculus and its embedding in Church allows us to make these ideas more formal as follows:

Probabilistic language of thought hypothesis (formal version):
Concepts are stochastic functions in a universal PPL.

In Church, knowledge is encoded in stochastic (and potentially recursive) function definitions. These functions describe elements of stochastic processes that can be composed together to describe various situations, to pose various questions and answer those questions with reasonable probabilistic guesses. Just as concepts are often seen as the stable and reusable units of human thought, stochastic functions are the units of knowledge encoded in a Church program. Identifying concepts with stochastic functions in Church (or any universal PPL) together with the construct of `query` immediately gives rise to all the desired features of concepts: their capacities to represent uncertainty about the world, to compose into an infinite set of more complex concepts, and to support statistical learning, generalization, and probabilistic inference, as well as causal and explanatory inferences, over unbounded spaces.

The formalism also lets us explore more precise answers to classic questions about the nature and origins of concepts: What constitutes meaning? How are the meanings of different concepts related? How are they acquired and used? Our answers to these questions can be subtle; for instance, on the face of it, the meaning of a stochastic function is simply its definition and the relation between concepts is determined by constituency—in our example earlier in the chapter, the meaning of `eyecolor` is its definition and it is related to other concepts only by its use of `mother` and `father` functions. However, when we consider the inferential relationships between concepts that come from conditional inference—`query`—we see additional aspects of meaning and conceptual relation. Conditioning on parentage can influence eye color, but also vice versa; conditioning on hair color may influence judgments about eye color indirectly, and so on. In the next section, we give an extended example in the domain of simple team games, illustrating these foundational issues as well as exploring the empirical adequacy of the PLoT hypothesis.

18.6 Putting the PLoT to Work: Bayesian Tug-Of-War and Ping Pong

As an example of how the PLoT has been developed to capture people's striking flexibility in understanding new situations and making productive probabilistic inferences, consider some of the inferences that you might make watching matches in a team game such as tug-of-war. In each match, we see two teams pitting their strength against each other by pulling on opposite ends of a heavy rope. If a team containing the first author of this chapter (NG) loses to a team containing the second author (TG), that might provide weak evidence that TG is the stronger of the two. If these teams contain only two members each, we might believe more in TG's greater strength than if the teams contain eight members each. If TG beats NG in a one-on-one tug-of-war, and NG goes on to beat three other individuals in similar one-on-one contests, we might believe that TG is not only stronger than NG, but strong in an absolute sense relative to the general population, even though we have only directly observed TG participating in a single match. However, if we later found out that NG did not try very hard in his match against TG but did try hard in his later matches, our convictions about TG's strength might subside.

This reasoning is clearly probabilistic. We may make good guesses about these propositions, but we will be far from certain; we will be more certain of TG's strength after seeing him play many matches. But our reasoning here is also highly abstract. It is not limited to a particular set of tug-of-war contestants. We can reason in analogous ways about matches between teams of arbitrary sizes and compositions and are unperturbed if a new player is introduced. We can also reason about the teams as collections: if team Alpha wins its first four matches but then loses to team Bravo, whom it has not faced before, we judge team Bravo very likely to be stronger than average. The smaller Bravo is, the more likely we are to judge a particular member of Bravo to be stronger than the average individual. And similar patterns of reasoning apply to inferences about skill and success in other kinds of team contests: we could be talking about math teams or teams for doubles Ping Pong and make analogous inferences for all the situations given here.

Our reasoning also supports inferences from complex combinations of evidence to complex conclusions. For example, suppose that participants have been divided into teams of two. If we learn that NG was lazy (not trying hard) whenever his team contested a match against TG's team, but NG's team nonetheless won each of these matches, it suggests both that NG is stronger than TG and that NG is often lazy. If we then learned that NG's teammate is stronger than any other individual in the population, we would probably revise the former belief (about NG's strength) but not the latter (about his laziness). If we learned that NG's team had won all its two-on-two matches but we were told nothing about NG's teammate, it is a good bet that the teammate—whomever that is—is stronger than average; this is all the more the case if we also learned that NG had lost several one-on-one matches while trying hard.

Finally, our reasoning in this one domain can be modularly combined with knowledge of other domains or manipulated based on subtle details of domain knowledge. If we observed TG lifting a number of very heavy boxes with apparent ease, we might reasonably expect his tug-of-war team to beat most others. But this would probably not raise our confidence that TG's math team (or even his Ping Pong team) is likely to be unusually successful. If we know that NG is trying to ingratiate himself to TG, perhaps to receive a favor, then we might not weight his loss very heavily in estimating strength. Likewise, if we knew that NG had received a distracting text message during the match.

How can we account for the wide range of flexible inferences people draw from diverse patterns of evidence such as these? What assumptions about the cognitive system are needed to explain the productivity and gradedness of these inferences? What kind of representations are abstract enough to extend flexibly to novel situations and questions, and yet concrete enough to support all the detailed predictions we can make about the world? The PLoT provides a way to approach all these questions, and its realization in a language such as Church lets us instantiate the approach in quantitatively predictive rational models of all the judgments that a person can make in this or any similarly rich domain.

18.6.1 Formalizing Tug-of-War

Let us use Church to formalize a simple intuitive theory of the tug-of-war domain, capturing some of the concepts described here and illustrating how they can be used for reasoning. The theory is built around matches, teams, and people. We observe a set of matches. A match is between two teams of people. Each person has a strength that is fixed, and each person may be lazy or not during a given match; being lazy reduces a person's pulling strength for that match. A match is won by the team that collectively pulls the hardest. Notice how we have a set of concepts that express our common-sense understanding of this domain—concepts like strength, laziness, pulling, and winning—and they can be used to reason about a wide range of situations and questions: about the likely outcomes of matches between different people, about different players' strengths or laziness, or about hypothetical outcomes of different kinds of matches between different kinds of players or teams.

In Church, we can use symbols as placeholders for unspecified individuals or entities of these types. This means that we do not need to define in advance how many people participate, what the size of the teams will be, or how many matches a tournament will have. Symbols can be constructed as needed by using a single quote, such as `'alice`, `'bob`, `'team-bayes`, and `'match-17`. In stochastic λ-calculus, symbols were used only for variables, but in Church, we can use symbols as values—and even values that serve no purpose other than to stand for themselves (so we can ask whether two symbols are equal) are remarkably useful for individuating situations and objects on the fly, as needed.

The only knowledge that we need to specify in advance is a lexicon of abstract concept definitions useful for reasoning in general about the entities in the domain—the people, teams, and matches—along with their properties and relations. We define an individual player's strength, `strength`, via a function that draws from a Gaussian distribution (with arbitrary mean $M = 10$ and standard deviation $SD = 3$):

```
(define strength (mem (lambda (person) (gaussian 10 3))))
```

This function is memoized to ensure that the strength value assigned to a person is persistent (as with our eye color function earlier in this chapter) and does not change between games. However, we assume that players are sometimes lazy. Suppose that the chance of a person being lazy in a particular match is 10 percent,

```
(define lazy (mem (lambda (person match) (flip 0.1))))
```

Suppose also that a lazy person pulls with only half of their actual strength. The overall strength of a team for a given match is the sum of the strength each person on the team contributes in that match, taking into account the possibility that any player could be lazy for that match:

```
(define teamstrength
  (mem (lambda (team match)
         (sum (map (lambda (person)
                     (if (lazy person match)
                         (/ (strength person) 2)
                         (strength person)))
                   team)))))
```

Note that this definition uses the function map, which applies a function to each element of a list; here, the function is the lambda expression encoding that being lazy on a given match divides a person's strength by 2, and the list is the variable `team`, which encodes the set of people comprising a team participating in the given match.

Finally, we specify how the winner of a game is determined. We simply say that the team wins that has the greater overall strength:

```
(define winner
  (mem (lambda (team1 team2 match)
         (if (> (teamstrength team1 match)
                (teamstrength team2 match))
             'team1 'team2))))
```

Taken together, this set of function definitions specifies a lexicon of concepts for reasoning about tug-of-war; it can also apply to many other team sports, as in the experiment we report next in which participants reason about a ping pong tournament.

With this set of concepts, we can ask many questions about many scenarios, taking advantage of multiple forms of compositionality in Church. For instance, we can ask how strong Bob is, given only the information that Bob played in one match on a team with Mary and beat a team of two other people, Sue and Bill:

```
(query
 ... CONCEPTS ...

 (strength 'bob)

 (equal? 'team1 (winner '(bob mary) '(sue bill) 'match-1))).
```

Here, `...CONCEPTS...` is shorthand for the definitions introduced in this discussion—the concepts that let us make inferences about a player's strength not only in this situation, but also in a multitude of possible situations with varying teams composed of any number of people, playing against each other with all thinkable combinations of game results in different tournament formats. With these concepts, we could just as easily ask about Bob's strength if we observed him participating on a team of three people in an asymmetric match against a team of six:

```
(query
 ... CONCEPTS ...
```

```
(strength 'bob)

(equal? 'team1
        (winner '(bob mary alice)
                '(sue bill frank joe henry nancy)
                'match-1))).
```

We encourage you to try these examples and others coming up in WebChurch,[4] a Church implementation that runs online in your browser, at http://v1.probmods.org/play-space.html. Think about what inferences you expect to see before you run each piece of code; this is a good way both to gain intuitions about probabilistic inference and to appreciate how the PLoT can capture the texture of your own intuitive thinking.

Begin by typing the following (with some WebChurch-specific syntax at the start and end):

```
(define samples
  (mh-query 1000 100

     ;; the concepts (definitions):
    (define strength (mem (lambda (person) (gaussian 10 3))))

    (define lazy (mem (lambda (person match) (flip 0.1))))

    (define teamstrength
       (mem (lambda (team match)
            (sum (map (lambda (person)
                     (if (lazy person match)
                         (/ (strength person) 2)
                         (strength person)))
                   team)))))

    (define winner
      (mem (lambda (team1 team2 match)
             (if (> (teamstrength team1 match)
                    (teamstrength team2 match))
                 'team1 'team2))))

    ;; the query:
     (strength 'bob)

    ;; conditioning on the evidence:
    (equal? 'team1 (winner '(bob mary) '(sue bill) 'match-1))
  )
)

(display (list``Expected strength:`` (mean samples)))
(density samples``Bob strength'' true).
```

Running this code in WebChurch generates 1,000 samples from the posterior for `(strength 'bob)` conditioned on the evidence that the team of Bob and Mary beat Sue

4. See also the dedicated section on reasoning in tug-of-war, which can be found at http://v1.probmods.org/conditioning.html#example-reasoning-about-the-tug-of-war.

and Bill; it will visualize the distribution of samples and also print out its mean (expectation). You should see a distribution that is roughly Gaussian, with a mean near 11 or slightly above the prior mean of 10.

Now try the same code, but with the conditioning line `(equal? team1 ... )` removed or commented out. This will generate samples from the prior on `(strength 'bob)`—a distribution that is not conditioned on any information. You should see a distribution that is Gaussian with the same standard deviation (equal to 3), but now with the prior mean of 10.

Finally, try the same code, but with the conditioning statement of the query changed to observe the very asymmetric match:

```
(equal? 'team1
        (winner '(bob mary alice)
                '(sue bill frank joe henry nancy)
                'match-1)),
```

but before running the code, think about how you would expect this piece of evidence from Bob winning a 3-on-6 match to affect your inference about Bob's strength differently from the evidence before, where Bob won a 2-on-2 match.

Intuitively, this will provide evidence that Bob is much stronger than before. Running the code shown here, you should see this effect in the posterior: the mean should now be around 13 (roughly a full standard deviation above the mean).

This capacity for productive extension over different possible situations, including different persons, different teams, and different winners of each game, renders the Church implementation a powerful model for general-purpose human reasoning. Contrast this way of defining domain knowledge, where specific individuals are represented with new symbols (`'bob`, etc.) only as needed, with more familiar forms of knowledge representation in probabilistic generative models, such as a Bayes net, where we require a fixed finite set of random variables (`strength-of-bob`, and so on) specified in advance; a Church program is like a specification from which infinitely many different Bayes nets can be built.

Another sense in which Church captures productive use of concepts comes from the ability to compose functions to form complex statements—with each new concept, there is a combinatorial increase in the number of queries that can be constructed. We can use this aspect of compositionality to capture complex patterns of data that we might observe, or complex questions that we can answer. Consider asking how strong Bob is, given that he and Mary won a match against Jane and Jim, and that Jane is more than twice as strong as Mary, by keeping the WebChurch code as before but replacing the conditioning line with

```
(and (equal? 'team1 (winner '(bob mary) '(jane jim) 'match-1))
     (> (strength 'jane) (* 2 (strength 'mary)))).
```

Or asking how strong Bob is, given that his team of three beat a team of six, and also the knowledge that he's stronger than his two teammates taken together, condition on the following:

```
(and
  (equal? 'team1
          (winner '(bob mary alice)
                  '(sue bill frank joe henry nancy)
                  'match-1))
  (> (strength 'bob)(+ (strength 'mary) (strength 'alice)))).
```

To ask how likely it is that Bob is stronger than both Jane and Jim, given the data that Bob and Mary beat Jane and Jim twice in a row (in matches 1 and 2), and the knowledge that Jane and Jim are each individually at least half again as strong as Mary, replace the query `(strength 'bob)` with

```
(> (strength 'bob) (max (strength 'jane) (strength 'jim)))
```

and the conditioning statement with

```
(and (equal? 'team1 (winner '(bob mary) '(jane jim) 'match-1))
     (equal? 'team1 (winner '(bob mary) '(jane jim) 'match-2))
     (> (strength 'jane) (* 1.5 (strength 'mary)))
     (> (strength 'jim) (* 1.5 (strength 'mary))))
```

You'll also want to replace the last two lines `(display ...)` and `(density ...)` with

```
(hist samples "Bob stronger?").
```

Or, given the same information, we can ask how likely it is that Bob's strength is more than one standard deviation above the mean of the population, by replacing the previous query with this line that estimates both mean and standard deviation from an imagined sample of 100 random individuals:

```
(> (strength 'bob) (+ (mean (map strength (iota 100)))
                      (sqrt (var (map strength (iota 100)))) )).
```

Again, we can see that the set of Church definitions lets us construct variables and relationships on the fly that would have to be prespecified in a more familiar Bayes net. Here, it is not just simple variables (like the strength of Bob) that are constructed, but quite complex conditions.

18.6.2 A Quantitative Experiment

The examples given thus far illustrate qualitatively how inference in probabilistic programs can model intuitive reasoning, but this account also performs well when assesed quantitatively in behavioral experiments. In Gerstenberg and Goodman (2012), participants were presented with information about a set of Ping Pong tournaments and asked to make judgments about the relative strengths of the players. Figure 18.1 shows an example of a single-player tournament (with all matches between teams of one player each). In this example, most people conclude that TG is relatively strong, while BL may be a little weaker than average.

This inference can be described with the following query in Church:

```
(query
  ...CONCEPTS...

  ;The query:
  (strength 'TG)

  ;The evidence:
  (and (equal? 'team1 (winner '(TG) '(NG) 'match-1))
       (equal? 'team1 (winner '(NG) '(AS) 'match-2))
       (equal? 'team1 (winner '(NG) '(BL) 'match-3))))
```

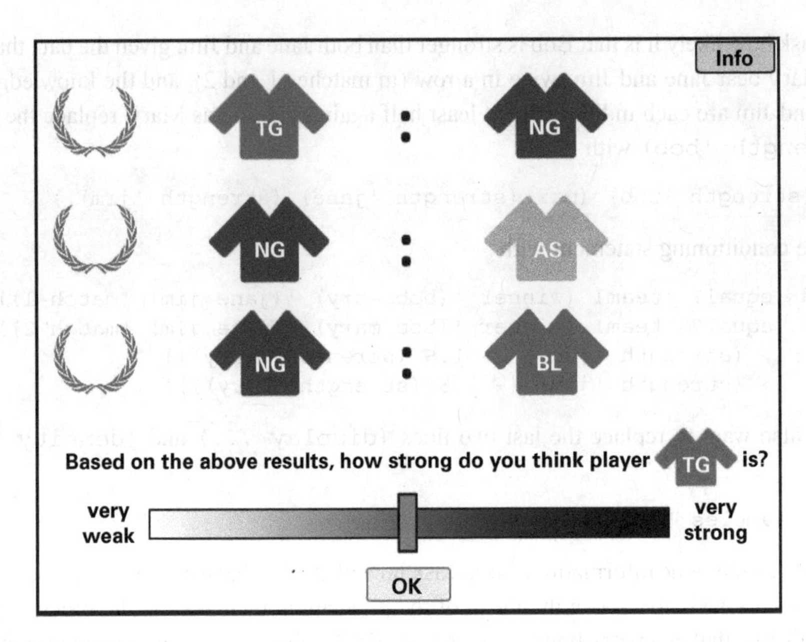

Figure 18.1
An example of a single-player tournament shown to human reasoners in Gerstenberg and Goodman (2012). A laurel wreath marks the winner of each match.

Participants in experiment 1 of Gerstenberg and Goodman (2012) saw both single-player tournaments and two-player tournaments (see tables 18.1 and 18.2 for the specific patterns of evidence shown). Participants were always asked to judge the strength of player A. They were also given some basic information about the scenarios corresponding to our modeling assumptions: that individual players have a fixed strength that does not vary between matches, and any player has a small chance of not playing as strongly as they can in any match, so even a strong player can sometimes lose to a weaker player. In experiment 2, participants saw only the single-player tournaments, but they also saw evidence in the form of a commentator (like a sportscaster, who always told the truth) reporting that a particular player (not player A) was lazy on a particular match. We asked for judgments about player A's strength in two stages: first after people saw only the matches, and then after they also heard the commentator's report, to assess how they updated their beliefs.

Across both experiments 1 and 2, our Church model accurately predicts variations in mean human strength estimates (see figures 18.2 and 18.3, respectively). Mean human participants' judgments and model estimates correlate extremely highly ($r = .98$, experiment 1; $r = .97$ experiment 2). We also find a high median correlation between the model and individual participants' judgments of ($r = .92$, experiment 1; $r = .86$, experiment 2), showing that the close fit is not merely an aggregation effect; we see it reflected in the untrained intuitions of almost every participant.

These results demonstrate quantitatively the strong form of generalization in human reasoning that we set out to capture with the PLoT. We presented people with a novel situation, bearing some similarity to familiar settings but also its own idiosyncratic features:

Table 18.1
Patterns of observation for the single-player tournaments in experiments 1 and 2 of Gerstenberg and Goodman (2012).

Confounded Evidence (Trials 1,2)	Strong Indirect Evidence (Trials 3,4)	Weak Indirect Evidence (Trials 5,6)	Diverse Evidence (Trials 7,8)
A > B	A > B	A > B	A > B
A > B	B > C	B < C	A > C
A > B	B > D	B < D	A > D
lazy,match: B,2	B,1	B,1	C,2

Note. Each tournament features three matches, and each match is between two players. The notation "A > B" means that player A won against player B in one match, and the numbers 1–8 indicate the trial types, with descriptive names for the patterns of evidence shown (e.g., "confounded evidence," "strong indirect evidence," etc.). Trials with odd numbers are shown explicitly, and an additional set of four patterns (even-numbered trials) were identical to these except that the outcomes of the matches were reversed. The bottom row shows the omniscient commentator's information in experiment 2. For example, in the confounded evidence case, the commentator reported that player B was lazy in the second match.

Table 18.2
Patterns of observation for the two-player tournaments in experiment 1 of Gerstenberg and Goodman (2012)

Confounded with Partner (Trials 9,10)			Confounded with Opponent (Trials 11,12)			Strong Indirect Evidence (Trials 13,14)		
AB	>	CD	AB	>	EF	AB	>	EF
AB	>	EF	AC	>	EG	BC	<	EF
AB	>	GH	AD	>	EH	BD	<	EF
Weak Indirect Evidence (Trials 15,16)			**Diverse Evidence (Trials 17,18)**			**Round Robin (Trials 19,20)**		
AB	>	EF	AB	>	EF	AB	>	CD
BC	>	EF	AC	>	GH	AC	>	BD
BD	>	EF	AD	>	IJ	AD	>	BC

Note. Trials with odd numbers are shown explicitly, and an additional set of six patterns (even-numbered trials) were identical to these except that the outcomes of the matches were reversed.

one- and two-player teams, multiple overlapping match-ups in a tournament, the potential for laziness, and information presented in both observed outcomes and linguistic commentary. With no additional training, people rationally and productively integrate multiple diverse pieces of evidence to form a reasonable overall estimate of a player's underlying strength. Additional studies are in progress to probe how far this ability goes—and we already know that there are questions where human and model predictions start to diverge. On balance, still, it is striking how well the patterns of intuitive human reasoning in a novel domain can be explained both qualitatively and quantitatively using only the very simple but very general machinery of a probabilistic language of thought.

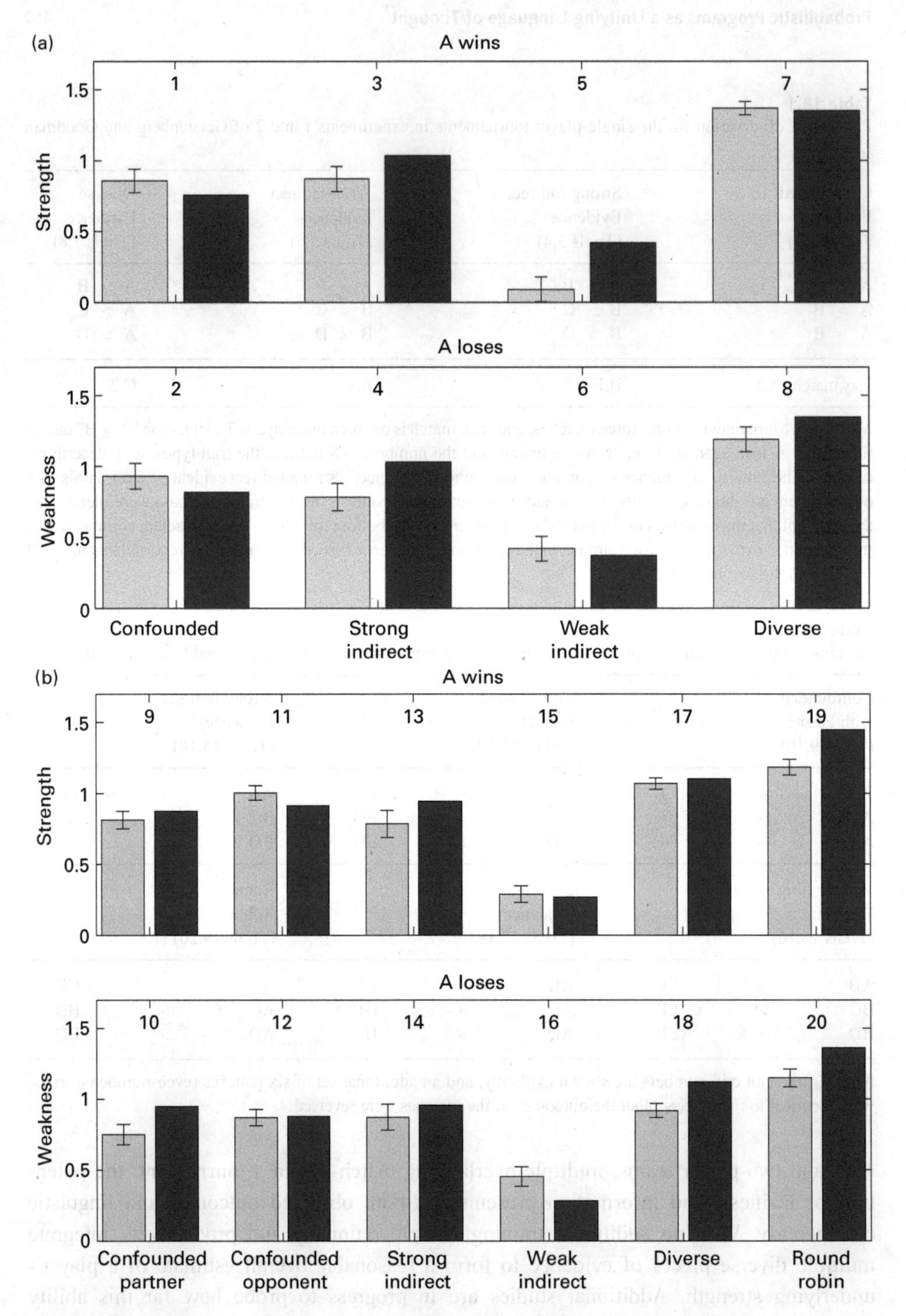

Figure 18.2
Mean strength estimates (gray bars) and model predictions (black bars) for (a) single-player and (b) two-player tournaments reported in experiment 1 of Gerstenberg and Goodman (2012). The top row shows strength judgments for cases in which the player won her games. The bottom row shows weakness judgments for cases in which the player lost. The numbers above the bars correspond to the trial patterns described in tables 18.1 and 18.2. Error bars are ± 1 *SEM*. *Note:* For ease of comparison, we z-scored both model predictions and each individual participants' judgments, and reverse-coded judgments and predictions for the situations in which the outcomes of the games were reversed (even-numbered trials) so both strength and weakness judgments go in the same direction.

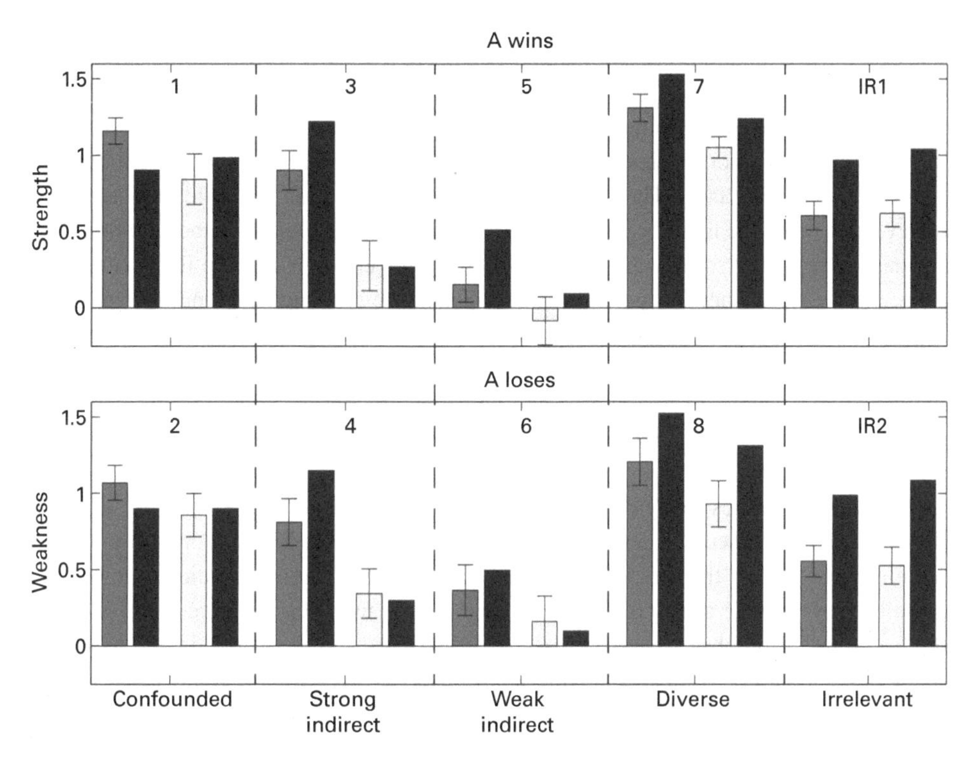

Figure 18.3
Mean strength estimates (gray bars) and model predictions (black bars) for estimates given tournament results only (dark gray) and after adding omniscient commentator information (light gray) in experiment 2 of Gerstenberg and Goodman (2012). *Note:* Situations IR1 and IR2 were cases in which the information given by the commentator was irrelevant and the model predicts no differences; see Goodman et al. (2014) for details.

18.7 Intuitive Theories

The examples given here provide concrete illustrations of how to represent concepts as functions in a probabilistic language of thought, how a system of such concepts supports inferences that are both productive and probabilistic, and how these inferences can capture the outputs of human reasoning at a high level of quantitative accuracy. But these examples are also quite limited in scope relative to the conceptual systems at the heart of human commonsense reasoning. In this section, we discuss how the same machinery can describe abstract concepts that are the backbone of thinking about everyday life and that have not fit easily into more traditional computational frameworks (Bayesian or otherwise).

Intuitive theories (Gopnik & Wellman, 2012; Wellman & Gelman, 1992; Carey, 1985, 2009; Gerstenberg & Tenenbaum, 2017), like their more familiar scientific counterparts, comprise a system of interrelated and interdefined concepts articulating a basic ontology of entities, the properties of and relations between those entities, and the causal laws that govern how these entities evolve over time and interact with each other. For instance, intuitive physics is a system for reasoning about physical objects and intuitive psychology for reasoning about intentional agents. These are called "theories" because, as in scientific theories, the essential constructs of intuitive theories are typically not directly observable. Yet intuitive theories also specify how unobservable states, properties, and processes do affect observable experience—and thus how they support competencies such as prediction, explanation, learning, and reasoning.

Intuitive theories can be found in some form in young infants and are also to some extent shared with many other species; they are arguably the earliest and oldest abstract concepts that we have (Carey, 2009). They provide the scaffolding for many of children's conceptual achievements over the first few years, as well as core building blocks for meaning in natural language. They are also transformed fundamentally as children develop their natural language abilities. Enriched, deepened, and broadened through language input, they continue to serve as the basis for adults' common-sense understanding of the world. For all these reasons, intuitive theories have long been a prime target for exploration by developmental and cognitive psychologists, linguists, and philosophers, but it is only recently—with the advent of probabilistic programs—that we have the machinery needed to represent them in computational terms.

Do intuitive theories really require the full power of probabilistic programs? In earlier chapters, we have discussed several approaches that were previously proposed to capture some aspects of folk theories. Bayesian networks are one natural candidate: interpreted causally, they provide a way to formalize how people use their intuitive theories to reason backwards from observed effects to unobserved causes in the world, as well as to learn about causal relations from sparse patterns of observations (Gopnik et al., 2004; Rehder, 2003; Goodman et al., 2006; Griffiths & Tenenbaum, 2005; Griffiths & Tenenbaum, 2009). These efforts were ultimately limited by the fact that Bayesian networks, like neural networks before them, fail to capture genuine productivity in thought. An intuitive theory of physics or psychology must be able to handle an infinite range of novel situations, differing in their specifics but not their abstract character, just as we illustrated on a much smaller scale in sketching a causal domain theory for tug-of-war or Ping Pong.

Hierarchical Bayesian models, especially those defined over structured relational representations such as first-order logic schemas, have been proposed as one way to increase the representational power of Bayesian networks, and they have given reasonable accounts of some aspects of abstract causal reasoning and learning with intuitive theories (e.g., Tenenbaum, Griffiths, & Kemp, 2006; Tenenbaum, Kemp, Griffiths, & Goodman, 2011; Goodman, Ullman, & Tenenbaum, 2011; Gopnik & Wellman, 2012). But these approaches still lack sufficiently fine-grained compositionality to capture our understanding of the physical and social worlds. The PLoT allows us to take a major step forward in this regard. Both Bayesian networks and hierarchical Bayesian models of intuitive theories can be naturally written as Church programs, preserving their insights into causal reasoning, abstraction, and learning, but Church programs go much further, letting us capture the essential representations of common-sense physics and psychology that have defied previous attempts at formalization within the probabilistic modeling tradition.

18.7.1 Mental Simulation, Sampling, and Intuitive Theories of Physics

There is a central link between the sampling semantics of Church programs, mental simulation, and the causality central to many intuitive theories. A Church program naturally expresses the causal, generative aspect of people's knowledge through the function dependencies in the program. The function dependencies dictate the causal flow of the sampling process: functions whose outputs serve as an input to another function must be evaluated first. Each run of a Church program can be interpreted as the dynamic generation of a possible world that is consistent with the causal laws as specified in the program (Chater & Oaksford, 2013). Because the sampling process is stochastic, a Church program specifies a probability distribution over possible worlds, and different modes of reasoning can be seen as different forms of mental simulation on top of this basic sampling process. While the notion of mental representation and simulation of possible worlds has had many advocates (Craik, 1943; Hegarty, 2004; Johnson-Laird, 1983), the PLoT view integrates this idea naturally into a view of mental models that is also probabilistic, causal, and sufficiently expressive to capture core intuitive theories.

Now we will illustrate the implications of this view via some concrete examples in an intuitive physics domain that was explored extensively in chapter 15. People can use their intuitive theories of physics to make many judgments about the trajectories of objects in motion, including predictions of where objects will go in the future, inferences about objects' latent physical properties or initial conditions, and explanations about *why* objects moved as they did, or judgments of "actual causation": namely, whether one specific physical event caused another specific event to happen. Consider an extension of the billiards scenarios in chapter 15: in addition to two billiard balls *A* and *B* and some solid walls with an opening gate, we also have a brick and a teleport gate that can be either active or inactive. Figure 18.4 shows diagrammatic illustrations of causal interactions between balls *A* and *B* in this world, assuming simple Newtonian elastic collisions between moving bodies. We might ask someone to predict how *B* will move at the point where *A* makes contact with it, or to judge where *A* must have come from if they see only *B*'s motion, or having seen the entirety of any of these four scenarios, to judge whether ball *A*'s collision with ball *B* caused

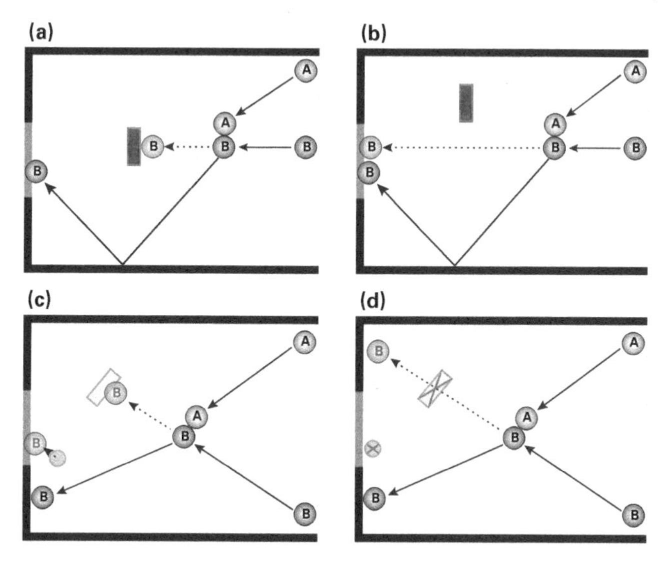

Figure 18.4
Diagrammatic illustrations of four collision events in a simple physics world (see Gerstenberg et al., 2021). *Note*: Solid arrows represent the actual trajectories of ball A before the collision and of ball B before and after the collision. Dashed arrows and faded circles represent the counterfactual trajectory of ball B. The brown rectangle, yellow rectangle, and blue circle represent a brick and the entry and exit of a teleport, respectively.

ball *B* to go through the red gate on the left of the screen, prevented it from going through, or did neither. The tools of probabilistic programs and sampling-based inference allow us to give a precise formal account of these causal judgments, which also accords well with intuitions of mental simulation and gives strong quantitative fits to behavioral experiments (see Gerstenberg, Goodman, Lagnado, & Tenenbaum, 2021).

To explain these judgments, we first need to be able to represent the relevant physical knowledge at the right level of abstraction. Despite its simplicity, our domain already affords an infinite number of interactions between *A* and *B*, and we want a model that yields a causal judgment for each possible situation. Rather than having to specify a new model for each causal interaction of interest (as we would have to do if we adopted a Bayesian network formulation; see Pearl, 2000), we want to represent the general laws that govern the interactions between the objects in our world. One way of representing people's knowledge of physical object motion in Church is by writing a probabilistic and approximate version of some aspects of Newtonian mechanics. Functions in the Church program compute the inertial time evolution and the outcome of collisions by taking as input the mass and velocity of objects, as well as more general aspects of the world such as friction and gravity.

So far, these are standard, deterministic simulation routines (so we omit the details). Critically, as in the models in chapter 15, we also assume that some noise in each object's momentum is inserted just after each collision, and perhaps at other times as well, resulting in trajectories that are noisy versions of their Newtonian counterparts. Research reviewed in chapter 15 has shown that people's intuitive physical judgments across many domains such as this are well described by such noisy Newtonian simulations (Sanborn, Mansinghka, & Griffiths, 2013; Battaglia, Hamrick, & Tenenbaum, 2013; Smith et al., 2013; Smith & Vul, 2013). Once we have a Church program that captures people's intuitive physics, we can model predictions about the future (e.g., will ball *B* go through the gate?) as simple forward simulations, and inferences about the past (e.g., where did ball *A* likely come from?) by a `query` of the past given the present, simulating possible histories that could have led up to the current state.

More subtly, a Church program can also be used to evaluate probabilistic counterfactuals (e.g., would ball *B* have gone through the gate if the collision with *A* hadn't happened?)—the key ingredients in judgments of actual causation as explained in chapter 15. In line with the theory in Pearl (2000) for counterfactual reasoning in Bayesian networks, the evaluation of counterfactuals in a Church program involves three steps. First, we condition all the random choices in the program based on what actually happened to estimate the unobserved values of the actual world. Second, we realize the truth of the counterfactual antecedent (e.g., that the collision did *not* happen) by intervening in the program execution that generated the actual world. This intervention breaks the normal flow of the program by setting some function inputs to desired values. For example, to model what would have happened if there had been no collision between *A* and *B*, we could set ball *A*'s velocity to zero or move ball *A* outside the scene shortly before the time of collision. Finally, to evaluate the truth of the counterfactual, we reevaluate all the functions downstream from the point at which we intervened in the program. This process generates a sample over counterfactual world states, and repeatedly running this process, allowing for different stochastic functions evaluations, can be used to express people's uncertainty over what would have happened in the relevant counterfactual world. Notice that the key feature of Church that allows this process to work is that it specifies a process for sampling particular situations and makes explicit the steps of the causal history that lead up to a situation (in the form of a program execution trace). Counterfactuals are then evaluated by a series of "simulation" steps that result in imagined counterfactual worlds, with a probability distribution over nearby possible worlds induced by the inherently probabilistic nature of the mental simulations.

As we explained in chapter 15, people's quantitative judgments of actual causation in scenarios like this are closely linked to such a probabilistic counterfactual analysis (Gerstenberg et al., 2021). When judging whether one event caused another event to happen, people appear to estimate something like the probability that the candidate cause was *necessary* to produce the outcome event: the probability that the outcome, which did in fact occur, would not have occurred in a counterfactual world where the candidate cause was absent. By implementing these computations in a PPL such as Church, they can be evaluated just as easily on simple and familiar billiards scenarios (like those in chapter 15) as on a wide range of novel and complex scenarios such as those shown in figure 18.4, featuring objects that rarely or never appear on a billiards table, but that we can still reason about through mental simulation.

Consider the top pair of diagrams shown in figure 18.4. In figure 18.4(a), ball *B* would have bounced off the brick if it hadn't collided with ball *A*. In figure 18.4(b), the motions of the balls are identical, but because the brick is in a slightly different position, we can infer that ball *B* would have gone through the gate even without the collision with ball *A*. As predicted, people's judgments about whether *A* caused *B* to go through the gate are significantly higher for figure 18.4(a) compared to figure 18.4(b) (Gerstenberg et al., 2021). In the bottom pair of cases, the contrast in the relevant counterfactual worlds was realized by comparing situations in which the teleport was either on or off (figures 18.4(c) and (d), respectively). Again, the motions of objects in these two cases were identical, but only when the teleport was on as shown in figure (18.4c) did people judge that *A* prevented *B* from going through the gate—because only in this case would *B* have gone through the gate (via first passing through the teleport) if it had not collided with *A*.

These examples demonstrate the flexibility of people's intuitive theories and the critical way that embodying a theory as a probabilistic generator for mental simulations supports modular counterfactual and causal thinking in novel situations. Once people have learned how balls bounce rigidly off solid objects, whether they are walls or other balls or bricks, the same simulation mechanisms apply. Once people have learned how the teleport works, they have no trouble imagining its effects either and incorporating them into their simulations and causal judgments. And because they are expressed at a high level of abstraction, the same simulation and inference programs can be used for a much wider range of intuitive physical judgments as well. Figure 18.5 shows several very different inferences that can all be implemented as queries in WebChurch using its built-in, rigid body simulator. The last example in figure 18.5(c) follows in the spirit of Gerstenberg et al. (2018), who present many more quantitative and systematic studies of people making these inverse physical inferences in a simulated "Plinko box." The histogram below the Plinko box shows an approximate posterior (based on 200 Markov chain Monte Carlo (MCMC) samples) of inferred horizontal locations for where a ball might have been dropped into the box along the top edge, given a final ball position along the bottom that is observed to be a normal distribution around the horizontal position 200 (on a range between 0 and 350), with standard deviation

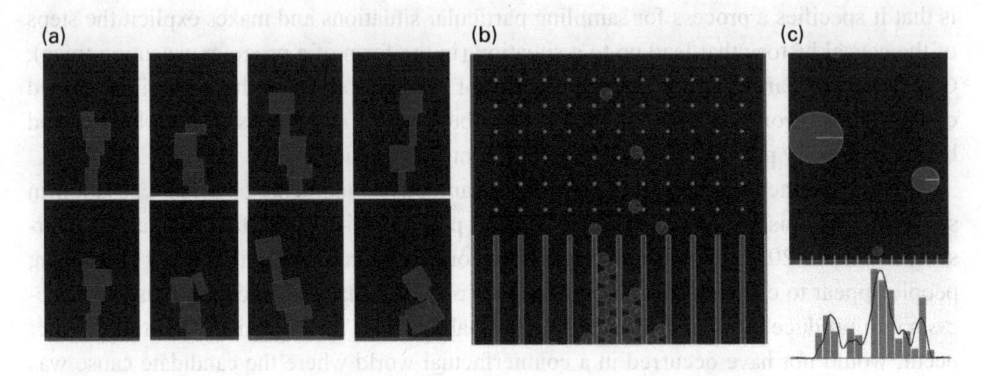

Figure 18.5
Examples of intuitive physical inferences that can be modeled using WebChurch with its built-in physics simulator, where green shapes indicate stationary or fixed objects and blue shapes indicate freely moveable objects (with uniformly distributed mass):

(a) Is a given tower stable (column 1), or slightly, moderately, or highly unstable (columns 2-4)? (http://v1.probmods.org/generative-models.html#example-intuitive-physics.)
(b) How will a large number of little balls dropped near the middle of this box tend to distribute themselves across the slots? (http://v1.probmods.org/generative-models.html#models-simulation-and-degrees-of-belief.)
(c) Given that a blue ball was dropped into the scene somewhere along the top edge and comes to rest at approximxately the location shown along the bottom edge, at what horizontal position along the top edge was the ball likely to have been dropped? (http://v1.probmods.org/conditioning.html#example-inverse-intuitive-physics.)

10. Note how the three peaks in the posterior reflect three qualitatively distinct hypotheses for where the ball could have been dropped: most likely, it was dropped in the center and fell straight down, but it could also have been dropped near either side of the box and landed in the middle after bouncing off the left or right circular obstacle—although if so, it had to have been dropped to the right side of the left obstacle or the left side of the right obstacle. This richly structured and fine-grained posterior belief state—which is also very intuitive to human reasoners—is easily computed in Church via a combination of general-purpose sampling—based approximate inference and general-purpose approximate physics simulation. Although the implementation details will surely differ, some analogously modular combination of sampling-based inference and simulation is both computationally and psychologically plausible as an account of how these inferences unfold in the human mind.

18.7.2 Recursive Inference and Intuitive Theories of Mind

This same general framework for flexible sampling-based inference and causal reasoning applies not only to the domain of intuitive physics, but to any domain for which we are able to represent people's intuitive theories in terms of a probabilistic program. For example, we could model people's judgments about whether agent *A*'s argument convinced agent *B* to try harder as a function of what actually happened and people's subjective degree of belief that *B* would still have tried harder had *A* not said anything. A probabilistic program that captures people's intuitive understanding of psychology looks different from a program that captures people's intuitive understanding of physics, but we can understand people's causal judgments in terms of the same process that compares the actual outcome with the outcomes of mental simulations of the relevant counterfactual worlds.

To illustrate how we might capture the core concepts of an intuitive psychology—a probabilistic model of how agents act rationally in response to their mental states, aiming to satisfy their desires as efficiently as possible given their beliefs—consider as an initial step an extension of the tug-of-war model discussed earlier in this chapter. Now we model laziness for a given player in a given game not simply as a random event, but as a rational choice (not necessarily conscious!) on the part of the player. That is, looking at both their team and the opposition in a given match, a player might estimate that the effective difference in team strengths is so great—either because their team is so strong or the other team so weak—that it is not worth the effort to try hard. We could express this in a Church model by imagining that a player asks themselves, "How should I act so my team will win?"; this translates into a `query`:

```
(define lazy (mem (lambda (person game)
  (query
    (define action (flip L))
    action
    (equal? (teamof person)
            (winner (team1of game) (team2of game) game))
  ))))
```

where we have helped ourselves to some innocuous helper functions such as one function for looking up the team of a player. Parameter `L` controls the a priori tendency to be lazy; this gives a simple way of including a principle of efficiency: a tendency to avoid undue effort. The condition statement of the query specifies the player's goal—for their team to win the game—hypothetically assuming that this goal will be achieved. The output of the query is an action (trying hard or not) that is a reasonable guess on the player's part for how that goal may be achieved. An inference about which team will win a match now leads to a subinference modeling each player's choice of whether to exert full effort, given the players on each team. We could further extend this model to take into account private evidence that each player might have about the strengths of the other players, expressing that player's process of belief formation about the total strengths of the two teams as an additional set of nested subinferences.

Because `query` can be defined within stochastic λ-calculus, it is formally a well-defined probabilistic inference to nest one query within another. In typical probabilistic models of cognition, we use `query` to model inferences of an agent (reasoning, learning, and so on), so a model with nested queries can be interpreted as *inference about inference*. The pattern of using an embedded `query` to capture the choices of another agent is a very general pattern for modeling intuitive psychology (Stuhlmüller & Goodman, 2014).

For instance, one general way to model an agent as approximately rational is to model their choice of actions as an inference about what action is likely to achieve their goal given their beliefs about the world's current state and dynamics (Goodman et al., 2012; Botvinick & Toussaint, 2012; Stuhlmüller & Goodman, 2014):

```
(define (choose-action goal? transition state)
  (query
    (define action (action-prior))
    action
    (goal? (transition state action))))
```

In this code, `state` is a variable encoding the agent's belief about the current state of the world, and `transition` is a function encoding the agent's beliefs about the transition model or environment dynamics: we can think of this as a causal model, either deterministic or stochastic, that outputs the next world state as a function of the previous state and (optionally) an action taken. Here, `goal?` is a function encoding what the agent desires to see happen in the world: we can think of this as a predicate that takes as input a state and returns `True` or `False` depending on whether the state satisfies the agent's desire.

This formulation gives a natural inferential role semantics (Block, 1986) for core concepts of belief-desire psychology. What is an agent's *goal*? It is a predicate that can be applied to states of the world and governs the agent's action choices: when the agent has that goal, they are expected to choose actions that they believe will lead to that predicate being true about the world's future state. What are the agent's *beliefs* about the world's causal structure? They are represented by a function that takes as input the agent's belief about the world's current state and any action that the agent could take, and returns as output an expected future state such that if that future state satsfies the agent's current goal, then the agent could reasonably be expected to take that action.

We can now model canonical Theory-of-Mind inferences by nesting this query representing a probabilistic rational planning process inside a query for goal inference. Conditioning on an agent having taken some action in a given state, this outer query asks which goal (drawn from some goal prior) is most likely to have given rise to that action choice via the inference process:

```
(define (inferred-goal? action transition state)
  (query
    (define goal? (goal-prior))
    goal?
    (equal? action (choose-action goal? transition state))))
```

This corresponds to a simple sampling-based version of the Bayesian *inverse planning* view of Theory-of-Mind, as described in chapter 14 and Baker, Tenenbaum, and Saxe (2007).

The nested query formalism allows planning and inverse planning computations to be flexibly interleaved to reason about how multiple agents will think and act in a wide range of social settings. Many examples can be found in the "Inference about inference" section of the Church-based online book (http://v1.probmods.org/inference-about-inference.html) or its successor using WebPPL. Two other books dive more deeply into modeling agents using probabilistic programs (both in WebPPL). *Modeling Agents with Probabilistic Programs* (Evans, Stuhlmüller, Salvatier, & Filan, 2017) (https://agentmodels.org/) develops PPL models of planning based on Markov decision processes (MDPs) and partially observable MDPs (POMDPs) (see chapter 7), and uses them for modeling intuitive reasoning about an agent's goals, preferences, and beliefs as Bayesian inverse planning (see chapter 14), as well as modeling the expectations about the actions of boundedly rational agents and simple multiagent interactions (e.g., playing simple games). *Probabilistic Language Understanding* (Scontras, Tessler, & Franke., 2017) takes a PPL approach to Rational Speech Act modeling of communication as a form of rational goal-directed action, and modeling language understanding as a Bayesian inference about a speaker's communicative intents (see chapter 16). Zhi-Xuan, Mann, Silver, Tenenbaum, and Mansinghka (2020) extend probabilistic programming for Bayesian inverse planning to more abstract symbolic planning settings, using the *Gen* PPL (Cusumano-Towner et al., 2019) and the *Planning Domain Description Language* to represent an agent's world models, goals, and plans using Gen also enables explicit probabilistic modeling of a boundedly rational agent's approximate planning algorithm to make goal inference robust when the agent might be prone to planning mistakes of various kinds.

Of course, many additional components will be needed to build fully adequate models of human intuitive psychology—for instance, how agents form higher-order beliefs about each other, experience emotions, or make moral judgments. Yet the ease with which researchers have already been able to express core aspects of Theory-of-mind in probabilistic programming terms, especially those within the broad framework of inverse planning, shows the potential of the PLoT approach both here and more generally for capturing core human common-sense reasoning. Extending the inverse planning approach to model inferences about agents' emotions (Houlihan, Kleiman-Weiner, Hewitt, Tenenbaum, & Saxe, 2023), communication about social roles and relationships (Radkani, Tenenbaum, & Saxe, 2022), and other rich social-cognitive inferences is one of the most active and exciting areas of research in probabilistic programming models of human cognition.

18.8 Concept Acquisition

Thus far, we have sketched a notion of concepts as stochastic functions in a PPL, and intuitive theories as systems (or libraries) of interrelated concepts. We have also described how such intuitive theories can be used to describe the complex causal knowledge that people use to reason about the world. If concepts constitute a library of useful (stochastic) functions, what is concept learning?

Forming new concepts from examples is fundamentally a problem of induction—in our case, the problem of program induction. This can be formulated as Bayesian inference of a set of concepts that best explain the experience we have in the world: conditioned on generating the examples we have seen, what is the likely new concept? Hypothesized concepts are formed in an effective language of thought based on the concepts learned so far—all the expressions that can be formed by composing the underlying PLoT and the already-defined function symbols. We can view these hypotheses as being generated by a higher-order *program-generating program*, a stochastic meta-program that generates candidate stochastic programs in our PLoT that might explain a given set of observed examples. Concept learning then reduces to a standard Church query: conditioning on the output of this process (the output of the stochastic program generated by the stochastic meta-program) being equal to the observed examples, we query for symbolic expressions in our PLoT that are both likely to be generated from the meta-program and that meet this condition.

While concept learning as probabilistic program induction is thus philosophically and mathematically well posed, inference over such a vast combinatorial space can be extremely challenging, and much more research is needed to fully realize this vision. The approach has already proved successful, however, in many settings of inductive learning that are simple enough to be studied in behavioral experiments: Boolean category learning (Goodman et al., 2008b); relational concept learning (Kemp, Goodman, & Tenenbaum, 2008a); learning action concepts, such as the drawing programs underlying handwritten characters (Lake, Salakhutdinov, & Tenenbaum, 2015); learning number concepts (Piantadosi et al., 2012b); learning list routines (Rule et al., in press); learning grammatical expressions (Yang & Piantadosi, 2022); and spatiotemporal sequence learning (Mills, Cheyette, & Tenenbaum, 2023). Chapter 19 presents an in-depth treatment of this work, including several detailed case studies of Bayesian program learning and concrete illustrations of the sampling-based inference algorithms that have been most successfully used to learn such expressive representations.

Of course, human concept learning is not always or merely a process of inducing a model that best accounts for a set of observed examples, and the PLoT view is well positioned to engage richer learning dynamics and mechanisms. In particular, social and communicative factors, with or without natural language experience, can play critical roles. Shafto, Goodman, and Frank (2012) show how inferences about the social context of examples in concept learning—for instance, examples generated communicatively by a helpful teacher—can strongly and rationally guide the generalizations that learners make beyond the examples. Natural language provides richly informative evidence for concept formation, especially in children's cognitive development—for instance, Piantadosi et al. (2012a) model how a natural-language count list provides children with a crucial input for learning concepts of natural numbers.

The PLoT offers a powerful way to think about these and other bootstrapping phenomena at the interface of social interaction, language acquisition, and concept acquisition, such as the contributions of syntactic and semantic bootstrapping in learning verbs and the contributions of pragmatic inference in learning quantifiers. Again, much further research is needed to understand how these social and linguistic factors integrate with inductive learning from examples in a full theory of concept acquisition; this will also need to build on general-purpose probabilistic programming tools for modeling pragmatic language understanding (Goodman & Stuhlmüller, 2013; Lassiter & Goodman, 2013) and grounding natural-language meaning representations compositionally and probabilistically in PLoT-based world models for reasoning (Goodman & Lassiter, 2015; Wong, Grand et al. 2023). What is important for us here is that the PLoT provides us with a theory of concepts, an approach to concept learning as probabilistic inference, and a unifying framework for modeling people's intuitive theories of social cognition and linguistic meaning that lets us explore these interactions in a productive way.

Note that this PLoT view of concept learning, although it emphasizes compositionality, is not incompatible with the view that many natural concepts begin life as little more than unanalyzed atoms or placeholder symbols in the mind, which only later come to acquire rich content. The impetus to add such a placeholder symbol may come from natural language (upon hearing a new word), from the interaction of knowledge about natural kinds and specific examples (as suggested by Margolis, 1998, and Carey, 2015), or from other explanatory pressures. Deeper causal content, knowledge about the causal relationships that instances of this concept enter into with other kinds of entities, and relationships to other concepts can be incorporated into the web of function definitions inductively and progressively, as the learner encounters additional examples and gains broader experience in a new domain.

An important implication of this richer inductive approach is that concept learning can change the effective language of thought in a domain: this happens when learned concepts are added as primitives to the language out of which new mental programs and new concepts will be composed. While the new effective language has the same mathematical expressivity as the underlying universal PLoT (which generates the hypotheses for all concepts, past and future), particular thoughts may be vastly simpler—expressed in shorter programs, and thus more computationally tractable—in the effective language. This has many cognitive implications. In particular, as the effective language of thought evolves, it gives rise to an evolving inductive bias for learning. Concepts that are initially complex and unlikely to be constructed by a naive learner may become simpler and more plausible later in the process of elaborating the learner's conceptual library. This process may be a critical driver of children's long-term cognitive development, as well as adults' development of expertise in a novel domain (Rule, Tenenbaum, & Piantadosi, 2020).

Although not directly implemented in a PPL, DreamCoder (Ellis et al., 2020, 2021) and its descendants (Wong, Ellis, Tenenbaum, & Andreas, 2022; Bowers et al., 2023) are recent AI approaches for Bayesian program learning inspired by Church and the PLoT hypothesis that show how such "library learning" can dramatically expand the repertoire of effectively learnable concepts, given a limited amount of computational resources. These systems work not only by adding newly learned concepts to the library, but also by abstracting out program components that are implicitly shared between previously learned concepts to identify the most explanatory and compact theory of a domain—effectively performing a heuristic version of hierarchical Bayesian inference to learn the prior for generating novel concepts. LAPS (Language for Abstraction and Program Search) (Wong et al., 2022) extends DreamCoder to learn a joint prior on programs and natural-language translations of those programs, and models how—consistent with studies in cognitive development (Carey, 2009)—experience with natural language in a doman can bootstrap a learner's theory acquisition and joint acquisition of word meanings and novel concepts, well beyond what could be gleaned merely from observed perceptual examples.

18.9 Future Directions

The tools and ideas in this chapter inherit much from earlier Bayesian modeling approaches that work with richly structured representations, but they frame them in new ways that also suggest important open directions for research in Bayesian cognitive science. This starts with the basic mechanics of reasoning under uncertainty, which gives a natural connection to inferential role notions of concept meaning (Block, 1986): It is not merely the proximal definitions, but also the complex ways that information flows under inference that determine the content of our concepts—much as we saw with the phenomena of explaining away in generative perception and causal reasoning. Rather than working at the level of monolithic probability distributions, the stochastic lambda calculus and Church allow us to work from the point of view of generative, sampling systems. As a form of knowledge representation, this makes a key connection to classic ways of studying mental simulation and mental imagery (Shepard, 1994) and more generally opens up rich avenues for modeling the roles of imagination, counterfactual reasoning, and other ways that human beings can use their conceptual systems to think flexibly about possible ways that the world could be. It also naturally brings in the tools of sampling-based approximate inference that can operate very generally on a wide range of symbolic structures—even the traces of Turing-complete programming languages—as candidate mechanistic models for thinking. And finally, by drawing on hierarchical Bayesian frameworks for learning inductive constraints and learning to learn, the PLoT lets us think in terms of programs as hypotheses for novel concepts, and probabilistic meta-programs that generate domain-specific languages of programs as powerful and dynamically evolvable hypothesis spaces and priors for inductive concept learning.

Looking ahead, we should start by acknowledging that Church models, and PLoT models more generally, are intended to capture the knowledge that people use to reason about the world and the inferences supported by this knowledge, but not in any precise way the algorithmic processes underlying inference, much less their neural instantiation. Connecting across these levels of analysis is one of the key future challenges for the PLoT hypothesis. Early implementations of Church as a programming language suggest one avenue of approach: Church's `query` works through various combinations of caching and Monte Carlo simulation, a very different view of computation than one might expect from a course on probability—not so much arithmetic tabulation as stochastic dynamical systems tuned to result in samples from the desired distributions, and not unlike the way that many neuroscientists think about computation in the brain. Long engineering practice shows that these algorithms can give efficient solutions to tough statistical inference problems, and recent work has provided initial connections between such sampling-based inference algorithms and the multiple levels at which human cognitive processes can be analyzed (see chapter 11).

But we are just at the beginning of this journey. The design and implementation of more recent PPLs suggest fascinating new ways to connect the computational, algorithmic, and neural levels of analysis that come from thinking about how to engineer probabilistic programs at scale: for example, via amortized inference in deep neural networks (Bingham et al., 2019), programmable inference that mixes symbolic and neural motifs (Cusumano-Towner et al., 2019), or, most intringuingly, generalizations of Monte Carlo inference that can be compiled into the network dynamics of biologically realistic spiking neurons (Bolton, Matheos et al., in prep).

18.10 Conclusion

As the other chapters in this book illustrate very well, advances in Bayesian models of cognition have the potential to affect almost every area of cognitive science. These models share the basic mechanics of probabilistic inference and a core philosophical commitment to what it means to reverse-engineer the mind, but they can bring a bewildering and heterogenous array of additional representational tools and claims. The universal probabilistic programming view presented in this chapter, as well as the underlying mathematics of stochastic λ-calculus, serves as a key unification showing that all these Bayesian models can be represented in, and hence reduced to, a simple system built from little more than function abstraction and random choice. This gives hope that future advances in probabilistic modeling of targeted domains will continue to be compatible with each other and can ultimately be combined into a broader architecture for modeling human knowledge, reasoning, and learning.

In a step toward that unification, we showed how embedding stochastic λ-calculus in a PPL such as Church offers a powerful way to model the structure and function of core concepts in higher-level cognition, along with the larger conceptual systems, or intuitive theories, that individual concepts contribute to and gain meaning from. Viewing concepts as the stable representations of a *probabilistic language of thought*—more formally, as functions in an enriched stochastic lambda calculus—builds on and unifies traditional approaches to concepts and mental representation that have often been cast as rivals. Like classical theories of concepts, the PLoT puts compositionality and symbolic scaffolding at center stage. Unlike these theories, however, but very much in the spirit of prototype, exemplar, and connectionist approaches to concepts, the PLoT views human reasoning as fundamentally graded. Compositionality has long been recognized as crucial to explaining the open-endedness and productivity of thought, while probabilistic inference explains why graded reasoning is necessary to make good guesses and good bets in an uncertain world. The PLoT seeks to explain how these various aspects of complex human cognition fit together and distinctively enable each other, making each work better in a way that previous formal theories have not. It also helps us understand many themes in everyday cognition with both new qualitative insights and quantitative accuracy.

Beyond the conceptual advances and unification that Church brings to Bayesian cognitive modeling, more recent PPLs such as Pyro and Gen focus on scaling up inference so that universal probabilistic programs can also be the basis for practical and state-of-the-art research in machine learning and AI. These tools are already letting data scientists and AI engineers work effectively with the kinds of richly structured probabilistic models that cognitive scientists have long seen as central to everyday human thinking—and that should be just as central in modern computational science and autonomous systems that we want to function in the human world.

As we finish writing this book, engineering efforts to scale up probabilistic programming for AI are just beginning to hit an inflection point of speed, cost-effectiveness, and generality. We look forward to following and building on the trajectories of these engineering developments, as well as to exploring the hypotheses that they will generate for reverse-engineering probabilistic inference in the mind and brain.

19

Learning as Bayesian Inference over Programs

Steven T. Piantadosi, Joshua S. Rule, and Joshua B. Tenenbaum

The previous chapters of this book show that Bayesian models form a powerful approach to understanding cognitive processes and illustrate how they can use diverse representations including continuous spaces, graphs, logical formulas, and grammars. This chapter covers foundational ideas in Bayesian inference over programs. The idea that people consider algorithmically rich hypothesis spaces has been found to match human performance on learning tasks. The theory of program learning also suggests one possible route for how human cognition might scale up, both developmentally in the life of a single child and culturally across generations. The idea that people do inference over programs can be motivated by the remarkably broad set of mental representations that human children acquire—from learning games, to social rules, or new formal systems like logic and programming languages. Our ability to interface with external dynamical or computational systems—a car engine, a pet, or a cash register—likely requires us to form mental models of these objects. Our successes across a diverse range of systems suggest that we are able to learn over hypothesis spaces that are computationally powerful, perhaps effectively the space of all computations but constrained by our limited memory.

If $\mathcal{H}$ is a space of computer programs and h is one program in that space which could be hypothesized to explain some observed data d, often program inputs and outputs, this chapter outlines how one typically can define a prior $P(h)$ and a likelihood $P(d \mid h)$, as well as some of the considerations that go into both. We then discuss typical statistical inference schemes for programs, focusing primarily on Markov chain Monte Carlo (MCMC) methods. We conclude with a discussion of advanced methods that learn subroutines, draw on a rich family of inferences, and learn to encode fundamentally new logical structures.

19.1 Background

Alan Turing's work in the mid-1900s to develop a mathematical model of computing (Turing, 1936, 1950) is widely recognized as foundational to computer science. It should also be celebrated as a profound discovery in cognitive psychology. At the time when Turing was working, the word "computer" referred to *people* who performed computations. Turing's formalization of computation (Turing machines) was thus an abstract model of what humans could do by combining a few basic abilities. What Turing discovered is staggering: operations like contingently updating an internal state and interfacing it with memory allow a human computer to simulate *any other* logical machine. With the right program, a person—or a mechanical/electronic computer—can simulate everything from physics to the stock market by combining elementary operations in new ways.[1]

This universal ability is articulated in the *Church-Turing thesis*, which holds that everything that can be computed can be computed on a Turing machine or equivalently programmed into a standard computer. The key is that basic operations can compose to determine more complex behavior, just as a complex novel can be composed only from a few dozen characters. Early in the history of computer science, logicians developed seemingly distinct models of computation (e.g., Turing machines, lambda calculus, combinatory logic, Post systems, and term rewriting) that were then found to be equal in computational power in that each system can simulate all the others. Today, remarkably diverse systems—arrays of cells that blink on and off based on the behavior of their neighbors, billiard balls bouncing, simple systems of rewrite rules, and neural networks (Wolfram, 2002; Abelson, Sussman, & Sussman, 1996; Fredkin & Toffoli, 1982; Pérez, Marinković, & Barceló, 2019)—have been found capable of universal computation. This means that there are many possible systems that could form the basis of human-like computational abilities.

People's cognitive capacity for universal computation appears to be distinctive in animal cognition. Human beings alone discover fundamentally new kinds of representations, internalize complex procedures, and revise and improve complex algorithms. Some of these processes are the bread and butter of cognitive psychology: people can learn to count, read, reason, and understand physical systems. But human abilities are much broader. People can also learn to fly the space shuttle, play the cello, perform a quadruple Lutz, or repair accordions. We learn to manipulate wholly new logical systems like calculus, Rubik's cube, and the rules of legal systems. We learn to reason about social entities and think strategically about competitors. When compared to other species, our ability to model the world and use those models in pursuit of our goals is virtually limitless.

This has led many to hypothesize that cognition must have the capacity to *infer* or *internalize* novel algorithmic processes—that is, programs. Much of human knowledge is well described using programs (figure 19.1). In turn, program induction models have been used to model many aspects of learning across a variety of domains. For example, Amalric et al. (2017) develop a model of mathematical reasoning that describes geometrical patterns as functions from step numbers (i.e., step 1 and step 2) to positions in space. Others have studied human induction of logical concepts (Goodman, Tenenbaum et al. 2008b; Kemp, 2009, 2012; Piantadosi, Tenenbaum, & Goodman, 2016) with models that take feature-based descriptions of objects as input and return rules that compute true/false judgments about whether each object belongs to the concept. Lake, Salakhutdinov, and Tenenbaum (2015) presented a model of hierarchical action planning that takes images of hand written characters, infers the probabilistic plan or program most likely to have generated them, and produces new exemplars of the same character. A number of models learn theories that take relational data as input and predict novel relations as output for domains like

1. Such scope is very much unlike early mechanical computers (Aspray, 2000). Babbage (1864), for example, viewed computation as merely a "mechanism for assisting the human mind in executing the operations of arithmetic." Babbage's colleague, Ada Lovelace, was perhaps the first to see the real potential for other forms of computation (Lovelace, 1842), but this vision was not realized until the creation of digital computers in the mid-twentieth century.

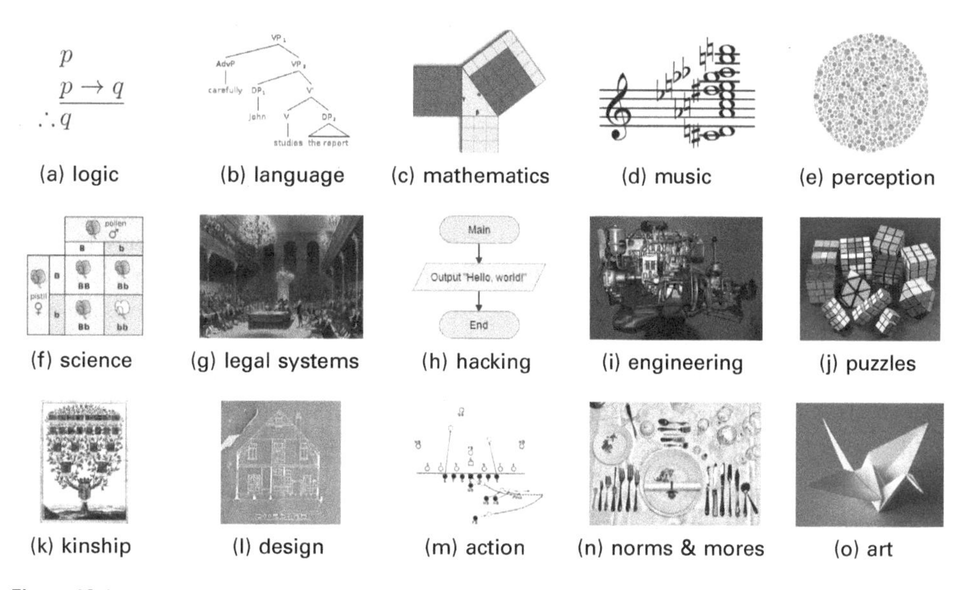

Figure 19.1
A few domains that can be described well using programs, usually because they involve multiple interacting or structural logical rules, or abstract processes.

magnetism, kinship, causality, biological taxonomies, and even spherical geometry (Kemp & Tenenbaum, 2008; Kemp, Tenenbaum, Niyogi, and Griffiths, 2010, Goodman, Ullman, & Tenenbaum, 2011; Ullman, Goodman, & Tenenbaum, 2012; Mollica & Piantadosi, 2021. Piantadosi et al. (2012a) and Piantadosi (2023) modeled the acquisition of the counting procedure as discovering a program for transforming sets of objects into number word labels.

A number of other models deal with more abstract mathematical relationships, including binary sequences (Planton et al., 2021), numerical concepts like even and odd (Tenenbaum & Griffiths, 2001a), fractals (Lake & Piantadosi, 2020), and functions over lists of natural numbers (Rule et al., in press). Siskind (1996) modeled lexical semantics, and other models have examined additional aspects of language, including quantifier semantics (Piantadosi, 2011), morphology (Ellis, 2020), syntax (Yang & Piantadosi, 2022), question answering (Rothe, Lake, & Gureckis, 2017), and pragmatics (Goodman & Lassiter, 2015; Goodman & Frank, 2016). Several others deal with visual reasoning over visual representations (Ellis, Morales, Sablé-Meyer, Solar-Lezama, & Tenenbaum, 2018), three-dimensional (3D) figures (Overlan, Jacobs, & Piantadosi, 2017), and even Bongard problems (Depeweg, Rothkopf, & Jäkel, 2024).

19.2 The Hypothesis Space

Most models of program learning do not work with Turing machines (cf. Graves, Wayne, & Danihelka, 2014; Kim & Bassett, 2022) because behavioral work in cognitive science makes a convincing case that human cognition is *compositional*. As discussed in chapter 18, Alonzo Church developed *lambda calculus* to formalize computation as function composition. In his framework, complex programs arise from gluing the input of one function to the output of another—for instance, composing f and g to form $f \circ g$. This kind of computation by composition survives today in functional programming languages like Haskell and Lisp, which express programs as a composition of primitive functions rather than a sequence of steps. A similar compositional approach is dominant in formalizing natural language semantics (Heim & Kratzer, 1998; Steedman, 2000; Blackburn & Bos, 2005). For example, the meaning of a sentence like "I found the concertina that Pietro dropped" can be captured by composing the individual function for each word in the sentence. Compositionality in language is likely deeply related to compositionality in thinking, leading many to argue for a "language of thought" (Fodor & Pylyshyn, 1988; Fodor, 1975; Piantadosi & Jacobs, 2016; Goodman, Tenenbaum, & Gerstenberg, 2014) in which thoughts themselves are comprised of constituent elements into sentence-like structures. This compositional approach has also been used in learning models, which operate over the space of possible compositions. Siskind (1996), for example, formulated a word-learning model where a word like "lift" would have the meaning

$$\texttt{lift(x,y)=CAUSE(x,GO(y,UP))}. \tag{19.1}$$

Here, `CAUSE`, `GO`, and `UP` are primitive operations that compose to express the meaning of `lift`. The expression in equation (19.1) means that "x lifts y" if `x` causes `y` to go up. This kind of model can be contrasted with models in which "lift" might be a conceptual primitive itself, so the only problem that learners face is determining the mapping from words to already-given meanings. In compositional models, learners face the much more difficult task of constructing a correct meaning from constituent parts and finding the correct composition out of the large (often infinite) space of possibilities.

19.2.1 A Grammar of Hypotheses Based on Types

Like the example in equation (19.1), we will typically think of programs as *functions* that map inputs to outputs. For example, `lift` in equation (19.1) takes two objects, `x` and `y`, and returns a Boolean value corresponding to whether `x` lifted `y`. When building a program learning model, it is our job to choose appropriate inputs and outputs, which often shapes what function we learn, as well as what primitives are necessary.[2] In typical program learning models, the hypothesis space consists of all semantically sensible ways of combining an assumed set of primitive operations. A learner couldn't even evaluate a hypothesis like `lift(x,y)=CAUSE(CAUSE,CAUSE)` or `lift(x,y)=GO(CAUSE,y)` because these make no sense.

The primary way that we ensure that programs are semantically coherent is to treat each primitive as having fixed input and output types. To illustrate, `lift` might take two things of type *object* as arguments. This means that the arguments `x` and `y` should be objects: you can lift a potato, but not an abstract thing like a joke; you also can't lift a whole sentence or proposition like, "Smoking cigarettes kills." The primitive `GO` might take an object and a direction, so `GO` can make a potato go left or right or up or down. Importantly, it cannot make a potato go "carrot" because it cannot take an object as its second argument. The primitive `CAUSE` is a little more interesting, in that it takes an object (a causer) `x` and an event

2. For example, we could alternatively imagine a version of `lift` that takes as an argument an object `X` and a context (e.g., set of objects) and returns the objects that `X` lifted.

that happens and asserts that x caused the event. Thus, CAUSE(x,GO(y,up)) means that x causes y to go up; this composition would evaluate to true only when the causal relationship were true.

The assumed types can be enforced using a context-free grammar (see chapter 16) to generate the full program composition. The nonterminals of this grammar specify the input and output types of each primitive function. For example, if GO takes an object and a direction and returns a Boolean truth value, we might write this in the grammar as BOOL → GO(OBJECT, DIRECTION). This means that when we are writing a program, any time we see a BOOL symbol, we may replace it with a call to the function GO applied to some OBJECT and some DIRECTION, both also derived from the grammar. Writing out more rules for this grammar gives the following:

```
   OBJECT → x | y | car | tree
DIRECTION → up | down | left | right | in | out
     BOOL → GO(OBJECT, DIRECTION) | CAUSE(OBJECT, BOOL)
          | or(BOOL, BOOL) | and(BOOL, BOOL) | not(BOOL).
```

This grammar specifies a vast hypothesis space. Even this toy example contains 24 possible GOing actions (each of 4 objects going in each of 6 directions), which then can be combined with conjunction (and), disjunction (or), and negation (not). Indeed, this combinatorial explosion is intentional. One of our chief design goals for program learning models is that a few built-in operations combine to express a huge variety of concepts. Such power is, as mentioned previously, motivated by people's ability to acquire such a rich diversity of concepts. For example, learners with this system could hypothesize that *x* lifts *y* if a car makes x go right or a car makes y go left:

```
lift(x, y) = or(
                 CAUSE(car, GO(x, right)),
                 CAUSE(car, GO(y,left))
                 ).                                  (19.2)
```

This hypothesis, though incorrect, is conceivable and possible—that is why we want the grammar to include it. That said, equation (19.2) is a bit unnatural, likely due to its complexity. This particular hypothesis uses 11 primitives and should perhaps be dispreferred a priori relative to simpler options, including the correct one. We can build a simplicity preference into our model by converting the grammar to a *probabilistic* grammar. This requires us to give a probability to each rule expansion (see chapter 16). We can then compute the probability of a full expression by multiplying the probabilities of each rule expansion it uses. Such a probabilistic grammar can then be used to define the prior for the program learning model. Because the prior typically favors short programs, Bayesian program learning models usually will tend to prefer programs that *compress* the observed data into a concise description. This compression is an effect of the broader generalization effect that is typical of Bayesian learners (see chapter 3).

19.2.2 A Simple Grammar of List Functions

We can see many Bayesian inference effects at work in the domain of *list functions*, where learners observe an unknown function $\mathcal{F}$ that transforms one list of numbers into another (Rule et al., in press). For example, we might be told only a single data point:

$$[3,6,1,7,3,2,9] \xrightarrow{\mathcal{F}} [9,7,6,3,3,2,1], \tag{19.3}$$

where the list on the left has been transformed to the list on the right by the unknown function $\mathcal{F}$. From this example, it is easy to guess that $\mathcal{F}$ sorts lists in descending order. After learning that, we could readily apply $\mathcal{F}$ to a new list. Other data sets would lead to different guesses about what $\mathcal{F}$ does, and consequently different generalizations:

$$[3,6,1,7,3,2,9] \xrightarrow{\mathcal{F}} [2,6]. \tag{19.4}$$

This is consistent with at least a few generalizations a learner might hypothesize: (1) remove odd numbers; (2) keep only 6s and 2s; (3) keep the second and sixth elements of the list; (4) keep only the second from the front and second from the back, etc. These options highlight a strength of probabilistic models–namely, that they can distribute belief across multiple hypotheses. These beliefs are sensitive to the amount and content of the data. For example, if we next saw two data points:

$$\begin{aligned} [3,6,1,7,3,2,9] &\xrightarrow{\mathcal{F}} [2,6] \\ [7,5,2] &\xrightarrow{\mathcal{F}} [2], \end{aligned} \tag{19.5}$$

the data would be best supported under hypotheses (1) and (2) but not (3) and (4), leading to revised beliefs. One more data point might give

$$\begin{aligned} [3,6,1,7,3,2,9] &\xrightarrow{\mathcal{F}} [2,6] \\ [7,5,2] &\xrightarrow{\mathcal{F}} [2] \\ [8,8,8] &\xrightarrow{\mathcal{F}} [8,8,8]. \end{aligned} \tag{19.6}$$

While it may feel at this point as though we have narrowed in on a single hypothesis, remember that, in the space of all possible computations, there are infinitely many possible hypotheses that are consistent with any data set. However, the prior will assign the highest probability to the shortest, simplest hypotheses that are consistent with the data, potentially leading to strong beliefs about the right answer.

Introspecting on this example motivates some basic operations that we might want in a formal model of the list domain, including operations to manipulate and select from lists:

```
     LIST → filter(PREDICATE, LIST)|pair(NUMBER, LIST)|
            ε|reverse(LIST)|rest(LIST)|lst
   NUMBER → first(LIST)|0|1|2|...|9
PREDICATE → is-even|is-odd|is-prime|...
```

Here, `lst` is the argument provided to $\mathcal{F}$; ϵ is an empty list; `pair` builds lists by prepending a number onto an existing list; `first` returns the first element of a list, while `rest` returns everything except the first element; `filter` loops over a list, keeping only elements that satisfy its predicate; `reverse` reverses a list; and we have several built-in predicates for operations that adults learning $\mathcal{F}$ might be expected to know. Predicates are themselves functions that can be applied to numbers (e.g., they map NUMBER to BOOL). Perhaps unsurprisingly, these operations are similar to those found in list-processing programming languages designed for human use (Abelson et al., 1996). Using these operations, we could formulate hypotheses like (1) as

$$\mathcal{F}(\texttt{lst}) = \texttt{filter(is-even, lst)} \tag{19.7}$$

or (4) as

$$\mathcal{F}(\texttt{lst}) = \texttt{pair(first(rest(lst)), first(rest(reverse(lst))))}. \tag{19.8}$$

Like the grammar for `lift`, this grammar allows us to express many other hypotheses. For example, $\mathcal{F}$`(lst) = pair(9, rest(lst))` would replace the first element of the list with 9.

19.2.3 Variables and Abstraction

The operations that we have discussed so far fit into a probabilistic, context-free grammar, but most programming-language grammars are not context free. One reason is that introducing a variable changes the programs that are valid: the expression `first(foo)` is valid only if `foo` has previously been declared as a variable. We face this problem if we introduce explicit variables into cognitive programs. For example, we might want to filter a list by whether a number is "even or prime," and in that case, we would have to include Boolean `or` and use it to define a subroutine:

$$\texttt{even-or-prime(y) := or(is-even(y), is-prime(y))}. \tag{19.9}$$

This subroutine is needed because of the types: `or(is-even(y), is-prime(y))` returns a BOOL, not a PREDICATE, and so it can't serve as the first argument of `filter`. If we just wrote `filter(or(is-even(y), is-prime(y)), lst`, it would not be clear that we intended `y` to be the variable mapped over each element of `lst`. Defining the function `even-or-prime` resolves these issues, so we can write

$$\mathcal{F}(\texttt{lst}) = \texttt{filter(even-or-prime, lst)}. \tag{19.10}$$

It is clunky to define an entire function `even-or-prime` if it is going to be used only once, particularly because learning means that we do not know ahead of time which functions are going to be necessary. Lambda calculus is useful for this because it allows us to define functions without having to give them names, often called *anonymous functions*. With lambda calculus, we would write equation (19.9) as

$$\lambda\ \texttt{y.or(is-even(y), is-prime(y))}, \tag{19.11}$$

where λ`y.` simply means "What comes next is a function of `y`." The whole thing can then be substituted into the first PREDICATE argument of `filter` because this lambda expression takes a number (`y`) and returns a BOOL, so it matches the type of PREDICATE:

$$\mathcal{F}(\texttt{lst}) = \texttt{filter(}\lambda\texttt{y.or(is-even(y),is-prime(y)),lst)}. \tag{19.12}$$

Many program induction libraries will allow lambda functions in their grammar.

Both equations (19.9) and (19.12) introduce the variable y, which is not in the grammar. It is only the "λ y" (or "even-or-prime(y)=...") that introduces y as a variable and a new symbol, and it could have been called z or symbol1244 without altering what the program computes. The way that variables dynamically change the underlying grammar is an annoying technical problem that gets handled in several ways across different implementations of cognitive models. The simplest solution is to build in a set number of variables and allow the learning model to sort out which ones go where, assuming an appropriate means of evaluating variables that are not defined. A more general approach that keeps the grammar context free is to introduce higher-order functions, an approach taken in the Fleet software package. For example, we might introduce a disjunction function (let's call it disj), which takes two PREDICATEs and returns a new PREDICATE of an argument x:

$$\texttt{disj(p1, p2)(x) := or(p1(x), p2(x)).}$$

Here, disj is called a *higher-order function* because instead of taking lists or numbers as arguments, it takes entire functions (PREDICATEs). Then we might form an expression like

$$\mathcal{F}\texttt{(lst) = filter(disj(is\_even, is\_prime), lst).} \qquad (19.13)$$

One downside to this approach is that it requires adding higher-order functions to the grammar that are conceptually close to other existing primitives. For example, disj is nearly identical to or, save for their types.[3] A third approach is to actually change the grammar and add rules for variables once they are introduced, an approach taken in the earlier LOTlib3 software package. The model must ensure that variables are given unique names, but there is a standard scheme for doing so known as *De Bruijn indexing*. While explicitly changing the grammar is flexible, it also increases the complexity of the type system. These kinds of engineering choices are typically far outside the resolution that we have for capturing human cognition—in fact, the priors for one approach can often be made similar or identical to priors for another.

Another reason that many programming languages are not context free is that the type system itself may not be context free. To see why this might happen, consider our current grammar. Expressions with the type NUMBER will always evaluate to numbers, and LIST expressions will always reduce to lists of numbers. Suppose, however, that we also want to include lists of *lists* of numbers. We cannot accommodate this type without extending our language and grammar:

$$\texttt{LIST}_\ell \rightarrow \texttt{filter(PREDICATE}_\ell\texttt{, LIST}_\ell\texttt{) | pair(LIST}_n\texttt{, LIST}_\ell\texttt{) | ...}$$

$$\texttt{LIST}_n \rightarrow \texttt{filter(PREDICATE}_n\texttt{, LIST}_n\texttt{) | pair(NUMBER, LIST}_n\texttt{) | ...}$$

$$\texttt{NUMBER} \rightarrow \texttt{first(LIST}_n\texttt{) | 0 | 1 | 2 | ... | 9}$$

$$\texttt{PREDICATE}_n \rightarrow \texttt{is-even | is-odd | is-prime | ...}$$

$$\texttt{PREDICATE}_\ell \rightarrow \texttt{is-empty | is-singleton | ...}$$

3. Some approaches include a *type-raising operation* T, such that T(or) was equivalent to disj (but T could be applied to any arbitrary function).

This grammar has both lists ($\texttt{LIST}_n$) and lists of lists ($\texttt{LIST}_\ell$) but it also contains many nearly redundant primitives, the same problem that we saw with `even-or-prime`. The $\texttt{LIST}_\ell$ and $\texttt{LIST}_n$ versions of `filter`, for example, behave nearly identically. They both apply a predicate to each element in a list and keep only those elements for which the predicate is true. This problem only grows worse if we want lists of characters, lists of lists of characters, lists of predicates, lists of lists of lists, and so on. It would be convenient to have a single primitive for `filter` that could apply to all lists.

We have looked so far at what are known as "simple types" (Pierce, 2002). Polymorphic type systems like the Hindley-Milner system[4] make this possible by adding variables to the type system itself. Instead of having $\texttt{LIST}_n$ and $\texttt{LIST}_\ell$, we have a single list type: `LIST x`. `x` is a variable that gets filled in with a concrete type like `NUMBER`. This approach allows us to provide a generic version of `filter`, for example, which takes a function from `x` to `BOOL` as its predicate and a `LIST x` as input. If `x` is `NUMBER`, it will filter a list of numbers. If `x` is `LIST NUM`, it will filter a list of list of numbers.

One reason that types are useful is that they allow us to encode information about how programs behave. When we have a program of type `NUMBER`, we know that it will evaluate to a number. As type systems grow more complex, they allow us to pack more information into the types. For example, consider the primitive `last`, which returns the last element of a list; and the primitive `length`, which returns the length of a list. Under simple types, these both take a `LIST` and return a `NUMBER`. With polymorphic types, however, we see that `length` takes a `LIST x` and returns a `NUMBER`, while `last` takes a `LIST x` and returns an `x`. The polymorphic types contain more information and thus differentiate `last` and `length`: they tell us that no matter what kind of list we give as input, `length` will return a `NUMBER`, while `last` will always return the kind of thing contained in the input list. Packing more information into types is useful because it allows a more complete description of what programs should be considered as potential solutions to a given problem. A more complete description is useful because it strengthens the learner's inductive bias, allowing it to rule out large groups of programs that it otherwise would have needed to consider.

Other kinds of type systems can add even more information than polymorphic types. Type systems with type classes, for instance, allow one to state that a primitive applies to any type (i.e., using a type variable), but then restricts that type to a class of types for which other specific primitives are guaranteed to be implemented. For example, you might say that `LIST x` can be sorted so long as type `x` has a comparison operator. Dependent types go so far as to encode entire first-order logical expressions in the type. The dependent type for a sorting function over `LIST x` not only can require that `x` have a comparison operator, but also can state that the output list will be a permutation of the input list such that each successive element is greater than or equal to its predecessor. The program-learning literature describes many techniques employing complex type systems to provide the inductive bias needed to learn complex programs efficiently. (Polikarpova, Kuraj, & Solar-Lezama, 2016; Osera & Zdancewic, 2015; Chlipala, 2022).

4. Hindley-Milner is a type system for what are technically known as "parameterically polymorphic types." Ad hoc polymorphism and subtyping are two other kinds of type polymorphism, with different abilities and benefits.

19.2.4 Recursion and Conditionals

Often, we will also be interested in modeling how people learn recursive functions. Recursion is hypothesized to be a core operation of human thought and language (Hauser, Chomsky, & Fitch, 2002; Corballis, 2014). It is also central in computer science because it allows complex problem to be solved with concise and elegant methods that break one problem into simpler subproblems. For example, if we wanted to include recursive sorting algorithms in our list example, we would need the ability to call the currently defined function $\mathcal{F}$ on new input.

We define the primitive `recurse`, which calls the function in which it is used. In this sense, `recurse` is a little unusual because the value that it returns is not determined beforehand by a defined primitive. Its return value must be computed on the fly by the current definition of $\mathcal{F}$, meaning that its value will depend on its context of use. Here, `recurse` will always have the same input and output types as $\mathcal{F}$. In our example, `recurse` will take and return a list, so we could write functions like

$$\mathcal{F}\texttt{(lst) = pair(first(lst), pair(1, recurse(rest(lst)))).} \tag{19.14}$$

This function is intended to insert a 1 in every other location, such as mapping [5, 6, 7] to [5, 1, 6, 1, 7, 1]. But it is actually not a correct implementation because there is no end to the recursion: even when given an empty list, the equation (19.14) function will try to call recurse on `rest(lst)`, which is another empty list. The recursion will never terminate, so the program will never return a value.

One solution is to introduce an `if` statement that recurses only when a condition is met. We can then write

$$\begin{aligned}\mathcal{F}\texttt{(lst) = pair(first(lst), pair(1, if(is-empty(lst),}\ \epsilon\texttt{,}\\ \texttt{recurse(rest(lst))))),}\end{aligned} \tag{19.15}$$

which recurses only when `lst` is not empty. The behavior of `if` is actually somewhat subtle. When we call a function in most programming languages, we evaluate its arguments first, so `filter(is-even, reverse(lst))` first calls `reverse(lst)` and then passes the value of that expression as the second argument to `filter`. If we do that with `if` in equation (19.15), we will *first* call `recurse(rest(lst))`, which will lead to another infinite recursion. The solution that programming languages have developed is that `if` is a special function that, unlike most other functions, does *not* evaluate all its arguments before being called. Instead, it evaluates the Boolean condition and then evaluates only one of its arguments, depending on the Boolean outcome. This is why `if` is a special keyword in most programming languages.[5]

5. Most languages also treat `and` and `or` in a special way via the related "short-circuit evaluation," where `and` evaluates its second argument only if the first is true; `or` evaluates its second argument only if the first is false. Lazy evaluation is an alternative way to execute programs that removes the need for short-circuit evaluation and other "special forms" (Abelson et al., 1996).

19.2.5 Stochastic Operations

Stochastic primitives are often useful when modeling the language of thought because we sometimes want to model learners who themselves conceive of the world as being partially random. For example, learners might look at a very long sequence like

$$11001001000011111101101010100010001000010110100011000010001101001100$$

and conceive of it as "random" when they cannot find a pattern in it, regardless of whether there actually is a pattern (like base 2 digits of π). Such a sequence, therefore, might have a very concise hypothesis ("concatenate random digits"), even though the data underlying it is complex or may look incompressible.

One easy way to introduce stochastic functions into the representation language is to add a function `flip` that "flips a coin" to determine whether to return "true" or "false." Here, `flip` is quite a different kind of operation from the ones listed before because now a program no longer necessarily has a single output. For example, the simple program

$$\mathcal{F}\texttt{(lst) = pair(if(flip(),1,2), lst)} \tag{19.16}$$

will return a distribution over two different lists: half the time a 1 appended onto the front of `lst`, and half the time a 2. Actually, we can think about the expression (19.16) in a few ways. We could think of it as a program that samples a random answer each time that it is run. We could also think of it as implicitly specifying or representing a particular probability distribution. Similarly, different implementations will handle stochastic operations like `flip` in different ways. For example, some might store partial evaluations of programs and their associated probabilities. When encountering a `flip`, they push *both* outcomes onto a queue of incomplete traces to continue evaluating later. Other examples of stochastic operations might sample an element from a discrete set (e.g., an alphabet) or a continuous set (a distribution).

Importantly, once stochastic operations interface with operations like recursion, things can become more complex. For example,

$$\mathcal{F}\texttt{(lst) = pair(if(flip(),1,head(lst)), recurse(rest(lst)))} \tag{19.17}$$

will take a list like $[1, 2, 3, 4]$ and flip a coin to determine whether each element should be replaced by 1. The distribution of outcomes, therefore, assigns nonzero probability to sequences like $[1, 2, 3, 4]$, $[1, 1, 3, 4]$, $[1, 2, 1, 4]$, $[1, 2, 3, 1]$, $[1, 1, 1, 4]$, $[1, 1, 1, 1]$, etc. In this kind of model, the probability of the data is tied to the probability that the program took a particular route in its evaluation. For example, if the program maps

$$[1, 2, 3, 4] \xrightarrow{\mathcal{F}} [1, 2, 1, 4], \tag{19.18}$$

then this data would have a probability of $(1/2)^3$ since it doesn't matter how the first coin flip comes out (either way will put a 1 at the start of the list), but the remaining three have to come out specific ways to generate the various outputs.

Many languages with stochastic operations, including Church (Goodman et al., 2012), also include *memoization*, which lets functions remember the outcome of specific coin flips or other random choices. To illustrate, imagine that we wanted to replace all of the 2s in a list with either 3 or 4 (but randomly one or the other). You might imagine writing a function like

```
F(lst) = if(flip(),
    pair(if(equals(first(lst),2), 3), recurse(rest(lst))),
    pair(if(equals(first(lst),2), 4), recurse(rest(lst)))).
```

This function would not be correct, though, because each time `recurse` is called, a *different* value of `flip` would be computed. So this would make a new random choice in each recursive step, potentially picking different ones each time. One solution is to have function mem, which "remembers" the random choices that were made. For example, `mem(flip)` would be a function that chooses a value the first time it is called and remembers that value the later times it is called, even across recursions. In this case, we could write

```
F(lst) = pair(
           if(
               first(lst) == 2,
               if(mem(flip)(), 2, 3),
               rest(lst)
           ),
           recurse(rest(lst))
         ).                                        (19.19)
```

19.3 Likelihoods for Program Models

To use a Bayesian model for program learning, we need more than the prior defined by the grammar over programs in our hypothesis space. We also need to specify a likelihood $P(d \mid h)$ that describes how probable the data d would be if generated using program h. One basic challenge arises when searching spaces of deterministic programs, which is that each program only returns a single output for a given input. Thus, most programs generate incorrect output. For example, we would have a very low probability of sampling a hypothesis from the prior and correctly predicting equation (19.3). One common way to address this is to smooth the likelihood by introducing noise into $P(d \mid h)$—namely, assuming that the output is computed by first running the program for h and then randomly altering the output in some way. Adding such noise is useful because hypotheses that are approximately correct can receive partial credit for getting some of the output correct. This in turn provides a signal that helps search algorithms move through the space of programs in a way that improves the fit to the data. Indeed, the choice of a helpful likelihood is often one of the most important choices for efficient inference.

Noise can be added to the likelihood in many ways. For example, we might assume in the domain of list functions that each element in an output list is corrupted by randomly sampling from $\{0, 1, 2, \ldots, 9\}$ with some small probability, perhaps 0.05. In this case, if the observed data list output were [9, 7, 6, 3, 3, 2, 1] but the program predicted [9, 7, 5, 3, 3, 2, 1], then the data would have the probability

$$\left(0.95 + \frac{0.05}{10}\right)^6 \cdot \frac{0.05}{10} \tag{19.20}$$

since six of seven data points could have been unchanged in the noise (probability 0.95) or changed via noise to their same values (probability 0.05/10), and one value, 5, had to be changed to its value. This computation illustrates that when computing the probability of noisy data, one must sum both the probability that it was generated by the hypothesis and the probability that it was generated by noise.

Noise in this likelihood has no way to change the length of the list. It would therefore assign zero probability to $[9, 7, 6, 3, 3, 2, 1]$ if the hypothesis predicted $[9, 7, 6, 3, 3, 2]$ or $[9, 7, 6, 3, 3, 2, 1, 8]$. This situation is problematic because it means that the hypothesis must get the length right on *every* output—which is unlikely to occur by chance—or else the entire data set is assigned zero probability. However, we can define other likelihoods that change the length. For example, Kashyap and Oommen (1984) define a probabilistic edit likelihood that allows one to compute the probability that one list is corrupted to another according to a series of insertions, deletions, and alterations, essentially a probabilistically coherent version of string edit (Levenshtein) distance. A faster but similar option, used in Yang and Piantadosi (2022), is to consider a noise model that first deletes from the end of the hypothesis's generated sequence and then appends randomly onto the end. Note that, as before, any string can come from any other by deleting everything, but hypotheses that share a prefix with the data will require fewer deletions and thus be assigned higher likelihood. This likelihood can not only be computed efficiently but also leads models to prefer to get the beginnings of lists correct. This ordering bias likely helps random search find hypotheses that match the initial stages of a recursive computation.

In settings other than lists or strings, the likelihood takes on a different form. For example, Piantadosi (2011) examined the learning of quantifier terms like "every" and "most" and showed that a program learning model could acquire programs for these terms. Quantifier terms are often formalized as functions on sets: "most A are B" (i.e., `most(A,B)`) is true if

$$|A \cap B| > |A \setminus B|. \tag{19.21}$$

In other words, "most barbers are happy" if the set of happy barbers (intersection of happy people and barbers) has greater cardinality than the set of nonhappy barbers (set difference between barbers and happy people). We therefore formulated a fairly standard hypothesis space with operations over sets and cardinalities. When we first considered how to set up the likelihood of this model, it seemed plausible to use a likelihood that gave high probability to utterances that were most often true—for example, the probability for a given data point (utterance in a context) would be high if the hypothesis evaluated it to be true, and false otherwise. We learned that this approach is doomed because the model will invariably learn meanings that are trivially true. For example, instead of learning a meaning like equation (19.21), the model might learn that $most(A, B) := true$.

While it is tempting to try to rule out these trivial meanings somehow in the grammar, it won't work because a model can always write expressions that are complex but trivially true (e.g., $|A \cup B| \geq |A|$). The example illustrates that a likelihood usually does need to be a proper probability distribution. Our initial likelihood was not quantifying the probability of the data in any sense. The problem was fixed by using a likelihood that evaluated all possible quantifiers in each context, and, with high probability, chose uniformly from those that were true. This size-principle likelihood (Tenenbaum 2000; Tenenbaum & Griffiths, 2001a; see also chapter 3) encourages the model to make quantifiers true in every context where they are possible, but it penalizes them for being true in situations where they do not occur.

19.4 Computability Concerns

When Turing first defined computability using Turing machines, he also showed that there is no algorithm to decide whether a given Turing machine will stop running and halt (Turing, 1936). There is, however, a much more powerful version of noncomputability known as *Rice's theorem* (Rice, 1953), which holds that essentially no question about what a program does is computable (more technically, nontrivial "semantic properties" of programs are undecidable). No algorithm, for example, can take an arbitrary program as input and always determine whether that program ever outputs the number 5. No algorithm can decide if a program outputs only ascending numbers or computes the function $f(x) = x^2$. That doesn't prevent us from gathering evidence about programs one way or the other—for example, we might run a program for a long time and see what it does—but Rice's theorem establishes that there is no procedure for definitely answering any questions like these that concern the output of the program.

Such theorems seem ominous for Bayesian inference over programs. When we hypothesize program h, we are unsure whether h will output the observed data because this question cannot generally be answered. Consequently, it seems as though we cannot compute the likelihood because no algorithm can even compute the outputs! But it is worth noting that psychologically, people may not face the same computability challenges in all domains where they learn programs. For example, if someone learns a program by observation—watching someone execute the steps of tying shoelaces or computing a derivative, for instance—then their task is substantially easier than if they were trying to induce an unseen algorithm. Cultural transmission of knowledge allows more complex representations—including programs (Thompson, Opheusden, Sumers, & Griffiths, 2022)—to be passed down with less learning difficulty, resulting in different forms of learning (from others versus from nature respectively) (Chater & Christiansen, 2010).

In learning models that discover programs from data, there are several ways around computability concerns. One is to carefully construct the hypothesis space so all programs are guaranteed to halt, but this can be difficult to do correctly. In this tradition, there are many computational or logical systems in which the key questions can be computed—these systems, though, necessarily have less power than Turing machines. The most common solution, however, is just to ignore the problem. While no algorithm can compute the likelihoods for all programs, many approximations seem to work fine in practice, likely because the programs in question tend to be relatively simple. If programs allow recursion or looping, then often when we run a program, it will not halt in a reasonable amount of time. If a program doesn't halt after, 1,024 steps, for instance, we just consider it to have output the empty string or a null output. Or we can solve this problem much more elegantly by using a prior that depends on the time and space resources that it uses. To illustrate the idea, if program h halts in time halt(h), we might choose a prior proportional to $2^{-\text{halt}(h)}$. This favors programs that evaluate quickly and gives zero prior probability to nonhalting ($\text{halt}(h) = \infty$) programs. In Bayesian inference—in particular, MCMC sampling—we could use this to reject nonhalting programs because, as they run, their prior would eventually drop below an acceptance threshold. There are several formal versions of this idea centering on optimal efficient search through programs (Levin, 1973; Schmidhuber, 2002; Hutter, 2002). One simple intuition for search is possible in these spaces is to consider Levin Search (Levin, 1973), in which one interleaves the computation of all programs. By dividing run-time resources between all possible programs, the nonhalting programs do not prevent the halting programs from being computed. While these ideas are theoretically elegant, we have yet to find a domain where experiments with humans empirically distinguish speed-based priors from grammar-based priors.

19.5 Markov Chain Monte Carlo for Programs

The Metropolis-Hastings algorithm, a kind of MCMC algorithm (see chapter 6), is the most common inference algorithm for Bayesian program learning in cognitive science. Metropolis-Hastings needs us to specify how to compute a proposed new sample from the current one. The simplest approach is to sample de novo from the underlying grammar, but doing so fails to preserve any information about the current hypothesis. This means that such proposals lose information about what has worked well so far. We can keep this information by making proposals conditioned on the existing program. A standard way to do this is to choose just a subexpression of the current program and regenerate it from the grammar (Goodman et al., 2008b). If the program is viewed as a tree, then this technique first chooses a subtree and then replaces that subtree with a new subtree of the same type generated by sampling from the grammar. This scheme typically preserves most of the existing program (hopefully the useful parts) and changes just a small part of it (hopefully something that can be improved).

Consider a program which duplicates the first element of an input list:

$$\mathcal{F}(\texttt{lst}) = \texttt{pair(first(lst), pair(first(lst), rest(lst)))}.$$

We can represent this program as the following tree structure:

The standard Goodman et al. (2008b) regeneration proposal would pick a subtree at random—we have denoted it here with $*$—and generate a new proposal from the grammar by sampling a new subtree with the same type. For example, we might generate

pair
first rest*
lst rest
lst; (19.23)

in this case, proposing a new program that skips the second element of the list. In this proposal, we have only changed the program under the $*$ node. Note that it is possible to regenerate the entire program (e.g., a proposal from the prior) by proposing to the root. But if subtrees are chosen uniformly, then most of the time, we will preserve some of the structure in the program.

The standard Metropolis-Hastings acceptance probability would then be

$$\min\left(1, \frac{P(h')P(d \mid h')}{P(h)P(d \mid h)} \cdot \frac{Q(h \mid h')}{Q(h' \mid h)}\right). \tag{19.24}$$

Some care must be taken in computing Q, the proposal probability, because it is asymmetric. If s is the subtree in h that is replaced with s' in h', then

$$Q(h' \mid h) = \frac{P_{\mathcal{G}}(s')}{N(h)},$$

where $P_{\mathcal{G}}(s')$ is the probability of s' under grammar $\mathcal{G}$ and $N(h)$ is the total number of subtrees in h. Q here has multiplied the $1/N(h)$ chance of picking s in h times the probability of replacing s with s' according to $\mathcal{G}$.[6]

Tree-regeneration proposals are fast and simple, but many other proposal schemes are worth considering. Many of these alternatives are motivated by the observation that regeneration proposals destroy potentially useful information in the subtree being replaced. One way to address this issue is to also include *insert moves*, which include the subtree being replaced as part of the newly generated subtree. For example, we might notice that `pair` takes a `NUMBER` and a `LIST` and returns a `LIST`. That means that if our tree had a `LIST` somewhere, like `𝓕(lst) = rest(rest(lst))`, we could select the `(rest(lst))` and insert a `pair` node that re-uses the existing structure, as in

```
𝓕(lst) = rest(pair(2,rest(lst))).
```
(19.25)

In equation (19.25), `rest(lst)` has been preserved from the initial hypothesis, and we have inserted `pair(2,LIST)`. The *delete move* implements the mirror transformation of replacing a subtree with an appropriately typed subtree of itself. Other proposals that preserve structure include those that swap arguments (e.g., the two branches of an `if` statement) or swap functions while preserving their arguments (e.g., `and` for `or`). Ensuring that the parts being swapped have the appropriate types is especially important in these kinds of moves. One other kind of important proposal extracts subroutines out of an existing program, but these are difficult to implement.

6. In computing $Q(h' \mid h)$, we must actually sum over *all* the possible ways that h' can be derived from h (and symmetrically for $Q(h \mid h')$). In general, that will be a complex sum because there are many possible subtrees that we could have proposed to in order to derive h', not just the one node that we did in fact choose. Conveniently, this "multiple path" problem can be ignored, thanks to an auxiliary variable argument in which we imagine augmenting our hypothesis space with a variable saying which subtree to replace, which is sampled uniformly on each MCMC iteration. Conditioned on the auxiliary variable, there is only one proposal path.

19.6 Example Model Runs

Figure 19.2 shows a sample of 25 runs of MCMC over programs for a version of the number learning model in Piantadosi et al. (2012a). This model captured children's rough developmental stages in learning to count, where they progress through a few intermediate levels of knowledge and behavioral ability before demonstrating full competence in saying what word goes with a given set of objects. The model captured the transitions between their intermediate knowledge states as the result of acquiring more data. With little data, the Bayesian inference would only justify short, simple programs that worked only partially, but as data accumulated, it would acquire a full counting procedure. Hypotheses in the model were set up as functions from sets to words, just like counting, and the model eventually discovered a recursive counting program:

```
F(s) = if(
          equals(cardinality(s,1)),
          ''one'',
          next(recurse( setminus(s,select(s))
          )                                        (19.26)
```

This program says the word "one" for sets containing one element, and otherwise, it computes the word after (`next`) as the word you get to by counting everything in s (the set to be counted) minus one element (`setminus(s,select(s))`). This emphasizes that in learning to count, children must come to understand how to update set s through recursive or iterative processes that move them along the list of number words that they know.

Figure 19.2 shows the dynamics of learning using MCMC with tree regeneration across 25 different models runs (subpanels), each run for 10 seconds using the Fleet software package. The horizontal axis in these plots shows the sample number (thinning by a factor of 1,000), and the vertical axis shows the posterior score $\log(P(h)P(d \mid h))$ for the sampled hypothesis. Chains are colored according to the behavioral pattern they exhibit or what numbers they get correct. This figure shows multiple runs to give a sense for what MCMC over programs is like. Roughly, within a run, the MCMC chain will tend to linger on a mode—typically, a program that behaves a certain way (same color). This is shown in the figure by the lines that jiggle up and down—these tend to be small, neutral proposals to a program that might slightly change their complexity but often won't affect the likelihood or functions that make a very minor difference to the likelihood, such as affecting performance on a rare data point. But sometimes a chain will propose a change that leads to a substantial improvement in either prior (complexity) or fit (simplicity). When this happens, the posterior score takes a large jump upward, and when this happens, the model often changes behavior (color).

A good picture to have in mind is that abstractly in "program space" there are modes of approximately equivalent programs that perform well, and the model improves by jumping from one mode to a better one. This jump is typically discrete because the programs are discrete, so the different performance levels (posterior scores, vertical axis) are discrete. Because of this, most of the interesting changes that happen while sampling over

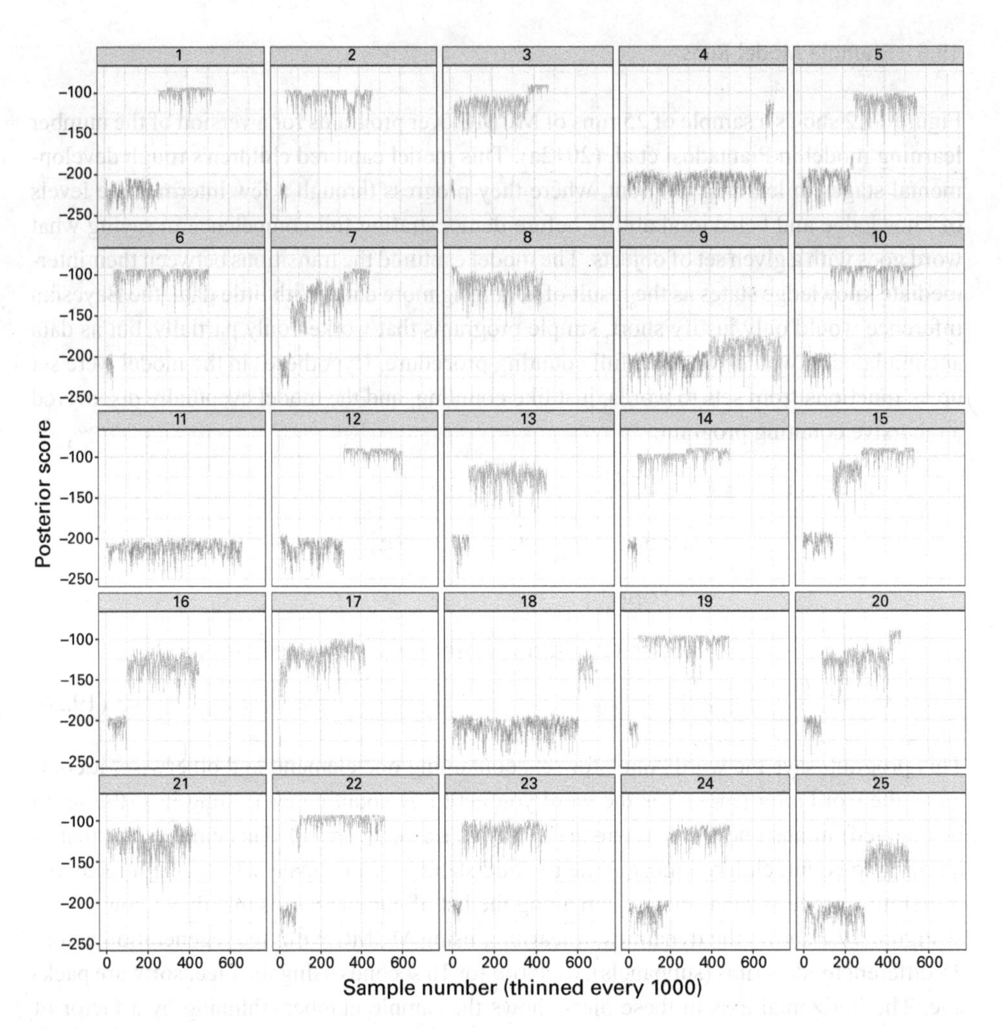

Figure 19.2
Example MCMC runs from a number-learning model.

program space tend to be large changes rather than small, incremental updates that one may be familiar with from MCMC over real-valued functions. Note, though, that a large change in performance or likelihood might actually correspond to a small change in the program itself. For example, we might replace `rest(lst)` with `recurse(rest(lst))` and change from a program that got nothing right to one that implements the correct recursion. Thus, the right thing to have in mind is that the modes in performance or likelihood may or may not be close to each other in program edit/proposal space. The mapping between the two is likely—possibly invariably—"brittle" such that little changes to a program can make or break it.

Random sampling is effective in this setup because a sampler, by definition, spends an amount of time on each hypothesis proportional to its posterior probability. This means that it proposes *more* from *better* hypotheses. In fact, the extent to which MCMC is better than enumeration tells us that there is some information about performance encoded into the proposal probabilities between programs.

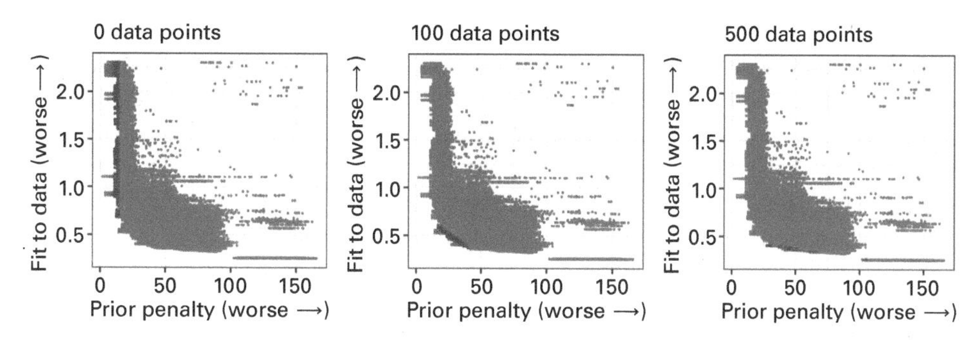

Figure 19.3
Priors versus likelihood for a counting program model illustrating the trade-off between the two and the distribution of hypotheses in this space. Good models will be near the left-to-lower edge, meaning those whose prior can't be improved without a worse likelihood or whose likelihood can't be improved without a worse prior.

Figure 19.3 shows a fancier version of this number model, from Piantadosi (2023). This version includes more primitives, such as ones for manipulating approximate quantity and noun types. This model also learns more complex programs because it derives operations on small sets (e.g., checking if a set has one element) from more basic operations like 1-to-1 matching. Despite this larger hypothesis space, the model is still able to sort out the computations involved in counting, including learning that numbers are not approximate and numbers don't refer to specific objects (e.g., "two" doesn't mean "two socks"). Visualizing this richer hypothesis space, figure 19.3 shows a scatterplot of different hypotheses (points) plotted at their negative log prior (horizontal axis) and negative log likelihood (vertical axis). The best programs are the ones near the origin, corresponding to having a high prior and a high likelihood. The hypotheses in this figure are colored according to their total posterior scores, which illustrates how probability mass moves as more and more data points are accumulated. Initially, the best hypotheses are the ones that are close to the origin primarily along the horizontal axis (prior); with moderate amounts of data, one might want the points to be close diagonally, minimizing both the prior and likelihood, and with lots of data, it primarily matters if the values on the vertical axis are close to zero. Thus, learners could be expected to move down the left edges, of these plots as data are accumulated.

It is important to remember, though, that the geometry of these plots are not themselves accessible to a learner. That is, from hypothesis h, it is not easy to determine which other hypotheses are approximately the same in terms of likelihood because those hypotheses that are nearby in likelihood might require a big change in terms of the underlying program.

19.6.1 Two Senses of Learning

These figures illustrate how MCMC sampling progressively searches a hypothesis space. This kind of search may map onto human learning in at least two distinct ways. In some studies, the program learning model is treated as an ideal observer, a statistically rational solution to the problem of deciding what hypothesis a learner should choose. This approach explains *learning as accumulation of evidence*: learners change hypotheses (as in figure 19.3) when they get more data. In this view, people perform approximate statistical inference at each amount of data; any non-adult-like hypotheses that early learners entertain are chosen because they represent the best solution given the available data. Several developmental effects have been modeled like this, including the shift from characteristic to defining features in conceptual development (Mollica & Piantadosi, 2021). This data-driven view also predicts other details of learning. For example, if learning is primarily about waiting for data, then that predicts a specific relationship between the mean and variance of the ages at which a given change occurs (Hidaka, 2013; Mollica & Piantadosi, 2017).

A potentially complementary approach treats *learning as search*. In this view, the primary limiting factor in development is how much of the hypothesis space learners have been able to explore. This idea seems to explain figure 19.2, where the sampling model transitions between discrete hypotheses as the learner finds new solutions to a problem that they have been considering. Ullman et al. (2012) argues for this point of view, pointing out that the drastic changes and reorganizations seen in searches, often without additional evidence, mirror qualitative phenomena in child development. Likely, both data and search are important to capturing how children learn and revise their theories over time.

19.6.2 Inference Hacks

Figure 19.2 shows 10 seconds of samples for each chain. In this time, for this model, Fleet will typically draw 400,000–600,000 samples. The variability occurs because larger programs take longer to evaluate: the chains that collect the most samples are the ones that did not achieve high posterior scores. In this case, low posterior-score hypotheses are typically simpler (e.g., don't involve recursion) and thus can be evaluated more quickly.

The discreteness of the hypothesis space has advantages and disadvantages. One primary disadvantage is that often, a single sampling chain will not be able to escape its local minima and thus not find the best programs. This can be seen in figure 19.2 as places where chains stay with low posterior scores. For programs, the best schemes seem to be parallel tempering schemes that run multiple chains at different temperatures and allow swapping between chains (see, e.g., Vousden, Farr, & Mandel, 2016; Earl & Deem, 2005). These methods are easy to implement, have low overhead, and allow MCMC sampling to mix between modes without requiring ad hoc decisions about how long to run chains for or when to restart them.

One advantage of discrete spaces is that summary statistics about the posterior can be computed by storing the best hypotheses found and then explicitly computing the desired statistical quantities over them. For example, if $h_1, h_2, \ldots, h_{100}$ are the top 100 programs found, we may estimate a quantity like whether a given primitive is used by re-normalizing the posterior score on this finite set of hypotheses. For example, if $I_{\text{recurse}}(h_i)$ indicates whether h_i contains a recursion primitive, then we could approximate the posterior probability of using a recursion primitive as

$$\sum_{i=1}^{100} \left(\frac{P(h_i)P(d \mid h_i)}{\sum_{j=1}^{100} P(h_j)P(d \mid h_j)} \right) \cdot I_{\text{recurse}}(h_i), \tag{19.27}$$

where the denominator of the first term is a normalizing term that is constant across all hypotheses h_i. The numerator is a hypothesis's posterior score, which is used to weight I_{recurse} (or equivalently adds up the posterior probability of all hypotheses for which I_{recurse} is true). Estimation from the top hypotheses works because typically, nearly all the posterior probability mass will be occupied by these highest-scoring hypotheses. By contrast, continuous spaces have no notion of the 100 highest posterior hypotheses and must use samples directly.

One general approach that has been successful is to run parallel tempering over programs and store a discrete set of the top hypotheses found by any chain. When we compute learning curves or make predictions, we do so following Bayes' rule, but treating these top hypotheses as a fixed, finite hypothesis space. If we are interested in modeling developmental change as accumulation of evidence, we will run parallel tempering at various amounts of data and compute the fixed hypothesis space as the union of the top hypotheses found at any amount of data. The result is often on the order of thousands or tens of thousands of hypotheses. These numbers are trivial to process as a discrete, finite space for Bayesian inference—for instance, plotting learning curves or computing posterior predictive distributions—but sampling good hypotheses for this set is still a hard inference problem. In other words, the discreteness of the space allows us to partially separate things that we want to compute from the posterior from the task of approximating that posterior with a finite set of hypotheses. Importantly, when we consider learning models with tens of thousands of hypotheses, we don't necessarily mean that learners represent them all. Learners with smart inference techniques (as discussed next) might represent only a handful of hypotheses. Often, we as modelers want to ensure that there aren't possible hypotheses that we have missed, so finding a large set of plausible hypotheses represents a kind of ideal observer model rather than a process model of learners' actual experiences.

19.7 Future Directions

This chapter has reviewed several key ideas that form the basis of how we can think about modeling developmental processes as program learning. We now give a flavor of how these ideas can be extended and combined, turning to a few ways that they have been explored in recent research.

19.7.1 Learning Reusable Components

The picture of program learning developed here focuses on acquiring a single program, typically a single function. One important extension of this idea is that of learning a library of functions that can be called as subroutines (Ellis et al., 2021; 2020; Dechter, Malmaud, Adams, & Tenenbaum, 2013; Talton et al., 2012; Cheyette & Piantadosi, 2017; Bowers et al., 2023). Intuitively, the ability to define subroutines extends the set of available primitives, which significantly affects the inductive bias of most program learners. For example, we might want to represent a concept like "every other element in a list" as follows:

```
every-other(lst) := if(
                        empty(lst),
                        ϵ,
                        pair(
                         first(lst),
                         recurse(rest(rest(lst)))
                        )
                      ).                              (19.28)
```

Having this primitive might allow us to easily learn a program like "all the odd elements followed by all the even elements" as follows:

$$\mathcal{F}(\texttt{lst}) = \texttt{append(every-other(lst),every-other(rest(lst)))}. \tag{19.29}$$

Without `every-other`, $\mathcal{F}$ it has a much longer description length (approximately as long as equation (19.29) plus twice the length of equation (19.28)).[7] Thus, it can be especially useful to learn subroutines in the *multitask* setting (i.e., when a learner is trying to acquire more than one program). For example, imagine a program learner trying to solve 30 distinct problems that all involved picking out subsets of a list. Assume that the initial grammar, unlike the grammars discussed earlier, did not include any sort of `filter` operation. In this case, it would be useful to define a shared library as including `filter` and perhaps a set of common predicates.

Library learning poses an interesting statistical problem of how learners should decide what to include in their library. One intuitive solution to this problem is to minimize total description length (i.e., the length of all the learned programs plus the length of each subroutine). Such a solution balances the cost of storing a subroutine against its potential to be reused many times. A number of mathematical models formalize this intuition (e.g., O'Donnell, 2015; Goodman et al., 2008b; Ziv & Lempel, 1978), but because there are so many possible libraries to consider, these solutions are often difficult to apply in practice. One approximate technique that works well in simple settings is to run multiple MCMC chains, each of which is constrained to a fixed library structure (e.g., one chain might have a library of three subroutines and require each one to be called, while another chain might have just one) (Yang & Piantadosi, 2022). The branch of program induction known as "inductive logic programming" has developed several other rule-based approaches to acquiring subroutines, which is known as *predicate invention* (e.g., Muggleton, Lin, & Tamaddoni-Nezhad, 2015).

An alternative approach that is particularly useful for multitask learners is to grow the library over time. For example, a learner might start with an initial grammar and solve 10 percent of the assigned problems with its initial grammar. It can examine these solutions to find repeated code, extract the repeated sections into subroutines, and add the new subroutines to its grammar. This newly extended grammar might now allow it to solve 20 percent of the problems, at which point it can look for new chunks of repeated code and repeat the process. By iterating in this way several times, a learner can develop an increasingly useful library of subroutines and eventually solve very difficult problems that would be impractical to solve using the original grammar. The DreamCoder algorithm (Ellis et al., 2020, 2021) applies a refined version of this idea and is perhaps the most successful library learning model available.

19.7.2 Learning Church Encodings

There is an important sense in which program learning models can create fundamentally new systems of knowledge, as is likely required for understanding human learning.

7. One word of warning: discussions about library learning often use confusing terminology. In one sense, it is correct to think about subroutines like `every-other` as new *primitives* since they are used in the grammar just like any other primitive. In another sense, `every-other` is a *derived* program that is itself defined only in terms of built-in operations.

Piantadosi (2021) considered program learning over logical systems of *Church encodings* (Pierce, 2002). A Church encoding is an idea from logic and computer science where one can construct something in one logical system to mirror the dynamics of another (note that a Church encoding is distinct from the Church programming language discussed in chapter 18). This work showed, for example, how to learn representations like number, Boolean logic, quantification, dominance hierarchies, and lists by encoding them into a minimalist underlying logic. To illustrate how this might work in lambda calculus, consider a mapping from the terms of Boolean logic (true, false, and, or, not) to pure lambda calculus (e.g., lambda calculus with only lambda terms):

$$\begin{aligned} true &:= \lambda a.\lambda b.a \\ false &:= \lambda c.\lambda d.d \\ and &:= \lambda p.\lambda q.pqp \\ or &:= \lambda r.\lambda s.rrs \\ not &:= \lambda t.\lambda u.\lambda v.tvu. \end{aligned} \tag{19.30}$$

The definitions in equation (19.30) should be thought of as encoding a program for each symbol in Boolean logic. But importantly, these programs aren't defined in terms of underlying primitives that have inherent content (like `CAUSE` or `GO`). These symbols get mapped to essentially just a syntactic construction in lambda calculus. However, they allow us to follow the rules of lambda calculus to compute the answer to a question, like "What is the result of `or(true,false)`?" To do this, we substitute the lambda terms into this expression and evaluate them according to the rules of lambda calculus:

$$\begin{aligned} \texttt{or(true,false)} &:= ((\lambda r.\lambda s.rrs)(\lambda a.\lambda b.a)(\lambda c.\lambda d.d)) \\ &= ((\lambda s.(\lambda a.\lambda b.a)(\lambda a.\lambda b.a)s)(\lambda c.\lambda d.d)) \\ &= ((\lambda a.\lambda b.a)(\lambda a.\lambda b.a)(\lambda c.\lambda d.d)) \\ &= ((\lambda b.(\lambda a.\lambda b.a))(\lambda c.\lambda d.d)) \\ &= (\lambda a.\lambda b.a). \end{aligned} \tag{19.31}$$

The result is the representation for `true`, which is the correct answer. Alternatively, we could compute `or(false,false)` as follows:

$$\begin{aligned} \texttt{or(false,false)} &:= ((\lambda r.\lambda s.rrs)(\lambda c.\lambda d.d)(\lambda c.\lambda d.d)) \\ &= ((\lambda s.(\lambda c.\lambda d.d)(\lambda c.\lambda d.d)s)(\lambda c.\lambda d.d)) \\ &= ((\lambda c.\lambda d.d)(\lambda c.\lambda d.d)(\lambda c.\lambda d.d))) \\ &= (\lambda d.d)(\lambda c.\lambda d.d))) \\ &= (\lambda c.\lambda d.d). \end{aligned} \tag{19.32}$$

The result is, correctly, `false`. The computations are tedious but profound: the dynamics of how lambda expressions evaluate have resulted in computing the correct outcome for an expression in Boolean logic. This is true for all propositions that we can compute in Boolean logic; for instance, `and(not(or(true,false)),not(true))` will evaluate to the correct answer when the corresponding lambda terms are substituted in for `and`, `or`, `not`, `true`, `false`.

This happens even though Boolean logic is not "built in" to the syntax of lambda calculus, at least not explicitly so. In other words, equation (19.30) provides a way to *encode* Boolean logic into lambda calculus to trick the default dynamics of lambda calculus into representing some other system. This is one way in which minimal programming systems might come to represent richer systems of knowledge. Success in getting lambda calculus—or any of dozens of other systems—to encode a new domain is essentially always possible: lambda calculus is a Turing-complete system, so any computation or computer program we want can be compiled into compositions of lambda calculus with no other primitives.

Learners learning Church-encoding programs, therefore, will create new logical systems without needing primitives beyond the syntax of lambda calculus. Piantadosi (2021) considers such learners as a developmental model using an even simpler system than lambda calculus, combinatory logic, which can also be encoded into a neural network in a more straightforward manner. Church encoding potentially answers deep questions about development because it shows how a Turing-complete system could be constructed without "building in" primitive concepts that human infants may not possess, like numbers, Boolean logic, family trees, and other key structures. All of these can be discovered as structured computations, allowing learners to build systems of logic and computation as models that explain their observations of the world.

19.7.3 Learning Like a Hacker

Rule et al. (2020)'s Hacker-like (HL) model has also explored resources other than simple sampling and search that learners might draw on in inference. The basic intuition is that when we observe data like

$$[3, 6, 1, 7, 3, 2, 9] \xrightarrow{\mathcal{F}} [3, 7, 1, 8, 3, 3, 9], \qquad (19.33)$$

we might notice some suspicious relationships between the input and the output. For example, the two are the same length, which suggests that operations like `map` may be likely operations. Second, we might notice that when the output numbers differ from the input, they only differ by 1. And then we might notice that the only ones that differ are every other element in the list. Learning models can formalize these inferences using a grammar of inference steps rather than primitives. Rule et al.'s model maintains not just a hypothesized program, but a series of transformations that convert the input into the output. For example, one transformation would be to simply memorize the correct output; another might abstract a lambda function that maps $+1$ over the appropriate list elements. Rule et al. show that this learning model dramatically out performs random sampling in terms of the number of hypotheses it must actively consider. It often requires on the order of 50–100 hypotheses, bringing it much closer to human scale in terms of search requirements. Notably, this model also provides a closer fit than alternatives to large-scale human experiments on list functions.

19.7.4 Learning over Term Rewriting

In addition to these search moves, the HL model learns over a different representation than standard program primitives or lambda calculus. It learns programs over *term rewriting systems*, which allow the learner to state explicit rules about how a structured computation proceeds. To illustrate the idea of term rewriting, consider a basic tranformation that children learn in algebra class, like

$$a \cdot (x+y) = a \cdot x + a \cdot y. \tag{19.34}$$

or

$$e^{x+y} = e^x \cdot e^y. \tag{19.35}$$

These are learned as an explicit rule about manipulating equations, which are themselves tree-structured. For example, learners who acquire these rules would be able to manipulating a symbolic structure:

$$e^{a \cdot (x+y)} \rightarrow e^{a \cdot x + a \cdot y} \rightarrow e^{a \cdot x} \cdot e^{a \cdot y}. \tag{19.36}$$

Term rewriting thinks of this sequence of steps as formalizing the computation itself, where the system of rules that are used are the program. This computation may halt at some point, when no more transformations can be applied, or loop forever, just as a program does. Typically, these systems allow nondeterminism, meaning that multiple rules can be applied at each point, and therefore evaluation requires exploring the tree of possible derivations (much as including `flip` requires exploring the space of execution paths). Importantly, term rewriting systems have the full power of Turing machines.

HL considers learners who have a grammar for writing such transformations and must find a collection of them that captures the knowledge required in a given domain. For example, you might construct a theory of numbers in this domain that explains how a number is formalized in mathematical logic (e.g., the Peano axioms; Peano, 1889; and the Robinson axioms; Robinson, 1949):

$$\begin{aligned} a + 0 &= a \\ a + S(b) &= S(a+b) \\ a \cdot 0 &= 0 \\ a \cdot S(b) &= a + (a \cdot b). \end{aligned} \tag{19.37}$$

These rules describe purely syntactic transformations of structures made out of symbols. On their own, no individual symbol has meaning. There is no sense, for example, in which $+$ symbolizes addition other than the role that it plays in the rules given here. A learner might acquire other rules, however, that place $+$ in different roles. For example, if we were to swap $\cdot$ and $+$ in these rules, the system would still describe a theory of numbers, but one using $\cdot$ for addition and $+$ for multiplication. If we were instead to include rules like $a + a = S(a)$ or $S(S(a)) = S(a)$, we would have an interesting computational system, but one that no longer captured a theory of natural number. The challenge for learners, then, is to find rules consistent with the dynamics of some domain (e.g., explaining data in that domain). When this happens, term rewriting systems serve to define both the key structures and processes in that domain.

Term rewriting can do everything that the Church encoding models can do because lambda expressions (or equivalent systems) can be evaluated according to term rewriting rules. But the theories and representations learned in term rewriting often feel more intuitive because their dynamics are defined transparently in the way that the terms rewrite, rather than obscurely encoded through the dynamics of another system like lambda calculus. One aspirational hope of these theories is that they will permit program learning models to acquire real systems of knowledge, like those required for reasoning in different domains and manipulating essentially arbitrary structures. One other useful feature of term-rewriting systems is that it is easy to change the set of symbols over which transformations are defined. Unlike Church encoding, symbols do not need to be defined in terms of some base system before they can be used. Simply by adding rules that use a new symbol, it begins to become part of the networks of computations that the system describes. Removing a symbol from the rules similarly removes it from the system. In this sense, a hypothesis space defined over the set of term rewriting systems is capable of genuine conceptual change.

The resulting systems of knowledge end up looking more like rich cognitive theories. As was thought for some time in artificial intelligence (AI) work, program learning over large sets of propositional rules may be a productive way toward humanlike knowledge. The programming language Prolog, for example, works in a fundamentally different way than most of the programs considered here, in that its programs are declarations of logical relationships. These programs determine the behavior of a backtracking search that, essentially, explores logical consequences of the initial statements. Term rewriting and Prolog-like representations may help program learning move toward acquiring richer systems of knowledge beyond single programs, and Prolog representations have been extensively considered in inductive inference (Muggleton, 1995; Muggleton & De Raedt, 1994).

19.8 Conclusion

This chapter discussed the essential ideas behind modeling learning as programming. This approach accounts for the breadth of conceptual structures that people appear to readily internalize and closely links studies of human learning to theories in AI, machine learning, and probabilistic inference. The computational ideas that we have outlined—including grammars over programs, random sampling, and Church encoding—provide a compelling tool set for refining debates about learning and development that have typically only relied on informal, philosophical notions. They are also empirically effective: program learning models explain many aspects of learning across a wide variety of domains. Together, these results show that Bayesian inference over programs is a powerful tool for understanding human learning.

20

Bayesian Models of Cognitive Development

Elizabeth Bonawitz and Tomer Ullman

How can a learner take ambiguous, noisy, and incomplete information and build causal, structured, and complete representations from it? This fundamental question has been viewed by symbolic, connectionist, and rationalist perspectives, each developing different, partial answers to this problem. As has been discussed in detail in previous chapters, probabilistic models naturally integrate strengths from each of these approaches toward creating a unifying framework that incorporates structured representations, learning, and a characterization of the goal of the cognitive system.

Classic approaches in cognitive science parallel areas of focus in developmental psychology that similarly grew out of a desire to answer the tension between representation and learning in a boundedly rational system. On one side, a purely nativist response was to deny learning and to focus on characterizing the detailed representations that were already in place. On another side, an empiricist response was to suggest that structured representations were not necessary, that learning (and the inferences that followed) was simply a bottom-up process of learning statistical associations. Other debates played out in cognitive development as well. Some research tended to characterized children as "noisy" or "irrational" adults, while other research sought to demonstrate that children are efficient and effective rational learners.

With the development of the probabilistic framework came a renewed interest in unifying developmental psychology, coined "rational constructivism" (e.g., Xu & Kushnir, 2013), as a nod toward a reinterpretation of Piagetian themes. Just as probabilistic models integrate themes from symbolic, connectionist, and rationalist approaches, did their emergence in cognitive development provided a framework to marry sides in nature-nurture and irrational-rational standoffs. As has been illustrated in previous chapters, theory-based probabilistic models of development provide a means to characterize core and intuitive beliefs in terms of representations. These models detail how these representations may be learned and revised in light of evidence. They operate at the rational level of analysis and depict what problems are being solved in the minds of young children. Their connection to algorithmic levels of analysis show how "boundedly rational" learning may resolve noisy behavior at the level of the individual while capturing optimal behavior at the level of the group. Furthermore, probabilistic models have helped researchers in developmental psychology find precision in both specifying the structure of internal representations and identifying the possible processes that drive their formation.

The problem can be broken into three related questions. First, how do children generalize from examples? Second, how do children learn the inductive biases that shape generalization from examples? Third, how do children learn the frameworks that shape the inductive biases that influence generalization from examples? For example, developmentalists have struggled with questions like how children learn to map words to their meanings so quickly, or how observing a single causal intervention supports inference about future outcomes. Inductive constraints, such as whole object biases (Markman, 1990) or core representations (e.g., Carey & Spelke, 1996), were proposed to explain these powerful inferences in early childhood.

Although these inductive biases appear early in development and are thus a natural candidate to support nativist theories, probabilistic frameworks can provide accounts of how these inductive biases may themselves be learned, pointing to overhypotheses (or framework theories, as they are often called in developmental psychology; Wellman & Gelman, 1992). Hierarchical Bayesian models (HBMs; see chapter 8) describe how frameworks can generate inductive biases that generate the evidence, and thus reciprocally speak to how the evidence can shape the biases which shape the frameworks, and so on. Although reminiscent of infinite regress arguments ("turtles all the way down"), most scientists who apply the probabilistic modeling approach to developmental questions will suggest that higher-level overhypotheses will eventually ground out in perceptual and conceptual core primitives, which we discuss in further detail in this chapter. Importantly, we are able to use the probabilistic modeling framework to help specify what must be built-in, given the evidence available to the developing mind,[1] what forms the representations may take at each level, what the consequences of these forms will be for learning, and whether the hallmarks found in empirical studies in development map to these consequents.

In this chapter, we touch on enduring questions for cognitive development, that mirror those discussed in past chapters: How is learning possible at multiple levels of abstraction? How can we reconcile computational/rational levels of analysis with the algorithms that children might be carrying out? How are new theories (hypotheses) generated? What representations are already in place to support learning? To find the answers, we focus on developments in probabilistic models of children's learning in causal domains. We then discuss empirical evidence for early emerging or core knowledge and how current probabilistic models are approaching the representational considerations.[2]

20.1 Causal Inference and Intuitive Theories

Children's cognitive development is characterized by conceptual revision of intuitive theories. At the core of these intuitive theories is the principle of causality; representations

1. Probabilistic models provide an account of what necessarily must be built into the system. However, they do not require that constraints "lower down" in the hierarchically are necessarily learned. That is, just because something can be learned does not mean that it is learned in development. Children may have more built in than is required by their experience, as a kind of redundancy in the learning system. See section 20.3 for more details.

2. Thanks to Lauren Leotti for help in preparing this chapter, and to Andy Perfors for discussions and feedback on an earlier draft.

in theories depends on one another. Reasoning about another's mind entails a notion that actions are driven by desires and beliefs. Reasoning about a physical system entails a notion of objects exerting forces on others. Reasoning about biology entails notions of the causal role of variables that result in a living organism's growth, illness, offspring, or death. As such, it becomes important to specify how causality is represented and learned in the developing mind.

As detailed in chapter 4, *causal graphical models* provide a representational language in the same lexicon of Bayesian probabilistic methods (Pearl, 1988). They were also one of the first types of representations used to characterize cognitive development in the early days of rational constructivism, demonstrating the power of children's learning from covariation that went beyond simple associative models (Gopnik, Sobel, Schulz, & Glymour, 2001; Gopnik et al., 2004; Schulz, Gopnik, & Glymour, 2007; Schulz, Goodman, Tenenbaum, & Jenkins, 2008; Schulz & Sommerville, 2006; Sobel, Tenenbaum, & Gopnik, 2004; Griffiths et al., 2011b). Since their use in early studies of children's causal reasoning, much as been learned about when and whether they best characterize children's behavior. In this section, we present a case study in preschoolers' causal reasoning, employing a simple framework theory over a bounded causal inference problem, and discuss some of the limitations of this model.

As briefly noted, the longstanding tension in developmental psychology rests on the nature-nurture debate. This debate has played out in children's causal reasoning as well. Some researchers have focused on whether children's causal beliefs are best understood as representations instantiated in domain-specific modules (School & Leslie, 1999) or innate concepts in core domains (Carey & Spelke, 1994; Keil, 1989), while other researchers have emphasized the role of domain-general learning mechanisms (e.g., Gopnik & Schulz, 2004). Research that has centered on the nature of the representation (giving less focus to the learning mechanisms) have suggested that children's causal reasoning respects domain boundaries (Carey, 1985; Estes, Wellman, & Woolley, 1989; Hatano & Inagaki, 1994; Wellman & Estes, 1986; Bloom, 2004; Shultz, 1982). For example, preschoolers deny that psychosomatic reactions are possible, rejecting the idea that (for example) being embarrassed can cause you to blush, or being worried can cause a stomachache (Notaro, Gelman, & Zimmerman, 2001).

Bayesian inference provides a natural framework in which to consider how prior knowledge and data interact in children's causal reasoning. (Schulz, Bonawitz, & Griffiths, 2007) gave children a storybook in which one variable co-occurred with an effect; in the Evidence condition, one cause recurred and the other causes were always novel (i.e., the evidence was in the form A & B → E; A & C → E; A & D → E . . .; see figure 20.1). In the Within-domain story, all the causes were domain-appropriate (e.g., a deer is running in the cattails, gardens, or other places and then gets itchy spots on its legs). Children were able to learn from the data as compared to a baseline; they were significantly more likely to infer that "A" was the cause in the Evidence condition. In the cross-domain story, the recurring cause (A, being worried) crossed domains from the effect (getting a tummy ache). Children disfavored this variable at baseline, as compared to domain-appropriate alternatives (e.g., eating a sandwich as the cause of the tummy ache). Following evidence, children learned and were significantly more likely to endorse cross-domain variable "A" as the cause than baseline, but their responses were tempered as compared to the within-domain story, suggesting that both the evidence and children's prior beliefs played a role in their causal learning.

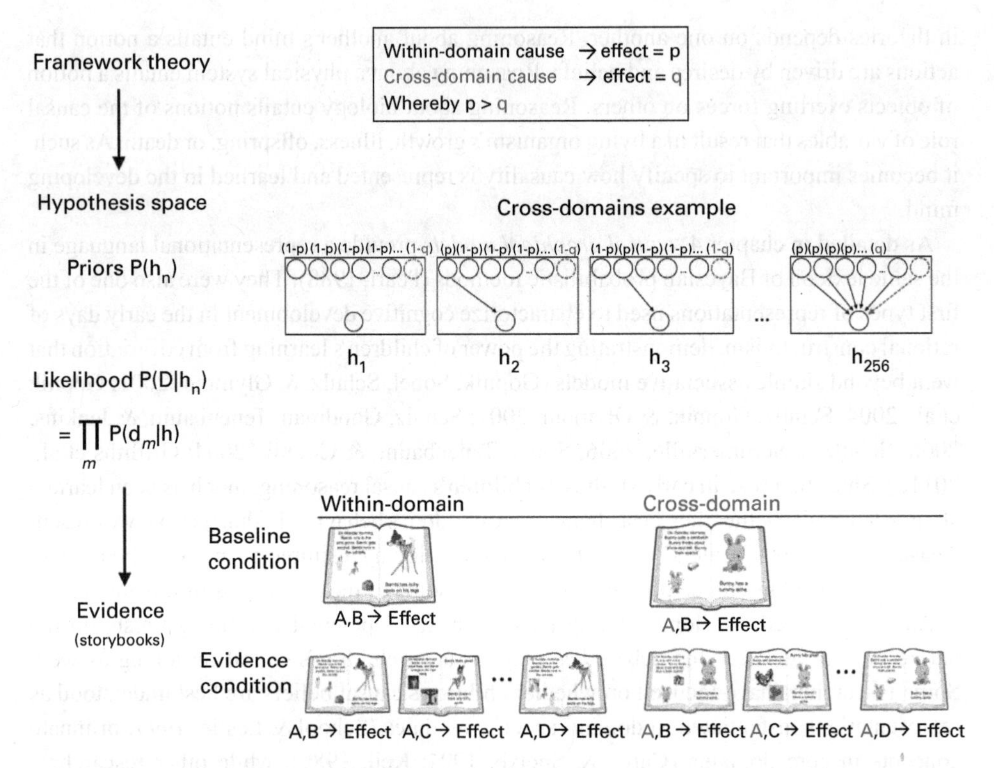

Figure 20.1
A framework theory in which within-domain causes are more likely than cross-domain causes, generating a space of possible causal models. Nodes in the models depict the causes and effects described in the storybook task from Schulz, Dayan, and Montague (2007). Priors over those models are informed by the framework theory such that the probability of each causal node containing a link to the effect is *p* or *q*, depending on whether it is a within- or cross-domain link. Storybooks read to children include all within-domain links (as in the *Bambi* books) or one cross-domain link (as in the *Bunny* book). Critically, the recurring cause co-occurs with other possible variables, which only provides probabilistic evidence in its favor.

This simplified causal learning task provides a nice case study of how probabilistic causal models can capture children's causal reasoning. In the task, children were provided with a force-choice alternative ("Was it variable A or variable B that is causing the effect"?). We can model the probability that children choose A as the correct explanation by directly contrasting it with the other possible explanation, B, given the observed data *d*:

$$P(A) = \frac{P(A|d)}{P(A|d) + P(B|d)}. \tag{20.1}$$

The explanation that includes A as a possible causal variable is consistent with many specific causal models linking the variables presented in the storybooks to the possible effects. As such, the probability of each particular candidate explanation, given the data, is computed by summing over all these possible causal model hypotheses that are consistent with the explanation:

$$P(A|d) = \sum_{h \in \mathcal{H}} P(A|h)P(h|d). \tag{20.2}$$

Here, h denotes a hypothesis about the specific causal model linking the variables in the story to the effect, and $\mathcal{H}$ represents the full hypothesis space of these models.

There are eight variables presented in the evidence books. A connection between a variable can either be present or absent, representing a causal link between the variable and effect (see chapter 4). Thus, for this simplified example with eight variables that can either be "on" or "off," the full hypothesis space includes 2^8 (or 256) possible different causal graphical models (h_n, for $n \in \{1, \ldots, 256\}$); see figure 20.1. $P(A|h)$ denotes the probability that the candidate explanation is correct, given the specific graphical model. When a causal link between A and the effect is present in the specific graphical model, this is simply 1; when no link is present, then the value is 0.

The second term, $P(h|d)$, represents the posterior probability of the hypothesis give the data. The probability of a particular causal structure given the data is expanded via Bayes' rule as

$$P(h|d) \propto P(d|h)P(h), \tag{20.3}$$

where $P(h)$ is the prior probability of a particular causal model, and $P(d|h)$ is the likelihood of observing data d, given the causal model h. Here, we are simplifying Bayes as a proportional equivalence because, due to the structure of the problem in which we are weighing two alternative explanations against each other, we do not require a normalizing constant.

The precise values of the prior and likelihood probabilities in Bayes' rule are determined by the intuitive causal theory entertained by the observer. As noted previously, past research has suggested that children have intuitive theories about the world (Gopnik, Meltzoff, & Kuhl, 1999; Carey & Spelke, 1994; Keil, 1995; Wellman & Gelman, 1992). These theories can be thought of as frameworks for guiding their causal reasoning. A hierarchical probabilistic framework provides a formalism for this intuition. Using a simple framework theory (as denoted in figure 20.1), within-domain variables are generated with probability p, where as cross-domain variables are generated with probability q. So long as p is greater than q (capturing the notion that within-domain causes are more likely that cross-domain causes), the qualitative predictions of the models hold across a range of values.

Critically, the framework theory gives us a model for how the causal models may be generated and how their prior probabilities and likelihood weights may be given. Specifically, the prior on any specific causal model is given by weighting the probability of generating each link (given by p or q depending on whether it is a within- or cross-domain variable, and given $1-p$ or $1-q$ when links are not present). The conditional probability distribution of the model provides a means to specify the probability of a causal link generating it's effect (specified by a noisy-OR parameterization, with weight ϵ; see chapter 4).

Model predictions are well captured by responses from older preschoolers, capturing the trade-offs between learning and strong prior beliefs for within-domain causes. The probabilistic framework provides a formal account for the interaction of children's intuitive theories (and beliefs in within-domain causes) and evidence at multiple levels.

Evidence shapes children's causal explanations in multiple domains (Bonawitz, Fischer, & Schulz, 2012; Bass et al., 2019; Bonawitz & Lombrozo, 2012; Goodman et al., 2006; Amsterlaw & Wellman, 2006). This work provides empirical support for the claim that even young children often act in ways consistent with optimal Bayesian models. However, just because it may be that on average, learners, responses look like the posterior distributions predicted by these rational models, it it not necessarily the case that learners are actually carrying out exact Bayesian inference at the algorithmic level in the mind. Given the computational complexity of exact Bayesian inference, the numerous findings that both children and adults struggle with explicit hypothesis testing, and the fact that children sometimes only slowly progress from one belief to the next, it becomes interesting to ask *how* learners might be behaving in a way that is consistent with Bayesian inference.

20.2 The Sampling Hypothesis

For most problems, the learner cannot actually consider every possible hypothesis; searching exhaustively through all the possible hypotheses rapidly becomes computationally intractable. So applications of Bayesian inference in computer science and statistics approximate these calculations using Monte Carlo methods, as discussed in chapter 6. In these methods, hypotheses are sampled from the appropriate distribution rather than being exhaustively evaluated. What's interesting is that a system that uses this sort of sampling will be variable—it will entertain different hypotheses, apparently at random. But this variability will be systematically related to the probability distribution of the hypotheses—more probable hypotheses will be sampled more frequently than less probable ones. This sampling method thus provides a way to reconcile rational reasoning with variable responding, a hallmark of early childhood. The *Sampling Hypothesis* is the idea that human learners may take a similar approach—an idea introduced in the context of adults in chapter 11.

There is growing empirical support for the idea that children are doing something that looks like sampling (Bonawitz et al., 2014b). For example, children provide explanations for causal events in proportion to their posterior probabilities (Denison, Bonawitz, Gopnik, & Griffiths, 2013). These studies show that children's responses are not simply noisily maximized, and further, that responses go beyond simple frequency tabulations in these causal learning tasks. These studies also raise questions about *how* a learner may sample hypotheses.

There are lots of ways in which a learner could sample hypotheses. They may resample every time they observe new data, they may take a hypothesis and stick with it, until they have impetus to re-evaluate (e.g., maybe data that is very unlikely given the current hypothesis). And, when they re-evaluate, they may make subtle changes to the hypothesis that they are currently entertaining, or go back and resample completely from the full posterior distribution. They may sample a few hypotheses or just one. All these ideas about how a learner "searches" through a space have analogs in computer science and machine learning. Here, we will contrast two: independent sampling and a modification of a classic algorithm, the *win-stay, lose-shift* (WSLS) algorithm.

The simplest idea, independent sampling, is that each time a learner observes new data, she recomputes the updated posterior and samples a guess from that updated distribution. This kind of approach to updating predicts that subsequent guesses from a single learner will be independent. That is, knowing that a learner predicts a specific hypothesis at a particular time tells you nothing about what hypothesis they are likely to have after the next observation of data.

However, another possibility is that a learner tends to maintain a hypothesis that makes a successful prediction and only tries a new hypothesis when the data weigh against the original choice. This means that an individual will tend toward "stickiness," being more likely to keep to the current hypothesis. This predicts dependency between responses. The central idea of maintaining a hypothesis, provided that it successfully accounts for the observed data, and otherwise shifting to a new hypothesis led to the name for this strategy: win-stay, lose-shift.

20.2.1 Win-Stay, Lose-Shift in Children's Causal Inferences

A general form of the WSLS algorithm has a long history in computer science and human concept learning (Robbins, 1952; Restle, 1962; Levine, 1975). However, it is possible to find specific classes of WSLS that approximate Bayesian inference. Specifically, there are different policies for when a hypothesis under current consideration should be rejected and different rules for how the next hypothesis should be drawn. It is possible to mathematically discover the specific instantiations of WSLS that provides a means to approximate Bayesian inference. That is, despite an individual's tendency toward stickiness, there are WSLS policies in which overall proportion of responses will reflect a sample from the full posterior distribution. Indeed, Bonawitz et al. (2014a) report two such specific instantiations of WSLS that approximate the posterior. These algorithms are based on stochastically deciding to draw a new hypothesis from the posterior with a probability determined by how well the current hypothesis accounts for each new piece of data is observed.

To compare the WSLS and independent sampling approaches, Bonawitz et al. (2014a) developed a mini-microgenetic method for investigating children's causal learning. In studies of children's learning, researchers typically look at responses at just one point—after all the evidence has been accumulated. However, an effective way to determine the actual algorithms that a learner uses to solve these problems is to examine how her behavior changes, trial by trial, as new evidence is accumulated. The differences between the dependency predictions for these algorithms can thus be used as a means to evaluate the process by which learners might update their beliefs.

In Bonawitz et al. (2014a), preschool-aged children were introduced to a machine and an experimenter demonstrated that each of these three kinds of blocks activate the machine with different probabilities when they are placed on it. The red blocks activate the machine on five out of six trials, the green blocks on three out of six trials, and the blue blocks just once out of six trials. Then children were shown a novel block that had lost its color, and children were asked to take an initial guess about what color (red, green, or blue) they thought it was supposed to be. After that, the block was placed on the machine once and children observed one of two possible outcomes as noted in figure 20.2.

To better understand the WSLS algorithm, we step through one of the particular instantiations in figure 20.2. A learner starts by sampling a hypothesis from the prior distribution before seeing any data about the mystery block. Here, the learner happens to choose red (as if a weighted die was rolled with this outcome). Then the block is set on the machine, and it turns out that it activates the toy. Because it is given in the demonstration phase of this experiment that the red block activates the machine 5/6 times, the likelihood is thus simply 5/6. Now the learner has to "decide" whether to stay or switch. The coin is now weighted 5/6 to stay. In this particular example, when it is flipped, it happens to come up in the more likely

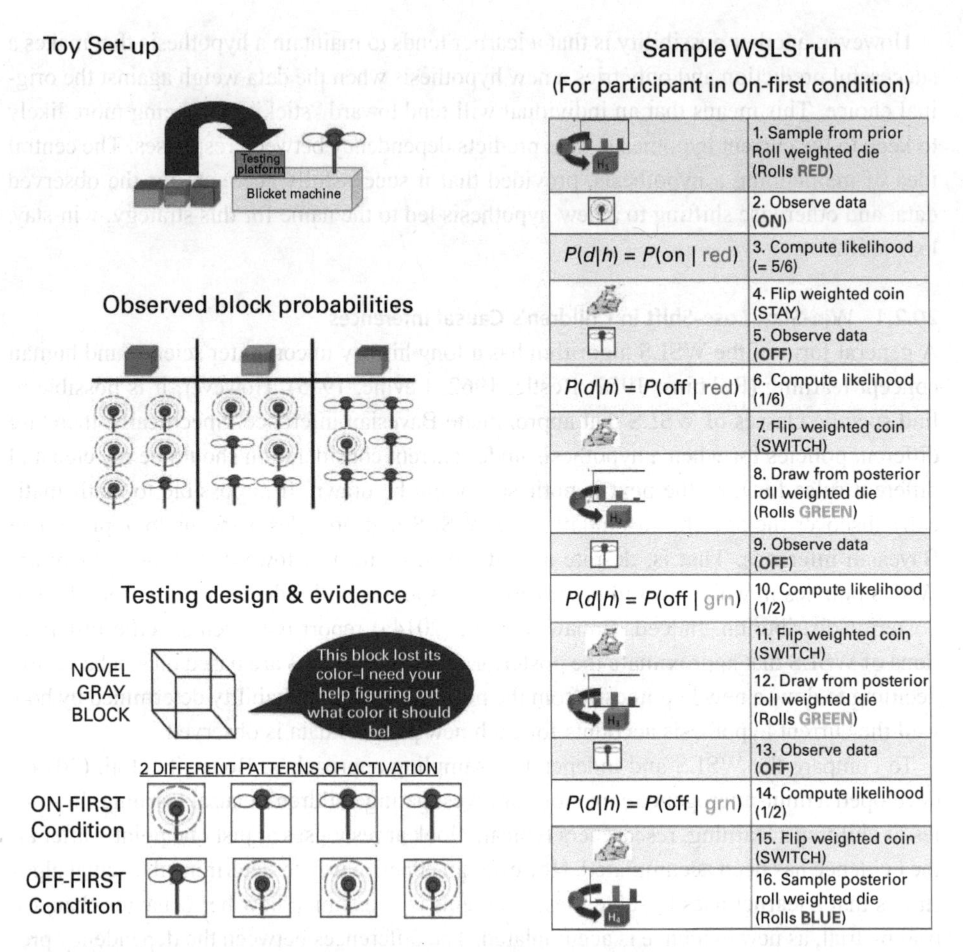

Figure 20.2
WSLS in children. The left column shows the method employed in experiment 2 of Bonawitz et al. (2014a). Children learned that blocks of different colors had difference causal affordances. Then a new block was introduced that lost its color. During the test, in the On-first condition, for example, children saw a pattern of evidence in which the mystery block first caused the toy to light, but then on subsequent trials, the toy failed to light. In the right column is an example run of a single individual carrying out the WSLS algorithm as evidence is observed over the four trials of the experiment.

case: to stay. Observing a second piece of data reveals that the machine does not activate. The likelihood is computed given *only this one piece of new evidence*. The currently hypothesized block, the red block, does not activate the machine with probability 1/6, as given in the demonstration phase. Thus, like likelihood is simply 1/6 and the weighted coin is flipped with probability "stay" at 1/6. In this example, the coin happens to come up "switch." At this point, the learner draws from the updated posterior, which includes all evidence observed so far. This time, the sampling die comes up green. So this is the new hypothesis that the learner holds in mind for the next observation.

An individual learner may look like they are randomly veering from one hypothesis to the next, starting with red and sticking with it, then switching to green, even though green may not be the most likely choice, etc. However, the lovely and surprising feature of WSLS is that

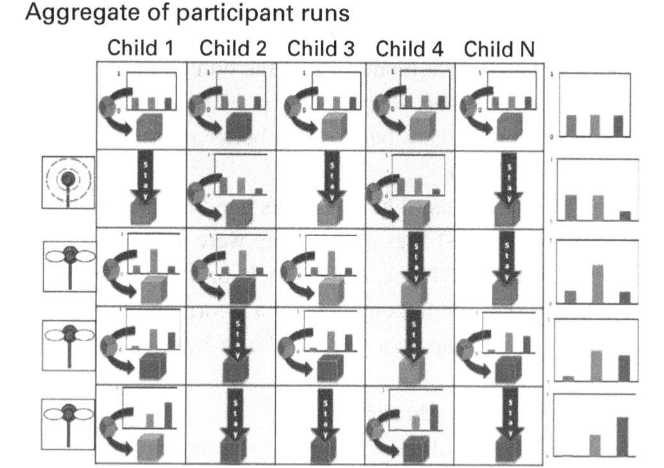

Figure 20.3
Example depiction of the patterns of several different child participants in the Bonawitz et al. (2014a) study after observing each new phase of evidence (with each new row). Although individual children appear to randomly veer from one hypothesis to the next (with a light "stickiness" to favor previously held hypotheses), on aggregate the distribution of children's responses capture the Bayesian posterior, as proved by the specific WSLS algorithm presented in Bonawitz et al. (2014a).

summing over enough participants, on aggregate, WSLS returns the posterior distribution on the aggregate. In particular, WSLS also helps solve the algorithmic problem of Bayesian inference because the learner can maintain just a single hypothesis in their working memory and need only recompute and resample from the posterior on occasion, but the responding of participants on aggregate still acts like a sample from a distribution (as in figure 20.3).

What is particularly nice about WSLS is that is provides a means for a young learner to make inferences given probabilistic information: the algorithm only considers a single hypothesis, but it acts like a sample from the distribution that makes it computationally more attractive than independent sampling because the learner need not compute and resample from the full posterior after each observation. Thinking about models at the algorithmic level can reveal important information about how children move from one belief to the next as in Bonawitz, Denison, Gopnik, and Griffiths (2014). These studies are important first steps in connecting the computational level and algorithmic level because they show how behavior can approximate Bayesian posterior distributions without requiring the learner to carry out exact Bayesian inference.

However, the study described here leaves open a number of interesting additional questions, such as whether there are dependencies even during sampling. For example, in WSLS, a resampled hypothesis is independent from the previously held hypothesis, but it is possible that the hypothesis that a current learner is entertaining is used as a kind of anchor for resampled hypotheses. There is reason to believe that adults at least show dependency even between switched hypotheses, though children may have weaker dependencies in the generation of subsequent examples (Bonawitz et al., in revision). This idea is consistent with the notion of simulated annealing in Markov chain Monte Carlo (MCMC) algorithms as discussed in chapter 6. Specifically, Gopnik and others have suggested that development may mirror the gradual cooling of an initially hot search. Childhood may be a time in which the mind employs wide and variable search of spaces at hot temperatures, supporting the notion that children are the creative innovators of the human race. In contrast, adults may employ colder-temperature searches, sticking with good enough hypotheses when they are discovered but missing out on unlikely, but possibly better alternatives (Gopnik, Griffiths, & Lucas, 2015; Gopnik et al., 2017).

A second open question pertains to how learners handle cases in which hypotheses are not easily enumerated. In the WSLS example, there were only three hypotheses available to learners (i.e., it was either a red block, a blue block, or a green block). However, in most learning problems, hypotheses are not enumerated a priori, and they may even be infinite. We turn now to a solution to this approach, as it informs developmental theories.

20.2.2 Framework Learning as Stochastic Search

In the studies described so far, children rapidly make inferences about the most likely causal models, probability matching responses in proportion to the posterior distribution. However, rich theory change as observed in typical development does not often appear so rapid, accurate, or linear. Learning takes time (Carey, 2009). Children move from one belief to something only slightly better; it is not often to see big conceptual jumps from beliefs that are conceptually incorrect to ones that are suddenly completely coherent. Instead, beliefs often gradually progress to capture more accurate and complete representations during development, (e.g., see Wellman & Liu, 2004 for a compelling case-study of the gradual development of a Theory of Mind). The transitions of children's beliefs have similarly been characterized as a somewhat stepwise process rather than suddenly all or none (e.g., Siegler, 1996). There are even apparent regressions in learning (e.g., Marcus et al., 1992; Thelen & Fisher, 1982). Why take a step backwards when a previously supported behavior has proved more accurate?

How can we understand this nonlinear, noisy, and slow progression of learning? One possibility is that children are performing a kind of stochastic search (as discussed in chapter 6). If so, we would expect that beliefs may appear to jump, as if randomly, from one point to the next. Similar evidence may lead children to different conclusions. Learning may unfold relatively slowly over time. These models can be applied to help explain belief revision in more sophisticated domains than those described earlier. Inspired by the stochastic search approach (detailed in Ullman, Goodman, & Tenenbaum, 2012), Bonawitz, Ullman, Bridgers, Gopnik, and Tenenbaum (2019) explored how preschool-aged children solve the chicken-and-egg problem of theory learning in the domain of magnetism, jointly identifying causal laws and the hidden categories that they are defined over. The hierarchical model employs stochastic search over logical laws and predicates (probabilistic context-free Horn clause grammar) that form a space of intuitive theories. To find the best theory, the search is carried out over the space using a grammar-based Metropolis-Hastings sampling method (also as in Goodman et al., 2008b). Simulating runs in this space revealed signatures of developmental belief revision; for example, the model revealed different convergence rates on individual runs capturing a similar phenomenon in development in which individual children may arrive at final, "correct" beliefs at different time points, despite similar evidence. In other work by Piantadosi et al. (2012a), a similar modeling approach captures the trademark progression of number knowledge, as children gradually build representations of the number concepts "1," then "2," then "3," and eventually "jump" to infer the cardinal principal of counting by applying a new sequential operator rule to the count list. These approaches critically demonstrate how core or primitive cognitive operations can combine to form more complex theories, and how a stochastic search process over this large space mirrors developmental trajectories.

Here, we have focused on Bayesian models, starting with how child learners may make causal inferences about specific data, moving up to models, and finally how abstract theories from those models may be learned. But in development, sometimes it appears as if children learn the abstract framework theories before they develop the specific ones (Wellman & Gelman, 1992; Simons & Keil, 1995). How could an abstract framework be inferred before a more specific-level model? The idea that framework theories (at least sometimes) seem to be in place before specific-level models provided tentative evidence that such knowledge is core (e.g., Spelke, Breinlinger, Macomber, & Jacobson, 1992). However, another answer comes again from HBMs. By applying computational modeling to developmental problems, it can be shown that there are cases in which the abstract learning happens at the same time—or sometimes even precedes—the specific level (Goodman, Ullman, & Tenenbaum, 2011). This has been called the *blessing of abstraction* by Goodman et al. (2011), and it provides an account of this surprising developmental phenomenon.

As noted previously, models show how and when specific framework theories could be learned (although such a demonstration does no provide proof that they are learned). In development, it is often taken for granted that there are domain-general concepts that must be built into the system as well, such as causality. The accounts suggesting that some learning happens have not provided a story about *how* such bootstrapping could occur. As in the case of early emerging framework theories, the lack of the story about how more domain-general concepts like causality could be learned have led many to assume that such concepts must be innate. Here again, HBMs provides a new proposal about how this kind of knowledge could be learned and about what must be built in given these models, such as perceptual input analyzers (Goodman et al., 2011). Some evidence suggests that children may develop causal theories piecemeal (Bonawitz et al., 2010), in the sense that associative and intervention information may not be spontaneously bound (at least in domains in which children have less familiarity).[3] Providing a story about how a domain-general principle like causality develops may also help to answer fundamental questions about how other domain-general theories can develop as well.

20.3 Core Knowledge

One of the themes of the last few chapters is that of program induction. The state of a mind at a given moment can be captured by a particular generative program, and the psychological process of thinking and learning can be seen as the mind executing program induction algorithms, leading to a new program that differs slightly or markedly in terms of parameters and structure. One could imagine all of development as proceeding from a very simple program with no structure, and discovering new routines, functions, and variables at different

3. See also Waismeyer, Meltzoff, and Gopnik (2015) and Meltzoff, Waismeyer, and Gopnik (2012).

levels of abstraction. This is not a new idea. Alan Turing, writing before the fields of artificial intelligence (AI) and cognitive science got their official start, proposed that the path to adult-level AI started with a child machine that learned new programs (Turing, 1950). Turing, like many others, pictured the starting point of a child's mind as something like a notebook with "rather little mechanism, and lots of blank sheets" (Turing, 1950, p. 546).

From an evolutionary and computational perspective, the notion of a blank notebook seems odd. Imagine, for example, the task of writing an algorithm to find a useful, short program that generates the sequence 101101110111101111 . . ., starting absolutely from scratch. One could try to order all possible programs on a universal Turing machine and search through them in various clever ways (Levin, 1973), but the search time for even a simple inversion problem is frightful. It is wasteful and unnecessary for each fresh organism to roam this landscape of possible programs starting from the same blank point. How much more useful it would be if evolution provided organisms with a "start-up library" of useful functions, variables, and routines (Lake, Ullman, Tenenbaum, & Gershman, 2017). Routines could be relatively hard-coded, allowing an okapi to get up and run soon after birth. Routines can also be relatively contentless, such as a routine for jumping back when detecting a scurrying motion without necessarily having a fully developed notion of what a spider is. But one could imagine, and in fact one would expect, built-in variables and functions and libraries that are far more general, abstract, and useful than hard-coded, contentless routines (Ullman & Tenenbaum, 2020; Baum, 2004).

From an empirical perspective, the notion of a blank notebook turns out to be wrong too. Research in cognitive development over the past decades has uncovered that infants have common-sense expectations about the workings of the world, present early on or innately (Spelke & Kinzler, 2007; Spelke, 1990; Woodward, 1998; Csibra, Bíró, Koós, & Gergely, 2003; Phillips & Wellman, 2005; Carey & Spelke, 1994). As might be expected from an evolutionary start-up library, these expectations are conserved across cultures, and they seem to be present in nonhuman animals as well. The expectations and principles are not all-encompassing and are modular in nature, focusing on several core domains, in particular number, space, agents, objects, and social relations. The principles are abstract and general, but with signature limits, and knowledge within core domains is acquired and developed throughout childhood.

To give a particular example, even young infants believe that solid bodies should not pass through one another (Spelke et al., 1992). This expectation holds true for all entities classified as objects. If an infant encounters a toy truck for the first time, they will expect the truck not to pass through a wall. Infants do not need separate and exhaustive re-training of their nervous system to form this expectation for a truck, and then a duck, and then a puck. This is what we mean by saying that the principle is abstract and general. However, it also has signature limits: infants cannot reason about more than several objects at once, and not all entities count as objects. Infants are lousy early on at reasoning about nonrigid bodies, and if the truck is perceived as an agent, then some of their physical expectations cease to hold. These principles can be exhibited empirically in different ways, but a common method is to show infants different displays and to measure the infants' looking time, which is a proxy for their surprise (but see Kidd, Piantadosi, & Aslin, 2012). For example, the infants may see display A, in which a rolling ball is stopped by a wall; and display B, in which the ball appears to roll through the wall. Infants will on average look longer at B than at A, indicating that they expected the ball not to pass through the wall.

We do not attempt a full specification of these principles of *core knowledge*, but highlight several such expectations (for a review, see Spelke & Kinzler, 2007). In all that follows, we emphasize that there is uncertainty both on the timeline itself (the earliest age at which infants demonstrate these expectations) and conceptual uncertainty over how best to characterize these expectations (and see the discussion that follows about formalization). For physics and objects, infants expect bodies to persist, cohere, follow smooth and continuous paths, not act at a distance, and not pass through one another. For agents and animate beings, even preverbal infants expect agents to act efficiently to achieve goals, and to trade off costs and rewards given environmental constraints (Spelke & Kinzler, 2007; Csibra et al., 2003; Csibra, 2008; Liu, Ullman, Tenenbaum, & Spelke, 2017). Infants also distinguish social and anti-social others (Hamlin & Wynn, 2011; Hamlin, Wynn, & Bloom, 2007; Hamlin, Ullman, Tenenbaum, Goodman, & Baker, 2013), although again, it is an open question exactly how early this distinction emerges. For spaces and places, young children and animals can use the layout of extended surfaces to reorient themselves and locate themselves and other objects with regard to the distances and directions to these surfaces, though navigation with respect to landmarks and small forms seems to rely on a separate process, with the two systems being integrated only at a later age (Hermer & Spelke, 1994; Dehaene, Izard, Pica, & Spelke, 2006; Spelke & Lee, 2012).

On their own, the principles of core knowledge do not directly specify how to implement them in a machine. Consider, for example, an engineer who accepts the core knowledge hypothesis but now wants to design a child machine with those principles. How should she build in things such as *The Principle of Continuity* or *The Principle of Efficiency*? This is an outstanding question in current computational cognitive science and machine learning, and different implementations amount to different answers to what core knowledge is, exactly. One possible route, at least for common-sense physics and psychology, is to assume that the generative models that capture adult reasoning exist from the beginning in some form. Or rather, as these are hierarchical models, the suggestion is that the top level of the hierarchy is built in or acquired early in infancy.

To use a rough analogy, imagine a game programmer who wants to design a new computer game. The programmer would likely not want to design the game from scratch—this would be onerous and replicate already-existing libraries of functions and routines. Sure, one particular game might involve jumping over alligators to collect diamonds, while another has you flinging boomerangs at killer bees, but both of them rely on the same basic physics engines and planners, just with different sprites and specific dynamics. On this picture, the infant is like a game programmer who is watching a game that they did not design, receiving the frames of the game and trying to figure out, given their libraries, what the underlying code is (Tsividis, Pouncy, Xu, Tenenbaum, & Gershman, 2017).

To be more specific, for intuitive physics, the generative model is that of the *Intuitive Physics Engine* (IPE; see chapter 15 and such sources as Battaglia, Hamrick, & Tenenbaum, 2013; Hamrick, Battaglia, Griffiths, & Tenenbaum, 2016). The idea would then be that the basic skeleton of a rough game engine is built in: we assume objects that have properties, and dynamics that update world states. The specific properties, object hierarchies,

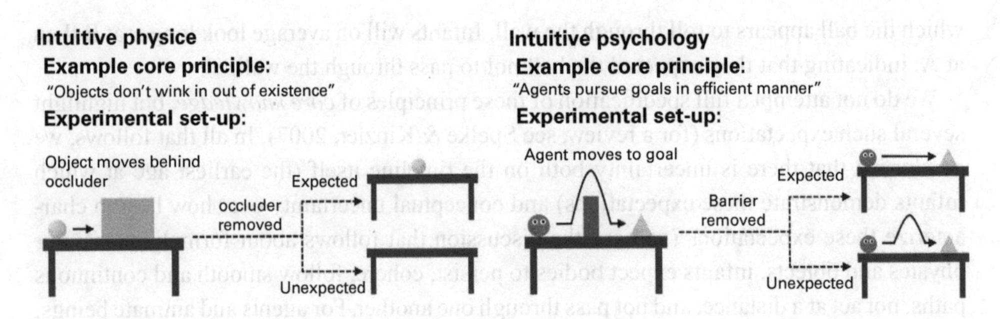

Figure 20.4
Example principles and supporting experiments in the core domains of intuitive physics and intuitive psychology. Intuitive physics: After observing an object moving behind an occluder and seeing the occluder being removed, young infants will express greater surprise at a display showing the object had disappeared (see, e.g., Spelke, 1990), leading to the formulation of the *Principle of Permanence*. Intuitive psychology: After observing an animate entity jump over a barrier to get to a target item, and then seeing the barrier removed, infants express surprise when the animate entity replicates its spatio-temporal trajectory (see, e.g., Csibra et al., 2003). This and similar experiments can be explained by positing that infants parse the scene in terms of goals and agents, and expect agents to pursue goals efficiently.

forces, and dynamic equations would then need to be learned (Ullman, Stuhlmüller, Goodman, & Tenenbaum, 2018; Lake et al., 2017). Various programs could then be learned on this basis, including simple ones that update object position based on velocity or random diffusion, along with simple collision resolvers as exist in nearly every game engine. Such programs are sufficient for quantitatively explaining infants' looking patterns to physical stimuli (Téglás et al., 2011), and embody core principles like permanence and cohesion without having to build them in explicitly. Such a program would predict that objects do not disappear or suddenly move in a discontinuous way, but *not* because such ideas exist explicitly in the code. Rather, it is because when the program simulates objects forward, the objects do not behave that way. Other principles are more explicit. For example, the *Principle of Solidity* can captured by the program for collision handling. Of course, such programs would be approximations to and simplifications of real-world physics, and such principled approximations can actually account for puzzling findings in infant reasoning about objects (Ullman, Spelke, Battaglia, & Tenenbaum, 2017).

For intuitive psychology, the generative model is that of planning or *Bayesian Theory of Mind* (see chapter 14 and, e.g., Baker, Saxe, & Tenenbaum, 2009; Baker, Jara-Ettinger, Saxe, & Tenenbaum, 2017). As with physics, the idea would be that the basic skeleton of a planner is built in: we assume agents that have utility functions with rewards and costs, as well as the ability to take actions to increase rewards and reduce costs. The learning process would then be to discover the types of agents, rewards, actions, skills, constraints, and costs that exist in different environments (Lake et al., 2017). Such a basic skeleton is sufficient for explaining many findings in preschool children (Jara-Ettinger, Gweon, Schulz, & Tenenbaum, 2016), and a similar simple model that trades off rewards and costs can explain several key findings in infants' reasoning about agents (Liu et al., 2017). In fact, the basic skeleton of a planner can be seen as embodying the *Principle of Efficiency*: agents are things with goals that act efficiently to achieve those goals within constraints.

Skeletal structured generative models are one route for embodying and implementing core knowledge, but other approaches are being developed as the AI and machine learning community re-engages with findings from cognitive development. One such promising direction, specifically in intuitive physics, is to combine artificial neural networks or graphs with different minimal notions of objects, and to let the network discover the dynamics and interactions between objects (for several recent examples, see Mrowca et al., 2018; Battaglia et al., 2016, 2018; Chang, Ullman, Torralba, & Tenenbaum, 2016). A very different computational approach, however, is try to recover the principles of core knowledge purely from vast amounts of empirical data (for recent examples in intuitive physics and psychology, respectively, see Piloto et al., 2018; Rabinowitz et al., 2018).

Such blank-slate models on their own do not yet generalize well, but it is too early to say whether this direction will turn out to be successful. Such blank-slate approaches (usually) do not claim to recover the trajectory of infant development, but rather that of evolution. This argument is common in other areas of machine learning, where humans do well without vast amounts of training data. For example, an feed-forward artificial neural network may require thousands of training images to recognize or the frame-equivalent of hundreds of hours to play a video game at a reasonable level (Mnih et al., 2015), whereas humans might need only two or three new examples of a new animal or several minutes with a new video game to reach the same level of performance. The claim is then that evolution has trained up the priors of the network, whereas the machine is training them anew. But this misses the dynamics of how evolutionary learning happens. Organisms do not perform the equivalent of being given several labeled horses, then passing on their trained neural state to offspring, this would be a folly on the level of the Lamarkian view that blacksmiths pass on their strong arms to their children. Re-discovering the priors of agents, objects, and other core knowledge would require reverse-engineering evolution, which has different dynamics than training and validating a network, and involves searching the uber-space of functions and variables, each set of which defines a subspace that an organism can explore over development (see, e.g., Baum, 2004).

In closing, the program induction view suggests that an organism would gain a serious leg up if it starts with a library of initial functions, routines, and variables. Initial built-in knowledge can involve specific feature detectors and hard-coded routines, but also more flexible and abstract concepts, and the recent decades of cognitive developmental research suggest that infants start with such a library for several core domains. Such functions and routines may take the form of structured generative models, which embody the principles discovered empirically, and they are similar in structure to later intuitive theories and are also the basis for the intuition that constructs scientific theories.

20.4 Future Directions

Cognitive development can be framed as a rational process: learners infer and build models of the world through an accumulation of data and time as they search through the possibilities. The progress and promise of probabilistic models are much like human development, in which more empirical work and time will lead to richer models of this process. Many areas remain open for exploration in the space of probabilistic models of development, and we highlight two particularly promising paths: the role of resource limitations in qualitative developmental shifts and the grounding of probabilistic developmental models with perceptually driven, bottom-up learning.

A once and future challenge for probabilistic models of development is the changing boundedness of the learning process. Cognitive changes in processing capacity can lead to quantitative and qualitative shifts in learning. For example, if your semantic memory capacity grows over development, you can store and retrieve more evidence during inference, a quantitative shift. However, it can also change the process that you use to store and retrieve evidence altogether, a qualitative shift. Changes in working memory and attention can affect not only the richness and content of mental simulations, but whether simulations are used as a means of prediction and inference at all. Changes in our representations of events can influence not only the speed of revising our beliefs, but also our assumptions of whether some observations count as evidence at all. Questions surrounding the cognitive effects of changing boundedness have a long history in the information-processing approach to cognitive development (Klahr & Wallace, 2022). In probabilistic models of cognition, they have a long future too.

A separate central challenge for probabilistic models of development concerns the need to ground this learning in perceptual representations. HBMs are presented as allowing researchers to unify ideas from both the nature and the nurture sides of the ongoing nature-nurture debate. However, the focus of much of the work in this area has advanced by supposing that central and difficult perceptual processes are largely solved, and then working with their output. For example, some models of intuitive physics assume a perceptual process that recognizes and distinguishes different objects. On the other hand, some of the best perceptual processing learning models currently make advances without paying much heed to learning programs over structured representations. Crucial breakthroughs are waiting for the right program learning models that take seriously the need for the last stages of a hierarchy to make contact with the basic input units of perception.

20.5 Conclusion

Children are the original learners, and most of the lessons in this book are inspired by their study. In this chapter, we provided a few examples that show how probabilistic models can speak to developmental questions. We presented cases in which probabilistic models similarly help solve open problems in the development of causal learning, demonstrating how children's prior beliefs and evidence interact to support causal inference, how approximation algorithms can explain dependencies in causal belief change, and even how causal search at abstract levels may unfold. Finally, we suggested that probabilistic models can be informed by the initial constraints observed in early infancy, operating as skeletal structures that afford rapid development of more complex, abstract structures over development.

The application of probabilistic models helps researchers to be precise about the content of early developing beliefs. These models illuminate early inductive biases that shape the learning process. They help reconcile debates regarding whether children are noisy learners or rational approximators. They provide a unifying framework for nature-nurture debates, demonstrating that structured representations can both drive inference and be inferred, even early in development.

The human experience weaves together social, emotional, and cognitive experiences to form a rich tapestry. Probabilistic models are a promising tool to help distinguish interlocking patterns, but we have only just begun to pull at the different threads. This approach is still in its infancy, and it is unlikely that we will develop general, unified theories of children's learning in the same time span that children take to develop their own abstract models of the world. But probabilistic models provide a framework that could make this goal possible, if we continue to let them develop.

21

The Limits of Inference and Algorithmic Probability

Nick Chater

Throughout this book, we have considered subjective probability to be our basic theoretical building block, and we have used Bayesian updating as a general recipe for adjusting beliefs in the light of new evidence. There are, indeed, good reasons to do so: a priori arguments that the laws of probability define a uniquely rational way of reasoning about uncertainty; and concrete illustrations of the productivity of this approach, in modeling a wide range of deep cognitive problems, as illustrated in this book and in the wider body of literature on Bayesian cognitive science, statistics, and machine learning.

How, though, does learning and reasoning get started? In analyzing any inferential problem (e.g., learning a grammar, a model of the perceptual world, or a theory of naive physics), the theorist starts by specifying a set of candidate generative models; and these models (and their parameters) are assigned initial prior probabilities. Only then can the confrontation with the data begin through the process of Bayesian updating.

But assigning prior probabilities might seem to provide both a perpetual challenge to the cognitive scientist and a problem for the brain whose computation is being modeled. What is the prior probability of a particular linguistic grammar, hierarchical model of visual scenes, or physical theory? How can such a probability be well defined, especially in view of the fact that most such models will never have been considered directly? To take an extreme example, a Bayesian model of scientific inference seems to require assigning a prior probability to general relativity, which is presumably supposed to predate any use of theory to explain physical data. If, by contrast, the prior is assigned after some of the data has been examined, the very idea of the prior as the starting point for Bayesian updating seems to be entirely violated.

Kolmogorov, the creator of the standard axiomatization of probability theory, was himself much concerned with whether probability can serve as a foundational idea (Kolmogorov, 1965)—and in his search for an alternative foundation (independently and in parallel with Chaitin (1969) and Solomonoff (1964, 1978)), he developed a theory that created deep links between computer programs and probability: *Kolmogorov complexity theory* (Li & Vitányi, 2008). This general approach has also been developed into a practical method for machine learning and statistics, most commonly known as the *minimum description length* (e.g., Grünwald, 2007; Rissanen, 1987, 1989) and the related *minimum message length* (Wallace & Boulton, 1968; Wallace & Freeman, 1987) principle.

The link between probabilities and programs connects with the broader idea that much inductive inference, in its most general form, concerns not fitting specific types of statistical model, but rather finding programs that can potentially generate the available data (see chapter 18). This viewpoint is given a concrete embodiment in a probabilistic programming language such as Church (Goodman et al., 2008a). And it provides a fascinating dual perspective on probability, according to which either programs or probabilities can be considered as basic.

This approach is, of course, of particular interest to cognitive science, given that the essence of the approach is to view the brain as a computational machine. Indeed, as we shall see, it is possible to see Bayesian inference in a new light—as searching for the shortest possible program that explains the available data. This connects Bayesian inference with a long and apparently distinct, intellectual tradition according to which perception, language acquisition, and science are driven by a search for simplicity, tracing to William of Ockham, Isaac Newton, Ernst Mach, and Albert Einstein (Baker, 2022), and more recently, to studies such as Attneave (1959), Blakemore, Adler, and Pointon (1990), Feldman (2000), Goldsmith (2001), Leeuwenberg (1971), and Mach (1883/1919), among many others. We shall see, moreover, that switching the focus to thinking in terms of programs rather than probabilities helps address the question of where priors come from and generates theoretical insights into a variety of specific areas of cognitive science. We begin with the problem of priors and focus, initially, on a specific programming language (namely, Church), for concreteness.

21.1 A Universal Recipe for Priors

Using a language like Church, we can encode a great many probability distributions can be encoded by probabilistic programs.[1] If the brain has a representational language to encode such programs, then we can envisage the possibility that, in principle at least, the brain may have a general-purpose way of representing and sampling probability distributions that might capture the structure of the external world.

It is interesting to consider the problem of learning the structure of the external world from this point of view. Perceptual, linguistic, or scientific data are observed; and an ideal learner can be presumed to attempt to find the program that is most probable in light of that data. This is, of course, a classic problem of Bayesian inference.

Working through the Bayesian calculation requires figuring out the probability of the data, given the probability distribution encoded by any specific probabilistic program. Calculating this likelihood value is, of course, computationally difficult in general—but it is at least well defined in principle and can be approximated by standard methods, such as Markov Chain Monte Carlo (MCMC) sampling (see chapter 6).

1. In fact, the space of probability distributions that can be encoded by a probabilistic program is an infinitesimal fraction of all probability distributions. The number of programs is countable, whereas the number of probability distributions is not—even the possible biases of a coin is a continuum between 0 and 1. From the point of view of cognitive science, though, probability distributions that cannot be represented at all can be set aside if we assume that the brain is restricted to the computable. Any distribution that cannot be represented certainly cannot be learned, for example. The probability distributions in practical use in statistics, machine learning and cognitive modeling are, of course, computable.

But what is the prior of the probability distribution specified by some arbitrary probabilistic program? Some probability distributions (e.g., independent coin tosses, linear statistical models in statistics, and so on) are intuitively much more probable than others (e.g., polynomials with hundreds of arbitrary parameters). However, putting intuition aside, how can we assign priors to probability distributions in a systematic and convincing way? While it is not immediately clear how to assign prior probabilities to probability distributions, it is much easier to see how priors might be assigned to the corresponding probabilistic programs. A simple, but powerful, line of thinking is the following. Suppose that we encode probabilistic programs in, for concreteness, binary strings of symbols (the choice of a binary alphabet is for convenience only). Suppose also that, like most conventional programming languages, and including probabilistic programming languages such as Church, these programs correspond to so-called prefix codes: that is, no program is the prefix of another program.

Now imagine generating programs using the simplest possible mechanism: flipping a coin to generate indefinitely long binary sequences. Most of the time, of course, no meaningful program is generated, just a meaningless binary string. Now and again, however, the binary sequence will happen to encode a probabilistic program, by sheer chance; and (because of the prefix property) each indefinitely long sequence of coin flips will generate no more than one such program, and hence no more than one probability distribution. Now we have an appealing generative model for probabilistic programs, and hence the probability distributions that they encode. A probabilistic program is generated with a probability that depends purely on its length. That is, a program that corresponds to a sequence of n binary symbols has a probability of 2^{-n}.

In short, probability distributions can neatly be assigned prior probabilities according to their simplicity, where simplicity is measured by the length of the programs that encode them. Hence, our intuition that a string of binary independent draws or a linear statistical model should be assigned fairly high prior probabilities corresponds to the observation that they can be encoded in a few lines of Church.

This line of thinking provides the prospect of a universal way of assigning prior probabilities to Church programs, and hence to the probability distributions that they encode. Moreover, it captures the widespread intuition that prior probability, whether in perception or science, should depend on simplicity: simple probability distributions, we suppose, have short Church programs, and hence high prior probabilities. This aligns with the widespread appeal, in science and everyday thought, of elegant ideas (see Baker, 2022); and our complementary dislike and suspicion of theories that appear overly complex.[2]

2. Here, we will be focusing on the simplicity of the codes used to express different models of the world. Another aspect of simplicity concerns the profusion of objects postulated in that model. To take an example, from science, Baker (2022) notes that the postulation of Neptune provided a simpler encoding of the orbits of the planets; but it did, after all, postulate an additional object. Similarly, inductive algorithms such as Dreamcoder (Ellis et al., 2021) infer general mathematical and scientific principles from examples by looking for simple regularities—but by doing so, these principles may imply the possible existence of infinite numbers of objects (e.g., generated by recursively building ever more complex programs). Merely recording the input data, without finding any patterns, would lead to a more complex representation, but it would be ontologically simpler. Whether "ontological" simplicity should be relevant to inference is by no means clear (e.g., in astronomy, it might seem to imply a strong prior for a very small universe)—but in any case, such questions are outside the scope of this discussion.

So far, we have sketched out a recipe for attaching prior probabilities to programs—but what we really need is a way of assigning priors to *models* of the world. And the very same model can be coded in endlessly many different programs of varying lengths. Not only that, ideally we don't want our prior probabilities to depend excessively on the particular details of Church, or whichever programming language we happen to choose (on the other hand, we might want our priors to be shaped to some extent by our representation language if we are viewing this language not merely as a useful technical tool, but as a hypothesis concerning mental representation). In this chapter, we will outline how this way of thinking can be extended and made precise using the theory of Kolmogorov complexity. We will see that it is possible to abstract away from the details of specific probabilistic (or indeed conventional) programming languages such as Church, and yet still arrive at interesting claims for cognitive science. The framework that we outline has, moreover, a number of interesting theoretical implications. We will highlight applications to theories in three areas of cognition: perception, similarity and language. First, though, we informally introduce and explore the key concepts in the context with which we began: that of defining a universal prior over probability distributions. For a comprehensive technical introduction to the underlying mathematics, we refer the reader to the textbook by Li and Vitányi (2008). While the mathematical framework is abstract and not tied to any specific representational language (and, indeed, the core mathematics was developed decades before modern probabilistic programming languages), it is useful to consider these ideas in a more concrete form.

21.2 From Programs to Priors

As we have noted, one of the most fundamental problems in Bayesian inference is assigning priors to hypotheses. It might seem that the most "neutral" prior (i.e., that which reflects the smallest amount of background knowledge), assigns each hypothesis the same probability. But this approach cannot work when, as in the most interesting cognitive models, there are an infinite number of hypotheses. This is because if we assign a particular probability δ to each hypothesis, then the sum of the probabilities of the hypotheses (which we assumed to be mutually exclusive) will be the sum of an infinite number of δs, which will be infinity, and, in particular, greater than 1, violating the rules of probability.

As before, we will limit ourselves to computable hypotheses, and even more concretely, to Church programs. Crucially, this restriction implies that the number of hypotheses is countable. This is because the number of computable hypotheses cannot exceed the number of Church programs, and the number of Church programs is clearly countable.[3] (While we often write Church programs and other models containing what we think of as real numbers, this is only meant as a useful simplification of the computable representations described earlier.) So what is the most neutral way to assign prior probabilities to computable hypotheses? Intuitively, it seems natural to give higher prior probabilities to simple hypotheses—perhaps those that correspond to short Church programs—and to assign low probabilities to

3. To see this, just note that the number of programs cannot exceed the number of strings in which they are encoded. The set of strings can easily be enumerated by enumerating strings of length 1, then 2, and so on.

hypotheses that can be expressed only using long Church programs. Following the logic outlined in section 21.1, without loss of generality, we can encode Church programs as binary sequences and then generate such sequences by repeatedly flipping a fair coin. As we noted, the probability that the fair coin will, by pure coincidence, specify a particular Church program depends, of course, on the length of that program. Specifically, for a program y, of length $\ell(y)$, the probability that the program will be generated by the fair coin is $2^{-\ell(y)}$ (of course, most sequences of coin flips will not generate valid Church programs at all).

But we don't really care, primarily, about assigning a prior probability to Church programs, but rather to the probabilistic models that those programs encode. Since many Church programs are equivalent, it is more natural to think of assigning a prior to equivalence classes of Church programs that generate the same computable probability distribution. To obtain the probability $\mathbf{m}_{\text{Church}}(h)$ that a computable probability distribution, h, is generated from a random sequence of coin flips interpreted as a Church program, we simply sum over all the programs that encode that distribution:

$$\mathbf{m}_{\text{Church}}(h) \propto \sum_{y\,:\,\mu_y=h} 2^{-\ell(y)}. \tag{21.1}$$

The prior, $\mathbf{m}_{\text{Church}}(h)$, is clearly one way of operationalizing a prior probability with a bias toward simple models. We may have, though, at least two reasons for skepticism concerning this choice of prior. First, surely no generally useful prior can depend excessively on the details of a specific programming language (i.e., on the details of Church). Second, we have given no justification concerning why this particular choice of prior is especially neutral. We next consider how far these concerns can be addressed in turn.

21.2.1 Choice of Programming Language

Suppose that instead of choosing Church, we defined a prior over computable distributions by using some other equally powerful probabilistic programming language, which we will call LangX. By the universality of Church, it must be possible to write an interpreter that translates any program in LangX into Church. Let us say that the length of this interpreter is $\log c$; if so, then

$$\mathbf{m}_{\text{Church}}(h) \le c\mathbf{m}_{\text{LangX}}(h). \tag{21.2}$$

Of course, if LangX is universal, this relationship applies in the opposite direction, too, giving us

$$\min\left(\frac{\mathbf{m}_{\text{Church}}(h)}{\mathbf{m}_{\text{LangX}}(h)}, \frac{\mathbf{m}_{\text{LangX}}(h)}{\mathbf{m}_{\text{Church}}(h)}\right) \le c_{\text{LangX,Church}} \tag{21.3}$$

for some constant $c_{\text{LangX,Church}}$ (which depends on the two languages, but not the hypothesis). So, up to a constant multiplicative factor, it turns out the choice of programming language is not crucial (see Li & Vitányi, 2008 for details). In the light of this observation, we can drop the subscript specifying a particular language and just refer to prior $\mathbf{m}(h)$ over computable distributions h. This is known as the *universal prior* (Solomonoff, 1978). While this abstraction is mathematically useful in some contexts, note that the choice of programming language nonetheless may be extremely important: indeed, the bounding constant, c, can be arbitrarily large!

21.2.2 Neutrality

Throughout this book, we have considered Bayesian models of increasing levels of complexity; and building such models requires that we provide some prior over their parameters. We have sketched how this may be done in specific cases, and there is a large and somewhat contested body of literature concerning general principles by which priors can be set, such as by choosing so-called conjugate priors (see chapter 3) or maximizing the entropy of the prior distribution, subject to constraints (Jaynes, 2003). But by limiting ourselves to a specific class of models, all models outside that class are ruled out (i.e., implicitly assign a prior probability of zero). Instead, we aim to provide a prior over all computable probabilistic models—and to do this in the most neutral possible way.

Note, first, that according to the universal prior, every computable probability distribution has a nonzero prior—for each of these hypotheses, there are (infinitely many) programs that represent it, and each of these has positive probability. But, more interestingly, it turns out that, in a certain sense, the universal prior gives almost as much prior probability to each hypothesis as any other computable prior does. Specifically, for any computable probability distribution, P, over *computable* probabilistic models, there is a constant c_P such that, for all h,

$$\frac{P(h)}{\mathbf{m}(h)} \leq c_P. \tag{21.4}$$

This means that if *any* computable prior assigned a particular hypothesis high probability, then the universal prior assigns that same hypothesis a "reasonably" high probability. In this somewhat limited but nonetheless interesting sense, the universal prior is neutral among all computable priors.

21.3 Kolmogorov Complexity and the Universal Prior

The universal prior, $\mathbf{m}(h)$, is derived by considering all the programs that can generate our probabilistic model h, weighted inversely by their lengths. But what happens if, instead, we consider only the *shortest* program and ignore the rest? According to this line of thought, a model should be given a high prior probability to the extent that it is simple from at least one point of view. Accordingly, the prior associated with a probabilistic model of the world will be given by the probability of generating the shortest binary Church program that encodes that probability distribution. Let us call the length of this shortest program for probabilistic model $h, K(h)$. The prior for h is then just $2^{-K(h)}$.

Fortunately, it turns out that we may not need to choose between these two different ways of defining priors: up to a multiplicative constant factor, each approach gives the same result. That is,

$$\sum_{i:Prog_i \text{encodes } h} 2^{-\ell(Prog_i)} \approx 2^{-K(h)}. \tag{21.5}$$

Roughly, then, the shortest code dominates the others in determining the prior for a particular probabilistic model. So we can focus our efforts on finding the shortest program that we can for a given set of data—and the length of that shortest code will specify the appropriate prior (with higher priors for probability distributions with short codes). But in general, of course, finding the shortest program that encodes a particular computable probability distribution is not computationally feasible. Indeed, there is not even a computer program that can decide which programs do, or do not, encode a specific probability distribution.[4] This observation is the interesting consequence that further computation (or, in the case of the mind, additional thought) may allow us to see that a particular hypothesis about the world can be represented more simply than we anticipated—and when we find an unexpectedly short code for a probabilistic generative model, or a hypothesis of any other form, we should therefore increase its prior appropriately.

A particularly attractive feature of this perspective is that it makes sense of the idea that general relativity, or a specific grammar or hierarchical visual model, can meaningfully be assigned a prior even before these hypotheses had even been formulated, let alone actively considered. If these hypotheses could be represented by the brain, then the brain has implicitly assigned a prior to those hypotheses based on the length of the shortest representation. But this shortest representation need not, and typically will not, be known.

Indeed, from this point of view, it makes sense to think of an agent finding out more about its own priors through reflection. Indeed, discovering one's own priors will typically be slow and partial—given any particular probabilistic model, M, it is always possible that there is a shorter code for M than has currently been formulated. So, for example, James Clerk Maxwell's original 20 or so equations formulating electrodynamics were later recast in spectacularly simpler form into the four Maxwell's equations taught in undergraduate physics classes: and this reformulation required enormously sophisticated analysis by the self-taught electrical engineer Oliver Heaviside (Hunt, 2005). If we think in terms of description length, we can see that by revealing a dramatically simpler formulation of Maxwell's theory, Heaviside has shown that theory had a much higher prior probability than had hitherto been suspected.

A coding perspective allows us to see the value of pure reflection in a new light. We do not need to suppose, implausibly, that prior probabilities are established for all relevant hypotheses before learning or scientific inquiry even begins. Instead, pure reflection (i.e., computation with existing information rather than employing new data) can estimate and refine priors for hypotheses as our inquiry proceeds. And the same goes for priors of theorists in cognitive science, of course: if we discover that a probabilistic model that we believed to be highly complex actually has an extremely short Church program, then we will revise its prior upward.

The coding viewpoint also allows us to appreciate the importance of the nature of mental representation. While the invariance theorem reassures us that the specific choice of programming language does not matter too much (i.e., for any two languages, the shortest code lengths for any computation can differ only by a fixed additive constant), it is clear that code lengths (and hence priors) may still differ considerably. After all, one of the rationales for developing specific programming languages is that they are especially well adapted for expressing certain kinds of programs. So while the Church program for, say, some standard linear statistical model, a mixture of Gaussians, or a hierarchical Bayesian model of some kind might correspond to a rather small number of lines of code, a considerably longer

4. This is a corollary of Rice's theorem (Rice, 1953), the remarkable result that all nontrivial semantic properties of programs are, in general, undecidable (see chapter 18).

code might be required to encode these distributions in a logic-programming language like Prolog. So our prior over probabilistic models is relative in a meaningful way to the programming language in which probability distributions are represented. Thus, from the point of view of cognitive science, this means that psychologically relevant priors will be relative to the system of representation used by the brain. So if Church, or some similar language, is a good analog of the "language of thought" (Piantadosi, Tenenbaum, & Goodman, 2016), then priors over probabilistic models induced by Church will be psychologically natural; but if the brain operates using some completely different formalism, it seems, this will not be the case. Of course, a great deal of cognitive science is focused on developing proposals concerning how the brain represents linguistic and sensory input, and with testing them using behavioral and neuroscientific methods. This type of evidence can potentially help constrain hypotheses concerning the cognitive simplicity (and hence the associated prior probability) of different hypotheses.

21.4 Bayes and Simplicity

We have sketched part of the link between probabilities and programs. In fact, the link can be extended in a variety of interesting directions. It turns out that, for example, under fairly general conditions, the problem of choosing the *most probable* program that generated a set of data, given some computable prior (i.e., a prior that can be encoded in a Church program), is equivalent to finding the *shortest* program that encodes that data (e.g., in Church, or some other programming language).

An intuitive sketch of why this makes sense runs as follows (see Vitányi & Li, 2000, for a rigorous analysis). According to standard information theory (Cover & Thomas, 1991), an optimal code should assign any state of affairs A, with probability $P(A)$, a code length of $C(A) = -\log_2(P(A))$.[5]

Suppose that we have some set of possible models M_i (these could be high-level descriptions of a scene, parses of a sentence, grammars for a language, or a scientific hypothesis, depending on the setting). These models are candidates to explain data d. As usual, a Bayesian approach to the choice of m_i applies Bayes' rule:

$$P(m_i|d) = \frac{P(d|m_i)P(m_i)}{P(d)}. \tag{21.6}$$

Suppose that we want to choose the most probable m_i, in light of the data d—that is, the model that maximizes $P(m_i|d)$, denoted as $\arg\max_i P(m_i|d)$.

The most probable model doesn't depend on $P(d)$, which is independent of i, which means that

$$\arg\max_i P(m_i|d) = \arg\max_i \frac{P(d|m_i)P(m_i)}{P(d)} = \arg\max_i P(d|m_i)P(m_i) \tag{21.7}$$

5. Here, purely for convenience, the code is usually assumed to be binary (e.g., consisting of 0s and 1s), which leads to base 2 logarithm. An optimal code is one that is as brief as possible but introduces no errors while communicating the outcome of samples from a probability distribution.

and the maximum won't be affected by taking logs (to base 2) as follows:

$$\arg\max_i P(m_i|d) = \arg\max_i \log P(d|m_i) + \log P(m_i). \tag{21.8}$$

And maximizing an expression is, of course, the same as minimizing the negative of that expression:

$$\arg\max_i \log P(d|m_i) + \log P(m_i) = \arg\min_i(-\log P(d|m_i) - \log P(m_i)). \tag{21.9}$$

Now we can see that these negative log probabilities are the optimal code lengths $C(.)$ according to information theory, so we can write this as

$$\arg\min_i(C(d|m_i) + C(m_i)). \tag{21.10}$$

So the upshot is that the m_i with the maximum posterior probability is the also the m_i that minimizes the code length for the data. This code length has two parts: one part encodes the model m_i itself (with length $C(m_i)$), and the other encodes the data, in terms of the model (with length $C(d|m_i)$). So using Bayes' rule to choose the model with the greatest probability is equivalent to choosing the model that provides the shortest encoding of the data.

This connection between probabilities and code lengths provides a crucial link between Bayesian cognitive science and the simplicity principle—that inference operates by choosing the model of the world that provides the shortest encoding of the data.[6] So we can see Bayesian and simplicity-based approaches to inference as not necessarily in conflict, but rather as alternative ways to express the same fundamental ideas.[7]

Note, of course, that a fully Bayesian approach to learning and inference would attempt to capture the full posterior distribution over possible models rather than merely identifying the single most probable model. In practice, though, sampling methods, such as MCMC and its many variants, focus on one (or a few) models at a time, as we saw in chapter 6—and such sampling can operate using code length as the quantity being minimized (e.g., Geman, Potter, & Chi, 2002).

Moreover, various psychological considerations suggest that we should at least take seriously the possibility that the brain can represent only one model of the environment at a time. So, for example, people can "see" an ambiguous figure as two faces in profile or as a single vase (where figure and ground have switched); but we can see only one interpretation at any time (Long & Toppino, 2004). Similarly, the brain appears to be able to interpret only one of two overlaid scenes at once (as when we look outside on a twilight evening and can focus either on the darkening scene outside or the window's reflection of the room, but not both, Neisser & Becklen, 1975); and when each eye is fed distinct and conflicting images, the information in one eye dominates and the other input is ignored (Tong, Meng, & Blake, 2006). Our difficulty in creating more than one interpretation of a

6. Note that this heuristic argument needs to be supplemented with a more rigorous analysis to make the link with a general notion of complexity in terms of the shortest program length (e.g., Chater, 1996; and for a formally rigorous analysis, see Vitányi and Li, 2000).

7. These approaches also generalize in different ways (e.g., Grünwald and Roos, 2019).

situation at a time seems to apply also to high-level cognition. Studies of real-world situations (e.g., Klein, 1993), as well as in the lab (e.g., Johnson-Laird, 1983) seem to imply that we reason by latching onto a particular interpretation of the information before us, and we find it difficult to conceive of other possible interpretations, let alone take such possibilities into account. Indeed, arguably one of the fundamental biases in human reasoning is our tendency to maintain our favored interpretation longer than we should, in the face of conflicting information.

So suppose that the brain is able to represent only one interpretation of a set of visual, linguistic, or scientific data at a time. Then, in the light of the equivalence outlined here, we can view a rational cognitive agent as searching for either the simplest explanation of the available data (where simplicity is measured according to some specific programming language) or the most probable explanation. By contrast, in perception, statistics, philosophy of science, and other fields, there has been a long history of opposition between Bayesian and simplicity- or program-based approaches (e.g., Leeuwenberg and Boselie, 1988; Rissanen, 1989). But the equivalence between the approaches suggests that we should instead be asking when it is theoretically useful to take each perspective.

Taking a probabilistic starting point is especially helpful when considering decision making (see chapter 7), whether concerning abstract plans or details of motor control. Here, the expected value or expected utility of a choice or action must be estimated—and calculating expectations requires thinking in terms of probabilities. Moreover, it is often useful to consider how far an organism is attuned to real-world frequencies in its natural environment by analyzing the statistics of natural visual images, auditory landscapes, linguistic corpora, and other data (Atick & Redlich, 1992; Jurafsky, 1996; Olshausen & Field, 1996; Parise, Knorre, & Ernst, 2014).

But when a theorist is building hypothetical generative models, whether in language, perception, or naive physics, it is often more natural to start with the following question: How would I write a program that would generate sentences, visual images, or natural scenes? The probability distribution created by the generative model arises as a side effect of the choices that must be made within the computer program—for example, in a generative model of scenes, there will be endless choices about precise locations of objects, their orientations, their colors, the position and nature of the lighting, and other factors. Of course, once we have a program that can generate sentences or scenes, we may be able to compare its output with statistics collected from real-world data; and perhaps even to tune the generative model in the light of such data (e.g., we can note that lighting tends to come from above; that some colors are more common than others, and so on). But the starting point is thinking in terms of codes rather than probabilities.

For the theorist, then, constructing a generative model is a matter of precisely specifying a generative process—and this is, in essence, a task of programming (e.g., by writing a Church program that implements that process). And this programming task is, like any programming task, guided by our need to create representations and algorithms that generate the appropriate material as elegantly and simply as possible. Indeed, the aesthetics of writing mathematical models and programs based on those models is powerfully biased in favor of simplicity (and this is evident in the generative models described throughout this book). The plethora of complex, messy models is not even contemplated by the theorist; it is not that they are explicitly considered and judged to have low probability. In sum, it is possible to think of Bayesian inference either in terms of probabilities (as we have throughout this book) or as a problem of minimizing code lengths in an internal representational language. Given that the nature of internal representation has always been one of the central topics in psychology and cognitive science, thinking in terms of codes is often particularly natural.

21.5 Applying a Code-Minimizing Perspective in Cognition

Why is a coding perspective on cognition a useful complement to the standard probabilistic Bayesian viewpoint? One reason is that we may sometimes be more confident in our assumptions about mental representations than about implicit subjective probabilities. After all, cognitive science is full of proposals about how shapes, categories, faces, transformations, or linguistic or musical structures are represented (whether using logical formulas, feature vectors, hierarchical representations of all kinds, and the contents of phonology, morphology, syntactic theory, and so on). These representations make some things easy to encode and other things difficult to encode—and by the equivalence between codes and probabilities, we can view this as a way of grounding prior probabilities. A second rationale arises from the mathematical methods that the coding perspective allows, ranging from classic information theory (Cover & Thomas, 1991) to the more specific machinery of Kolmogorov complexity theory (Li & Vitányi, 2008) that we discuss here. In any area of science, different frameworks for looking at the same phenomenon are often useful because claims may be much easier to state, justify, or prove in one framework than in another. Different perspectives also suggest different generalizations, approximations, and experimental tests. But ultimately, the question of which framework is the most useful is a practical one—we can use whichever perspective provides the most useful way to consideer the aspect of cognition that we are exploring. To get a sense of the variety of ways in which it can be useful to take a code minimization perspective, we briefly illustrate how thinking in terms of minimizing code length can lead to useful insights across three areas of cognition: perception; similarity and; language processing and acquisition.

21.5.1 Perception

A simplicity approach to perception starts with the assumption that the brain constructs representations of the environment that provide the briefest possible codes that can reconstruct sensory input—that is, the aim is to compress the sensory input as much as possible. This viewpoint on perception traces back at least to the physicist and philosopher Ernst Mach (1919/1883), who saw the goal of simplification of sensory data as applying equally to perception and science alike, and this is a theme among the Gestalt psychologists (e.g., Koffka, 2013), as well as theories inspired by information theory in psychology and neurosciences (Attneave, 1959; Blakemore et al., 1990) and approaches based on developing specific coding languages (e.g., Van Der Helm & Leeuwenberg, 1996).

Rather than focusing on specific models, here we focus on the types of qualitative phenomena that are naturally explained by the simplicity approach. To begin with, note that any kind of structure or pattern in the sensory world can potentially be recovered through searching for the simplest representation. So, for example, consider two sets of random dots overlaid upon each other (as if 'marked on separate transparent sheets of glass). If both sets of dots are still, then there is no advantage, from a coding point of view, to distinguishing the two sets of dots—and they will be perceived as a single, slightly denser random dot pattern. If, on the other hand, the sets of dots move independently (as if the imaginary glass sheets are translated or rotated in different ways), then it is much more efficient to encode this motion by separating the dots into two sheets and representing the motion of each sheet separately. Without this grouping, the motion of each individual dot has to be explained independently, which will be a very inefficient code. The result of this grouping is the phenomenon of transparent motion (e.g., Braddick, Wishart, & Curran, 2002): the dots are perceived as transparent sheets, moving independently.

The same logic applies very broadly. Parts of an image that move coherently or have the same color or texture will tend to be grouped together—because this grouping allows a single representation of the relevant motion, color, or texture to be applied to at a stroke, rather than having to encode this information for each part separately. Similarly, perceiving the world in terms of hierarchically structured objects has a natural justification from the point of view of creating short codes. Thus, for example, representing an entire object, such as a dog, provides an efficient way of encoding its overall color, texture, and motion; representing body, legs, head, and other parts as distinct entities allows an efficient coding of the location of those parts and their tendency to move coherently (e.g., the head may tilt to one side, legs may be extended, and so on). Moreover, it would be inefficient to encode the properties of each leg, or each tooth, independently, as they share a great many properties.

Or consider the case of depth from stereo fusion (e.g., Julesz, 1986). If the brain can represent the images in each eye as displaced versions of one another, then a single code for both images will suffice, alongside an auxiliary code for the relative displacement. Now, of course, the displacement is determined by depth—so this auxiliary code will implicitly reveal a map of the depths of each location in the visual field. The same point extends to structure from motion, where explaining the rate of displacement in the visual image as the eye moves in space implicitly reveals the depth of the surrounding surfaces (and, indeed, this indicates that the information will itself be relatively stable, at least if the environment is relatively static).

Real images, of course, are often complex and cluttered. The simplicity principle applies here, too. It will typically be more efficient to encode a cluttered scene as an overlapping pattern of whole objects, with a particular three-dimensional (3D) layout rather than encoding patches of the image independently. Or to take a simple example, consider figure 21.1, created by the Italian psychologist and artist Gaetano Kanizsa. We see a square containing portions of black circles surrounded by a square frame. A rather lengthy code would be required to specify the particular pattern by which parts of some of the circles are "snipped" off. But postulating that these are bounded by a square, whose size and location can be represented briefly, provides a shorter, and hence preferred, explanation.

Or consider figure 21.2. On the left half of the picture, we see white "worms" against a uniform black background. On the right side of the image, we see the reverse. But why do we not also see the opposite figure-ground relationship? That is, why do we not see a white background, with irregularly shaped blacked shapes superimposed upon it, on the left? According to the simplicity principles, the answer lies precisely in that irregularity.

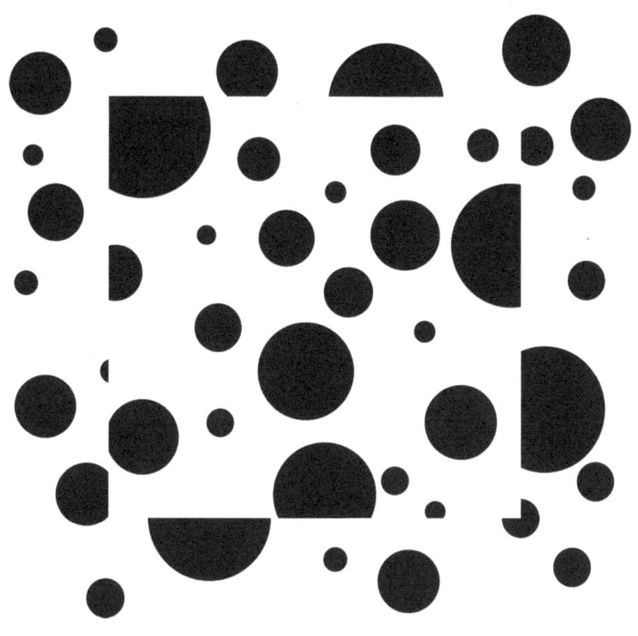

Figure 21.1
Kanizsa's image, redrawn here, can be concisely encoded by postulating an "invisible" inner square, which bounds the black circles within it (Kanizsa, 1985). One natural, though not inevitable, perceptual interpretation is that the inner square lies behind a square white frame (also with black circles). The invisible square provides a brief (and hence probable) encoding of where the black circles have been "cut off." The same is true of the equally invisible "outer square" that frames the image.

Figure 21.2
Figure-ground as determined by a simplicity principle. In this and similar examples, the "figure" corresponds to the structure that is simplest to encode (here, the wormlike forms with approximately constant width). By contrast, encoding the areas between the worms requires specifying an irregular and hence more complex shape.

The worms, by contrast, have a roughly uniform width and are all roughly the same width and spacing as each other. If the worms are treated as the figure, then they can concisely encoded. Indeed, it is focusing on encoding the worms as objects (and the background as mere spaces between neighboring worms) that allows the brief encoding.

Three points about the interpretation of the simplicity principle are worth noting. First, the simplicity principle provides a metric for evaluating candidate representations of perceptual input, such that there is a preference for representations corresponding to short codes. But the perceptual system cannot be expected to alight on the shortest possible code—indeed, the task of finding the shortest possible code for an arbitrary set of data, *D*, is known to be, in general, uncomputable (e.g., Chaitin, 1998). Second, the perceptual system cannot provide a reconstruction of *all* current sensory data—indeed, the psychology of visual attention seems to indicate that the brain is able to reconstruct at most only a few objects at a given moment, and the rest of the sensory input in largely ignored. Indeed, the computational challenges of the required reconstructive analysis-by-synthesis process may be one explanation for such attentional limitations (e.g., Chater, 2018). Third, if the brain seeks short codes in a local fashion, as reconstructing the entire sensory input is infeasible, then a crucial challenge is finding heuristics that suggest which data should be grouped together so that a common explanation can reasonably be attempted. Low-level factors, some of which can perhaps be computed in parallel, like common motion, texture, and timing, are likely to be important, but top-down factors are likely to be important as well (witness our tendency to "project" particularly familiar and important objects such as faces even in accidental patterns; Yuille & Kersten, 2006).

But these low-level factors can be misleading and can generate intriguing illusions. So, for example, in the rubber hand illusion (Botvinick & Cohen, 1998), a visibly "fake" rubber hand is repeatedly stroked with a brush in synchrony in both time and location with stroking the participant's real (but not visible) hand. The alignment between vision and touch suggests that these signals must be linked—and this linkage allows them to be represented more briefly. To create such a link, the brain generates the bizarre phenomenology that the disembodied rubber hand "belongs" to the participant. Indeed, when asked to shut their eyes and to point to the "target" hand with their other hand, people frequently point to the rubber hand, not their actual hand. This hypothesis provides a simple coding of the immediate visual and tactile input—even though it is globally a poor explanation in light of a person's entire history of sensory experience, world knowledge, and so on.

Finally, it is interesting to ask whether a simplicity principle operating over codes may offer an interesting perspective on how people infer causal structure in the world. Human perception does not just tell us how the world is; it also tells us a lot about how it could be. Surveying a tabletop, we can "see" that moving the table will simultaneously move the objects on it; that if we tip the table sufficiently, the objects will slide off (i.e., they are not glued down); that if a glass is upended, the water that it contains would spread across the table, perhaps drip onto the floor, and so on.

One viewpoint, explored in chapter 19, is that vision can be viewed as inverse computer graphics, and this computer graphics is partly driven by a "physics engine" that embodies causal principles (Wu, Yildirim, Lim, Freeman, & Tenenbaum, 2015). A pure simplicity principle appears focused purely on finding whatever structure compresses the data most effectively. Does this principle clash with the aim of causally reconstructing the origin of the perceptual input?

This question touches on one of the deepest questions in inductive reasoning and statistics. Compression seems to involve finding patterns of association (of whatever complexity) that we actually have, but not uncovering a causal structure that will help tell us what patterns of data we would obtain if the world were modified in various ways (i.e., the table were moved, the glass overturned, and so on). But mere association does not imply causation (e.g., Pearl, 2000). There are many deep and open issues here, which seem likely to be productive to explore.

One intriguing observation is that the simplicity-based account of perception aims to reconstruct data via *programs*; and programs can be viewed not merely as computing a single function (i.e., taking an input and generating an output), but as divided into an algorithm and a data structure over which that algorithm operates. This implies that, as computation proceeds, we can ask what would happen if there were some external intervention that modified the contents of the data structure (and over which the algorithm would continue to operate as before). That is, a program itself is a causally rich entity, defining counterfactuals, not just a single function. Thus, we can define counterfactuals based on the results of hypothetical interventions on the data structure. For example, we can ask how a chess engine would play out a game after making a bizarre first move, even though it would never spontaneously make that move. Or we can see that a recursive program for factorials would calculate 5! to be 85, if it had (wrongly) concluded that 4! had the value 17 (rather than 24). We can obtain these counterfactuals by considering interventions in the data structures as the computation proceeds. From this point of view, the separation between algorithm (which is treated as fixed) and data structure (which, hypothetically, can be modified during the computation) becomes crucial (see Chater & Oaksford, 2013)—what is represented as data can be modified, while the algorithm is invariant. The central question, from the point of view of the operation of the perceptual system, is whether the causal structure implicit in the program that reconstructs perceptual data maps onto the causal structure of the external world. For example, are chunks of sensory input that are assigned to be part of a single object (e.g., a presumed animal shrouded by foliage) actually associated with parts of the world that do cohere together (as will be true of an actual animal, which might hypothetically move coherently in any number of ways) or not (as where meaningless patterns of light and shadow have been spuriously grouped together)?

A simplicity principle alone seems inadequate to reliably reconstruct a causal structure using this viewpoint because the same function can be encoded (and, indeed, encoded in programs of approximately the same length) with many divisions between algorithm and data structure. But perhaps an additional constraint may help. One possible proposal is that, controlling for program length, the brain should prefer the program that has splits algorithm and data so that the encoding of the *algorithm* is as short as possible; another possibility is that the algorithm should be chosen so that the data structure can be as modifiable as possible to give the richest possible set of counterfactuals (indeed, these criteria may themselves be closely related).

The intuition, here, is that the perceptual system should impose the loosest causal structure (thus making the fixed algorithm as minimal as possible and allowing the greatest flexibility in the data structure) required to account for the perceptual input. Thus, an object sitting on a tabletop should be assumed not to be fixed to, or part of the table unless this is required by the data (e.g., the object moves with the table even when it is tipped up dramatically); by contrast, a tiled pattern on the floor should be assumed to be glued down by default; otherwise, how they maintain their geometric alignment is unexplained. The question of whether it is possible to turn these, or similar, intuitions into a rigorous formal account of how causal structure can reliably be inferred from perceptual input is an question for future research.

21.5.2 Similarity and Categorization

A coding viewpoint also provides a simple and general perspective on some aspects of cognition that are traditionally viewed as more abstract. For example, suppose that we consider mental representations in some coding language, which could be a language like Church, which can represent hierarchical structures of all kinds, rather than just a list of features (Tversky, 1977) or a location in a mental space (Shepard, 1957).

We can then ask: What is the length of the code required to transform one representation into the other? This idea can be spelled out in Kolmogorov complexity terms using the notion of *conditional* Kolmogorov complexity, $K(y|x)$, defined as the length of the shortest program that can transform input, x, into output, y. There is a rich mathematical theory of conditional Kolmogorov complexity, which appears to capture an abstract notion of the similarity between two representations (Li, Chen, Li, Ma, & Vitányi, 2004)—roughly, two representations are similar if one can be transformed into the other by a short code. This is a general formulation of a popular type of measure in computer science: edit distance, which is defined in terms of the number of operations required to transform one string, graph, tree, or other representation into another using a specified set of operations (e.g., for strings, insertion, replacement, and deletion defines the widely used Levenshtein distance (Levenshtein et al., 1966), and its many generalizations, including transpositions (Damerau, 1964) and movable substrings (Cormode & Muthukrishnan, 2007)). Edit distances are widely used in engineering, computational linguistics, and computational biology—for example, for correcting typing errors, matching strings of deoxyribonucleic acid (DNA), or comparing representations of images (e.g., Gao, Xiao, Tao, & Li, 2010). This approach can be turned into a psychological account, by seeing psychological similarity as captured by the complexity of the transformation from one mental representation into another (Imai, 1977; Hahn, Chater, & Richardson, 2003).

Figure 21.3 provides a simple illustration: one simple geometric stimulus (a square and a triangle) can be transformed into another (a triangle and a circle) by a series of transformational steps (swapping the objects in the pair, creating a new object, and replacing one object with another). The empirical prediction is that the longer transformational sequence is required, the less similar the stimuli will seem (see Hodgetts, Hahn, & Chater, 2009, for data and experimental details).

The generality of this transformational approach to similarity is particularly helpful when we consider representations of complex, structured objects, such as sentences. Thus, it seems intuitively natural that "The cat sat on the mat" is rather similar to "The mat was sat on by the cat," and somewhat similar to "The cat with green eyes and a bushy tail sat on the mat" because there are linguistically natural changes (e.g., a passive transformation,

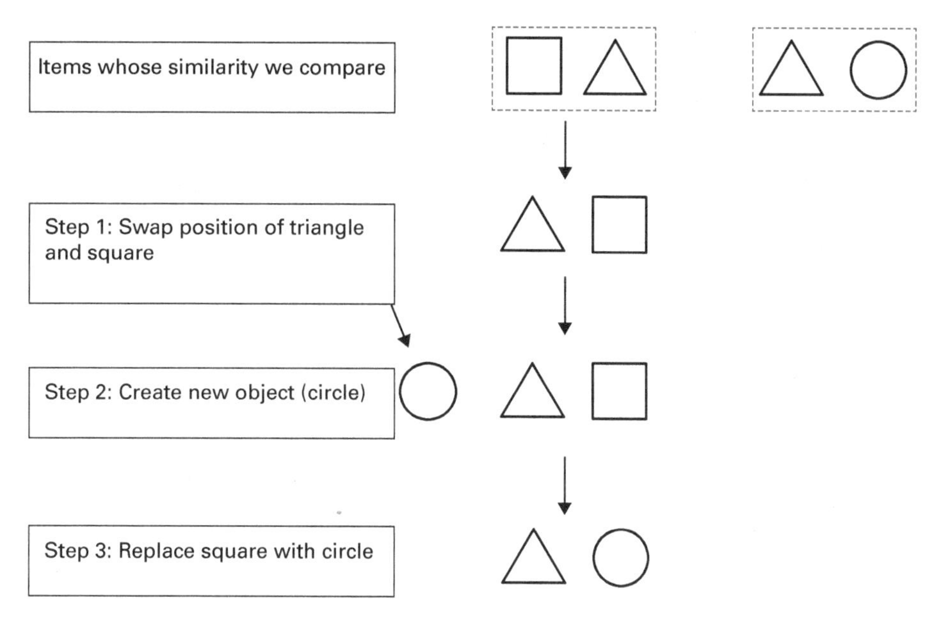

Figure 21.3
Similarity as transformation. A series of steps allows one stimulus (the square and triangle) to be turned into another (the triangle and circle). Experimental data show that, for elementary stimuli such as those shown here, the greater the number of transformational steps between two stimuli, the less similar they are judged to be. Of course, a more sophisticated approach would allow that steps may themselves differ in complexity, and people's numerical similarity judgments will also depend on many contextual factors. Figure redrawn from Hodgetts et al. (2009).

adding a relative clause) that turn one into the other. For that matter, "The cat sat on the mat" seems rather similar to the speech error "The sat cat on the mat," where two initial phonemes (or perhaps whole words) have inadvertently been swapped—presumably a psychologically natural transformation because it is often arises spontaneously even in fluent speech (e.g., Cutler, 1982). On the other hand, a random scrambling of words to "on cat the mat sat" seems much less similar—indeed, it is hard for us to imagine the transformation from one to the other. In the same way, when considering structured representations of the visual world, it is easy to imagine, say, a human body changing position through the transformation of joint angles. For this reason, a person moving from sitting to standing may seem highly similar. Likewise, a person with or without a sweater or a pair of glasses may be perceived as highly similar—at an abstract level, the transformation is fairly simple, though at the level of the pixels of the image itself, the changes may be substantial.

If this viewpoint is correct, then making a particular transformation more available (e.g., through recent exposure or longer-term learning) should make items connected by that transformation seem more similar. Hahn, Close, and Graf (2009) found that showing people sequences of morphs from one item into another in a particular direction seems to prime that direction of transformation, and thus shifts similarity judgments in an asymmetrical way. Relatedly, Langsford, Hendrickson, Perfors, and Navarro (2017) familiarized people with rotations and arbitrary color-swapping rules for abstract patterns, and they found that this prior experience increased the perceived similarity of pairs of items that could be related by these rules more easily. Mathematical results, moreover, have shown how the transformational approach to similarity (expressed in terms of conditional Kolmogorov complexity) naturally generates Shepard's (1987) celebrated "universal law of generalization," which relates similarity with confusability in a wide range of experimental contexts (Chater & Vitányi, 2003b).

It is natural to imagine that a psychological theory of similarity should map naturally into a theory of categorization, on the assumption that categories that should put similar things together. Indeed, one psychological model takes this approach directly: Pothos and Chater (2002) treat similarity data as the starting point and use a simplicity principle to find categories that encode that similarity data as efficiently as possible (where the key assumption is that, other things being equal, items in the same category will tend to be more similar than items in different categories). But an alternative approach, more closely aligned with the Bayesian approaches to categorization described in this book (see chapters 5 and 9), would be to use a simplicity principle to find categories that most efficiently describe the items themselves, rather than similarity relations between them. Indeed, one of the very earliest applications of code minimization in machine learning took this approach (e.g., Wallace & Boulton, 1968).

Feldman (2000, 2003, 2006) applies a coding approach to model classic experiments on human categorization (Shepard, Hovland, & Jenkins, 1961): looking at how people learn so-called Boolean concepts (e.g., categories defined by logical rules). People are given positive and negative examples of a category, where the items have three binary dimensions (e.g., shape [say, triangle versus circle], size [say, large versus small] and color [say, dark versus light]). Feldman measures the complexity of a code by the number of nonlogical terms in the most economical formula describing the category (so the not-triangle would have a complexity of 1; $[\neg(B \wedge C) \wedge \neg A]$ has a complexity of 3; $[(\neg A \wedge \neg B) \vee (A \wedge B \wedge \neg C)]$ has a complexity of 5; and $[(\neg A \wedge (\neg B \wedge \neg C)) \vee (B \wedge C) \vee A \wedge \neg B \wedge C]$ has a complexity of 8 (where $A, B,$ and C are particular binary features). Feldman finds that the complexity of these formulas successfully predicts how easily they are learned from examples by human participants.

21.5.3 Simplicity, Learning, and Language

A simplicity-based viewpoint also provides an interesting perspective on how language acquisition is possible. We discussed specific Bayesian models of language acquisition and processing in chapter 16. Here, instead, we consider the general question of the sense in which learning, and in particular language learning, is possible at all. This issue has been particularly controversial, as there have been influential claims that very strong innate constraints are required for learning, and in particular, for learning language.

Chomsky (1965, 1980) has argued that the problem of learning a grammar from a partial, finite, and often noisy corpus of linguistic input is not possible without innate language-specific constraints that are so strong as to embody almost a complete blueprint for language, aside from specific "parameters" that differentiate one language from another and the accumulation of inventories of words and idioms. This line of thinking seems to be strengthened by some "negative" results in formal learning theory, pioneered by Gold (1967) and extensively developed since, as outlined in Jain, Osherson, Royer, and Sharma (1999), which seem to imply that learning from experience can almost never be guaranteed not to go badly awry (see chapter 16).

A distinct, though related, line of argument is that it is inherently impossible to recover from overgeneral grammars (or overgeneral models in other domains), as these can be perfectly consistent with the observed empirical data—and typically, we observe only what is possible in language and receive little or no negative evidence about what cannot be said (Bowerman, 1988). This line of argument has been influential in motivating some nativist approaches to language acquisition (Crain & Lillo-Martin, 1999), who consider examples such as:

a) Which man did Fred want to see?

b) Which man did Fred want to win?

c) Which man did Fred wanna see?

d) *Which man did Fred wanna win?

where the final ungrammatical case (marked with a "star") is presumed to be disallowed because of rather deep linguistic reasons (roughly, that there is a "gap" in (b) *Which man did Fred want _ to win*, which is implicitly filled by *that man* as in *Which man did Fred want [that man] to win)*. And the idea is that contractions can't occur over such gaps—so "want to" can be contracted to "wanna," but "want _ to" cannot be. According to Crain and Lillo-Martin (1999), this provides evidence that the relevant linguistic principles concerning gaps and contraction are innate.

It is natural to suspect that there must be something wrong with this line of argument—it appears far too strong. The problem of overgeneralization arises not just for aspects of language that some linguists see as arising from deep universal principles, but just about everywhere else in language: language is a mixture of regularities, subregularities, and downright exceptions (Culicover, 1999), and these exceptions are often highly idiosyncratic and not credibly viewed as innately specified. For example, as we noted in chapter 16, "give" and "donate" have very similar meanings;

a) Ali gave the book to Eva.

b) Ali donated the book to Eva.

c) Could Ali give Eva the book?

d) * Could Ali donate Eva the book?

We can say "Ali gave the book to Eva" and "Ali donated the book to Eva." But while "Could you give me the book?" is perfectly acceptable, there is something distinctly odd about "*Could you donate me the book?" But how does the learner know this? If recovery from overgeneralization is inherently problematic, then once the learner has conjectured that sentence d) might be possible, there may be no way back. And, indeed, were this the case, we should expect such exceptions gradually to be eliminated throughout language over time—quite the opposite from what is observed across the world's languages (Culicover, 1999).

Thus, if overgeneralizations are difficult to recover from without direct negative evidence (i.e., specifying that certain things cannot occur), then the problem should be ubiquitous not merely in learning language, but learning about the natural world. After all, children, and for that matter, scientists, can only ever observe data concerning what is possible according to the laws of nature, and cannot directly obtain data concerning what is not possible. But children and scientists do seem, in practice, to be able to learn about the world successfully without getting stuck with overgeneral hypotheses. Indeed, many of the most fundamental scientific principles concern specifying what is not possible: for instance, heat cannot flow from a cooler to a hotter body; energy cannot be created or destroyed; and so on.

A Bayesian framework, whether viewed in terms of probabilities or codes, provides a simple explanation of how recovery from overgeneralization need not be problematic. Suppose that we consider two models, one of which predicts (m_1) just the data that we actually obtain; and another, overgeneral, model with much less precise predictions (m_2). Suppose, for example, that our data are simply a sequence of heads and tails; and one model is that the data come from a coin that has heads on both sides; and the other assumes a fair coin (so that, on each trial, heads and tails are equally probable). So, on a single trial, m_1 predicts heads with probability 1, whereas m_2 just predicts heads with probability $\frac{1}{2}$. Thus, applying Bayes' rule to the ratio of probabilities between the two theories, we have

$$\frac{P(m_1)|\text{heads})}{P(m_2)|\text{heads}} = \frac{P(m_1)}{P(m_2)} \frac{P(\text{heads}|m_1)}{P(\text{heads}|m_2)} = \frac{P(m_1)}{P(m_2)} \frac{1}{1/2} = \frac{P(m_1)}{P(m_2)}.2. \tag{21.11}$$

In the same way, a string of n heads will boost the ratio of posterior probabilities in favor of m_1 by 2^n. As n increases, this is likely to overwhelm any differences in priors between the theories rapidly. Thus, the overgeneral theory is eliminated by Bayes' rule because it assigns a lower probability to the observed data (conversely, of course, observing a single tails will set the posterior probability of m_1 to zero).

Given the equivalence between the probabilistic and simplicity perspectives, of course, we can make the same point in terms of code lengths. The simplicity viewpoint is that we should choose whichever of m_1 and m_2 provides the shortest code for the data, where the code length for theory m_i is

$$\begin{aligned} C(d, m_i) &= C(m_i) + C(d|m_i) \\ &= C(m_i) + \log_2\left(\frac{1}{P(d|m_i)}\right). \end{aligned} \tag{21.12}$$

The $\log_2(\frac{1}{P(d|m_i)})$ term is simply the amount of information, in bits, required to encode the data, D (here, one or more heads or tails) in terms of the theory. A theory that makes more precise predictions will require less information to reconstruct the data. So, at least with independent and identically distributed (i.i.d.) data, such as our coin flips, an overgeneral model will eventually generate a higher $C(d, m_i)$ value and will hence be rejected. In our simple example, m_1, predicts that each successive heads has a probability 1, which can be encoded with no information at all: $\log_2(\frac{1}{P(\text{heads}|m_1)}) = \log_2(\frac{1}{1}) = 0$. And m_2, by contrast, assigns each successful heads a probability of $\frac{1}{2}$, and hence requires 1 bit of information for each observation $\log_2(\frac{1}{P(\text{heads}|m_1)}) = \log_2(\frac{1}{1/2}) = \log_2(2) = 1$. So after n successive heads, the overgeneral hypothesis is penalized by n bits, which will rapidly overwhelm any small differences in encoding the codes for models themselves, i.e., between $C(m_1)$ and $C(m_2)$.

The same logic applies to eliminating overgeneral models in language, and it has been illustrated both from a Bayesian (Dowman, 2000) and a simplicity perspective (Onnis, Roberts, & Chater, 2002) using simple artificial languages. This logic can be extended in various ways. Here, we first consider how this approach can be generalized to create a formal framework for learnability, broad enough to create useful results for language acquisition, as a counterpoint to the often negative results that follow the assumptions of formal learnability theory. We then ask how this framework can be scaled down to deal with the learning of specific aspects of language.

First, then, let us look at the general framework. Kolmogorov complexity theory provides a framework within which very minimal assumptions still allow a learner to converge on knowing what is, and is not, acceptable in a language, purely from exposure to a sufficiently large corpus of language. The assumptions required are remarkably modest: roughly, that the infinite corpus (which we can, without loss of generality, encode as a string of 0s and 1s) is generated by a computable probability distribution—which, roughly, means an arbitrary computer program that accepts a source of randomness as input (this could, for example, be encoded in Church). Notice that we need make no assumptions that sentences are chosen i.i.d., or that sentences are generated in a particular grammatical formalism. Indeed, the linguistic material can have any computable structure whatever. But this computability assumption is crucial: it rules out the overwhelming number of corpora that are uncomputable; but assuming a standard computational view of the mind, such corpora could not be generated by a human in any case. Let us call the "true" distribution, over the infinite sequences of 0s and 1s, μ, such that it assigns a probability to every corpus—that is, a finite sequence of 0s and 1s, however long.

Suppose that we want to predict the next element in this sequence. Then it is possible to apply a remarkable theorem by Solomonoff (1978), which considers a universal prior, λ, over these binary strings, following roughly the recipe we described earlier in this chapter (applied to infinite strings of 0s and 1s). So suppose that we use λ to predict μ, in the following way. For the very first item in the sequence, we consider the squared difference between $\lambda(0)$ and $\mu(0)$. Call this error s_1. On the nth trial, we consider the squared difference between $\lambda(0|x)$ and $\mu(0|x)$, weighted by the probability $\mu(x)$, of each $n-1$ length sequence, x, being generated by μ. This is the expected error on the nth piece of data, which we call s_n. And then we can sum up these expected errors, across the infinite sequence of data. If λ does not converge on μ, then this sum will be infinite. But Solomonoff shows that, remarkably, the sum has a finite bound:

$$\sum_{j=n}^{\infty} s_n \leq \frac{\log_e 2}{2} K(\mu), \tag{21.13}$$

where $K(\mu)$ is the Kolmogorov complexity of the computable probability distribution μ. This means that the total summed expected squared error to infinity, using the universal distribution λ to predict the specific μ, is determined by the length of the shortest Church program (or whatever language we choose) for μ. So λ converges onto μ; and the mistakes that it makes in doing so depends on the complexity of specifying the target μ. So if the language to be learned is generated by i.i.d sampling from a small phrase structure grammar, λ will learn that language with few mistakes en route (as it makes provisional predictions at every step); specifying the full complexity of a natural language, including its statistical patterns, discourse structure, and other elements, will be much greater, and hence learning will be corresponding slow. Still, if target μ is computable at all, λ will converge to it eventually, with errors bounded by the target distribution's complexity.

It may seem mysterious, even impossible, that a single distribution λ can simultaneously approximate all computable distributions μ—after all, the values of different μ can wildly diverge from each other. To get an intuition for how this is possible, imagine a process that, given any string of binary data of length n, finds the shortest program that generates that sequence and predicts element $n+1$ by following whatever that program would predict (or, better, would aggregate the predictions of multiple possible programs, appropriately weighted by their length). So as soon as a pattern begins to emerge in a sequence, this process would rapidly find, and follow, that pattern; and it will eventually lock onto the program that actually generated the data (or, more precisely, will gradually assign that program an ever-higher probability). So a single process can model all possible computable distributions by learning to mimic them as rapidly as possible.

It turns out that Solomonoff's remarkable result has implications, in the context of natural language, for judgments of grammaticality—the ability to make such judgments has been viewed as an important measure of linguistic knowledge. And, in particular, let us focus on the challenge of recovery from overgeneralization—that a learner needs to find some way to prune back a grammar that allows unacceptable sentences, but with access only to positive examples. To frame this problem, we need to switch from focusing on binary codes to considering our data as a sequence of words; and the challenge of the learner is, given the $j-1$ words encountered so far, to conjecture the probabilities of possible jth words. If the learner has an overgeneral model of the language, then it will assign a positive probability to continuations that are not possible (i.e., have zero probability because they are inconsistent with the grammar). Let us call the probability of making such an overgeneralization error $\Delta_j(x)$. The expected value of overgeneralization on the jth word, weighted by the probabilities of the preceding strings of $j-1$ words, can then be written as $\langle \Delta_j \rangle$.

Now, by using Solomonoff's result, it turns out to be possible to prove that the expected value of generating ungrammatical continuations is bounded by the Kolmogorov complexity of the probabilistic process, μ, that actually generates the data:

$$\sum_{j=1}^{\infty} \langle \Delta_j \rangle \leq \frac{K(\mu)}{\log_e 2}. \tag{21.14}$$

This means that, using an ideal simplicity-based learning method, it is possible to recover from overgeneralization. Specifically, the expected number of overgeneralization errors that the learner makes is bounded over an infinite corpus of language. To see that this addresses the problem of overgeneralization, suppose that, on the contrary, the learner failed to recover from the possible overgeneralization concerning "donate" discussed previously—so the learner would attach a positive probability to "Could Ali donate Eva the book?" "Donate me the book," and so on. A learner that did not eliminate this overgeneralization would have a small, but roughly constant, probability of making such a mistake on the jth word of the corpus, implying that $\langle \Delta_j \rangle$ is above some constant, c, for all j. But if this were true, then the sum, $\sum_{j=1}^{\infty} \langle \Delta_j \rangle$, would be infinite, because an infinite number of any finite quantity, however small, will sum to infinity. So, by contradiction, any such overgeneralizations must be eliminated by our simplicity-based "ideal learner." So overgeneralization cannot provide a logical problem for language acquisition—a sufficiently large corpus does contain enough information to allow overgeneralizations to be ruled out.

These positive results have their limitations, however. One of these is uncomputability—because it is impossible to compute the shortest code for a given set of data, as noted already, Solomonoff's results and their extensions cannot be used to build an algorithm that could capture how people actually learn language. A partial response to this limitation is a result by Vitányi and Chater (2017), which provides a computable algorithm that recovers any computable probability distribution exactly in the limit from an infinitely long i.i.d. sample. So, given a sufficient i.i.d. sample of, say, a stochastic phrase structure grammar, or any other grammar that can generated by a combination of random noise and a computable process (e.g., can be encoded in the Church program), then there is algorithm that conjectures precisely that stochastic phrase structure grammar and sticks to that conjecture forever, with probability 1. But the algorithm proposed by Vitányi and Chater (2017), while computable, is extraordinarily slow.

Another limitation is that these abstract results use arbitrarily large amounts of data—but what really matters is whether recovery from overgeneralizations can occur from realistic amounts of data, given the language input received by the child. Dowman (2000) offers an early Bayesian analysis. Anne Hsu and colleagues have proposed simplicity-based and traditional Bayesian analyses for a range of linguistic data, and experimental paradigms (Hsu & Chater, 2010; Hsu et al., 2011; Hsu, Chater, & Vitányi, 2013; Hsu, Horng, Griffiths, & Chater, 2017). Hsu and Chater (2010) use a simplicity-based framework to predict the learnability of linguistic patterns that have been viewed as problematic in the field of language acquisition: contractions of "want to," "going to," "what is," and "who is"; the "dative alternation" for the verbs "donate" (mentioned earlier), "whisper," "shout," "suggest," "create," and "pour"; and transitivity for the verbs "disappear," "vanish," "arrive," "come," and "fall." Interestingly, in the light of the number examples encountered in real language, some constructions appeared learnable within a few years, whereas others required many lifetimes of linguistic input.

Do the latter cases indicate that the relevant regularities must be innate? No, not necessarily in the traditional sense of the term. Note that languages are shaped through cultural evolution to be easy to learn and process—so that languages will embody cognitive biases that predate language (just as written scripts are adapted to be easy to create and to recognize, given that constraints of the human motor and visual systems). So unlearned aspects of language need not be language-specific, or in any way adapted through biological evolution for language processing. The mesh between languages and learning biases may arise from languages evolving around our cognitive systems rather than vice versa (e.g., Chater & Christiansen, 2010; Christiansen & Chater, 2008, 2022; Zuidema, 2002).

21.6 Future Directions

The dual relationship between probability and coding is a fascinating one for cognitive science—and flipping back and forth between these perspectives is likely to continue to be a valuable source of ideas and insight. Here, we briefly highlight some of the many issues that seem to be tractable and exciting topics for future research.

Consider, for example, the nature of objects (windows, people, numbers, tunes, voices) and their properties (colors, heights, sizes, timbres). Symbolic representations of the world typically *start* with some set of objects and properties and build up a logical language consisting of formulas expressing relationships between them—and such logical formulas are also often assumed to capture the meaning of expressions in natural language. But where do objects and their properties come from? Why do we group together the parts of an animal into a single entity, rather than grouping together an animal's legs and the floor that it is standing on, or any other arbitary arrangement (Quine, 1960)? The preference for simple codes provides one line of attack: for one thing, the animal moves in unison, but the floor doesn't; for another, many of its properties (perhaps furriness, color, and so on) are naturally defined over most of the whole animal; and, of course, defining animals as objects allows the huge simplifications generated by similarities across animals, both within and across species (if they arise from a common biological history). Yet the same approach may be applicable to more abstract objects such as tunes, where grouping together a particular set of notes may make it particularly easy to encode their sequence and rhythm (Bregman, 1994). Perhaps properties similarly provide a repertoire of descriptions that make regularities in the world particularly easy to encode briefly. For example, it seems reasonable to say that properties of an object are likely to be useful if they are often invariant over transformations of that object (e.g., the mass, symmetry, or color of a bird is preserved as it takes flight). More broadly, in both perception and science, it is interesting to ask how far our conceptions of objects and properties can arise from the broader challenge of providing brief encodings of sensory data.

It is interesting, too, to consider how coding may help address questions of innateness in our understanding of the perceptual world, as discussed in this chapter in the context of language acquisition. Building in useful representational primitives is likely to assist the process of finding simple patterns in the world; but this bias toward particular primitives need not constitute a rigid theory of how the world works—it may be more analogous to having a useful set of functions to make a particular operation easy to code in a universal programming language (e.g., multiplying matrices, appending lists, capturing logical inferences, and so on). But the representational language may still be able to capture any computable process—the representational bias simply makes some processes more natural than others (and, in probabilistic terms, corresponds to a prior that favors these natural options, while not strictly ruling out alternative possibilities).

Finally, a crucial challenge for the coding approach is to deal with the fact that merely compressing the available data, without paying attention to the process by which that data is gathered and selected, may lead the cognitive system systematically awry. To borrow an example from the philosopher Gilbert Harman, after we have seen a sufficiently large number of white swans, we might tentatively and reasonably conclude that all, or at least the great majority, of swans are white—wrongly, as it happens, as black swans are rather common in Australia (Harman, 1965). But, Harman points out, the observation that all swans that we have seen so far have been within one mile of a human being would be very bad evidence for the generalization that most or all swans are within one mile of a human being—because there is a better explanation for this regularity, which is little to do with swans at all, and mainly depends on the limits of human vision (and, of course, could immediately be falsified as soon as we use remote cameras or drones).

From a Bayesian perspective, it seems natural that we view our data as a joint product of the data sensing and gathering machinery (and the active way in which that machinery is deployed), as well as the structure of the world—and the many patterns in the data that are explicable due to our data-processing and gathering methods (e.g., the degree of blur in the visual image across different retinal locations and wavelengths; the "shadows" of fixed structures in the eye, such as the retina's own blood supply, should be factored out; and, of course, sensory data should be interpreted in the light of feedback from our own eye and head movements). And active learning (see chapter 7) will causes biases in how we sample the world, which we should attribute to our own choices, not to the world itself. Thus, if we are relentlessly searching for apples, and hence focus our attention on applelike visual inputs, we need to avoid interpreting sensory data as confirming that the world is itself well-nigh full of apples, merely because our sensory input is full of apple images. It is by no means obvious how this factorization of the explanation of the data into properties of the world, rather than properties of the perceiver, should be achieved.

21.7 Conclusion

A Bayesian viewpoint on cognition can be re-framed in terms of coding: rather than seeking probable models of observed data, the mind is viewed as searching for short codes for that data. This provides a connection with simplicity-based models in perception (which have often been viewed as a rival, rather than a complementary, perspective on Bayesian models), the basis for a transformation-based view of similarity, and a framework for providing positive results about language learnability. Indeed, thinking in terms of codes as well as probabilities may be a useful perspective for cognitive science more broadly.

22

A Bayesian Conversation

Nick Chater, Thomas L. Griffiths, and Joshua B. Tenenbaum

What are the prospects for Bayesian cognitive science? Is it too specific, making claims about the mind that are already known to be false? Or too general, providing a framework so capacious that any empirical data can fit comfortably within it without danger of falsification? How does the Bayesian approach relate to other work in cognitive science? In this chapter, we discuss these issues in the form of a conversation between a skeptical (S) and a Bayesian (B) cognitive scientist.

To do so, we imagine our hypothetical skeptic having concerns arising from several different perspectives, such as traditional cognitive psychology and connectionist cognitive modeling. We draw upon questions from actual skeptics (McClelland et al., 2010; Jones & Love, 2011; Bowers & Davis, 2012) and our responses to them (Griffiths, Chater, Kemp, Perfors, & Tenenbaum, 2010; Chater et al., 2011; Griffiths, Chater, Norris, & Pouget, 2012a). In some cases, our responses serve to clarify potential misunderstandings or weaken what may be perceived as overly strong or confident claims. In other cases, the challenges raise important open questions for the Bayesian approach that future work should seek to address. We hope the ideas given here will stimulate further debate and research by readers of this book to push forward the development of cognitive science, whether within the Bayesian framework or outside it.

So let the debate begin!

S: The premise of this book is that cognitive science is an exercise in reverse-engineering. Let's suppose that is right. Then we should be looking for inspiration for the cutting-edge engineering methods that are driving forward contemporary machine learning and artificial intelligence (AI). But surely many of the real breakthroughs in AI and machine learning aren't Bayesian at all. The biggest game-changer in the past decade or two in machine learning has been the rise of deep learning (Goodfellow, Bengio, & Courville, 2016). This is the technology that underpins Google images, autonomous driving, and some incredible recent results in natural-language processing, and has helped computers finally beat the best human at Go and a wide range of other games (Brown et al., 2020; Silver et al., 2017). The Bayesian approach is philosophically well justified and mathematically elegant, perhaps. But isn't it just blown away by the general-purpose power of deep learning from really large amounts of data?

B: The results of deep learning are undoubtedly remarkable and have led to incredibly useful practical applications. But these methods don't work by modeling human intelligence. Instead, they are solving specific, and practically important, problems by methods that appear to sidestep the need for human intelligence.

What is remarkable about people is their ability to learn rapidly, reason flexibly, and think creatively about entirely novel problems. To do this, our brains are able to make wild inferential leaps from minimal amounts of data. By contrast, deep learning typically involves making comparatively small generalizations from extremely large amounts of data. Now, of course, deep learning is not just table lookup; it is more like a very sophisticated table lookup with extremely clever interpolation. This type of mechanism may very well be important in understanding many aspects of cognition, particularly concerning problems of pattern recognition that arise in motor control. But surely much more is required to help engineer machines that learn like humans (Muggleton & Chater, 2021).

Lake, Ullman, Tenenbaum, and Gershman (2017) stress the importance of three key elements, which have all figured heavily in this book: the ability to build causal models of the external world rather than purely finding patterns in data (chapters 4, 5, and 9), the capacity to learn and apply rich background theories of how the world works (including naive physics and folk psychology; chapters 14 and 15); and the ability to construct compositional representations (chapters 17–20) and to learn how to learn (chapter 8).

In short, the response to your question is: yes, deep learning is terrific engineering, but it doesn't reverse-engineer human cognition (though it might be part of the story). It just doesn't solve the deep problems of human intelligence, and it doesn't operate in a humanlike way (or at least not so far).

Consider a parallel with the body. Understanding the heart as a pump was a spectacularly successful piece of reverse-engineering, which would have been impossible without the invention of pumps for moving water. And the valves in the circulatory system would be hard to understand without the engineer's prior conception of a valve. But the complex set of dynamically contracting and relaxing chambers that make up the heart work very differently from any pumps known in engineering. And the most efficient pumps in engineering certainly do not mimic the solution found by natural selection. Or, to take a more extreme example, the problem of reverse-engineering the biomechanics of human locomotion depends on very complex background engineering knowledge; but it does not depend on the technology of internal combustion engines, gears, or wheels.

S: That all sounds plausible enough—but are you sure that deep learning isn't more humanlike than you think? Look at the incredible results obtained by GPT-3, GPT-4, and other large language models that were discussed briefly in chapter 16, ranging from generating pastiche chunks of period novels, "fake" but strangely plausible philosophical conversations on consciousness, decent-looking computer code, answers to questions on a huge variety of topics, and much more. Who is to say what is going on deep inside the 175 billion parameters in GPT-3's neural network? Maybe GPT-3, GPT-4, or one of their many rival models have already developed causal models of the world, figured out rudimentary naive physics and psychology, and created compositional representations from scratch. Or maybe some future even-more-powerful deep neural network model will soon do so? Perhaps the Bayesian approach to cognition, the importance of structured representations, and the entire rational analysis approach are perfectly fine, but really no more than a distraction.

The real action is going to be building the ever-more-powerful learning algorithms, trained on ever-richer sources of data (perhaps not just words and images but something closer to the rich sensorimotor input available to a human child) that will explain how all this complexity can be created from scratch.

B: Well, even if that turns out to be right, the Bayesian story will be still be important: We will want to understand how giant neural network models are working, just as we will want to understand the operation of the human brain. But the evidence so far strongly suggests that at least early large language models using deep learning, like GPT-3, work in profoundly non-humanlike ways. Remember how, in chapter 16, we saw that GPT-3 seems to do language processing by something like a highly sophisticated table lookup, with incredibly clever cut-and-paste interpolation, rather than having the remotest conception that language has meaning or is used in communication, or that there is an external world to be communicated *about*. GPT-3 knows a lot of "facts" in domains in which its training corpus (a large fraction of the entire internet) contains relevant sentences—so it knows that spiders have eight eyes even though many people don't. But, as we saw, it has no idea that a foot has no eyes because the internet is not full of discussions of this rather nonsensical topic. A human, with even the most rudimentary knowledge of biology, knows that feet have no eyes; but GPT-3 does not have this rudimentary understanding—indeed, this might suggest that GPT-3, at least, has no real understanding at all. Or, to put the point more neutrally, the understanding of such models is at least hotly contested (Mitchell & Krakauer, 2023).

S: But the new generation of deep learning systems, or the generations after these, might spontaneously develop such understanding. GPT-4's flexibiliy and intelligence may not merely be papering over the cracks of earlier models, but learning to work in a fundamentally different, and more human, way. And who knows what will be possible with large neural networks in a decade, or perhaps a century.

B: Time will, of course, tell. For now, there are at least good reasons to believe that the apparent "sparks of general intelligence" (Bubeck, et al., 2023) that even the most sophisticated large language models exhibit may have a more prosaic stastical basis and may fail in ways that seem unnatural from the perspective of human cognition (McCoy, Yao, Friedman, Hardy & Griffiths, 2023). Indeed, it may well turn out that relying on deep learning or similar approaches to solve the fundamental problems of cognition is like betting that the next generation of automotive technology will miraculously explain the details of human biomechanics. For reverse-engineering to succeed, we can't just rely on finding good solutions to engineering problems. We must find engineering solutions to the actual challenges faced by human intelligence (Griffiths, 2020); and we also need to check that any successful system is solving these problems in a humanlike way.

Suppose, though, that future generations of neural network model do seem not merely to be emulating the results of human intelligence, but also can build a real understanding of how the world, and other humans, work—or at least can build as much of an understanding as we humans have. Then there is still the question of how these results are achieved. And the Bayesian reverse-engineering approach may turn out to be crucial to answering this question. Just specifying the learning algorithm, the architecture of the network, and its sources of data isn't enough to give us much insight into how such networks are working. We still wouldn't know what knowledge it represents and in what form. In fact, the challenge of figuring out how a giant artificial network works is not so different from the challenge of understanding the biological neural network that is the human brain. Of course, the artificial neural network is in principle far more "transparent"—we can see its activity in minute detail rather than relying on the relatively crude results of brain imagining. But these details may not help much anyway, at least without some theory of the possible computations that the system is carrying out—and to make this possible, we will most likely need the Bayesian analysis after all.

S: Well, I can see we're not going to agree on this—and perhaps that's OK: having different research strategies running in parallel is usually a good approach in science, after all. But all this discussion of solving problems in a humanlike way seems a double-edged sword for the Bayesian cognitive scientist. After all, Bayesian cognitive science sees the brain as an exquisite probabilistic reasoning machine. But surely this isn't a humanlike model at all, but rather an assumption of hyperrationality—what Gigerenzer and Goldstein (1996, p. 650) have termed the "Laplacean demon."

Doesn't this fly in the face of developments in rational choice explanation in the social sciences? Surely, the whole point of behavioral economics, for example, is that such rational idealizations are incorrect—or at least incomplete; and these disciplines look to the cognitive sciences to build more realistic, and hence less hyperrational, models of behavior. This viewpoint seems to fit, too, with many and diverse traditions in psychology, the cognitive sciences, and AI, which sees intelligence as more like a flexible bag of tricks rather than the result of fully rational calculation (Agre & Chapman, 1987; Brooks, 1991; Gigerenzer & Todd, 1999; Ramachandran, 1990).

B: But the Bayesian approach does not require that people necessarily carry out enormously complex Bayesian calculations. The Bayesian approach may, in some instances, specify the "right" pattern of inference or course of action; but the agent need not necessarily recapitulate this calculation to justify its thoughts and behavior, any more than the bird needs a knowledge of fluid dynamics to "justify" the shape of its wings (Marr, 1982). Instead, the cognitive system may come upon the right solution by evolution, through trial-and-error learning, or by some other means.

One of the striking contributions of approaches in cognitive science and AI based on simple heuristics is to show that in some environments, such approaches can be surprisingly successful. But why? A Bayesian rational explanation is surely required—spelling out an optimal solution, given the structure of typical environments, and showing that simple heuristics often come close to this optimum. It seems likely that when such shortcuts are available, the cognitive system may use them—and a Bayesian analysis is helpful in identifying those situations.

Yet the "bag of tricks" perspective is fundamentally incomplete as a reverse-engineering account in at least two ways. A complete solution would indicate, first, how the cognitive system knows which heuristic to apply in which circumstances; and second, and more important, how the cognitive system adjusts these heuristics in the light of new information. In a nutshell, an "adaptive toolbox" or "bag of tricks" model of cognition may be appropriate if there is a well-defined and stable "bag of problems" to which the tricks naturally correspond. But if the agent has to respond deftly and appropriately to a continually changing set of problems, which is surely the case with human cognition, then it needs to be guided by some principles concerning how to think or act. If these principles are to work effectively, then presumably they must have some justification—and hence we are forced back to at least some approximation of a rational approach (indeed, the rational solutions to these problems are discussed at length in chapter 13).

An analogy may be helpful. Given a fixed set of arithmetic problems, a good strategy may be simply to store the answers to each by rote, so the questions can be answered rapidly, without calculation (Logan, 1988). But if the set of problems is broad and unpredictable and the scope of those problems is continually liable to shift (e.g., if negative numbers, fractions, real numbers, or imaginary numbers are successively introduced), then it is vital to understand the general principles of arithmetic.

S: But what about computational complexity limitations? Surely we've known for a long time that Bayesian calculations cannot possibly scale up to deal with real-world inference.

B: As we have just noted, complex Bayesian calculations may sometimes end up explaining why the cognitive system sometimes sticks to simple heuristics—for which no complexity issues arise. But, in any case, complexity results can be misleading, as they typically consider inference over arbitrary probability distributions. The magic of graphical models is that they provide a way of factorizing probability distributions into a particularly workable form—and a form that lends itself to parallel implementation: each node can be viewed as a processor, which makes calculations entirely locally, depending on its own inputs; but the resulting collective behavior may correspond to a globally "good" solution. More precisely, the computational cost of naive probabilistic inference is exponential in the number of variables involved, but inference in a Bayesian network is exponential in the size of the largest clique in the underlying graph (Cooper, 1990). Performing exact inference in such a network may thus, nonetheless, be costly—but then numerical methods, such as Markov chain Monte Carlo (MCMC; Gilks, Richardson, & Spiegelhalter, 1996) and other approximation methods that we met in chapter 6 may allow good approximate solutions.

But, of course, issues of computational tractability are hugely important, from the point of view of both engineering and reverse-engineering. Cognitive processes must run fast on the slow parallel hardware of the brain (Feldman & Ballard, 1982). But over the past few decades, there has been great progress in finding ways in which seemingly intractable probabilistic problems can be approximated (see chapters 4, 5, and 6). We suggest that these, and future, developments concerning tractable, approximate, Bayesian inference will be an important source of hypotheses in reverse-engineering how the brain deals effectively with what may appear to be unmanageably severe computational challenges.

S: But doesn't more than a half-century of research in judgment and decision-making tell us just the opposite? Rather than finding elegant approximations to Bayesian calculations, people seem to deviate drastically from them, even for the very simplest problems. Endless experiments have found that human probabilistic reasoning seems to be riddled with systematic errors and biases. People persistently have the wrong qualitative intuitions too: the conjunction fallacy, ignoring base rates, the gambler's fallacy, Simpson's paradox, the Monty Hall problem, and many more surely make it clear that the human mind is not Bayesian. Surely this is the elephant in the room for Bayesian cognitive science: it is completely incompatible with the ubiquitous inadequacies of human probabilistic reasoning.

B: I'm glad you raised this point! It is so easy to think: "I'm sure I've heard that Kahneman and Tversky have shown people aren't Bayesian. So this whole Bayesian cognitive science movement has got to be on the wrong track."

But this would be a serious mistake! There are several points to make. To start with, note that the Bayesian cognitive models considered in this book, and similar models in computational neuroscience, perception, and motor control, have absolutely no direct implications for human probability judgments.

Remember that cognitive models of all kinds involve complex mathematics. Now it is true that the mathematics underpin the operation of the models. But it would be a complete misunderstanding to imagine that such models imply that people have to understand and be proficient in the relevant branch of mathematics. This would be a mistake akin to thinking that the deep reinforcement learning can't possibly explain how rats learn because rats don't know even basic algebra and calculus, let alone how to back-propagate a gradient; or that the rabbit retina can't be convolving the image with its receptive fields because rabbits haven't learned about convolution; or, for that matter, that complex biochemistry can't underpin your own digestion unless you have a deep understanding of biochemistry. Indeed, Marr (1982) made this point 40 years ago, noting that a bird does not need to know the principles of aerodynamics that make flight possible. Likewise, Bayesian models of cognition remain a useful tool even if we seek to understand a mind that is utterly ignorant of probability theory. So the elephant in the room turns out not to loom quite so large after all. The worry is primarily based on a conceptual confusion.

But that's not quite the end of the story. As we've seen throughout this book, the Bayesian computations involved in any cognitively interesting problem tend to be far too complex to solve precisely. So in practice, Bayesian models, in cognitive science just as in statistics and machine learning, generally use approximation methods of various types. It would particularly neat if it turned out that the biases and errors that people fall into when they are reasoning about probabilities turn out to be side effects of this process of approximation. Now, of course, a lot of the probabilistic reasoning tasks that people are given in laboratory experiments do have simple analytic solutions, so no approximation methods are required. But, of course, the probabilistic machinery of the brain is not adapted to simple verbal or numerical probability problems, so it is likely that the approximation algorithms used to solve complex problems will be used here too.

As we saw in chapter 11 (see also, e.g., Dasgupta, Schulz, & Gershman, 2017; Sanborn & Chater, 2016), one approach assumes that people make probability estimates through sampling, possibly with some process of correction for sample size (Zhu, Sanborn, & Chater, 2020). Then, as we saw, a variety of biases will occur, potentially due to biases concerning where sampling begins, the fact that samples are autocorrelated rather than independent, and so on. And some of these biases map neatly onto observed probabilistic reasoning "biases" including conservatism, anchoring, and representativeness (for a survey, see Chater et al., 2020). So it is possible that understanding the mind as approximate Bayesian inference may provide an elegant and unifying explanation of apparently disparate quirks in human probabilistic reasoning. Thus, the frailties of human probabilistic reasoning may fit neatly within a Bayesian framework rather than being a counterexample to it.

This viewpoint is part of a broader perspective on Bayesian cognitive science: that empirical predictions need not directly be drawn from pure Bayesian analysis of the probabilistic problems faced by the brain, but rather through understanding how these problems are approximated, given the computational machinery of the brain. This type of resource-rationality perspective can be developed in a number of ways (Griffiths, Lieder, & Goodman,

2015; Lieder & Griffiths, 2020). Sticking with sampling models, one approach is to assume that, if computationally expensive, samples will be rationally biased to where they are likely to be most valuable (utility-based sampling; see chapter 13). This will lead to an overrepresentation of rare but significant events, providing an explanation for the excessive influence of such events (e.g., plane crashes, terrorist attacks) on probability judgments in a wide range of contexts (Lieder et al. 2018a). Relatedly, resource-bounded rational models can provide a basis for rank-based models of sampling (Stewart, Chater, & Brown, 2006), on the assumption that psychoeconomic scales are unavoidably noisy (Bhui & Gershman, 2018).

S: But doesn't it still seem just a bit weird that Bayesian cognitive scientists are postulating that the brain solves really hard probability problems easily (involving vision, language, categorization, and so on) and yet falls down completely on trivial probability problems?

B: Really, it shouldn't sound weird at all! Consider vision. Suppose that it turns out that human, and perhaps more broadly animal, vision, can be viewed as involving (suitably approximated) Bayesian inference over images, represented in a hierarchy of levels of representation. An agent with such a visual system would not thereby be expected to be able to engage in general probabilistic reasoning, over verbally stated problems—indeed, nonhuman species would inevitably lack this ability.

S: Well that just brings me to another problem. Probability theory is all about quantifying uncertainty in numerical form—and making calculations that link different numerical uncertainties with each other. But most of the time, people don't think about numerical probability at all*—instead, we mostly reason about how the world works, and what will, might, or certainly won't happen in purely qualitative terms. The brain is like a lawyer arguing that the defendant is, or is not, guilty—and our conclusion is determined by the qualitative structure of the arguments one way or the other.*[1] *Lawyers don't talk about probabilities of guilt; and jurors and judges wouldn't like it if they did. That's not the way the mind works. Of course, in the last few centuries, human have developed mathematics for dealing with numerical probability—but the very recency of probability theory, and the trouble each new generation of students has figuring out how to use it, surely testify to how cognitively unnatural it is.*

B: I couldn't agree more. But it is a mistake to think that the Bayesian approach to cognition, or indeed inference in general, is about probabilities represented as numbers. It's really much better viewed as a set of tools for thinking about the structure of reasoning, and specifically reasoning about uncertainty.

Think about graphical models, which encode dependence relations between propositions (and, for causal graphical models, causal links that have implications about counterfactuals; see chapter 4). The structure of models is really where the action is: it is this structure that underpins qualitative patterns of reasoning, showing which propositions depend on which others, what causes what, and other questions.

Common-sense reasoning is incredibly rich and flexible, and dependent on rich background understanding—which, we've noted, is naturally modeled within a Bayesian framework. Thus, seeing a window smashed, and footprints in the flower bed below it, the reasoner may use rich causal and cultural knowledge to infer that a burglary has taken place. Given

1. In fact, a good case can be made that verbal arguments between people may even underpin reasoning and argumentation within an individual mind (Mercier & Sperber, 2011).

the additional knowledge that a camera crew is nearby, the most likely hypothesis readily switches to the assumption that no real crime has been committed, and that instead a murder mystery is being filmed. Still further information, such as that the camera crew is carefully hidden and using long-range lenses, might flip the reasoner's most probable interpretation back again: the camera crew are police who have been tipped off about the burglary and want to capture it on film. Such reasoning involves drawing on extremely rich knowledge of the world; indeed, the knowledge that may be engaged seems entirely open-ended.

The engineering project of attempting to capture such inferences, although initially framed in terms of logical reasoning, is now most commonly addressed in AI in probabilistic terms. Graphical models have been particularly fruitful ways of representing people's knowledge of the world (Pearl, 1988); and, in AI, there have been important steps toward understanding how people can reason about causal relationships using probabilistic methods (Pearl, 2000). The Bayesian perspective is a powerful framework for such analysis—so, in this example, a boost to the probability of one hypothesis ("fictional crime reconstruction") reduces the probability assigned to an alternative hypothesis ("genuine crime being committed"), a pattern of inference known variously as "causal discounting" (in psychology; Einhorn & Hogarth, 1986; Kelley, 1987) or "explaining away" (in AI; Pearl, 1988), which we have encountered in various places in this book (particularly in chapter 4). Note, moreover, that patterns of common-sense reasoning, both in adult and infant cognition, is often viewed as analogous to and perhaps continuous with, scientific reasoning—where again the Bayesian approach is now a dominant mode of explanation (Horwich, 1982; Howson & Urbach, 1993; Bovens & Hartmann, 2004; though see Gelman & Shalizi, 2013, for an important dissenting perspective). Moreover, many laboratory studies of human reasoning, once interpreted as indicating that people violate logical patterns of reasoning, can be reinterpreted as indicating that patterns of human reasoning neatly accord with Bayesian principles (e.g., Hahn & Oaksford, 2007; Oaksford & Chater, 1994, 1998b, 2007, 2020), given appropriate assumptions about the background knowledge that people possess about the domain about which they are reasoning.

S: So let me take a different tack. Even from the earliest developments in probability theory, it was assumed that the calculus of probability should help to clarify the nature of everyday human reasoning about uncertain events. Indeed, Bernoulli's book, The Art of Conjecture, explicitly aims both to characterize and to improve human reasoning (Bernoulli, 2006/1713). And after all, the subjective interpretation of probability as "degree of belief," which is the starting point of the Bayesian approach, surely should have some relationship with the intuitive notion of belief—perhaps the most fundamental notion in folk psychology.

So this raises two issues. First, to what extent is the Bayesian approach them attached to the folk psychological conception of the mind is driven by beliefs and desires? And surely, since Nisbett and Wilson (1977), and through the rich tradition in social and cognitive psychology including Johansson, Hall, Sikstrom, and Olsson (2005), it has become apparent that our explanations of our thoughts and behavior in terms of beliefs and desires are at best highly suspect and at worst complete fiction (Chater, 2018; Churchland, 1981). But if it turns out that this folk psychological perspective is unworkable, where does this leave the Bayesian approach?

Second, if we do want to maintain a direct connection between the concept of belief in the Bayesian approach to cognition and the folk psychological notion of belief, is it really possible to do so? That is, can one credibly claim that probability theory is rich enough to describe patterns of everyday inference? Or the entire machinery of logic and probability? Surely the philosophy of the later-period Wittgenstein (1953), the apparently unruly nature of real common-sense and scientific inference (Feyerabend, 1993; Lakatos, 1970), the frame problem in AI (Ford & Pylyshyn, 1996; McCarthy & Hayes, 1969), and many related lines of thought should undermine the idea that human reasoning can be reconstructed in a formal framework of any kind.

B: These are deep issues, and the question of the relationship between degree of belief and the everyday notion of belief is indeed of fundamental importance. The mathematical and computational resources of the Bayesian approach are, of course, still available even if one rejects any such relationship. So, for example, Bayesian models in computer vision and computational linguistics make reference to low-level visual or linguistic features that have absolutely no counterparts in "folk psychology," and hence are not the kind of thing about which we are typically assumed to have beliefs in the full-blown sense. The "rules of belief" propagation in a Bayesian network can be applied, after all, regardless of the interpretation that we assigns to the nodes of the network, if any. This means that the notion of belief for the application of formal Bayesian methods is very minimal. So if the folk psychological enterprise were to turn out to be no more scientifically respectable than folk physics or folk biology, as some have argued (Chater, 2018; Churchland, 1981; Rosenberg, 2019), Bayesian cognitive science would in no way be threatened.

Still, though, we see that the potential connection with common-sense beliefs, typically expressed in natural language, is one of the most important selling points for the Bayesian approach: forging close links between cognitive modelling and everyday notions of belief, desire, reasoning, and knowledge. In many areas of psychology, the dominant style of explanation is not computational, but folk-psychological: social psychologists explain people's behavior in terms of consonance or dissonance between their beliefs or attitudes; and clinical psychologists aim to explore the abnormal belief systems formed by people with mental disorders. But the same applies within the cognitive science of high-level thought, which is similarly steeped in talk of categories, inferences, and knowledge, which are notions of transparently folk-psychological origin. Equally, developmental psychologists characterize many aspects of cognitive development as involving the acquisition of increasingly rich beliefs and concepts, perhaps organized into theories of the world. This has allowed Bayesian models to capture a wide range of phenomena in cognitive, social, and developmental psychology, which are usually conceived of in folk psychological terms.

This is, moreover, an important virtue of the Bayesian approach for another reason: that models of human intelligence need to capture the human ability to fluently generate and understand language and to relate this understanding to our internal models of the world. If our internal models happen to be interpretable in intuitively meaningful terms (e.g., as capturing claims about what causes what), then these internal models can, in principle, be reported by the cognitive system and modified in response to linguistic input from others.

The contrast with neural network models and earlier and more primitive associative models of learning is interesting. A series of now rather little-known studies (reviewed in Brewer, 1974) showed that associative learning turns out to be surprisingly easily manipulated by verbal information. A typical study might train people to expect an electric shock in response to a cue and measure the strength of this association implicitly by monitoring skin conductance after the stimulus was presented. The experimenter would then explicitly tell people that the contingency would no longer hold (e.g., by saying something like "I'm turning the shock machine off now"). People's skin conductance responses declined sharply in proportion to how much the participants reported that they believed the experimenter's reassurance. More recent work similarly suggests that associations are formed and modified by general processes of knowledge revision rather than by a distinct, mechanistic associative mechanism (Mitchell, De Houwer, & Lovibond, 2009). This makes sense if people are constructing and training graphical models, with links that they can abruptly disable when they hear that a previously active causal link has been severed. But it hard to understand how this can work if, for example, knowledge of causal connections is distributed in an impenetrable way throughout the weights in a vast deep neural network.

A probabilistic "language of thought" perspective (chapters 16–18; see also Piantadosi & Jacobs, 2016) takes this approach a step further. If the kinds of symbolic knowledge that are expressed in natural language can be captured in a rich, internal, and symbolic language, then the interface between that knowledge and natural language becomes potentially more transparent. But there need be no necessary commitment that the "sentences" in such a language refer to the objects and categories of natural language (see, e.g., Dennett, 1978). More broadly, the less direct the connection between internal representations and natural language is, the more work will be required in building a computational theory of communication (see, e.g., Christiansen & Chater, 2022; Fodor, 1975; Goodman & Frank, 2016, for contrasting perspectives).

S: But even suppose that this is right—what about my second concern: that probability theory just isn't up to the job of capturing the complexity of everyday inference.

B: Well, probability theory is only part of the story—although we believe it to be an important part. What probability theory does is to help map from a current knowledge state to an appropriate future knowledge date in the light of new data that the system has received. But the challenge of characterizing this knowledge (i.e., describing the content of the naive theories of the natural and social world that guide our behavior and representing it in some formal machinery) remains an enormous challenge.

S: That may turn out to be a considerable understatement! The project of developing classical symbolic AI foundered on this very problem, after all. Consider the notorious frame problem (McCarthy & Hayes, 1969; Ford & Pylyshyn, 1996), which is roughly that whenever an agent updates a belief or takes an action, it cannot be sure which of its other beliefs it needs to update. In short, the problem is that each individual belief can have implications throughout an entire network of beliefs, and these implications can be very far-reaching, but also very unexpected. So I might, for example, operate with a particular intuitive model of the functioning of, say, the appliances in my house, but a single unexpected observation (e.g., seeing the streetlight outside my house snap "off" without warning) might lead me to suppose that there may be a power outage, and hence that my freezer or television will cease to function.

So I may have a model of the functioning of my appliances that includes the possibility of power failure and its consequences (the contents of my freezer begin to thaw, and my television screen goes blank). Moreover, one might suppose that the inference from the streetlight going out, to there being a power failure can itself be captured in probabilistic terms. But the problem is not that each extra complication cannot be handled piecemeal, but rather that the open-ended character of world knowledge requires that the process of extending and elaborating the probabilistic model seems to have no obvious end. Noticing the distant glow of light in another room, I might conclude that the power supply to the house must be intact after all; however, recalling that one of my lamps is powered by a battery, I may consider whether to retract this inference; noticing a repair van pulling up next to the streetlight might lead me to suspect that the bulb has blown, and hence that there is no reason to suppose that the power supply is not intact; but noticing that the man emerging from the van lives next door may undercut this conclusion because I suspect that he has not arrived to repair the bulb but has simply returned home. To be sure, each of these inferential steps can be reconstructed, perhaps quite neatly, in a Bayesian framework; but surely the implication is that our general intuitive knowledge of the world cannot be reconstructed as a set of stable and well-defined probabilistic models, such as those described in this book.

Indeed, this seems hopelessly implausible. After all, Leonard Savage, one of the key pioneers of the foundations of Bayesian statistics, explicitly distinguished between "small worlds," which are simple enough for a person to have consistent degrees of belief (and hence a probabilistic model capturing these degrees of belief can be constructed), and "large worlds," where we will inevitably be inconsistent, and no probabilistic model is possible (Savage, 1972). Savage thought that the Bayesian approach was only appropriate for small worlds; yet Bayesian cognitive science seems blithely, and implausibly, to assume that the same approach applies quite generally (Binmore, 2008).

At best, the Bayesian approach seems to be left attempting to make sense of one gigantically complex and impossibly unwieldy probabilistic model of all of world knowledge, which captures the entire "web of belief" (Quine, 1960; Quine & Ullian, 1978). At worst, we might see the probabilistic approach to capturing world knowledge as sinking into the same quicksand that devoured symbolic approaches to knowledge representation in AI in the 1970s and 80s.

B: These sorts of problems certainly raise serious challenges for cognitive science. But they certainly don't weigh against the Bayesian approach specifically. After all, according to just about any theory, the brain is somehow able to construct a rich and highly flexible model of the natural and social worlds. But the natural and social worlds are of course, of enormous complexity, and they do not come divided into neatly presealed units, each governed by its own set of entirely independent principles. The formal and conceptual tools required to capture the structure of our environment will be many and varied; and this must surely be embodied, to some extent at least, in the complexity of our mental representations of this information. And the inferences that we draw will be largely dependent on the contents of these mental representations, and only secondarily on general inference principles such as the laws of probability. One of the central challenges in AI and cognitive science is to understand the nature of such knowledge.

Indeed, as many influential theoretical models of cognitive development suppose, the child (and indeed the adult) should be thought of as engaging in an activity analogous to science (e.g., Gopnik, Meltzoff, & Kuhl, 1999). Then the contribution of the Bayesian approach to understanding cognition is analogous to the contribution of the Bayesian movement in the philosophy of science—that is, helping to explain general patterns of reasoning. Only indirectly can it assist with the problem of characterizing the content of the intuitive "science" that underlies cognition.

The problem of understanding human knowledge is, therefore, not specifically a challenge to the Bayesian approach, but rather a challenge for the cognitive sciences in general. Probability theory, therefore, no more promises to solve problems of knowledge representation than it promises to solve the problems of the natural and social sciences. What the Bayesian approach does give us, however, is a framework for helping to draw certain types of inferences from such knowledge—and perhaps some tools for helping to represent limited aspects of this knowledge as well. But the vast bulk of the problem of understanding human knowledge, and the inferences that can be drawn from it, concern the nature of their knowledge itself.

These disclaimers aside, it may be that Bayesian methods may be helpful in characterizing important aspects of human knowledge. We have particularly stressed the importance of defining probabilistic knowledge and reasoning over rich, structured representations, including graphs, grammars, logics, and programming languages. Related developments in the probabilistic representation of knowledge may also prove to be important. For example, Pearl (2000) and Spirtes, Glymour, and Schienes (1993) have made important steps toward the understanding of reasoning about causality, and in particular, toward understanding the difference between an observation and an intervention, working within a probabilistic framework, and some aspects of this work may be directly relevant to cognitive science (see also Gerstenberg, Goodman, Lagnado, & Tenenbaum, 2021; Glymour, 2001; Sloman, 2005).

Bayesian cognitive science, moreover, provides deep general ideas about how knowledge can be acquired, represented, and applied, regardless of topic. Ideas such as graphical models, belief propagation, and MCMC may prove to be fruitful metaphors for understanding cognition. The Bayesian framework may also be essential to capture the general inferential principles by which different semiautonomous domains of knowledge can be linked together into a functioning reasoning system.

S: In a well-known article, Gould and Lewontin (1979) point out one of the potential pitfalls of any style of explanation based on optimality. We may inadvertently begin to follow Voltaire's Professor Pangloss to reframe the domain of study so that it becomes the "best of all possible worlds." Gould and Lewontin (1979) were concerned about possible excesses in adaptationist explanation in biology—that the biologist may start to see all and every feature of a living organism as exquisitely adapted to some environmental challenge. They argued that, by contrast, many aspects of the structure of an organism are not adaptations at all—but arise as side effects of structures that may be adaptations. Doesn't the Bayesian cognitive scientist falls into precisely this trap? Almost any aspect of thought and behavior can, with sufficient ingenuity, be interpreted as arising from an optimal Bayesian calculation if we can freely vary assumptions about the goals and knowledge of the agent. As with adaptationist explanation in biology, any apparent counterexample (i.e., deviation from optimal design or behavior) is surely all too easy to explain away—leaving the Bayesian proposal unfalsifiable and empirically empty.

B: This is a real concern—but no more for Bayesian explanation than for any other theory (Dennett, 1983). Any style of theorizing must be judged by the standard criteria of scientific explanation—not merely to fit with the data, but, for example, to provide, as far as possible, a simple explanation with accurate predictions. And any scientific hypothesis can be made to fit the data by suitable ad hoc adjustments. Thus, as Putnam (1974) notes, Newton's laws could be saved, even if we observed square planetary orbits, if suitable additional forces, of unknown origin, are assumed to be operating in addition to gravity (see Stanford, 2017). So the Bayesian cognitive scientist is in a no better, and no worse, position than advocates of any other approach, with respect to being able to "save the theory," come what may; and, as with other approaches, such attempts will ultimately be fruitless because the theory will become increasingly complex and less predictively successful.

So the question is a pragmatic one: How useful is the Bayesian approach for understanding cognition in comparison with (and alongside) other approaches? Many of the chapters in this book outline encouraging developments; and the literature in Bayesian cognitive science has been expanding rapidly, suggesting that the approach is often productive. My guess is that this trend is likely to continue, and attempting to understand cognition without Bayesian ideas in the theoretical toolbox is somewhat perverse. But let's see what happens over the coming decades.

S: Well, indeed, the ultimate usefulness of any approach can be determined only by trying it out—and perhaps we can agree that the jury is still out on this one, however promising some of the progress so far may be. But let me end with a final worry, which makes me doubtful that the Bayesian approach can really be on the right track: and this comes from thinking about neuroscience.

Cognition is generated by the brain. Yet this entire book scarcely mentions a single fact of neuroscience. The approach is relentlessly top-down. Try to analyze the nature of the problem that the brain is facing (Marr's computational level) and how it might be solved in principle; perhaps consider algorithms that might approximate the solution (the algorithmic level); and only then worry about how such calculations might be implemented in neural hardware (the implementation level). But let's face it—we never got to the implementation level, did we?

And isn't this a crucial step? Aren't we ultimately reverse-engineering the brain, *not merely the mind. And surely we can't reverse-engineer the brain without paying a lot of attention to the computational machinery of the brain. Here, neural networks seem more promising because their basic mode of operation (computing with densely interconnected networks of numerical processing units and learning through adjusting the strengths of connections between those units) seems a lot closer than more abstract Bayesian calculations to how the brain operates.*

B: This is more a question of research strategy than substantive debate. Building a link between high-level cognition and low-level neuroscience is inevitably going to be incredibly difficult; and it likely that there will be strong constraints between levels of analysis. So, for example, it has long been pointed out that purely sequential symbolic computations seem incompatible with the slow, parallel hardware of the brain (Feldman & Ballard, 1982).

Trying to build in all constraints at once is terrific, if it is doable. In some areas, such as aspects of perception and motor control, and perhaps a basic process of reinforcement learning (if these can usefully be distinguished from higher-level cognition; Chater, 2009; Mitchell, De Houwer, & Lovibond, 2009), a multilevel analysis seems at least possible. Indeed, this is the project of computational neuroscience, much of which takes an explicitly Bayesian perspective (Doya, Ishii, Pouget, & Rao, 2007; Knill & Pouget, 2004).

When we consider areas of higher-level cognition, as is the focus of this book, this goal seems to be out of reach at this time. But to get there, we should surely be working piecemeal at each level of analysis, and linking levels where we can, to try to piece together a complete story. As it happens, progress has been easiest at Marr's computational and algorithmic levels so far—connections with neural implementation are not yet well sufficiently developed to provide neural constraints on reasoning about the physical or social worlds; we don't yet have a clear sense of how to neurally implement complex symbolic representations; and many other challenges remain. For now, let's make progress where we can and keep channels of communication across levels as open as possible.

To end on an optimistic note, it is worth stressing that the challenge of meeting constraints from multiple levels of analysis, where we can figure out how to do it, is actually likely to speed rather than slow our progress (Christiansen & Chater, 2016b). If we are attempting to solve a puzzle of any kind, more clues are better than fewer—more clues will guide us toward good approaches and eliminate bad ones. So when it comes to reverse-engineering the brain/mind, we should use constraints wherever we can find them. Bayesian cognitive science, when applied to high-level cognition, is not always able to account for constraints from human biology because it is not clear how neural implementation connects with quite abstract computation problems and algorithms. But the more we can join together insights from different levels of analysis—and from diverse fields, from AI to cognitive psychology to linguistics, philosophy, and neuroscience—the more rapidly we will progress. Bayesian cognitive science is, after all *cognitive science*. And it is this interdisciplinary approach to providing an integrated account of the brain as a computational machine that gives us our best hope of understanding the astonishing phenomenon of human intelligence.

Conclusion

We began this book with a big question: How do our minds get so much from so little? The intervening pages have sketched an answer using Bayesian inference as a tool for reverse-engineering the mind. Thinking about the ideal solution to the computational problems that the mind faces provides a framework for exploring people's inductive biases, expressed in the form of prior distributions. The challenge then becomes one of finding the right language for describing these prior distributions (drawing on structured representations such as graphical models, logic, grammars, and programs), understanding how they can in turn be learned and deployed when relevant (via hierarchical Bayes, nonparametric Bayes, and metalearning), and how they can be used effectively and efficiently by minds and brains (through sampling and optimization). The chapters that you have read are the result of more than two decades pursuing these threads, but they all join back to that first idea of Bayes as a paradigm for inductive inference.

Consistent with the reverse-engineering approach outlined in chapter 1, the book has worked through multiple levels of analysis from the top down. At the level of computational theory, we offered Bayesian inference as an account of the ideal solution to the problems faced by human minds. At the algorithmic level, we explored Monte Carlo and variational inference as mechanisms for approximating Bayes and considered how they might be best deployed as a bridge back to the computational level. At the implementational level, we highlighted connections to neural networks and ways in which different algorithms might be expressed in neural circuits. And across all levels, we considered what kinds of representations might be needed to effectively and efficiently express the hypothesized operations, with all the adaptiveness and robustness that human intelligence displays. The result is a fully integrated picture of how our minds could work—as machines for approximating Bayesian inferences and decisions over hierarchies of flexibly structured representations, carried out via neurally implementable, resource-rational algorithms.

This picture of human minds is equally useful as a lens for understanding artificial intelligence (AI) systems. While our focus has been on human cognition, our stated goal is for the ideas that we present here to be equally useful to practitioners of AI. By contrast to the approach we have taken in this book, much of contemporary AI has adopted a more bottom-up approach, in which modeling begins by specifying a neural network architecture that is differentiable, together with sources of training data and loss functions, and then intelligence is left to emerge via gradient-based optimization of the parameters of the architecture, given the data and loss function. This approach has been remarkably successful in machine learning, but it also has exposed increasingly wide gaps between AI systems and human minds. We would suggest that those gaps are in exactly the capacities that are highlighted by studying those human minds from the top down: inductive biases, representational modularity, hierarchy and flexibility, and rationally adaptive ways of approximating rational Bayesian inferences and decisions.

Our formula for reverse-engineering the mind—Bayesian inference over structured representations approximated sensibly via neurally plausible algorithms—also provides a way to close those gaps. Many of today's AI systems that are built by training large artificial neural networks, loosely inspired by the brain and without explicit reference to either structured representations or probabilistic inference, can be shown to be implicitly doing something like this—forming representations that can be decoded to reveal analogs of posterior distributions over structured representations that have been learned to make sense of their data.

In some cases, given enough data and the appropriate architectures, good-enough approximations to structured probabilistic models and inference algorithms might be learned without explicitly building in these components. The more this program is successful, the more we expect that the concepts and tools in this book will be useful in understanding how those large-scale neural network systems actually work—so they do not just have to be treated as black boxes. And to the extent that this approach to AI continues to fall short of what we expect from intelligence, if AI designers want to create systems that learn and reason more like people—from less data, in a broader range of settings, more robustly and coherently, and with a more robust and coherent sense of their own uncertainty and ability to use their own computational resources rationally—we also expect that they will find value in the tools and ideas presented here, as an explicit guide for engineering, not just reverse-engineering intelligence. Intelligent designers of intelligent machines owe it to themselves, to their creations, and to all of us to be as thoughtful as possible about how they might build into those systems the right machinery for thinking—the right inductive biases, representations, and computations.

In the preceding chapters, we have laid out some of our ideas for how to achieve this synthesis of AI and cognitive science. By better understanding human inductive biases, we are better able to construct the targets for our AI systems. By exploring methods such as hierarchical Bayes and metalearning, we can see how those targets might be reached partly through experience. By discovering how people efficiently approximate complex problems of probabilistic reasoning, we obtain clues that can be transferred to machines. And by exploring the potential of probabilistic inference over representations like graphs, grammars, logic, and programs, we have the components for describing human inductive biases clearly, and in a form that we can use to create AI systems that come closer to instantiating those inductive biases.

In the second part of the book, we also laid out some of the most important ongoing and future directions for this research program. Perhaps the most significant theme is reunifying AI and cognitive science. There are still going to be significant differences between the fields—human intelligence is shaped by a specific set of computational constraints that are the product of our biological and cultural evolution, and not the same as those that shape AI, in a continually changing economic, technological, and cultural landscape that AI itself is reshaping. Yet there remains a significant amount that these fields can continue to learn from one another. Reverse-engineering can provide insights that can inform engineering, and vice versa, so long as the scientists and engineers involved speak a common language. We hope that this book provides a dictionary and a grammar—or at least a phrase book—for such a language.

Acknowledgments

This book is the result of more than a decade of work by a large group of people, and we want to take a moment to acknowledge their efforts. The book would not exist without them.

That descriptor—as a necessary cause—is particularly true of Roger Shepard, to whom the book is dedicated. Roger was not just necessary, but in some sense the "prime mover" of the entire causal system on which the book depends. There is a clear causal chain from his groundbreaking work on generalization through rational analysis to everything that appears in the book. He also moved us through conversations, mentorship, and other interactions over the years. We continue to draw inspiration from his work and appreciate the time that he took to share it with us.

A more proximal cause is the summer school on probabilistic models of cognition at the Institute for Pure and Applied Mathematics at the University of California, Los Angeles, in 2007 and 2011. The core faculty for those summer schools—the three of us, plus Charles Kemp, Noah Goodman, and Alan Yuille—started talking about putting the materials together in a book at that point, with more detailed plans emerging at a subsequent meeting on Bayesian models of cognitive development at the Banff International Research Station. We are grateful to both of these institutions for supporting these meetings. Consistent with this history, various chapters have grown out of other pieces of writing and reproduce some of that material. We have noted some of these places in the text—chapter 1 builds on (or grows from) Tenenbaum, Kemp, Griffiths, and Goodman (2011); and chapters 5, 9, 11, and 18 draw on material from Griffiths et al. (2011a), Lucas, Griffiths, Williams, and Kalish (2015), Griffiths et al. (2012b), and Goodman, Tenenbaum, and Gerstenberg (2014), respectively. In addition, chapters 3–6 are based on lecture notes that were used in various classes and subsequently were developed into published tutorials. As a consequence, these chapters have overlapping content in places with Griffiths and Yuille (2006) and Griffiths et al. (2008).

Bob Prior at the MIT Press has been a constant champion of this book and has helped us feel bad about not finishing it every time we have seen him for the last decade. Maya Malaviya and Logan Nelson provided a variety of editorial assistance, from disentangling bibTeX bugs to helping with figures.

We would also like to thank the funding agencies that have supported this work. Science is just as dependent on program managers who can appreciate a vision as it is on the scientists who propose it, and we appreciate the funding agencies that have been willing to invest in Bayesian models of cognition over the years. In particular, aspects of the work presented here were supported by the National Science Foundation through many individual grants, as well as the Science and Technology Center (STC) Grant to the Center for Brains, Minds, and Machines; the Air Force Office of Scientific Research; the Office of Naval Research; and the Army Research Office, through individual grants as well as a number of Multidisciplinary University Research Initiative (MURI) projects; the Defense Advanced Research Projects Agency (DARPA) via the CALO, BOLT, L2M, PPAML, MCS, COLTRANE, and NGS2 programs, among others; the Intelligence Advanced Research Projects Agency (IARPA); and private foundations such as the John Templeton Foundation, the Templeton World Charity Foundation, Siegel Family Endowment, and the NOMIS Foundation. Tom Griffiths was partially supported by the Guggenheim Fellowship, and Josh Tenenbaum was partially supported by the MacArthur Fellowship while working on this book. Nick Chater was supported by the ESRC Network for Integrated Behavioural Science (grant number ES/P008976/1), the Leverhulme Trust (grant number RP2012-V- 022). The support of the Economic and Social Research Council (ESRC), via the Rebuilding Macroeconomics Network (grant number ES/R00787X/1), is gratefully acknowledged.

Many friends (a few of them also partners) have encouraged, critiqued, or contributed to the efforts represented in this book through joint teaching, tutorials, and workshops, or by organizing and participating in those activities, or just innumerable casual conversations. In addition to those who coauthored chapters, we would like to thank especially Mira Bernstein, Susan Carey, Louie Fooks, Alison Gopnik, Keith Holyoak, John Kruschke, Tania Lombrozo, Kevin Murphy, Mike Oaksford, Rebecca Saxe, Laura Schulz, Rich Shiffrin, Elizabeth Spelke, and Fei Xu. Our children—Orli Griffiths and Anica Griffiths, Maya Fooks and Caitlin Fooks, and Abi Tenenbaum—have made their mark as well, whether through contributing pilot data or philosophical insights, finding confounds in experimental designs or arguments, or pointing out connections that we had never thought to see. We know we're biased, but we are quite certain that their everyday inductive leaps are the best.

Our families deserve additional thanks for tolerating us being distracted thinking about mathematics at dinner, or spending late nights or weekends plugging away at chapters. We realize it doesn't make up for it, but we hope they know how much we appreciate their willingness to let us spend time frolicking in abstract fields.

Finally, the current form of the book has provided a chance to work with our academic families—our current and former students, postdocs, and collaborators—on many of the chapters. Every single one of these people is another necessary cause. None of the work we discussed would have been possible without this wonderful, talented crew of scientists. In addition to those who are represented as coauthors, we received feedback on the chapters of this book or the material therein from a vast number of former students and collaborators. We thank Josh Abbott, Samy Abdel-Ghaffar, Joshua Aduol, Mayank Agrawal, Kelsey Allen, Alex Aronovich, Somya Arora, Zoe Ashwood, Xuechunzi Bai, Halely Balaban, Ruairidh Battleday, Mario Belledonne, Gianluca Bencomo, Natalia Bilenko, Damon Binder, Colton Bishop, Andrew Bolton, David Bourgin, Matt Bowers, Tim Brady, Tyler

Brooke-Wilson, Daphna Buchsbaum, David Burkett, Branson Byers, Alex Carstensen, Kartik Chandra, Colby Chang, Julia Chang, Michael Chang, Alicia Chen, Howard Chen, Sihan Chen, Tony Chen, Sam Cheyette, Ashley Chung, Thomas Clark, Cedric Colas, James Coleman, Matthew Coleman, Katie Collins, Carlos Correa, Sholei Croom, Maddie Cusimano, Marco Cusumano-Towner, Emily Dale, Thurston Dang, Wendy de Heer, Eyal Dechter, Bhishma Dedhia, Kaiyuan Deng, Arjun Devraj, Jack Dewey, Sarah Dillender, Rachit Dubey, Daniel Duckworth, Maria Eckstein, Bernhard Egger, Kevin Ellis, Francesco Fabbri, Naomi Feldman, Sammy Floyd, Mike Frank, Dan Friedman, Jaime Fisac, Yoni Friedman, Vael Gates, Tao Gao, Jon Gauthier, Sam Gershman, Brett Goldstein, Gabriel Grand, Ekin Gurgen, Ulrike Hahn, Allison Hamilos, Mathew Hardy, Addele Hargenrader, Sev Harootonian, Josh Hartshorne, Sylvia Herbert, Luke Hewitt, Michelle Ho, Matthias Hofer, Chris Holdgraf, Dae Houlihan, Konstantin Howard, Haimin Hu, Michael Hu, Peter Hu, Allison Huang, Andrew Hutchinson, Kai-Chieh Hsu, Shu-Yuan Hsueh, Frank Jakel, Rachel Jansen, Andy Jeon, Kyle Jennings, Aditi Jha, Laila Johnston, Jiu Jin, Daniel Jubas, Skylar Jung, Gili Karni, Yarden Katz, Hope Kean, Max Kleiman-Weiner, Konrad Koerding, Eliza Kosoy, Paul Krueger, Tevye Krynski, Nastasia Klevak, Sreejan Kumar, Alexander La Cour, Brenden Lake, Sayeri Lala, Michelle Lamar, Tuan-Anh Le, Brian Leahy, Rengye Lee, Matthew Leung, Alex Lew, Casey Lewry, Sam Liang, Theresa Lin, Emily Liquin, Grace Liu, Colton Loftus, Joao Loula, Qihong Lu, Paul Stefan Lunis, Jean Luo, Andrew Mains, Shayna Maleson, Rivka Mandelbaum, Vikash Mansinghka, Ioana Marinescu, Jay Martin, Raja Marjieh, George Matheos, Ralf Mayrhofer, Noah McGuinness, John McCoy, Tom McCoy, Katie McLaughlin, Stephan Meylan, Smitha Milli, Tracey Mills, Joshua Moller-Mara, Ken Nakamura, Rebecca Neumann, Sean-Wyn Ng, Vivien Nguyen, Anwar Nunez-Elizalde, Max Nye, Tim O'Donnell, Kerem Oktar, Danny Oppenheimer, Begum Ortaoglu, Sinan Ozbay, Etiosa Omeike, M Pacer, Vivian Paulun, Josh Peterson, Hien Pham, Thomas Pouncy, Peng Qian, Frederick Qiu, Shiyun Qiu, Anna Rafferty, Setayesh Radkani, Jacob Raghoobar, Sunayan Rane, Natalie Reptak, Eno Reyes, Dan Roy, Maya Rozenshteyn, Evan Russek, Feras Saad, Basil Saeed, Russ Salakhutdinov, Lauren Schmidt, Suganda Sharma, Anoopkumar Sonar, Sophia Sanborn, Sara Schwartz, Sarah Schwettmann, Simon Segert, Pat Shafto, Benjamin Shapiro, Hye-Young Shin, Max Siegel, Aditi Singh, Jake Snell, Olga Solodova, Felix Sosa, Mark Steyvers, Sean Stromsten, Andreas Stuhlmuller, Erik Sudderth, Ellen Su, Ted Sumers, Skylar Sutherland, M. H. Tessler, Prerit Terway, Kevin Tham, Lucas Tian, Pedro Tsividis, Shikhar Tuli, Yoko Urano, Camila Vasquez, Ed Vul, Caren Walker, Greg Weaving, Alex Wettig, Andrew Whalen, Joseph Williams, Samarie Wilson, Lionel Wong, Jiajun Wu, Shunyu Yao, Soohwang Yeem, Shelley Xia, Jing Xu, Ilker Yildirim, Lance Ying, Matei Zaharia, Byron Zhang, Ced Zhang, Olivia Zhang, Charles Zhao, Tan Zhi-Xuan, Adam Ziff, and Yoonseo Zoh.

References

Abadi, M., Agarwal, A., Barham, P., et al. (2015). *TensorFlow: Large-scale machine learning on heterogeneous systems.* (Software available from tensorflow.org).

Abbott, J., Hamrick, J., & Griffiths, T. (2013). Approximating Bayesian inference with a sparse distributed memory system. In *Proceedings of the 35th Annual Meeting of the Cognitive Science Society.*

Abbott, J. T., & Griffiths, T. L. (2011). Exploring the influence of particle filter parameters on order effects in causal learning. In L. Carl- son (Ed.), *Proceedings of the 33rd Annual Meeting of the Cognitive Science Society.*

Abel, D., Dabney, W., Harutyunyan, A., Ho, M. K., Littman, M., Precup, D., & Singh, S. (2021). On the expressivity of Markov reward. In *Advances in Neural Information Processing Systems 34* (pp. 7799–7812).

Abelson, H., Sussman, G. J., & Sussman, J. (1996). *Structure and interpretation of computer programs.* MIT Press.

Abend, O., Kwiatkowski, T., Smith, N., Goldwater, S., & Steedman, M. (2017). Bootstrapping language acquisition. *Cognition*, *164*, 116–143.

Ackley, D. H., Hinton, G. E., & Sejnowski, T. J. (1985). A learning algorithm for Boltzmann machines. *Cognitive Science*, *9*, 147–169.

Agre, P. E., & Chapman, D. (1987). Pengi: An implementation of a theory of activity. In *Proceedings of the AAAI Conference on Artificial Intelligence* (pp. 286–272).

Ahuja, A., & Sheinberg, D. L. (2019). Behavioral and oculomotor evidence for visual simulation of object movement. *Journal of Vision*, *19*(6), 13.

Alais, D., & Burr, D. (2004). The ventriloquist effect results from near-optimal bimodal integration. *Current Biology*, *14*(3), 257–262.

Alayrac, J.-B., Donahue, J., Luc, P., et al. (2022). Flamingo: A visual language model for few-shot learning. In *Advances in Neural Information Processing Systems 35* (pp. 23716–23736).

Aldous, D. (1985). Exchangeability and related topics. In *École d'été de probabilités de Saint-Flour, XIII—1983* (pp. 1–198). Springer.

Ali, A., Kolter, J. Z., & Tibshirani, R. J. (2019). A continuous-time view of early stopping for least squares regression. In *The 22nd International Conference on Artificial Intelligence and Statistics* (pp. 1370–1378).

Alishahi, A., & Stevenson, S. (2008). A computational model for early argument structure acquisition. *Cognitive Science*, *32*(5), 789–834.

Allais, M. (1953). Le comportement de l'homme rationnel devant le risque: Critique des postulats et axiomes de l'école américaine. *Econometrica*, *21*(4), 503–546.

Allen, K. R., Smith, K. A., & Tenenbaum, J. B. (2020). Rapid trial-and- error learning with simulation supports flexible tool use and physical reasoning. *Proceedings of the National Academy of Sciences*, *117*(47), 29302–29310.

Amalric, M., Wang, L., Pica, P., Figueira, S., Sigman, M., & Dehaene, S. (2017). The language of geometry: Fast comprehension of geometrical primitives and rules in human adults and preschoolers. *PLoS Computational Biology*, *13*(1), e1005273.

Amsterlaw, J., & Wellman, H. (2006). Theories of mind in transition: A microgenetic study of the development of false belief understanding. *Journal of Cognition and Development*, *7*, 139–172.

Anderson, J. R. (1990). *The adaptive character of thought.* Erlbaum.

Anderson, J. R. (1991a). The adaptive nature of human categorization. *Psychological Review*, *98*(3), 409–429.

Anderson, J. R. (1991b). Is human cognition adaptive? *Behavioral and Brain Sciences*, *14*(3), 471–485.

Anderson, J. R., & Milson, R. (1989). Human memory: An adaptive perspective. *Psychological Review*, *96*, 703–719.

Anderson, J. R., & Schooler, L. J. (1991). Reflections of the environment in memory. *Psychological Science*, *2*, 396–408.

Antoniak, C. (1974). Mixtures of Dirichlet processes with applications to Bayesian nonparametric problems. *Annals of Statistics*, *2*, 1152– 1174.

Ashby, F. G., & Alfonso-Reese, L. A. (1995). Categorization as probability density estimation. *Journal of Mathematical Psychology*, *39*, 216–233.

Aspray, W. (2000). *Computing before computers.* Iowa State University Press.

Atick, J. J., & Redlich, A. N. (1992). What does the retina know about natural scenes? *Neural Computation*, *4*(2), 196–210.

Atran, S. (1998). Folk biology and the anthropology of science: Cognitive universals and cultural particulars. *Behavioral and Brain Sciences*, *21*, 547–609.

Attneave, F. (1959). *Applications of information theory to psychology: A summary of basic concepts, methods and results.* Holt-Dryden.

Austerweil, J. L., & Griffiths, T. L. (2011). A rational model of the effects of distributional information on feature learning. *Cognitive Psychology*, *63*(4), 173–209.

Austerweil, J. L., & Griffiths, T. L. (2013). A nonparametric Bayesian framework for constructing flexible feature representations. *Psychological Review*, *120*(4), 817–851.

Austerweil, J. L., Sanborn, S., & Griffiths, T. L. (2019). Learning how to generalize. *Cognitive Science*, *43*(8), e12777.

Babaeizadeh, M., Finn, C., Erhan, D., Campbell, R. H., & Levine, S. (2018). Stochastic variational video prediction. In *International Conference on Learning Represenations.*

Babbage, C. (1864). *Passages from the life of a philosopher.* Longman.

Badham, S. P., Sanborn, A. N., & Maylor, E. A. (2017). Deficits in category learning in older adults: Rule-based versus clustering accounts. *Psychology and Aging*, *32*(5), 473–488.

Baillargeon, R. (1994). How do infants learn about the physical world? *Current Directions in Psychological Science*, *3*(5), 133–140.

Baillargeon, R. (2002). The acquisition of physical knowledge in infancy: A summary in eight lessons. In U. Goswami (Ed.), *Blackwell handbook of childhood cognitive development* (pp. 46–83). Blackwell.

Baillargeon, R. (2004). Infants' physical world. *Current Directions in Psychological Science*, *13*(3), 89–94.

Baker, A. (2022). Simplicity. In E. N. Zalta (Ed.), *Stanford encyclopedia of philosophy* (Summer 2022 ed.). Metaphysics Research Lab, Stanford University.

Baker, C. (1979). Syntactic theory and the projection problem. *Linguistic Inquiry*, *10*(4), 533–581.

Baker, C. L., Jara-Ettinger, J., Saxe, R., & Tenenbaum, J. B. (2017). Rational quantitative attribution of beliefs, desires and percepts in human mentalizing. *Nature Human Behaviour*, *1*, 0064.

Baker, C. L., Saxe, R., & Tenenbaum, J. B. (2009). Action understanding as inverse planning. *Cognition*, *113*(3), 329–349.

Baker, C. L., Tenenbaum, J. B., & Saxe, R. R. (2007). Goal inference as inverse planning. In *Proceedings of the 29th Annual Meeting of the Cognitive Science Society.*

Bannard, C., Lieven, E., & Tomasello, M. (2009). Modeling children's early grammatical knowledge. *Proceedings of the National Academy of Sciences*, *106*(41), 17284–17289.

Bapst, V., Sanchez-Gonzalez, A., Doersch, C., et al. (2019). Structured agents for physical construction. *arXiv preprint arXiv:1904.03177.*

Barker, A. A. (1965). Monte Carlo calculations of the radial distribution functions for a proton-electron plasma. *Australian Journal of Physics*, *18*, 119–133.

Baron-Cohen, S. (1997). *Mindblindness: An essay on autism and theory of mind.* MIT Press.

Barreto, A., Dabney, W., Munos, R., et al. (2017). Successor features for transfer in reinforcement learning. In I. Guyon, U. V. Luxburg et al. (Eds.), *Advances in Neural Information Processing Systems 30.*

Barrett, H. C., Todd, P. M., Miller, G. F., & Blythe, P. W. (2005). Accurate judgments of intention from motion cues alone: A cross-cultural study. *Evolution and Human Behavior*, *26*(4), 313–331.

Bartlett, F. C. (1932). *Remembering: a study in experimental and social psychology.* Cambridge University Press.

Bass, L., Gopnik, A., Hanson, M., et al. (2019). Children's developing theory of mind and pedagogical evidence selection. *Developmental Psychology*, *55*, 286–302.

Bass, I., Smith, K., Bonawitz, E., & Ullman, T. (2021). Partial mental simulation explains fallacies in physical reasoning. *Cognitive Neuropsychology*, *38*, 413–424.

Bates, C. J., Yildirim, I., Tenenbaum, J. B., & Battaglia, P. (2019). Modeling human intuitions about liquid flow with particle-based simulation. *PLoS Computational Biology*, *15*(7), e1007210.

Battaglia, P., Hamrick, J., & Tenenbaum, J. B. (2013). Simulation as an engine of physical scene understanding. *Proceedings of the National Academy of Sciences*, *110*(45), 18327–18332.

Battaglia, P. W., Hamrick, J. B., Bapst, V., et al. (2018). Relational inductive biases, deep learning, and graph networks. *arXiv preprint arXiv:1806.01261.*

Battaglia, P., Jacobs, R. A., & Aslin, R. N. (2003). Bayesian integration of visual and auditory signals for spatial localization. *Journal of the Optical Society of America A*, *20*(7), 1391–1397.

Battaglia, P., Pascanu, R., Lai, M., Jimenez Rezende, D., & Kavukcuoglu, K. (2016). Interaction networks for learning about objects, relations and physics. In *Advances in Neural Information Processing Systems 29.*

Baum, E. B. (2004). *What is thought?* MIT Press.

Baum, L. E., & Petrie, T. (1966). Statistical inference for probabilistic functions of finite state Markov chains. *Annals of Mathematical Statistics*, *37*, 1554–1563.

Bayes, T. (1763/1958). Studies in the history of probability and statistics: IX. Thomas Bayes's essay towards solving a problem in the doctrine of chances. *Biometrika*, *45*, 296–315.

Beaumont, M. A., Zhang, W., & Balding, D. J. (2002). Approximate Bayesian computation in population genetics. *Genetics*, *162*(4), 2025–2035.

Becker, F., Skirzynski, J., van Opheusden, B., & Lieder, F. (2022). Boosting human decision-making with AI-generated decision aids. *Computational Brain & Behavior*, *5*(3).

Behrens, T. E., Muller, T. H., Whittington, J. C., Mark, S., Baram, A. B., Stachenfeld, K. L., & Kurth-Nelson, Z. (2018). What is a cognitive map? Organizing knowledge for flexible behavior. *Neuron*, *100*(2), 490–509.

Bell, D. E. (1985). Disappointment in decision making under uncertainty. *Operations Research*, *33*(1), 1–27.

Bellemare, M. G., Dabney, W., & Munos, R. (2017). A distributional perspective on reinforcement learning. In *Proceedings of the International Conference on Machine Learning* (pp. 449–458).

Bellemare, M. G., Dabney, W., & Rowland, M. (2023). *Distributional reinforcement learning.* MIT Press.

Bellman, R. (1957). *Dynamic programming.* Princeton University Press.

Bem, D. J. (1972). Self-perception theory. In *Advances in experimental social psychology* (Vol. 6, pp. 1–62). Elsevier.

Berg, R. van den, Anandalingam, K., Zylberberg, A., Kiani, R., Shadlen, M. N., & Wolpert, D. M. (2016). A common mechanism underlies changes of mind about decisions and confidence. *eLife*, *5*, e12192.

Berg, R. Van den, & Ma, W. J. (2018). A resource-rational theory of set size effects in human visual working memory. *eLife*, *7*, e34963.

Bergen, L., Gibson, E., & O'Donnell, T. (2013). Arguments and modifiers from the learner's perspective. In *Proceedings of the 51st Annual Meeting of the Association for Computational Linguistics.*

Bergen, L., Levy, R., & Goodman, N. (2016). Pragmatic reasoning through semantic inference. *Semantics and Pragmatics*, *9*.

Berger, J. O. (1993). *Statistical decision theory and Bayesian analysis.* Springer.

Bernardo, J. M., & Smith, A. F. M. (1994). *Bayesian theory.* Wiley.

Bernoulli, J. (2006/1713). *The art of conjecturing, together with Letter to a friend on sets in court tennis* (Sylla, E. D., Translator). Johns Hopkins University Press, Baltimore (Original work published in 1713, Thurneysen Brothers Press, Basel, Switzerland).

Berwick, R. C., Pietroski, P., Yankama, B., & Chomsky, N. (2011). Poverty of the stimulus revisited. *Cognitive Science, 35*(7), 1207–1242.

Berwick, R. C., & Pilato, S. (1987). Learning syntax by automata induction. *Machine Learning, 2*(1), 9–38.

Bever, T. G. (1970). The cognitive basis for linguistic structures. In J. R. Hayes (Ed.), *Cognition and the development of language* (pp. 279–362).Wiley.

Bhui, R., & Gershman, S. J. (2018). Decision by sampling implements efficient coding of psychoeconomic functions. *Psychological Review, 125*(6), 985–1001.

Bingham, E., Chen, J. P., Jankowiak, M., et al. (2019). Pyro: Deep universal probabilistic programming. *Journal of Machine Learning Research, 20*(1), 973–978.

Binmore, K. (2008). *Rational decisions.* Princeton University Press.

Blackburn, P., & Bos, J. (2005). *Representation and inference for natural language: A first course in computational semantics.* Center for the Study of Language and Information, Stanford, CA.

Blackwell, D., & MacQueen, J. (1973). Ferguson distributions via Polyaurn schemes. *Annals of Statistics, 1*, 353–355.

Blakemore, C., Adler, K., & Pointon, M. (1990). *Vision: Coding and efficiency.* Cambridge University Press.

Blei, D. M., Griffiths, T. L., & Jordan, M. I. (2010). The nested Chinese restaurant process and Bayesian nonparametric inference of topic hierarchies. *Journal of the ACM, 57*(2), 1–30.

Blei, D. M., & Jordan, M. I. (2006). Variational inference for Dirichlet process mixtures. *Bayesian Analysis, 1*(1), 121–143.

Blei, D. M., Kucukelbir, A., & McAuliffe, J. D. (2017). Variational inference: A review for statisticians. *Journal of the American Statistical Association, 112*(518), 859–877.

Blei, D. M., Ng, A. Y., & Jordan, M. I. (2003). Latent Dirichlet allocation. *Journal of Machine Learning Research, 3*, 993–1022.

Block, N. (1986). Advertisement for a semantics for psychology. *Midwest Studies in Philosophy, 10*, 615–678.

Block, N. (1998). Conceptual role semantics. In *Routledge Encyclopedia of Philosophy* (pp. 242–256). Routledge.

Bloom, P. (2000). *How children learn the meanings of words.* MIT Press.

Bloom, P. (2004). *Descartes' baby.* Basic Books.

Blundell, C., Sanborn, A. N., & Griffiths, T. L. (2012). Look-ahead Monte Carlo with people. In *Proceedings of the 34th Annual Meeting of the Cognitive Science Society* (pp. 1356–1361).

Boas, M. L. (1983). *Mathematical methods in the physical sciences* (2nd ed.). Wiley.

Bogacz, R., Brown, E., Moehlis, J., Holmes, P., & Cohen, J. D. (2006). The physics of optimal decision making: A formal analysis of models of performance in two-alternative forced-choice tasks. *Psychological Review, 113*(4), 700–765.

Bonawitz, E., Denison, S., Gopnik, A., & Griffiths, T. (2014a). Win- stay, lose-sample: A simple sequential algorithm for approximating Bayesian inference. *Cognitive Psychology, 74*, 35–65.

Bonawitz, E., Denison, S., Griffiths, T. L., & Gopnik, A. (2014b). Probabilistic models, learning algorithms, and response variability: sampling in cognitive development. *Trends in Cognitive Sciences, 18*(10), 497–500.

Bonawitz, E., Ferranti, D., Saxe, R., Gopnik, A., Meltzoff, A., Woodard, J., & Schulz, L. (2010). Just do it? Investigating the gap between prediction and action in toddlers' causal inferences. *Cognition, 115*, 104–117.

Bonawitz, E., Fischer, A., & Schulz, L. (2012). Teaching three-and-a-half- year-olds to revise their beliefs given ambiguous evidence. *Journal of Cognition and Development, 13*, 266–280.

Bonawitz, E., & Lombrozo, T. (2012). Occam's rattle: Children's use of simplicity and probability to constrain inference. *Developmental Psychology, 48*, 1156–1164.

Bonawitz, E., Ullman, T. D., Bridgers, S., Gopnik, A., & Tenenbaum, J. B. (2019). Sticking to the evidence? a behavioral and computational case study of micro-theory change in the domain of magnetism. *Cognitive Science, 43*(8), e12765.

Bonawitz, E., Walker, C., Hemmer, P., Abbot, J., Griffiths, T., & Gopnik, A. (in revision). Variability in preschoolers' cognitive search.

Borschinger, B., Demuth, K., & Johnson, M. (2012). Studying the effect of input size for Bayesian word segmentation on the Providence corpus. In *Proceedings of COLING* (pp. 325–340).

Borschinger, B., & Johnson, M. (2011). A particle filter algorithm for Bayesian word segmentation. In *Proceedings of the Australasian Language Technology Association Workshop* (pp. 10–18).

Borschinger, B., & Johnson, M. (2014). Exploring the role of stress in Bayesian word segmentaiton using adaptor grammars. In S. Riezler (Ed.), *Transactions of the Association for Computational Linguistics* (Vol. 2, pp. 93–104).

Botvinick, M., & Cohen, J. (1998). Rubber hands "feel" touch that eyes see. *Nature*, *391*(6669), 756–756.

Botvinick, M., & Toussaint, M. (2012). Planning as inference. *Trends in Cognitive Sciences*, *16*(10), 485–488.

Botvinick, M. M., Niv, Y., & Barto, A. C. (2009). Hierarchically organized behavior and its neural foundations: a reinforcement learning perspective. *Cognition*, *113*(3), 262–280.

Bouman, K. L., Xiao, B., Battaglia, P., & Freeman, W. T. (2013). Estimating the material properties of fabric from video. In *Proceedings of the IEEE International Conference on Computer Vision* (pp. 1984–1991).

Bourgin, D., Abbott, J., Griffiths, T. L., Smith, K., & Vul, E. (2014). Empirical evidence for Markov chain Monte Carlo in memory search. In *Proceedings of the 36th Annual Meeting of the Cognitive Science Society*.

Bovens, L., & Hartmann, S. (2004). *Bayesian epistemology*. Oxford University Press.

Bowerman, M. (1988). The "no negative evidence" problem: How do children avoid constructing an overly general grammar? In J. Hawkins (Ed.), *Explaining language universals* (pp. 73–101). Blackwell.

Bowers, J., & Davis, C. (2012). Bayesian just-so stories in psychology and neuroscience. *Psychonomic Bulletin and Review*, *138*(3), 389–414.

Bowers, M., Olausson, T. X., Wong, L., et al. (2023). Top-down synthesis for library learning. *Proceedings of the ACM on Programming Languages*, *7*(POPL), 1182–1213.

Box, G. E. P., & Tiao, G. C. (1992). *Bayesian inference in statistical analysis*. Wiley.

Boyen, X., & Koller, D. (1998). Tractable inference for complex stochastic processes. In *Proceedings of the Fourteenth Conference on Uncertainty in Artificial Intelligence* (pp. 33–42).

Braddick, O. J., Wishart, K. A., & Curran, W. (2002). Directional performance in motion transparency. *Vision Research*, *42*(10), 1237–1248.

Bradley, R. (2017). *Decision theory with a human face*. Cambridge University Press.

Bramley, N. R., Dayan, P., Griffiths, T. L., & Lagnado, D. A. (2017). For- malizing Neurath's ship: Approximate algorithms for online causal learning. *Psychological Review*, *124*(3), 301–338.

Bratman, M. (1987). *Intention, plans, and practical reason*. University of Chicago Press.

Braun, D. A., Waldert, S., Aertsen, A., Wolpert, D. M., & Mehring, C. (2010). Structure learning in a sensorimotor association task. *PLoS One*, *5*(1), 1–8.

Bregman, A. S. (1994). *Auditory scene analysis: The perceptual organization of sound*. MIT Press.

Brehmer, B. (1974). Hypotheses about relations between scaled variables in the learning of probabilistic inference tasks. *Organizational Behavior and Human Decision Processes*, *11*, 1–27.

Brent, M. (1999). An efficient, probabilistically sound algorithm for segmentation and word discovery. *Machine Learning*, *34*, 71–105.

Brewer, W. F. (1974). There is no convincing evidence for operant or classical conditioning in adult humans. In W. B. Weimer & D. S. Palermo (Eds.), *Cognition and the symbolic processes* (pp. 1–42). Erlbaum.

Britten, K. H., Shadlen, M. N., Newsome, W. T., & Movshon, J. A. (1992). The analysis of visual motion: a comparison of neuronal and psychophysical performance. *Journal of Neuroscience*, *12*(12), 4745–4765.

Brooks, R. A. (1991). Intelligence without representation. *Artificial Intelligence*, *47*(1–3), 139–159.

Brown, R., & Hanlon, C. (1970). Derivational complexity and the order of acquisition in child speech. In *Cognition and the Development of Language* (pp. 11–53). Wiley.

Brown, S. D., & Heathcote, A. (2008). The simplest complete model of choice response time: Linear ballistic accumulation. *Cognitive Psychology*, *57*(3), 153–178.

Brown, S. D., & Steyvers, M. (2009). Detecting and predicting changes. *Cognitive Psychology*, *58*, 49–67.

Brown, T., Mann, B., Ryder, N., et al. (2020). Language models are few-shot learners. In *Advances in Neural Information Processing Systems 33* (pp. 1877–1901).

Bruner, J. S., Goodnow, J. J., & Austin, G. A. (1956). *A study of thinking*. Wiley.

Bubeck, S., Chandrasekaran, V., Eldan, R., et al. (2023). Sparks of artificial general intelligence: Early experiments with GPT-4. *arXiv preprint arXiv:2303.12712*.

Buch, P. (1994). Future prospects discussed. *Nature*, *368*, 107–108.

Buehner, M., & Cheng, P. W. (1997). Causal induction: The Power PC theory versus the Rescorla-Wagner theory. In *Proceedings of the 19th Annual Conference of the Cognitive Science Society* (pp. 55–61).

Buehner, M. J., Cheng, P. W., & Clifford, D. (2003). From covariation to causation: A test of the assumption of causal power. *Journal of Experimental Psychology: Learning, Memory, and Cognition*, *29*, 1119–1140.

Buesing, L., Bill, J., Nessler, B., & Maass, W. (2011). Neural dynamics as sampling: A model for stochastic computation in recurrent networks of spiking neurons. *PLoS Computational Biology*, *7*(11), e1002211.

Burger, L., & Jara-Ettinger, J. (2020). Mental inference: Mind perception as Bayesian model selection. In *Proceedings of the 42nd Annual Meeting of the Cognitive Science Society*.

Busemeyer, J. R. (1985). Decision making under uncertainty: A comparison of simple scalability, fixed-sample, and sequential-sampling models. *Journal of Experimental Psychology: Learning, Memory, and Cognition*, *11*(3), 538.

Busemeyer, J. R., Byun, E., DeLosh, E. L., & McDaniel, M. A. (1997). Learning functional relations based on experience with input-output pairs by humans and artificial neural networks. In K. Lamberts D. Shanks (Eds.), *Concepts and categories* (pp. 405–437). MIT Press.

Bush, C. A., & MacEachern, S. N. (1996). A semi-parametric Bayesian model for randomized block designs. *Biometrika*, *83*, 275–286.

Bush, R., & Mosteller, F. (1955). *Stochastic models of learning*. Wiley.

Callaway, F., Hamrick, J., & Griffiths, T. L. (2017). *Discovering simple heuristics from mental simulation*. In G. Gunzelmann, A. Howes, T. Tenbrink, & E. J. Davelaar (Eds.), *Proceedings of the 39th Annual Conference of the Cognitive Science Society*. Austin, TX: Cognitive Science Society.

Callaway, F., Hardy, M., & Griffiths, T. L. (2020). Optimal nudging. In *Proceedings of the 42nd Annual Conference of the Cognitive Science Society*.

Callaway, F., Hardy, M., & Griffiths, T. L. (2023). Optimal nudging for cognitively bounded agents: A framework for modeling, predicting, and controlling the effects of choice architectures. *Psychological Review*, *130*(6), 1457–1491.

Callaway, F., Jain, Y. R., van Opheusden, B., et al. (2022a). Leveraging artificial intelligence to improve people's planning strategies. *Proceedings of the National Academy of Sciences*, *119*(12), e2117432119.

Callaway, F., Rangel, A., & Griffiths, T. L. (2021). Fixation patterns in simple choice reflect optimal information sampling. *PLoS Computational Biology*, *17*(3), e1008863.

Callaway, F., van Opheusden, B., Gul, S., et al. (2022b). Rational use of cognitive resources in human planning. *Nature Human Behaviour*, 1–14.

Canini, K. R., Griffiths, T. L., Vanpaemel, W., & Kalish, M. L. (2014). Revealing human inductive biases for category learning by simulating cultural transmission. *Psychonomic Bulletin & Review*, *21*(3), 785– 793.

Canini, K. R., Shashkov, M. M., & Griffiths, T. L. (2010). Modeling transfer learning in human categorization with the hierarchical Dirichlet process. In *Proceedings of the 27th International Conference on Machine Learning* (pp. 151–158).

Caramazza, A., McCloskey, M., & Green, B. (1981). Naive beliefs in "sophisticated" subjects: Misconceptions about trajectories of objects. *Cognition*, *9*(2), 117–123.

Carey, S. (1985). *Conceptual change in childhood*. MIT Press.

Carey, S. (2009). *The origin of concepts*. Oxford University Press.

Carey, S. (2015). Why theories of concepts should not ignore the problem of acquisition. In *The conceptual mind: New directions in the study of concepts* (pp. 113–163). MIT Press.

Carey, S., & Bartlett, E. (1978). Acquiring a single new word. *Papers and Reports on Child Language Development, 15*, 17–29.

Carey, S., & Spelke, E. (1994). Domain-specific knowledge and conceptual change. In L. Hirschfeld & S. Gelman (Eds.), *Mapping the mind: Domain specificity in cognition and culture* (pp. 169–200). Cambridge University Press.

Carey, S., & Spelke, E. (1996). Science and core knowledge. *Philosophy of Science, 63*, 515–533.

Carnie, A. (2013). *Syntax: A generative introduction.* Wiley.

Caron, F. (2012). Bayesian nonparametric models for bipartite graphs. *Advances in Neural Information Processing Systems 25.*

Caron, F., & Teh, Y. (2012). Bayesian nonparametric models for ranked data. *Advances in Neural Information Processing Systems, 25.*

Carpenter, B., Gelman, A., Hoffman, M. D., et al. (2017). Stan: A probabilistic programming language. *Journal of Statistical Software, 76*(1), 1–32.

Carroll, J. D. (1963). *Functional learning: The learning of continuous functional mappings relating stimulus and response continua.* Educational Testing Service.

Castillo, L., León Villagrá, P., Chater, N., & Sanborn, A. (2021). Local sampling with momentum accounts for human random sequence generation. In *Proceedings of the 43rd Annual Meeting of the Cognitive Science Society.*

Ceccarelli, F., La Scaleia, B., Russo, M., et al. (2018). Rolling motion along an incline: Visual sensitivity to the relation between acceleration and slope. *Frontiers in Neuroscience, 12.*

Chaitin, G. J. (1969). On the length of programs for computing finite binary sequences: Statistical considerations. *Journal of the ACM, 16*, 145–159.

Chaitin, G. J. (1998). *The limits of mathematics: A course on information theory and the limits of formal reasoning.* Springer.

Chang, C. C., & Keisler, H. J. (1973). *Model theory.* Elsevier.

Chang, M. B., Ullman, T., Torralba, A., & Tenenbaum, J. B. (2016). A compositional object-based approach to learning physical dynamics. *arXiv preprint arXiv:1612.00341.*

Charniak, E. (1993). *Statistical language learning.* MIT Press.

Charniak, E., Hendrickson, C., Jacobson, N., & Perkowitz, M. (1993). Equations for part-of-speech tagging. In *Proceedings of the Tenth National Conference on Artificial Intelligence (AAAI-93)* (pp. 784–789).

Chater, N. (1996). Reconciling simplicity and likelihood principles in perceptual organization. *Psychological Review, 103*, 566–581.

Chater, N. (2009). Rational and mechanistic perspectives on reinforcement learning. *Cognition, 113*(3), 350–364.

Chater, N. (2018). *The mind is flat: The illusion of mental depth and the improvised mind.* Penguin.

Chater, N. (2023). How could we make a social robot? A virtual bargaining approach. *Philosophical Transactions of the Royal Society A, 381*(2251), 20220040.

Chater, N., & Brown, G. D. (2008). From universal laws of cognition to specific cognitive models. *Cognitive Science, 32*(1), 36–67.

Chater, N., & Christiansen, M. H. (2010). Language acquisition meets language evolution. *Cognitive Science, 34*(7), 1131–1157.

Chater, N., Clark, A., Goldsmith, J., & Perfors, A. (2015). *Empiricism and language learnability.* Oxford University Press.

Chater, N., Felin, T., Funder, D. C., et al. (2018). Mind, rationality, and cognition: An interdisciplinary debate. *Psychonomic Bulletin & Review, 25*, 793–826.

Chater, N., Goodman, N., Griffiths, T. L., Kemp, C., Oaksford, M., & Tenenbaum, J. B. (2011). The imaginary fundamentalists: The unshocking truth about Bayesian cognitive science. *Behavioral and Brain Sciences, 34*(4), 194–196.

Chater, N., & Loewenstein, G. (2016). The under-appreciated drive for sense-making. *Journal of Economic Behavior & Organization, 126*, 137–154.

Chater, N., & Manning, C. D. (2006). Probabilistic models of language processing and acquisition. *Trends in Cognitive Sciences, 10*, 335–344.

Chater, N., McCauley, S. M., & Christiansen, M. H. (2016). Language as skill: Intertwining comprehension and production. *Journal of Memory and Language*, *89*, 244–254.

Chater, N., & Misyak, J. (2021). Spontaneous communicative conventions through virtual bargaining. In S. Muggleton & N. Chater (Eds.), *Human-like machine intelligence* (pp. 52–67). Oxford University Press.

Chater, N., & Oaksford, M. (1999). The probability heuristics model of syllogistic reasoning. *Cognitive Psychology*, *38*(2), 191–258.

Chater, N., & Oaksford, M. (2012). Normative systems: Logic, probability, and rational choice. In K. J. Holyoak & R. G. Morrison (Eds.), *Oxford handbook of thinking and reasoning* (pp. 11–21). Oxford University Press.

Chater, N., & Oaksford, M. (2013). Programs as causal models: Speculations on mental programs and mental representation. *Cognitive Science*, *37*(6), 1171–1191.

Chater, N., & Vitányi, P. M. (2003a). The generalized universal law of generalization. *Journal of Mathematical Psychology*, *47*(3), 346–369.

Chater, N., & Vitányi, P. M. (2003b). Simplicity: A unifying principle in cognitive science. *Trends in Cognitive Science*, *7*, 19–22.

Chater, N., & Vitányi, P. M. (2007). "Ideal learning" of natural language: Positive results about learning from positive evidence. *Journal of Mathematical Psychology*, *51*(3), 135–163.

Chater, N., Zeitoun, H., & Melkonyan, T. (2022). The paradox of social interaction: Shared intentionality, we-reasoning and virtual bargaining. *Psychological Review, 129*(3), 415–437.

Chater, N., Zhu, J.-Q., Spicer, J., Sundh, J., León-Villagrá, P., & Sanborn, A. (2020). Probabilistic biases meet the Bayesian brain. *Current Directions in Psychological Science*, *29*(5), 506–512.

Chen, S. F., & Goodman, J. (1996). An empirical study of smoothing techniques for language modeling. In *Proceedings of the 34th Annual Meeting of the Association for Computational Linguistics* (pp. 310–318).

Cheng, P. (1997). From covariation to causation: A causal power theory. *Psychological Review*, *104*(2), 367–405.

Cheng, P. W., & Holyoak, K. J. (1985). Pragmatic reasoning schemas. *Cognitive Psychology*, *17*(4), 391–416.

Cheyette, S., & Piantadosi, S. (2017). Knowledge transfer in a probabilistic language of thought. In *Proceedings of the 39th Annual Meeting of the Cognitive Science Society.*

Chlipala, A. (2022). *Certified programming with dependent types: A pragmatic introduction to the Coq proof assistant.* MIT Press.

Chomsky, N. (1956). Three models for the description of language. *IRE Transactions on Information Theory*, *2*(3), 113–124.

Chomsky, N. (1957). *Syntactic structures.* Mouton.

Chomsky, N. (1959). A review of B. F. Skinner's verbal behavior. *Language*, *35*, 26–58.

Chomsky, N. (1965). *Aspects of the theory of syntax.* MIT Press.

Chomsky, N. (1969). Quine's empirical assumptions. In *Words and objections* (pp. 53–68). Springer.

Chomsky, N. (1980). Rules and representations. *Behavioral and Brain Sciences*, *3*(1), 1–15.

Chomsky, N. (1986). *Language and problems of knowledge: The Managua lectures.* MIT Press.

Chomsky, N. (1995). *The minimalist program.* MIT Press.

Chomsky, N. (2015). *What kind of creatures are we?* Columbia University Press.

Chopin, N., & Papaspiliopoulos, O. (2020). *An introduction to sequential Monte Carlo.* Springer.

Chouinard, M. M., & Clark, E. V. (2003). Adult reformulations of child errors as negative evidence. *Journal of Child Language*, *30*(3), 637–669.

Chouinard, P. A., Leonard, G., & Paus, T. (2005). Role of the primary motor and dorsal premotor cortices in the anticipation of forces during object lifting. *Journal of Neuroscience*, *25*(9), 2277–2284.

Christiansen, M. H., & Chater, N. (1999). Toward a connectionist model of recursion in human linguistic performance. *Cognitive Science*, *23*(2), 157–205.

Christiansen, M. H., & Chater, N. (2008). Language as shaped by the brain. *Behavioral and Brain Sciences*, *31*(5), 489–509.

Christiansen, M. H., & Chater, N. (2016a). *Creating language: Integrating evolution, acquisition, and processing.* MIT Press.

Christiansen, M. H., & Chater, N. (2016b). The now-or-never bottle- neck: A fundamental constraint on language. *Behavioral and Brain Sciences*, *39*, e62.

Christiansen, M. H., & Chater, N. (2022). *The language game: How improvisation created language and changed the world.* Hachette.

Christianson, K., Hollingworth, A., Halliwell, J. F., & Ferreira, F. (2001). Thematic roles assigned along the garden path linger. *Cognitive Psychology*, *42*(4), 368–407.

Churchland, P. M. (1981). Eliminative materialism and propositional attitudes. *Journal of Philosophy*, *78*(2), 67–90.

Clark, A. (2013). Whatever next? Predictive brains, situated agents, and the future of cognitive science. *Behavioral and Brain Sciences*, *36*(03), 181–204.

Clark, E. V. (2009). *First language acquisition.* Cambridge University Press.

Clark, H. H. (1996). *Using language.* Cambridge University Press.

Clocksin, W. F., & Mellish, C. S. (2003). *Programming in PROLOG.* Springer Science & Business Media.

Cohen, A., & Ross, M. (2009). Exploring mass perception with Markov chain Monte Carlo. *Journal of Experimental Psychology: Human Perception and Performance*, *35*, 1833–1844.

Cohen, A. L., Sidlowski, S., & Staub, A. (2017). Beliefs and Bayesian reasoning. *Psychonomic Bulletin & Review*, *24*(3), 972–978.

Collins, A. M., & Loftus, E. F. (1975). A spreading activation theory of semantic processing. *Psychological Review*, *82*, 407–428.

Collins, A. M., & Quillian, M. R. (1969). Retrieval time from semantic memory. *Journal of Verbal Learning and Verbal Behavior*, *8*, 240–247.

Consul, S., Heindrich, L., Stojcheski, J., & Lieder, F. (2022). Improving human decision-making by discovering efficient strategies for hierarchical planning. *Computational Brain & Behavior*, *5*(2), 185–216.

Contreras Kallens, P., Kristensen-McLachlan, R. D., & Christiansen, M. H. (2023). Large language models demonstrate the potential of statistical learning in language. *Cognitive Science*, *47*(3), e13256.

Cooper, G. F. (1990). The computational complexity of probabilistic inference using Bayesian belief networks. *Artificial Intelligence*, *42*(2–3), 393–405.

Cooper, G., & Herskovits, E. (1992). A Bayesian method for the induction of probabilistic networks from data. *Machine Learning*, *9*, 308–347.

Cooter, R., & Rappoport, P. (1984). Were the ordinalists wrong about welfare economics? *Journal of Economic Literature*, *22*(2), 507–530.

Corballis, M. C. (2014). *The recursive mind.* Princeton University Press.

Cormode, G., & Muthukrishnan, S. (2007). The string edit distance matching problem with moves. *ACM Transactions on Algorithms (TALG)*, *3*(1), 1–19.

Correa, C. G., Ho, M. K., Callaway, F., & Griffiths, T. L. (2020). Resource-rational task decomposition to minimize planning costs. In *Proceedings of the 42nd Annual Meeting of the Cognitive Science Society.*

Cosmides, L. (1989). The logic of social exchange: Has natural selection shaped how humans reason? Studies with the Wason selection task. *Cognition*, *31*(3), 187–276.

Costello, F., & Watts, P. (2014). Surprisingly rational: probability theory plus noise explains biases in judgment. *Psychological Review*, *121*(3), 463–480.

Costello, F., & Watts, P. (2017). Explaining high conjunction fallacy rates: The probability theory plus noise account. *Journal of Behavioral Decision Making*, *30*(2), 304–321.

Costello, F., & Watts, P. (2019). The rationality of illusory correlation. *Psychological Review*, *126*(3), 437–450.

Cover, T., & Thomas, J. (1991). *Elements of information theory.* Wiley.

Cox, R. (1946). Probability, frequency, and reasonable expectation. *American Journal of Physics*, *14*, 1–13.

Cox, R. T. (1961). *The algebra of probable inferencce.* Johns Hopkins University Press.

Craik, K. J. W. (1943). *The nature of explanation.* Cambridge University Press.

Crain, S., & Lillo-Martin, D. C. (1999). *An introduction to linguistic theory and language acquisition.* Oxford University Press.

Crupi, V., Chater, N., & Tentori, K. (2013). New axioms for probability and likelihood ratio measures. *British Journal for the Philosophy of Science*, *64*(1), 189–204.

Csibra, G. (2008). Goal attribution to inanimate agents by 6.5-month-old infants. *Cognition*, *107*(2), 705–717.

Csibra, G., Bíró, S., Koós, O., & Gergely, G. (2003). One-year-old infants use teleological representations of actions productively. *Cognitive Science*, *27*(1), 111–133.

Culicover, P. W. (1999). *Syntactic nuts: Hard cases, syntactic theory, and language acquisition.* Oxford University Press.

Cushman, F., & Morris, A. (2015). Habitual control of goal selection in humans. *Proceedings of the National Academy of Sciences*, *112*(45), 13817–13822.

Cusumano-Towner, M. F., Saad, F. A., Lew, A. K., & Mansinghka, V. K. (2019). Gen: A general-purpose probabilistic programming system with programmable inference. In *Proceedings of the 40th ACM Sigplan Conference on Programming Design and Implementation* (pp. 221–236). ACM.

Cutler, A. (1982). *Slips of the tongue.* Mouton.

Dabney, W., Kurth-Nelson, Z., Uchida, N., et al. (2020). A distributional code for value in dopamine-based reinforcement learning. *Nature*, *577*(7792), 671–675.

Dabney, W., Rowland, M., Bellemare, M., & Munos, R. (2018). Distributional reinforcement learning with quantile regression. In *Proceedings of the 32nd AAAI Conference on Artificial Intelligence.*

Dahl, D. B. (2003). *An improved merge-split sampler for conjugate Dirichlet process mixture models* (Tech. Rep. No. 1086). Department of Statistics, University of Wisconsin.

Damerau, F. J. (1964). A technique for computer detection and correction of spelling errors. *Communications of the ACM*, *7*(3), 171–176.

Dasgupta, I., & Gershman, S. J. (2021). Memory as a computational resource. *Trends in Cognitive Sciences*, *25*(3), 240–251.

Dasgupta, I., & Griffiths, T. L. (2022). Clustering and the efficient use of cognitive resources. *Journal of Mathematical Psychology*, *109*, 102675.

Dasgupta, I., Schulz, E., & Gershman, S. J. (2017). Where do hypotheses come from? *Cognitive Psychology*, *96*, 1–25.

Dasgupta, I., Schulz, E., Tenenbaum, J. B., & Gershman, S. J. (2020). A theory of learning to infer. *Psychological Review*, *127*(3), 412–441.

Dasgupta, I., Smith, K. A., Schulz, E., Tenenbaum, J. B., & Gershman, S. J. (2018). Learning to act by integrating mental simulations and physical experiments. *bioRxiv*, 321497.

Dasgupta, I., Wang, J., Chiappa, S., et al. (2019). Causal reasoning from meta-reinforcement learning. *arXiv preprint arXiv:1901.08162.*

Davidson, P. R., & Wolpert, D. M. (2005). Widespread access to predictive models in the motor system: A short review. *Journal of Neural Engineering*, *2*(3), S313.

Davis, E. (2014). *Representations of commonsense knowledge.* Morgan Kaufmann.

Davis, E., Marcus, G., & Frazier-Logue, N. (2017). Commonsense reasoning about containers using radically incomplete information. *Artificial Intelligence*, *248*, 46–84.

Davis, T., Love, B. C., & Maddox, W. T. (2012). Age-related declines in the fidelity of newly acquired category representations. *Learning & Memory*, *19*(8), 325–329.

Davis, Z. J., Bramley, N. R., & Rehder, B. (2020). Causal structure learning in continuous systems. *Frontiers in Psychology*, *11*, 244.

Daw, N., & Courville, A. (2008). The pigeon as particle filter. *Advances in Neural Information Processing Systems 20*, 369–376.

Daw, N. D., Courville, A. C., & Dayan, P. (2008). Semi-rational models of conditioning: The case of trial order. In N. Chater & M. Oaksford (Eds.), *The probabilistic mind* (pp. 431–452). Oxford University Press.

Daw, N. D., & Dayan, P. (2014). The algorithmic anatomy of model- based evaluation. *Philosophical Transactions of the Royal Society B: Biological Sciences*, *369*(1655), 20130478.

Daw, N. D., Gershman, S. J., Seymour, B., Dayan, P., & Dolan, R. J. (2011). Model-based influences on humans' choices and striatal prediction errors. *Neuron*, *69*(6), 1204–1215.

Daw, N. D., Niv, Y., & Dayan, P. (2005). Uncertainty-based competition between prefrontal and dorsolateral striatal systems for behavioral control. *Nature Neuroscience*, *8*(12), 1704–1711.

Dawkins, R. (1978). *The selfish gene.* Oxford University Press.

Dayan, P. (1993). Improving generalization for temporal difference learning: The successor representation. *Neural Computation*, *5*(4), 613–624.

Dayan, P., & Abbott, L. (2001). *Theoretical neuroscience.* MIT Press.

Dayan, P., & Daw, N. D. (2008). Decision theory, reinforcement learning, and the brain. *Cognitive, Affective, & Behavioral Neuroscience*, *8*(4), 429–453.

Dechter, E., Malmaud, J., Adams, R. P., & Tenenbaum, J. B. (2013). Bootstrap learning via modular concept discovery. In *Proceedings of the 29th International Joint Conference in Artificial Intelligence (IJCAI)* (pp. 1302–1309).

Deeb, A.-R., Cesanek, E., & Domini, F. (2021). Newtonian predictions are integrated with sensory information in 3D motion perception. *Psychological Science*, *32*(2), 280–291.

de Finetti, B. (1937). "Foresight: Its logical laws, its subjective sources" in *Annales de l'Institut Henri Poincaré*, *7*, 1–68.

Dehaene, S., Al Roumi, F., Lakretz, Y., Planton, S., & Sablé-Meyer, M. (2022). Symbols and mental programs: A hypothesis about human singularity. *Trends in Cognitive Sciences*, 751–766.

Dehaene, S., Izard, V., Pica, P., & Spelke, E. (2006). Core knowledge of geometry in an Amazonian indigene group. *Science*, *311*(5759), 381–384.

De Lafuente, V., Jazayeri, M., & Shadlen, M. N. (2015). Representation of accumulating evidence for a decision in two parietal areas. *Journal of Neuroscience*, *35*(10), 4306–4318.

DeLosh, E. L., Busemeyer, J. R., & McDaniel, M. A. (1997). Extrapolation: The sine qua non of abstraction in function learning. *Journal of Experimental Psychology: Learning, Memory, and Cognition*, *23*, 968–986.

Dempster, A. P., Laird, N. M., & Rubin, D. B. (1977). Maximum likelihood from incomplete data via the EM algorithm. *Journal of the Royal Statistical Society, B*, *39*, 1–38.

Denić, M., & Szymanik, J. (2022). Reverse-engineering the language of thought: A new approach. In *Proceedings of the 44th Annual Meeting of the Cognitive Science Society.*

Denison, S., Bonawitz, E., Gopnik, A., & Griffiths, T. (2013). Rational variability in children's causal inferences: The sampling hypothesis. *Cognition*, *126*, 285–300.

Dennett, D. C. (1978). *Brainstorms: Philosophical essays on mind and psychology.* MIT Press.

Dennett, D. C. (1983). Intentional systems in cognitive ethology: The "Panglossian paradigm" defended. *Behavioral and Brain Sciences*, *6*(3), 343–90.

Depeweg, S., Rothkopf, C. A., & Jäkel, F. (2024). Solving Bongard problems with a visual language and pragmatic constraints. *Cognitive Science, 48*(5), e13432.

DeRaedt, L., Frasconi, P., & Muggleton, S. H. (Eds.). (2008). *Probabilistic inductive logic programming.* Springer.

Desmet, T., De Baecke, C., Drieghe, D., Brysbaert, M., & Vonk, W. (2006). Relative clause attachment in Dutch: On-line comprehension corresponds to corpus frequencies when lexical variables are taken into account. *Language and Cognitive Processes*, *21*(4), 453–485.

Desmet, T., & Gibson, E. (2003). Disambiguation preferences and corpus frequencies in noun phrase conjunction. *Journal of Memory and Language*, *49*(3), 353–374.

Dezfouli, A., & Balleine, B. W. (2013). Actions, action sequences and habits: Evidence that goal-directed and habitual action control are hierarchically organized. *PLoS Computational Biology*, *9*(12), e1003364.

Diacu, F. (1996). The solution of the n-body problem. *The Mathematical Intelligencer*, *18*(3), 66–70.

Dietterich, T. G. (2000). Hierarchical reinforcement learning with the MAXQ value function decomposition. *Journal of Artificial Intelligence Research*, *13*, 227–303.

DiSessa, A. A. (1993). Toward an epistemology of physics. *Cognition and Instruction*, *10*(2–3), 105–225.

Dolan, R. J., & Dayan, P. (2013). Goals and habits in the brain. *Neuron*, *80*(2), 312–325.

Doll, B. B., Simon, D. A., & Daw, N. D. (2012). The ubiquity of model- based reinforcement learning. *Current Opinion in Neurobiology*, *22*(6), 1075–1081.

Doucet, A., de Freitas, & Gordon, N. (2001). *Sequential Monte Carlo methods in practice.* Springer.

Dowe, P. (2000). *Physical causation.* Cambridge University Press.

Dowman, M. (2000). Addressing the learnability of verb subcategorizations with Bayesian inference. In *Proceedings of the 22nd Annual Meeting of the Cognitive Science Society* (pp. 107–112).

Doya, K., Ishii, S., Pouget, A., & Rao, R. P. (2007). *Bayesian brain: Probabilistic approaches to neural coding.* MIT Press.

Doyle, G., & Levy, R. (2013). Combining multiple information types in Bayesian word segmentation. In *Proceedings of NAACL-HIT* (pp. 117–126).

Duncker, K. (1945). On problem-solving. *Psychological Monographs, 58*(5), i–113.

Dunn, J. (2017). Computational learning of construction grammars. *Language and cognition, 9*(2), 254–292.

Earl, D. J., & Deem, M. W. (2005). Parallel tempering: Theory, applications, and new perspectives. *Physical Chemistry Chemical Physics, 7*(23), 3910–3916.

Earman, J. (1992). *Bayes or bust? A critical examination of Bayesian confirmation theory.* MIT Press.

Eckstein, M. K., & Collins, A. G. (2020). Computational evidence for hierarchically structured reinforcement learning in humans. *Proceedings of the National Academy of Sciences, 117*(47), 29381–29389.

Edwards, W. (1954). The theory of decision making. *Psychological Bulletin, 51*(4), 380.

Efron, B. (1975). The efficiency of logistic regression compared to normal discriminant analysis. *Journal of the American Statistical Association, 70*(352), 892–898.

Einhorn, H. J., & Hogarth, R. M. (1986). Judging probable cause. *Psychological Bulletin, 99*(1), 3.

Ellis, K. (2020). *Algorithms for learning to induce programs.* Unpublished doctoral dissertation, Massachusetts Institute of Technology.

Ellis, K., Morales, L., Sablé-Meyer, M., Solar-Lezama, A., & Tenenbaum, J. (2018). Learning libraries of subroutines for neurally–guided Bayesian program induction. In *Advances in Neural Information Processing Systems* (pp. 7815–7825).

Ellis, K., Wong, L., Nye, M., (2023). Dreamcoder: Growing generalizable, interpretable knowledge with wake-sleep Bayesian program learning. *Philosophical Transactions of the Royal Society A, 381*(2251) 20220050.

Ellis, K., Wong, C., Nye, M.,. et al. (2021). Dreamcoder: Bootstrapping inductive program synthesis with wake-sleep library learning. In *Proceedings of the 42nd ACM SIGPLAN International Conference on Programming Language Design and Implementation* (pp. 835–850).

Elman, J. L. (1990). Finding structure in time. *Cognitive Science, 14*, 179–211.

Elman, J. L., Bates, E. A., Johnson, M. H., Karmiloff-Smith, A., Parisi, D., & Plunkett, K. (1996). *Rethinking innateness: A connectionist perspective.* MIT Press.

Ennis, D. M. (1988). Toward a universal law of generalization. *Science, 242*(4880), 944.

Erdogan, G., Chen, Q., Garcea, F. E., Mahon, B. Z., & Jacobs, R. A. (2016). Multisensory part-based representations of objects in human lateral occipital cortex. *Journal of Cognitive Neuroscience, 28*(6), 869–881.

Erdogan, G., Yildirim, I., & Jacobs, R. A. (2015). From sensory signals to modality-independent conceptual representations: A probabilistic language of thought approach. *PLoS Computational Biology, 11*(11), e1004610.

Ernst, M. O., & Banks, M. S. (2002). Humans integrate visual and haptic information in a statistically optimal fashion. *Nature, 415*(6870), 429.

Escobar, M. D., & West, M. (1995). Bayesian density estimation and inference using mixtures. *Journal of the American Statistical Association, 90*, 577–588.

Estes, D., Wellman, H., & Woolley, J. (1989). Children's understanding of mental phenomena. *Advances in Child Development and Behavior, 22*, 41–87.

Evans, J. S. B., Handley, S. J., Over, D. E., & Perham, N. (2002). Background beliefs in Bayesian inference. *Memory & Cognition, 30*(2), 179–190.

Evans, J. S. B., Newstead, S. E., & Byrne, R. M. (1993). *Human reasoning: The psychology of deduction.* Erlbaum.

Evans, J., & Over, D. (2004). *If.* Oxford University Press.

Evans, O., Stuhlmüller, A., Salvatier, J., & Filan, D. (2017). *Modeling Agents with Probabilistic Programs.* Accessed September 20, 203, from http://agentmodels.org.

Favaro, S., Nipoti, B., & Teh, Y. W. (2016). Rediscovery of Good-Turing estimators via Bayesian nonparametrics. *Biometrics, 72*(1), 136–145.

Fazeli, N., Oller, M., Wu, J., Wu, Z., Tenenbaum, J., & Rodriguez, A. (2019). See, feel, act: Hierarchical learning for complex manipulation skills with multisensory fusion. *Science Robotics*, *4*(26), eaav3123.

Fearnhead, P. (2002). Markov chain Monte Carlo, sufficient statistics, and particle filters. *Journal of Computational and Graphical Statistics*, *11*(4), 848–862.

Feldman, J. (2000). Minimization of Boolean complexity in human concept learning. *Nature*, *407*, 630–633.

Feldman, J. (2001). Bayesian contour integration. *Perception & Psychophysics*, *63*(7), 1171–1182.

Feldman, J. (2003). A catalog of Boolean concepts. *Journal of Mathematical Psychology*, *47*, 98–112.

Feldman, J. (2006). An algebra of human concept learning. *Journal of Mathematical Psychology*, *50*, 339–368.

Feldman, J. A., & Ballard, D. H. (1982). Connectionist models and their properties. *Cognitive Science*, *6*(3), 205–254.

Feldman, J., & Singh, M. (2005). Information along contours and object boundaries. *Psychological Review*, *112*(1), 243–252.

Feldman, N., Goldwater, S., Griffiths, T. L., & Morgan, J. (2013). A role for the developing lexicon in phonetic category acquisition. *Psychological Review*, *120*(4), 751–778.

Feldman, N. H., & Griffiths, T. L. (2007). A rational account of the perceptual magnet effect. In *Proceedings of the 29th Annual Meeting of the Cognitive Science Society* (pp. 257–262). Cognitive Science Society.

Feldman, N. H., Griffiths, T. L., & Morgan, J. L. (2009). The influence of categories on perception: Explaining the perceptual magnet effect as optimal statistical inference. *Psychological Review*, *116*(4), 752–782.

Ferguson, T. S. (1983). Bayesian density estimation by mixtures of normal distributions. In M. Rizvi, J. Rustagi, & D. Siegmund (Eds.), *Recent advances in statistics* (pp. 287–302). Academic Press.

Ferreira, F., Christianson, K., & Hollingworth, A. (2001). Misinterpretations of garden-path sentences: Implications for models of sentence processing and reanalysis. *Journal of Psycholinguistic Research*, *30*(1), 3–20.

Festinger, L. (1957). *A theory of cognitive dissonance.* Stanford University Press.

Feyerabend, P. (1993). *Against method.* Verso.

Finn, C., Abbeel, P., & Levine, S. (2017). Model-agnostic meta- learning for fast adaptation of deep networks. *arXiv preprint arXiv:1703.03400.*

Fischer, J., Mikhael, J. G., Tenenbaum, J. B., & Kanwisher, N. (2016). Functional neuroanatomy of intuitive physical inference. *Proceedings of the National Academy of Sciences*, *113*(34), E5072–E5081.

Fiser, J., Berkes, P., Orbán, G., & Lengyel, M. (2010). Statistically optimal perception and learning: from behavior to neural representations. *Trends in Cognitive Sciences*, *14*(3), 119–130.

Fisher, J. C. (2006). Does simulation theory really involve simulation? *Philosophical Psychology*, *19*(4), 417–432.

Fisher, R. A. (1930). Inverse probability. *Mathematical Proceedings of the Cambridge Philosophical Society*, *26*(4), 528–535.

Flanagan, J. R., & Wing, A. M. (1997). The role of internal models in motion planning and control: Evidence from grip force adjustments during movements of hand-held loads. *Journal of Neuroscience*, *17*(4), 1519–1528.

Fodor, J. A. (1975). *The language of thought.* Harvard University Press.

Fodor, J. A. (1983). *The modularity of mind.* MIT Press.

Fodor, J. A. (1998). *Concepts: Where cognitive science went wrong.* Oxford University Press.

Fodor, J. A., Bever, T. G., & Garrett, M. F. (1974). *The psychology of language: An introduction to psycholinguistics and generative grammar.* McGraw-Hill.

Fodor, J. A., & Pylyshyn, Z. W. (1988). Connectionism and cognitive architecture: A critical analysis. *Cognition*, *28*(1–2), 3–71.

Foraker, S., Regier, T., Khetarpal, N., Perfors, A., & Tenenbaum, J. (2009). Indirect evidence and the poverty of the stimulus: The case of anaphoric *one*. *Cognitive Science*, *33*, 287–300.

Ford, K. M., & Pylyshyn, Z. W. (1996). *The robot's dilemma revisited: The frame problem in artificial intelligence.* Greenwood Publishing Group.

Forstmann, B. U., Ratcliff, R., & Wagenmakers, E.-J. (2016). Sequential sampling models in cognitive neuroscience: Advantages, applications, and extensions. *Annual Review of Psychology*, *67*, 641–666.

Frank, M. C., Goldwater, S., Griffiths, T. L., & Tenenbaum, J. B. (2010). Modeling human performance in statistical word segmentation. *Cognition, 117*(2), 107–125.

Frank, M. C., & Goodman, N. D. (2012). Predicting pragmatic reasoning in language games. *Science, 336*(6084), 998–998.

Frank, M. C., & Tenenbaum, J. B. (2011). Three ideal observer models for rule learning in simple languages. *Cognition, 120*, 360-371.

Frazier, L., & Fodor, J. D. (1978). The sausage machine: A new two-stage parsing model. *Cognition, 6*(4), 291–325.

Fredkin, E., & Toffoli, T. (1982). Conservative logic. *International Journal of Theoretical Physics, 21*(3), 219–253.

Fried, L. S., & Holyoak, K. J. (1984). Induction of category distributions: A framework for classification learning. *Journal of Experimental Psychology: Learning, Memory and Cognition, 10*, 234–257.

Friedman, N. (1997). Learning belief networks in the presence of missing values and hidden variables. In D. Fisher (Ed.), *Fourteenth International Conference on Machine Learning* (pp. 125–133). Morgan Kaufmann.

Friedman, N., & Koller, D. (2000). Being Bayesian about network structure. In *Proceedings of the 16th Annual Conference on Uncertainty in AI* (pp. 201–210). Stanford, CA.

Friston, K. (2009). The free-energy principle: A rough guide to the brain? *Trends in Cognitive Sciences, 13*(7), 293–301.

Futrelle, R., Albright, P., Graff, P. & O'Donnell, T. (2013). A generative model of phonotactics. In *Transactions of the Association for Computational Linguistics.*

Gabaix, X. (2014). A sparsity-based model of bounded rationality. *Quarterly Journal of Economics, 129*(4), 1661–1710.

Gal, Y., & Ghahramani, Z. (2016). Dropout as a Bayesian approximation: Representing model uncertainty in deep learning. In *Proceedings of the International Conference on Machine Learning* (pp. 1050–1059).

Galantucci, B. (2005). An experimental study of the emergence of human communication systems. *Cognitive Science, 29*(5), 737–767.

Gao, X., Xiao, B., Tao, D., & Li, X. (2010). A survey of graph edit distance. *Pattern Analysis and Applications, 13*(1), 113–129.

Garrett, A. J. M., & Coles, P. (1993). Bayesian inductive inference and the anthropic principles. *Comments on Astrophysics and Space Physics, 17*, 23–47.

Garthwaite, P. H., Kadane, J. B., & O'Hagan, A. (2005). Statistical methods for eliciting probability distributions. *Journal of the American Statistical Association, 100*(470), 680–701.

Ge, H., Xu, K., & Ghahramani, Z. (2018). Turing: A language for flexible probabilistic inference. In *International Conference on Artificial Intelligence and Statistics* (pp. 1682–1690).

Gelman, A., Carlin, J. B., Stern, H. S., & Rubin, D. B. (1995). *Bayesian data analysis.* Chapman & Hall.

Gelman, A., & Shalizi, C. R. (2013). Philosophy and the practice of Bayesian statistics. *British Journal of Mathematical and Statistical Psychology, 66*(1), 8–38.

Gelman, S. A. (2003). *The essential child: Origins of essentialism in everyday thought.* Oxford University Press.

Geman, S., & Geman, D. (1984). Stochastic relaxation, Gibbs distributions, and the Bayesian restoration of images. *IEEE Transactions on Pattern Analysis and Machine Intelligence, 6*, 721–741.

Geman, S., Potter, D. F., & Chi, Z. (2002). Composition systems. *Quarterly of Applied Mathematics, 60*(4), 707–736.

Genesereth, M. R., & Nilsson, N. J. (1987). *Logical foundations of artificial intelligence.* Springer.

Gentner, D., & Stevens, A. L. (1983). *Mental models.* Erlbaum.

Gergely, G., Nádasdy, Z., Csibra, G., & Bíró, S. (1995). Taking the intentional stance at 12 months of age. *Cognition, 56*(2), 165–193.

Gershman, S. J. (2018). The successor representation: Its computational logic and neural substrates. *Journal of Neuroscience, 38*(33), 7193–7200.

Gershman, S. J. (2021). The rational analysis of memory. In A. D. Wagner & M. J. Kahana (Eds.), *Oxford handbook of human memory.* Oxford University Press.

Gershman, S. J., Blei, D. M., & Niv, Y. (2010). Context, learning, and extinction. *Psychological Review, 117*(1), 197.

Gershman, S. J., Horvitz, E. J., & Tenenbaum, J. B. (2015). Computational rationality: A converging paradigm for intelligence in brains, minds, and machines. *Science*, *349*(6245), 273–278.

Gershman, S. J., & Niv, Y. (2015). Novelty and inductive generalization in human reinforcement learning. *Topics in Cognitive Science*, *7*(3), 391–415.

Gershman, S. J., Vul, E., & Tenenbaum, J. B. (2012). Multistability and perceptual inference. *Neural Computation*, *24*(1), 1–24.

Gerstenberg, T., & Goodman, N. D. (2012). Ping Pong in Church: Productive use of concepts in human probabilistic inference. In *Proceedings of the 34th Annual Meeting of the Cognitive Science Society* (pp. 1590–1595).

Gerstenberg, T., Goodman, N. D., Lagnado, D. A., & Tenenbaum, J. B. (2021). A counterfactual simulation model of causal judgments for physical events. *Psychological Review*,*128*(5), 936–975.

Gerstenberg, T., Peterson, M. F., Goodman, N. D., Lagnado, D. A., & Tenenbaum, J. B. (2017). Eye-tracking causality. *Psychological Science*, *28*(12), 1731–1744.

Gerstenberg, T., Siegel, M. H., & Tenenbaum, J. B. (2018). What happened? Reconstructing the past from vision and sound. In *Proceedings of the 40th Annual Meeting of the Cognitive Science Society.*

Gerstenberg, T., & Stephan, S. (2021). A counterfactual simulation model of causation by omission. *Cognition*, *216*, 104842.

Gerstenberg, T., & Tenenbaum, J. B. (2016). Understanding "almost": Empirical and computational studies of near misses. In *Proceedings of the 38th Annual Meeting of the Cognitive Science Society* (pp. 2777–2782). Cognitive Science Society.

Gerstenberg, T., & Tenenbaum, J. B. (2017). Intuitive theories. In M. Waldmannn (Ed.), *Oxford handbook of causal reasoning* (pp. 515–548). Oxford University Press.

Gerstenberg, T., Zhou, L., Smith, K. A., & Tenenbaum, J. B. (2017). Faulty towers: A hypothetical simulation model of physical support. In *Proceedings of the 39th Annual Meeting of the Cognitive Science Society* (pp. 409–414).

Gettier, E. L. (1963). Is justified true belief knowledge? *Analysis*, *23*(6), 121–123.

Geweke, J. (1989). Bayesian inference in econometric models using Monte Carlo integration. *Econometrica*, *57*(6), 1317–1339.

Ghahramani, Z. (2004). Unsupervised learning. In O. Bousquet, G. Raetsch, & U. von Luxburg (Eds.), *Advanced lectures on machine learning* (pp. 72–112). Springer-Verlag.

Ghallab, M., Nau, D., & Traverso, P. (2016). *Automated planning and acting.* Cambridge University Press.

Gibson, E., Bergen, L., & Piantadosi, S. T. (2013). Rational integration of noisy evidence and prior semantic expectations in sentence interpretation. *Proceedings of the National Academy of Sciences*, *110*(20), 8051–8056.

Gibson, E., & Thomas, J. (1999). Memory limitations and structural forgetting: The perception of complex ungrammatical sentences as grammatical. *Language and Cognitive Processes*, *14*(3), 225–248.

Gibson, E., & Wexler, K. (1994). Triggers. *Linguistic Inquiry*, *25*(3), 407–454.

Gibson, J. J. (1950). *The perception of the visual world.* Houghton Mifflin.

Gibson, J. J. (1979). *The ecological approach to visual perception.* Erlbaum.

Gigerenzer, G., & Goldstein, D. G. (1996). Reasoning the fast and frugal way: Models of bounded rationality. *Psychological Review*, *103*(4), 650–669.

Gigerenzer, G., & Hug, K. (1992). Domain-specific reasoning: Social contracts, cheating, and perspective change. *Cognition*, *43*(2), 127–171.

Gigerenzer, G., Swijtink, Z., Porter, T., Daston, L., Beatty, J., & Kruger, L. (1989). *The empire of chance.* Cambridge University Press.

Gigerenzer, G., & Todd, P. M. (1999). *Simple heuristics that make us smart.* Oxford University Press.

Gilden, D. L., & Proffitt, D. R. (1989). Understanding collision dynamics. *Journal of Experimental Psychology: Human Perception and Performance*, *15*(2), 372.

Gilks, W. R., Richardson, S., & Spiegelhalter, D. J. (1996). *Markov chain Monte Carlo in practice.* Chapman & Hall.

Gläscher, J., Daw, N., Dayan, P., & O'Doherty, J. P. (2010). States versus rewards: Dissociable neural prediction error signals underlying model-based and model-free reinforcement learning. *Neuron*, *66*(4), 585–595.

Glassen, T., & Nitsch, V. (2016). Hierarchical Bayesian models of cognitive development. *Biological Cybernetics, 110*(2), 217–227.

Glimcher, P. W. (2011). Understanding dopamine and reinforcement learning: The dopamine reward prediction error hypothesis. *Proceedings of the National Academy of Sciences, 108*(supplement_3), 15647–15654.

Glymour, C. N. (2001). *The mind's arrows: Bayes nets and graphical causal models in psychology.* MIT Press.

Godfrey-Smith, P. (2003). *Theory and reality.* University of Chicago Press.

Gold, E. M. (1967). Language identification in the limit. *Information and Control, 10*(5), 447–474.

Gold, J. I., & Shadlen, M. N. (2002). Banburismus and the brain: Decoding the relationship between sensory stimuli, decisions and reward. *Neuron, 36*, 299–308.

Gold, J. I., & Shadlen, M. N. (2007). The neural basis of decision making. *Annual Review of Neuroscience, 30*(1), 535–574.

Goldberg, A. E. (2006). *Constructions at work: The nature of generalization in language.* Oxford University Press.

Goldenberg, G., & Spatt, J. (2009). The neural basis of tool use. *Brain, 132*(6), 1645–1655.

Goldsmith, J. (2001). Unsupervised learning of the morphology of a natural language. *Computational linguistics, 27*(2), 153–198.

Goldwater, S., & Griffiths, T. L. (2007). A fully Bayesian approach to unsupervised part-of-speech tagging. In *Proceedings of the 45th Annual Meeting of the Association of Computational Linguistics* (pp. 744–751).

Goldwater, S., Griffiths, T. L., & Johnson, M. (2006a). Contextual dependencies in unsupervised word segmentation. In *Proceedings of COLING/ACL 2006.*

Goldwater, S., Griffiths, T. L., & Johnson, M. (2006b). Interpolating between types and tokens by estimating power-law generators. *Advances in Neural Information Processing Systems 18*, 459–466.

Goldwater, S., Griffiths, T. L., & Johnson, M. (2009). A Bayesian framework for word segmentation: Exploring the effects of context. *Cognition, 112*, 21–54.

Gomez, R. L., & Gerken, L. (1999). Artificial grammar learning by 1-year-olds leads to specific and abstract knowledge. *Cognition, 70*, 109–135.

Good, I. J. (1950). *Probability and the weighing of evidence.* C. Griffin.

Good, I. J. (1979). A. M. Turing's statistical work in World War II. *Biometrika, 66*, 393–396.

Goodfellow, I., Bengio, Y., & Courville, A. (2016). *Deep learning.* MIT Press.

Goodman, J., & Weare, J. (2010). Ensemble samplers with affine invariance. *Communications in Applied Mathematics and Computational Science, 5*(1), 65–80.

Goodman, N., Baker, C., Bonawitz, E., et al. (2006). Intuitive theories of mind: A rational approach to false belief. In *Proceedings of the 28th Annual Meeting of the Cognitive Science Society.*

Goodman, N., Mansinghka, V., Roy, D. M., Bonawitz, K., & Tenenbaum, J. B. (2012). Church: A language for generative models. *arXiv preprint arXiv:1206.3255.*

Goodman, N. D., Mansinghka, V. K., Roy, D. M., Bonawitz, K., & Tenenbaum, J. B. (2008a). Church: A language for generative models. In *Proceedings of the 24th Conference on Uncertainty in Artificial Intelligence.*

Goodman, N. D., & Frank, M. C. (2016). Pragmatic language interpretation as probabilistic inference. *Trends in Cognitive Sciences, 20*(11), 818–829.

Goodman, N. D., & Lassiter, D. (2015). Probabilistic semantics and pragmatics: Uncertainty in language and thought. In S. Lappin & C. Fox (Eds.), *Handbook of contemporary semantic theory* (2nd ed., pp. 655–686). Wiley-Blackwell.

Goodman, N. D., & Stuhlmüller, A. (2013). Knowledge and implicature: Modeling language understanding as social cognition. *Topics in Cognitive Science, 5*(1), 173–184.

Goodman, N. D., Tenenbaum, J. B., Feldman, J., & Griffiths, T. L. (2008b). A rational analysis of rule-based concept learning. *Cognitive Science, 32*(1), 108–154.

Goodman, N. D., Tenenbaum, J. B., & Gerstenberg, T. (2014). *Concepts in a probabilistic language of thought* (Tech. Rep.). Center for Brains, Minds and Machines (CBMM).

Goodman, N. D., Ullman, T. D., & Tenenbaum, J. B. (2011). Learning a theory of causality. *Psychological Review, 118*(1), 110–119.

Gopnik, A. (2003). The theory as an alternative to the innateness hypothesis. In L. M. Antony & N. Hornstein (Eds.), *Chomsky and his critics* (pp. 238–254). Wiley.

Gopnik, A., Glymour, C., Sobel, D., Schulz, L., Kushnir, T., & Danks, D. (2004). A theory of causal learning in children: Causal maps and Bayes nets. *Psychological Review*, *111*, 1–31.

Gopnik, A., Griffiths, T. L., & Lucas, C. G. (2015). When younger learners can be better (or at least more open-minded) than older ones. *Current Directions in Psychological Science*, *24*(2), 87–92.

Gopnik, A., & Meltzoff, A. N. (1997). *Words, thoughts, and theories.* MIT Press.

Gopnik, A., Meltzoff, A. N., & Kuhl, P. K. (1999). *The scientist in the crib: Minds, brains, and how children learn.* William Morrow & Co.

Gopnik, A., O'Grady, S., Lucas, C. G., et al. (2017). Changes in cognitive flexibility and hypothesis search across human life history from childhood to adolescence to adulthood. *Proceedings of the National Academy of Sciences*, *114*(30), 7892–7899.

Gopnik, A., & Schulz, L. (2004). Mechanisms of theory formation in young children. *Trends in Cognitive Science*, *8*, 371–377.

Gopnik, A., Sobel, D. M., Schulz, L. E., & Glymour, C. (2001). Causal learning mechanisms in very young children: Two, three, and four-year-olds infer causal relations from patterns of variation and covariation. *Developmental Psychology*, *37*, 620–629.

Gopnik, A., & Tenenbaum, J. B. (2007). Bayesian networks, Bayesian learning and cognitive development. *Developmental Science*, *10*(3), 281–287.

Gopnik, A., & Wellman, H. M. (2012). Reconstructing constructivism: Causal models, Bayesian learning mechanisms, and the theory theory. *Psychological Bulletin*, *138*(6), 1085–1108.

Gott, J. R. III. (1993). Implications of the Copernican principle for our future prospects. *Nature*, *363*, 315–319.

Gott, J. R. III. (1994). Future prospects discussed. *Nature*, *368*, 108.

Gould, S. J., & Lewontin, R. C. (1979). The spandrels of San Marco and the Panglossian paradigm: A critique of the adaptationist programme. *Proceedings of the Royal Society of London. Series B. Biological Sciences*, *205*(1161), 581–598.

Grant, E., Finn, C., Levine, S., Darrell, T., & Griffiths, T. L. (2018). Recasting gradient-based meta-learning as hierarchical Bayes. In *Proceeedings of the International Conference on Learning Representations (ICLR).*

Graves, A., Wayne, G., & Danihelka, I. (2014). Neural Turing machines. *arXiv preprint arXiv:1410.5401.*

Green, D. M., & Swets, J. A. (1966). *Signal detection theory and psychophysics.* Wiley.

Green, P., & Richardson, S. (2001). Modelling heterogeneity with and without the Dirichlet process. *Scandinavian Journal of Statistics*, *28*, 355–377.

Gregory, J. (2018). *Game engine architecture, third edition.* CRC Press.

Grice, H. P. (1957). Meaning. *Philosophical Review*, *66*(3), 377–388.

Grice, H. P. (1975). Logic and conversation. In D. Davidson & G. Harman (Eds.), *The logic of grammar* (pp. 64–75). Dickenson.

Griffiths, T. L. (2020). Understanding human intelligence through human limitations. *Trends in Cognitive Sciences*, *24*(11), 873–883.

Griffiths, T. L., Callaway, F., Chang, M. B., Grant, E., Krueger, P. M., & Lieder, F. (2019). Doing more with less: Meta-reasoning and meta- learning in humans and machines. *Current Opinion in Behavioral Sciences*, *29*, 24–30.

Griffiths, T. L., Canini, K. R., Sanborn, A. N., & Navarro, D. J. (2007). Unifying rational models of categorization via the hierarchical Dirichlet process. In *Proceedings of the Twenty-Ninth Annual Meeting of the Cognitive Science Society.*

Griffiths, T. L., Chater, N., Kemp, C., Perfors, A., & Tenenbaum, J. (2010). Probabilistic models of cognition: Exploring the laws of thought. *Trends in Cognitive Sciences*, *14*, 357–364.

Griffiths, T. L., Chater, N., Norris, D., & Pouget, A. (2012a). How the Bayesians got their beliefs (and what those beliefs actually are): Comment on Bowers and Davis (2012). *Psychological Bulletin*, *138*, 415–422.

Griffiths, T. L., Christian, B. R., & Kalish, M. L. (2008a). Using category structures to test iterated learning as a method for identifying inductive biases. *Cognitive Science*, *32*, 68–107.

Griffiths, T. L., Daniels, D., Austerweil, J. L., & Tenenbaum, J. B. (2018). Subjective randomness as statistical inference. *Cognitive Psychology*, *103*, 85–109.

Griffiths, T., Lucas, C., Williams, J., & Kalish, M. (2008b). Modeling human function learning with Gaussian processes. In *Advances in Neural Information Processing Systems 21*.

Griffiths, T. L., & Ghahramani, Z. (2005). *Infinite latent feature models and the Indian buffet process* (Tech. Rep. No. 2005-001). Gatsby Computational Neuroscience Unit.

Griffiths, T. L., & Ghahramani, Z. (2006). Infinite latent feature models and the Indian buffet process. In *Advances in Neural Information Processing Systems 18*.

Griffiths, T. L., & Kalish, M. L. (2005). A Bayesian view of language evolution by iterated learning. In *Proceedings of the Twenty-Seventh Annual Conference of the Cognitive Science Society* (pp. 827–832).

Griffiths, T. L., & Kalish, M. L. (2007). Language evolution by iterated learning with Bayesian agents. *Cognitive Science*, *31*(3), 441–480.

Griffiths, T. L., Kemp, C., & Tenenbaum, J. B. (2008c). Bayesian models of cognition. In R. Sun (Ed.), *Cambridge handbook of computational cognitive modeling* (pp. 59–100). Cambridge University Press.

Griffiths, T. L., Lieder, F., & Goodman, N. D. (2015). Rational use of cognitive resources: Levels of analysis between the computational and the algorithmic. *Topics in Cognitive Science*, *7*(2), 217–229.

Griffiths, T. L., & Pacer, M. (2011). A rational model of causal inference with continuous causes. In *Advances in Neural Information Processing Systems* (pp. 2384–2392).

Griffiths, T. L., Sanborn, A. N., Canini, K. R., & Navarro, D. J. (2008c). Categorization as nonparametric Bayesian density estimation. In N. Chater & M. Oaksford (Eds.), *The probabilistic mind* (pp. 303–328). Oxford University Press.

Griffiths, T. L., Sanborn, A. N., Canini, K. R., Navarro, D. J., & Tenenbaum, J. B. (2011a). Nonparametric Bayesian models of categorization. In E. M. Pothos & A. J. Wills (Eds.), *Formal approaches in categorization* (pp. 173–198). Cambridge University Press.

Griffiths, T. L., Sobel, D. M., Tenenbaum, J. B., & Gopnik, A. (2011b). Bayes and blickets: Effects of knowledge on causal induction in children and adults. *Cognitive Science*, *35*, 1407–1455.

Griffiths, T. L., & Steyvers, M. (2002). A probabilistic approach to semantic representation. In *Proceedings of the Twenty-Fourth Annual Conference of the Cognitive Science Society*.

Griffiths, T. L., & Steyvers, M. (2003). Prediction and semantic association. In *Advances in Neural Information Processing systems 15*. MIT Press.

Griffiths, T. L., & Steyvers, M. (2004). Finding scientific topics. *Proceedings of the National Academy of Science*, *101*, 5228–5235.

Griffiths, T. L., Steyvers, M., Blei, D. M., & Tenenbaum, J. B. (2005). Integrating topics and syntax. In *Advances in Neural Information Processing Systems 17*.

Griffiths, T. L., Steyvers, M., & Tenenbaum, J. B. (2007). Topics in semantic association. *Psychological Review*, *114*, 211–244.

Griffiths, T. L., & Tenenbaum, J. B. (2000). Teacakes, trains, toxins, and taxicabs: A Bayesian account of predicting the future. In *Proccedings of the 22nd Annual Conference of the Cognitive Science Society* (pp. 202–207).

Griffiths, T. L., & Tenenbaum, J. B. (2001). Randomness and coincidences: Reconciling intuition and probability theory. In *Proceedings of the Twenty-Third Annual Conference of the Cognitive Science Society*.

Griffiths, T. L., & Tenenbaum, J. B. (2005). Structure and strength in causal induction. *Cognitive Psychology*, *51*(4), 334–384.

Griffiths, T. L., & Tenenbaum, J. B. (2006). Optimal predictions in everyday cognition. *Psychological Science*, *17*(9), 767–773.

Griffiths, T. L., & Tenenbaum, J. B. (2007a). From mere coincidences to meaningful discoveries. *Cognition*, *103*, 180–226.

Griffiths, T. L., & Tenenbaum, J. B. (2007b). Two proposals for causal grammars. In A. Gopnik & L. Schulz (Eds.), *Causal learning: Psychology, philosophy, and computation* (pp. 323–345). Oxford University Press.

Griffiths, T. L., & Tenenbaum, J. B. (2009). Theory-based causal induction. *Psychological Review*, *116*(4), 661–716.

Griffiths, T. L., Vul, E., & Sanborn, A. N. (2012b). Bridging levels of analysis for probabilistic models of cognition. *Current Directions in Psychological Science*, *21*(4), 263–268.

Griffiths, T. L., & Yuille, A. (2006). A primer on probabilistic inference. *Trends in Cognitive Sciences*, *10*(7), 33–58.

Grossberg, S. (1987). Competitive learning: From interactive activation to adaptive resonance. *Cognitive Science*, *11*(1), 23–63.

Grünwald, P. (2007). *The minimum description length principle.* MIT Press.

Grünwald, P., & Roos, T. (2019). Minimum description length revisited. *International Journal of Mathematics for Industry*, *11*(1), 1930001.

Grush, R. (2004). The emulation theory of representation: Motor control, imagery, and perception. *Behavioral and Brain Sciences*, *27*(3), 377–396.

Grzeszczuk, R., Terzopoulos, D., & Hinton, G. (1998, July 19–24). Neuroanimator: Fast neural network emulation and control of physics-based models. *In the Proceedings of SIGGRAPH 98.*

Guevara, T., Pucci, R., Taylor, N., Gutmann, M., Ramamoorthy, S., & Subr, K. (2018). To stir or not to stir: Online estimation of liquid properties for pouring actions. In *Modeling the physical world: Perception, learning, and control workshop.*

Hacking, I. (1975). *The emergence of probability.* Cambridge University Press.

Hagmayer, Y., & Mayrhofer, R. (2013). Hierarchical Bayesian models as formal models of causal reasoning. *Argument & Computation*, *4*(1), 36–45.

Hagmayer, Y., Sloman, S. A., Lagnado, D. A., & Waldmann, M. R. (2007). Causal reasoning through intervention. In A. Gopnik & L. Schulz (Eds.), *Causal learning: Psychology, philosophy, and computation* (pp. 86–100). Oxford University Press.

Hahn, U., Chater, N., & Richardson, L. B. (2003). Similarity as transformation. *Cognition*, *87*(1), 1–32.

Hahn, U., Close, J., & Graf, M. (2009). Transformation direction influences shape-similarity judgments. *Psychological Science*, *20*(4), 447–454.

Hahn, U., & Oaksford, M. (2007). The rationality of informal argumentation: A Bayesian approach to reasoning fallacies. *Psychological Review*, *114*(3), 704–732.

Hall, K., Hume, E., Jaeger, F., & Wedel, A. (2018). The role of predictability in shaping phonological patterns. *Linguistics Vanguard*, *4*, 20170027.

Halle, M., & Stevens, K. (1962). Speech recognition: A model and a program for research. *IRE Transactions on Information Theory*, *8*(2), 155–159.

Hamlin, J. K., & Wynn, K. (2011). Young infants prefer prosocial to antisocial others. *Cognitive Development*, *26*(1), 30–39.

Hamlin, J. K., Wynn, K., & Bloom, P. (2007). Social evaluation by preverbal infants. *Nature*, *450*, 557–559.

Hamlin, K., Ullman, T., Tenenbaum, J., Goodman, N., & Baker, C. (2013). The mentalistic basis of core social cognition: Experiments in preverbal infants and a computational model. *Developmental Science*, *16*(2), 209–226.

Hammersley, D. C., & Handscomb, J. M. (1964). *Monte Carlo methods.* Methuen.

Hamrick, J. B. (2019). Analogues of mental simulation and imagination in deep learning. *Current Opinion in Behavioral Sciences*, *29*, 8–16.

Hamrick, J. B., Battaglia, P. W., Griffiths, T. L., & Tenenbaum, J. B. (2016). Inferring mass in complex scenes by mental simulation. *Cognition*, *157*, 61–76.

Hamrick, J. B., Smith, K. A., Griffiths, T. L., & Vul, E. (2015). Think again? The amount of mental simulation tracks uncertainty in the outcome. In *Proceedings of the 37st Annual Meeting of the Cognitive Science Society.*

Hanes, D. P., & Schall, J. D. (1996). Neural control of voluntary movement initiation. *Science*, *274*(5286), 427–430.

Hansen, E. A., & Zilberstein, S. (2001). LAO*: A heuristic search algorithm that finds solutions with loops. *Artificial Intelligence*, *129*(1–2), 35–62.

Hanson, S., & Pratt, L. (1988). Comparing biases for minimal network construction with back-propagation. In *Advances in Neural Information Processing Systems 1.*

Härkönen, E., Hertzmann, A., Lehtinen, J., & Paris, S. (2020). GANspace: Discovering interpretable GAN controls. *Advances in Neural Information Processing Systems 33*, 9841–9850.

Harman, G. H. (1965). The inference to the best explanation. *Philosophical Review*, *74*(1), 88–95.

Harrison, P., Marjieh, R., Adolfi, F., et al. (2020). Gibbs sampling with people. *Advances in Neural Information Processing Systems 33*, 10659–10671.

Hart, P. E., Nilsson, N. J., & Raphael, B. (1968). A formal basis for the heuristic determination of minimum cost paths. *IEEE Transactions on Systems Science and Cybernetics*, *4*(2), 100–107.

Hastie, T., Tibshirani, R., & Friedman, J. (2009). *The elements of statistical learning* (2nd ed.). Springer.

Hastings, W. K. (1970). Monte Carlo methods using Markov chains and their applications. *Biometrika*, *57*, 97–109.

Hatano, G., & Inagaki, K. (1994). Young children's naïve theory of biology. *Cognition*, *50*, 171–188.

Hauf, P., Paulus, M., & Baillargeon, R. (2012). Infants use compression information to infer objects' weights: Examining cognition, exploration, and prospective action in a preferential-reaching task. *Child Development*, *83*(6), 1978–1995.

Hauser, M. D., Chomsky, N., & Fitch, W. T. (2002). The faculty of language: What is it, who has it, and how did it evolve? *Science*, *298*(5598), 1569–1579.

Hawkins, R. D., Franke, M., Frank, M. C., et al. (2023). From partners to populations: A hierarchical Bayesian account of coordination and convention. *Psychological Review*, *130*(4), 977–1016.

Hay, N., Russell, S., Tolpin, D., & Shimony, S. (2012). Selecting computations: Theory and applications. In N. de Freitas & K. Murphy (Eds.), *Proceedings of the 28th Conference on Uncertainty in Artificial Intelligence*. AUAI Press.

Hayes, P. J. (1979). The naive physics manifesto. In D. Michie (Ed.), *Expert systems in the microelectronic age* (pp. 242—270).Edinburgh University Press.

He, R., Jain, Y. R., & Lieder, F. (2021). Measuring and modelling how people learn how to plan and how people adapt their planning strategies to the structure of the environment. In *Proceedings of the International Conference on Cognitive Modeling*.

He, R., & Lieder, F. (under review). *Where do adaptive planning strategies come from?* Under review. Preprint, doi:10.13140/RG.2.2.28966.60487

Hecht, H., & Bertamini, M. (2000). Understanding projectile acceleration. *Journal of Experimental Psychology: Human Perception and Performance*, *26*(2), 730–746.

Heckerman, D. (1998). A tutorial on learning with Bayesian networks. In M. I. Jordan (Ed.), *Learning in graphical models* (pp. 301–354). MIT Press.

Hegarty, M. (1992). Mental animation: Inferring motion from static dis- plays of mechanical systems. *Journal of Experimental Psychology: Learning, Memory, and Cognition*, *18*(5), 1084–1102.

Hegarty, M. (2004). Mechanical reasoning by mental simulation. *Trends in Cognitive Sciences*, *8*(6), 280–285.

Heider, F. (1958). *The psychology of interpersonal relations*. Wiley.

Heider, F., & Simmel, M. (1944). An experimental study of apparent behavior. *American Journal of Psychology*, *57*(2), 243–259.

Heim, I., & Kratzer, A. (1998). *Semantics in generative grammar*. Wiley- Blackwell.

Heit, E., & Rubinstein, J. (1994). Similarity and property effects in inductive reasoning. *Journal of Experimental Psychology: Learning, Memory and Cognition*, *20*(2), 411–422.

Heller, K. A., Sanborn, A., & Chater, N. (2009). Hierarchical learning of dimensional biases in human categorization. *Advances in Neural Information Processing Systems 22*, 727–735.

Helmholtz, H. von (1866/1962). Concerning the perceptions in general. In J. P. C. Southall (Ed.), *Treatise on physiological optics* (Vol. 3, pp. 1–37). Dover.

Henderson, L. (2022). The problem of induction. In E. N. Zalta & U. Nodelman (Eds.), *Stanford encyclopedia of philosophy* (Winter 2022 ed.). Metaphysics Research Lab, Stanford University.

Hermer, L., & Spelke, E. S. (1994). A geometric process for spatial reorientation in young children. *Nature*, *370*(6484), 57.

Hernández-Lobato, J. M., Hoffman, M. W., & Ghahramani, Z. (2014). Predictive entropy search for efficient global optimization of black-box functions. In *Advances in Neural Information Processing Systems* (pp. 918–926).

Hespos, S. J., Ferry, A. L., Anderson, E. M., Hollenbeck, E. N., & Rips, L. J. (2016). Five-month-old infants have general knowledge of how nonsolid substances behave and interact. *Psychological Science*, *27*(2), 244–256.

Hidaka, S. (2013). A computational model associating learning process, word attributes, and age of acquisition. *PLoS One*, *8*(11), e76242.

Hinton, G. E. (1986). Learning distributed representations of concepts. In *Proceedings of the 8th Annual Conference of the Cognitive Science Society.*

Hinton, G., & Anderson, J. (1981). *Parallel models of associative memory.* Erlbaum.

Hinton, G. E., & Salakhutdinov, R. R. (2006). Reducing the dimensionality of data with neural networks. *Science*, *313*(5786), 504–507.

Hinton, G. E., & Sejnowski, T. J. (1983). Optimal perceptual inference. In *Proceedings of the IEEE Conference on Computer Vision and Pattern Recognition* (Vol. 448, pp. 448–453).

Hinton, G. E., & Sejnowski, T. J. (1986). Learning and relearning in Boltzmann machines. In J. McClelland & D. Rumelhart (Eds.), *Parallel distributed processing: Explorations in the microstructure of cognition* (Vol. 1, pp. 282–317). MIT Press.

Hirsh-Pasek, K., Treiman, R., & Schneiderman, M. (1984). Brown & Hanlon revisited: Mothers' sensitivity to ungrammatical forms. *Journal of Child Language*, *11*(1), 81–88.

Hjort, N. L. (1990). Nonparametric Bayes estimators based on beta processes in models for life history data. *Annals of Statistics*, *18*, 1259–1294.

Hjort, N. L., Holmes, C., Müller, P., & Walker, S. G. (2010). *Bayesian nonparametrics.* Cambridge University Press.

Ho, M. K., Abel, D., Correa, C. G., Littman, M. L., Cohen, J. D., & Griffiths, T. L. (2022). People construct simplified mental representations to plan. *Nature*, *606*(7912), 129–136.

Ho, M. K., Abel, D., Griffiths, T. L., & Littman, M. L. (2019). The value of abstraction. *Current Opinion in Behavioral Sciences*, *29*, 111–116.

Ho, M. K., Cushman, F., Littman, M. L., & Austerweil, J. L. (2019). People teach with rewards and punishments as communication, not reinforcements. *Journal of Experimental Psychology: General*, *148*(3), 520–549.

Hodgetts, C. J., Hahn, U., & Chater, N. (2009). Transformation and alignment in similarity. *Cognition*, *113*(1), 62–79.

Hofmann, T. (1999). Probablistic latent semantic indexing. In *Proceedings of the Twenty-Second Annual International SIGIR Conference.*

Holtzen, S., Zhao, Y., Gao, T., Tenenbaum, J. B., & Zhu, S.-C. (2016). Inferring human intent from video by sampling hierarchical plans. In *IEEE/RSJ International Conference on Intelligent Robots and Systems (IROS),* (pp. 1489–1496).

Holyoak, K. J. (2012). Analogy and relational reasoning. In K. J. Holyoak & R. G. Morrison (Eds.), *Oxford handbook of thinking and reasoning* (pp. 234–259). Oxford University Press.

Hommel, B., Chapman, C. S., Cisek, P., Neyedli, H. F., Song, J.-H., & Welsh, T. N. (2019). No one knows what attention is. *Attention, Perception, & Psychophysics*, *81*(7), 2288–2303.

Hopcroft, J. E., Motwani, R., & Ullman, J. D. (2001). Introduction to automata theory, languages, and computation. *Acm Sigact News*, *32*(1), 60–65.

Hopper, P. J., & Traugott, E. C. (2003). *Grammaticalization.* Cambridge University Press.

Horvitz, E. J. (1987). Reasoning about beliefs and actions under computational resource constraints. In J. F. Lemmer, T. Levitt, & L. N. Kanal (Eds.), *Proceedings of the Third Conference on Uncertainty in Artificial Intelligence* (pp. 301–324). AUAI Press.

Horwich, P. (1982). *Probability and evidence.* Cambridge University Press.

Hoshen, Y. (2017). Vain: Attentional multi-agent predictive modeling. In *Advances in Neural Information Processing Systems* (pp. 2701–2711).

Houlihan, S. D., Kleiman-Weiner, M., Hewitt, L. B., Tenenbaum, J. B., & Saxe, R. (2023). Emotion prediction as computation over a generative theory of mind. *Philosophical Transactions of the Royal Society A*, *381*(2251), 20220047.

Howes, A., Lewis, R. L., & Vera, A. (2009). Rational adaptation under task and processing constraints: Implications for testing theories of cognition and action. *Psychological Review*, *116*(4), 717–751.

Howson, C., & Urbach, P. (1993). *Scientific reasoning: The Bayesian approach.* Open Court.

Hsu, A. S., & Chater, N. (2010). The logical problem of language acquisition: A probabilistic perspective. *Cognitive Science, 34*(6), 972–1016.

Hsu, A. S., Chater, N., & Vitányi, P. M. (2011). The probabilistic analysis of language acquisition: Theoretical, computational, and experimental analysis. *Cognition, 120*(3), 380–390.

Hsu, A. S., Chater, N., & Vitányi, P. M. (2013). Language learning from positive evidence, reconsidered: A simplicity-based approach. *Topics in Cognitive Science, 5*(1), 35–55.

Hsu, A., & Griffiths, T. L. (2009). Differential use of implicit negative evidence in generative and discriminative language learning. *Advances in Neural Information Processing Systems 22,* 754–762.

Hsu, A. S., Horng, A., Griffiths, T. L., & Chater, N. (2017). When absence of evidence is evidence of absence: Rational inferences from absent data. *Cognitive Science, 41,* 1155–1167.

Hsu, A. S., Martin, J. B., Sanborn, A. N., & Griffiths, T. L. (2019). Identifying category representations for complex stimuli using discrete Markov chain Monte Carlo with people. *Behavior Research Methods, 51*(4), 1706–1716.

Huang, Y., & Rao, R. P. (2014). Neurons as Monte Carlo samplers: Bayesian inference and learning in spiking networks. In Z. Ghahramani, M. Welling, C. Cortes, N. Lawrence, & K. Weinberger (Eds.), *Advances in Neural Information Processing Systems* (pp. 1943–1951).

Huelsenbeck, J. P., & Ronquist, F. (2001). MRBAYES: Bayesian inference of phylogenetic trees. *Bioinformatics, 17*(8), 754–755.

Hume, D. (1739/1978). *A treatise of human nature.* Oxford University Press.

Hunt, B. J. (2005). *The Maxwellians.* Cornell University Press.

Huttenlocher, J., Hedges, L. V., & Duncan, S. (1991). Categories and particulars: prototype effects in estimating spatial location. *Psychological Review, 98*(3), 352.

Huttenlocher, J., Hedges, L. V., & Vevea, J. L. (2000). Why do categories affect stimulus judgment? *Journal of Experimental Psychology: General, 129,* 220–241.

Hutter, M. (2002). The fastest and shortest algorithm for all well-defined problems. *International Journal of Foundations of Computer Science, 13*(03), 431–443.

Hutter, M. (2005). *Universal artificial intelligence.* Springer.

Huys, Q. J., Eshel, N., O'Nions, E., Sheridan, L., Dayan, P., & Roiser, J. P. (2012). Bonsai trees in your head: How the Pavlovian system sculpts goal-directed choices by pruning decision trees. *PLoS Computational Biology, 8*(3), e1002410.

Huys, Q. J. M., Lally, N., Faulkner, P., et al. (2015). Interplay of approximate planning strategies. *Proceedings of the National Academy of Sciences, 112*(10), 3098–3103.

Icard, T., & Goodman, N. D. (2015). A resource-rational approach to the causal frame problem. In *Proceedings of the Annual Meeting of the Cognitive Science Society.*

Icarte, R. T., Klassen, T., Valenzano, R., & McIlraith, S. (2018). Using reward machines for high-level task specification and decomposition in reinforcement learning. In *International Conference on Machine Learning* (pp. 2107–2116).

Imai, S. (1977). Pattern similarity and cognitive transformations. *Acta Psychologica, 41*(6), 433–447.

Ishwaran, H., & James, L. F. (2001). Gibbs sampling methods for stick- breaking priors. *Journal of the American Statistical Association, 96,* 1316–1332.

Isik, L., Koldewyn, K., Beeler, D., & Kanwisher, N. (2017). Perceiving social interactions in the posterior superior temporal sulcus. *Proceedings of the National Academy of Sciences,* 201714471.

Iverson, P., & Kuhl, P. K. (1995). Mapping the perceptual magnet effect for speech using signal detection theory and multidimensional scaling. *Journal of the Acoustical Society of America, 97*(1), 553–562.

Jabbar, Y., Majed, A., Hsu, A., et al. (2013). Decision-making in proximal humeral fractures. *Shoulder & Elbow, 5*(2), 78–83.

Jack, A. I., Dawson, A. J., Begany, K. L., et al. (2013). fMRI reveals reciprocal inhibition between social and physical cognitive domains. *NeuroImage, 66,* 385–401.

Jacoby, N., & McDermott, J. H. (2017). Integer ratio priors on musical rhythm revealed cross-culturally by iterated reproduction. *Current Biology, 27*(3), 359–370.

Jacoby, N., Polak, R., Grahn, J., et al. (2024). Commonality and variation in mental representations of music revealed by a cross-cultural comparison of rhythm priors in 15 countries. *Nature Human Behavior.* https://doi.org/10.1038/s41562-023-01800-9

Jäger, G., & Rogers, J. (2012). Formal language theory: Refining the Chomsky hierarchy. *Philosophical Transactions of the Royal Society B: Biological Sciences, 367*(1598), 1956–1970.

Jain, S., & Neal, R. M. (2004). A split-merge Markov chain Monte Carlo procedure for the Dirichlet Process mixture model. *Journal of Computational and Graphical Statistics, 13*, 158–182.

Jain, S., Osherson, D., Royer, J. S., & Sharma, A. (1999). *Systems that learn: An introduction to learning theory.* MIT Press.

Jain, Y. R., Callaway, F., Griffiths, T. L., et al. (2022). A computational process-tracing method for measuring people's planning strategies and how they change over time. *Behavior Research Methods 55*(4), 2037–2079.

James, W. (1890). *Principles of psychology.* Holt.

Janner, M., Levine, S., Freeman, W. T., Tenenbaum, J. B., Finn, C., & Wu, J. (2019). Reasoning about physical interactions with object-oriented prediction and planning. In *Proceedings of the 7th International Conference on Learning Representations.*

Jara-Ettinger, J., Baker, C., & Tenenbaum, J. (2012). Learning what is where from social observations. In *Proceedings of the 34th Annual Meeting of the Cognitive Science Society.*

Jara-Ettinger, J., Floyd, S., Tenenbaum, J. B., & Schulz, L. E. (2017). Children understand that agents maximize expected utilities. *Journal of Experimental Psychology: General, 146*(11), 1574.

Jara-Ettinger, J., Gweon, H., Schulz, L. E., & Tenenbaum, J. B. (2016). The naïve utility calculus: Computational principles underlying commonsense psychology. *Trends in Cognitive Sciences, 20*(8), 589–604.

Jara-Ettinger, J., Schulz, L. E., & Tenenbaum, J. B. (2020). The naive utility calculus as a unified, quantitative framework for action understanding. *Cognitive Psychology, 123*, 101334.

Jaynes, E. T. (2003). *Probability theory: The logic of science.* Cambridge University Press.

Jazayeri, M., & Shadlen, M. N. (2010). Temporal context calibrates interval timing. *Nature Neuroscience, 13*(8), 1020–1026.

Jeffreys, H. (1961). *Theory of probability.* Oxford University Press.

Jeffreys, W. H., & Berger, J. O. (1992). Ockham's razor and Bayesian analysis. *American Scientist, 80*(1), 64–72.

Jenkins, H. M., & Ward, W. C. (1965). Judgment of contingency between responses and outcomes. *Psychological Monographs: General and Applied, 79*(1), 1–17.

Jerfel, G., Grant, E., Griffiths, T. L., & Heller, K. A. (2019). Reconciling meta-learning and continual learning with online mixtures of tasks. In *Advances in Neural Information Processing Systems.*

Jern, A., Lucas, C. G., & Kemp, C. (2017). People learn other people's preferences through inverse decision-making. *Cognition, 168*, 46–64.

Johansson, P., Hall, L., Sikstrom, S., & Olsson, A. (2005). Failure to detect mismatches between intention and outcome in a simple decision task. *Science, 310*(5745), 116–119.

Johnson, E. J., Häubl, G., & Keinan, A. (2007a). Aspects of endowment: A query theory of value construction. *Journal of Experimental Psychology: Learning, Memory, and Cognition, 33*(3), 461–474.

Johnson, K. (2004). Gold's theorem and cognitive science. *Philosophy of Science, 71*(4), 571–592.

Johnson, M., Demuth, K., & Frank, M. C. (2012). Exploiting social information in grounded language learning via grammatical reductions. In *Proceedings of the 50th Annual Meeting of the Association for Computational Linguistics.*

Johnson, M., Griffiths, T. L., & Goldwater, S. (2007b). Adaptor grammars: A framework for specifying compositional nonparametric Bayesian models. In *Advances in Neural Information Processing Systems.*

Johnson-Laird, P. N. (1983). *Mental models: Towards a cognitive science of language, inference, and consciousness.* Harvard University Press.

Jones, M., & Love, B. (2011). Bayesian fundamentalism or enlightenment? On the explanatory status and theoretical contributions of Bayesian models of cognition. *Behavioral and Brain Sciences, 34*(4), 169–231.

Jones, S. S., & Smith, L. B. (2002). How children know the relevant properties for generalizing object names. *Developmental Science, 5*(2), 219–232.

Jordan, M. I. (1986). An introduction to linear algebra in parallel distributed processing. In J. McClelland, D. Rumelhart, & the PDP Research Group (Eds.), *Parallel distributed processing: Explorations in the microstructure of cognition* (Vol. 1, pp. 365–422). MIT Press.

Jordan, M. I., Ghahramani, Z., Jaakkola, T. S., & Saul, L. K. (1999). An introduction to variational methods for graphical models. *Machine Learning, 37*(2), 183–233.

Joshi, A. K. (1985). Tree adjoining grammars: How much context- sensitivity is required to provide reasonable structural descriptions? In D. R. Dowty, L. Karttunen, & A. M. Zwicky (Eds.), *Natural language parsing* (pp. 206–250). Cambridge University Press.

Julesz, B. (1986). Stereoscopic vision. *Vision Research, 26*(9), 1601–1612.

Jurafsky, D. (1996). A probabilistic model of lexical and syntactic access and disambiguation. *Cognitive Science, 20*(2), 137–194.

Jurafsky, D. (2003). Probabilistic modeling in psycholinguistics: Linguistic comprehension and production. In R. Bod, J. Hay, & S. Jannedy (Eds.), *Probabilistic linguistics* (pp. 39–96). MIT Press.

Jurafsky, D., & Martin, J. H. (2000). *Speech and language processing: An introduction to natural language processing, computational linguistics, and speech recognition.* Prentice Hall.

Jurafsky, D., & Martin, J. H. (2008). *Speech and language processing: An introduction to natural language processing, computational linguistics, and speech recognition* (2nd ed.). Prentice Hall.

Juslin, P., Winman, A., & Hansson, P. (2007). The naïve intuitive statistician: A naïve sampling model of intuitive confidence intervals. *Psychological Review, 114*(3), 678.

Kaelbling, L., Littman, M., & Cassandra, A. (1998). Planning and acting in partially observable stochastic domains. *Artificial Intelligence, 101*(1–2), 99–134.

Kahneman, D., & Tversky, A. (1979). Prospect theory: An analysis of decision under risk. *Econometrica, 47*(2), 263–292.

Kahneman, D., & Tversky, A. (1982). The simulation heuristic. In D. Kahneman & A. Tversky (Eds.), *Judgment under uncertainty: Heuristics and biases* (pp. 201–208). Cambridge University Press.

Kahneman, D., & Tversky, A. (1984). Choices, values, and frames. *American Psychologist, 39*(4), 341–350.

Kalichman, S. C. (1988). Individual differences in water-level task per- formance: A component-skills analysis. *Developmental Review, 8*(3), 273–295.

Kalish, M. L., Griffiths, T. L., & Lewandowsky, S. (2007). Iterated learning: Intergenerational knowledge transmission reveals inductive biases. *Psychonomic Bulletin and Review, 14*, 288–294.

Kalish, M., Lewandowsky, S., & Kruschke, J. (2004). Population of linear experts: Knowledge partitioning and function learning. *Psychological Review, 111*, 1072–1099.

Kalman, R. E. (1960). A new approach to linear filtering and prediction problems. *Journal of Basic Engineering, 82*, 35–45.

Kanerva, P. (1988). *Sparse distributied memory.* MIT Press.

Kanizsa, G. (1985). Seeing and thinking. *Acta Psychologica, 59*(1), 23–33.

Kant, I. (1781/1964). *Critique of pure reason.* Macmillan.

Kanwisher, N., McDermott, J., & Chun, M. M. (1997). The fusiform face area: A module in human extrastriate cortex specialized for face perception. *Journal of Neuroscience, 17*(11), 4302–4311.

Kao, J. T., Wu, J. Y., Bergen, L., & Goodman, N. D. (2014). Nonliteral understanding of number words. *Proceedings of the National Academy of Sciences, 111*(33), 12002–12007.

Kashyap, R. L., & Oommen, B. J. (1984). Spelling correction using probabilistic methods. *Pattern Recognition Letters, 2*(3), 147–154.

Kass, R. E., & Raftery, A. E. (1995). Bayes factors. *Journal of the American Statistical Association, 90*, 773–795.

Katz, L. N., Yates, J. L., Pillow, J. W., & Huk, A. C. (2016). Dissociated functional significance of decision-related activity in the primate dorsal stream. *Nature, 535*(7611), 285–288.

Katz, Y., Goodman, N. D., Kersting, K., Kemp, C., & Tenenbaum, J. B. (2008). Modeling semantic cognition as logical dimensionality reduction. In *Proceedings of the 30th Annual Meeting of the Cognitive Science Society* (pp. 71–76).

Kawato, M. (1999). Internal models for motor control and trajectory planning. *Current Opinion in Neurobiology*, *9*(6), 718–727.

Kay, P., Berlin, B., Maffi, L., Merrifield, W. R., & Cook, R. (2009). *The World Color Survey.* CSLI Publications.

Keil, F. C. (1989). *Concepts, kinds, and cognitive development.* MIT Press.

Keil, F. (1995). The growth of causal understanding of natural kinds. In D. Sperber & D. Premack (Eds.), *Causal cognition: A multidisciplinary debate* (pp. 234–267). Clarendon Press/Oxford University Press.

Kelley, H. H. (1987). Attribution in social interaction. In E. E. Jones, D. E. Kanouse, H. H. Kelley, R. E. Nisbett, S. Valins, & B. Weiner (Eds.), *Attribution: Perceiving the causes of behavior* (pp. 1–26). Lawrence Erlbaum Associates, Inc.

Kemp, C. (2009). Quantification and the language of thought. In *Advances in Neural Information Processing Systems 21.*

Kemp, C. (2012). Exploring the conceptual universe. *Psychological Review*, *119*(4), 685–722.

Kemp, C., Goodman, N. D., & Tenenbaum, J. B. (2008a). Learning and using relational theories. *Advances in Neural Information Processing Systems 20* (pp. 753–760).

Kemp, C., Goodman, N. D., & Tenenbaum, J. B. (2008b). Theory acquisition and the language of thought. In *Proceedings of the 30th Annual Meeting of the Cognitive Science Society* (pp. 1606–1611).

Kemp, C., Goodman, N. D., & Tenenbaum, J. B. (2010a). Learning to learn causal models. *Cognitive Science*, *34*(7), 1185–1243.

Kemp, C., Han, F., & Jern, A. (2011). Concept learning and modal reasoning. In *Proceedings of the 33rd Annual Meeting of the Cognitive Science Society* (pp. 513–518).

Kemp, C., & Jern, A. (2009). Abstraction and relational learning. In *Advances in Neural Information Processing Systems 21* (pp. 934–942).

Kemp, C., Perfors, A., & Tenenbaum, J. B. (2004). Learning domain structures. In *Proceedings of the 26th Annual Conference of the Cognitive Science Society.*

Kemp, C., Perfors, A., & Tenenbaum, J. B. (2007). Learning overhypotheses with hierarchical Bayesian models. *Developmental Science*, *10*(3), 307–321.

Kemp, C., & Regier, T. (2012). Kinship categories across languages reflect general communicative principles. *Science*, *336*(6084), 1049–1054.

Kemp, C., & Tenenbaum, J. B. (2003). Theory-based induction. In *Proceedings of the Twenty-Fifth Annual Conference of the Cognitive Science Society* (pp. 658–663). Erlbaum.

Kemp, C., & Tenenbaum, J. B. (2008). The discovery of structural form. *Proceedings of the National Academy of Sciences*, *105*(31), 10687– 10692.

Kemp, C., & Tenenbaum, J. B. (2009). Structured statistical models of inductive reasoning. *Psychological Review*, *116*(1), 20–58.

Kemp, C., Tenenbaum, J. B., Griffiths, T. L., Yamada, T., & Ueda, N. (2006). Learning systems of concepts with an infinite relational model. In *Proceedings of the 21st National Conference on Artificial Intelligence.*

Kemp, C., Tenenbaum, J. B., Niyogi, S., & Griffiths, T. L. (2010b). A probabilistic model of theory formation. *Cognition*, *114*, 165–196.

Keramati, M., Smittenaar, P., Dolan, R. J., & Dayan, P. (2016). Adaptive integration of habits into depth-limited planning defines a habitual goal-directed spectrum. *Proceedings of the National Academy of Sciences*, *113*(45), 12868–12873.

Kersten, D., & Yuille, A. (2003). Bayesian models of object perception.*Current Opinion in Neurobiology*, *13*(2), 150–158.

Kessler, S., Nguyen, V., Zohren, S., & Roberts, S. J. (2021). Hierarchical Indian buffet neural networks for Bayesian continual learning. In *Proceedings of the 37th Conference on Uncertainty in Artificial Intelligence* (pp. 749–759).

Kidd, C., Piantadosi, S. T., & Aslin, R. N. (2012). The Goldilocks effect: Human infants allocate attention to visual sequences that are neither too simple nor too complex. *PLoS One*, *7*(5), e36399.

Kim, J. Z., & Bassett, D. S. (2022). A neural programming language for the reservoir computer. *arXiv preprint arXiv:2203.05032.*

Kinzler, K. D., & Spelke, E. S. (2007). Core systems in human cognition. *Progress in Brain Research*, *164*, 257–264.

Kipf, T., Elsayed, G. F., Mahendran, A., et al. (2021). Conditional object-centric learning from video. *arXiv preprint arXiv:2111.12594.*

Kirby, S. (2001). Spontaneous evolution of linguistic structure: An iterated learning model of the emergence of regularity and irregularity. *IEEE Journal of Evolutionary Computation, 5*, 102–110.

Kirby, S., & Tamariz, M. (2021). Cumulative cultural evolution, population structure, and the origin of combinatoriality in human language. *Philosophical Transactions of the Royal Society B: Biological Sciences, 377*, 20200319.

Klahr, D., & Wallace, J. G. (2022). *Cognitive development: An information-processing view.* Routledge.

Klayman, J., & Ha, Y.-W. (1987). Confirmation, disconfirmation, and information in hypothesis testing. *Psychological Review, 94*(2), 211–228.

Klein, G. A. (1993). A recognition-primed decision (RPD) model of rapid decision making. *Decision Making in Action: Models and Methods, 5*(4), 138–147.

Kleinschmidt, D., & Jaeger, F. (2015). Robust speech perception: Recognize the familiar, generalize to the similar, and adapt to the novel. *Psychological Review, 122*(2), 148–203.

Klingeman, M. (2020). *An imaginary Jerome K. Jerome writes about Twitter.* (https://twitter.com/quasimondo/status/1284509525500989445).

Knight, F. H. (1921). *Risk, uncertainty and profit.* Hart, Schaffner, & Marx.

Knill, D. C., & Pouget, A. (2004). The Bayesian brain: The role of uncertainty in neural coding and computation. *Trends in Neurosciences, 27*(12), 712–719.

Knill, D. C., & Richards, W. A. (1996). *Perception as Bayesian inference.* Cambridge University Press.

Kocsis, L., & Szepesvári, C. (2006). Bandit based Monte-Carlo planning. In *European Conference on Machine Learning* (pp. 282–293).

Koffka, K. (1925). *The growth of the mind: An introduction to child- psychology.* Harcourt Brace & Company.

Koffka, K. (2013). *Principles of gestalt psychology.* Routledge.

Koh, K., & Meyer, D. E. (1991). Function learning: Induction of continuous stimulus-response relations. *Journal of Experimental Psychology: Learning, Memory, and Cognition, 17*, 811–836.

Kohler, W. (2018). *The mentality of apes.* Routledge.

Kok, S., & Domingos, P. (2007). Statistical predicate invention. In *Proceedings of the 24th International Conference on Machine Learning.*

Koller, D., & Friedman, N. (2009). *Probabilistic graphical models: principles and techniques.* MIT Press.

Kolmogorov, A. N. (1965). Three approaches to the quantitative definition of information. *Problems of Information Transmission, 1*, 1–7.

Kool, W., Cushman, F. A., & Gershman, S. J. (2016). When does model-based control pay off? *PLoS Computational Biology, 12*(8), e1005090.

Kool, W., Cushman, F. A., & Gershman, S. J. (2018). Competition and cooperation between multiple reinforcement learning systems. In R. Morris, A. Bornstein, & A. Shenhav (Eds.), *Goal-directed decision making* (pp. 153–178). Academic Press.

Kool, W., Gershman, S. J., & Cushman, F. A. (2017). Cost-benefit arbitration between multiple reinforcement-learning systems. *Psychological Science, 28*(9), 1321–1333.

Korb, K., & Nicholson, A. (2003). *Bayesian artificial intelligence.* Chapman and Hall/CRC.

Körding, K. P., Beierholm, U., Ma, W. J., Quartz, S., Tenenbaum, J. B., & Shams, L. (2007). Causal inference in multisensory perception. *PLoS One, 2*(9), e943.

Körding, K., & Wolpert, D. M. (2004). Bayesian integration in sensorimotor learning. *Nature, 427*, 244–247.

Körding, K., & Wolpert, D. (2006). Bayesian decision theory in sensorimotor control. *Trends in Cognitive Sciences, 10*(7), 319–326.

Kosslyn, S. M., Ball, T. M., & Reiser, B. J. (1978). Visual images preserve metric spatial information: evidence from studies of image scanning. *Journal of Experimental Psychology: Human Perception and Performance, 4*(1), 47–60.

Kouh, M., & Poggio, T. (2008). A canonical neural circuit for cortical nonlinear operations. *Neural Computation, 20*(6), 1427–1451.

Kowalski, R. (1974). Predicate logic as programming language. In *Proceedings of the IFIP Congress* (pp. 569–544).

Krajbich, I., Armel, C., & Rangel, A. (2010). Visual fixations and the computation and comparison of value in simple choice. *Nature Neuroscience*, *13*(10), 1292–1298.

Krajbich, I., & Rangel, A. (2011). Multialternative drift-diffusion model predicts the relationship between visual fixations and choice in value- based decisions. *Proceedings of the National Academy of Sciences*, *108*(33), 13852–13857.

Kreps, D. M. (1988). *Notes on the theory of choice.* Westview Press.

Kreps, D. M. (1990). *Game theory and economic modelling.* Oxford University Press.

Kronrod, Y., Coppess, E., & Feldman, N. (2016). A unified account of categorical effects in phonetic perception. *Psychonomic Bulletin and Review*, *23*(6), 1681–1712.

Krueger, P. M., Lieder, F., & Griffiths, T. L. (2017). Enhancing metacognitive reinforcement learning using reward structures and feedback. In *Proceedings of the 39th Annual Meeting of the Cognitive Science Society.*

Kruschke, J. K. (1992). ALCOVE: An exemplar-based connectionist model of category learning. *Psychological Review*, *99*, 22–44.

Kruschke, J. K. (2006). Locally Bayesian learning with applications to retrospective revaluation and highlighting. *Psychological Review*, *113*, 677–699.

Krynski, T. R., & Tenenbaum, J. B. (2007). The role of causality in judgment under uncertainty. *Journal of Experimental Psychology: General*, *136*(3), 430–450.

Kubricht, J., Jiang, C., Zhu, Y., Zhu, S.-C., Terzopoulos, D., & Lu, H. (2016). Probabilistic simulation predicts human performance on viscous fluid-pouring problem. In *Proceedings of the 38th Annual Meeting of the Cognitive Science Society* (pp. 1805–1810).

Kubricht, J., Zhu, Y., Jiang, C., Terzopoulos, D., Zhu, S.-C., & Lu, H. (2017). Consistent probabilistic simulation underlying human judgment in substance dynamics. In *Proceedings of the 39th Annual Meeting of the Cognitive Science Society* (pp. 700–705).

Kuhl, P. (2004). Early language acquisition: Cracking the speech code. *Nature Reviews Neuroscience*, *5*, 831–843.

Kulkarni, T. D., Kohli, P., Tenenbaum, J. B., & Mansinghka, V. (2015). Picture: A probabilistic programming language for scene perception. In *IEEE Conference on Computer Vision and Pattern Recognition (cvpr)* (pp. 4390–4399).

Kumar, S., Dasgupta, I., Cohen, J. D., Daw, N. D., & Griffiths, T. L. (2021). Meta-learning of structured task distributions in humans and machines. In *Proceedings of the 9th International Conference on Learning Representations (ICLR).*

Kumar, S., Dasgupta, I., Marjieh, R., Daw, N. D., Cohen, J. D., & Griffiths, T. L. (2022). Disentangling abstraction from statistical pattern matching in human and machine learning. *arXiv preprint arXiv:2204.01437.*

Kurumada, C., Meylan, S., & Frank, M. C. (2013). Zipfian frequency distributions facilitate word segmentation in context. *Cognition*, *127*, 439–453.

Kutschireiter, A., Surace, S. C., Sprekeler, H., & Pfister, J.-P. (2015). The neural particle filter. *arXiv preprint arXiv:1508.06818.*

Lacker, K. (2020). *Giving GPT-3 a Turing Test.* Kevin Lacker's blog. https://lacker.io/ai/2020/07/06/giving-gpt-3-a-turing-test.html

Lagnado, D., & Sloman, S. A. (2004). The advantage of timely intervention. *Journal of Experimental Psychology: Learning, Memory, and Cognition*, *30*, 856–876.

Lakatos, I. (1970). Falsification and the methodology of scientific research programmes. In I. Lakatos & A. Musgrave (Eds.), *Criticism and the growth of knowledge* (pp. 91–196). Cambridge University Press.

Lake, B. M., & Piantadosi, S. T. (2020). People infer recursive visual concepts from just a few examples. *Computational Brain & Behavior*, *3*(1), 54–65.

Lake, B. M., Salakhutdinov, R., & Tenenbaum, J. B. (2015). Human-level concept learning through probabilistic program induction. *Science*, *350*(6266), 1332–1338.

Lake, B. M., Ullman, T. D., Tenenbaum, J. B., & Gershman, S. J. (2017). Building machines that learn and think like people. *Behavioral and Brain Sciences*, *40*, e253.

Lamberts, K. (2000). Information-accumulation theory of speeded categorization. *Psychological Review*, *107*(2), 227–260.

Landauer, T. K., & Dumais, S. T. (1997). A solution to Plato's problem: The latent semantic analysis theory of acquisition, induction, and representation of knowledge. *Psychological Review*, *104*(2), 211–240.

Langley, P., Simon, H. A., Bradshaw, G. L., & Zytkow, J. M. (1987). *Scientific discovery: Computational explorations of the creative process.* MIT Press.

Langlois, T. A., Jacoby, N., Suchow, J. W., & Griffiths, T. L. (2021). Serial reproduction reveals the geometry of visuospatial representations. *Proceedings of the National Academy of Sciences*, *118*(13), e2012938118.

Langsford, S., Hendrickson, A., Perfors, A., & Navarro, D. J. (2017). When do learned transformations influence similarity and categorization? In *Proceedings of Annual Meeting of the Cognitive Science Society.*

Lassiter, D., & Goodman, N. (2013). Context, scale structure, and statistics in the interpretation of positive-form adjectives. *Semantics and Linguistic Theory*, *23*, 587–610.

Le, T. A., Baydin, A. G., & Wood, F. (2016). Inference compilation and universal probabilistic programming. *arXiv preprint arXiv:1610.09900.*

LeCun, Y. (2022). A path towards autonomous machine intelligence version 0.9. *OpenReview preprint* https://openreview.net/forum?id=BZ5a1r-kVsf

LeCun, Y., Bengio, Y., & Hinton, G. (2015). Deep learning. *Nature*, *521*(7553), 436–444.

Lee, D. N., & Aronson, E. (1974). Visual proprioceptive control of standing in human infants. *Perception & Psychophysics*, *15*(3), 529–532.

Lee, S. A., & Spelke, E. S. (2008). Children's use of geometry for reorientation. *Developmental Science*, *11*(5), 743–749.

Lee, T. S., & Mumford, D. (2003). Hierarchical Bayesian inference in the visual cortex. *Journal of the Optical Society of America A*, *20*(7), 1434–1448.

Leeuwenberg, E. L. (1971). A perceptual coding language for visual and auditory patterns. *The American journal of psychology*, *84*, 307–349.

Leeuwenberg, E. L., & Boselie, F. (1988). Against the likelihood principle in visual form perception. *Psychological Review*, *95*(4), 485–491.

Legenstein, R., & Maass, W. (2014). Ensembles of spiking neurons with noise support optimal probabilistic inference in a dynamically changing environment. *PLoS Computational Biology*, *10*(10), e1003859.

Legge, G. E., Klitz, T. S., & Tjan, B. S. (1997). Mr. Chips: An ideal- observer model of reading. *Psychological Review*, *104*(3), 524–553.

Leong, Y. C., Radulescu, A., Daniel, R., DeWoskin, V., & Niv, Y. (2017). Dynamic interaction between reinforcement learning and attention in multidimensional environments. *Neuron*, *93*(2), 451–463.

Leslie, A. (1994). ToMM, ToBy and Agency: Core architecture and domain specificity. In L. Hirschfeld & S. Gelman (Eds.), *Domain specificity and cultural knowledge* (pp. 119–148). Cambridge University Press.

Levenshtein, V. I. (1966). Binary codes capable of correcting deletions, insertions, and reversals. *Soviet Physics Doklady*, *10*(8), 707–710.

Levesque, H. J. (2012). *Thinking as computation: A first course.* MIT Press.

Levin, L. A. (1973). Universal sequential search problems. *Problemy Peredachi Informatsii*, *9*(3), 115–116.

Levine, M. (1975). *A cognitive theory of learning: Research on hypothesis testing.* Erlbaum.

Levinson, S. C. (2016). Turn-taking in human communication–origins and implications for language processing. *Trends in Cognitive Sciences*, *20*(1), 6–14.

Levy, R. (2008). Expectation-based syntactic comprehension. *Cognition*, *106*(3), 1126–1177.

Levy, R., Reali, F., & Griffiths, T. (2008). Modeling the effects of memory on human online sentence processing with particle filters. *Advances in Neural Information Processing Systems 21.*

Lewandowsky, S., Griffiths, T. L., & Kalish, M. L. (2009). The wisdom of individuals: Exploring people's knowledge about everyday events using iterated learning. *Cognitive Science*, *33*(6), 969–998.

Lewis, R. L., Howes, A., & Singh, S. (2014). Computational rationality: Linking mechanism and behavior through bounded utility maximization. *Topics in Cognitive Science*, *6*(2), 279–311.

Li, F., Fergus, R., & Perona, P. (2006). One-shot learning of object categories. *IEEE Transactions on Pattern Analysis and Machine Intelligence*, *28*(4), 594–611.

Li, M., Chen, X., Li, X., Ma, B., & Vitányi, P. M. (2004). The similarity metric. *IEEE Transactions on Information Theory*, *50*(12), 3250–3264.

Li, M. Y., Grant, E., & Griffiths, T. L. (2021). Meta-learning inductive biases of learning systems with Gaussian processes. In *Fifth Workshop on Meta-learning at the Conference on Neural Information Processing Systems*.

Li, M., & Vitányi, P. (2008). *An introduction to Kolmogorov complexity and its applications* (3rd ed.). Springer.

Li, S., Sun, Y., Liu, S., Wang, T., Gureckis, T., & Bramley, N. (2019). Active physical inference via reinforcement learning.

Li, Y., Wu, J., Tedrake, R., Tenenbaum, J. B., & Torralba, A. (2018). Learning particle dynamics for manipulating rigid bodies, deformable objects, and fluids. *arXiv preprint arXiv:1810.01566*.

Lieder, F., Chen, O. X., Krueger, P. M., & Griffiths, T. L. (2019). Cognitive prostheses for goal achievement. *Nature Human Behaviour*, *3*(10), 1096–1106.

Lieder, F., & Griffiths, T. L. (2017). Strategy selection as rational metareasoning. *Psychological Review*, *124*(6), 762–794.

Lieder, F., & Griffiths, T. L. (2020). Resource-rational analysis: Understanding human cognition as the optimal use of limited computational resources. *Behavioral and Brain Sciences*, *43*, e1.

Lieder, F., Griffiths, T. L., & Goodman, N. D. (2012). Burn-in, bias, and the rationality of anchoring. In *Advances in Neural Information Processing Systems 26* (pp. 2690–2798).

Lieder, F., Griffiths, T. L., & Hsu, M. (2018a). Overrepresentation of extreme events in decision making reflects rational use of cognitive resources. *Psychological Review*, *125*(1), 1–32.

Lieder, F., Griffiths, T. L., Huys, Q. J. M., & Goodman, N. D. (2018b). The anchoring bias reflects rational use of cognitive resources. *Psychonomic Bulletin & Review*, *25*(1), 322–349.

Lieder, F., Griffiths, T. L., Huys, Q. J. M., & Goodman, N. D. (2018c). Empirical evidence for resource-rational anchoring and adjustment. *Psychonomic Bulletin & Review*, *25*, 775–784.

Lieder, F., & Iwama, G. (2021). Toward a formal theory of proactivity. *Cognitive, Affective, & Behavioral Neuroscience*, *21*(3), 490–508.

Lieder, F., Shenhav, A., Musslick, S., & Griffiths, T. L. (2018c). Rational metareasoning and the plasticity of cognitive control. *PLoS Computational Biology*, *14*, e1006043.

Lindley, D. V. (1956). On a measure of the information provided by an experiment. *Annals of Mathematical Statistics*, *27*(4), 986–1005.

Liu, S., Ullman, T. D., Tenenbaum, J. B., & Spelke, E. S. (2017). Ten-month-old infants infer the value of goals from the costs of actions. *Science*, *358*(6366), 1038–1041.

Lloyd, K., Sanborn, A., Leslie, D., & Lewandowsky, S. (2019). Why higher working memory capacity may help you learn: sampling, search, and degrees of approximation. *Cognitive Science*, *43*(12), e12805.

Loewenstein, G., & Prelec, D. (1992). Anomalies in intertemporal choice: Evidence and an interpretation. *Quarterly Journal of Economics*, *107*(2), 573–597.

Logan, G. D. (1988). Toward an instance theory of automatization. *Psychological Review*, *95*(4), 492–527.

Long, G. M., & Toppino, T. C. (2004). Enduring interest in perceptual ambiguity: Alternating views of reversible figures. *Psychological Bulletin*, *130*(5), 748.

Loomes, G., & Sugden, R. (1984). The importance of what might have been. In O. Hagen & F. Wenstøp (Eds.), *Progress in utility and risk theory* (Vol. 42, pp. 219–235). Springer.

Loomes, G., & Sugden, R. (1986). Disappointment and dynamic consistency in choice under uncertainty. *Review of Economic Studies*, *53*(2), 271–282.

Lovelace, A. A. (1842). Sketch of the analytical engine invented by Charles Babbage, by lf Menabrea, officer of the military engineers, with notes upon the memoir by the translator. *Taylor's Scientific Memoirs*, *3*, 666–731.

Lowder, M. W., Choi, W., Ferreira, F., & Henderson, J. M. (2018). Lexical predictability during natural reading: Effects of surprisal and entropy reduction. *Cognitive Science*, *42*, 1166–1183.

Lu, H., Rojas, R. R., Beckers, T., & Yuille, A. L. (2016). A Bayesian theory of sequential causal learning and abstract transfer. *Cognitive Science*, *40*(2), 404–439.

Lu, H., Yuille, A., Liljeholm, M., Cheng, P. W., & Holyoak, K. J. (2008). Bayesian generic priors for causal learning. *Psychological Review*, *115*, 955–984.

Lucas, C. G., Bridgers, S., Griffiths, T. L., & Gopnik, A. (2014a). When children are better (or at least more open-minded) learners than adults: Developmental differences in learning the forms of causal relationships. *Cognition, 131*(2), 284–299.

Lucas, C. G., & Griffiths, T. L. (2010). Learning the form of causal relationships using hierarchical Bayesian models. *Cognitive Science, 34*(1), 113–147.

Lucas, C. G., Griffiths, T. L., Williams, J. J., & Kalish, M. L. (2015). A rational model of function learning. *Psychonomic Bulletin & Review, 22*(5), 1193–1215.

Lucas, C. G., Griffiths, T. L., Xu, F., et al. (2014b). The child as econometrician: A rational model of preference understanding in children. *PLoS One, 9*(3), e92160.

Luce, R. D. (1959). *Individual choice behavior.* Wiley.

Ludwin-Peery, E., Bramley, N. R., Davis, E., & Gureckis, T. M. (2020). Broken physics: A conjunction-fallacy effect in intuitive physical reasoning. *Psychological Science, 31*(12), 1602–1611.

Lund, K., & Burgess, C. (1996). Producing high-dimensional semantic spaces from lexical co-occurrence. *Behavior Research Methods, Instrumentation, and Computers, 28*, 203–208.

Luong, M.-T., Frank, M. C., & Johnson, M. (2013). Parsing entire discourses as very long strings: Capturing topic continuity in grounded language learning. In *Transactions of the Association for Computational Linguistics* (pp. 315–326).

Luong, M.-T., O'Donnell, T., & Goodman, N. (2015). Evaluating models of computation and storage in human sentence processing. *Proceedings of the Sixth Workshop on Cognitive Aspects of Computational Language Learning* (pp. 14–21).

Ma, W. J., Beck, J., Latham, P., & Pouget, A. (2006). Bayesian inference with probabilistic population codes. *Nature Neuroscience, 9*, 1432– 1438.

Mach, E. (1883/1919). *The science of mechanics: A critical account of its development* (4th ed.). T. J. McCormack, trans. Open Court.

Machina, M. J., & Siniscalchi, M. (2014). Ambiguity and ambiguity aversion. In *Handbook of the economics of risk and uncertainty* (Vol. 1, pp. 729–807). Elsevier.

MacKay, D. J. C. (1992a). Bayesian interpolation. *Neural Computation, 4*, 415–447.

Mackay, D. J. C. (1992b). *Bayesian methods for adaptive models.* California Institute of Technology.

MacKay, D. J. (1995). Bayesian neural networks and density networks. *Nuclear Instruments and Methods in Physics Research Section A: Accelerators, Spectrometers, Detectors and Associated Equipment, 354*(1), 73–80.

Mackay, D. J. C. (1998). Introduction to Monte Carlo methods. In M. I. Jordan (Ed.), *Learning in graphical models* (pp. 175–204). MIT Press.

Mackay, D. J. C. (2003). *Information theory, inference, and learning algorithms.* Cambridge University Press.

Madan, C. R., Ludvig, E. A., & Spetch, M. L. (2014). Remembering the best and worst of times: Memories for extreme outcomes bias risky decisions. *Psychonomic Bulletin & Review, 21*(3), 629–636.

Maddox, J. (1994). Star masses and Bayesian probability. *Nature, 371*, 649.

Mandt, S., Hoffman, M. D., & Blei, D. M. (2017). Stochastic gradient descent as approximate Bayesian inference. *Journal of Machine Learning Research, 18*, 1–35.

Manning, C., & Schütze, H. (1999). *Foundations of statistical natural language processing.* MIT Press.

Mansinghka, V. K., Kemp, C., Tenenbaum, J. B., & Griffiths, T. L. (2006). Structured priors for structure learning. In *Proceedings of the Twenty-Second Conference on Uncertainty in Artificial Intelligence* (pp. 324–331).

Mansinghka, V., Selsam, D., & Perov, Y. (2014). Venture: A higher-order probabilistic programming platform with programmable inference. *arXiv preprint arXiv:1404.0099.*

Mansinghka, V., Shafto, P., Jonas, E., Petschulat, C., Gasner, M., & Tenenbaum, J. B. (2016). CrossCat: A fully Bayesian nonparametric method for analyzing heterogeneous, high dimensional data. *Journal of Machine Learning Research, 17*(1), 4760–4808.

Marcus, G. F., & Davis, E. (2013). How robust are probabilistic models of higher-level cognition? *Psychological Science, 24*(12), 2351–2360.

Marcus, G., & Davis, E. (2019). *Rebooting AI: Building artificial intelligence we can trust.* Vintage.

Marcus, G. F., Pinker, S., Ullman, M., et al. (1992). Overregularization in language acquisition. *Monographs of the Society for Research in Child Development, 57*(4), 1–182.

Marcus, G. F., Vijayan, S., Bandi Rao, S., & Vishton, P. M. (1999). Rule learning by seven-month-old infants. *Science, 283*(5398), 77–80.

Margolis, E. (1998). How to acquire a concept. *Mind & Language, 13*(3), 347–369.

Marjieh, R., Harrison, P. M., Lee, H., Deligiannaki, F., & Jacoby, N. (2022). Reshaping musical consonance with timbral manipulations and massive online experiments. *bioRxiv*.

Markman, E. S. (1989). *Categorization and naming in children.* MIT Press.

Markman, E. (1990). Constraints children place on word meanings. *Cognitive Science, 14*, 57–77.

Markov, A. A. (1913). Primer statisticheskogo issledovaniya nad tekstom "Evgeniya Onegina", illyustriruyuschij svyaz ispytanij v cep. *Izvestiya Akademii Nauk, Ser. 6*(3), 153–162.

Marr, D. (1982). *Vision.* W. H. Freeman.

Martin, J. B., Griffiths, T. L., & Sanborn, A. N. (2012). Testing the efficiency of Markov chain Monte Carlo with people using facial affect categories. *Cognitive Science, 36*(1), 150–162.

Mattar, M. G., & Daw, N. D. (2018). Prioritized memory access explains planning and hippocampal replay. *Nature Neuroscience, 21*(11), 1609–1617.

Mattys, S. L., Davis, M. H., Bradlow, A. R., & Scott, S. K. (2012). Speech recognition in adverse conditions: A review. *Language and Cognitive Processes, 27*(7–8), 953–978.

McAllester, D. A. (1998). Some PAC-Bayesian theorems. In *Proceedings of the Eleventh Annual Conference on Computational Learning Theory* (pp. 230–234).

McCarthy, J. (1959). Programs with common sense. In *Mechanization of Thought Processes: Symposium at the National Physical Laboratory, 24–27 November, 1958.* Cambridge University Press.

McCarthy, J. (1960). Recursive functions of symbolic expressions and their computation by machine, Part I. *Communications of the ACM, 3*(4), 184–195.

McCarthy, J. (1980). Circumscription—a form of non-monotonic reasoning. *Artificial Intelligence, 13*(1–2), 27–39.

McCarthy, J., & Hayes, P. J. (1969). Some philosophical problems from the standpoint of artificial intelligence. In B. Meltzer & D. Michie (Eds.), *Machine intelligence* (Vol. 4, pp. 463–502). Edinburgh University Press.

McClelland, J. L. (1998). Connectionist models and Bayesian inference. In M. Oaksford & N. Chater (Eds.), *Rational models of cognition* (pp. 21–53). Oxford University Press.

McClelland, J. L., Botvinick, M. M., Noelle, D. C., et al. (2010). Letting structure emerge: connectionist and dynamical systems approaches to cognition. *Trends in Cognitive Sciences, 14*(8), 348–356.

McClelland, J., & Rumelhart, D. (Eds.). (1986). *Parallel distributed processing: Explorations in the microstructure of cognition.* MIT Press.

McCloskey, M., Caramazza, A., & Green, B. (1980). Curvilinear motion in the absence of external forces: Naive beliefs about the motion of objects. *Science, 210*(5), 1139–1141.

McCloskey, M., & Cohen, N. J. (1989). Catastrophic interference in connectionist networks: The sequential learning problem. In *Psychology of learning and motivation* (Vol. 24, pp. 109–165). Academic Press.

McCoy, R. T., Grant, E., Smolensky, P., Griffiths, T. L., & Linzen, T. (2020). Universal linguistic inductive biases via meta-learning. In *Proceedings of the 42nd Annual Meeting of the Cognitive Science Society.*

McCoy, R. T., & Griffiths, T. L. (2023). Modeling rapid language learning by distilling Bayesian priors into artificial neural networks. *arXiv preprint arXiv:2305.14701*.

McCoy, R. T., Yao, S., Friedman, D., Hardy, M., & Griffiths, T. L. (2023). Embers of autoregression: Understanding large language models through the problem they are trained to solve. *arXiv preprint arXiv:2309.13638*.

McCulloch, W. S., & Pitts, W. (1943). A logical calculus of the ideas immanent in nervous activity. *Bulletin of Mathematical Biophysics, 5*(4), 115–133.

McDaniel, M. A., & Busemeyer, J. R. (2005). The conceptual basis of function learning and extrapolation: Comparison of rule-based and associative-based models. *Psychonomic Bulletin and Review, 12*, 24– 42.

McDuff, D. (2010). A human-Markov chain Monte Carlo method for investigating facial expression categorization. In *Proceedings of the 10th International Conference on Cognitive Modeling* (pp. 151–156).

McFarland, D., & Bösser, T. (1993). *Intelligent behavior in animals and robots.* MIT Press.

McNamee, D., & Wolpert, D. M. (2019). Internal models in biological control. *Annual Review of Control, Robotics, and Autonomous Systems*, *2*, 339–364.

Medin, D. L., & Schaffer, M. M. (1978). Context theory of classification learning. *Psychological Review*, *85*, 207–238.

Mednick, S. (1962). The associative basis of the creative process. *Psychological Review*, *69*(3), 220.

Mehta, A., Jain, Y. R., Kemtur, A., et al. (2022). Leveraging machine learning to automatically derive robust decision strategies from imperfect knowledge of the real world. *Computational Brain & Behavior*, *5*, 343–377.

Meltzoff, A., Waismeyer, A., & Gopnik, A. (2012). Learning about causes from people: observational causal learning in 24-month old infants. *Developmental Psychology*, *48*, 1215–1228.

Mengersen, K. L., & Tweedie, R. L. (1996). Rates of convergence of the Hastings and Metropolis algorithms. *Annals of Statistics*, *24*(1), 101–121.

Mercier, H., & Sperber, D. (2011). Why do humans reason? Arguments for an argumentative theory. *Behavioral and Brain Sciences*, *34*(2), 57–74.

Metropolis, A. W., Rosenbluth, A. W., Rosenbluth, M. N., Teller, A. H., & Teller, E. (1953). Equations of state calculations by fast computing machines. *Journal of Chemical Physics*, *21*, 1087–1092.

Miall, R. C., & Wolpert, D. M. (1996). Forward models for physiological motor control. *Neural networks*, *9*(8), 1265–1279.

Mikhail, J. (2008). The poverty of the moral stimulus. In W. Sinnott- Armstrong (Ed.), *Moral psychology* (Vol. 1, pp. 353–359). MIT Press.

Mikolov, T., Sutskever, I., Chen, K., Corrado, G. S., & Dean, J. (2013). Distributed representations of words and phrases and their compositionality. *Advances in Neural Information Processing Systems 26*.

Mikulik, V., Delétang, G., McGrath, T., et al. (2020). Meta-trained agents implement bayes- optimal agents. *Advances in Neural Information Processing Systems*, *33*, 18691–18703.

Milch, B., Marthi, B., & Russell, S. (2004). BLOG: Relational modeling with unknown objects. In T. Dietterich, L. Getoor, & K. Murphy (Eds.), *ICML 2004 Workshop on Statistical Relational Learning and its Connections to Other Fields* (pp. 67–73).

Milch, B., & Russell, S. (2006). General-purpose MCMC inference over relational structures. In *Proceedings of the 22nd Conference on Uncertainty in Artificial Intelligence (UAI)* (pp. 349–358).

Mill, J. S. (1843). *A system of logic, ratiocinative and inductive: Being a connected view of the principles of evidence, and methods of scientific investigation*. J. W. Parker.

Miller, E. K., & Cohen, J. D. (2001). An integrative theory of prefrontal cortex function. *Annual Review of Neuroscience*, *24*(1), 167–202.

Miller, G. A. (1951). *Language and communication*. McGraw-Hill.

Miller, G. A. (1995). WordNet: A lexical database for English. *Communications of the ACM*, *38*(11), 39–41.

Miller, K. J., Shenhav, A., & Ludvig, E. A. (2019). Habits without values. *Psychological Review*, *126*(2), 292–311.

Mills, T., Cheyette, S. J., & Tenenbaum, J. B. (2023). Spatiotemporal sequence learning as probabilistic program induction. In *Advances in Neural Information Processing Systems*.

Minka, T. (2001). *Inferring a Gaussian distribution*.(http://www.stat.cmu.edu/~minka/papers/gaussian.html).

Minka, T. (2005). *Divergence measures and message passing* (Tech. Rep. No. MSR-TR-2005-173). Microsoft Research Ltd.

Minka, T., & Lafferty, J. (2002). Expectation-propagation for the generative aspect model. In *Proceedings of the 18th Conference on Uncertainty in Artificial Intelligence (UAI)*.

Minsky, M. L. (1977). Frame theory. In P. Johnson-Laird & P. C.Wason (Eds.), *Thinking: Readings in cognitive science*. Cambridge University Press.

Minsky, M. (1982). *Semantic information processing*. MIT Press.

Minsky, M. L., & Papert, S. A. (1969). *Perceptrons*. MIT Press.

Misyak, J., Noguchi, T., & Chater, N. (2016). Instantaneous conventions: The emergence of flexible communicative signals. *Psychological Science*, *27*(12), 1550–1561.

Mitchell, C. J., De Houwer, J., & Lovibond, P. F. (2009). The propositional nature of human associative learning. *Behavioral and Brain Sciences*, *32*(2), 183–198.

Mitchell, M., & Krakauer, D. C. (2023). The debate over understanding in AI's large language models. *Proceedings of the National Academy of Sciences*, *120*(13), e2215907120.

Mitchell, T. M. (1997). *Machine learning*. McGraw Hill.

Mnih, V., Kavukcuoglu, K., Silver, D., et al. (2015). Human-level control through deep reinforcement learning. *Nature*, *518*(7540), 529–533.

Mollica, F., & Piantadosi, S. T. (2017). How data drive early word learning: A cross-linguistic waiting time analysis. *Open Mind*, *1*(2), 67–77.

Mollica, F., & Piantadosi, S. T. (2021). Logical word learning: The case of kinship. *Psychonomic Bulletin & Review*, 1–34.

Momennejad, I., Otto, A. R., Daw, N. D., & Norman, K. A. (2017). Offline replay supports planning: fMRI evidence from reward revaluation. *bioRxiv*, 196758.

Momennejad, I., Russek, E. M., Cheong, J. H., Botvinick, M. M., Daw, N. D., & Gershman, S. J. (2017). The successor representation in human reinforcement learning. *Nature Human Behaviour*, *1*(9), 680–692.

Moreno-Bote, R., Knill, D. C., & Pouget, A. (2011). Bayesian sampling in visual perception. *Proceedings of the National Academy of Sciences*, *108*(30), 12491–12496.

Morrill, G. (2019). Parsing/theorem-proving for logical grammar "CatLog3." *Journal of Logic, Language and Information*, *28*, 183–216.

Moulton, S. T., & Kosslyn, S. M. (2009). Imagining predictions: Mental imagery as mental emulation. *Philosophical Transactions of the Royal Society of London B: Biological Sciences*, *364*(1521), 1273–1280.

Mrowca, D., Zhuang, C., Wang, E., et al. (2018). Flexible neural representation for physics prediction. In *Advances in Neural Information Processing Systems* (pp. 8813–8824).

Muggleton, S. (1995). Inverse entailment and progol. *New Generation Computing*, *13*(3), 245–286.

Muggleton, S. (1996). Stochastic logic programs. In L. D. Raedt (Ed.), *Advances in Inductive Logic Programming* (pp. 254–264).

Muggleton, S., & Chater, N. (2021). *Human-like machine intelligence*. Oxford University Press.

Muggleton, S., & De Raedt, L. (1994). Inductive logic programming: Theory and methods. *Journal of Logic Programming*, *19*, 629– 679.

Muggleton, S. H., Lin, D., & Tamaddoni-Nezhad, A. (2015). Meta-interpretive learning of higher-order dyadic datalog: Predicate invention revisited. *Machine Learning*, *100*(1), 49–73.

Mulder, M. J., Wagenmakers, E.-J., Ratcliff, R., Boekel, W., & Forstmann, B. U. (2012). Bias in the brain: a diffusion model analysis of prior probability and potential payoff. *Journal of Neuroscience*, *32*(7), 2335–2343.

Muller, P., & Quintana, F. A. (2004). Nonparametric Bayesian data analysis. *Statistical Science*, *19*, 95–110.

Murphy, G. L. (2002). *The big book of concepts*. MIT Press.

Murphy, K. P. (2012). *Machine learning: A probabilistic perspective*. MIT Press.

Myung, I. J., Forster, M. R., & Browne, M. W. (2000). Model selection [special issue]. *Journal of Mathematical Psychology*, *44*(1-2), 37.

Myung, I. J., & Pitt, M. A. (1997). Applying Occam's razor in modeling cognition: A Bayesian approach. *Psychonomic Bulletin and Review*, *4*, 79–95.

Myung, I. J., & Shepard, R. N. (1996). Maximum entropy inference and stimulus generalization. *Journal of Mathematical Psychology*, *40*(4), 342–347.

Nair, V., & Hinton, G. E. (2010). Rectified linear units improve restricted Boltzmann machines. In *Proceedings of the International Conference on Machine Learning*.

Narens, L., & Luce, R. D. (1986). Measurement: The theory of numerical assignments. *Psychological Bulletin*, *99*(2), 166–180.

Natarajan, R., Murray, I., Shams, L., & Zemel, R. (2009). Characterizing response behavior in multisensory perception with conflicting cues. In *Advances in Neural Information Processing Systems*.

Nau, R. F., & McCardle, K. F. (1991). Arbitrage, rationality, and equilibrium. *Theory and Decision*, *31*, 199–240.

Navarro, D. (2006). From natural kinds to complex categories. In R. Sun & N. Miyake (Eds.), *Proceedings of the 28th Annual Conference of the Cognitive Science society* (pp. 621–626). Psychology Press.

Navarro, D. J., & Perfors, A. F. (2011). Hypothesis generation, sparse categories, and the positive test strategy. *Psychological Review, 118*(1), 120–134.

Neal, R. M. (1993). *Probabilistic inference using Markov chain Monte Carlo methods* (Tech. Rep. No. CRG-TR-93-1). University of Toronto.

Neal, R. M. (1998). *Markov chain sampling methods for Dirichlet process mixture models* (Tech. Rep. No. 9815). Department of Statistics, University of Toronto.

Neal, R. M., & Hinton, G. E. (1998). A view EM algorithm that justifies incremental, sparse, and other variants. In M. I. Jordan (Ed.), *Learning in graphical models* (pp. 355–368). MIT Press.

Neisser, U. (1967). *Cognitive psychology.* Appleton-Century-Crofts.

Neisser, U., & Becklen, R. (1975). Selective looking: Attending to visually specified events. *Cognitive Psychology, 7*(4), 480–494.

Nelson, D. L., McEvoy, C. L., & Schreiber, T. A. (1998). *The University of South Florida word association, rhyme, and word fragment norms.* (http://w3.usf.edu/FreeAssociation/).

Nematzadeh, A., Meylan, S. C., & Griffiths, T. L. (2017). Evaluating vector-space models of word representation, or the unreasonable effectiveness of counting words near other words. In *Proceedings of the 39th Annual Meeting of the Cognitive Science Society.*

Neumann, J. von, & Morgenstern, O. (1944). *The theory of games and economic behavior.* Princeton University Press.

Newell, A., Shaw, J. C., & Simon, H. A. (1959). *Report on a general problem solving program* (Tech. Rep. No. P-1584). RAND Corporation. version. Also in *Proceedings of the International Conference on Information Processing* (pp. 256–264)

Newell, A., & Simon, H. (1956). The logic theory machine: A complex information processing system. *IRE Transactions on Information Theory, IT-2*, 61–79.

Newell, A., & Simon, H. A. (1972). *Human problem solving.* Prentice-Hall.

Newell, A., & Simon, H. A. (1976). Computer science as empirical inquiry: Symbols and search. *Communications of the ACM, 19*, 11–26.

Newell, B. R., Lagnado, D. A., & Shanks, D. R. (2022). *Straight choices: The psychology of decision making.* Psychology Press.

Newman, M. E. J., & Barkema, G. T. (1999). *Monte Carlo methods in statistical physics.* Clarendon Press.

Newsome, W. T., & Pare, E. B. (1988). A selective impairment of motion perception following lesions of the middle temporal visual area (MT). *Journal of Neuroscience, 8*(6), 2201–2211.

Ng, A., & Jordan, M. (2001). On discriminative vs. generative classifiers: A comparison of logistic regression and naive Bayes. *Advances in Neural Information Processing Systems.*

Ng, A. Y., Harada, D., & Russell, S. (1999). Policy invariance under reward transformations: Theory and application to reward shaping. In *Proceedings of the 16th Annual International Conference on Machine Learning* (pp. 278–287).

Nisbett, R. E., & Wilson, T. D. (1977). Telling more than we can know: Verbal reports on mental processes. *Psychological Review, 84*(3), 231–259.

Niv, Y. (2009). Reinforcement learning in the brain. *Journal of Mathematical Psychology, 53*(3), 139–154.

Niv, Y. (2019). Learning task-state representations. *Nature Neuroscience, 22*(10), 1544–1553.

Niv, Y., Daniel, R., Geana, A., Gershman, S. J., Leong, Y. C., Radulescu, A., & Wilson, R. C. et al. (2015). Reinforcement learning in multidimensional environments relies on attention mechanisms. *Journal of Neuroscience, 35*(21), 8145–8157.

Niyogi, P., & Berwick, R. C. (1996). A language learning model for finite parameter spaces. *Cognition, 61*, 161–193.

Norman, D. A. (1972). Memory, knowledge, and the answering of questions. In R. L. Solso (Ed.), *Contemporary issues in cognitive psychology: The Loyola Symposium.* V. H. Winston & Sons.

Norris, D. (2006). The Bayesian reader: Explaining word recognition as an optimal Bayesian decision process. *Psychological Review, 113*, 327–357.

Norris, J. R. (1997). *Markov chains.* Cambridge University Press.

Norvig, P. (2012). Colorless green ideas learn furiously: Chomsky and the two cultures of statistical learning. *Significance, 9*(4), 30–33.

Nosofsky, R. M. (1986). Attention, similarity, and the identification- categorization relationship. *Journal of Experimental Psychology: General, 115*, 39–57.

Nosofsky, R. M., Palmeri, T. J., & McKinley, S. C. (1994). Rule-plus-exception model of classification learning. *Psychological Review, 101*(1), 53–79.

Notaro, P., Gelman, S., & Zimmerman, M. (2001). Children's understanding of psychogenic bodily reactions. *Child Development, 72*, 444–459.

Oaksford, M., & Chater, N. (1994). A rational analysis of the selection task as optimal data selection. *Psychological Review, 101*(4), 608–631.

Oaksford, M., & Chater, N. (1998a). *Rationality in an uncertain world: Essays on the cognitive science of human reasoning.* Psychology Press.

Oaksford, M., & Chater, N. (Eds.). (1998b). *Rational models of cognition.* Oxford University Press.

Oaksford, M., & Chater, N. (2001). The probabilistic approach to human reasoning. *Trends in Cognitive Sciences, 5*, 349–357.

Oaksford, M., & Chater, N. (2003). Optimal data selection: Revision, review, and reevaluation. *Psychonomic Bulletin & Review, 10*(2), 289–318.

Oaksford, M., & Chater, N. (2007). *Bayesian rationality: The probabilistic approach to human reasoning.* Oxford University Press.

Oaksford, M., & Chater, N. (2020). New paradigms in the psychology of reasoning. *Annual Review of Psychology, 71*(1), 305–330.

Oberauer, K., Wilhelm, O. IV, & Diaz, R. R. (1999). Bayesian rationality for the Wason selection task? A test of optimal data selection theory. *Thinking & Reasoning, 5*(2), 115–144.

O'Donnell, T. J. (2015). *Productivity and reuse in language: A theory of linguistic computation and storage.* MIT Press.

O'Hagan, A., Buck, C. E., Daneshkhah, A., et al. (2006). *Uncertain judgements: Eliciting experts' probabilities.* Wiley.

Ohlsson, S. (2012). The problems with problem solving: Reflections on the rise, current status, and possible future of a cognitive research paradigm. *Journal of Problem Solving, 5*(1), 7.

Ólafsdóttir, H. F., Barry, C., Saleem, A. B., Hassabis, D., & Spiers, H. J. (2015). Hippocampal place cells construct reward related sequences through unexplored space. *eLife, 4*, e06063.

Olshausen, B. A., & Field, D. J. (1996). Emergence of simple-cell receptive field properties by learning a sparse code for natural images. *Nature, 381*(6583), 607–609.

Onnis, L., Roberts, M., & Chater, N. (2002). Simplicity: A cure for overgeneralizations in language acquisition? In *Proceedings of the 24th Annual Meeting of the Cognitive Science Society* (pp. 720-725).

Opheusden, B. van, Galbiati, G., Bnaya, Z., Li, Y., & Ma, W. J. (2017). A computational model for decision tree search. In *Proceedings of the Annual Meeting of the Cognitive Science Society.*

O'Reilly, R. C., & Munakata, Y. (2000). *Computational explorations in cognitive neuroscience: Understanding the mind by simulating the brain.* MIT Press.

Osera, P.-M., & Zdancewic, S. (2015). Type- and example-directed program synthesis. *ACM SIGPLAN Notices, 50*(6), 619–630.

Osherson, D. N., Smith, E. E., Wilkie, O., Lopez, A., & Shafir, E. (1990). Category-based induction. *Psychological Review, 97*(2), 185–200.

Otto, A. R., Gershman, S. J., Markman, A. B., & Daw, N. D. (2013). The curse of planning: dissecting multiple reinforcement-learning systems by taxing the central executive. *Psychological Science, 24*(5), 751–761.

Ouyang, L., Wu, J., Jiang, X., et al. (2022). Training language models to follow instructions with human feedback. *Advances in Neural Information Processing Systems* (pp. 27730–27744).

Overlan, M., Jacobs, R., & Piantadosi, S. T. (2017). Learning abstract visual concepts via probabilistic program m induction in a language of thought. *Cognition*, *168*, 320–334.

Paccanaro, A., & Hinton, G. E. (2002). Learning hierarchical structures with linear relational embedding. In *Advances in Neural Information Processing Systems* (pp. 857–864).

Pacer, M., & Griffiths, T. L. (2012). Elements of a rational framework for continuous-time causal induction. In *Proceedings of the 34th Annual Meeting of the Cognitive Science Society.*

Pacer, M., & Griffiths, T. L. (2015). Upsetting the contingency table: Causal induction over sequences of point events. In *Proceedings of the 37th Annual Meeting of the Cognitive Science Society.*

Paivio, A. (1991). Dual coding theory: Retrospect and current status. *Canadian Journal of Psychology/Revue canadienne de psychologie*, *45*(3), 255.

Pajak, B., Fine, A. B., Kleinschmidt, D. F., & Jaeger, T. F. (2016). Learning additional languages as hierarchical probabilistic inference: Insights from first language processing. *Language Learning*, *66*(4), 900–944.

Pajak, B., & Levy, R. (2014). The role of abstraction in non-native speech perception. *Journal of Phonetics*, *46*, 147–160.

Parise, C. V., Knorre, K., & Ernst, M. O. (2014). Natural auditory scene statistics shapes human spatial hearing. *Proceedings of the National Academy of Sciences*, *111*(16), 6104–6108.

Parr, R., & Russell, S. (1998). Reinforcement learning with hierarchies of machines. In *Advances in Neural Information Processing Systems.*

Paszke, A., Gross, S., Massa, F., et al. (2019). Pytorch: An imperative style, high-performance deep learning library. In *Advances in Neural Information Processing Systems* (pp. 8024–8035).

Pavlick, E. (2023). Symbols and grounding in large language models. *Philosophical Transactions of the Royal Society A*, *381*(2251), 20220041.

Payne, J. W., Bettman, J. R., & Johnson, E. J. (1993). *The adaptive decision maker.* Cambridge University Press.

Peano, G. (1889). *Arithmetices principia: Nova methodo.* Fratres Bocca.

Pearl, J. (1988). *Probabilistic reasoning in intelligent systems.* Morgan Kaufmann.

Pearl, J. (2000). *Causality: Models, reasoning and inference.* Cambridge University Press.

Pearl, L. (2021). Modeling syntactic acquisition. In J. Sprouse (Ed.), *Oxford handbook of experimental syntax* (pp. 209–270). Oxford University Press.

Pearl, L., & Sprouse, J. (2013). Syntactic islands and learning biases: Combining experimental syntax and computational modeling to investigate the language acquisition problem. *Language Acquisition*, *20*(1), 23–68.

Pearl, L., & Sprouse, J. (2019). Comparing solutions to the linking problem using an integrated quantitative framework of language acquisition. *Language*, *95*(4), 583–611.

Pearson, J., & Kosslyn, S. M. (2015). The heterogeneity of mental representation: Ending the imagery debate. *Proceedings of the National Academy of Sciences*, *112*(33), 10089–10092.

Pennington, J., Socher, R., & Manning, C. D. (2014). GloVe: Global vectors for word representation. In *Proceedings of the 2014 Conference on Empirical Methods in Natural Language Processing (EMNLP)* (pp. 1532–1543).

Pereira, F. (2000). Formal grammar and information theory: Together again? *Philosophical Transactions of the Royal Society A*, *358*(1769), 1239–1253.

Pérez, J., Marinković, J., & Barceló, P. (2019). On the Turing completeness of modern neural network architectures. *arXiv preprint arXiv:1901.03429.*

Perfors, A., & Tenenbaum, J. B. (2009). Learning to learn categories. In *Proceedings of the 31st Annual Meeting of the Cognitive Science Society* (pp. 136–141).

Perfors, A., Tenenbaum, J. B., Griffiths, T. L., & Xu, F. (2011). A tutorial introduction to Bayesian models of cognitive development. *Cognition*, *120*(3), 302–321.

Perfors, A., Tenenbaum, J. B., & Regier, T. (2011). The learnability of abstract syntactic principles. *Cognition*, *118*(3), 306–338.

Perfors, A., Tenenbaum, J. B., & Wonnacott, E. (2010). Variability, negative evidence, and the acquisition of verb argument constructions. *Journal of Child Language*, *37*, 607–642.

Pericliev, V., & Valdés-Peréz, R. E. (1998). Automatic componential analysis of kinship semantics with a proposed structural solution to the problem of multiple models. *Anthropological Linguistics*, *40*(2), 272–317.

Peterson, C. R., & Beach, L. R. (1967). Man as an intuitive statistician. *Psychological Bulletin*, *68*, 29–46.

Pfaff, T., Fortunato, M., Sanchez-Gonzalez, A., & Battaglia, P. W. (2021). Learning mesh-based simulation with graph networks. In *International Conference on Learning Representations.*

Pfeiffer, B. E., & Foster, D. J. (2013). Hippocampal place-cell sequences depict future paths to remembered goals. *Nature*, *497*(7447), 74–79.

Phillips, A. T., & Wellman, H. M. (2005). Infants' understanding of object- directed action. *Cognition*, *98*, 137–155.

Phillips, L., & Pearl, L. (2015). The utility of cognitive plausibility in language acquisition modeling: Evidence from word segmentation. *Cognitive Science*, *39*, 1–31.

Piaget, J. (1954). *The construction of reality in the child.* Basic Books.

Piantadosi, S. T. (2011). *Learning and the language of thought.* Unpublished doctoral dissertation, Massachusetts Institute of Technology.

Piantadosi, S. T. (2021). The computational origin of representation.*Minds and Machines*, *31*(1), 1–58.

Piantadosi, S. T. (2023). The algorithmic origins of counting. *Child Development, 94*(6), 1472–1490.

Piantadosi, S. T., & Jacobs, R. A. (2016). Four problems solved by the probabilistic language of thought. *Current Directions in Psychological Science*, *25*(1), 54–59.

Piantadosi, S. T., Tenenbaum, J. B., & Goodman, N. D. (2012a). Bootstrapping in a language of thought: A formal model of numerical concept learning. *Cognition*, *123*(2), 199–217.

Piantadosi, S. T., Tenenbaum, J. B., & Goodman, N. D. (2016). The logical primitives of thought: Empirical foundations for compositional cognitive models. *Psychological Review*, *123*(4), 392.

Piantadosi, S. T., Tily, H., & Gibson, E. (2012b). The communicative function of ambiguity in language. *Cognition*, *122*(3), 280–291.

Pickering, M. J., & Garrod, S. (2013). An integrated theory of language production and comprehension. *Behavioral and Brain Sciences*, *36*(4), 329–347.

Pickering, M. J., & Garrod, S. (2021). *Understanding dialogue: Language use and social interaction.* Cambridge University Press.

Pierce, B. C. (2002). *Types and programming languages.* MIT Press.

Piloto, L., Weinstein, A., Dhruva T. B., et al. (2018). Probing physics knowledge using tools from developmental psychology. *arXiv:1804.01128 [cs].*

Pinker, S. (1979). Formal models of language learning. *Cognition*, *7*(3), 217–283.

Pinker, S. (1994). *The language instinct: How the mind creates language.* Harper Collins.

Pinker, S. (1997). *How the mind works.* W. W. Norton & Company.

Pinker, S. (1999). *Words and rules: The ingredients of language.* Basic Books.

Pinker, S. (2003). *The blank slate: The modern denial of human nature.* Penguin.

Pinker, S., & Prince, A. (1988). On language and connectionism: Analysis of a parallel distributed processing model of language acquisition. *Cognition*, *28*(1–2), 73–193.

Pitman, J. (1993). *Probability.* Springer-Verlag.

Pitman, J. (2002). *Combinatorial stochastic processes.* Notes for Saint Flour Summer School.

Planton, S., van Kerkoerle, T., Abbih, L., et al. (2021). A theory of memory for binary sequences: Evidence for a mental compression algorithm in humans. *PLoS Computational Biology*, *17*(1), e1008598.

Platt, J. R. (1964). Strong inference: Certain systematic methods of scientific thinking may produce much more rapid progress than others. *Science*, *146*(3642), 347–353.

Pleskac, T. J., & Busemeyer, J. R. (2010). Two-stage dynamic signal detection: A theory of choice, decision time, and confidence. *Psychological Review*, *117*(3), 864–901.

Polikarpova, N., Kuraj, I., & Solar-Lezama, A. (2016). Program synthesis from polymorphic refinement types. *ACM SIGPLAN Notices, 51*(6), 522–538.

Pollard, C., & Sag, I. A. (1994). *Head-driven phrase structure grammar.* University of Chicago Press.

Popper, K. R. (1959/1990). *The logic of scientific discovery.* Unwin Hyman.

Pothos, E. M., & Chater, N. (2002). A simplicity principle in unsupervised human categorization. *Cognitive Science, 26*(3), 303–343.

Pouget, A., Beck, J. M., Ma, W. J., & Latham, P. E. (2013). Probabilistic brains: Knowns and unknowns. *Nature Neuroscience, 16*(9), 1170–1178.

Pramod, R., Cohen, M., Tenenbaum, J., & Kanwisher, N. (2022). *Invariant representation of physical stability in the human brain.* eLife, *11*, e71736.

Prystawski, B., Mohnert, F., Tošić, M., & Lieder, F. (2020). Resource-rational models of human goal pursuit. *Topics in Cognitive Science, 14*(3), 528–549.

Pullum, G. K. (2020). Theorizing about the syntax of human language: A radical alternative to generative formalisms. *Cadernos de Linguística, 1*(1), 1–33.

Puterman, M. L. (1994). *Markov decision processes: Discrete stochastic dynamic programming.* Wiley.

Putnam, H. (1974). The 'corroboration' of theories. In P. A. Schilpp (Ed.), *The philosophy of Karl Popper* (pp. 221–240). Open Court.

Pylyshyn, Z. W. (2002). Mental imagery: In search of a theory. *Behavioral and Brain Sciences, 25*(2), 157–238.

Quine, W. V. O. (1960). *Word and object.* MIT Press.

Quine, W. V. O., & Ullian, J. S. (1978). *The web of belief* (Vol. 2). Random House New York.

Quinlan, J. R. (1990). Learning logical definitions from relations. *Machine Learning, 5*(3), 239–266.

Rabi, R., & Minda, J. P. (2016). Category learning in older adulthood: A study of the Shepard, Hovland, and Jenkins (1961) tasks. *Psychology and Aging, 31*(2), 185–197.

Rabiner, L. (1989). A tutorial on hidden Markov models and selected applications in speech recognition. *Proceedings of the IEEE, 77.*

Rabinowitz, N. C., Perbet, F., Song, H. F., Zhang, C., Eslami, S., & Botvinick, M. (2018). Machine theory of mind. *arXiv preprint arXiv:1802.07740.*

Radkani, S., Tenenbaum, J. B., & Saxe, R. (2022). Modeling punishment as a rational communicative social action. In *Proceedings of the Annual Meeting of the Cognitive Science Society* (Vol. 44).

Radulescu, A., Niv, Y., & Ballard, I. (2019). Holistic reinforcement learning: The role of structure and attention. *Trends in Cognitive Sciences, 23*(4), 278–292.

Ramachandran, V. (1990). Interactions between motion, depth, color and form: The utilitarian theory of perception. *Vision: Coding and efficiency*, 346–360.

Ramesh, A., Dhariwal, P., Nichol, A., Chu, C., & Chen, M. (2022). Hierarchical text-conditional image generation with clip latents. *arXiv preprint arXiv:2204.06125.*

Ramlee, F., Sanborn, A., & Tang, N. (2017). What sways people's judgement of sleep quality? A quantitative choice-making study with good and poor sleepers. *Sleep, 40*(7), zsx091.

Ramsey, F. P. (1926/1931). Truth and probability. In R. B. Braithwaite (Ed.), *The foundations of mathematics and other logical essays* (pp. 156–198). Harcourt, Brace and Company.

Ranganath, R., Gerrish, S., & Blei, D. (2014). Black box variational inference. In *Proceedings of the Seventeenth International Conference on Artificial Intelligence and Statistics* (pp. 814–822).

Rasmussen, C. E. (2000). The infinite Gaussian mixture model. In *Advances in Neural Information Processing Systems.*

Rasmussen, C. E., & Ghahramani, Z. (2003). Bayesian Monte Carlo. In *Advances in Neural Information Processing Systems* (pp. 505–512).

Ratcliff, R. (1978). A theory of memory retrieval. *Psychological Review, 85*(2), 59–108.

Ratcliff, R., Smith, P. L., Brown, S. D., & McKoon, G. (2016). Diffusion decision model: Current issues and history. *Trends in Cognitive Sciences, 20*(4), 260–281.

Redington, M., Chater, N., & Finch, S. (1998). Distributional information: A powerful cue for acquiring syntactic categories. *Cognitive Science, 22*(4), 425–469.

Reed, S. K. (1972). Pattern recognition and categorization. *Cognitive Psychology*, *3*, 393–407.

Reed, S., Zolna, K., Parisotto, E., et al. (2022). A generalist agent. *arXiv preprint arXiv:2205.06175*.

Rehder, B. (2003). A causal-model theory of conceptual representation and categorization. *Journal of Experimental Psychology: Learning, Memory, and Cognition*, *29*, 1141–1159.

Rehder, B. (2014). Independence and dependence in human causal reasoning. *Cognitive Psychology*, *72*, 54–107.

Rehder, B. (2018). Beyond Markov: Accounting for independence violations in causal reasoning. *Cognitive Psychology*, *103*, 42–84.

Rehder, B., & Burnett, R. C. (2005). Feature inference and the causal structure of categories. *Cognitive Psychology*, *50*(3), 264–314.

Rehder, B., & Waldmann, M. R. (2017). Failures of explaining away and screening off in described versus experienced causal learning scenarios. *Memory & cCgnition*, *45*, 245–260.

Reiter, R. (1980). A logic for default reasoning. *Artificial Intelligence*, *13*(1–2), 81–132.

Rescorla, R. A., & Wagner, A. R. (1972). A theory of Pavlovian conditioning: Variations on the effectiveness of reinforcement and non-reinforcement. In A. H. Black & W. F. Prokasy (Eds.), *Classical conditioning II: Current research and theory* (pp. 64–99). Appleton-Century-Crofts.

Restle, F. (1962). The selection of strategies in cue learning. *Psychological Review*, *69*(4), 329–343.

Ribas-Fernandes, J. J., Solway, A., Diuk, C., et al. (2011). A neural signature of hierarchical reinforcement learning. *Neuron*, *71*(2), 370–379.

Ricardo, D. (1817). *On the principles of political economy and taxation.* John Murray.

Rice, H. G. (1953). Classes of recursively enumerable sets and their decision problems. *Transactions of the American Mathematical society*, *74*(2), 358–366.

Rice, J. A. (1995). *Mathematical statistics and data analysis* (2nd ed.). Duxbury.

Richards, W. E. (1988). *Natural computation.* MIT Press.

Richardson, M., & Domingos, P. (2006). Markov logic networks. *Machine Learning*, *62*(1–2), 107–136.

Riochet, R., Castro, M. Y., Bernard, M., et al. (2018). IntPhys: A framework and benchmark for visual intuitive physics reasoning. *arXiv:1803.07616 [cs]*.

Rips, L. J. (1975). Inductive judgments about natural categories. *Journal of Verbal Learning and Verbal Behavior*, *14*, 665–681.

Rips, L. J. (1994). *The psychology of proof: Deductive reasoning in human thinking.* MIT Press.

Rish, I. (2001). An empirical study of the naive Bayes classifier. In *IJCAI 2001 Workshop on Empirical Methods in Artificial Intelligence* (Vol. 3, pp. 41–46).

Rissanen, J. (1987). Stochastic complexity. *Journal of the Royal Statistical Society: Series B (Methodological)*, *49*(3), 223–239.

Rissanen, J. (1989). *Stochastic complexity in statistical inquiry.* World Scientific Press.

Robbins, H. (1952). Some aspects of the sequential design of experiments. *Bulletin of the American Mathematical Society*, *58*, 527–535.

Robert, C. P. (2007). *The Bayesian choice: From decision-theoretic foundations to computational implementation.* Springer.

Robert, C. P., & Casella, G. (1999). *Monte Carlo statistical methods* (Vol. 2). Springer.

Robert, C. P., & Casella, G. (2009). *Introducing Monte Carlo methods with R.* Springer Science & Business Media.

Robinson, J. (1949). Definability and decision problems in arithmetic. *The Journal of Symbolic Logic, 14*(2), 98–114.

Rogers, T., & McClelland, J. (2004). *Semantic cognition: A parallel distributed processing approach.* MIT Press.

Rohe, T., Ehlis, A.-C., & Noppeney, U. (2019). The neural dynamics of hierarchical Bayesian causal inference in multisensory perception. *Nature Communications*, *10*(1), 1907.

Rosenberg, A. (2019). *How history gets things wrong: The neuroscience of our addiction to stories.* MIT Press.

Rosenblatt, F. (1958). The perceptron: A probabilistic model for information storage and organization in the brain. *Psychological Review*, *65*, 386–408.

Rosseel, Y. (2002). Mixture models of categorization. *Journal of Mathematical Psychology*, *46*, 178–210.

Rothe, A., Lake, B. M., & Gureckis, T. (2017). Question asking as program generation. In *Advances in Neural Information Processing Systems* (pp. 1046–1055).

Rule, J. S., Piantadosi, S. T., Cropper, A., Ellis, K., Nye, M., & Tenenbaum, J. B. (in press). Symbolic meta-program search improves learning efficiency and explains rule learning in humans. *Nature Communications*.

Rule, J. S., Tenenbaum, J. B., & Piantadosi, S. T. (2020). The child as hacker. *Trends in Cognitive Sciences*, *24*(11), 900–915.

Rumelhart, D., & McClelland, J. (1986). On learning the past tenses of English verbs. In J. McClelland, D. Rumelhart, & the PDP Research Group (Eds.), *Parallel distributed processing: Explorations in the microstructure of cognition* (Vol. 2, pp. 535–551). MIT Press.

Rumelhart, D. E., Hinton, G. E., & Williams, R. J. (1985). *Learning internal representations by error propagation* (Technical Report). California University San Diego La Jolla Institute for Cognitive Science.

Rumelhart, D. E., Hinton, G. E., & Wilson, R. J. (1986a). Learning representations by back-propagating errors. *Nature*, *323*, 533–536.

Rumelhart, D. E., Smolensky, P., McClelland, J. L., & Hinton, G. (1986b). Sequential thought processes in PDP models. *Parallel Distributed Processing: Explorations in the Microstructures of Cognition*, *2*, 3–57.

Russek, E. M., Momennejad, I., Botvinick, M. M., Gershman, S. J., & Daw, N. D. (2017). Predictive representations can link model- based reinforcement learning to model-free mechanisms. *PLoS Computational Biology*, *13*(9), e1005768.

Russell, S. J., & Norvig, P. (2021). *Artificial intelligence: A modern approach* (4th ed.). Prentice Hall.

Russell, S. J., & Subramanian, D. (1994). Provably bounded-optimal agents. *Journal of Artificial Intelligence Research*, *2*, 575–609.

Russell, S. J., & Wefald, E. (1991). Principles of metareasoning. *Artificial Intelligence*, *49*(1–3), 361–395.

Saffran, J. R., Aslin, R. N., & Newport, E. L. (1996). Statistical learning by 8-month old infants. *Science*, *274*, 1926–1928.

Salakhutdinov, R., Tenenbaum, J. B., & Torralba, A. (2013). Learning with hierarchical-deep models. *IEEE Transactions on Pattern Analysis and Machine Intelligence*, *35*(8), 1958–1971.

Samuelson, P. A. (1938). A note on the pure theory of consumer's behaviour. *Economica*, *5*(17), 61–71.

Sanborn, A. N. (2014). Testing Bayesian and heuristic predictions of mass judgments of colliding objects. *Frontiers in Psychology*, *5*, 1–7.

Sanborn, A. N., & Chater, N. (2016). Bayesian brains without probabilities. *Trends in Cognitive Sciences*, *20*(12), 883–893.

Sanborn, A. N., & Griffiths, T. L. (2008). Markov chain Monte Carlo with people. In *Advances in Neural Information Processing Systems*.

Sanborn, A. N., Griffiths, T. L., & Navarro, D. J. (2006). A more rational model of categorization. In *Proceedings of the 28th Annual Conference of the Cognitive Science Society*.

Sanborn, A. N., Griffiths, T. L., & Navarro, D. J. (2010a). Rational approximations to rational models: Alternative algorithms for category learning. *Psychological Review*, *117*(4), 1144–1167.

Sanborn, A. N., Griffiths, T. L., & Shiffrin, R. M. (2010b). Uncovering mental representations with Markov chain Monte Carlo. *Cognitive Psychology*, *60*(2), 63–106.

Sanborn, A. N., Heller, K., Austerweil, J. L., & Chater, N. (2021). Refresh: A new approach to modeling dimensional biases in perceptual similarity and categorization. *Psychological Review*, *128*(6), 1145–1186.

Sanborn, A. N., Mansinghka, V. K., & Griffiths, T. L. (2013). Reconciling intuitive physics and Newtonian mechanics for colliding objects. *Psychological Review*, *120*(2), 411–437.

Sanborn, A. N., & Silva, R. (2013). Constraining bridges between levels of analysis: A computational justification for locally Bayesian learning. *Journal of Mathematical Psychology*, *57*(3–4), 94–106.

Sanchez-Gonzalez, A., Godwin, J., Pfaff, T., Ying, R., Leskovec, J., & Battaglia, P. (2020). Learning to simulate complex physics with graph networks. In *International Conference on Machine Learning* (pp. 8459–8468).

Santos, R. J. (1996). Equivalence of regularization and truncated iteration for general ill-posed problems. *Linear Algebra and Its Applications*, *236*, 25–33.

Savage, L. J. (1972). *The foundations of statistics* (2nd ed.). Courier Corporation.

Saxe, R., & Kanwisher, N. (2003). People thinking about thinking people: The role of the temporo-parietal junction in "theory of mind". *Neuroimage*, *19*(4), 1835–1842.

Scarfe, P., & Glennerster, A. (2014). Humans use predictive kinematic models to calibrate visual cues to three-dimensional surface slant. *Journal of Neuroscience*, *34*(31), 10394–10401.

Schachter, R. (1998). Bayes-ball: The rational pastime (for determining irrelevance and requisite information in belief networks and influence diagrams. In *Proceedings of the Fourteenth Annual Conference on Uncertainty in Artificial Intelligence (UAI 98).*

Schank, R. C., & Abelson, R. P. (1977). *Scripts, plans, goals, and understanding.* Erlbaum.

Schmidhuber, J. (1987). *Evolutionary principles in self-referential learning.* Unpublished doctoral dissertation, Institut für Informatik, Technische Universität München.

Schmidhuber, J. (2002). The speed prior: A new simplicity measure yielding near-optimal computable predictions. In *International Conference on Computational Learning Theory* (pp. 216–228).

Schmidt, M., & Lipson, H. (2009). Distilling free-form natural laws from experimental data. *Science*, *324*(5923), 81–85.

Scholl, B. J., & Tremoulet, P. D. (2000). Perceptual causality and animacy. *Trends in Cognitive Sciences*, *4*(8), 299–309.

School, B., & Leslie, A. (1999). Explaining the infant's object concept: Beyond the perception/cognition dichotomy. In E. Lepore & Z. Pylyshyn (Eds.), *What is cognitive science?* (pp. 26–73). Blackwell.

Schooler, L. J., & Anderson, J. R. (1997). The role of process in the rational analysis of memory. *Cognitive Psychology*, *32*, 219–250.

Schubotz, R. I. (2007). Prediction of external events with our motor system: Towards a new framework. *Trends in Cognitive Sciences*, *11*(5), 211–218.

Schulz, E., Tenenbaum, J. B., Duvenaud, D., Speekenbrink, M., & Gershman, S. J. (2017). Compositional inductive biases in function learning. *Cognitive Psychology*, *99*, 44–79.

Schulz, L. E., Bonawitz, E. B., & Griffiths, T. L. (2007). Can being scared make your tummy ache? Naive theories, ambiguous evidence, and preschoolers' causal inferences. *Developmental Psychology*, *43*, 1124–1139.

Schulz, L., Goodman, N., Tenenbaum, J., & Jenkins, C. (2008). Going beyond the evidence: Abstract laws and preschoolers' responses to anomalous data. *Cognition*, *109*(2), 211–233.

Schulz, L., Gopnik, A., & Glymour, C. (2007). Preschool children learn about causal structure from conditional interventions. *Developmental Science*, *10*, 322–332.

Schulz, L. E., & Sommerville, J. (2006). God does not play dice: Causal determinism and children's inferences about unobserved causes. *Child Development*, *77*, 427–442.

Schultz, T., & Frank, M. C. (2023). Turning large language models into cognitive models. *arXiv preprint arXiv:2306.03917.*

Schultz, W., Dayan, P., & Montague, P. R. (1997). A neural substrate of prediction and reward. *Science*, *275*(5306), 1593–1599.

Schwartz, D. L., & Black, J. B. (1996a). Analog imagery in mental model reasoning: Depictive models. *Cognitive Psychology*, *30*(2), 154–219.

Schwartz, D. L., & Black, J. B. (1996b). Shuttling between depictive models and abstract rules: Induction and fallback. *Cognitive Science*, *20*(4), 457–497.

Schwartz, D. L., & Black, T. (1999). Inferences through imagined actions: Knowing by simulated doing. *Journal of Experimental Psychology: Learning, Memory, and Cognition*, *25*(1), 116–136.

Schwettmann, S., Tenenbaum, J. B., & Kanwisher, N. (2019). Invariant representations of mass in the human brain. *eLife*, *8*, e46619.

Scontras, G., Tessler, M. H., & Franke., M. (2017). *Probabilistic language understanding: An introduction to the Rational Speech Act framework.* Accessed September 20, 2023, from http://problang.org.

Sethuraman, J. (1994). A constructive definition of Dirichlet priors. *Statistica Sinica*, *4*, 639–650.

Shafto, P., Goodman, N. D., & Frank, M. C. (2012). Learning from others: The consequences of psychological reasoning for human learning. *Perspectives on Psychological Science*, *7*(4), 341–351.

Shafto, P., Goodman, N., & Griffiths, T. L. (2014). A rational account of pedagogical reasoning: Teaching by, and learning from, examples. *Cognitive Psychology*, *71*, 55–89.

Shafto, P., Kemp, C., Bonawitz, E. B., Coley, J. D., & Tenenbaum, J. B. (2008). Inductive reasoning about causally transmitted properties. *Cognition*, *109*(2), 175–192.

Shafto, P., Kemp, C., Mansinghka, V., Gordon, M., & Tenenbaum, J. B. (2006). Learning cross-cutting systems of categories. In *Proceedings of the 28th Annual Meeting of the Cognitive Science Society* (pp. 2151–2156).

Shafto, P., Kemp, C., Mansinghka, V., & Tenenbaum, J. B. (2011). A probabilistic model of cross-categorization. *Cognition*, *120*(1), 1–25.

Shanks, D. R., Tunney, R. J., & McCarthy, J. D. (2002). A re-examination of probability matching and rational choice. *Journal of Behavioral Decision Making*, *15*(3), 233–250.

Shannon, C. E. (1948). The mathematical theory of communication. *Bell System Technical Journal*, *27*, 379–423, 623–656.

Shastri, L., & Ajjanagadde, V. (1993). From simple associations to systematic reasoning: A connectionist representation of rules, variables and dynamic bindings using temporal synchrony. *Behavioral and Brain Sciences*, *16*(3), 417–451.

Shenhav, A., Musslick, S., Lieder, F., et al. (2017). Toward a rational and mechanistic account of mental effort. *Annual Review of Neuroscience*, *40*, 99–124.

Shepard, R. N. (1957). Stimulus and response generalization: A stochastic model relating generalization to distance in psychological space. *Psychometrika*, *22*(4), 325–345.

Shepard, R. N. (1980). Multidimensional scaling, tree-fitting, and clustering. *Science*, *210*, 390–398.

Shepard, R. N. (1987). Toward a universal law of generalization for psychological science. *Science*, *237*, 1317–1323.

Shepard, R. N. (1994). Perceptual-cognitive universals as reflections of the world. *Psychonomic Bulletin & Review*, *1*, 2–28.

Shepard, R. N., & Feng, C. (1972). A chronometric study of mental paper folding. *Cognitive Psychology*, *3*(2), 228–243.

Shepard, R. N., Hovland, C. I., & Jenkins, H. M. (1961). Learning and memorization of classifications. *Psychological Monographs*, *75* (13, No. 517).

Shepard, R. N., & Metzler, J. (1971). Mental rotation of three-dimensional objects. *Science*, *171*, 701–703.

Shevlin, H. (2020). A #gpt3 interview about AI with an imaginary David Chalmers. (https://twitter.com/dioscuri/status/1285385825971245058?lang=en).

Shi, L., & Griffiths, T. (2009). Neural implementation of hierarchical Bayesian inference by importance sampling. In *Advances in Neural Information Processing Systems* (pp. 1669–1677).

Shi, L., Griffiths, T. L., Feldman, N. H., & Sanborn, A. N. (2010). Exemplar models as a mechanism for performing Bayesian inference. *Psychological Bulletin and Review*, *17*, 443–464.

Shiffrin, R. M., & Steyvers, M. (1997). A model for recognition memory: REM: Retrieving Effectively from Memory. *Psychonomic Bulletin & Review*, *4*, 145–166.

Shultz, T. R. (1982). Causal reasoning in the social and non-social realms. *Canadian Journal of Behavioural Science*, *14*, 307–322.

Shultz, T. R., Mareschal, D., & Schmidt, W. C. (1994). Modeling cognitive development on balance scale phenomena. *Machine Learning*, *16*(1), 57–86.

Siegel, M., Magid, R., Tenenbaum, J., & Schulz, L. (2014). Black boxes: Hypothesis testing via indirect perceptual evidence. In *Proceedings of the 36th Annual Meeting of the Cognitive Science Society*.

Siegler, R. (1996). *Emerging minds: The process of change in children's thinking.* Oxford University Press.

Siegler, R. S. (1976). Three aspects of cognitive development. *Cognitive Psychology*, *8*(4), 481–520.

Silver, D., Huang, A., Maddison, C. J., et al. (2016). Mastering the game of Go with deep neural networks and tree search. *Nature*, *529*(7587), 484.

Silver, D., Schrittwieser, J., Simonyan, K., et al. (2017). Mastering the game of Go without human knowledge. *Nature*, *550*(7676), 354–359.

Silver, D., Singh, S., Precup, D., & Sutton, R. S. (2021). Reward is enough. *Artificial Intelligence*, *299*, 103535.

Silverman, B. W. (1986). *Density estimation for statistics and data analysis*. Chapman and Hall.

Simon, H. A. (1956). Rational choice and the structure of the environment. *Psychological Review*, *63*(2), 129–138.

Simon, H. A. (1982). *Models of bounded rationality: Empirically grounded economic reason* (Vol. 3). MIT Press.

Simons, D. J., & Keil, F. C. (1995). An abstract to concrete shift in the development of biological thought: The insides story. *Cognition*, *56*(2), 129–163.

Sims, C. R., Jacobs, R. A., & Knill, D. C. (2012). An ideal observer analysis of visual working memory. *Psychological Review*, *119*(4), 807–830.

Singh, S., Lewis, R. L., & Barto, A. G. (2009). Where do rewards come from? In *Proceedings of the Annual Meeting of the Cognitive Science Society* (pp. 2601–2606).

Siskind, J. (1996). A computational study of cross-situational techniques for learning word-to-meaning mappings. *Cognition*, *61*, 31–91.

Skinner, B. F. (1953). *Science and human behavior.* Simon and Schuster.

Skinner, B. F. (1957). *Verbal behavior.* Prentice Hall.

Skirzynński, J., Becker, F., & Lieder, F. (2021). Automatic discovery of interpretable planning strategies. *Machine Learning*, *110*, 2641–2683.

Sloman, S. (2005). *Causal models: How people think about the world and its alternatives*. Oxford University Press.

Sloman, S. A., & Lagnado, D. (2015). Causality in thought. *Annual Review of Psychology*, *66*, 223–247.

Sloman, S. A., & Lagnado, D. (2005). Do we "do"? *Cognitive Science*, *29*(1), 5–39.

Smith, K. A., Battaglia, P. W., & Vul, E. (2018). Different physical intuitions exist between tasks, not domains. *Computational Brain & Behavior*, *1*(2), 101–118.

Smith, K. A., Dechter, E., Tenenbaum, J. B., & Vul, E. (2013). Physical predictions over time. In *Proceedings of the 35th Annual Meeting of the Cognitive Science Society.*

Smith, K. A., Huber, D. E., & Vul, E. (2013). Multiply-constrained semantic search in the remote associates test. *Cognition*, *128*(1), 64–75.

Smith, K. A., Mei, L., Yao, S., et al. (2019). Modeling expectation violation in intuitive physics with coarse probabilistic object representations. In *Advances in Neural Information Processing Systems.*

Smith, K. A., de Peres, F. A. B., Vul, E., & Tenenbaum, J. B. (2017). Thinking inside the box: Motion prediction in contained spaces using simulation. In *Proceedings of the 39th Annual Meeting of the Cognitive Science Society* (pp. 3209–3214).

Smith, K. A., & Vul, E. (2013). Sources of uncertainty in intuitive physics. *Topics in Cognitive Science*, *5*(1), 185–199.

Smith, K. A., & Vul, E. (2015). Prospective uncertainty: The range of possible futures in physical prediction. In *Proceedings of the 37th Annual Meeting of the Cognitive Science Society.*

Smith, L. B. (2000). How to learn words: An associative crane. In K. H.-P. R. Golinkoff (Ed.), *Breaking the word learning barrier* (pp. 51–80). Oxford University Press.

Smith, L. B., Jones, S. S., Landau, B., Gershkoff-Stowe, L., & Samuelson, L. (2002). Object name learning provides on-the-job training for attention. *Psychological Science*, *13*(1), 13–19.

Smith, P. L., & Ratcliff, R. (2004). Psychology and neurobiology of simple decisions. *Trends in Neurosciences*, *27*(3), 161–168.

Smolensky, P. (1990). Tensor product variable binding and the representation of symbolic structures in connectionist systems. *Artificial Intelligence*, *46*(1–2), 159–216.

Sobel, D. M., Tenenbaum, J. B., & Gopnik, A. (2004). Children's causal inferences from indirect evidence: Backwards blocking and Bayesian reasoning in preschoolers. *Cognitive Science*, *28*, 303–333.

Solomonoff, R. (1978). Complexity-based induction systems: Comparisons and convergence theorems. *IEEE Transactions on Information Theory*, *24*(4), 422–432.

Solomonoff, R. J. (1964). A formal theory of inductive inference. Part I. *Information and Control*, *7*(1), 1–22.

Solway, A., & Botvinick, M. M. (2015). Evidence integration in model- based tree search. *Proceedings of the National Academy of Sciences*, *112*(37), 11708–11713.

Solway, A., Diuk, C., Córdova, N., et al. (2014). Optimal behavioral hierarchy. *PLoS Computational Biology*, *10*(8), e1003779.

Sorg, J., Singh, S. P., & Lewis, R. L. (2010). Internal rewards mitigate agent boundedness. In *Proceedings of the 27th International Conference on Machine Learning* (pp. 1007–1014).

Sosa, F. A., Ullman, T., Tenenbaum, J. B., Gershman, S. J., & Gerstenberg, T. (2021). Moral dynamics: Grounding moral judgment in intuitive physics and intuitive psychology. *Cognition*, *217*, 104890.

Spelke, E. S. (1990). Principles of object perception. *Cognitive Science*, *14*(1), 29–56.

Spelke, E. S. (2022). *What babies know: Core knowledge and composition.* Oxford University Press.

Spelke, E. S., Breinlinger, K., Jacobson, K., & Phillips, A. (1993). Gestalt relations and object perception: A developmental study. *Perception*, *22*(12), 1483–1501.

Spelke, E. S., Breinlinger, K., Macomber, J., & Jacobson, K. (1992). Origins of knowledge. *Psychological Review*, *99*(4), 605–632.

Spelke, E. S., & Kinzler, K. D. (2007). Core knowledge. *Developmental Science*, *10*(1), 89–96.

Spelke, E. S., & Lee, S. A. (2012). Core systems of geometry in animal minds. *Philosophical Transactions of the Royal Society B: Biological Sciences*, *367*, 2784–2793.

Sperber, D. (1985). Anthropology and psychology: Towards an epidemi- ology of representations. *Man*, *20*, 73–89.

Spirtes, P., Glymour, C., & Schienes, R. (1993). *Causation, prediction, and search.* Springer-Verlag.

Srivastava, N., Hinton, G., Krizhevsky, A., Sutskever, I., & Salakhutdinov, R. (2014). Dropout: A simple way to prevent neural networks from overfitting. *Journal of Machine Learning Research*, *15*(1), 1929–1958.

Stachenfeld, K. L., Botvinick, M. M., & Gershman, S. J. (2017). The hippocampus as a predictive map. *Nature Neuroscience*, *20*(11), 1643–1653.

Stacy, S., Li, C., Zhao, M., Yun, Y., Zhao, Q., Kleiman-Weiner, M., & Gao, T. (2021). Modeling communication to coordinate perspectives in cooperation. *arXiv preprint, arXiv:2106.02164.*

Stanford, K. (2017). Underdetermination of scientific theory. In E. N. Zalta (Ed.), *The Stanford Encyclopedia of Philosophy* (Winter 2017 ed.). Metaphysics Research Lab, Stanford University. https://plato.stanford.edu/archives/sum2023/entries/scientific-underdetermination/

Steedman, M. (2000). *The syntactic process.* MIT Press.

Stephens, D. W., & Krebs, J. R. (1986). *Foraging theory.* Princeton University Press.

Stewart, N., Chater, N., & Brown, G. D. (2006). Decision by sampling. *Cognitive Psychology*, *53*(1), 1–26.

Steyvers, M., Griffiths, T. L., & Dennis, S. (2006). Probabilistic inference in human semantic memory. *Trends in Cognitive Sciences*, *10*, 327– 334.

Steyvers, M., Tenenbaum, J. B., Wagenmakers, E. J., & Blum, B. (2003). Inferring causal networks from observations and interventions. *Cognitive Science*, *27*, 453–489.

Stigler, S. M. (1986). *The history of statistics: The measurement of uncertainty before 1900.* Harvard University Press.

Stocker, A. A., & Simoncelli, E. P. (2006). Noise characteristics and prior expectations in human visual speed perception. *Nature Neuroscience*, *9*(4), 578–585.

Strange, W. (2011). Automatic selective perception (ASP) of first and second language speech: A working model. *Journal of Phonetics*, *39*(4), 456–466.

Stuhlmüller, A., & Goodman, N. D. (2014). Reasoning about reasoning by nested conditioning: Modeling theory of mind with probabilistic programs. *Cognitive Systems Research*, *28*, 80–99.

Stuhlmüller, A., Taylor, J., & Goodman, N. (2013). Learning stochastic inverses. In *Advances in Neural Information Processing Systems* (pp. 3048–3056).

Suchow, J. W., & Griffiths, T. L. (2016a). Deciding to remember: Memory maintenance as a Markov decision process. In *Proceedings of the 38th Annual Meeting of the Cognitive Science Society* (pp. 2063–2068).

Suchow, J. W., & Griffiths, T. L. (2016b). Rethinking experiment design as algorithm design. In *CrowdML: Workshop on Crowdsourcing and Machine Learning.*

Sudderth, E. B., Torralba, A., Freeman, W. T., & Willsky, A. S. (2005). Learning hierarchical models of scenes, objects, and parts. In *10th IEEE International Conference on Computer Vision (ICCV 2005)* (pp.1331–1338).

Sukhbaatar, S., Szlam, A., & Fergus, R. (2016). Learning multiagent communication with backpropagation. In *Advances in Neural Information Processing Systems* (pp. 2244–2252).

Sun, C., Karlsson, P., Wu, J., Tenenbaum, J. B., & Murphy, K. (2019). Stochastic prediction of multi-agent interactions from partial observations. *arXiv preprint arXiv:1902.09641.*

Sundh, J., Zhu, J., Chater, N., & Sanborn, A. (2023). A unified explanation of variability and bias in human probability judgments: How computational noise explains the mean–variance signature. *Journal of Experimental Psychology: General, 152*(10), 2842–2860.

Sutton, R. S., & Barto, A. G. (1987). A temporal-difference model of classical conditioning. In *Proceedings of the Ninth Annual Meeting of the Cognitive Science Society* (pp. 355–378).

Sutton, R. S., & Barto, A. G. (1998). *Reinforcement learning: An introduction* (1st ed.). MIT Press.

Sutton, R. S., & Barto, A. G. (2018). *Reinforcement learning: An introduction* (2nd ed.). MIT Press.

Sutton, R. S., Precup, D., & Singh, S. (1999). Between MDPs and semi-MDPs: A framework for temporal abstraction in reinforcement learning. *Artificial Intelligence*, *112*(1–2), 181–211.

Synnaeve, G., Dautriche, I., Börschinger, B., Johnson, M., & Dupoux, E. (2014). Unsupervised word segmentation in context. In *Proceedings of COLING* (pp. 2326–2334).

Tacchetti, A., Song, H. F., Mediano, P. A., et al. (2018). Relational forward models for multiagent learning. *arXiv preprint arXiv:1809.11044.*

Talton, J., Yang, L., Kumar, R., Lim, M., Goodman, N. D., & Me¡ch, R. (2012). Learning design patterns with Bayesian grammar induction. In *Proceedings of the 25th Annual ACM Symposium on User Interface Software and Technology* (pp. 63–74).

Taylor, K. I., Moss, H. E., Stamatakis, E. A., & Tyler, L. K. (2006). Binding cross-modal object features in perirhinal cortex. *Proceedings of the National Academy of Sciences*, *103*(21), 8239–8244.

Téglás, E., Vul, E., Girotto, V., Gonzalez, M., Tenenbaum, J. B., & Bonatti, L. L. (2011). Pure reasoning in 12-month-old infants as probabilistic inference. *Science*, *332*(6033), 1054–1059.

Teh, Y. W. (2006). A hierarchical Bayesian language model based on Pitman-Yor processes. In *Proceedings of COLING/ACL* (pp. 985–992).

Tenenbaum, J. B. (1999). *A Bayesian framework for concept learning.* Unpublished doctoral dissertation, Massachussetts Institute of Technology.

Tenenbaum, J. B. (2000). Rules and similarity in concept learning. In *Advances in Neural Information Processing Systems.*

Tenenbaum, J. B., & Griffiths, T. L. (2001a). Generalization, similarity, and Bayesian inference. *Behavioral and Brain Sciences*, *24*, 629–641.

Tenenbaum, J. B., & Griffiths, T. L. (2001b). Structure learning in human causal induction. In *Advances in Neural Information Processing Systems* (pp. 59–65).

Tenenbaum, J. B., Griffiths, T. L., & Kemp, C. (2006). Theory-based Bayesian models of inductive learning and reasoning. *Trends in Cognitive Sciences*, *10*, 309–318.

Tenenbaum, J. B., Kemp, C., Griffiths, T. L., & Goodman, N. D. (2011). How to grow a mind: Statistics, structure, and abstraction. *Science*, *331*(6022), 1279–1285.

Thagard, P. (2002). *Coherence in thought and action.* MIT Press.

Thelen, E., & Fisher, D. M. (1982). Newborn stepping: An explanation for a "disappearing" reflex. *Developmental Psychology*, *18*(5), 760.

Thompson, B., van Opheusden, B., Sumers, T., & Griffiths, T. (2022). Complex cognitive algorithms preserved by selective social learning in experimental populations. *Science*, *376*(6588), 95–98.

Thorndike, E. L. (1898). Animal intelligence: An experimental study of the associative processes in animals. *Psychological Review: Monograph Supplements*, *2*(4), i.

Thrun, S., Burgard, W., & Fox, D. (2005). *Probabilistic robotics.* MIT Press.

Thrun, S., & Pratt, L. (2012). *Learning to learn.* Springer.

Todd, J. T., & Warren, W. H. Jr. (1982). Visual perception of relative mass in dynamic events. *Perception*, *11*(3), 325–335.

Tomasello, M. (2009). *Constructing a language.* Harvard University Press.

Tomov, M. S., Yagati, S., Kumar, A., Yang, W., & Gershman, S. J. (2020). Discovery of hierarchical representations for efficient planning. *PLoS Computational Biology*, *16*(4), e1007594.

Tong, F., Meng, M., & Blake, R. (2006). Neural bases of binocular rivalry. *Trends in Cognitive Sciences*, *10*(11), 502–511.

Traer, J., & McDermott, J. H. (2016). Statistics of natural reverberation enable perceptual separation of sound and space. *Proceedings of the National Academy of Sciences*, *113*(48), E7856–E7865.

Traxler, M. J. (2014). Trends in syntactic parsing: Anticipation, Bayesian estimation, and good-enough parsing. *Trends in Cognitive Sciences*, *18*(11), 605–611.

Trommershäuser, J., Maloney, L. T., & Landy, M. S. (2003). Statistical decision theory and the selection of rapid, goal-directed movements. *Journal of the Optical Society of America A*, *20*(7), 1419–1433.

Tsividis, P. A., Loula, J., Burga, J., et al. (2021). Human-level reinforcement learning through theory-based modeling, exploration, and planning. *arXiv preprint arXiv:2107.12544*.

Tsividis, P. A., Pouncy, T., Xu, J. L., Tenenbaum, J. B., & Gershman, S. J. (2017). Human learning in Atari. In *AAAI Spring Symposium Series, Science of Intelligence: Computational Principles of Natural and Artificial Intelligence.*

Turing, A. M. (1936). On computable numbers, with an application to the Entscheidungsproblem. *Journal of Math*, *58*(345–363), 5.

Turing, A. M. (1950). Computing machinery and intelligence. *Mind*, *59*(236), 433–460.

Tversky, A. (1977). Features of similarity. *Psychological Review*, *84*, 327–352.

Tversky, A., & Kahneman, D. (1974). Judgment under uncertainty: Heuristics and biases. *Science*, *185*(4157), 1124–1131.

Tversky, A., & Kahneman, D. (1983). Extensional vs. intuitive reasoning: The conjunction fallacy in probability judgment. *Psychological Review*, *90*, 293–315.

Ullman, T., Baker, C., Macindoe, O., Evans, O., Goodman, N., & Tenenbaum, J. B. (2009). Help or hinder: Bayesian models of social goal inference. In *Advances in Neural Information Processing Systems* (pp. 1874–1882).

Ullman, T. D., Goodman, N. D., & Tenenbaum, J. B. (2012). Theory learning as stochastic search in the language of thought. *Cognitive Development*, *27*(4), 455–480.

Ullman, T. D., Spelke, E., Battaglia, P., & Tenenbaum, J. B. (2017). Mind games: Game engines as an architecture for intuitive physics. *Trends in Cognitive Sciences*, *21*(9), 649–665.

Ullman, T. D., Stuhlmüller, A., Goodman, N. D., & Tenenbaum, J. B. (2018). Learning physical parameters from dynamic scenes. *Cognitive Psychology*, *104*, 57–82.

Ullman, T. D., & Tenenbaum, J. B. (2020). Bayesian models of conceptual development: Learning as building models of the world. *Annual Review of Developmental Psychology*, *2*, 533–558.

Usher, M., & McClelland, J. L. (2001). The time course of perceptual choice: The leaky, competing accumulator model. *Psychological Review*, *108*(3), 550–592.

Vallabha, G., & McClelland, J. (2007). Success and failure of new speech category learning in adulthood: Consequences of learned Hebbian attractors in topographic maps. *Cognitive, Affective, and Behavioral Neuroscience*, *7*, 53–73.

Van Assen, J. J. R., Barla, P., & Fleming, R. W. (2018). Visual features in the perception of liquids. *Current Biology*, *28*(3), 452–458.

Van Der Helm, P. A., & Leeuwenberg, E. L. (1996). Goodness of visual regularities: A nontransformational approach. *Psychological Review*, *103*(3), 429–456.

Vanpaemel, W., Storms, G., & Ons, B. (2005). A varying abstraction model for categorization. In *Proceedings of the 27th Annual Meeting of the Cognitive Science Society.*

Van Rijn, P., Mertes, S., Schiller, D., et al. (2021). Exploring emotional prototypes in a high dimensional TTS latent space. *arXiv preprint arXiv:2105.01891.*

Vasta, R., & Liben, L. S. (1996). The water-level task: An intriguing puzzle. *Current Directions in Psychological Science*, *5*(6), 171–177.

Vazquez-Chanlatte, M., Jha, S., Tiwari, A., Ho, M. K., & Seshia, S. (2018). Learning task specifications from demonstrations. In *Advances in Neural Information Processing Systems*.

Veerapaneni, R., Co-Reyes, J. D., Chang, M., et al. (2020). Entity abstraction in visual model-based reinforcement learning. In *Conference on Robot Learning* (pp. 1439–1456).

Velez-Ginorio, J., Siegel, M., Tenenbaum, J. B., & Jara-Ettinger, J. (2017). Interpreting actions by attributing compositional desires. In *Proceedings of the 39th Annual Meeting of the Cognitive Science Society*.

Verbrugge, R. (2009). Logic and social cognition. *Journal of Philosophical Logic*, *38*(6), 649–680.

Vitányi, P. M., & Chater, N. (2017). Identification of probabilities. *Journal of Mathematical Psychology*, *76*, 13–24.

Vitányi, P. M., & Li, M. (2000). Minimum description length induction, Bayesianism, and Kolmogorov complexity. *IEEE Transactions on Information Theory*, *46*(2), 446–464.

Viterbi, A. (1967). Error bounds for convolutional codes and an asymptotically optimum decoding algorithm. *Transactions on Information Theory*, *13*(2), 260–269.

Von Helmholtz, H. (1867). *Handbuch der physiologischen optik* (Vol. 9). Voss.

Vousden, W., Farr, W. M., & Mandel, I. (2016). Dynamic temperature selection for parallel tempering in Markov chain Monte Carlo simulations. *Monthly Notices of the Royal Astronomical Society*, *455*(2), 1919–1937.

Vul, E., Alvarez, G., Tenenbaum, J., & Black, M. (2009). Explaining human multiple object tracking as resource-constrained approximate inference in a dynamic probabilistic model. *Advances in Neural Information Processing Systems*.

Vul, E., Goodman, N., Griffiths, T. L., & Tenenbaum, J. B. (2014). One and done? Optimal decisions from very few samples. *Cognitive Science*, *38*(4), 599–637.

Vulkan, N. (2000). An economist's perspective on probability matching. *Journal of Economic Surveys*, *14*, 101–118.

Waismeyer, A., Meltzoff, A., & Gopnik, A. (2015). Causal learning from probabilistic events in 24-month-olds: An action measure. *Developmental Science*, *18*, 175–182.

Wald, A. (1947). *Sequential analysis*. Wiley.

Waldmann, M. R. (Ed.). (2017). *Oxford handbook of causal reasoning*. Oxford University Press.

Wallace, C. S., & Boulton, D. M. (1968). An information measure for classification. *Computer Journal*, *11*(2), 185–194.

Wallace, C. S., & Freeman, P. R. (1987). Estimation and inference by compact coding. *Journal of the Royal Statistical Society: Series B (Methodological)*, *49*(3), 240–252.

Wallace, M. T., Roberson, G., Hairston, W. D., Stein, B. E., Vaughan, J. W., & Schirillo, J. A. (2004). Unifying multisensory signals across time and space. *Experimental Brain Research*, *158*(2), 252–258.

Wang, J. X., Kurth-Nelson, Z., Tirumala, D., et al. (2016). Learning to reinforcement learn. *arXiv preprint arXiv:1611.05763*.

Wang, R., Wu, S., Evans, J., Tenenbaum, J., Parkes, D., & Kleiman- Weiner, M. (2021). Too many cooks: Coordinating multi-agent collaboration through inverse planning. In S. Muggleton & N. Chater (Eds.), *Human-like machine intelligence* (pp. 152–170). Oxford University Press.

Wason, P. C. (1966). Reasoning. In B. Foss (Ed.), *New horizons in psychology*. Penguin.

Wason, P. C. (1968). Reasoning about a rule. *Quarterly Journal of Experimental Psychology*, *20*(3), 273–281.

Watters, N., Zoran, D., Weber, T., Battaglia, P., Pascanu, R., & Tacchetti, A. (2017). Visual interaction networks: Learning a physics simulator from video. In *Advances in Neural Information Processing Systems* (pp. 4539–4547).

Weber, E. U., Johnson, E. J., Milch, K. F., Chang, H., Brodscholl, J. C., & Goldstein, D. G. (2007). Asymmetric discounting in intertemporal choice: A query-theory account. *Psychological Science*, *18*(6), 516–523.

Wedell, D. H., Fitting, S., & Allen, G. L. (2007). Shape effects on memory for location. *Psychonomic Bulletin & Review*, *14*(4), 681–686.

Wei, J., Wang, X., Schuurmans, D., et al. (2022). Chain of thought prompting elicits reasoning in large language models. *arXiv preprint arXiv:2201.11903*.

Weiss, Y. (1996). Interpreting images by propagating Bayesian beliefs. *Advances in Neural Information Processing Systems 9*, 908–915.

Wellman, H., & Estes, D. (1986). Early understanding of mental entities: A reexamination of childhood realism. *Child Development, 57*, 910–923.

Wellman, H. M., & Gelman, S. A. (1992). Cognitive development: Foundational theories of core domains. *Annual Review of Psychology, 43*, 337–375.

Wellman, H. M., & Liu, D. (2004). Scaling of theory-of-mind tasks. *Child Development, 75*(2), 523–541.

West, M., Muller, P., & Escobar, M. (1994). Hierarchical priors and mixture models, with application in regression and density estimation. In P. Freeman & A. Smith (Eds.), *Aspects of uncertainty* (pp. 363–386). Wiley.

Wexler, K., & Culicover, P. (1983). *Formal principles of language acquisition.* MIT Press.

Wiederholt, M. (2010). Rational inattention. In *The new Palgrave Dictionary of Economics (online ed.)* (pp. 1–8). https://doi.org/10.1057/978-1-349-95121-5_2901-1

Williams, D. (2018). Hierarchical Bayesian models of delusion. *Consciousness and Cognition, 61*, 129–147.

Wilson, A., Fern, A., Ray, S., & Tadepalli, P. (2007). Multi-task reinforcement learning: A hierarchical Bayesian approach. In *Proceedings of the 24th International Conference on Machine Learning* (pp. 1015–1022).

Wilson, A. G., Dann, C., Lucas, C., & Xing, E. P. (2015). The human kernel. In *Advances in Neural Information Processing Systems.*

Winston, P. H., Horn, B., Minsky, M., Shirai, Y., & Waltz, D. (1975). *The psychology of computer vision.* McGraw-Hill.

Wittgenstein, L. (1953). *Philosophical investigations.* MacMillan.

Wolff, P. (2007). Representing causation. *Journal of Experimental Psychology: General, 136*(1), 82–111.

Wolff, P., Barbey, A. K., & Hausknecht, M. (2010). For want of a nail: How absences cause events. *Journal of Experimental Psychology: General, 139*(2), 191–221.

Wolfram, S. (2002). *A new kind of science.* Wolfram Media.

Wolpert, D. H., & Macready, W. G. (1997). No free lunch theorems for optimization. *IEEE Transactions on Evolutionary Computation, 1*(1), 67–82.

Wolpert, D. M. (2007). Probabilistic models in human sensorimotor control. *Human Movement Science, 26*(4), 511–524.

Wolpert, D. M., & Kawato, M. (1998). Multiple paired forward and inverse models for motor control. *Neural Networks, 11*(7–8), 1317–1329.

Wolpert, D. M., Miall, R. C., & Kawato, M. (1998). Internal models in the cerebellum. *Trends in Cognitive Sciences, 2*(9), 338–347.

Wong, L., Ellis, K., Tenenbaum, J. B., & Andreas, J. (2022). Leveraging language to learn program abstractions and search heuristics. *arXiv preprint arXiv:2106.11053.*

Wong, L., Grand, G., Lew, A. K., et al. (2023). From word models to world models: Translating from natural language to the probabilistic language of thought. *arXiv preprint arXiv:*2306.12672.

Wong, S.-M., Dras, M., & Johnson, M. (2012). Exploring adaptor grammars for native language identification. In *Proceedings of EMNLP/CoNLL* (pp. 699–709).

Wood, F., Meent, J. W., & Mansinghka, V. (2014). A new approach to probabilistic programming inference. In *International Conference on Artificial Intelligence and Statistics* (pp. 1024–1032).

Wood, W., & Rünger, D. (2016). Psychology of habit. *Annual Review of Psychology, 67*(1), 289–314.

Woodward, A. L. (1998). Infants selectively encode the goal object of an actor's reach. *Cognition, 69*(1), 1–34.

Woodward, J. (2003). *Making things happen: A theory of causal explanation.* Oxford University Press.

Wu, J., Yildirim, I., Lim, J. J., Freeman, B., & Tenenbaum, J. (2015). Galileo: Perceiving physical object properties by integrating a physics engine with deep learning. In *Advances in Neural Information Processing Systems* (pp. 127–135).

Xu, F. (2005). Categories, kinds, and object individuation in infancy. In *Building object categories in developmental time* (pp. 81–108). Psychology Press.

Xu, F. (2019). Towards a rational constructivist theory of cognitive development. *Psychological Review*, *126*(6), 841–864.

Xu, F., & Carey, S. (1996). Infants' metaphysics: The case of numerical identity. *Cognitive Psychology*, *30*(2), 111–153.

Xu, F., & Kushnir, T. (2013). Infants are rational constructivist learners. *Current Directions in Psychological Science*, *22*, 28–32.

Xu, F., & Tenenbaum, J. B. (2007). Word learning as Bayesian inference. *Psychological Review*, *114*(2), 245–272.

Xu, J., Dowman, M., & Griffiths, T. L. (2013). Cultural transmission results in convergence towards colour term universals. *Proceedings of the Royal Society B: Biological Sciences*, *280*(1758), 20123073.

Xu, J., & Griffiths, T. L. (2010). A rational analysis of the effects of memory biases on serial reproduction. *Cognitive Psychology*, *60*(2), 107–126.

Yang, Y., & Piantadosi, S. T. (2022). One model for the learning of language. *Proceedings of the National Academy of Sciences*, *119*(5), e2021865119.

Yeung, S., & Griffiths, T. L. (2015). Identifying expectations about the strength of causal relationships. *Cognitive Psychology*, *76*, 1–29.

Yildirim, I., Belledonne, M., Freiwald, W., & Tenenbaum, J. (2020). Efficient inverse graphics in biological face processing. *Science Advances*, *6*(10), eaax5979.

Yildirim, I., Gerstenberg, T., Saeed, B., Toussaint, M., & Tenenbaum, J. (2017). Physical problem solving: Joint planning with symbolic, geometric, and dynamic constraints. *arXiv preprint arXiv:1707.08212*.

Yildirim, I., & Jacobs, R. A. (2013). Transfer of object category knowledge across visual and haptic modalities: Experimental and computational studies. *Cognition*, *126*(2), 135–148.

Yildirim, I., Saeed, B., Bennett-Pierre, G., Gerstenberg, T., Tenenbaum, J., & Gweon, H. (2019). Explaining intuitive difficulty judgments by modeling physical effort and risk. *Proceedings of the 41st Annual Meeting of the Cognitive Science Society*. arXiv preprint arXiv:1905.04445

Yildirim, I., Smith, K. A., Belledonne, M., Wu, J., & Tenenbaum, J. B. (2018). Neurocomputational modeling of human physical scene understanding. In *2018 Conference on Cognitive Computational Neuroscience.*

Yoo, A. H., Klyszejko, Z., Curtis, C. E., & Ma, W. J. (2018). Strategic allocation of working memory resource. *Scientific Reports*, *8*(1), 1–8.

Yuille, A., & Kersten, D. (2006). Vision as Bayesian inference: Analysis by synthesis? *Trends in Cognitive Sciences*, *10*(7), 301–308.

Yuille, A. L., Burgi, P.-Y., & Grzywacz, N. M. (1998). Visual motion estimation and prediction: A probabilistic network model for temporal coherence. In *Sixth International Conference on Computer Vision* (pp. 973–978).

Zabaras, N. (2010). *Importance sampling* (Technical Report). Cornell University.

Zago, M., & Lacquaniti, F. (2005). Cognitive, perceptual and action- oriented representations of falling objects. *Neuropsychologia*, *43*(2), 178–188.

Zhang, C., Xie, S., Jia, B., Wu, Y. N., Zhu, S.-C., & Zhu, Y. (2022). Learning algebraic representation for systematic generalization in abstract reasoning. In *European Conference on Computer Vision* (pp. 692–709).

Zhi-Xuan, T., Mann, J., Silver, T., Tenenbaum, J., & Mansinghka, V. (2020). Online Bayesian goal inference for boundedly rational planning agents. *Advances in Neural Information Processing Systems*, 19238–19250.

Zhou, L., Smith, K. A., Tenenbaum, J. B., & Gerstenberg, T. (2023). Mental Jenga: A counterfactual simulation model of causal judgments about physical support. *Journal of Experimental Psychology: General*, *152*(8), 2237–2269.

Zhu, J., Sanborn, A., & Chater, N. (2018). Mental sampling in multimodal representations. In *Advances in Neural Information Processing Systems.*

Zhu, J.-Q., Sanborn, A. N., & Chater, N. (2020). The Bayesian sampler: Generic Bayesian inference causes incoherence in human probability judgments. *Psychological Review*, *127*(5), 719–748.

Zhu, J.-Q., Sundh, J., Chater, N., & Sanborn, A. (2023). The auto- correlated Bayesian sampler for estimation, choice, confidence, and response times. *Psychological Review*. https://doi.org/10.1037/rev0000427

Zhu, L., Chen, Y., Torralba, A., Freeman, W. T., & Yuille, A. L. (2010). Part and appearance sharing: Recursive compositional models for multi-view. In *Twenty-Third IEEE Conference on Computer Vision and Pattern Recognition, CVPR 2010* (pp. 1919–1926).

Zipf, G. (1932). *Selective studies and the principle of relative frequency in language.* Harvard University Press.

Ziv, J., & Lempel, A. (1978). Compression of individual sequences via variable-rate coding. *IEEE Transactions on Information Theory*, *24*(5), 530–536.

Zuidema, W. (2002). How the poverty of the stimulus solves the poverty of the stimulus. In *Advances in Neural Information Processing Systems.*

Index

Note: Photos, figures, and tables are indicated by *italicized* page numbers.